Methods in Microbiology
Volume 42

Recent titles in the series

Methods in Microbiology
Volume 42

Current and Emerging Technologies for the Diagnosis of Microbial Infections

Edited by

Andrew Sails
Public Health England Newcastle Laboratory,
National Infection Service,
The Medical School, Royal Victoria Infirmary,
Newcastle, UK

Yi-Wei Tang
Department of Laboratory Medicine,
Memorial Sloan-Kettering Cancer Center, and
Department of Pathology and Laboratory Medicine,
Weill Medical College of Cornell University,
New York, USA

ELSEVIER

AMSTERDAM • BOSTON • HEIDELBERG • LONDON
NEW YORK • OXFORD • PARIS • SAN DIEGO
SAN FRANCISCO • SINGAPORE • SYDNEY • TOKYO
Academic Press is an imprint of Elsevier

Academic Press is an imprint of Elsevier
125 London Wall, London, EC2Y 5AS, UK
The Boulevard, Langford Lane, Kidlington, Oxford OX5 1GB, UK
525 B Street, Suite 1800, San Diego, CA 92101–4495, USA
225 Wyman Street, Waltham, MA 02451, USA

First edition 2015

ISBN: 978-0-12-803297-8
ISSN: 0580-9517 (Series)

For information on all Academic Press publications
visit our website at www.store.elsevier.com

Working together
to grow libraries in
developing countries

www.elsevier.com • www.bookaid.org

Contents

**CHAPTER 8   Technical and Software Advances in Bacterial
Pathogen Typing**...**289**
Linda Chui, Vincent Li

**CHAPTER 9   Molecular Strain Typing and Characterisation
of Toxigenic *Clostridium difficile***...................**329**
Tanis C. Dingle, Duncan R. MacCannell

CHAPTER 12 Gene Amplification and Sequencing for Bacterial Identification 433

Susanna K.P. Lau, Jade L.L. Teng, Chi-Chun Ho, Patrick C.Y. Woo

Contributors

Michel G. Bergeron
Centre de recherche en infectiologie de l'Université Laval, Axe maladies infectieuses et immunitaires, Centre de recherche du CHU de Québec, and Département de microbiologie-infectiologie et d'immunologie, Faculté de médecine, Université Laval, Quebec City, Quebec, Canada

Kimberly A. Bishop-Lilly
Henry M. Jackson Foundation for the Advancement of Military Medicine, Bethesda, and Naval Medical Research Center—Frederick, Fort Detrick, Frederick, Maryland, USA

Luc Bissonnette
Centre de recherche en infectiologie de l'Université Laval, Axe maladies infectieuses et immunitaires, Centre de recherche du CHU de Québec, and Département de microbiologie-infectiologie et d'immunologie, Faculté de médecine, Université Laval, Quebec City, Quebec, Canada

Lori Bourassa
Department of Laboratory Medicine, University of Washington, Seattle, Washington, USA

Jessica N. Brazelton de Cárdenas
St. Jude Children's Research Hospital, Memphis, Tennessee, USA

Eoin Brodie
Earth Sciences Division, Lawrence Berkeley National Laboratory, Berkeley, California, USA

Susan M. Butler-Wu
Department of Laboratory Medicine, University of Washington, Seattle, Washington, USA

Linda Chui
Provincial Laboratory for Public Health, Walter Mackenzie Health Sciences Centre, and Department of Laboratory Medicine and Pathology, University of Alberta Hospital, Edmonton, Alberta, Canada

Martin D. Curran
Public Health England, Cambridge University Hospitals NHS, Cambridge, United Kingdom

Tanis C. Dingle
Icahn School of Medicine at Mount Sinai, New York, USA

Sherry Dunbar
Luminex Corporation, Austin, Texas, USA

Julia Engstrom-Melnyk
Medical and Scientific Affairs, Roche Diagnostic Corporation, Indianapolis, Indiana, USA

Kenneth G. Frey
Henry M. Jackson Foundation for the Advancement of Military Medicine, Bethesda, and Naval Medical Research Center—Frederick, Fort Detrick, Frederick, Maryland, USA

Nicholas S. Gleadall
Public Health England, Cambridge University Hospitals NHS, Cambridge, United Kingdom

Jane S. Greatorex
Public Health England, Cambridge University Hospitals NHS, Cambridge, United Kingdom

Randall T. Hayden
St. Jude Children's Research Hospital, Memphis, Tennessee, USA

Jude Heaney
Department of Medicine, University of Cambridge, Cambridge, and Institute of Hepatology, Foundation for Liver Research, London, United Kingdom

Raymond C. Hein
Medical and Scientific Affairs, Roche Diagnostic Corporation, Indianapolis, Indiana, USA

Chi-Chun Ho
Department of Microbiology, The University of Hong Kong, Queen Mary Hospital, Hong Kong

Ulas Karaoz
Earth Sciences Division, Lawrence Berkeley National Laboratory, Berkeley, California, USA

Markus Kostrzewa
Bruker Daltonik GmbH, Bremen, Germany

Susanna K.P. Lau
Department of Microbiology, and State Key Laboratory of Emerging Infectious Diseases, The University of Hong Kong, Queen Mary Hospital, Hong Kong

Vincent Li
Provincial Laboratory for Public Health, Walter Mackenzie Health Sciences Centre, University of Alberta Hospital, Edmonton, Alberta, Canada

Duncan R. MacCannell
Centers for Disease Control and Prevention (CDC), Atlanta, Georgia, USA

Elizabeth M. Marlowe
TPMG Regional Reference Laboratories, Berkeley, California, USA

Steve Miller
Clinical Microbiology Laboratory, University of California, San Francisco, California, USA

Susan M. Novak-Weekley
SCPMG Regional Reference Laboratories, North Hollywood, California, USA

Olivier Peraud
Medical and Scientific Affairs, Roche Diagnostic Corporation, Indianapolis, Indiana, USA

Michael A. Pfaller
T2Biosystems, Lexington, Massachusetts, and University of Iowa College of Medicine and College of Public Health, Iowa City, Iowa, USA

Hannes Pouseele
Applied Maths, Sint-Martens-Latem, Belgium

Bobbi S. Pritt
Mayo Clinic, Rochester, Minnesota, USA

Pedro L. Rodriguez
Medical and Scientific Affairs, Roche Diagnostic Corporation, Indianapolis, Indiana, USA

Kathryn Rolfe
Public Health England, Cambridge University Hospitals NHS, Cambridge, United Kingdom

Sören Schubert
Max von Pettenkofer Institute, Ludwig-Maximilians-Universität (LMU), Munich, Germany

Alon Singer
HelixBind, Inc., Marlborough, Massachusetts, USA

Philip Supply
Center for Infection and Immunity of Lille; INSERM U1019; CNRS UMR 8204; Université Lille; Institut Pasteur de Lille, and Genoscreen, Lille, France

Yi-Wei Tang
Department of Laboratory Medicine, Memorial Sloan-Kettering Cancer Center, and Department of Pathology and Laboratory Medicine, Weill Medical College of Cornell University, New York, USA

Jade L.L. Teng
Department of Microbiology, The University of Hong Kong, Queen Mary Hospital, Hong Kong

Ephraim L. Tsalik
Center for Applied Genomics & Precision Medicine; Division of Infectious Diseases, Department of Medicine, Duke University Medical Center and Emergency Medicine Service, Durham VAMC, Durham, North Carolina, USA

Richard Allen White III
Department of Microbiology and Immunology at the University of British Columbia, Vancouver, British Columbia, Canada, and Pacific Northwest National Laboratory, Richland, Washington, USA

Patrick C.Y. Woo
Department of Microbiology, and State Key Laboratory of Emerging Infectious Diseases, The University of Hong Kong, Queen Mary Hospital, Hong Kong

Christopher W. Woods
Medicine Service, Durham VAMC; Center for Applied Genomics & Precision Medicine, and Division of Infectious Diseases, Department of Medicine, Duke University Medical Center, Durham, North Carolina, USA

William E. Yang
Duke University School of Medicine, Center for Applied Genomics and Precision Medicine, Durham, North Carolina, USA

Preface

As microbiologists we live in exciting times. In the last 20 years, we have witnessed revolutionary changes in the way we diagnose microbial infections which have been driven by the emergence and adoption of new technologies. Techniques such as *in vitro* nucleic acid amplification and MALDI-TOF mass spectroscopy have become common place in most clinical microbiology laboratories. The way we characterise and type or fingerprint microbial pathogens has also undergone revolutionary changes. Next-generation sequencing (NGS) technologies facilitate the sequencing of whole bacterial genomes in days, potentially making these technologies suitable for routine use in clinical practice. Further developments in NGS technologies leading to longer read lengths, in combination with further improvements in data analysis and bioinformatics, will facilitate their use in diagnostics directly on patient samples in the near future. Finally, the use of panomic techniques such as genomics, transcriptomics, proteomics, and metabolomics at both hosts and bugs as an integrated approach will further substantiate and broaden the microbial infection diagnostic armamentarium.

This volume of *Methods in Microbiology* is intended to be a comprehensive review of current and emerging technologies in the field of clinical microbiology and features a wide variety of state-of-the art methods and techniques for the diagnosis and management of microbial infections with chapters authored by internationally renowned experts. The audience for this volume includes pathologists and clinical microbiologists working in hospital laboratories, public health laboratories, national reference laboratories, academic and research microbiologists in universities, students studying clinical microbiology or biomedical science, and industrial microbiologists working in the *in vitro* diagnostic industry. The editors are indebted to the contributors, all of whom have produced cutting edge reviews of diagnostic technologies.

The editors would also like to thank Colin Harwood for inviting them to edit this volume of *Methods in Microbiology* and for supporting us throughout the project. Finally, we would like to thank our colleagues at Elsevier, in particular Helene Kabes and Omer Mukthar, and Mary Ann Zimmerman for overseeing the project from its initial concept to the finished volume.

Andrew David Sails
Yi-Wei Tang

Total Laboratory Automation in Clinical Bacteriology

1

Susan M. Novak-Weekley*,[1], Elizabeth M. Marlowe[†]

*SCPMG Regional Reference Laboratories, North Hollywood, California, USA
†TPMG Regional Reference Laboratories, Berkeley, California, USA
[1]Corresponding author: e-mail address: susan.m.novak@kp.org

1 CLINICAL BACTERIOLOGY AND AUTOMATION: BACKGROUND

The primary mission of the clinical bacteriology laboratory is to assist the health care provider in the diagnosis of infectious diseases. Due to the variety of specimens submitted to the bacteriology laboratory, many of the steps related to the processing and workup of a specimen have remained manual. The specimen is inoculated onto an agar medium (with plating protocols typically driven by the source of the specimen), the plates transferred manually to an incubator, the plates removed after a defined period of time and the culture examined by a technologist to look for potential pathogens. The cost of healthcare in many countries and the Affordable Care Act in the United States are collectively driving institutions to explore new and novel ways to provide continuous, quality care in a more affordable, efficient fashion. One of those options to enhance efficiency and affordability is automation.

Notwithstanding the pressure to maintain affordability and quality, there are many other pressures health care institutions face. It is a fact that fewer clinical laboratorians are entering the workplace and this parallels the decline in medical technology training programmes in the United States (Microbiology, 2008). The American Society for Clinical Pathology (ASCP) vacancy survey indicated that in 2012, 9% of the microbiology staff will most likely retire in the next 2 years (Garcia, Ali, & Choudhry, 2013). In the state of California alone, the average age of a medical technologist is approximately 55 years of age. Again, with fewer programmes producing licenced, trained technologists and the anticipation of the growing need for laboratory services, automation is a potential solution to mitigate the decreased staffing situation (Garcia et al., 2013). In addition to automation, some institutions are investigating the training and implementation of lean within the laboratory section. Although other industries have seen the adoption of lean impact efficiency and bottom line, the healthcare setting has been slow to adopt the lean philosophy. Seminars are now being offered in the context of the laboratory and lean management. Some health care institutions have published on lean in the laboratory space and offer

Methods in Microbiology, Volume 42, ISSN 0580-9517, http://dx.doi.org/10.1016/bs.mim.2015.09.002

In addition to the pre-analytical plating instruments, there are "smart incubators" which include not only the incubator to incubate the culture plates at the appropriate temperature but also a sophisticated digital camera to take images of the culture plates. The digital camera can detect colonies that are not able to be visualised by the human eye, therefore potentially allowing a specimen to be worked on much earlier than if the plates were examined manually with the naked eye. Digital imaging has the potential to impact in the areas of competency, training and quality assurance. Images of organisms on culture plates can be stored and subsequently retrieved if there is a quality problem in the laboratory or for training or competency testing. Digital plate reading (DPR) allows the technologist to work up the cultures without handling the plates. This impacts on the ergonomics of opening and closing plates in addition to decreasing unnecessary pathogen exposure as mentioned above.

To date, there is limited literature on TLA and the components that make up these systems (Bourbeau & Swartz, 2009; Mutters et al., 2014; Novak & Marlowe, 2013). Once these types of instruments are placed in more clinical laboratories additional performance data will be forthcoming. This chapter will summarise the currently available automation systems from a modular and TLA perspective.

2 SPECIMEN COLLECTION: LIQUID MICROBIOLOGY

A wide variety of specimen types are submitted to the bacteriology department for processing. In order to fully automate the plating of samples for microbiology the specimen (for several of the instruments mentioned below) must be in a liquid form. There is one pre-analytical plating instrument that can process the liquid from a transport swab or inoculate the media using the swab in semi-solid transport medium. Regardless of the specimen collection device, the importance of adequate specimen collection is always paramount when submitting microbiology samples for laboratory analysis. For many years, rayon swabs have been used for routine microbiology specimen collection. The rayon swab contains tightly round rayon fibres placed on the end of the collection shaft (Figure 1). To improve on specimen collection, Copan Diagnostics (Brescia, Italy) developed the flocked swab or ESwab™ which allows for better collection and subsequently more homogenous dispersal of the specimen into a liquid Amies transport medium within the transport tube (Buchan, Olson, Mackey, & Ledeboer, 2014; Trotman-Grant, Raney, & Dien Bard, 2012; Van Horn, Audette, Sebeck, & Tucker, 2008). The flocked swab is unique in the design with nylon fibres arranged perpendicular to the head of the swab thus increasing the surface area for collection and trapping of the specimen and associated organisms (Figure 2). Figure 3 shows various types of ESwabs from Copan Diagnostics available today on the market. In the case of the nylon swab, the net that is created traps the sample preventing release of the specimen during plating. After the sample is collected with the ESwab, approximately 90% of the specimen is released into the liquid Amies transport medium present in the transport tube. This liquid Amies transport medium is suitable for plating using an automated instrument such as those

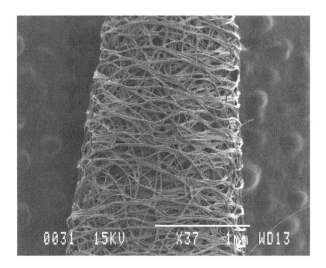

FIGURE 1

Electron micrograph image of the tip of a rayon specimen collection swab.

Image courtesy of Copan Diagnostics.

FIGURE 2

Electron micrograph image of the tip of a flocked swab showing the increased surface area.

Image courtesy of Copan Diagnostics.

described in this chapter. Several studies have demonstrated equivalent or superior performance of the ESwab compared to conventional swabs. In one study, there was a 3.6-fold increase in the recovery of viable methicillin-resistant *Staphylococcus aureus* with the ESwab versus the Venturi swab (Smismans, Verhaegen,

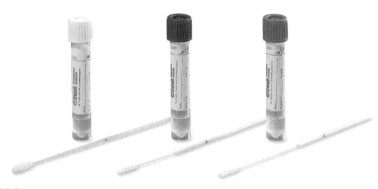

FIGURE 3

Various ESwab™ vials. Each tube contains a different size swab, which correlates to the colour cap of the tube.

Image courtesy of Copan Diagnostics.

Schuermans, & Frans, 2009). In another study, the ESwab was equivalent or better than (respectively) to other swabs tested (BD CultureSwab MaxV and the Remel BactiSwab (Lenexa, KS)) when using Clinical Laboratory Standards Institute (CLSI) criteria for the evaluation of swab systems (Van Horn et al., 2008). The liquid specimen is a more homogenous specimen than a swab in semi-solid media which should result in all plates receiving the same amount of specimen. This is advantageous when multiple plates are inoculated for specific specimen types. There are other manufacturers of flocked swabs in addition to Copan Diagnostics such as Puritan (Guilford, ME) and Millipore (Billerica, MA) (Buchan & Ledeboer, 2014).

3 PRE-ANALYTICAL AUTOMATION

In regards to specimen processing and inoculation of media, there are several instruments that can fully automate the pre-analytical processing of the majority of clinical samples submitted for routine bacteriology testing. It is important to note that because laboratories are very different in terms of the volume and types of specimens processed there will seldom be a "one size fits all" solution for automation in the clinical bacteriology laboratory. Many of the systems described below have features that are complimentary to one another yet differences do exist. Briefly, all fully automated pre-analytical plating instruments will read the barcoded specimen and can query the Laboratory Information System (LIS) to determine the type of specimen and the plating protocol. This is a very important feature because it allows for less intervention from the personnel that are managing the upfront processing of the specimens and in addition allows for traceability and positive identification throughout the pre-analytical process. After the barcode is read, the systems will then apply a barcode label to each plate linking that plate to a specific patient specimen. Specimen tracking and traceability are paramount in today's clinical laboratory where staff are continuously being challenged to work very efficiently yet maintain excellent quality. Enhanced procedures should be put in place to minimise labelling errors and specimen mix-up (Wagar, Stankovic, Raab, Nakhleh, & Walsh, 2008). Plating automation can significantly reduce or mitigate these errors from occurring in the context of pre-analytical specimen processing.

The following sections will describe the automated plating systems available today, but it is important to note that the functionality of these systems and associated software is constantly evolving (Bourbeau & Ledeboer, 2013; Buchan & Ledeboer, 2014; Novak & Marlowe, 2013). If a laboratory is interested in pursuing automation, due to the dynamic nature of the software and functionality of the instrumentation, it is important to obtain up to date information from the manufacturer.

4 PRE-ANALYTICAL BACTERIOLOGY SPECIMEN PLATING INSTRUMENTS

4.1 ISOPLATER

One of the first semi-automated plating instruments on the market for use in the clinical laboratory was the Isoplater (Vista Technology Inc., Edmonton, Alberta, Canada; www.vistatechnology.com). The initial version of the instrument was developed almost 24 years ago and continues to be used in clinical laboratories today. Unlike newer instrumentation, the Isoplater is a semi-automated plating instrument which means the sample must be manually inoculated onto the agar media prior to the placement of the plate on the instrument for streaking. A newer version of the instrument on the market today is the Isoplater 180i (Figure 4). The specimen must be inoculated over the standard ¼″ Frosted Mark found on most Petri dishes, so the Isoplater can properly orient the plates before streaking. A carousel can hold up to four stacks of 20 Petri dishes with a total of 80 plates resulting in a 30-min walk-away cycle. The instrument has dimensions of 30″ (width) × 25″ (depth) × 30″ (height) and can fit on the table or bench top within the laboratory. The Isoplater streaking pattern is the standard four quadrant overlapping pattern (the streaks lines are curved for maximum dish usage) which is very standardised and can easily be quantitated by the technologist. By using four wire loops in succession, the throughput is 180 plates per hour producing isolated colonies Figure 5. The manufacturer states that the life span of the wire loop is approximately 20,000 inoculated plates. Four "S"-shaped loops (Figure 6), designed for maximum streak coverage and improved isolation are used to streak a single plate. The instrument can streak one plate in 18 s with a total of 180 plates per hour. The newer instrument offers a touch screen interactive control panel as shown in Figure 4. Streaking of an inoculated plate occurs in a negative pressure area within the instrument that is HEPA filtered to provide safety to the laboratorians using the instrument and those personnel in the surrounding laboratory space. The negative pressure and HEPA filter also remove smoke when heating the

FIGURE 4

Isoplater 180i pre-analytical specimen processor, Vista Technology.

Image courtesy of Vista Technology.

FIGURE 5

Isoplater 180i streaking pattern.

Image courtesy of Vista Technology.

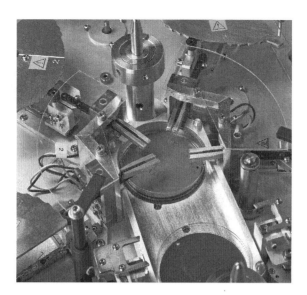

FIGURE 6

Inside view of Isoplater 180i instrument loop mechanism.

Image courtesy of Vista Technology.

loops and reduces the risk when a hazardous specimen containing potential pathogenic organisms is being processed. The HEPA filter is a feature present in all of the instruments described below as well. The Isoplater 180i cannot be interfaced with the LIS, and patient labels must be applied to the plates manually when the Petri dish is inoculated. The four individual S-shaped streaking loops the Isoplater allow effective streaking of sputum, stool and other thick specimens without any special specimen pre-treatment.

Generally, prepping, labelling and inoculating represent about 50% of the labour of set up, and streaking takes the other 50%. Therefore, while there are some manual steps associated with preparing the media for placement on the Isoplater, this instrument may be an ideal solution for those laboratories that process a lower volume of samples but want to enhance the quality of plating and remove the manual task of streaking the plates which can free up staff for other essential departmental tasks.

4.2 INNOVA

The number of manufacturers developing bacteriology pre-analytical automation have been limited to date. Dynacon, manufactured one of the first fully automated liquid plating instruments in 2002, called the Inoculab (Novak & Marlowe, 2013). This instrument set the stage for newer fully automated pre-analytical plating instruments and is still used in many laboratories today although as of 2015 this instrument will no longer be manufactured. The next-generation instrument from Dynacon was the Innova, and at the time of introduction to the clinical market, Dynacon was acquired by Becton Dickinson in 2010.

The Innova (Figure 7) was introduced around 2010 and had enhanced features compared to the Inoculab and could accommodate various sizes and shapes of specimen containers. The instrument can be interfaced to the LIS, has a capacity to hold up to 200 specimens and contains six plate silos that can hold six different types of media with a total plate capacity of 270. The instrument measures 60″ (width) × 49.5″ (depth) × 71″ (height) and access to the instrument is through the front. As shown in Table 1, the Innova is similar to other plating instruments in that a re-useable loop is used for inoculation. Loop sizes can vary depending on the need and range from 1, 10 and 30 µl. To increase the homogeneity of the specimen plated, the Innova has an agitator/shaker so that the specimen is mixed prior to plating. An internal camera takes a picture of the loop to ensure loop alignment since a misaligned loop (on any system) can impact the quality of plating and the isolation and recovery of important organisms. To ensure an adequate volume is present in the original specimen, an ultrasonic level sensor is present and if adequate volume is not available for plating, the specimen will not be processed, and the instrument

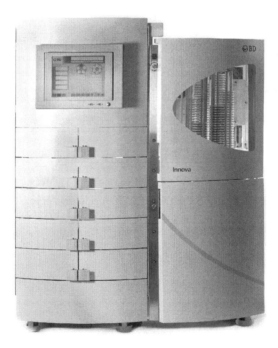

FIGURE 7

Innova™ pre-analytical specimen processor, Becton, Dickinson and Company.

Image courtesy of BD.

Table 1 Comparison of the Fully Automated Plating Instruments Currently on the Market

	PREVI Isola	Innova	WASP	InoqulA+	PreLUD
De-cap/cap containers	No	Yes	Yes	Yes	Yes
Number of different media at once	5	6	9	12	8
Number of samples at once (max)	114	200	72	288	300
Number of plates streaked at once	1	1	1	Up to 5 at once	1
Streak only mode	No	Yes	Yes	Yes—MI module	Yes
Inoculate Gram Slide	No	No	Yes	Yes	Future
Inoculate broth tube	No	No	Yes	Yes	Future
Detect Eswab presence	No	Yes	No	No	Yes
Method of inoculation	Pipette	Re-useable loop	Re-useable loop	Pipette	Pipette, re-usable loop, primary swab
Throughput[a]	~180 inoculations/h	~130 inoculations/h	~180 inoculations/h	~220 inoculations/h	~120 inoculations/h
Integrate into track system	No	No	Yes	Yes	Future
Sample vortex/agitation	No	Yes	Yes	Yes	Yes
Streaking method	Spiral-plastic comb	Custom loop	Custom loop	Custom-rolling bead	Custom loop, primary swab, custom bead
Sort plates by incubator	Yes	Yes	Yes	Yes	Yes—Custom—8 stacks
Consumables/waste	Streaking comb, pipette tip, extra cap	Re-useable loop	Re-useable loop	Re-useable bead, pipette tip	Re-useable loop, pipette tip, re-useable bead

[a]Varies depending on streak pattern.

flag will notify the operator that the specimen was not plated. The Innova was the first next-generation plating instrument to have a universal specimen de-capper/ capper which can adjust to the container type when the instrument scans the barcode on the individual specimen prior to plating the specimen. After the specimen is de-capped and plated, the instrument re-caps the specimen so the sample can be stored depending on the departmental protocols. This de-capping and re-capping feature is integral for all the newer plating instruments. Specimens are placed in metal racks or "canoes" within the Innova that are located in five individual drawers, with a total of 40 specimens in each drawer. The metal racks are flexible in that the rack can accommodate different size specimen containers such as a boric acid tube or a stool Culture & Sensitivity vial (C&S, Meridian Bioscience, Inc., Cincinnati, OH). An advanced feature of this instrument, again similar to the others that will be described below, is that the Innova can be programmed to inoculate media using various plating protocols that can be configured based on the specimen source or laboratory procedures. The instrument can obtain this information from the LIS which is a very important feature for small- to medium-size laboratories that might need to place a variety of specimens on the plating run. The Innova does not have a track for constant feeding of specimens into the unit nor does it have a track system exiting the system as do many of the other instruments described below. Consequently, this instrument would not be part of a TLA suite of instruments but could be used as a standalone instrument to automate the pre-analytical plating process. If necessary, certain specimens can be plated on the media within a biological safety cabinet (BSC) and a streak only mode used on the Innova similar to the Isoplater. Positive identification of the sample is ensured throughout the plating process since the instrument queries the LIS for instructions as to the patient information, type of media to plate (based on the specimen source or protocol) and attaches the correct patient label to the plate. In an era of increased awareness in terms of patient safety, positive patient identification throughout the journey of the specimen as it travels through the laboratory is critical for most laboratories in today's healthcare environment.

In 2012, BD acquired KIESTRA Lab Automation (Drachten, the Netherlands) which had designed and developed another specimen processing system—the InoqulA. The InoqulA, which will be described in more detail below, offers several enhanced features. As the InoqulA is the latest specimen-processing platform offered by BD, the Innova will eventually no longer be marketed for sale but will be serviced and supported for those that are currently using this technology.

4.3 BD KIESTRA™ InoqulA+™

As mentioned above in 2012, BD acquired Kiestra™, a Dutch company that specialises in microbiology automation. Kiestra has been in the microbiology domain for 17 years and installed the first automated instrumentation in 2006 in Europe. The InoqulA™ pre-analytical automated plating instrument has been in the laboratory setting since 2011 in Western Europe. In 2013, BD Kiestra introduced the InoqulA+ (Figure 8) which is the first system on the market to include an optional

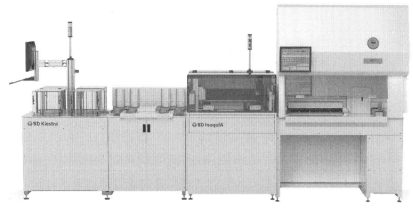

FIGURE 8

InoculA+™ pre-analytical specimen processor, Becton, Dickinson and Company.

Image courtesy of BD.

BSC, which was designed with the intent to assist the laboratorian in processing samples that cannot otherwise be processed with the full laboratory automation. This would include specimens such as tissues, catheter tips and sputum, for example, and may obviate the need for a standalone BSC in the laboratory. The previous configuration of the InoqulA did not have the BSC feature. The InoqulA+ is interfaceable and has a dimension of 174″ (width) × 37″ (depth). The instrument has a fully loaded plate capacity of 612 plates and can be loaded with 12 different media types depending on the protocols of the individual laboratory. Compared with the other instruments described in this chapter, this system has the largest plating capacity and is well suited for higher-volume clinical laboratories. The system has customisable container racks and a universal de-capper (as described with previous instrumentation) which enables the de-capping of different-sized sample containers. A calibrated pipette which samples the liquid specimen is used to inoculate plates, broth tubes and slides according to the sample protocol set by the end user. The pipetting mechanism is unique to the InoqulA+; however, the instrument has a unique streaking technology that uses a magnetic rolling bead (shown in Figure 9) to streak the plate using customisable patterns (zig zag, four quadrants, bi-plate, antimicrobial susceptibility testing (AST) lawn and semi-quantitative patterns) based on operator need. When comparing the magnetic bead streak pattern and colony isolation, it appears that the number of single or discrete colonies is enhanced compared to manual methods. From a capacity perspective, the instrument can streak up to five plates at once using the rolling bead technology. From a throughput perspective if a sample requires seven plates to be inoculated, the InoqulA+ can streak 220 plates per hour. The instrument has similar features already mentioned in that the specimen is vortex-mixed prior to inoculation and a picture is taken of the pipette tip prior to inoculation to ensure adequate specimen delivery. The InoqulA+ also has a barcode-

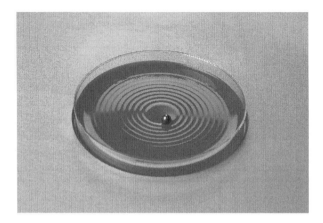

FIGURE 9

Petri dish streaked with magnetic bead technology using the InoqulA+™, Becton, Dickinson and Company.

Image courtesy of BD.

driven semi-automated mode that is designed for continuous processing of specimens that are not suitable for fully automated plating, such as tissues, catheter tips and other non-liquid samples types. In this mode, even though the specimen is not liquid, plates can be selected by the instrument and barcoded. Other plating instruments do not have the BSC feature, though laboratories are likely to have a standalone BSC in the laboratory. The attached BSC might be advantageous to those labs that do not have a BSC or the additional space for both plating automation and a BSC. This also places all necessary components for specimen processing in a line which compliments a lean process flow for pre-analytical specimen processing. The InoqulA+ contains a module that will label a slide with patient identifier information and prepares the slide for a Gram stain offline. It is important to note that consumables are needed in the form of pipette tips and magnetic beads for specimen inoculation with the InoqulA+. The bead and tip will be biological waste and must be disposed of accordingly. These costs should be figured into the overall cost analysis if this instrument is introduced into the clinical laboratory.

4.4 PREVI® ISOLA

The Previ® Isola (Figure 10) is bioMerieux's solution for pre-analytical microbiology specimen processing. The instrument is interfaceable, has footprint of 66.5″ (width) × 35.7″ (depth) × 58.7″ (height) and can be loaded with five different types of media at once with a total of 150 plates. The Previ Isola is similar to the Isoplater in that the streak pattern is spiral. Each plate is inoculated using a single-use disposable specimen applicator (comb). The spiral applicator comb is unique in that it mimics

FIGURE 10

PREVI™ Isola pre-analytical specimen processor, bioMerieux.

Image courtesy of bioMerieux.

16 loops streaking simultaneously which covers more surface area on the plate versus conventional quadrant streaking patterns. The applicator can be changed with every plate or can be used for each specimen regardless of the number of plates inoculated. This feature is defined by the end user. One additional feature of the Previ Isola is the ability to streak a bi-plate, in the case of a urine culture, with the comb applicator. The throughput of the instrument is approximately 180 plates per hour streaking the full plate in a spiral configuration.

To date, the Previ Isola does not have the capability of de-capping and re-capping specimens which is a feature present in some of the other plating instrumentation described in this chapter. The impact of the lack of de-capping/capping within the instrument may be varied depending on the size of the laboratory and the number of specimens processed per day. This impact might be minimal in a low- to medium-volume lab but may be more significant in a higher-volume lab where thousands of specimens are being processed throughout the day. Having to move the specimens back and forth between the de-capper/capper and the plating instrument will be an added step adding labour and reducing the overall efficiency. As with other instruments, the Previ Isola can query the LIS and can be programmed to separate plates into individual canisters based on the atmospheric environment the plates will be incubated in. One study compared the Previ Isola to manual methods and demonstrated decreased hands-on time and improved efficiency in the laboratory space with integration of the instrumentation (Mischnik, Mieth, Busch, Hofer, &

Zimmermann, 2012). The same study demonstrated that the Previ Isola reduced the need for re-incubation in regard to obtaining isolated colonies compared to a manual plating method (0.8–1.1% with Previ Isola compared to 5–15% with manual streaking) (Mischnik et al., 2012).

It is important to note that after sometime in 2016, the Previ Isola is likely to be discontinued as a product by bioMerieux. bioMerieux will be partnering with Copan Diagnostics in the future to enhance the TLA solutions offered by Copan Diagnostics using the strengths of both companies to enhance the features of the WASP and WASPLab™ TLA.

4.5 COPAN WASP®

Early on as described above, Copan Diagnostics recognised the value of liquid microbiology and flocked swabs from both a pathogen recovery perspective and from the ability to automate swab specimens using plating instrumentation. Complimentary to the flocked swab, Copan Diagnostics has developed the WASP® (Walk Away Specimen Processor) for the automation of bacteriology plating (Figures 11 and 12; Bourbeau & Swartz, 2009). As described with the other systems, the WASP comes

FIGURE 11

Copan WASP® pre-analytical specimen processor. Front view.

Image courtesy of Copan Diagnostics.

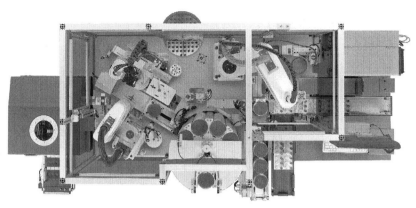

FIGURE 12

Copan WASP® pre-analytical specimen processor. Top view.

Image courtesy of Copan Diagnostics.

equipped with HEPA filtration to mitigate exposure of staff to potential pathogens in the specimen. The footprint of the WASP is 43.5″ (depth) × 81.5″ (width) × 76″ (height) and can accommodate tracking both to load specimens onto the system and to move the plates to a smart incubator which will be discussed below in more detail (WASPLab™). The agar plates are inoculated using custom loops (Table 1) that are available in sizes of 1, 10 and 30 µl. The universal de-capper/capper can accommodate various specimen containers while relying on the LIS to direct the plating protocol. The WASP tool belt can carry five loop devices, and each loop device is comprised of two individual loops, each of which can inoculate approximately 15,000 plates. Two loops allows for faster plating since one loop can be cooling post sterilisation, while the other loop is plating a specimen. A dual streaking tool is also available which comprises two double-loop heads. Therefore, both sides of a bi-plate can be streaked simultaneously. The throughput of the WASP varies depending on the specimen type and plating protocol that is being used. When streaking a bi-plate, for example, 180 plates/h can be inoculated. With the LIS-driven process, the WASP can select the appropriate protocol based on the specimen type. This will drive which loop is automatically selected for plating and which streak pattern is used. Similar to other plating systems, the specimen is vortex mixed, and a camera takes a picture of the loop prior to each inoculation ensuring ample specimen is present on the loop. This is a very important quality control feature of the WASP system.

An additional feature on the WASP is a module which prepares a slide for subsequent Gram staining. The Gram SlidePrep Module is a barcode-driven process as well, which prepares the smear using the liquid sample while automatically labelling the slide with the patient information via permanent inkjet printing, again facilitating traceability with the specimen. Adding to the functionality of the WASP is an additional area within the system called the warehouse carousel. The warehouse carousel (Figure 13) can hold supplies for Kirby–Bauer susceptibility testing in addition to

FIGURE 13

Copan WASP® warehouse carousel.

Image courtesy of Copan Diagnostics.

other disks such as bacitracin or optochin since some laboratories might apply these disks directly to the primary plate to aid in initial organism identification. Lastly, within the warehouse carousel, there is a Broth Inoculation Module which automatically applies a barcode to sterile tubes and can prepare up to four different broth tubes per specimen. The WASP has what is termed the Sort Out Stacker module that automatically sorts inoculated plates into four different categories depending on the incubation protocol of the culture plates. This facilitates placement of the inoculated media into the appropriate incubator without adding steps in sorting the plates. There are limited studies comparing automated plating instruments to manual processes. One study found the WASP to be comparable to manual methods in the clinical laboratory setting (Bourbeau & Swartz, 2009).

Since liquid microbiology is a prerequisite for many plating instruments, Copan Diagnostics has also developed other products to facilitate the plating of specimens such as sputa. Since these specimens are not in a completely fluid form and often not homogenous in nature (for optimal plating), they are incompatible with most plating systems on the market today. Copan Diagnostics has developed the Snot Buster™ which is a sputum-liquefying agent. The Snot Buster consists of dithiothreitol

(DTT) premeasured in instrument (WASP) ready tubes. The laboratorian uses a small plastic disposable sampling tool (Sputum Dipper) to collect approximately 0.5 g or 0.5 ml of sputum which facilitates manual specimen placement into the Snot Buster™. The shaft of the dipper is broken off and the cap placed on the tube (capturing the swab in the cap head). The sputum to DTT is at a 1:1 ratio which is optimal for liquefying the sample. The tube can be labelled and after mixing and standing for 15 min the sample is ready to be placed on the WASP instrument for plating. The liquid sample is inoculated to the designated media as is described above for other liquid samples.

Because mass spectrometry is rapidly being integrated into most laboratories, Copan has recognised the need for accurate inoculum application on the mass spectrometry slide. This can be a tedious task, and in addition, the technologist has to remember which MALDI spot the specific patient organism needs to be applied to, based on the corresponding mass spec slide map set up for patient tracking. Copan has developed a standalone instrument called the C-Tracer™ that can assist the technologist in identifying the precise colony and the location on the mass spec plate where the colony needs to be placed (Figure 14). This is a stand alone piece of equipment separate from the WASP.

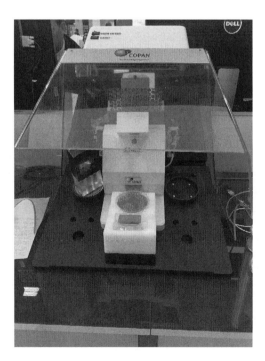

FIGURE 14

Copan Diagnostics MALDI C-Tracer™.

Image courtesy of Copan Diagnostics.

4.6 PreLUDE: i2a

i2a (Montpellier, France) is a French company with its own unique solution for pre-analytical automation for the microbiology laboratory. The pre-analytical specimen processor is called the PreLUD™ (Pre-Analytical Laboratory Universal Device) Figure 15. The PreLUD™ has a capacity of 480 plates with 60 plates in eight individual stacks. The throughput depends on the plating protocol, but the instrument can process up to 120 plates/hour. The PreLUD™ has a footprint of $90'' \times 40'' \times 75''$ and can be placed up against the wall unlike other instruments that need access from behind. When loading the instrument, the plates are loaded upside down to avoid condensation if they are just removed from the refrigerator (other systems the Petri dishes are loaded top-side up). Plates are front loaded into the instrument into the stacking unit. The instrument has the ability to track the specimen automatically with an integrated reader or by a barcode scanner, while the instrument is being loaded. Lot numbers of the media can be tracked based on expiration date of the media. What is unique about the PreLUD™ is that the instrument will process for inoculation a wide variety of tubes, even a swab in a gel or semi-solid medium. This is different than the other processors that need the sample in the form of a liquid. The PreLUD™ therefore has the ability to use the primary swab for direct streaking, without needing a pipette or loop to perform the initial plate inoculation. The instrument contains a vortex for liquid specimen mixing and a camera to take pictures of the inoculum preparation before and after the plating occurs. The system will alert the user if the specimen was not inoculated onto the media. The instrument can handle tubes with or without caps and also has the ability to recognise the sample

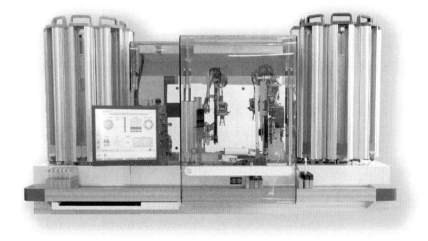

FIGURE 15

i2a PreLUD™, pre-analytical specimen processor.

Image courtesy of i2a.

rack which is barcoded. Information on a rack barcode can be read which will direct the plating protocol.

Another unique feature of the PreLUD™ is that there are three different methods for streaking or inoculating plates; (1) via an inoculating loop, (2) with a pipette and the Trigalski bead and (3) with a primary swab. The Trigalski bead is used to streak the plate after the pipette has released the inoculum. The process is similar to the rolling bead used by the InoqulA+ except the Trigalski bead is maintained by the robot arm. The loop and the bead are sterilised between each plate. Streaking methods are customisable and can be spiral or quadrant. After inoculation, the plates are stacked and stored upside down. To enhance traceability, the plates are labelled with the barcode imprinted directly on the side (or anywhere else) of the Petri dish, which eliminates the expense of labels. Since there are no paper labels being used, this eliminates label jamming that does occur with other systems that utilise paper labels. All printing on the plate can be customisable by the end user which adds to the flexibility of this feature.

Integrated into the instrument is a module that automates the preparation of the Kirby–Bauer disk susceptibility test. The proprietary middleware SIRweb™ linked with the LIS can determine the susceptibility testing set up for a customisable panel of antibiotic disks; each disk is dispensed by the instrument onto the agar plate (Figure 16). The plate can then be placed in the AST incubator/reader SIRscan™ and then automatically read after the appropriate incubation protocol. The instrument comes equipped with HEPA filtration similar to other pre-analytical plating instruments to mitigate any exposure to the staff working with the equipment.

4.7 deltalab AUTOPLAK

The deltalab AUTOPLAK (NTE Healthcare, Barcelona, Spain) (Figure 17) is a newer instrument to the market that is similar to other pre-analytical plating instruments described above. The instrument has a footprint of $72''$ (width) $\times 33''$ (depth) $\times 78''$ (height) and can be interfaced with the LIS. The instrument can process liquid samples and accommodates a wide range of tube sizes. Similar to the other instrumentation described, there is a mechanism to remove and re-cap tubes. Two independent drawers allow continuous loading of samples, and each rack can accommodate 10 specimen tubes. The loading capacity is 120 samples. Inoculation is performed with re-usable loops, and no extra consumables are required. Streak patterns are customisable by the end user, and both solid and liquid media can be inoculated. The instrument can streak 90 mm plates and up to 240 plates can be loaded with the ability to extend to 480 plates. This translates into six media silos, extendable to 12. The plates can be sorted by incubation atmosphere, and on average, 140 plates can be streaked per hour. Inoculated media are labelled by the instrument for traceability, and the user interaction is performed via a touch screen and pop-up keyboard. There is a HEPA filter within the instrument and a station to prepare a smear for a Gram stain that is labelled with the patient information directly on the slide. As this

FIGURE 16

i2a PreLUD Kirby–Bauer susceptibility test preparation module.

Image courtesy of i2a.

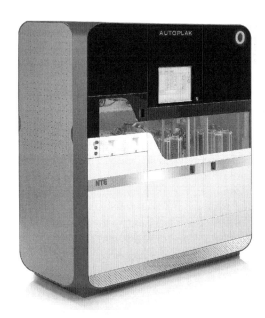

FIGURE 17

AUTOPLAK pre-analytical plating processor, deltalab.

Image courtesy of deltalab.

instrument is introduced into the clinical laboratory market, more information on performance and functional characteristics will be forthcoming.

5 DIGITAL PLATE READING

The manual reading and workup of bacterial cultures is a process that all clinical microbiologists are familiar with. To date, there is little in the literature regarding DPR as it relates to microbiology (Rhoads, Novak, & Pantanowitz, 2015). With the advent of the "smart incubator" as part of the TLA suite of instruments, DPR can now become a reality. The smart incubator (a term used generically to describe an incubator with a camera) consists of an incubator, and a sophisticated digital camera juxtaposed to the instrument that can take an image of the growth on an agar Petri dish over the course of a defined incubation period for a specimen or specimen type. Several manufacturers are incorporating this technology into their TLA systems. The digital camera works collectively with the middleware or software solution, depending on the vendor, so that images can be presented to the technologist for reading (Figures 18–20). Figures 18–20 are an example of COPAN screen displays showing how plates are visualised using the Synapse Pro (Copan Diagnostics middleware software). As mentioned previously, bioMerieux and Copan will be partnering collectively to enhance the middleware system for the Copan WASPLab. Gone are the days where the technologist handles the plates directly, digital microbiology allows for culture workup by observing the plates and the associated bacterial growth on the computer screen. Digital cameras are part of the newer incubator systems, and user-defined protocols can be set as to when the individual plates should be observed and at what intervals. DPR allows the plates to be kept under the appropriate incubation conditions compared to the process currently where technologists take out stacks of plates to be read and left on the bench top. This decreases the growth time associated with the culture because plates are held under the appropriate conditions at all times.

Incubators that are available in combination with TLA consist of a compartment for each individual plate, a robotic mechanism to move the plates in an out of the incubator or to the camera for image analysis and a camera. The sophistication of the camera is increasing with a range of 9–27 megapixels present today in the variety of systems on the market. Various images are presented to the laboratory technologist using different lighting conditions. Backlighting can be used to enhance haemolysis or tangential lighting to enhance colony texture. Composite images are used within DPR to render the best image possible for presentation to the technologist. Composite images consist of several images within one image. The time required to photograph an individual plate is seconds, but protocols will need to be developed so that imaging of all plates within an incubator can occur within a defined period of time. Depending on the types of cultures, images can be taken (and stored) more frequently (e.g. sterile site cultures), or less frequently (Chromagar that requires a defined incubation time prior to reading). These individual configurations for plate reading can be incorporated into the software. These user-defined capabilities allow

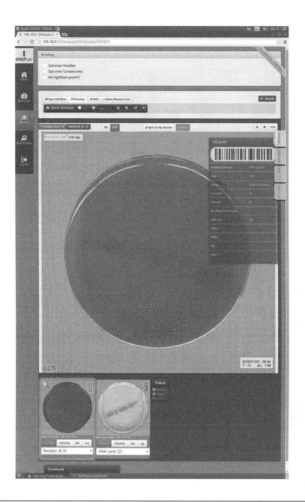

FIGURE 20

COPAN WASPLab middleware screen shot shows an example of digital plate image.

Image courtesy of Copan Diagnostics.

As mentioned previously, DPR and archiving routine cultures can allow laboratories to manage quality assurance issues that arise. There are instances where an organism is missed on culture workup or misidentified, and in these cases, the laboratory can go back to the archived picture of the plate to determine where the problem may have arisen. From a competency or educational perspective, pictures of unique or rare organisms could be saved to the server for access at a future time. Sharing unique results and organisms isolated in the clinical laboratory is a powerful way to train and keep technologists competent at the bench level. Libraries of unique and rare or hard to identify organisms can be used for plate rounds and case studies. There are advantages and perhaps challenges associated with DPR (Novak & Marlowe, 2013) since the technology is very new to the bacteriology space. Studies are lacking due to the infancy of DPR within microbiology, and there are no professional guidelines for implementation or quality management to date.

6 TLA SYSTEMS

There are many drivers towards automation in the clinical microbiology laboratory as there are in other clinical laboratory testing areas. Automation can result in increased efficiency and personnel cost savings, can reduce repetitive motion injuries and even mitigate pathogen exposure in the workplace in the case of the smart incubators described above. Automation allows for positive patient identification since the barcode on the initial patient specimen is read prior to plating, and then all of the subsequent plates that are inoculated are tied to the original patient barcode. Quality and reproducibility can be enhanced by the automation as well. Many of the above preanalytical systems mentioned above are meant to be part of a suite of instrumentation called TLA for microbiology. To refresh, these systems can include a number of modular components, but the full suite would consist of a plating system, a track leading from the plating instrument to the "smart incubator" which contains the digital camera for the imaging of all culture and Kirby–Bauer susceptibility plates. Some systems come complete with track systems emanating from the smart incubators and are connected to ergonomic workstations where the technologist will reside to analyse the cultures. The incubators can either be configured as an O_2 or CO_2 incubator no different than today. When a specimen is inoculated, the system delivers the plate directly to a defined spot within the incubator via a track system. When a plate is called up by a technologist, retrieval of the plate can usually occur within less than a minute. This is accomplished via sophisticated software that tracks the location of the plated media and a robotic arm that will manipulate the plates inside the incubator cabinet.

Depending on the size of a laboratory, the complete needs for a full TLA system will vary. Some laboratories might need more than one complete system due to the high volume nature of their laboratory. The reader is encouraged to reach out to the manufacturer of all the systems described in this chapter since the instruments and functionality are changing at a very fast rate. Middleware systems are constantly being adapted to provide more information and functionality to enhance what TLA can offer the clinical microbiology laboratory.

Given the flexibility of digital microbiology, technologists would in theory not have to be physically present within the lab to work up specimens. Future technologists might be working up cultures from a reading room, away from the busy, active laboratory, so they could concentrate on reading the cultures. As TLA is adopted in the microbiology laboratory, we will see the unique ways the systems can be used to enhance the care that we provide the patient.

6.1 BD KIESTRA

The BD Kiestra TLA offers total microbiology lab automation that maximises throughput and staff efficiency to reduce TAT and improve workflow in the microbiology laboratory (Figure 21). Modular and flexible with integrated workbenches,

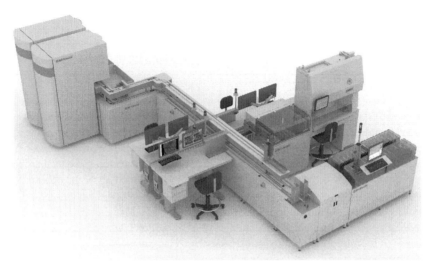

FIGURE 21

BD/Kiestra Total Laboratory Automation. Consists of InoculA plating system, track system, BSC and incubator with digital camera.

Image courtesy of BD.

the TLA can be configured to meet the unique space and capacity requirements in each laboratory. This system includes automated specimen processing, plate transportation, incubation and digital imaging systems as well as integrated work benches designed to improve workflow and the efficiency of the laboratory technologist. The modules that make up the Kiestra TLA consist of the module for the loading and sorting of the plates (SorterA), barcoding (BarcodeA), and as discussed above the InoqulA+ for automated plating of liquid specimens.

The BD Kiestra incubation and digital imaging system is the ReadA Compact (Figures 22 and 23). This second generation system has a capacity of 1150 plates and requires a floor space of about four square feet. The unit can operate in either O_2 or CO_2 conditions and is designed to optimise the growth conditions with a stable and measured temperature and humidity control. The track-based solutions of the BD Kiestra WCA work cell automation (WCA) and TLA Systems allow for plates to move in and out of the ReadA Compact incubator based on request or protocol. Each ReadA Compact includes four output stackers. In the WCA configuration, these stackers are used to facilitate follow-up work, waste handling and request for visual inspection of plates. The TLA configuration will present plates to the user at the work bench (ErgonmicA) needed for follow-up inspection or work. The ReadA Compact automatically takes plates into the incubation section and places them in designated racks. Each plate will have a unique location enabling direct access, ensuring fast retrieval and short delivery times. When a plate is requested by a user or finished its protocol and is assigned to go to the waste, the system will move that plate to

FIGURE 22

BD Kiestra ReadA Compact Incubator.

Image courtesy of BD.

FIGURE 23

BD Kiestra ReadA Compact Incubator—inside close up view.

Image courtesy of BD.

the output lane. The ReadA Compact has the unique capability of parallel processing plates that come into the incubator, plates that go out of the incubator and plates that need to be imaged, due to the three different plate tracks/lanes designed for this purpose. The incubator uses a laboratory-defined protocol to image plates at set intervals allowing for the monitoring of growth. The HD industrial camera creates a realistic image of the plates using different light sources to create multiple image scenarios. The system is capable of imaging 300 plates per hour (approximately 4 h to image a full incubator). Once all plates of a sample have been imaged, the sample will be flagged as ready for reading and updated on the BD Kiestra dashboard and in the ReadA Browser work list.

The ReadA Browser software enables plate reading and follow-up work selection. The software will indicate plates that are ready for reading. The user will access these files and imaged plates. Users can see and compare growth over time and have a complete patient overview over multiple samples. Per plate the software can show various scenarios and light conditions. When growth is detected and follow-up work is required, the user will mark a colony and select a specific follow-up task. This sets up a follow-up action in the system which is again visible on the dashboard.

One study showed that for pathogen detection, the Kiestra TLA combined with mass spectrometry resulted in approximately 30 h time gained per isolate compared to conventional methods used in this laboratory (Mutters et al., 2014). Most TLA systems are equipped with a "dashboard" (which is a large display screen) that can be positioned in the laboratory which shows the cultures that are ready to be analysed by the technologist. This dashboard allows timely workup of cultures and movement away from reading cultures in a batch mode. This is a paradigm shift from how bacteriology laboratories are operated today where batches of specimens are usually read which consists of a number of cultures. Some of those cultures might not be ready for analysis and culture workup due to the level or amount of growth being insufficient. The images taken by the camera and corresponding middleware display the cultures that are ready for analysis on the dashboard in the laboratory.

6.2 COPAN WASPLab™

The TLA solution for Copan Diagnostics is the WASPLab (Figures 24 and 25). Inoculated media are transferred from the WASP to the WASPLab smart incubator via a track or conveyor system so that the inoculated plate can be incubated and imaged with the digital camera. The WASPLab Image Acquisition technology uses sophisticated lighting associated with a camera. The camera has a sensor that acquires the image as the plate sweeps laterally beneath the camera. The WASPLab camera produces a 27 megapixel image. The camera and WASPLab software can detect and differentiate colonies as small as 0.1 m in diameter. The WASPLab Image Acquisition technology uses a variety of different types of lighting which varies depending on the media colour or opacity. Top light with background can be used for transparent agar simulating viewing the plate at the bench. Top light without background can be used for opaque agar simulating viewing the plate at the bench. Bottom light can

FIGURE 24

WASPLab™. Consists of WASP® plating system, track and incubator with digital camera.

Image courtesy of Copan Diagnostics.

FIGURE 25

WASPLab™ smart incubator, inside close up view.

Image courtesy of Copan Diagnostics.

be used without background to see haemolysis. The WASPLab features telecentric camera optics and software which aids in visualising three-dimensional objects. A telecentric lens uses constant magnification so that the image on the screen is not distorted to the user. The depth of field on the camera is 9 mm which ensures that all colonies (regardless of height, shape, colour or texture) are captured for visualisation. As with most digital imagers and associated software, an image of the plate will be taken at "time 0", so all subsequent images can be compared back to the original plate. WASPLab unique discriminative image analysis software uses the plate image taken at time 0 and compares it to the images taken after incubation. WASPLab software is able to discriminate artifacts present on the plate at time 0, focus on the growth and even recognise small colonies. The software then groups the plates according to the estimated number of colonies. The system then sorts the plates from the most estimated colonies to the least estimated colonies and presents it to the technologist for interpretation and analysis. The technologist can then decide which plates represent significant growth and choose to work those up first. This new technology helps speed up the workup of positive cultures, by presenting them to the technologist first, leaving no growth cultures for last. In the case of no growth cultures, the technologist, after reviewing the plates, can batch result them in groups.

The WASPLab solution is flexible in that large specimen managers can be used for higher-volume laboratories that process thousands of specimens per day and can be used to feed samples into the WASP system. These specimen manager type systems are manufactured by other automation companies such as Inpeco (Lugano, Switzerland) and are being configured to work with WASPLab systems being placed today. As shown Figure 18, the technologist will analyse and work up the culture by visualising the images on the computer screen. If the technologist needs to identify a colony on the pate, the plate can be called up and sent to the colony picking station (in development) which will inoculate the mass spectrometry template and prepare a McFarland standard for susceptibility testing. If plates need to be manually visualised or handled by the technologist, the plates can be called from the incubator into silos along the track system for manual retrieval.

6.3 i2a: ECITALS™

i2a offers a TLA solution for the clinical laboratory illustrated in Figure 26. The suite of instrument is called RECITALS™ and contains a number of modular instruments that are part of the fully automated solution. RECITALS is defined as number of analysers that can be used as standalone pieces of equipment or with a conveyor track system depending on the needs and size of the laboratory. As mentioned above, the PreLUD™ system is the pre-analytical analyser including automatic culture plates streaking, as well as automatic preparation of AST (Mueller Hinton agar plate streaking and antibiotic disks dispensing). The MAESTRO™ is the smart incubator–reader module (with a throughput of up to 240 plates/h) where the digital plating reading station is housed and all plate reading occurs. MAESTRO™ includes incubation of culture plates and Mueller Hinton plates for AST under different conditions

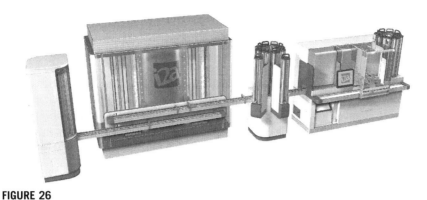

FIGURE 26

i2a Recitals, Total Laboratory Automation.

Image courtesy of i2a.

(atmosphere, temperature and hygrometry), automatic reading of plates, results interpretation that are customisable (with interpretation rules set by the user) and sorting. MAESTRO includes integrated software, a microbiology safety cabinet and a remote validation workstation. The SIRSCAN™ is a unique solution for automatic incubation/reading of AST (disk diffusion on round or square plates that holds up to 16 antibiotic disks each) that allows for rapid results for many microorganisms (5–7 h when used with i2a's Rapid Mueller Hinton media). The middleware is called SIRWEB™ which is a proprietary middleware that enables the complete management and centralisation in one single database of all bacteriology results. This is an open system which can be bi-directionally interfaced with other analysers as well as having full compatibility with various LIS and HIS. The system has an expert system for customised results according to CLSI guidelines. There is a data management component as well which contains features for generating epidemiology reports, detecting and following multi-resistant bacteria and managing hospital-associated infections. This system is relatively new to the market with no placements in the United States to date but the manufacturer is partnering with one of the mass spectrometry vendors to integrate that technology into the TLA solution as well.

7 CHANGE MANAGEMENT: A HOLISTIC APPROACH TO AUTOMATION IN BACTERIOLOGY

The landscape within the microbiology laboratory is changing due to automation. Automation will standardise processes, improve quality and TAT and reduce workplace injury due to repetitive motion associated with tasks that are currently performed manually. The anticipated gain from a clinical perspective is that inoculated media can be imaged based on custom protocols, and software programs can alert the end user to when the cultures are ready to be analysed. This will reduce TAT to result for many cultures. As most of the microbiology laboratory are aware, automating culture workup is new and will require adoption from the staff. This change management is important to recognises and includes not only training staff members but also includes a robust vendor/laboratory partnership that can help the laboratory make the most of the new automation. Technologists in the clinical laboratory have been used to working up cultures using conventional methods for years and need to be given the support to understand and adopt the newer automation at a pace that is comfortable for them. Technicians in reality might be able to process specimens in a more timely fashion than the pre-analytical automation can, but it is important to recognise that automating the plating function will allow those staff members to be freed up to perform other important laboratory tasks, with subsequent efficiency gain in the laboratory overall.

As mentioned at the beginning of this chapter, automation is here but a one size fits all solution does not exist for all laboratories. There will be a significant investment in workflow analysis with the vendor to understand each laboratories individual needs, given the diversity of available automation. Each facility will have to investigate space constraints, and monetary investment may be necessary for remodelling, in addition to the dollars needed for capital equipment expenditure. In today's competitive market, a business case is often necessary to engage upper administration about new technology and the placement of that new technology in the laboratory setting. The reader can refer to other published articles that illustrate how to formulate a robust business case for TLA (Novak & Marlowe, 2013).

REFERENCES

American Academy of Microbiology (2008). *Clinical microbiology in the 21st century—Keeping the pace.* Washington, D.C.: American Academy of Microbiology.

Barenfanger, J., Drake, C., & Kacich, G. (1999). Clinical and financial benefits of rapid bacterial identification and antimicrobial susceptibility testing. *Journal of Clinical Microbiology, 37*(5), 1415–1418.

Bourbeau, P. P., & Ledeboer, N. A. (2013). Automation in clinical microbiology. *Journal of Clinical Microbiology, 51*(6), 1658–1665.

Bourbeau, P. P., & Swartz, B. L. (2009). First evaluation of the WASP, a new automated microbiology plating instrument. *Journal of Clinical Microbiology, 47*(4), 1101–1106.

Buchan, B. W., & Ledeboer, N. A. (2014). Emerging technologies for the clinical microbiology laboratory. *Clinical Microbiology Reviews, 27*(4), 783–822.

Buchan, B. W., Olson, W. J., Mackey, T. L., & Ledeboer, N. A. (2014). Clinical evaluation of the walk-away specimen processor and ESwab for recovery of *Streptococcus agalactiae* isolates in prenatal screening specimens. *Journal of Clinical Microbiology, 52*(6), 2166–2168.

Burnham, C. A., Dunne, W. M., Jr., Greub, G., Novak, S. M., & Patel, R. (2013). Automation in the clinical microbiology laboratory. *Clinical Chemistry, 59*(12), 1696–1702.

Centers for Disease Control and Prevention (2003). Laboratory exposure to *Burkholderia pseudomallei*—Los Angeles, CA, 2003. In *Morbidity and mortality weekly report: Vol. 53*, (pp. 989–990). Department of Health and Human Services (October 29, 2004 edition).

Ellison, J., & Jensen, K. (2011). Six ways to manage lab ergonomics more efficiently. *Professional Safety,* 76–78.

Garcia, E., Ali, A., & Choudhry, S. (2013). The American Society for Clinical Pathology's 2012 vacancy survey of clinical laboratories in the United States. *Laboratory Medicine*, *44*, e1–e18.

Ledeboer, N. A., & Dallas, S. D. (2014). The automated clinical microbiology laboratory: Fact or fantasy? *Journal of Clinical Microbiology*, *52*(9), 3140–3146.

Mischnik, A., Mieth, M., Busch, C. J., Hofer, S., & Zimmermann, S. (2012). First evaluation of automated specimen inoculation for wound swab samples by use of the Previ Isola system compared to manual inoculation in a routine laboratory: Finding a cost-effective and accurate approach. *Journal of Clinical Microbiology*, *50*(8), 2732–2736.

Mutters, N. T., Hodiamont, C. J., de Jong, M. D., Overmeijer, H. P., van den Boogaard, M., & Visser, C. E. (2014). Performance of Kiestra total laboratory automation combined with MS in clinical microbiology practice. *Annals of Laboratory Medicine*, *34*(2), 111–117.

Novak, S. M., & Marlowe, E. M. (2013). Automation in the clinical microbiology laboratory. *Clinics in Laboratory Medicine*, *33*(3), 567–588.

Rhoads, D. D., Novak, S. M., & Pantanowitz, L. (2015). A review of the current state of digital plate reading of cultures in clinical microbiology. *Journal of Pathology Informatics*, *6*, 23.

Sayin-Kutlu, S., Kutlu, M., Ergonul, O., Akalin, S., Guven, T., Demiroglu, Y. Z., et al. (2012). Laboratory-acquired brucellosis in Turkey. *The Journal of Hospital Infection*, *80*(4), 326–330.

Smismans, A., Verhaegen, J., Schuermans, A., & Frans, J. (2009). Evaluation of the Copan ESwab transport system for the detection of methicillin-resistant *Staphylococcus aureus*: A laboratory and clinical study. *Diagnostic Microbiology and Infectious Disease*, *65*(2), 108–111.

Traxler, R. M., Lehman, M. W., Bosserman, E. A., Guerra, M. A., & Smith, T. L. (2013). A literature review of laboratory-acquired brucellosis. *Journal of Clinical Microbiology*, *51*(9), 3055–3062.

Trotman-Grant, A., Raney, T., & Dien Bard, J. (2012). Evaluation of optimal storage temperature, time, and transport medium for detection of group B Streptococcus in StrepB carrot broth. *Journal of Clinical Microbiology*, *50*(7), 2446–2449.

Van Horn, K. G., Audette, C. D., Sebeck, D., & Tucker, K. A. (2008). Comparison of the Copan ESwab system with two Amies agar swab transport systems for maintenance of microorganism viability. *Journal of Clinical Microbiology*, *46*(5), 1655–1658.

Wagar, E. A., Stankovic, A. K., Raab, S., Nakhleh, R. E., & Walsh, M. K. (2008). Specimen labeling errors: A Q-probes analysis of 147 clinical laboratories. *Archives of Pathology & Laboratory Medicine*, *132*(10), 1617–1622.

MALDI-TOF Mass Spectrometry for Microorganism Identification

2

Lori Bourassa, Susan M. Butler-Wu[1]

Department of Laboratory Medicine, University of Washington, Seattle, Washington, USA
[1]Corresponding author: e-mail address: butlerwu@uw.edu

1 INTRODUCTION

The use of mass spectrometry to characterise bacteria was first proposed in 1975 when it was observed that small molecules from lyophilised bacteria produce distinct spectra for different bacterial genera and species (Anhalt & Fenselau, 1975). However, it was not until the advent of so-called soft ionisation techniques such as matrix-assisted laser desorption ionisation time-of-flight mass spectrometry (MALDI-TOF MS) that intact proteins could also be analysed. While initial studies used cellular extraction and purification techniques prior to analysis by MALDI-TOF MS, later studies showed that whole bacterial cells could also be analysed (Carbonnelle et al., 2011; Holland et al., 1996; Lavigne et al., 2013). Since then, there have been an ever-increasing number of studies demonstrating that MALDI-TOF MS produces characteristic and reproducible spectra that can be used to differentiate microorganisms to the genus and species levels (Clark, Kaleta, Arora, & Wolk, 2013; Croxatto, Prod'hom, & Greub, 2012; Dingle & Butler-Wu, 2013). With the development of easy to use, commercially available systems, identification of organisms by MALDI-TOF MS has become the standard of practice in many clinical microbiology laboratories worldwide. While MALDI-TOF MS has also been applied to the detection of antimicrobial resistance and strain typing, these applications are unfortunately beyond the scope of this chapter. The interested reader is referred to the following articles for further information on additional applications of MALDI-TOF MS to clinical microbiology (Clark et al., 2013; DeMarco & Ford, 2013; Hrabak, Chudackova, & Walkova, 2013).

1.1 MALDI-TOF MS: PRINCIPLES AND PROCESSES

The simplest application of MALDI-TOF MS to the identification of microorganisms is the analysis of intact microbial cells without extensive preparation. This is known as 'intact cell' or 'direct cell' analysis and is shown in Figure 1. Testing begins with application of a freshly grown bacterial or yeast colony from a culture

Methods in Microbiology, Volume 42, ISSN 0580-9517, http://dx.doi.org/10.1016/bs.mim.2015.07.003

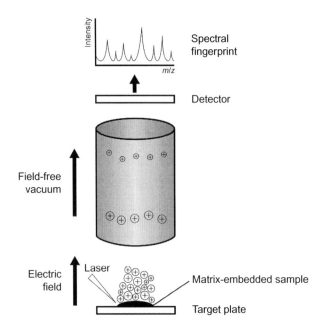

FIGURE 1

Principle of MALDI-TOF MS for microbial identification. Microbial samples are deposited on a conductive target plate/slide and overlaid with a matrix solution composed of an organic acid. Samples are ionised by short laser pulses once placed in the instrument, followed by acceleration of variably charged particles through an electric field. The amount of time it takes for each particle to pass through a field-free tube, or time-of-flight (TOF), is dependent on their mass and charge. This is measured by a detector resulting in the production of an organism-specific spectral fingerprint that is then compared to known microorganism spectral profiles in a database to obtain a reliable identification of the organism.

Figure reproduced from Dingle and Butler-Wu (2013) with permission of the publisher.

plate directly onto a target plate or slide using a wooden stick, toothpick or a sterile loop. Use of swabs should be avoided to prevent build-up of fibres within the mass spectrometer. Between 10^4 and 10^6 colony-forming units (CFU) of the test microorganism is generally required for analysis by MALDI-TOF MS using the intact cell method (Wieser, Schneider, Jung, & Schubert, 2012). The spotted microorganism is then overlaid with a matrix solution composed of soluble, small acid molecules. Commonly used matrix solutions include α-cyano-4-hydroxycinnamic acid (HCCA or CHCA) and 2,5-dihydroxybenzoic acid (DBA). As the matrix dries, the microorganism and the matrix co-crystallise embedding the microorganism into the matrix. Most intact bacteria are lysed from exposure to the water, organic solvent or strong acid in the matrix solution (Buchan & Ledeboer, 2014; Clark et al., 2013; Patel, 2013; Randell, 2014; Wojewoda, 2013).

Not all microorganisms are amenable to analysis by the intact cell method and thus require extraction prior to analysis by MALDI-TOF MS. The requirement for prior extraction may be multifactorial: certain microorganisms may not be completely lysed by the matrix solution, e.g., yeast cells (Croxatto et al., 2012). Secondly, testing of microorganisms such as mycobacteria and filamentous fungi by direct cell analysis also presents a safety hazard. This includes *Mycobacterium tuberculosis*, which is not inactivated by the matrix solution, as well as dimorphic mould species such as *Coccidiodes* and *Histoplasma* species (Centers for Disease Control and Prevention & National Institutes of Health, 2009). The extraction methods described in the literature vary widely, ranging from manufacturer-recommended to custom, user-developed extraction methods. However, the most commonly reported extraction methods for routine bacteria and yeast isolates include ethanol–formic acid tube extraction and on-target extraction (Figure 2). Tube extraction increases the ability of MALDI-TOF MS to identify some microorganisms but is time-consuming and can disrupt clinical microbiology workflow. On-target extraction, performed by applying formic acid directly to the spotted microorganism, also leads to increased identification rates compared to standard intact cell analysis (Clark et al., 2013; Haigh, Degun, Eydmann, Millar, & Wilks, 2011; McElvania Tekippe, Shuey, Winkler, Butler, & Burnham, 2013; Theel et al., 2012). This method has only a minimal impact on clinical workflow, with use of a variety of formic acid concentrations reported (25–100%).

After the matrix and microbial specimen are dry, the target plate or slide is loaded into the ionisation chamber of the MALDI-TOF MS instrument. A nitrogen laser applies short pulses of energy to each sample, leading to sublimation of the matrix and microbial specimen from solid to gas phase with resulting ionisation of the sample. The ionised molecules are accelerated by an electric charge and travel through a vacum tube towards a detector. As they travel, the ionised molecules are separated based on their mass-to-charge ratio (i.e. m/z). The mass analyser of the instrument precisely measures the time required for each ion to reach the detector and records the TOF, generating a mass spectrum (Clark et al., 2013; Patel, 2013). MALDI-TOF MS detects highly abundant proteins, predominately ribosomal proteins and other abundant cytosolic proteins (Holland et al., 1999; Ryzhov & Fenselau, 2001; Sun et al., 2006). Because protein composition differs between different bacterial genera and species, unique spectra are produced (Clark et al., 2013; Dingle & Butler-Wu, 2013). The spectrum must then be compared to a database of reference spectra in order to identify the unknown microorganism. The entire process is extremely rapid, usually taking less than 60 s to complete from the time of laser pulsation to completion of database interrogation.

1.2 COMMERCIALLY AVAILABLE MALDI-TOF MS PLATFORMS

There are currently three commercial MALDI-TOF MS platforms being used for microbial identification: the MALDI Biotyper (Bruker Daltonics), the VITEK MS (bioMérieux) and the Andromas MS (Andromas SAS) systems. The latter system is predominately used in Europe and is not available in the United States. There

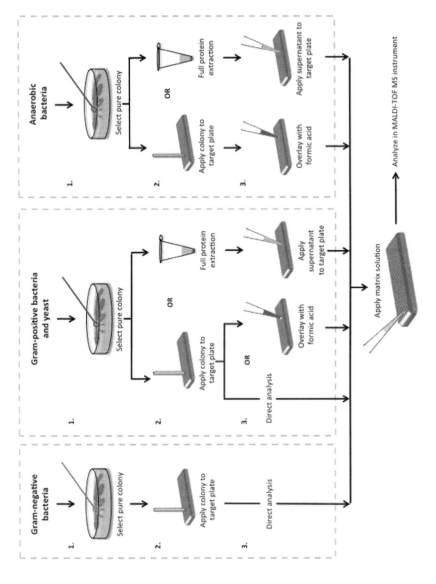

FIGURE 2

MALDI-TOF MS sample preparation methods utilised for routine clinical bacterial and yeast isolates. Most Gram-negative organisms can be reliably identified by 'direct cell' analysis, whereas many Gram-positive bacteria, anaerobic bacteria and yeast may require extraction prior to analysis by MALDI-TOF MS. The requirement for extraction and the method used vary by instrument manufacturer. Published extraction methods for these microorganisms include on-target extraction with formic acid as well as ethanol–formic acid tube extraction. For the latter method, bacteria are resuspended in a mixture of ethanol and distilled water, centrifuged, and the bacterial is pellet resuspended in formic acid and acetonitrile. Following additional centrifugation, the supernatant is then applied to the target plate/slide.

are differences between the three systems with respect to instrumentation, software, identification algorithms and the proprietary reference databases used for identification (summarised in Table 1).

The Bruker MALDI Biotyper system compares the spectra of the test microorganism to a database of reference spectra which are generated by repeated measurements of a particular strain consolidated into a 'main spectral profile' (MSP) (Bruker, 2008). Spectral comparisons are made based on the position, intensity and frequency of the component peaks. The database can also be supplemented with additional MSPs by the user for the Research Use Only (RUO) system (Bruker, 2011). Separate RUO databases are also available from the manufacturer for mycobacteria and filamentous fungi, respectively. The MALDI Biotyper produces log score values ranging from 0.000 to 3.000. A score ≥ 2.0 indicates probable identification to the species level; scores between 1.700 and 1.999 indicate probable genus-level identification and scores <1.700 are not considered reliable for identification (Clark et al., 2013; Patel, 2013). For the MALDI Biotyper CA system, which is approved for *in vitro* diagnostic (IVD) use in the United States, a score value ≤ 1.999 represents no identification. In the United States, the database composition varies between RUO and IVD systems, although this is not the case in Europe. Importantly, some laboratories have validated the use of lower score thresholds for the identification of certain microorganisms. Some published studies documenting performance of the system for certain microorganism groups therefore use alternative score thresholds rather than those specifically recommended by the manufacturer.

As with the MALDI Biotyper, the bioMérieux VITEK MS system is available in both RUO- and IVD-cleared formats (VITEK MS RUO and VITEK MS, respectively). For the former, the SARAMIS database (previously part of the Shimadzu Axima-iD Plus system) is queried using one of two algorithms. While the first algorithm matches spectra, the second converts the spectra into a list of peaks and intensities that are then compared against 'SuperSpectra', derived from at least 15 individual isolates grown under a variety of culture media and growth conditions. The peaks are weighted according to their specificity for organism identification with the output being a confidence value that ranges from 0% to 100%. Scores >90% are indicative of species-level identification (Clark et al., 2013; Kok, Chen, Dwyer, & Iredell, 2013; Patel, 2013). The VITEK MS uses an 'Advanced Spectra Classifier', whereby the spectral peaks of the test microorganism are detected and sorted by mass and intensity into a series of 1300 bins that are weighted according to their importance for identification. Output is colour coded to reflect high (green), medium (yellow) or low (red) confidence, respectively. Entries from both mycobacteria and filamentous fungi are present in the Saramis and in the upcoming IVD v3.0 databases (Mather, Rivera, & Butler-Wu, 2014; Panagea et al., 2015).

The Andromas system uses three different databases (bacteria, yeasts + *Aspergillus* species and mycobacteria) and is predominantly used in Europe. There are multiple spectral profiles corresponding to each microorganism in the database with the comprising spectra generated after growth on different types of

Table 1 Comparison of the Three Commercially Available MALDI-TOF MS Systems

MALDI-TOF MS System	Manufacturer	Instrumentation	Identification Algorithm	Currently Available Databases	Organisms Included in Database	Depth of Spectral Coverage	Regulatory Status
Bruker MALDI Biotyper MS	Bruker Daltonics, Billerica, MA	Desktop system	Algorithm: Spectral comparisons between the unknown organism and reference spectra	CA System	Aerobic and anaerobic bacteria, yeast	Spectra for >280 species	FDA approved
				Biotyper—RUO	Gram-positive and Gram-negative bacterial spectra (non-mycobacterial) anaerobes and yeast	Spectra for 2200 species of >300 genera	Not FDA approved
			Output: Log score values ranging from 0.000 to 3.000	Mycobacteria Library v2.0-RUO	Mycobacteria	131 species	Not FDA approved
				Filamentous Fungi Library 1.0—RUO	Fungi	124 species, 30 genera	Not FDA approved
				Security-Relevant Database	Select agents		Not FDA approved
VITEK MS	bioMérieux, Durham, NC	Floor model	Algorithm: Advanced Spectral Classifier	VITEK MS	Aerobic and anaerobic bacteria, yeast	193 species	FDA approved
			Output: Confidence value that ranges from 0% to 100%.	SARAMIS v4.12 database—RUO	Gram-positive and Gram-negative bacterial spectra, anaerobes, yeast, mycobacteria, some select agents and moulds	>1300 species	Not FDA approved
Andromas	Andromas SAS, Paris, France	Floor or desktop system	Algorithm: Spectra from unknown organism queried against database	Bacteria			Not FDA approved
				Yeast, *Aspergillus* species			Not FDA approved
			Output: Conserved species-specific peaks. Minimum relative intensity must be above threshold	Mycobacteria			Not FDA approved

RUO, research use only.

culture media without extraction. After query of the appropriate database with spectra from an unknown microorganism, a percent similarity between the reference spectra in the database and the unknown organism is produced. Results are then interpreted as a 'good identification', 'identification to be confirmed' or 'no identification' based on manufacturer-defined thresholds (Bille et al., 2012; Kok et al., 2013).

1.3 FACTORS INFLUENCING THE PERFORMANCE OF MALDI-TOF MS

As mentioned previously, MALDI-TOF MS predominately detects ribosomal proteins as well as other abundant cytosolic proteins (Holland et al., 1999; Ryzhov & Fenselau, 2001; Sun et al., 2006). Because bacterial ribosomal proteins are ubiquitously expressed throughout the growth cycle, the spectral patterns produced by analysis of these proteins are not generally affected by incubation temperature or atmosphere (Clark et al., 2013; Jarman et al., 1999; McElvania Tekippe et al., 2013). Similarly, extended exposure to oxygen does not appear to significantly impact identification rates for most anaerobic bacteria (Hsu & Burnham, 2014; Veloo, Elgersma, Friedrich, Nagy, & van Winkelhoff, 2014). However, one study reported decreased rates of identification of non-fermenting/fastidious Gram-negative bacilli when isolates were incubated at room temperature rather than at 37 °C (Ford & Burnham, 2013).

MALDI-TOF MS performance can, however, be influenced by culture media composition. In particular, differences in spectral profiles can be observed with specific culture media (Anderson et al., 2012; Walker, Fox, Edwards-Jones, & Gordon, 2002). Growth on certain selective media, such as Hektoen Enteric and MacConkey Lactose agars, can negatively impact the rates of identification although misidentifications are generally not observed. The reduced rates of bacterial identification can be overcome by performing microorganism extraction prior to MALDI-TOF MS analysis (Anderson et al., 2012; Bizzini, Durussel, Bille, Greub, & Prod'hom, 2010; Ford & Burnham, 2013; Saffert et al., 2011). Other pre-analytical variables that can affect performance include a mucoid phenotype (e.g. that expressed by *Pseudomonas aeruginosa*), the density of organism spotted (heavy vs. light spotting), selection of an impure 'colony' (i.e. more than one bacterial species present in the spot) and frequent subculture (Anderson et al., 2012; Ferreira, Sanchez-Juanes, et al., 2010; Ford & Burnham, 2013; Haigh et al., 2011; McElvania Tekippe et al., 2013; Pinto et al., 2011; Sogawa et al., 2011). Suboptimal performance can also be observed when the extraction method used to test microorganisms of interest differs from that used to generate the database, particularly in the case of filamentous fungi (Lau, Drake, Calhoun, Henderson, & Zelazny, 2013).

As identification of microorganisms by MALDI-TOF MS depends on comparison to a reference database, identification is dependent on the depth and breadth of microorganism coverage in the database. Failure of MALDI-TOF MS to identify a microbial isolate is most commonly due to the absence of reference spectra corresponding to that species, or poor representation of a particular species, within the database (Bizzini et al., 2011; Croxatto et al., 2012; Fournier et al., 2013). Similarly, MALDI-TOF MS may be unable to adequately distinguish the spectra of certain closely related bacterial species leading to species misidentification (e.g. *Shigella* species vs. *Escherichia coli*) (Croxatto et al., 2012; Patel, 2013; Wieser et al., 2012).

As a result of poor spectral representation for certain microorganism groups or species in commercially available databases, some laboratories have chosen to supplement existing databases with spectra from additional microorganisms of interest. Indeed, supplementation of early database versions with spectra from additional clinical isolates improved the performance of MALDI-TOF MS (Sogawa et al., 2012). There are a large number of publications reporting the performance of these supplemented or so-called custom databases. While these publications show the ability of MALDI-TOF MS to identify microorganisms specifically targeted by the custom database of interest, laboratories should be aware that these results may differ from those obtained with the manufacturer-provided databases. Similarly, use of different manufacturer database versions varies widely between studies (e.g. RUO vs. IVD). Moreover, manufacturers regularly update databases, making it challenging to compare inter-study performance even within the same system. Nevertheless, MALDI-TOF MS is highly reproducible when performed across multiple laboratories using the same MS system with the same identification database (Westblade et al., 2015).

2 MICROBIAL IDENTIFICATION BY MALDI-TOF MS

Clinical microbiology laboratories require accurate and reliable methods to identify clinically significant microorganisms. Conventional microbiological methods rely on isolating potential pathogens in culture prior to microorganism identification. The same requirement is also true for MALDI-TOF MS, although the method has also been applied directly to select clinical samples (discussed later in this chapter). However, many conventional microbiological identification systems (e.g. Phoenix, Vitek 2, Microscan, etc.) require additional growth of the microorganism *in vitro* following isolation in culture. MALDI-TOF MS does not have this requirement. Because only 10^4–10^6 CFU are required for testing, direct identification from the primary culture plate is therefore also possible without the need to perform subculture. Furthermore, not all microorganisms can be reliably identified using traditional biochemical-based methods and may require molecular-based methods for identification (e.g. 16S rRNA sequencing). Molecular methods are generally expensive to perform, require specialised technical expertise and may not be available in all clinical microbiology laboratories (Clark et al., 2013; Dingle & Butler-Wu, 2013; Wojewoda, 2013).

2.1 COMPARISON OF MALDI-TOF MS AND CONVENTIONAL METHODS FOR THE IDENTIFICATION OF ROUTINE ISOLATES

There is a relatively large body of the literature comparing the performance of MALDI-TOF MS to conventional methods for identification of routine clinical isolates of bacteria and yeast. In a landmark study by Seng and colleagues, MALDI-TOF identified 84.1% of prospectively encountered routine bacterial isolates to the species level (Seng et al., 2009). Since then, a number of additional prospective evaluation studies have been published (Benagli, Rossi, Dolina, Tonolla, & Petrini, 2011; Bille et al., 2012; Bizzini et al., 2010; Carbonnelle et al., 2012; Cherkaoui et al., 2010; Deak et al., 2015; Dubois et al., 2012; Eigner et al., 2009; El-Bouri et al., 2012; Jamal, Albert, & Rotimi, 2014; Martiny, Busson, et al., 2012; Neville et al., 2011; van Veen, Claas, & Kuijper, 2010). Of those studies using manufacturer-recommended identification thresholds, correct species-level identification rates varied between 84.5% and 97.3%, with rates of 86.7% and 99.2% observed among studies using alternative interpretive thresholds. While these studies generally include a large number of isolates (often ≥1000), it is important to note that the performance of MALDI-TOF MS for the identification of yeast, anaerobes and less-frequently encountered bacteria is not necessarily robustly evaluated. However, the prospective nature of these studies provides important information regarding the overall expected performance of MALDI-TOF MS in a clinical microbiology laboratory.

A number of prospective evaluation studies have directly compared the performance of MALDI-TOF MS to conventional biochemical identification methods. Overall, these show that MALDI-TOF MS performs equivalently to or even outperforms conventional identification methods for most routine isolates of bacteria and yeast (Bizzini et al., 2010; Carbonnelle et al., 2012; Cherkaoui et al., 2010; El-Bouri et al., 2012; Martiny, Busson, et al., 2012; Seng et al., 2009; Sogawa et al., 2011; van Veen et al., 2010). When discrepancies between MALDI-TOF MS and conventional identification methods occur, 16S rDNA sequencing tends to favour the MALDI-TOF MS identification in the majority of cases (Benagli et al., 2011).

Not surprisingly, MALDI-TOF MS has demonstrated clear utility in the identification of bacteria that are difficult to identify using conventional identification methods. In one study, nearly half of the difficult-to-identify bacterial isolates tested (49.8%) that could previously only be identified by 16S rDNA sequencing were correctly identified to the species level by MALDI-TOF MS. Failure of MALDI-TOF MS to identify a given isolate was due to lack of adequate reference spectra in the database at that time (58.6%) or failure to obtain an adequate spectrum from the isolate (41.4%). While only half of the isolates were identified, results were available 48 h earlier and at a fraction of the cost of sequencing (Bizzini et al., 2011). More recent studies have also reported similar findings (McElvania TeKippe & Burnham, 2014; Rodriguez-Sanchez et al., 2014).

2.2 COMPARISON OF COMMERCIALLY AVAILABLE MALDI-TOF MS SYSTEMS

The majority of studies published to date have investigated the performance of the MALDI Biotyper system using either proprietary manufacturer- or user-built custom databases. There are also now an increasing number of publications documenting the performance of the VITEK MS system. Nevertheless, only a handful of studies have directly compared the performance of both systems for routine bacterial isolates. Initial studies showed slightly better performance for the MALDI Biotyper compared with the VITEK MS (or the preceding Shimadzu system) with identification rates of 94.4–95.3% and 88.8–93.4%, respectively. Despite using identification thresholds lower than those recommended by the manufacturers, low misidentification rates were observed for both systems (<1%) (Carbonnelle et al., 2012; Cherkaoui et al., 2010).

More recently, the overall rates of identification have been shown to be similar for routine isolates using the manufacturer-recommended identification thresholds (92.7–93.9% and 93.2–93.7%, respectively) (Deak et al., 2015; Martiny, Busson, et al., 2012). Lower identification rates were observed for anaerobes (61.3% and 75.3%, respectively). Interestingly, there appear to be differences between the systems with regard to which anaerobic species are challenging to identify; *Bacteroides* and *Fusobacterium* species were problematic for the MALDI Biotyper and VITEK MS, respectively. Identical identification rates (85.1%) between both systems were observed for bacteria considered to be either unusual of difficult to identify by conventional methods (McElvania TeKippe & Burnham, 2014).

2.3 LABORATORY COST SAVINGS ASSOCIATED WITH USING MALDI-TOF MS FOR ISOLATE IDENTIFICATION

Despite the relatively high instrumentation costs associated with MALDI-TOF MS (approximately USD $200,000 at the time of writing), implementation of MALDI-TOF MS for routine isolate identification decreases the number of consumables required for identification, often replacing conventional identification systems. Studies have estimated that reagents cost only USD $0.35–0.79 per test (Dhiman, Hall, Wohlfiel, Buckwalter, & Wengenack, 2011; Tan et al., 2012). The reduced reagent costs and reduced 'hands-on' technician time associated with MALDI-TOF MS can lead to an overall cost savings for the laboratory. Early studies estimated cost savings of 68–78% over conventional methods (Seng et al., 2009) with a per-test cost savings of approximately GB £2.00 (El-Bouri et al., 2012). More recently, one laboratory reported a 59.3% reduction in costs over the course of 12 months for a savings of USD $102,413 for almost 48,000 isolates identified (Tan et al., 2012). Additionally, implementation of MALDI-TOF MS for yeast identification decreased supply costs by over 80% and decreased turnaround time by 1–5 days, depending on the microorganism (Patel, 2013). Thus, despite the relatively high instrumentation costs, the time to return on investment can be fairly short for many laboratories.

In addition to reduced costs, one of the primary advantages of MALDI-TOF MS over conventional methods is its speed. Where conventional identification systems can take from 5 to 48 h depending on the method used (Seng et al., 2009), an early study estimated a possible reduction in turnaround time by more than 8 h using MALDI-TOF MS (Cherkaoui et al., 2010). A prospective study of MALDI-TOF MS identification in parallel with conventional methods reported a turnaround time to identification of 6–8.5 min per isolate by MALDI-TOF MS, which included time for specimen preparation and analysis. More recently, a prospective study with 952 microbial isolates found a reduction in the average time to identification by 1.45 days using MALDI-TOF MS compared to conventional identification methods. Over 85% of isolates were identified on the first day of analysis compared with only 9.4% identified using conventional methods (Tan et al., 2012). Target plate/ slide design allows for high throughput with the ability to batch analyse microorganisms leading to enhanced efficiency. The actual run time for MALDI-TOF MS depends on the system used, with 45 min required for the Bruker Biotyper system (96 spots) and 15 min (for 16 spots, >1 h for 96 spots) required for the VITEK MS system, respectively (Martiny, Busson, et al., 2012).

Not surprisingly, the reduced turnaround times associated with the use of MALDI-TOF MS for microbial identification can in turn also positively impact patient outcomes. Use of MALDI-TOF MS to identify cultured isolates recovered from positive blood cultures resulted in reduced time to initiation of more appropriate antimicrobial therapy, and reduced mortality and length of intensive care stay compared with identifications performed using conventional methods (Huang et al., 2013). Even greater reductions in the time to the initiation of more directed antimicrobial therapy and mortality are observed when MALDI-TOF is applied directly to the positive blood cultures (Perez et al., 2013, 2014). In all cases, the communication of these identification results with an antimicrobial stewardship practitioner is essential to ensure the benefits of more rapid microorganism identification from patients with bacteraemia are realised.

3 PERFORMANCE OF MALDI-TOF MS FOR THE IDENTIFICATION OF ROUTINE CLINICAL ISOLATES

Numerous studies have demonstrated that MALDI-TOF MS identification of routine aerobic bacterial isolates performs as well as and often exceeds identification using conventional methods. An overview of the performance of MALDI-TOF MS for the identification of common microorganism groupings is discussed in this section (summarised in Table 2).

Table 2 Percent Correct Species- or Genus-Level Identification of the Indicated Organisms Using the Manufacturer-Provided Databases

Organisms Evaluated	Percent Correct Identification by MALDI-TOF MS					
	Andromas		MALDI Biotyper (Bruker)		Vitek MS (bioMérieux)	
Gram-Positive Bacteria	**% Correct ID**	**Reference (s)**	**% Correct ID**	**Reference(s)**	**% Correct ID**	**Reference(s)**
Enterococcus faecalis	99.5%	Bille et al. (2012)	100%	Bizzini et al. (2010), Deak et al. (2015), Fang, Ohlsson, Ullberg, and Ozenci (2012), Schulthess et al. (2013)	97–100%	Deak et al. (2015), Fang et al. (2012), Moon, Lee, Chung, Lee, and Lee (2013), Rychert et al. (2013), Wang et al. (2014)
Enterococcus faecium	96.9%	Bille et al. (2012)	100%	Bizzini et al. (2010), Deak et al. (2015), Fang et al. (2012), Schulthess et al. (2013)	100%	Deak et al. (2015), Fang et al. (2012), Moon et al. (2013), Rychert et al. (2013), Wang et al. (2014)
Staphylococcus aureus	100%	Bille et al. (2012)	99.6–100%	Bizzini et al. (2010), Deak et al. (2015), Schulthess et al. (2013), Szabados, Woloszyn, Richter, Kaase, and Gatermann (2010)	98–100%	Deak et al. (2015), Moon et al. (2013), Rychert et al. (2013), Wang et al. (2014)
Coagulase-negative staphylococci	99.8%	Bille et al. (2012)	87.7–100%	Bizzini et al. (2010), Cherkaoui et al. (2010), Deak et al. (2015), Dubois et al. (2010), Dupont et al. (2010), Loonen, Jansz, Bergland, et al. (2012), Schulthess et al. (2013), Seng et al. (2009)	85–100%	Cherkaoui et al. (2010), Deak et al. (2015), Moon et al. (2013), Rychert et al. (2013), Wang et al. (2014)
Streptococcus agalactiae	98.3%	Bille et al. (2012)	100%	Bizzini et al. (2010), Cherkaoui, Emonet, Fernandez, Schorderet, and Schrenzel (2011), Deak et al. (2015), Schulthess et al. (2013)	100%	Moon et al. (2013), Rychert et al. (2013)
Streptococcus dysgalactiae			25–100%	Bizzini et al. (2010), Cherkaoui et al. (2011), Deak et al. (2015), Schulthess et al. (2013)	51–90.9%	Moon et al. (2013), Rychert et al. (2013)
Mitis group streptococci	98%	Bille et al. (2012)	0–66%	Deak et al. (2015), Neville et al. (2011), Stevenson, Drake, and Murray (2010), van Veen et al. (2010)	86–100%	Moon et al. (2013), Rychert et al. (2013)
Streptococcus pneumoniae	100%	Bille et al. (2012)	86.4–100%	Bizzini et al. (2010), Deak et al. (2015), Schulthess et al. (2013), van Veen et al. (2010)	96–100%	Angeletti et al. (2015), Branda, Markham, Garner, Rychert, and Ferraro (2013), Dubois, Segonds, Prere, Marty, and Oswald (2013), Rychert et al. (2013)
Streptococcus pyogenes	100%	Bille et al. (2012)	89–100%	Bizzini et al. (2010), Cherkaoui et al. (2011), Deak et al. (2015), Schulthess et al. (2013)	96–100%	Moon et al. (2013), Rychert et al. (2013)
Streptococcus bovis/equius group			100%	Angeletti et al. (2015), Romero et al. (2011)	100%	Angeletti et al. (2015), Rychert et al. (2013)
Micrococcus spp.			66.6%[a]	Deak et al. (2015)	94%	Rychert et al. (2013)
Abiotrophia defectiva			100%	Alatoom, Cunningham, Ihde, Mandrekar, and Patel (2011)	100%[a]	Moon et al. (2013), Ratcliffe et al. (2013), Rychert et al. (2013)
Aerococcus spp.			95–100%	Alatoom et al. (2011), Deak et al. (2015)	100% (*A. viridans*)	Rychert et al. (2013)
Gemella spp.			100%	Alatoom et al. (2011)	100%[a]	Rychert et al. (2013)

Table 2 Percent Correct Species- or Genus-Level Identification of the Indicated Organisms Using the Manufacturer-Provided Databases—cont'd

Organisms Evaluated	Percent Correct Identification by MALDI-TOF MS					
	Andromas		MALDI Biotyper (Bruker)		Vitek MS (bioMérieux)	
Gram-Positive Bacteria	% Correct ID	Reference(s)	% Correct ID	Reference(s)	% Correct ID	Reference(s)
Actinomyces spp.	98.7%	Farfour et al. (2012)	68.9–100%[b]	Barberis et al. (2014), Schulthess, Bloemberg, Zbinden, Bottger, and Hombach (2014)	75.8%	Garner et al. (2014)
Bacillus spp.	100%	Farfour et al. (2012)	100%[a]	Bizzini et al. (2010)	90–100%	Navas et al. (2014)
Corynebacterium spp.	98.8–100%	Bille et al. (2012), Farfour et al. (2012)	72.0–100%[b]	Barberis et al. (2014), Deak et al. (2015), Konrad et al. (2010), Schulthess, Bloemberg, et al. (2014)	50–100%	Deak et al. (2015), Navas et al. (2014), Rychert et al. (2013)
Lactobacillus spp.	100%	Bille et al. (2012), Farfour et al. (2012)	100%[a,b]	Barberis et al. (2014)		
Listeria monocytogenes	92.6%	Farfour et al. (2012)	88–100%[b]	Barberis et al. (2014), Hsueh et al. (2014)	76%	Rychert et al. (2013)
Nocardia spp.	91.3%	Farfour et al. (2012)	14.9–90.6%[b]	Hsueh et al. (2014), Segawa et al. (2015), Verroken et al. (2010)		
Propionibacterium spp.	100%	Farfour et al. (2012)	66.7–100%[b]	Deak et al. (2015), Schulthess, Bloemberg, et al. (2014)	82.7%	Garner et al. (2014)
Gram-negative bacteria						
Acinetobacter baumannii complex	100%	Bille et al. (2012)	100%	Deak et al. (2015)	100%	Deak et al. (2015), Wang et al. (2014)
Burkholderia cepacia complex	100%[a] (n=2)	Bille et al. (2012)	60–76.9%[a]	Alby, Gilligan, and Miller (2013), Deak et al. (2015), Fehlberg et al. (2013)	50–100%[a]	Alby et al. (2013), Deak et al. (2015), Wang et al. (2014)
Campylobacter spp.	100%	Bille et al. (2012)	67–100%	Bizzini et al. (2010), Deak et al. (2015), He, Li, Lu, Stratton, and Tang (2010)	94–100%	Branda et al. (2014), Deak et al. (2015), Deng et al. (2014)
Citrobacter freundii complex	100%	Bille et al. (2012)	100%	Deak et al. (2015), Saffert et al. (2011)	100%	Deak et al. (2015), Wang et al. (2014)
Escherichia coli	99.4%	Bille et al. (2012)	99–100%	Bizzini et al. (2010), Cherkaoui et al. (2010), Deak et al. (2015), He et al. (2010), Saffert et al. (2011)	95–100%	Deak et al. (2015), Richter et al. (2013), Wang et al. (2014)
Haemophilus influenzae	100%	Bille et al. (2012)	100%	Bizzini et al. (2010), Cherkaoui et al. (2010), Deak et al. (2015)	96–100%	Branda et al. (2014), Deak et al. (2015), Wang et al. (2014)
Klebsiella oxytoca	100%	Bille et al. (2012)	90–100%	Bizzini et al. (2010), Cherkaoui et al. (2010), Deak et al. (2015)	100%	Cherkaoui et al. (2010), Deak et al. (2015), Wang et al. (2014)
Klebsiella pneumoniae	99.5%	Bille et al. (2012)	97–100%	Bizzini et al. (2010), Cherkaoui et al. (2010), Deak et al. (2015), He et al. (2010), Saffert et al. (2011)	87–100%	Cherkaoui et al. (2010), Deak et al. (2015), Wang et al. (2014)
Moraxella catarrhalis	25%	Bille et al. (2012)	100%	Deak et al. (2015)	100%	Branda et al. (2014), Deak et al. (2015)
Neisseria gonorrhoeae			100%[a]	Deak et al. (2015)	90–100%	Branda et al. (2014), Deak et al. (2015), Wang et al. (2014)
Proteus mirabilis	100%	Bille et al. (2012)	100%	Bizzini et al. (2010), Deak et al. (2015), He et al. (2010), Saffert et al. (2011)	100%	Deak et al. (2015), Wang et al. (2014)
Proteus vulgaris	100%	Bille et al. (2012)	85–100%	Bizzini et al. (2010), Deak et al. (2015), He et al. (2010)	0–100%	Deak et al. (2015), Wang et al. (2014)

Continued

Table 2 Percent Correct Species- or Genus-Level Identification of the Indicated Organisms Using the Manufacturer-Provided Databases—cont'd

Organisms Evaluated	Percent Correct Identification by MALDI-TOF MS					
	Andromas		**MALDI Biotyper (Bruker)**		**Vitek MS (bioMérieux)**	
Gram-Positive Bacteria	**% Correct ID**	**Reference (s)**	**% Correct ID**	**Reference(s)**	**% Correct ID**	**Reference(s)**
Pseudomonas aeruginosa	100%	Bille et al. (2012)	99–100%	Bizzini et al. (2010), Cherkaoui et al. (2010), Deak et al. (2015), Marko et al. (2012), Saffert et al. (2011)	100%	Deak et al. (2015), Marko et al. (2012), Wang et al. (2014)
Salmonella group	100%[a]	Bille et al. (2012)	100%	Bizzini et al. (2010), Deak et al. (2015)	100%	Deak et al. (2015), Deng et al. (2014)
Serratia marcescens	100%	Bille et al. (2012)	100%	Bizzini et al. (2010), Deak et al. (2015), He et al. (2010), Saffert et al. (2011)	100%	Deak et al. (2015), Wang et al. (2014)
Shigella spp.			0%	Bizzini et al. (2010), He et al. (2010), Seng et al. (2009)	0%	Deng et al. (2014)
Stenotrophomonas maltophilia	100%	Bille et al. (2012)	75–100%	Deak et al. (2015), Marko et al. (2012), Saffert et al. (2011)	100%	Deak et al. (2015), Marko et al. (2012)
Mycobacteria						
Mycobacterium abscessus	100%	Bille et al. (2012)	42–92%	Buchan, Riebe, Timke, Kostrzewa, and Ledeboer (2014), Chen, Yam, et al. (2013), Mather et al. (2014)	96–100%	Mather et al. (2014)
Mycobacterium avium complex			70–100%	Buchan et al. (2014), Chen, Yam, et al. (2013), Mather et al. (2014)	93–100%	Mather et al. (2014)
Mycobacterium chelonae	100%[a]	Bille et al. (2012)	0–100%[a]	Buchan et al. (2014), Chen, Yam, et al. (2013), Mather et al. (2014)	79–100%	Mather et al. (2014)
Mycobacterium fortuitum complex			83–100%	Chen, Yam, et al. (2013), Mather et al. (2014)	83–100%	Mather et al. (2014)
Mycobacterium gordonae			42–100%[a]	Buchan et al. (2014), Chen, Yam, et al. (2013), Mather et al. (2014)	75–100%	Mather et al. (2014)
Mycobacterium kansasii			29–55%[a]	Chen, Yam, et al. (2013), Mather et al. (2014)	71–86%	Mather et al. (2014)
Mycobacterium marinum			75–100%[a]	Chen, Yam, et al. (2013), Mather et al. (2014)	100%	Mather et al. (2014)
Mycobacterium tuberculosis complex	100%[a]	Bille et al. (2012)	50–100%[a,b]	Balada-Llasat, Kamboj, and Pancholi (2013), Buchan et al. (2014), Chen, Yam, et al. (2013), Mather et al. (2014)	100%	Mather et al. (2014)
Select agents						
Brucella spp.			66.7%[c]	Cunningham and Patel (2013)		
Burkholderia pseudomallei			50.0%[c]	Cunningham and Patel (2013)		
Francisella tularensis			77.8%[c]	Cunningham and Patel (2013)		

Table 2 Percent Correct Species- or Genus-Level Identification of the Indicated Organisms Using the Manufacturer-Provided Databases—cont'd

Organisms Evaluated	Percent Correct Identification by MALDI-TOF MS					
	Andromas		MALDI Biotyper (Bruker)		Vitek MS (bioMérieux)	
Gram-Positive Bacteria	% Correct ID	Reference (s)	% Correct ID	Reference(s)	% Correct ID	Reference(s)
Yeasts						
Candida spp.	96.3–100%	Bille et al. (2012), Lacroix et al. (2014)	50–100%	Bizzini et al. (2010), Chao et al. (2014), Deak et al. (2015), Goyer et al. (2012), Jamal, Ahmad, Khan, and Rotimi (2014), Lacroix et al. (2014), Mancini et al. (2013), Pence, McElvania TeKippe, Wallace, and Burnham (2014)	75–100%	Chao et al. (2014), Deak et al. (2015), Jamal, Ahmad, et al. (2014), Mancini et al. (2013), Pence et al. (2014), Wang et al. (2014), Westblade et al. (2013)
Cryptococcus spp.			85–100%[a]	Chao et al. (2014), Deak et al. (2015), Firacative, Trilles, and Meyer (2012), Mancini et al. (2013), McTaggart et al. (2011)	70–100%[a]	Chao et al. (2014), Deak et al. (2015), Mancini et al. (2013), Wang et al. (2014), Westblade et al. (2013)

Genus-level identification is indicated by spp.
[a]*Some studies analysed a small number of isolates (<5).*
[b]*Some studies used a score cut-off of 1.7 for species-level identification.*
[c]*Bruker's Security-Relevant Library used. See text for performance with custom libraries.*

3.1 IDENTIFICATION OF GRAM-POSITIVE BACTERIA

Multiple studies have shown that MALDI-TOF MS provides accurate species-level identification for the majority of commonly isolated Gram-positive cocci. These include *Staphylococcus aureus* (100%) (Bille et al., 2012; Bizzini et al., 2010; Deak et al., 2015; Rajakaruna et al., 2009; Schulthess et al., 2013; Szabados et al., 2010; Wang et al., 2014), coagulase-negative staphylococci (87.7–100%) (Bille et al., 2012; Bizzini et al., 2010; Cherkaoui et al., 2010; Deak et al., 2015; Dubois et al., 2010; Dupont et al., 2010; Loonen, Jansz, Bergland, et al., 2012; Schulthess et al., 2013; Seng et al., 2009; Wang et al., 2014), enterococci (≥96.9%) (Bille et al., 2012; Bizzini et al., 2010; Deak et al., 2015; Fang et al., 2012; Moon et al., 2013; Schulthess et al., 2013; Wang et al., 2014) and most beta-haemolytic streptococci (Fang et al., 2012; Moon et al., 2013; Schulthess et al., 2013; Wang et al., 2014). Among the beta-haemolytic streptococci, MALDI-TOF MS appears to have difficulty identifying *Streptococcus dysgalactiae* to the species level (25–51% correctly identified) (Rychert et al., 2013; Schulthess et al., 2013). This is believed to result from spectra overlapping with *Streptococcus pyogenes*.

The identification of alpha-haemolytic streptococci by MALDI-TOF MS is problematic for a number of species. Several studies have shown that the Bruker Biotyper frequently misidentifies members of the *Streptococcus mitis* group as *Streptococcus pneumoniae* (Deak et al., 2015; McElvania Tekippe et al., 2013; Neville et al., 2011; Schulthess et al., 2013; Seng et al., 2009; van Veen et al., 2010). This issue appears to be system specific as approximately 96% of *S. pneumoniae* isolates were correctly identified by the VITEK MS system in multiple studies (Branda et al., 2013; Dubois et al., 2013). Correct identification of non-pneumococcal *S. mitis* group isolates to the species level by the VITEK MS system ranged between 88% and 90.8% (Branda et al., 2013; Dubois et al., 2013).

While an early study reported 100% accuracy for the identification of viridans group streptococci isolates using a custom database and data analysis (Friedrichs, Rodloff, Chhatwal, Schellenberger, & Eschrich, 2007), this has not been fully realised using manufacturer-provided databases. However, both the MALDI Biotyper and VITEK MS systems were recently shown to correctly identify viridans streptococci to group levels (Angeletti et al., 2015). Among the *Streptococcus anginosus* group, rates of correct identification for *S. anginosus* and *Streptococcus constellatus* appear to be higher than *Streptococcus intermedius* for both the VITEK MS and Bruker Biotyper (Rychert et al., 2013; Woods, Beighton, & Klein, 2014). It is important to note that the recently identified *Streptococcus tigurinus*, which has been implicated as an important agent of infective endocarditis (Zbinden et al., 2012) is not present in either of the commercially available databases (Isaksson et al., 2015). The ability of MALDI-TOF MS to identify *Streptococcus suis*, which is an important cause of meningitis in Asia, has also not been evaluated (Huong et al., 2014).

The ability of MALDI-TOF MS to accurately identify members of the *Streptococcus bovis* group has also been studied. Within this group, bacteraemia with certain members, in particular *Streptococcus gallolyticus* subsp. *gallolyticus*, appears to be more strongly associated with the presence of colonic malignancy (Boleij, van Gelder, Swinkels, & Tjalsma, 2011; Burnett-Hartman, Newcomb, & Potter, 2008). Even though MALDI-TOF MS is less discriminatory than sequencing of the *sodA* gene (Romero et al., 2011), it does appear to be able to identify these bacteria to the species level (Romero et al., 2011; Rychert et al., 2013). The identification of *S. bovis* group to the subspecies level by the manufacturer-provided databases has not been extensively evaluated, although one study reported misidentification of *S. gallolyticus* subspecies *gallolyticus* as *S. gallolyticus* subspecies *pasteurianus* by the MALDI Biotyper but not the VITEK MS system (Angeletti et al., 2015). Subspecies-specific mass peak biomarkers have been noted, suggesting the potential for improvement using enhanced databases and software (Hinse et al., 2011).

While the majority of studies published to date have focused on staphylococci, streptococci and enterococci, only a handful of studies have addressed the performance of MALDI-TOF MS for the identification of less commonly encountered Gram-positive cocci, e.g., *Abiotrophia* spp., *Aerococcus* spp., and *Gemella* spp. Lower species-level identification rates are observed with these bacteria (56.3–69%) even when extraction is performed (Alatoom et al., 2011; Schulthess et al., 2013). Much of this stems from poor representation in the database as supplementation of the database with additional strains dramatically improved performance (Christensen et al., 2012). More recently, 90% of other genera of Gram-positive cocci tested by the VITEK MS system were correctly identified to the species level albeit with a relatively low number of isolates tested (Rychert et al., 2013). Further studies are required to thoroughly evaluate this technology for these bacteria as most studies to date has been very limited in the number of isolates tested (Ratcliffe et al., 2013).

While many studies have assessed the performance of MALDI-TOF MS for the identification of routinely encountered Gram-positive cocci, relatively few have assessed the performance of MALDI-TOF MS for the identification of Gram-positive bacilli. The ability of MALDI-TOF MS to identify Gram-positive bacilli including *Corynebacterium* spp., *Actinomyces* spp., *Nocardia* spp. and *Listeria* spp. has been analysed. Whilst Gram-positive cocci do not necessarily appear to require extraction prior to analysis; extraction appears to be required for Gram-positive bacilli (Schulthess, Bloemberg, et al., 2014). In one study, species-level identification rates of 36.8% were observed with direct cell analysis compared with approximately 62% for on-target and ethanol-formic acid tube extractions (Schulthess, Bloemberg, et al., 2014). However, there appear to be system-specific differences with respect to this requirement: 98.5% of non-*Listeria* Gram-positive bacilli isolates were correctly identified to the species level by the Andromas MS system with a supplemented reference database and using the direct cell analysis method. This study used 659 isolates including *Corynebacterium* spp., *Listeria* spp. and *Nocardia* spp. (Farfour et al., 2012). The rates of identification of these organisms appear to be substantially lower than those of common Gram-positive cocci using the manufacturer-recommended identification thresholds for the Bruker Biotyper system. Multiple studies have reported improved performance of the Bruker Biotyper system for these organisms by reducing the species-level score threshold from 2.0 to 1.7, which resulted in a 20% increase in the rate of correct species-level identification (Barberis et al., 2014; Schulthess, Bloemberg, et al., 2014). Despite the use of lower identification thresholds, very low rates of species-level identifications were observed for members of the *Kocuria* and *Nocardia* genera (26.7% and 14.9%, respectively). Similarly, *Gordonia* spp. ($n=7$) and *Tsukamurella* spp. ($n=2$) isolates were unable to be identified to the species level (Hsueh et al., 2014). Nevertheless, correct species-level identifications can be obtained by MALDI-TOF MS for organisms that typically require 16S sequencing for identification such as *Nocardia* spp., *Rhodococcus equi*, *Erysipelothrix rhusiopathiae*, *Actinomyces* spp. and *Bacillus* and *Lactobacillus* species. Use of modified extraction methods that include cellular disruption (e.g. bead-beating) as well as the development of custom databases improved the species-level identification rates of *Nocardia* spp. (79–90% correctly identified) (Segawa et al., 2015; Verroken et al., 2010).

3.2 IDENTIFICATION OF GRAM-NEGATIVE BACTERIA

Numerous studies have evaluated the use of MALDI-TOF MS for the identification of Gram-negative bacteria routinely encountered in the clinical microbiology laboratory, outperforming traditional identification methods (e.g. Vitek 2) (Saffert et al.,

2011). High rates of correct identification to the species and/or genus level (>95%) are observed for commonly isolated *Enterobacteriaceae* (e.g. *K. pneumoniae*) using both the MALDI Biotyper and VITEK MS systems (Richter et al., 2013). Species-level identification rates to the species level are somewhat lower due to the current inability of MALDI-TOF MS to identify certain complex members to the species level (e.g. *Enterobacter cloacae* complex, *Citrobacter freundii* complex, etc.). For example, *Enterobacter asburiae*, *Enterobacter hormaechei*, *Enterobacter kobei* and *Enterobacter ludwigii* cannot be delineated from *E. cloacae* (Pavlovic et al., 2012). This inability to accurately resolve complex members may preclude the discovery of novel disease associations by laboratories using this methodology, e.g., *Klebsiella variicola* may be more virulent but is unlikely to be differentiated from *K. pneumoniae* by MALDI-TOF MS currently (Maatallah et al., 2014). However, the molecular gold standard for the identification of bacteria (i.e. 16S rDNA sequencing) is similarly unable to speciate these complex members.

One significant limitation of the identification of *Enterobacteriaceae* by MALDI-TOF MS is the inability of either the MALDI Biotyper or VITEK MS system to reliably discriminate between *E. coli* and *Shigella* spp. (Deng et al., 2014; He et al., 2010; Saffert et al., 2011; Seng et al., 2009). It is possible that MALDI-TOF MS may be able to do so in the future as use of cluster analysis and alternative classification algorithm software showed promise for differentiating these bacteria (Schaumann et al., 2013). However, in the interim, clinical microbiologists must continue to use additional phenotypic characteristics (lactose fermentation, motility) and perform additional tests (indole, LIA, TSI) to distinguish *E. coli* from *Shigella* spp. (Patel, 2013). Similarly, MALDI-TOF MS is thus far unable to differentiate *Salmonella enterica* subspecies *enterica* to the serovar level (i.e. Typhi, Typhimurium, etc.), (Dieckmann & Malorny, 2011) although one study reported being able to differentiate Typhi by VITEK MS (Martiny, Busson, et al., 2012). However, *Salmonella* are accurately identified to the genus level by both systems. Identification to the serovar level may be possible using alternative classification schemes (Dieckmann & Malorny, 2011; Kuhns et al., 2012; Schaumann et al., 2013). With respect to other bacteria isolated from stool cultures, 96.6% of normal flora isolates were correctly identified to the species level by the Bruker MS system (He et al., 2010). *Yersinia* spp. and *Campylobacter* spp. from stool cultures are also correctly identified by MALDI-TOF MS (He et al., 2010). Commonly encountered *Vibrio* species (including *Vibrio vulnificus*) also appear to be accurately identified to the species level (>90%) by MALDI-TOF MS and correctly differentiated from *Aeromonas* spp. (Cheng et al., 2015; Lamy, Kodjo, & Laurent, 2011; Manji et al., 2014; Rychert et al., 2015).

The identification of non-fermenters by MALDI-TOF MS has also been extensively investigated, although many of these studies have used isolates recovered from patients with cystic fibrosis. The use of older bacterial nomenclature led to higher misidentification rates in early studies (Seng et al., 2009). For example, *Stenotrophomonas maltophilia* was misidentified as *Pseudomonas beteli*, *Pseudomonas geniculata* and *Pseudomonas hibisciola*. This issue has since been resolved and more recently published studies have shown high identification rates for these bacteria with both the MALDI Biotyper and VITEK MS systems (Alby et al., 2013; Fernandez-Olmos et al., 2012; Manji et al., 2014; Marko et al., 2012). Both systems show excellent performance of the identification of *P. aeruginosa, S. maltophilia* and most *Acinetobacter* species, among others. However, as with enteric bacteria, MALDI-TOF MS struggles to identify certain very closely related bacterial species. Thus, members of the *Pseudomonas putida* and *Pseudomonas fluorescens* groups cannot be differentiated. Similarly, members of the *Achromobacter* genus cannot be identified to the species level with the exception of *Achromobacter xylosoxidans.* Variable performance between MALDI-TOF MS systems has been observed for *Ochrobactrum anthropi, Elizabethkingia meningoseptica* and the emerging pathogen *Inquilinus limosus* (Almuzara et al., 2015; Jamal, Albert, et al., 2014).

MALDI-TOF MS outperforms traditional identification methods and has lower misidentification rates for the identification of fastidious Gram-negative bacteria (Powell, Blecker-Shelly, Montgomery, & Mortensen, 2013). High identification rates have been observed with both the VITEK MS and MALDI Biotyper systems for these bacteria (Branda et al., 2014). Nevertheless, *Haemophilus haemolyticus* was frequently misidentified as *Haemophilus influenzae* although this issue was abrogated through the use of custom databases (Bruin et al., 2014; Zhu et al., 2013). Similarly, low identification rates of *Aggregatibacter aphrophilus* (35%) and *Cardiobacterium hominis* with the MALDI Biotyper could be overcome using a custom database that included these bacteria (Couturier, Mehinovic, Croft, & Fisher, 2011). *Kingella* spp. appear to be accurately identified to the genus level with both MALDI-TOF MS platforms (Couturier et al., 2011; Powell et al., 2013). Even though *Neisseria* species show high identification rates, laboratories should be aware of the possibility for misidentification of *Neisseria polysaccharea* as *Neisseria meningitidis* by the MALDI Biotyper (Cunningham, Mainella, & Patel, 2014). This is believed to result from a relative over-representation of spectra from *Neisseria gonorrhoeae* and *N. meningitidis* relative to other species in the database.

3.3 IDENTIFICATION OF ANAEROBES

The identification of anaerobic bacteria by conventional phenotypic methods can produce ambiguous and even inaccurate results. As a result, 16S rRNA sequencing is often required to accurately identify anaerobic isolates, increasing the cost and time to identification. Early studies suggested that MALDI-TOF MS was less reliable for the identification of anaerobic bacteria, mainly due to inadequate spectral representation in the databases (Fedorko, Drake, Stock, & Murray, 2012; Fournier et al., 2012; Justesen et al., 2011; Nagy, Becker, Kostrzewa, Barta, & Urban, 2012; Veloo, Knoester, Degener, & Kuijper, 2011). Initial publications comparing the Bruker MALDI Biotyper and Shimadzu MS systems identified 63–67% and 49.0–61% of isolates, respectively, with higher rates of incorrect identifications for the MALDI Biotyper (7.9% vs. 1.4%) (Justesen et al., 2011; Veloo et al., 2011). More recently, similar identification and misidentification rates (<1%) were observed when these systems were directly compared (Jamal, Shahin, & Rotimi, 2013). MALDI-TOF MS has now been demonstrated to be more accurate than phenotypic identification methods (e.g. RapID 32A) (Barba et al., 2014; Chean et al., 2014; Kierzkowska, Majewska, Kuthan, Sawicka-Grzelak, & Mlynarczyk, 2013).

Bacterial extraction, both by on-target and by tube methods, led to improvements in the species-level identification rates for anaerobes (Fedorko et al., 2012; Schmitt, Cunningham, Dailey, Gustafson, & Patel, 2013). However, this effect is mostly due to improved performance with Gram-positive anaerobes (Hsu & Burnham, 2014). As a result, in an early study evaluating the performance of the MALDI Biotyper for *Bacteroides* spp., 97.5% of isolates were correctly identified to the species level (Nagy, Maier, Urban, Terhes, & Kostrzewa, 2009). More recently, a number of large studies have reported higher species-level identification rates than had been reported previously. The VITEK MS (v2.0 IVD) correctly identified 91.2% of 651 anaerobic isolates to the species level (Garner et al., 2014). Lower rates of species-level identification were observed for some anaerobic species in this study including certain *Bacteroides* spp. (*Bacteroides uniformis*) as well as *Fusobacterium nucleatum*. Lower identification rates for *Peptoniphilus* spp. and *Veilonella* spp. were also observed with an older version VITEK MS system (v1.1 IVD) (Lee et al., 2015).

Similarly, 92.5% of 1325 anaerobic isolates tested were correctly identified to the species level using the MALDI Biotyper (Barreau, Pagnier, & La Scola, 2013). Reduced rates of identification were observed for *Eggerthella* spp., *Finegoldia* spp., *Anaerococcus* spp. and *Fusobacterium* spp. Importantly, an alternative score threshold was used (≥ 1.9) in this study. Had the manufacturer-recommended score threshold been used, the species-level identification rate would have dropped to 77.7%. The use of lower species-level score thresholds, rather than the ≥ 2.0 threshold recommended by the manufacturer, has been shown in a number of studies to improve performance of the MALDI Biotyper system for anaerobic isolates (Hsu & Burnham, 2014; Schmitt et al., 2013). Interestingly, this did not appear to be required for the identification of *Clostridium* species (Chean et al., 2014). In contrast to the impact of score threshold, identification of anaerobic bacteria by MALDI-TOF MS does not appear to be influenced by culture media, incubation time (up to 96 h) or exposure to oxygen (Hsu & Burnham, 2014; Veloo et al., 2014). Future improvements in the composition of the databases will aid in further improving the identification of anaerobic bacteria by MALDI-TOF MS.

3.4 IDENTIFICATION OF MYCOBACTERIA

The identification of *Mycobacterium* species has traditionally relied on the performance of biochemical methods as well as growth properties such as pigmentation and temperature preferences. More recently, laboratories in the United States and Europe have become increasingly reliant on molecular-based methods to accurately identify these organisms on account of the poor accuracy and slow-turnaround time associated with traditional identification methodologies. MALDI-TOF MS therefore holds great promise for mycobacteriology if it can provide rapid and accurate results.

Unlike other bacteria, mycobacteria are not inactivated by matrix solution (Hettick et al., 2004). Making suspensions of the isolate to be tested in 70% ethanol prior to analysis by MALDI-TOF MS has also been attempted (Lotz et al., 2010), although differences in the ability of ethanol to inactivate *M. tuberculosis* have been noted (Best, Sattar, Springthorpe, & Kennedy, 1990). Because of the low infectious dose associated with *M. tuberculosis*, slowly growing mycobacteria must be manipulated in a laminar flow biosafety cabinet with testing personnel required to wear fit-tested respirators with an N-95 rating (Singh, 2009). A number of different extraction protocols have been published, with heat most commonly used to inactivate organisms. An extraction procedure based on bead-beating in the presence of ethanol followed by additional incubation at room temperature has also been developed. Inactivation of *M. tuberculosis* has been documented with both heat-based and bead-beating in ethanol extraction protocols (Balada-Llasat et al., 2013; Dunne et al., 2014; Machen, Kobayashi, Connelly, & Wang, 2013; Mather et al., 2014). Failure to extract mycobacteria prior to analysis by MALDI-TOF MS results in an overabundance of molecules with a mass-to-charge ratio (m/z) of <2000, most likely correlating with cell wall lipid molecules (Hettick et al., 2004; Pignone, Greth, Cooper, Emerson, & Tang, 2006).

A number of studies have been published using custom mycobacterial databases (El Khechine, Couderc, Flaudrops, Raoult, & Drancourt, 2011; Machen et al., 2013; Saleeb, Drake, Murray, & Zelazny, 2011). More recently, the performance of manufacturer-provided databases targeting these organisms has been investigated. The Bruker Mycobacteria Library 1.0 contains 173 spectra from 94 species of *Mycobacterium*, while v2.0 contains 313 spectra from 131 species of *Mycobacterium*. Initial evaluations of this database showed low rates of identification of *Mycobacterium* spp. from colonies (50–62.8%) (Buchan et al., 2014; Chen, Yam, et al., 2013) although a later study showed that 93.8% of 178 isolates tested were correctly identified to the species level (Balada-Llasat et al., 2013). Each of these studies used the manufacturer-recommended extraction protocol, which involves a number of technical manipulations to ensure organism inactivation. Using a simplified extraction method, however, 79.3% of 198 clinical isolates tested with the Bruker v2.0 library were correctly identified (Mather et al., 2014). Lowering the score threshold to ≥ 1.7 increased the species-identification rate to 93.9%. The SARMIS v4.12 RUO database covers 47 *Mycobacterium* species with the inclusion of mycobacteria in the future IVD v3.0 database. Using the SARAMIS v4.12 database, the VITEK MS correctly identified 94.4% of isolates to the species level (Mather et al., 2014). The strains used in the aforementioned study were among the most commonly encountered in a clinical laboratory; thus, performance of the VITEK MS for more rarely encountered organisms has not yet been established. Finally, there are no published studies detailing the performance of the VITEK MS system directly from liquid broth cultures.

In published studies to date, 100% of all *M. tuberculosis* complex strains tested are accurately identified to the complex level. As with other bacterial genera, MALDI-TOF MS appears to be unable to differentiate between members of certain complexes (e.g. *Mycobacterium abscessus* complex). Nevertheless, such differentiation may be possible in the future as a number of subspecies-specific spectra or biomarkers appear to be discernible by cluster analysis (Fangous et al., 2014; Panagea et al., 2015). MALDI-TOF MS reliably distinguishes *Mycobacterium chelonae* from the *M. abscessus* group, which cannot be accomplished using 16S rRNA gene sequencing alone. This particular finding is important clinically, enabling earlier selection of optimal antimicrobial therapy. In contrast, other closely related organisms such as *Mycobacterium chimaera* and *Mycobacterium intracellulare* (both members of the *Mycobacterium avium–intracellularae* complex) could not be distinguished by MALDI-TOF MS.

Spectral differences have been observed between older and younger isolates (Lotz et al., 2010). In one study, rates of identification for certain rapidly growing mycobacteria decreased by 20% at 10 days (Mather et al., 2014). For the slow-growing mycobacteria tested (*Mycobacterium gordonae* and *Mycobacterium kansasii*), dramatically reduced rates of identification were observed at 28 days compared with 21 days of growth (Mather et al., 2014). These data highlight the need for consistency between the age of the organisms used for database generation and those used for testing. Higher rates of automated mass spectra acquisition of are observed from mycobacteria growing in broth media (7H9 broth) compared with agar (7H10), although this difference could be overcome using manual spectra acquisition (Balazova et al., 2014). This may contribute to the higher rates of identification that have been previously observed for broth media (Buchan et al., 2014).

3.5 IDENTIFICATION OF YEAST

Early studies on the application of MALDI-TOF MS to the identification of clinical yeast isolates showed higher identification rates for *Candida* species compared with non-*Candida* species (96–100% vs. 41.2–61.9%, respectively) (Bader et al., 2011; Buchan & Ledeboer, 2013; Dhiman et al., 2011). More recently, that has been shown not to be the case with similar overall identification rates observed between *Candida* spp. and non-*Candida* species (Westblade et al., 2013). Initial studies comparing the performance of different MALDI-TOF systems showed higher misidentification rates with the VITEK MS compared with the Bruker MALDI Biotyper (12.1% vs. 1%) (Mancini et al., 2013). Importantly, all misidentifications with the VITEK MS were reported with a high confidence score. A more recent comparison showed similar identification and misidentification rates (Jamal, Ahmad, et al., 2014). However, variability in performance has been observed between both platforms in other studies (Chao et al., 2014; Pence et al., 2014).

The Andromas MS system has also been shown to be highly accurate for species-level identification of yeasts, with 96.3% correct identifications given after the first acquisition of spectra and 98.8% after a second acquisition (Bille et al., 2012;

Lacroix et al., 2014). Only the Bruker MS system could identify *Prototheca wickerhamii*, *Lodderomyces elongisporus* and *Sporobolomyces salmonicolor*. However, all systems appear to have difficulty identifying *Candida orthopsilosis* and differentiating between species of the *Candida guilliermondii* complex (Chao et al., 2014; Jamal, Ahmad, et al., 2014). Nevertheless, identification of yeast by MALDI-TOF MS outperforms traditional identification methods and is associated with far lower rates of misidentification (e.g. 10% vs. <1.1%) (Goyer et al., 2012; Lacroix et al., 2014).

MALDI-TOF MS has been shown to perform well for the identification of *Cryptococcus* species. Whereas conventional identification systems (e.g. Vitek 2) fail to accurately distinguish *Cryptococcus gattii* from *Cryptococcus neoformans*, these species can be differentiated by MALDI-TOF MS (Firacative et al., 2012; McTaggart et al., 2011; Posteraro et al., 2012). However, these studies used custom databases. More recently, spectra from *C. gattii* have been included in the MALDI Biotyper database although the performance of this updated database for *C. gattii* is unknown. Although the VITEK MS correctly identified 70–100% of *Cryptococcus* isolates tested (Chao et al., 2014; Mancini et al., 2013; Westblade et al., 2013), *C. gattii* was misidentified as *C. neoformans* in one study using database IVD v1.2. The performance of more updated database versions is thus currently unknown.

Unlike bacterial cells, fungal cells contain chitin and thus extraction is required to analyse yeast cells by MALDI-TOF MS. While on-target extraction is sufficient for accurate identifications with the VITEK MS and Andromas systems, Bruker recommends a formic acid extraction for the identification of yeast (Bille et al., 2012; Chao et al., 2014; Lacroix et al., 2014; Mancini et al., 2013). Similar identification rates were observed between on target and tube extractions for yeast on the MALDI Biotyper system (Theel et al., 2012), although some studies found improved performance with tube extraction (Cunningham & Patel, 2013; Vlek et al., 2014). As was discussed previously with anaerobes, using a species-level score threshold of ≥ 1.7 improves performance compared with the manufacturer-recommended threshold (Van Herendael et al., 2012). Culture age (up to 72 h) did not affect identification rates and colonies can be taken directly from chromogenic culture media (Goyer et al., 2012). High inter-laboratory accuracy was observed for the identification of medically relevant yeasts collected at distinct sites across North America (Westblade et al., 2013). In summary, MALDI-TOF MS performs as well as and often exceeds conventional methods for the identification of medically relevant yeast and can delivers results days to weeks earlier than conventional methods (Marcos & Pincus, 2013).

3.6 IDENTIFICATION OF FILAMENTOUS FUNGI

Invasive fungal infections are frequently devastating in immunocompromised individuals, with reported mortality rates often in excess of 50% (Pfaller & Diekema, 2010). Accurate identification of filamentous fungi is essential due to species-specific differences in antifungal susceptibility patterns between members of the certain genera (e.g. *Fusarium solani* complex) (Alastruey-Izquierdo, Cuenca-Estrella,

Monzon, Mellado, & Rodriguez-Tudela, 2008). The identification of filamentous fungi by MALDI-TOF MS is more complicated than bacterial identification for several reasons: (1) fungal taxonomy can be highly complicated, (2) filamentous fungi frequently produce both hyphae and conidia, each of which produce different spectral patterns, (3) these microorganisms must be extracted due to the presence of a robust, chitinous cell wall and (4) they must be manipulated in a laminar flow cabinet to both protect the operator and prevent environmental contamination.

Extraction methods for filamentous fungi vary widely, but most use some variation of an ethanol-based formic acid and acetonitrile extraction. Cellular disruption by bead-beating with glass beads has also been used (Cassagne et al., 2011; Lau et al., 2013). Despite the hazardous nature of these microorganisms, only one study to date has confirmed organism inactivation after extraction (Lau et al., 2013). Because of the relatively recent availability of commercially available databases that include spectra from filamentous fungi, most published studies have used custom databases that have overwhelmingly targeted only limited fungal groups (e.g. dermatophytes) (Alshawa et al., 2012; Coulibaly et al., 2011; Schrodl et al., 2012; Sitterle et al., 2014; Theel, Hall, Mandrekar, & Wengenack, 2011). A handful of more inclusive custom reference databases have been also been described that include hyaline and dematiaceous moulds, as well as zygomycetes, dermatophytes and dimorphic isolates. Using these databases, excellent performance has been observed for the identification of routine clinical mould isolates (88.9–96.8% identification to the species level) (Cassagne et al., 2011; Lau et al., 2013; Normand et al., 2013).

In contrast, evaluation of manufacturer-provided databases has thus far been relatively limited. Using the Bruker RUO Filamentous Fungi Library 1.0, 72% of isolates were correctly identified to the species level ($n = 200$ isolates) (Schulthess, Ledermann, et al., 2014). However, use of a simpler, user-developed extraction protocol resulted in only 16.2% of isolates being correctly identified with almost half remaining unidentified (48.4%) (Lau et al., 2013). The same extraction method appears to be required for both testing and database construction for filamentous fungi. Readers should be aware of the limitations of evaluation studies where the isolates tested may be biased towards isolates of certain genera (e.g. *Aspergillus* spp.). The VITEK MS IVD database (version 2.0) covers a relatively limited number of filamentous fungi (8 genera, 34 species) although additional fungal genera and species will be present in a future version (IVD v3.0). Examination of the performance of this database has thus far been limited to the identification to *Aspergillus* species. One hundred percent of *Aspergillus* isolates tested were correctly identified including *Aspergillus lentulus* (Verwer et al., 2014), a cryptic species of *Aspergillus fumigatus* that is more resistant to voriconazole. Future studies will be required to more fully assess the performance of this system for the identification of filamentous fungi.

Unlike most bacteria, significant spectral differences are observed between earlier and later stages of fungal growth (Alanio et al., 2011; De Carolis et al., 2012). Subcultures of the same strain also show spectral heterogeneity (Normand et al., 2013). Although extraction following overnight in broth culture has been proposed to reduce spectral heterogeneity, such an approach increases the turnaround time for analysis. In contrast, culture media does not appear to affect performance of MALDI-TOF MS (De Carolis et al., 2012; Lau et al., 2013) although some studies have reported this as being an issue (Coulibaly et al., 2011). Finally, while early studies reported poor performance of MALDI-TOF MS for the analysis of pigmented filamentous fungi such as *Aspergillus niger* (Hettick et al., 2008; Valentine, Wahl, Kingsley, & Wahl, 2002), a subsequent study showed that melanin (which is produced by dematiaceous moulds) suppresses protein and peptide ion signals during the acquisition of spectra (Buskirk et al., 2011). However, this phenomenon has not been observed in more recently published studies where more extensive extraction procedures were used (Lau et al., 2013; Ranque et al., 2014).

3.7 IDENTIFICATION OF SELECT AGENTS

The identification of select agents isolated in culture (e.g. *Brucella* spp., *Francisella* spp., etc.) represents arguably one of the most serious limitations of MALDI-TOF MS for microbial identification. This stems from a dearth to complete absence of select agent representation within commercially available databases depending on the MALDI-TOF MS system used. In the case of the MALDI Biotyper, no spectra corresponding to any select agent are present in the regular manufacturer-provided databases. As a result, *Brucella* spp. and *Francisella* spp. remain unidentifiable (Cunningham & Patel, 2013). In contrast, use of custom databases containing spectra from these organisms resulted in correct identification of these organisms (Ferreira, Vega Castano, et al., 2010; Inglis, Healy, Fremlin, & Golledge, 2012; Lista et al., 2011; Seibold, Maier, Kostrzewa, Zeman, & Splettstoesser, 2010). Use of Bruker's security relevant (SR) library, which contains spectra for some select agents, correctly identified that all *Brucella* spp. and *Francisella* spp. isolates tested to the genus level with correct species-level identifications were given for 70% of the 20 isolates tested (Cunningham & Patel, 2013). Nevertheless, the SR library does not contain spectra for either *Bacillus anthracis* or *Yersinia pestis*.

Laboratories should also be aware of the potential for misidentification of certain select agents by MALDI-TOF MS. Multiple studies have shown that *Burkholderia pseudomallei* can be misidentified as the closely related *Burkholderia thailandensis* when using the Biotyper reference library (Cunningham & Patel, 2013; Dingle, Butler-Wu, & Abbott, 2014). In both publications concerning this observation, the score obtained on the Bruker system was less than that required for species-level identification per the manufacturer's recommendations (score range 1.864–1.962, manufacturer-recommended species-level ID score value ≥ 2.0). Similarly, *Y. pestis* can be misidentified as *Yersinia pseudotuberculosis* (Ayyadurai, Flaudrops, Raoult, & Drancourt, 2010). It is therefore imperative that clinical laboratories continue to follow sentinel guidelines for selected agents.

4 DIRECT IDENTIFICATION FROM CLINICAL SPECIMENS

As a result of its ability to accurately and rapidly identify microorganisms, there is much interest in using MALDI-TOF MS to identify microorganisms directly from patient specimens. MALDI-TOF MS is most commonly performed on pure colonies isolated in culture. Clinical specimens, in contrast, can contain variable numbers of microorganisms or may be polymicrobial (i.e. have multiple microorganisms present). Furthermore, host proteins are also present in clinical specimens and can result in interference with the protein spectra. Thus, clinical specimens must generally be purified and concentrated prior to analysis by MALDI-TOF MS. (Buchan & Ledeboer, 2014).

4.1 IDENTIFICATION OF MICROORGANISMS DIRECTLY FROM POSITIVE BLOOD CULTURES

Bloodstream infections (BSI) are associated with increased mortality and increased length of hospital stay and concomitant costs (Pittet, Tarara, & Wenzel, 1994). Additionally, mortality increases nearly 8% for every hour of inappropriate antimicrobial therapy administration in septic patients (Afshari, Schrenzel, Ieven, & Harbarth, 2012). Blood cultures remain the standard of care for diagnosing the aetiologic agent of BSI on account of the low concentration of bacteria present in patients with BSI; ≤ 10 and 1.0 CFU/mL in adults and children, respectively (Croxatto et al., 2012; Kellogg, Manzella, & Bankert, 2000). The definitive identification of the microorganism present using traditional methodologies typically takes between 24 and 72 h from the time the culture is flagged as positive for growth (Klouche & Schroder, 2008). Although rapid, nucleic acid-based identification methods can be performed directly on positive blood cultures (e.g. PNA-FISH), these are costly and limited in the number of microorganisms they can identify (Buchan & Ledeboer, 2014). Because the concentration of bacteria present in the blood culture bottle at the time of positivity typically ranges between 10^6 and 10^9 CFU/mL, MALDI-TOF MS has been proposed as a method to rapidly identify microorganisms directly from positive blood cultures (Christner et al., 2010; Croxatto et al., 2012).

Despite the relatively high concentration of microorganisms present, positive blood cultures cannot be directly analysed by MALDI-TOF MS: only 47% of positive blood cultures were correctly identified to the species level when directly analysed by MALDI-TOF MS (Ferreira, Sanchez-Juanes, Munoz-Bellido, & Gonzalez-Buitrago, 2011). Blood cultures must therefore be extracted prior to analysis by MALDI-TOF MS to remove the non-bacterial proteins present, e.g., haemoglobin, serum proteins, etc. A variety of extraction methods have been published including differential centrifugation and washing, serum separator tubes, filtration, selective lysis of blood cells and use of commercially available processing methods such as the Sepsityper Kit (Bruker Daltonics) (Buchan, Riebe, & Ledeboer,

2012; Christner et al., 2010; Ferroni et al., 2010; Gray, Thomas, Olma, Iredell, & Chen, 2013; Leli et al., 2013; Machen, Drake, & Wang, 2014; March-Rossello, Munoz-Moreno, Garcia-Loygorri-Jordan de Urries, & Bratos-Perez, 2013; Schubert et al., 2011; Stevenson et al., 2010; Tadros & Petrich, 2013). While extraction methods add a pre-analytical step to the procedure, these typically take less than an hour to complete (Lagace-Wiens et al., 2012; Wojewoda, 2013).

The performance of MALDI-TOF MS for the direct identification of pathogens from positive blood cultures ranges from 67% to 100% (Buchan et al., 2012; Chen, Ho, et al., 2013; Kok, Thomas, Olma, Chen, & Iredell, 2011; Lagace-Wiens et al., 2012; Loonen, Jansz, Stalpers, Wolffs, & van den Brule, 2012; Martiny, Dediste, & Vandenberg, 2012; Meex et al., 2012; Nonnemann, Tvede, & Bjarnsholt, 2013; Yan et al., 2011). Even though performance is affected by the extraction method used, >95% of the MALDI-TOF MS identifications obtained agree with conventional methods (Christner et al., 2010; Clark et al., 2013; Stevenson et al., 2010). It is important to note that published studies vary widely in the extraction procedures and identification thresholds used, making inter-study comparison difficult to interpret. Use of blood culture-specific parameters for MALDI-TOF MS analysis by the MALDI Biotyper appears to improve results (Buchan et al., 2012). This includes the exclusion of mass peaks with m/z ratios of <4000, as well as the use of lower identification thresholds (≥ 1.8 and ≥ 1.6–<1.8 for species- and genus-level, respectively).

A number of studies have reported superior performances for Gram-negative bacteria compared with blood cultures positive for Gram-positive bacteria (>95% vs. 57–63%, respectively) (Buchan et al., 2012; Christner et al., 2010; La Scola & Raoult, 2009). Nevertheless, one study reported superior rates of Gram-positive identification by inclusion of an additional wash step during Sepsityper processing suggesting the possibility for further optimisation of extraction protocols (Martinez, Bauerle, Fang, & Butler-Wu, 2014). Interestingly, similar rates of identification of Gram-negative and Gram-positive bacteria were observed using the VITEK MS system (90.1% and 83.7%, respectively) (Foster, 2013). In a direct comparison of the VITEK MS and MALDI Biotyper, both systems identified >80% of the bacteria at the species level and >90% at the genus level from mono-microbial blood cultures (Chen, Ho, et al., 2013).

In experiments with blood culture broths spiked with *E. coli* and *S. aureus*, spectra produced from extracted blood cultures with $>10^8$ CFU/mL resembled spectra obtained from pure colonies (Christner et al., 2010). However, spectra were indistinguishable from the spectra from sterile blood culture media at concentrations of 10^6 CFU/mL. The performance of direct identification by MALDI-TOF MS also appears to be influenced by the blood culture medium used. Identification rates were much lower from charcoal containing broths compared to non-charcoal containing broths from the BacT/Alert system (Romero-Gomez & Mingorance, 2011; Schmidt et al., 2012). Earlier studies showed that BACTEC blood culture media appears to produce higher rates of identification than other blood culture media

(74–76% vs. 45.6–69%) (Romero-Gomez & Mingorance, 2011; Schmidt et al., 2012). However, a more recent study found no difference between two commercially available resin-containing blood culture media, with >90% of the bacteria isolated from both the BacT/Alert Plus and Bactec Plus bottles correctly identified by MALDI-TOF MS (Fiori et al., 2014).

One major limitation of the use of MALDI-TOF MS for direct identification of bacteria from blood cultures is its poor performance with polymicrobial specimens (Chen, Ho, et al., 2013; Christner et al., 2010; Fiori et al., 2014; La Scola & Raoult, 2009; Vlek, Bonten, & Boel, 2012). Regardless of the commercial system used, MALDI-TOF MS tends to only identify one of the microorganisms present in the majority of cases (Chen, Ho, et al., 2013; Ferroni et al., 2010; Kok et al., 2011). Thus, to ensure that polymicrobial infections are recognised, Gram stains of all positive blood cultures are recommended prior to MALDI-TOF MS analysis although laboratories should be aware of the relative insensitivity of the Gram stain for detecting these infections (La Scola & Raoult, 2009).

An alternative MALDI-TOF-MS-based strategy for improving the time to identification of positive blood cultures involves incubation of the blood culture subcultures for several hours with the resulting growth or 'scum' tested by MALDI-TOF MS (Kroumova et al., 2011; Ngan, Lin, & Teo, 2012; Patel, 2013). While Gram-negative bacteria appear to be adequately identified after an incubation period of 2–4 h, Gram-positive bacteria require 6 h of incubation (Bhatti, Boonlayangoor, Beavis, & Tesic, 2014; Idelevich et al., 2014; Verroken et al., 2014). However, use of an ethanol/formic acid extraction was shown to reduce the incubation required for Gram-positive microorganisms to 3 h (Idelevich et al., 2014). Yeast and anaerobic bacteria were not readily identified after a short incubation on culture media (Verroken et al., 2014). Even though an additional incubation step is required, these methods allow for same-day identification of bacteria in positive blood culture broths. As discussed previously, pathogen identification directly from positive blood cultures has been shown to improve patient outcomes in a number of studies, most likely due to dramatic reductions in the time to initiation of optimal appropriate antimicrobial therapy (Perez et al., 2014; Vlek et al., 2012).

4.2 IDENTIFICATION OF BACTERIA DIRECTLY FROM URINE SPECIMENS

A handful of studies have investigated the use of MALDI-TOF MS for identification of bacteria directly from urine specimens. Urine specimens are considered ideal candidates for direct identification by MALDI-TOF MS because urine typically lacks host proteins and normal flora that could confound protein spectra. Additionally, spectra for pathogens most frequently associated with urinary tract infection (e.g. *E. coli*, *Staphylococcus saprophyticus* and other *Enterobacteriaceae*) are present in all commercially available MALDI-TOF MS databases (Demarco & Burnham, 2014). When an infection is present, urine often contains a high concentration of a single pathogen and should, in theory, have a sufficient quantity of bacteria present for analysis by MALDI-TOF MS (Ferreira, Sanchez-Juanes, et al., 2010; Ferreira et al., 2011). However, direct identification of bacteria from urine specimens by MALDI-TOF MS has been hindered by the limited volume of urine that can be deposited on the MALDI-TOF MS target plate (Croxatto et al., 2012; DeMarco & Ford, 2013). Therefore, urine specimens must be concentrated prior to analysis by MALDI-TOF MS. Processing of urine specimens by centrifugation, filtration and diafiltration prior to MALDI-TOF MS analysis have been described (Demarco & Burnham, 2014; Ferreira, Sanchez-Juanes, et al., 2010; Kohling et al., 2012; Wang et al., 2013).

Regardless of the method used for specimen concentration, urine specimens with higher concentrations of bacteria (i.e. $>10^5$ CFU/mL) are more likely to yield species-level identifications than those with lower concentrations of organisms present (Croxatto et al., 2012; Ferreira, Sanchez-Juanes, et al., 2010; Wang et al., 2013). The inclusion of a filtration step appears to improve performance for urines with lower concentrations of bacteria with correct bacterial identifications in 58% of urine specimens containing 10^2 to $\geq 10^5$ CFU/mL of bacteria (Kohling et al., 2012). Interestingly, 37% of specimens that failed to be identified by MALDI-TOF MS were found to have intense peaks corresponding to human α-defensins that likely suppressed the bacterial spectra. As with positive blood culture identification, MALDI-TOF MS struggles to identify polymicrobial urinary tract infections (Kohling et al., 2012; Wang et al., 2013). Recently, DeMarco and colleagues used a diafiltration method to desalt, fractionate and concentrate urine specimens prior to MALDI-TOF MS analysis. Using this method, the uropathogen was identified in 67% of the UTI-positive specimens and all specimens without clinically relevant bacteriuria were identified. The limit of detection for this assay was 10^5–10^6 CFU/mL of urine. The turnaround time was reduced from 24–48 to 2–3 h (Demarco & Burnham, 2014).

4.3 IDENTIFICATION OF BACTERIA DIRECTLY FROM CEREBROSPINAL FLUID

A preliminary diagnosis of the aetiology of bacterial meningitis is traditionally made by performing a Gram stain of a cerebrospinal fluid (CSF) specimen. However, definitive identification requires isolation in culture. Identification of microorganisms directly form CSF specimens therefore has the potential to improve clinical outcome by allowing faster diagnosis of bacterial meningitis and more timely initiation of optimal antimicrobial therapy (Clark et al., 2013). To date, there has been no systematic analysis of the ability to perform direct pathogen identification from CSF using MALDI-TO-MS. However, two case reports have preliminarily reported its utility. In both cases, the CSF was processed using a two-step double-spin centrifugation protocol and the identification matched that subsequently obtained when the cultured isolate was tested (Nyvang Hartmeyer et al., 2010; Segawa et al., 2014). The direct identification of bacteria in CSF by MALDI-TOF MS is therefore a promising tool that warrants further investigation.

LIMITATIONS AND CONCLUSIONS

The use of MALDI-TOF MS for the identification of microorganisms in the clinical microbiology laboratory has several limitations (summarised in Table 3). Because MALDI-TOF MS identification of microorganisms is based on analysis of spectra produced by abundant proteins (e.g. ribosomal proteins), highly related organisms with well-conserved ribosomal proteins cannot be distinguished by MALDI-TOF MS. As discussed earlier, MALDI-TOF MS cannot distinguish between *E. coli* and *Shigella* species, has difficulty in distinguishing *S. pneumoniae* from *S. mitis* group and cannot always resolve complex members to the species level (e.g. *E. cloacae* complex). Clinical microbiology laboratories must therefore continue to rely on conventional methods to identify certain organisms (e.g. *Salmonella* Typhi). Critically, laboratories cannot assume that a select agent will or will not be identified by MALDI-TOF MS as a result of the lack of reference spectra in both the MALDI Biotyper and VITEK MS databases (Cunningham & Patel, 2013).

Identification of a microorganism by MALDI-TOF MS relies on proprietary databases of reference spectra. Advantageously, the databases are adaptable and open source; the manufacturer or user may improve the database by adding mass spectral for additional strains or microorganisms. However, ensuring the accuracy of spectra added to the database may beyond the capabilities of many laboratories (Dingle & Butler-Wu, 2013; Patel, 2013). IVD-cleared databases in the United States lack this functionality. While the breadth of microorganisms present in the MALDI-TOF MS databases can aid in the identification of rare organisms, reporting of esoteric names may lead to clinician confusion. Guidelines should be in place to use common names or groups when reporting identifications.

Additionally, as noted above, MALDI-TOF MS struggles with polymicrobial specimens, complicating the identification of microorganisms directly from clinical specimens. Other limitations of note include negative effects on workflows and identification if the MALDI-TOF MS instrument is down, particularly if only one instrument is available in the lab. Additionally, certain organisms require extraction to be identified by MALDI-TOF MS. While such extraction procedures do not significantly increase turnaround time, they require the use of strong acids that must be disposed of (Dingle & Butler-Wu, 2013). While generally reproducible, inter- and intra-laboratory variability can be introduced by a number of factors including culture conditions, biological variability, brand of mass spectrometer used, matrix composition and the individual user (Patel, 2013).

Microorganism identification by MALDI-TOF MS has truly revolutionised the field of clinical microbiology by enabling rapid and highly accurate identification of most cultured microorganisms. Nevertheless, the method is neither infallible

Table 3 Advantages and Limitations of MALDI-TOF MS for Microbial Identification

	Advantages	Limitations

Performance	• Identification of a broad range of microorganisms (bacteria, yeast) on a single platform
	• Performs well or better than conventional methods for the correct identification of microorganisms
	• Identifies microorganisms that previously required molecular methods for identification
	• High inter-laboratory reproducibility
	• Rapid turnaround time
	• FDA-approved databases are available for both the Biotyper and VITEK MS platforms
		• Poor discrimination of certain highly related species can lead to misidentifications, including:
		  - *Shigella* and *E. coli*
		  - *S. pneumoniae* and *S. mitis/ S. oralis* (Biotyper only)
		  - *Haemophilus influenzae* and other non-influenzae *Haemophilus species*
		• Accurate identification is critically dependent on the breadth and depth of database
		• Absence of spectra for select agents in commercially available databases
		• Polymicrobial specimens problematic
		• FDA-cleared databases lack the depth and breadth of research use only databases

Specimen requirements and processing	• Minimal quantity of isolate required (10^4–10^6 CFU)
	• Ease of use—simple sample preparation, assay set-up, and result interpretation
	• High throughput
		• Some microorganisms require protein extraction prior to analysis
		  - Increases turnaround time
		  - Use of hazardous chemicals necessary for extraction
		• Requires pure or well-isolated colonies
		• Requires sufficient growth of an microorganism before identification

Cost	• Low cost per test
	• Cost savings for laboratories
	  - Less technician time necessary for identification
	  - Replaces many proprietary identification systems
	  - Few consumables required
		• High initial instrument cost
		• Maintenance costs

nor immune from operator error. Even with the continued development an improvement of the identification databases, unanticipated misidentifications will continue to be encountered (Alby, Glaser, & Edelstein, 2015). Thus, it is imperative that laboratory staff continue to retain key skills such as recognition of colony morphology and knowledge of basic biochemical reactions.

REFERENCES

Afshari, A., Schrenzel, J., Ieven, M., & Harbarth, S. (2012). Bench-to-bedside review: Rapid molecular diagnostics for bloodstream infection—A new frontier? *Critical Care*, *16*(3), 222.

Alanio, A., Beretti, J. L., Dauphin, B., Mellado, E., Quesne, G., Lacroix, C., et al. (2011). Matrix-assisted laser desorption ionization time-of-flight mass spectrometry for fast and accurate identification of clinically relevant *Aspergillus* species. *Clinical Microbiology and Infection*, *17*(5), 750–755.

Alastruey-Izquierdo, A., Cuenca-Estrella, M., Monzon, A., Mellado, E., & Rodriguez-Tudela, J. L. (2008). Antifungal susceptibility profile of clinical *Fusarium* spp. isolates identified by molecular methods. *The Journal of Antimicrobial Chemotherapy*, *61*(4), 805–809.

Alatoom, A. A., Cunningham, S. A., Ihde, S. M., Mandrekar, J., & Patel, R. (2011). Comparison of direct colony method versus extraction method for identification of gram-positive cocci by use of Bruker Biotyper matrix-assisted laser desorption ionization-time of flight mass spectrometry. *Journal of Clinical Microbiology*, *49*(8), 2868–2873.

Alby, K., Gilligan, P. H., & Miller, M. B. (2013). Comparison of matrix-assisted laser desorption ionization-time of flight (MALDI-TOF) mass spectrometry platforms for the identification of gram-negative rods from patients with cystic fibrosis. *Journal of Clinical Microbiology*, *51*(11), 3852–3854.

Alby, K., Glaser, L. J., & Edelstein, P. H. (2015). *Kocuria rhizophila* misidentified as *Corynebacterium jeikeium* and other errors caused by the Vitek MS system call for maintained microbiological competence in the era of matrix-assisted laser desorption ionization-time of flight mass spectrometry. *Journal of Clinical Microbiology*, *53*(1), 360–361.

Almuzara, M., Barberis, C., Traglia, G., Famiglietti, A., Ramirez, M. S., & Vay, C. (2015). Evaluation of matrix-assisted laser desorption ionization-time-of-flight mass spectrometry for species identification of nonfermenting gram-negative bacilli. *Journal of Microbiological Methods*, *112*, 24–27.

Alshawa, K., Beretti, J. L., Lacroix, C., Feuilhade, M., Dauphin, B., Quesne, G., et al. (2012). Successful identification of clinical dermatophyte and *Neoscytalidium* species by matrix-assisted laser desorption ionization-time of flight mass spectrometry. *Journal of Clinical Microbiology*, *50*(7), 2277–2281.

Anderson, N. W., Buchan, B. W., Riebe, K. M., Parsons, L. N., Gnacinski, S., & Ledeboer, N. A. (2012). Effects of solid-medium type on routine identification of bacterial isolates by use of matrix-assisted laser desorption ionization-time of flight mass spectrometry. *Journal of Clinical Microbiology*, *50*(3), 1008–1013. http://dx.doi.org/10.1128/jcm.05209-11.

Angeletti, S., Dicuonzo, G., Avola, A., Crea, F., Dedej, E., Vailati, F., et al. (2015). Viridans group streptococci clinical isolates: MALDI-TOF mass spectrometry versus gene sequence-based identification. *PLoS One*, *10*(3), e0120502.

Anhalt, J. P., & Fenselau, C. (1975). Identification of bacteria using mass spectrometry. *Analytical Chemistry*, *45*, 219–225.

Ayyadurai, S., Flaudrops, C., Raoult, D., & Drancourt, M. (2010). Rapid identification and typing of *Yersinia pestis* and other *Yersinia* species by matrix-assisted laser desorption/ionization time-of-flight (MALDI-TOF) mass spectrometry. *BMC Microbiology*, *10*, 285.

Bader, O., Weig, M., Taverne-Ghadwal, L., Lugert, R., Gross, U., & Kuhns, M. (2011). Improved clinical laboratory identification of human pathogenic yeasts by matrix-assisted laser desorption ionization time-of-flight mass spectrometry. *Clinical Microbiology and Infection, 17*(9), 1359–1365.

Balada-Llasat, J. M., Kamboj, K., & Pancholi, P. (2013). Identification of mycobacteria from solid and liquid media by matrix-assisted laser desorption ionization-time of flight mass spectrometry in the clinical laboratory. *Journal of Clinical Microbiology, 51*(9), 2875–2879. http://dx.doi.org/10.1128/JCM.00819-13.

Balazova, T., Makovcova, J., Sedo, O., Slany, M., Faldyna, M., & Zdrahal, Z. (2014). The influence of culture conditions on the identification of *Mycobacterium* species by MALDI-TOF MS profiling. *FEMS Microbiology Letters, 353*(1), 77–84.

Barba, M. J., Fernandez, A., Oviano, M., Fernandez, B., Velasco, D., & Bou, G. (2014). Evaluation of MALDI-TOF mass spectrometry for identification of anaerobic bacteria. *Anaerobe, 30*, 126–128.

Barberis, C., Almuzara, M., Join-Lambert, O., Ramirez, M. S., Famiglietti, A., & Vay, C. (2014). Comparison of the Bruker MALDI-TOF mass spectrometry system and conventional phenotypic methods for identification of Gram-positive rods. *PLoS One, 9*(9), e106303.

Barreau, M., Pagnier, I., & La Scola, B. (2013). Improving the identification of anaerobes in the clinical microbiology laboratory through MALDI-TOF mass spectrometry. *Anaerobe, 22*, 123–125.

Benagli, C., Rossi, V., Dolina, M., Tonolla, M., & Petrini, O. (2011). Matrix-assisted laser desorption ionization-time of flight mass spectrometry for the identification of clinically relevant bacteria. *PLoS One, 6*(1), e16424.

Best, M., Sattar, S. A., Springthorpe, V. S., & Kennedy, M. E. (1990). Efficacies of selected disinfectants against *Mycobacterium tuberculosis*. *Journal of Clinical Microbiology, 28*(10), 2234–2239.

Bhatti, M. M., Boonlayangoor, S., Beavis, K. G., & Tesic, V. (2014). Rapid identification of positive blood cultures by matrix-assisted laser desorption ionization-time of flight mass spectrometry using prewarmed agar plates. *Journal of Clinical Microbiology, 52*(12), 4334–4338.

Bille, E., Dauphin, B., Leto, J., Bougnoux, M. E., Beretti, J. L., Lotz, A., et al. (2012). MALDI-TOF MS Andromas strategy for the routine identification of bacteria, mycobacteria, yeasts, Aspergillus spp. and positive blood cultures. *Clinical Microbiology and Infection, 18*(11), 1117–1125.

Bizzini, A., Durussel, C., Bille, J., Greub, G., & Prod'hom, G. (2010). Performance of matrix-assisted laser desorption ionization-time of flight mass spectrometry for identification of bacterial strains routinely isolated in a clinical microbiology laboratory. *Journal of Clinical Microbiology, 48*(5), 1549–1554.

Bizzini, A., Jaton, K., Romo, D., Bille, J., Prod'hom, G., & Greub, G. (2011). Matrix-assisted laser desorption ionization-time of flight mass spectrometry as an alternative to 16S rRNA gene sequencing for identification of difficult-to-identify bacterial strains. *Journal of Clinical Microbiology, 49*(2), 693–696.

Boleij, A., van Gelder, M. M., Swinkels, D. W., & Tjalsma, H. (2011). Clinical importance of *Streptococcus gallolyticus* infection among colorectal cancer patients: Systematic review and meta-analysis. *Clinical Infectious Diseases, 53*(9), 870–878.

Branda, J. A., Markham, R. P., Garner, C. D., Rychert, J. A., & Ferraro, M. J. (2013). Performance of the Vitek MS v2.0 system in distinguishing *Streptococcus pneumoniae* from nonpneumococcal species of the *Streptococcus mitis* group. *Journal of Clinical Microbiology*, *51*(9), 3079–3082.

Branda, J. A., Rychert, J., Burnham, C. A., Bythrow, M., Garner, O. B., Ginocchio, C. C., et al. (2014). Multicenter validation of the VITEK MS v2.0 MALDI-TOF mass spectrometry system for the identification of fastidious gram-negative bacteria. *Diagnostic Microbiology and Infectious Disease*, *78*(2), 129–131.

Bruin, J. P., Kostrzewa, M., van der Ende, A., Badoux, P., Jansen, R., Boers, S. A., et al. (2014). Identification of *Haemophilus influenzae* and *Haemophilus haemolyticus* by matrix-assisted laser desorption ionization-time of flight mass spectrometry. *European Journal of Clinical Microbiology & Infectious Diseases*, *33*(2), 279–284.

Bruker. (2008). *Microorganism identification and classification based on MALDI-TOF MS fingerprinting with MALDI Biotyper*. Application Note #MT-80. 2015.

Bruker. (August 2011). *MALDI Biotyper clinical microbiology. Fast and accurate identification of microorganisms*. http://www.bruker.com/products/mass-spectrometry-and-separations/literature/literature-room-mass-spec.html. Bruker Daltonics.

Buchan, B. W., & Ledeboer, N. A. (2013). Advances in identification of clinical yeast isolates by use of matrix-assisted laser desorption ionization-time of flight mass spectrometry. *Journal of Clinical Microbiology*, *51*(5), 1359–1366.

Buchan, B. W., & Ledeboer, N. A. (2014). Emerging technologies for the clinical microbiology laboratory. *Clinical Microbiology Reviews*, *27*(4), 783–822.

Buchan, B. W., Riebe, K. M., & Ledeboer, N. A. (2012). Comparison of the MALDI Biotyper system using Sepsityper specimen processing to routine microbiological methods for identification of bacteria from positive blood culture bottles. *Journal of Clinical Microbiology*, *50*(2), 346–352.

Buchan, B. W., Riebe, K. M., Timke, M., Kostrzewa, M., & Ledeboer, N. A. (2014). Comparison of MALDI-TOF MS with HPLC and nucleic acid sequencing for the identification of *Mycobacterium* species in cultures using solid medium and broth. *American Journal of Clinical Pathology*, *141*(1), 25–34.

Burnett-Hartman, A. N., Newcomb, P. A., & Potter, J. D. (2008). Infectious agents and colorectal cancer: A review of *Helicobacter pylori*, *Streptococcus bovis*, JC virus, and human papillomavirus. *Cancer Epidemiology, Biomarkers & Prevention*, *17*(11), 2970–2979.

Buskirk, A. D., Hettick, J. M., Chipinda, I., Law, B. F., Siegel, P. D., Slaven, J. E., et al. (2011). Fungal pigments inhibit the matrix-assisted laser desorption/ionization time-of-flight mass spectrometry analysis of darkly pigmented fungi. *Analytical Biochemistry*, *411*(1), 122–128.

Carbonnelle, E., Grohs, P., Jacquier, H., Day, N., Tenza, S., Dewailly, A., et al. (2012). Robustness of two MALDI-TOF mass spectrometry systems for bacterial identification. *Journal of Microbiological Methods*, *89*(2), 133–136.

Carbonnelle, E., Mesquita, C., Bille, E., Day, N., Dauphin, B., Beretti, J. L., et al. (2011). MALDI-TOF mass spectrometry tools for bacterial identification in clinical microbiology laboratory. *Clinical Biochemistry*, *44*(1), 104–109.

Cassagne, C., Ranque, S., Normand, A. C., Fourquet, P., Thiebault, S., Planard, C., et al. (2011). Mould routine identification in the clinical laboratory by matrix-assisted laser desorption ionization time-of-flight mass spectrometry. *PLoS One*, *6*(12), e28425.

Centers for Disease Control and Prevention, & National Institutes of Health. (2009). In Biosafety in Microbiological and Biomedical Laboratories (BMBL). L. C. Chosewood & D. Wilson (Eds.), *21–1112*(5th ed.): Atlanta, GA, USA.

Chao, Q. T., Lee, T. F., Teng, S. H., Peng, L. Y., Chen, P. H., Teng, L. J., et al. (2014). Comparison of the accuracy of two conventional phenotypic methods and two MALDI-TOF MS systems with that of DNA sequencing analysis for correctly identifying clinically encountered yeasts. *PLoS One*, *9*(10), e109376.

Chean, R., Kotsanas, D., Francis, M. J., Palombo, E. A., Jadhav, S. R., Awad, M. M., et al. (2014). Comparing the identification of Clostridium spp. by two matrix-assisted laser desorption ionization-time of flight (MALDI-TOF) mass spectrometry platforms to 16S rRNA PCR sequencing as a reference standard: A detailed analysis of age of culture and sample preparation. *Anaerobe*, *30*, 85–89.

Chen, J. H., Ho, P. L., Kwan, G. S., She, K. K., Siu, G. K., Cheng, V. C., et al. (2013). Direct bacterial identification in positive blood cultures by use of two commercial matrix-assisted laser desorption ionization-time of flight mass spectrometry systems. *Journal of Clinical Microbiology*, *51*(6), 1733–1739.

Chen, J. H., Yam, W. C., Ngan, A. H., Fung, A. M., Woo, W. L., Yan, M. K., et al. (2013). Advantages of using matrix-assisted laser desorption ionization-time of flight mass spectrometry as a rapid diagnostic tool for identification of yeasts and mycobacteria in the clinical microbiological laboratory. *Journal of Clinical Microbiology*, *51*(12), 3981–3987.

Cheng, W. C., Jan, I. S., Chen, J. M., Teng, S. H., Teng, L. J., Sheng, W. H., et al. (2015). Evaluation of the Bruker Biotyper matrix-assisted laser desorption ionization-time of flight mass spectrometry system for identification of blood isolates of *Vibrio* species. *Journal of Clinical Microbiology*, *53*(5), 1741–1744.

Cherkaoui, A., Emonet, S., Fernandez, J., Schorderet, D., & Schrenzel, J. (2011). Evaluation of matrix-assisted laser desorption ionization-time of flight mass spectrometry for rapid identification of beta-hemolytic streptococci. *Journal of Clinical Microbiology*, *49*(8), 3004–3005.

Cherkaoui, A., Hibbs, J., Emonet, S., Tangomo, M., Girard, M., Francois, P., et al. (2010). Comparison of two matrix-assisted laser desorption ionization-time of flight mass spectrometry methods with conventional phenotypic identification for routine identification of bacteria to the species level. *Journal of Clinical Microbiology*, *48*(4), 1169–1175.

Christensen, J. J., Dargis, R., Hammer, M., Justesen, U. S., Nielsen, X. C., & Kemp, M. (2012). Matrix-assisted laser desorption ionization-time of flight mass spectrometry analysis of Gram-positive, catalase-negative cocci not belonging to the *Streptococcus* or *Enterococcus* genus and benefits of database extension. *Journal of Clinical Microbiology*, *50*(5), 1787–1791.

Christner, M., Rohde, H., Wolters, M., Sobottka, I., Wegscheider, K., & Aepfelbacher, M. (2010). Rapid identification of bacteria from positive blood culture bottles by use of matrix-assisted laser desorption-ionization time of flight mass spectrometry fingerprinting. *Journal of Clinical Microbiology*, *48*(5), 1584–1591.

Clark, A. E., Kaleta, E. J., Arora, A., & Wolk, D. M. (2013). Matrix-assisted laser desorption ionization-time of flight mass spectrometry: A fundamental shift in the routine practice of clinical microbiology. *Clinical Microbiology Reviews*, *26*(3), 547–603.

Coulibaly, O., Marinach-Patrice, C., Cassagne, C., Piarroux, R., Mazier, D., & Ranque, S. (2011). *Pseudallescheria/Scedosporium* complex species identification by matrix-assisted laser desorption ionization time-of-flight mass spectrometry. *Medical Mycology*, *49*(6), 621–626.

Couturier, M. R., Mehinovic, E., Croft, A. C., & Fisher, M. A. (2011). Identification of HA-CEK clinical isolates by matrix-assisted laser desorption ionization-time of flight mass spectrometry. *Journal of Clinical Microbiology*, *49*(3), 1104–1106.

Fernandez-Olmos, A., Garcia-Castillo, M., Morosini, M. I., Lamas, A., Maiz, L., & Canton, R. (2012). MALDI-TOF MS improves routine identification of non-fermenting Gram negative isolates from cystic fibrosis patients. *Journal of Cystic Fibrosis, 11*(1), 59–62.

Ferreira, L., Sanchez-Juanes, F., Gonzalez-Avila, M., Cembrero-Fucinos, D., Herrero-Hernandez, A., Gonzalez-Buitrago, J. M., et al. (2010). Direct identification of urinary tract pathogens from urine samples by matrix-assisted laser desorption ionization-time of flight mass spectrometry. *Journal of Clinical Microbiology, 48*(6), 2110–2115.

Ferreira, L., Sanchez-Juanes, F., Munoz-Bellido, J. L., & Gonzalez-Buitrago, J. M. (2011). Rapid method for direct identification of bacteria in urine and blood culture samples by matrix-assisted laser desorption ionization time-of-flight mass spectrometry: Intact cell vs. extraction method. *Clinical Microbiology and Infection, 17*(7), 1007–1012.

Ferreira, L., Vega Castano, S., Sanchez-Juanes, F., Gonzalez-Cabrero, S., Menegotto, F., Orduna-Domingo, A., et al. (2010). Identification of *Brucella* by MALDI-TOF mass spectrometry. Fast and reliable identification from agar plates and blood cultures. *PLoS One, 5*(12), e14235.

Ferroni, A., Suarez, S., Beretti, J. L., Dauphin, B., Bille, E., Meyer, J., et al. (2010). Real-time identification of bacteria and *Candida* species in positive blood culture broths by matrix-assisted laser desorption ionization-time of flight mass spectrometry. *Journal of Clinical Microbiology, 48*(5), 1542–1548.

Fiori, B., D'Inzeo, T., Di Florio, V., De Maio, F., De Angelis, G., Giaquinto, A., et al. (2014). Performance of two resin-containing blood culture media in detection of bloodstream infections and in direct matrix-assisted laser desorption ionization-time of flight mass spectrometry (MALDI-TOF MS) broth assays for isolate identification: Clinical comparison of the BacT/Alert Plus and Bactec Plus systems. *Journal of Clinical Microbiology, 52*(10), 3558–3567.

Firacative, C., Trilles, L., & Meyer, W. (2012). MALDI-TOF MS enables the rapid identification of the major molecular types within the *Cryptococcus neoformans/C. gattii* species complex. *PLoS One, 7*(5), e37566.

Ford, B. A., & Burnham, C. A. (2013). Optimization of routine identification of clinically relevant Gram-negative bacteria by use of matrix-assisted laser desorption ionization-time of flight mass spectrometry and the Bruker Biotyper. *Journal of Clinical Microbiology, 51*(5), 1412–1420.

Foster, A. G. (2013). Rapid identification of microbes in positive blood cultures by use of the Vitek MS matrix-assisted laser desorption ionization-time of flight mass spectrometry system. *Journal of Clinical Microbiology, 51*(11), 3717–3719.

Fournier, P. E., Drancourt, M., Colson, P., Rolain, J. M., La Scola, B., & Raoult, D. (2013). Modern clinical microbiology: New challenges and solutions. *Nature Reviews. Microbiology, 11*(8), 574–585.

Fournier, R., Wallet, F., Grandbastien, B., Dubreuil, L., Courcol, R., Neut, C., et al. (2012). Chemical extraction versus direct smear for MALDI-TOF mass spectrometry identification of anaerobic bacteria. *Anaerobe, 18*(3), 294–297.

Friedrichs, C., Rodloff, A. C., Chhatwal, G. S., Schellenberger, W., & Eschrich, K. (2007). Rapid identification of viridans streptococci by mass spectrometric discrimination. *Journal of Clinical Microbiology, 45*(8), 2392–2397.

Garner, O., Mochon, A., Branda, J., Burnham, C. A., Bythrow, M., Ferraro, M., et al. (2014). Multi-centre evaluation of mass spectrometric identification of anaerobic bacteria using the VITEK(R) MS system. *Clinical Microbiology and Infection, 20*(4), 335–339.

Goyer, M., Lucchi, G., Ducoroy, P., Vagner, O., Bonnin, A., & Dalle, F. (2012). Optimization of the preanalytical steps of matrix-assisted laser desorption ionization-time of flight mass spectrometry identification provides a flexible and efficient tool for identification of clinical yeast isolates in medical laboratories. *Journal of Clinical Microbiology*, *50*(9), 3066–3068.

Gray, T. J., Thomas, L., Olma, T., Iredell, J. R., & Chen, S. C. (2013). Rapid identification of Gram-negative organisms from blood culture bottles using a modified extraction method and MALDI-TOF mass spectrometry. *Diagnostic Microbiology and Infectious Disease*, *77*(2), 110–112.

Haigh, J., Degun, A., Eydmann, M., Millar, M., & Wilks, M. (2011). Improved performance of bacterium and yeast identification by a commercial matrix-assisted laser desorption ionization-time of flight mass spectrometry system in the clinical microbiology laboratory. *Journal of Clinical Microbiology*, *49*(9), 3441.

He, Y., Li, H., Lu, X., Stratton, C. W., & Tang, Y. W. (2010). Mass spectrometry Biotyper system identifies enteric bacterial pathogens directly from colonies grown on selective stool culture media. *Journal of Clinical Microbiology*, *48*(11), 3888–3892.

Hettick, J. M., Green, B. J., Buskirk, A. D., Kashon, M. L., Slaven, J. E., Janotka, E., et al. (2008). Discrimination of *Aspergillus* isolates at the species and strain level by matrix-assisted laser desorption/ionization time-of-flight mass spectrometry fingerprinting. *Analytical Biochemistry*, *380*(2), 276–281.

Hettick, J. M., Kashon, M. L., Simpson, J. P., Siegel, P. D., Mazurek, G. H., & Weissman, D. N. (2004). Proteomic profiling of intact mycobacteria by matrix-assisted laser desorption/ionization time-of-flight mass spectrometry. *Analytical Chemistry*, *76*(19), 5769–5776.

Hinse, D., Vollmer, T., Erhard, M., Welker, M., Moore, E. R., Kleesiek, K., et al. (2011). Differentiation of species of the *Streptococcus bovis/equinus*-complex by MALDI-TOF mass spectrometry in comparison to sodA sequence analyses. *Systematic and Applied Microbiology*, *34*(1), 52–57.

Holland, R. D., Duffy, C. R., Rafii, F., Sutherland, J. B., Heinze, T. M., Holder, C. L., et al. (1999). Identification of bacterial proteins observed in MALDI TOF mass spectra from whole cells. *Analytical Chemistry*, *71*(15), 3226–3230.

Holland, R. D., Wilkes, J. G., Rafii, F., Sutherland, J. B., Persons, C. C., Voorhees, K. J., et al. (1996). Rapid identification of intact whole bacteria based on spectral patterns using matrix-assisted laser desorption/ionization with time-of-flight mass spectrometry. *Rapid Communications in Mass Spectrometry*, *10*(10), 1227–1232.

Hrabak, J., Chudackova, E., & Walkova, R. (2013). Matrix-assisted laser desorption ionization-time of flight (MALDI-TOF) mass spectrometry for detection of antibiotic resistance mechanisms: From research to routine diagnosis. *Clinical Microbiology Reviews*, *26*(1), 103–114.

Hsu, Y. M., & Burnham, C. A. (2014). MALDI-TOF MS identification of anaerobic bacteria: Assessment of pre-analytical variables and specimen preparation techniques. *Diagnostic Microbiology and Infectious Disease*, *79*(2), 144–148.

Hsueh, P. R., Lee, T. F., Du, S. H., Teng, S. H., Liao, C. H., Sheng, W. H., et al. (2014). Bruker Biotyper matrix-assisted laser desorption ionization-time of flight mass spectrometry system for identification of *Nocardia*, *Rhodococcus*, *Kocuria*, *Gordonia*, *Tsukamurella*, and *Listeria* species. *Journal of Clinical Microbiology*, *52*(7), 2371–2379.

Huang, A. M., Newton, D., Kunapuli, A., Gandhi, T. N., Washer, L. L., Isip, J., et al. (2013). Impact of rapid organism identification via matrix-assisted laser desorption/ionization time-of-flight combined with antimicrobial stewardship team intervention in adult patients with bacteremia and candidemia. *Clinical Infectious Diseases, 57*(9), 1237–1245.

Huong, V. T., Ha, N., Huy, N. T., Horby, P., Nghia, H. D., Thiem, V. D., et al. (2014). Epidemiology, clinical manifestations, and outcomes of *Streptococcus suis* infection in humans. *Emerging Infectious Diseases, 20*(7), 1105–1114.

Idelevich, E. A., Schule, I., Grunastel, B., Wullenweber, J., Peters, G., & Becker, K. (2014). Rapid identification of microorganisms from positive blood cultures by MALDI-TOF mass spectrometry subsequent to very short-term incubation on solid medium. *Clinical Microbiology and Infection, 20*(10), 1001–1006.

Inglis, T. J., Healy, P. E., Fremlin, L. J., & Golledge, C. L. (2012). Use of matrix-assisted laser desorption/ionization time-of-flight mass spectrometry analysis for rapid confirmation of *Burkholderia pseudomallei* in septicemic melioidosis. *The American Journal of Tropical Medicine and Hygiene, 86*(6), 1039–1042.

Isaksson, J., Rasmussen, M., Nilson, B., Stadler, L. S., Kurland, S., Olaison, L., et al. (2015). Comparison of species identification of endocarditis associated viridans streptococci using rnpB genotyping and 2 MALDI-TOF systems. *Diagnostic Microbiology and Infectious Disease, 81*(4), 240–245.

Jamal, W. Y., Ahmad, S., Khan, Z. U., & Rotimi, V. O. (2014). Comparative evaluation of two matrix-assisted laser desorption/ionization time-of-flight mass spectrometry (MALDI-TOF MS) systems for the identification of clinically significant yeasts. *International Journal of Infectious Diseases, 26*, 167–170.

Jamal, W., Albert, M. J., & Rotimi, V. O. (2014). Real-time comparative evaluation of bioMerieux VITEK MS versus Bruker Microflex MS, two matrix-assisted laser desorption-ionization time-of-flight mass spectrometry systems, for identification of clinically significant bacteria. *BMC Microbiology, 14*, 289.

Jamal, W. Y., Shahin, M., & Rotimi, V. O. (2013). Comparison of two matrix-assisted laser desorption/ionization-time of flight (MALDI-TOF) mass spectrometry methods and API 20AN for identification of clinically relevant anaerobic bacteria. *Journal of Medical Microbiology, 62*(Pt. 4), 540–544.

Jarman, K. H., Daly, D. S., Petersen, C. E., Saenz, A. J., Valentine, N. B., & Wahl, K. L. (1999). Extracting and visualizing matrix-assisted laser desorption/ionization time-of-flight mass spectral fingerprints. *Rapid Communications in Mass Spectrometry, 13*(15), 1586–1594.

Justesen, U. S., Holm, A., Knudsen, E., Andersen, L. B., Jensen, T. G., Kemp, M., et al. (2011). Species identification of clinical isolates of anaerobic bacteria: A comparison of two matrix-assisted laser desorption ionization-time of flight mass spectrometry systems. *Journal of Clinical Microbiology, 49*(12), 4314–4318.

Kellogg, J. A., Manzella, J. P., & Bankert, D. A. (2000). Frequency of low-level bacteremia in children from birth to fifteen years of age. *Journal of Clinical Microbiology, 38*(6), 2181–2185.

Kierzkowska, M., Majewska, A., Kuthan, R. T., Sawicka-Grzelak, A., & Mlynarczyk, G. (2013). A comparison of Api 20A vs MALDI-TOF MS for routine identification of clinically significant anaerobic bacterial strains to the species level. *Journal of Microbiological Methods, 92*(2), 209–212.

Klouche, M., & Schroder, U. (2008). Rapid methods for diagnosis of bloodstream infections. *Clinical Chemistry and Laboratory Medicine, 46*(7), 888–908.

Kohling, H. L., Bittner, A., Muller, K. D., Buer, J., Becker, M., Rubben, H., et al. (2012). Direct identification of bacteria in urine samples by matrix-assisted laser desorption/ionization time-of-flight mass spectrometry and relevance of defensins as interfering factors. *Journal of Medical Microbiology, 61*(Pt. 3), 339–344.

Kok, J., Chen, S. C., Dwyer, D. E., & Iredell, J. R. (2013). Current status of matrix-assisted laser desorption ionisation-time of flight mass spectrometry in the clinical microbiology laboratory. *Pathology, 45*(1), 4–17.

Kok, J., Thomas, L. C., Olma, T., Chen, S. C., & Iredell, J. R. (2011). Identification of bacteria in blood culture broths using matrix-assisted laser desorption-ionization Sepsityper and time of flight mass spectrometry. *PLoS One, 6*(8), e23285.

Konrad, R., Berger, A., Huber, I., Boschert, V., Hormansdorfer, S., Busch, U., et al. (2010). Matrix-assisted laser desorption/ionisation time-of-flight (MALDI-TOF) mass spectrometry as a tool for rapid diagnosis of potentially toxigenic *Corynebacterium* species in the laboratory management of diphtheria-associated bacteria. *Euro Surveillance, 15*(43), pii: 19699.

Kroumova, V., Gobbato, E., Basso, E., Mucedola, L., Giani, T., & Fortina, G. (2011). Direct identification of bacteria in blood culture by matrix-assisted laser desorption/ionization time-of-flight mass spectrometry: A new methodological approach. *Rapid Communications in Mass Spectrometry, 25*(15), 2247–2249.

Kuhns, M., Zautner, A. E., Rabsch, W., Zimmermann, O., Weig, M., Bader, O., et al. (2012). Rapid discrimination of *Salmonella enterica* serovar Typhi from other serovars by MALDI-TOF mass spectrometry. *PLoS One, 7*(6), e40004.

Lacroix, C., Gicquel, A., Sendid, B., Meyer, J., Accoceberry, I., Francois, N., et al. (2014). Evaluation of two matrix-assisted laser desorption ionization-time of flight mass spectrometry (MALDI-TOF MS) systems for the identification of *Candida* species. *Clinical Microbiology and Infection, 20*(2), 153–158.

Lagace-Wiens, P. R., Adam, H. J., Karlowsky, J. A., Nichol, K. A., Pang, P. F., Guenther, J., et al. (2012). Identification of blood culture isolates directly from positive blood cultures by use of matrix-assisted laser desorption ionization-time of flight mass spectrometry and a commercial extraction system: Analysis of performance, cost, and turnaround time. *Journal of Clinical Microbiology, 50*(10), 3324–3328.

Lamy, B., Kodjo, A., & Laurent, F. (2011). Identification of *Aeromonas* isolates by matrix-assisted laser desorption ionization time-of-flight mass spectrometry. *Diagnostic Microbiology and Infectious Disease, 71*(1), 1–5.

La Scola, B., & Raoult, D. (2009). Direct identification of bacteria in positive blood culture bottles by matrix-assisted laser desorption ionisation time-of-flight mass spectrometry. *PLoS One, 4*(11), e8041.

Lau, A. F., Drake, S. K., Calhoun, L. B., Henderson, C. M., & Zelazny, A. M. (2013). Development of a clinically comprehensive database and a simple procedure for identification of molds from solid media by matrix-assisted laser desorption ionization-time of flight mass spectrometry. *Journal of Clinical Microbiology, 51*(3), 828–834.

Lavigne, J. P., Espinal, P., Dunyach-Remy, C., Messad, N., Pantel, A., & Sotto, A. (2013). Mass spectrometry: A revolution in clinical microbiology? *Clinical Chemistry and Laboratory Medicine, 51*(2), 257–270.

Lee, W., Kim, M., Yong, D., Jeong, S. H., Lee, K., & Chong, Y. (2015). Evaluation of VITEK mass spectrometry (MS), a matrix-assisted laser desorption ionization time-of-flight MS system for identification of anaerobic bacteria. *Annals of Laboratory Medicine, 35*(1), 69–75.

Leli, C., Cenci, E., Cardaccia, A., Moretti, A., D'Alo, F., Pagliochini, R., et al. (2013). Rapid identification of bacterial and fungal pathogens from positive blood cultures by MALDI-TOF MS. *International Journal of Medical Microbiology, 303*(4), 205–209.

Lista, F., Reubsaet, F. A., De Santis, R., Parchen, R. R., de Jong, A. L., Kieboom, J., et al. (2011). Reliable identification at the species level of *Brucella* isolates with MALDI-TOF-MS. *BMC Microbiology, 11,* 267.

Loonen, A. J., Jansz, A. R., Bergland, J. N., Valkenburg, M., Wolffs, P. F., & van den Brule, A. J. (2012). Comparative study using phenotypic, genotypic, and proteomics methods for identification of coagulase-negative staphylococci. *Journal of Clinical Microbiology, 50*(4), 1437–1439.

Loonen, A. J., Jansz, A. R., Stalpers, J., Wolffs, P. F., & van den Brule, A. J. (2012). An evaluation of three processing methods and the effect of reduced culture times for faster direct identification of pathogens from BacT/ALERT blood cultures by MALDI-TOF MS. *European Journal of Clinical Microbiology & Infectious Diseases, 31*(7), 1575–1583.

Lotz, A., Ferroni, A., Beretti, J. L., Dauphin, B., Carbonnelle, E., Guet-Revillet, H., et al. (2010). Rapid identification of mycobacterial whole cells in solid and liquid culture media by matrix-assisted laser desorption ionization-time of flight mass spectrometry. *Journal of Clinical Microbiology, 48*(12), 4481–4486.

Maatallah, M., Vading, M., Kabir, M. H., Bakhrouf, A., Kalin, M., Naucler, P., et al. (2014). *Klebsiella variicola* is a frequent cause of bloodstream infection in the Stockholm area, and associated with higher mortality compared to *K. pneumoniae. PLoS One, 9*(11), e113539.

Machen, A., Drake, T., & Wang, Y. F. (2014). Same day identification and full panel antimicrobial susceptibility testing of bacteria from positive blood culture bottles made possible by a combined lysis-filtration method with MALDI-TOF VITEK mass spectrometry and the VITEK2 system. *PLoS One, 9*(2), e87870.

Machen, A., Kobayashi, M., Connelly, M. R., & Wang, Y. F. (2013). Comparison of heat inactivation and cell disruption protocols for identification of mycobacteria from solid culture media by use of vitek matrix-assisted laser desorption ionization-time of flight mass spectrometry. *Journal of Clinical Microbiology, 51*(12), 4226–4229.

Mancini, N., De Carolis, E., Infurnari, L., Vella, A., Clementi, N., Vaccaro, L., et al. (2013). Comparative evaluation of the Bruker Biotyper and Vitek MS matrix-assisted laser desorption ionization-time of flight (MALDI-TOF) mass spectrometry systems for identification of yeasts of medical importance. *Journal of Clinical Microbiology, 51*(7), 2453–2457.

Manji, R., Bythrow, M., Branda, J. A., Burnham, C. A., Ferraro, M. J., Garner, O. B., et al. (2014). Multi-center evaluation of the VITEK(R) MS system for mass spectrometric identification of non-*Enterobacteriaceae* Gram-negative bacilli. *European Journal of Clinical Microbiology & Infectious Diseases, 33*(3), 337–346.

March-Rossello, G. A., Munoz-Moreno, M. F., Garcia-Loygorri-Jordan de Urries, M. C., & Bratos-Perez, M. A. (2013). A differential centrifugation protocol and validation criterion for enhancing mass spectrometry (MALDI-TOF) results in microbial identification using blood culture growth bottles. *European Journal of Clinical Microbiology & Infectious Diseases, 32*(5), 699–704.

Marcos, J. Y., & Pincus, D. H. (2013). Fungal diagnostics: Review of commercially available methods. *Methods in Molecular Biology*, *968*, 25–54.

Marko, D. C., Saffert, R. T., Cunningham, S. A., Hyman, J., Walsh, J., Arbefeville, S., et al. (2012). Evaluation of the Bruker Biotyper and Vitek MS matrix-assisted laser desorption ionization-time of flight mass spectrometry systems for identification of nonfermenting gram-negative bacilli isolated from cultures from cystic fibrosis patients. *Journal of Clinical Microbiology*, *50*(6), 2034–2039.

Martinez, R. M., Bauerle, E. R., Fang, F. C., & Butler-Wu, S. M. (2014). Evaluation of three rapid diagnostic methods for direct identification of microorganisms in positive blood cultures. *Journal of Clinical Microbiology*, *52*(7), 2521–2529.

Martiny, D., Busson, L., Wybo, I., El Haj, R. A., Dediste, A., & Vandenberg, O. (2012). Comparison of the Microflex LT and Vitek MS systems for routine identification of bacteria by matrix-assisted laser desorption ionization-time of flight mass spectrometry. *Journal of Clinical Microbiology*, *50*(4), 1313–1325.

Martiny, D., Dediste, A., & Vandenberg, O. (2012). Comparison of an in-house method and the commercial Sepsityper kit for bacterial identification directly from positive blood culture broths by matrix-assisted laser desorption-ionisation time-of-flight mass spectrometry. *European Journal of Clinical Microbiology & Infectious Diseases*, *31*(9), 2269–2281.

Mather, C. A., Rivera, S. F., & Butler-Wu, S. M. (2014). Comparison of the Bruker Biotyper and Vitek MS matrix-assisted laser desorption ionization-time of flight mass spectrometry systems for identification of mycobacteria using simplified protein extraction protocols. *Journal of Clinical Microbiology*, *52*(1), 130–138.

McElvania TeKippe, E., & Burnham, C. A. (2014). Evaluation of the Bruker Biotyper and VITEK MS MALDI-TOF MS systems for the identification of unusual and/or difficult-to-identify microorganisms isolated from clinical specimens. *European Journal of Clinical Microbiology & Infectious Diseases*, *33*(12), 2163–2171.

McElvania Tekippe, E., Shuey, S., Winkler, D. W., Butler, M. A., & Burnham, C. A. (2013). Optimizing identification of clinically relevant Gram-positive organisms by use of the Bruker Biotyper matrix-assisted laser desorption ionization-time of flight mass spectrometry system. *Journal of Clinical Microbiology*, *51*(5), 1421–1427.

McTaggart, L. R., Lei, E., Richardson, S. E., Hoang, L., Fothergill, A., & Zhang, S. X. (2011). Rapid identification of *Cryptococcus neoformans* and *Cryptococcus gattii* by matrix-assisted laser desorption ionization-time of flight mass spectrometry. *Journal of Clinical Microbiology*, *49*(8), 3050–3053.

Meex, C., Neuville, F., Descy, J., Huynen, P., Hayette, M. P., De Mol, P., et al. (2012). Direct identification of bacteria from BacT/ALERT anaerobic positive blood cultures by MALDI-TOF MS: MALDI Sepsityper kit versus an in-house saponin method for bacterial extraction. *Journal of Medical Microbiology*, *61*(Pt. 11), 1511–1516.

Moon, H. W., Lee, S. H., Chung, H. S., Lee, M., & Lee, K. (2013). Performance of the Vitek MS matrix-assisted laser desorption ionization time-of-flight mass spectrometry system for identification of Gram-positive cocci routinely isolated in clinical microbiology laboratories. *Journal of Medical Microbiology*, *62*(Pt. 9), 1301–1306.

Nagy, E., Becker, S., Kostrzewa, M., Barta, N., & Urban, E. (2012). The value of MALDI-TOF MS for the identification of clinically relevant anaerobic bacteria in routine laboratories. *Journal of Medical Microbiology*, *61*(Pt. 10), 1393–1400.

Nagy, E., Maier, T., Urban, E., Terhes, G., & Kostrzewa, M. (2009). Species identification of clinical isolates of *Bacteroides* by matrix-assisted laser-desorption/ionization time-of-flight mass spectrometry. *Clinical Microbiology and Infection, 15*(8), 796–802.

Navas, M., Pincus, D. H., Wilkey, K., Sercia, L., LaSalvia, M., Wilson, D., et al. (2014). Identification of aerobic Gram-positive bacilli by use of Vitek MS. *Journal of Clinical Microbiology, 52*(4), 1274–1277.

Neville, S. A., Lecordier, A., Ziochos, H., Chater, M. J., Gosbell, I. B., Maley, M. W., et al. (2011). Utility of matrix-assisted laser desorption ionization-time of flight mass spectrometry following introduction for routine laboratory bacterial identification. *Journal of Clinical Microbiology, 49*(8), 2980–2984.

Ngan, G. J., Lin, R. T., & Teo, J. W. (2012). Utility of the Bruker Biotyper matrix-assisted laser desorption ionisation time-of-flight mass spectrometer in a clinical microbiology laboratory. *Pathology, 44*(5), 493–496.

Nonnemann, B., Tvede, M., & Bjarnsholt, T. (2013). Identification of pathogenic microorganisms directly from positive blood vials by matrix-assisted laser desorption/ionization time of flight mass spectrometry. *APMIS: Acta Pathologica, Microbiologica, et Immunologica Scandinavica, 121*(9), 871–877.

Normand, A. C., Cassagne, C., Ranque, S., L'Ollivier, C., Fourquet, P., Roesems, S., et al. (2013). Assessment of various parameters to improve MALDI-TOF MS reference spectra libraries constructed for the routine identification of filamentous fungi. *BMC Microbiology, 13*, 76.

Nyvang Hartmeyer, G., Kvistholm Jensen, A., Bocher, S., Damkjaer Bartels, M., Pedersen, M., Engell Clausen, M., et al. (2010). Mass spectrometry: Pneumococcal meningitis verified and *Brucella* species identified in less than half an hour. *Scandinavian Journal of Infectious Diseases, 42*(9), 716–718.

Panagea, T., Pincus, D. H., Grogono, D., Jones, M., Bryant, J., Parkhill, J., et al. (2015). *Mycobacterium abscessus* complex identification with matrix-assisted laser desorption ionization time of flight (MALDI-TOF) mass spectrometry. *Journal of Clinical Microbiology, 53*(7), 2355–2358.

Patel, R. (2013). Matrix-assisted laser desorption ionization-time of flight mass spectrometry in clinical microbiology. *Clinical Infectious Diseases, 57*(4), 564–572.

Pavlovic, M., Konrad, R., Iwobi, A. N., Sing, A., Busch, U., & Huber, I. (2012). A dual approach employing MALDI-TOF MS and real-time PCR for fast species identification within the *Enterobacter cloacae* complex. *FEMS Microbiology Letters, 328*(1), 46–53.

Pence, M. A., McElvania TeKippe, E., Wallace, M. A., & Burnham, C. A. (2014). Comparison and optimization of two MALDI-TOF MS platforms for the identification of medically relevant yeast species. *European Journal of Clinical Microbiology & Infectious Diseases, 33*(10), 1703–1712.

Perez, K. K., Olsen, R. J., Musick, W. L., Cernoch, P. L., Davis, J. R., Land, G. A., et al. (2013). Integrating rapid pathogen identification and antimicrobial stewardship significantly decreases hospital costs. *Archives of Pathology & Laboratory Medicine, 137*(9), 1247–1254.

Perez, K. K., Olsen, R. J., Musick, W. L., Cernoch, P. L., Davis, J. R., Peterson, L. E., et al. (2014). Integrating rapid diagnostics and antimicrobial stewardship improves outcomes in patients with antibiotic-resistant Gram-negative bacteremia. *The Journal of Infection, 69*(3), 216–225.

Pfaller, M. A., & Diekema, D. J. (2010). Epidemiology of invasive mycoses in North America. *Critical Reviews in Microbiology, 36*(1), 1–53.

Pignone, M., Greth, K. M., Cooper, J., Emerson, D., & Tang, J. (2006). Identification of mycobacteria by matrix-assisted laser desorption ionization-time-of-flight mass spectrometry. *Journal of Clinical Microbiology, 44*(6), 1963–1970.

Pinto, A., Halliday, C., Zahra, M., van Hal, S., Olma, T., Maszewska, K., et al. (2011). Matrix-assisted laser desorption ionization-time of flight mass spectrometry identification of yeasts is contingent on robust reference spectra. *PLoS One, 6*(10), e25712.

Pittet, D., Tarara, D., & Wenzel, R. P. (1994). Nosocomial bloodstream infection in critically ill patients. Excess length of stay, extra costs, and attributable mortality. *JAMA, 271*(20), 1598–1601.

Posteraro, B., Vella, A., Cogliati, M., De Carolis, E., Florio, A. R., Posteraro, P., et al. (2012). Matrix-assisted laser desorption ionization-time of flight mass spectrometry-based method for discrimination between molecular types of *Cryptococcus neoformans* and *Cryptococcus gattii*. *Journal of Clinical Microbiology, 50*(7), 2472–2476.

Powell, E. A., Blecker-Shelly, D., Montgomery, S., & Mortensen, J. E. (2013). Application of matrix-assisted laser desorption ionization-time of flight mass spectrometry for identification of the fastidious pediatric pathogens *Aggregatibacter, Eikenella, Haemophilus*, and *Kingella*. *Journal of Clinical Microbiology, 51*(11), 3862–3864.

Rajakaruna, L., Hallas, G., Molenaar, L., Dare, D., Sutton, H., Encheva, V., et al. (2009). High throughput identification of clinical isolates of *Staphylococcus aureus* using MALDI-TOF-MS of intact cells. *Infection, Genetics and Evolution, 9*(4), 507–513.

Randell, P. (2014). It's a MALDI but it's a goodie: MALDI-TOF mass spectrometry for microbial identification. *Thorax, 69*(8), 776–778.

Ranque, S., Normand, A. C., Cassagne, C., Murat, J. B., Bourgeois, N., Dalle, F., et al. (2014). MALDI-TOF mass spectrometry identification of filamentous fungi in the clinical laboratory. *Mycoses, 57*(3), 135–140.

Ratcliffe, P., Fang, H., Thidholm, E., Borang, S., Westling, K., & Ozenci, V. (2013). Comparison of MALDI-TOF MS and VITEK 2 system for laboratory diagnosis of *Granulicatella* and *Abiotrophia* species causing invasive infections. *Diagnostic Microbiology and Infectious Disease, 77*(3), 216–219.

Richter, S. S., Sercia, L., Branda, J. A., Burnham, C. A., Bythrow, M., Ferraro, M. J., et al. (2013). Identification of *Enterobacteriaceae* by matrix-assisted laser desorption/ionization time-of-flight mass spectrometry using the VITEK MS system. *European Journal of Clinical Microbiology & Infectious Diseases, 32*(12), 1571–1578.

Rodriguez-Sanchez, B., Marin, M., Sanchez-Carrillo, C., Cercenado, E., Ruiz, A., Rodriguez-Creixems, M., et al. (2014). Improvement of matrix-assisted laser desorption/ionization time-of-flight mass spectrometry identification of difficult-to-identify bacteria and its impact in the workflow of a clinical microbiology laboratory. *Diagnostic Microbiology and Infectious Disease, 79*(1), 1–6.

Romero, B., Morosini, M. I., Loza, E., Rodriguez-Banos, M., Navas, E., Canton, R., et al. (2011). Reidentification of *Streptococcus bovis* isolates causing bacteremia according to the new taxonomy criteria: Still an issue? *Journal of Clinical Microbiology, 49*(9), 3228–3233.

Romero-Gomez, M. P., & Mingorance, J. (2011). The effect of the blood culture bottle type in the rate of direct identification from positive cultures by matrix-assisted laser desorption/ionisation time-of-flight (MALDI-TOF) mass spectrometry. *The Journal of Infection, 62*(3), 251–253.

Rychert, J., Burnham, C. A., Bythrow, M., Garner, O. B., Ginocchio, C. C., Jennemann, R., et al. (2013). Multicenter evaluation of the Vitek MS matrix-assisted laser desorption ionization-time of flight mass spectrometry system for identification of Gram-positive aerobic bacteria. *Journal of Clinical Microbiology, 51*(7), 2225–2231.

Rychert, J., Creely, D., Mayo-Smith, L. M., Calderwood, S. B., Ivers, L. C., Ryan, E. T., et al. (2015). Evaluation of matrix-assisted laser desorption ionization-time of flight mass spectrometry for identification of *Vibrio cholerae*. *Journal of Clinical Microbiology, 53*(1), 329–331.

Ryzhov, V., & Fenselau, C. (2001). Characterization of the protein subset desorbed by MALDI from whole bacterial cells. *Analytical Chemistry, 73*(4), 746–750.

Saffert, R. T., Cunningham, S. A., Ihde, S. M., Jobe, K. E., Mandrekar, J., & Patel, R. (2011). Comparison of Bruker Biotyper matrix-assisted laser desorption ionization-time of flight mass spectrometer to BD Phoenix automated microbiology system for identification of gram-negative bacilli. *Journal of Clinical Microbiology, 49*(3), 887–892.

Saleeb, P. G., Drake, S. K., Murray, P. R., & Zelazny, A. M. (2011). Identification of mycobacteria in solid-culture media by matrix-assisted laser desorption ionization-time of flight mass spectrometry. *Journal of Clinical Microbiology, 49*(5), 1790–1794.

Schaumann, R., Knoop, N., Genzel, G. H., Losensky, K., Rosenkranz, C., Stingu, C. S., et al. (2013). Discrimination of *Enterobacteriaceae* and non-fermenting Gram negative bacilli by MALDI-TOF mass spectrometry. *The Open Microbiology Journal, 7*, 118–122.

Schmidt, V., Jarosch, A., Marz, P., Sander, C., Vacata, V., & Kalka-Moll, W. (2012). Rapid identification of bacteria in positive blood culture by matrix-assisted laser desorption ionization time-of-flight mass spectrometry. *European Journal of Clinical Microbiology & Infectious Diseases, 31*(3), 311–317.

Schmitt, B. H., Cunningham, S. A., Dailey, A. L., Gustafson, D. R., & Patel, R. (2013). Identification of anaerobic bacteria by Bruker Biotyper matrix-assisted laser desorption ionization-time of flight mass spectrometry with on-plate formic acid preparation. *Journal of Clinical Microbiology, 51*(3), 782–786.

Schrodl, W., Heydel, T., Schwartze, V. U., Hoffmann, K., Grosse-Herrenthey, A., Walther, G., et al. (2012). Direct analysis and identification of pathogenic *Lichtheimia* species by matrix-assisted laser desorption ionization-time of flight analyzer-mediated mass spectrometry. *Journal of Clinical Microbiology, 50*(2), 419–427.

Schubert, S., Weinert, K., Wagner, C., Gunzl, B., Wieser, A., Maier, T., et al. (2011). Novel, improved sample preparation for rapid, direct identification from positive blood cultures using matrix-assisted laser desorption/ionization time-of-flight (MALDI-TOF) mass spectrometry. *The Journal of Molecular Diagnostics, 13*(6), 701–706.

Schulthess, B., Bloemberg, G. V., Zbinden, R., Bottger, E. C., & Hombach, M. (2014). Evaluation of the Bruker MALDI Biotyper for identification of Gram-positive rods: Development of a diagnostic algorithm for the clinical laboratory. *Journal of Clinical Microbiology, 52*(4), 1089–1097.

Schulthess, B., Brodner, K., Bloemberg, G. V., Zbinden, R., Bottger, E. C., & Hombach, M. (2013). Identification of Gram-positive cocci by use of matrix-assisted laser desorption ionization-time of flight mass spectrometry: Comparison of different preparation methods and implementation of a practical algorithm for routine diagnostics. *Journal of Clinical Microbiology, 51*(6), 1834–1840.

Schulthess, B., Ledermann, R., Mouttet, F., Zbinden, A., Bloemberg, G. V., Bottger, E. C., et al. (2014). Use of the Bruker MALDI Biotyper for the identification of molds in the clinical mycology laboratory. *Journal of Clinical Microbiology, 52*(8), 2797–2803.

Segawa, S., Nishimura, M., Sogawa, K., Tsuchida, S., Murata, S., Watanabe, M., et al. (2015). Identification of Nocardia species using matrix-assisted laser desorption/ionization–time-of-flight mass spectrometry. *Clinical Proteomics*, *12*(1), 6.

Segawa, S., Sawai, S., Murata, S., Nishimura, M., Beppu, M., Sogawa, K., et al. (2014). Direct application of MALDI-TOF mass spectrometry to cerebrospinal fluid for rapid pathogen identification in a patient with bacterial meningitis. *Clinica Chimica Acta*, *435*, 59–61.

Seibold, E., Maier, T., Kostrzewa, M., Zeman, E., & Splettstoesser, W. (2010). Identification of *Francisella tularensis* by whole-cell matrix-assisted laser desorption ionization-time of flight mass spectrometry: Fast, reliable, robust, and cost-effective differentiation on species and subspecies levels. *Journal of Clinical Microbiology*, *48*(4), 1061–1069.

Seng, P., Drancourt, M., Gouriet, F., La Scola, B., Fournier, P. E., Rolain, J. M., et al. (2009). Ongoing revolution in bacteriology: Routine identification of bacteria by matrix-assisted laser desorption ionization time-of-flight mass spectrometry. *Clinical Infectious Diseases*, *49*(4), 543–551.

Singh, K. (2009). Laboratory-acquired infections. *Clinical Infectious Diseases*, *49*(1), 142–147.

Sitterle, E., Giraud, S., Leto, J., Bouchara, J. P., Rougeron, A., Morio, F., et al. (2014). Matrix-assisted laser desorption ionization-time of flight mass spectrometry for fast and accurate identification of *Pseudallescheria/Scedosporium* species. *Clinical Microbiology and Infection*, *20*(9), 929–935.

Sogawa, K., Watanabe, M., Sato, K., Segawa, S., Ishii, C., Miyabe, A., et al. (2011). Use of the MALDI BioTyper system with MALDI-TOF mass spectrometry for rapid identification of microorganisms. *Analytical and Bioanalytical Chemistry*, *400*(7), 1905–1911.

Sogawa, K., Watanabe, M., Sato, K., Segawa, S., Miyabe, A., Murata, S., et al. (2012). Rapid identification of microorganisms by mass spectrometry: Improved performance by incorporation of in-house spectral data into a commercial database. *Analytical and Bioanalytical Chemistry*, *403*(7), 1811–1822.

Stevenson, L. G., Drake, S. K., & Murray, P. R. (2010). Rapid identification of bacteria in positive blood culture broths by matrix-assisted laser desorption ionization-time of flight mass spectrometry. *Journal of Clinical Microbiology*, *48*(2), 444–447.

Sun, L., Teramoto, K., Sato, H., Torimura, M., Tao, H., & Shintani, T. (2006). Characterization of ribosomal proteins as biomarkers for matrix-assisted laser desorption/ionization mass spectral identification of *Lactobacillus plantarum*. *Rapid Communications in Mass Spectrometry*, *20*(24), 3789–3798.

Szabados, F., Woloszyn, J., Richter, C., Kaase, M., & Gatermann, S. (2010). Identification of molecularly defined *Staphylococcus aureus* strains using matrix-assisted laser desorption/ionization time of flight mass spectrometry and the Biotyper 2.0 database. *Journal of Medical Microbiology*, *59*(Pt. 7), 787–790.

Tadros, M., & Petrich, A. (2013). Evaluation of MALDI-TOF mass spectrometry and Sepsityper Kit™ for the direct identification of organisms from sterile body fluids in a Canadian pediatric hospital. *The Canadian Journal of Infectious Diseases & Medical Microbiology*, *24*(4), 191–194.

Tan, K. E., Ellis, B. C., Lee, R., Stamper, P. D., Zhang, S. X., & Carroll, K. C. (2012). Prospective evaluation of a matrix-assisted laser desorption ionization-time of flight mass spectrometry system in a hospital clinical microbiology laboratory for identification of bacteria and yeasts: A bench-by-bench study for assessing the impact on time to identification and cost-effectiveness. *Journal of Clinical Microbiology*, *50*(10), 3301–3308.

Theel, E. S., Hall, L., Mandrekar, J., & Wengenack, N. L. (2011). Dermatophyte identification using matrix-assisted laser desorption ionization-time of flight mass spectrometry. *Journal of Clinical Microbiology*, *49*(12), 4067–4071.

Theel, E. S., Schmitt, B. H., Hall, L., Cunningham, S. A., Walchak, R. C., Patel, R., et al. (2012). Formic acid-based direct, on-plate testing of yeast and *Corynebacterium* species by Bruker Biotyper matrix-assisted laser desorption ionization-time of flight mass spectrometry. *Journal of Clinical Microbiology*, *50*(9), 3093–3095.

Valentine, N. B., Wahl, J. H., Kingsley, M. T., & Wahl, K. L. (2002). Direct surface analysis of fungal species by matrix-assisted laser desorption/ionization mass spectrometry. *Rapid Communications in Mass Spectrometry*, *16*(14), 1352–1357.

Van Herendael, B. H., Bruynseels, P., Bensaid, M., Boekhout, T., De Baere, T., Surmont, I., et al. (2012). Validation of a modified algorithm for the identification of yeast isolates using matrix-assisted laser desorption/ionisation time-of-flight mass spectrometry (MALDI-TOF MS). *European Journal of Clinical Microbiology & Infectious Diseases*, *31*(5), 841–848.

van Veen, S. Q., Claas, E. C., & Kuijper, E. J. (2010). High-throughput identification of bacteria and yeast by matrix-assisted laser desorption ionization-time of flight mass spectrometry in conventional medical microbiology laboratories. *Journal of Clinical Microbiology*, *48*(3), 900–907.

Veloo, A. C., Elgersma, P. E., Friedrich, A. W., Nagy, E., & van Winkelhoff, A. J. (2014). The influence of incubation time, sample preparation and exposure to oxygen on the quality of the MALDI-TOF MS spectrum of anaerobic bacteria. *Clinical Microbiology and Infection*, *20*(12), O1091–O1097.

Veloo, A. C., Knoester, M., Degener, J. E., & Kuijper, E. J. (2011). Comparison of two matrix-assisted laser desorption ionisation-time of flight mass spectrometry methods for the identification of clinically relevant anaerobic bacteria. *Clinical Microbiology and Infection*, *17*(10), 1501–1506.

Verroken, A., Defourny, L., Lechgar, L., Magnette, A., Delmee, M., & Glupczynski, Y. (2014). Reducing time to identification of positive blood cultures with MALDI-TOF MS analysis after a 5-h subculture. *European Journal of Clinical Microbiology & Infectious Diseases*, *34*(2), 405–413.

Verroken, A., Janssens, M., Berhin, C., Bogaerts, P., Huang, T. D., Wauters, G., et al. (2010). Evaluation of matrix-assisted laser desorption ionization-time of flight mass spectrometry for identification of *Nocardia* species. *Journal of Clinical Microbiology*, *48*(11), 4015–4021.

Verwer, P. E., van Leeuwen, W. B., Girard, V., Monnin, V., van Belkum, A., Staab, J. F., et al. (2014). Discrimination of *Aspergillus lentulus* from *Aspergillus fumigatus* by Raman spectroscopy and MALDI-TOF MS. *European Journal of Clinical Microbiology & Infectious Diseases*, *33*(2), 245–251.

Vlek, A. L., Bonten, M. J., & Boel, C. H. (2012). Direct matrix-assisted laser desorption ionization time-of-flight mass spectrometry improves appropriateness of antibiotic treatment of bacteremia. *PLoS One*, *7*(3), e32589.

Vlek, A., Kolecka, A., Khayhan, K., Theelen, B., Groenewald, M., Boel, E., et al. (2014). Interlaboratory comparison of sample preparation methods, database expansions, and cutoff values for identification of yeasts by matrix-assisted laser desorption ionization-time of flight mass spectrometry using a yeast test panel. *Journal of Clinical Microbiology*, *52*(8), 3023–3029.

Walker, J., Fox, A. J., Edwards-Jones, V., & Gordon, D. B. (2002). Intact cell mass spectrometry (ICMS) used to type methicillin-resistant *Staphylococcus aureus*: Media effects and inter-laboratory reproducibility. *Journal of Microbiological Methods, 48*(2–3), 117–126.

Wang, W., Xi, H., Huang, M., Wang, J., Fan, M., Chen, Y., et al. (2014). Performance of mass spectrometric identification of bacteria and yeasts routinely isolated in a clinical microbiology laboratory using MALDI-TOF MS. *Journal of Thoracic Disease, 6*(5), 524–533.

Wang, X. H., Zhang, G., Fan, Y. Y., Yang, X., Sui, W. J., & Lu, X. X. (2013). Direct identification of bacteria causing urinary tract infections by combining matrix-assisted laser desorption ionization-time of flight mass spectrometry with UF-1000i urine flow cytometry. *Journal of Microbiological Methods, 92*(3), 231–235.

Westblade, L. F., Garner, O. B., MacDonald, K., Bradford, C., Pincus, D. H., Mochon, A. B., et al. (2015). Assessment of the reproducibility of MALDI-TOF mass spectrometry for bacterial and yeast identification. *Journal of Clinical Microbiology, 53*(7), 2349–2352.

Westblade, L. F., Jennemann, R., Branda, J. A., Bythrow, M., Ferraro, M. J., Garner, O. B., et al. (2013). Multicenter study evaluating the Vitek MS system for identification of medically important yeasts. *Journal of Clinical Microbiology, 51*(7), 2267–2272.

Wieser, A., Schneider, L., Jung, J., & Schubert, S. (2012). MALDI-TOF MS in microbiological diagnostics—Identification of microorganisms and beyond (mini review). *Applied Microbiology and Biotechnology, 93*(3), 965–974.

Wojewoda, C. (2013). Pathology consultation on matrix-assisted laser desorption ionization-time of flight mass spectrometry for microbiology. *American Journal of Clinical Pathology, 140*(2), 143–148.

Woods, K., Beighton, D., & Klein, J. L. (2014). Identification of the 'Streptococcus anginosus group' by matrix-assisted laser desorption ionization-time-of-flight mass spectrometry. *Journal of Medical Microbiology, 63*(Pt. 9), 1143–1147.

Yan, Y., He, Y., Maier, T., Quinn, C., Shi, G., Li, H., et al. (2011). Improved identification of yeast species directly from positive blood culture media by combining Sepsityper specimen processing and Microflex analysis with the matrix-assisted laser desorption ionization Biotyper system. *Journal of Clinical Microbiology, 49*(7), 2528–2532.

Zbinden, A., Mueller, N. J., Tarr, P. E., Eich, G., Schulthess, B., Bahlmann, A. S., et al. (2012). *Streptococcus tigurinus*, a novel member of the *Streptococcus mitis* group, causes invasive infections. *Journal of Clinical Microbiology, 50*(9), 2969–2973.

Zhu, B., Xiao, D., Zhang, H., Zhang, Y., Gao, Y., Xu, L., et al. (2013). MALDI-TOF MS distinctly differentiates nontypable *Haemophilus influenzae* from *Haemophilus haemolyticus*. *PLoS One, 8*(2), e56139.

POC Tests in Microbial Diagnostics: Current Status

3

Luc Bissonnette*,†, Michel G. Bergeron*,†,1

*Centre de recherche en infectiologie de l'Université Laval, Axe maladies infectieuses et immunitaires, Centre de recherche du CHU de Québec, Québec City, Québec, Canada
†Département de microbiologie-infectiologie et d'immunologie, Faculté de médecine, Université Laval, Québec City, Québec, Canada
1Corresponding author: e-mail address: michel.g.bergeron@crchul.ulaval.ca

1 INTRODUCTION

In healthcare, the management of infectious diseases remains a high-level activity since infections still claim tens of millions of lives annually (Khabbaz, Moseley, Steiner, Levitt, & Bell, 2014; Lozano et al., 2012; Table 1). Bacteria such as *Staphylococcus aureus*, *Klebsiella pneumoniae*, or *Mycobacterium tuberculosis*, which were controllable 10–20 years ago with the antibiotic arsenal of the time have become highly or extensively resistant to drugs, and contributed to the expansion of life-threatening healthcare-acquired/associated infections (HAIs) (Laxminarayan et al., 2013; Munoz-Price et al., 2013).

Compared to other diagnostic disciplines, clinical microbiology is still disadvantaged by the time (minimum of 2–3 days) required to detect, isolate, and/or identify the microorganism(s) suspected of causing an infection. Molecular microbiology, in the form of nucleic acid-based tests (NATs) derived from the polymerase chain reaction (PCR), molecular hybridisation and isothermal amplification techniques, as well as newer approaches derived from mass spectrometry technologies have certainly contributed to shortening the time required to perform a diagnostic procedure, provide more specific microbial identification, and inform the physician of the potential for antimicrobial resistance, but the overall turnaround time (TAT) of the test has become highly dependent on the number and length of steps of the diagnostic process required to forward a clinical sample to the lab and to return the result to the physician (Bissonnette & Bergeron, 2010, 2012a, 2012b; Bissonnette et al., 2015).

Needless to say but, upon entrance in a healthcare facility, more efficient management could occur if a non-invasive clinical sample from a patient would be subjected, at the point of care/need (POC), to a simple (single step), affordable, and rapid (less than 1 h) test contained in a simple device. Furthermore, the POC test done in a decentralised setting by (healthcare) personnel with no or minimal training should

Methods in Microbiology, Volume 42, ISSN 0580-9517, http://dx.doi.org/10.1016/bs.mim.2015.09.003

Table 1 Global Mortality, Overall Rank, and Disability-Adjusted Life Years (DALYs) of Important Infectious Diseases During Year 2010

	Infectious Disease(s)	Deaths	Global Rank	DALYs
All Causes	**Not Applicable**	**52,769,700**		**2,490,385,000**
	Lower respiratory tract infections	2,814,400	4	115,227,000
	Influenza	507,900		19,244,000
	Respiratory syncytial virus pneumonia	235,500		20,472,000
	Streptococcus pneumoniae pneumonia	827,300		26,906,000
	Haemophilus influenzae type B pneumonia	379,900		21,315,000
	Other lower respiratory tract infections	845,800		27,289,000
	Upper respiratory tract infections	3000		1,866,000
Lower and upper respiratory tract infections	*Total*	*5,613,800*		*232,319,000*
	HIV/AIDS	1,465,400	6	81,547,000
	HIV disease leading to mycobacterial infection	256,900		14,948,000
	HIV disease leading to other disease(s)	1,208,400		66,600,000
	Tuberculosis	1,196,000	10	49,396,000
	Syphilis	113,300		9,578,000
	Chlamydial diseases	1200		714,000
	Gonococcal infection	900		282,000
	Trichomoniasis	Not available		167,000
	Other sexually transmitted infections	2900		236,000
HIV/AIDS, tuberculosis, and sexually transmitted infections	*Total*	*4,245,000*		*223,468,000*
	Rotaviral enteritis	250,900		18,650,000
	E. coli infections and shigellosis	332,300		21,488,000
	Campylobacter enteritis	109,700		7,541,000
	Salmonella infections	81,300		4,847,000

Table 1 Global Mortality, Overall Rank, and Disability-Adjusted Life Years (DALYs) of Important Infectious Diseases During Year 2010—cont'd

	Infectious Disease(s)	Deaths	Global Rank	DALYs
All Causes	Not Applicable	52,769,700		2,490,385,000
	Cholera	58,100		4,463,000
	Cryptosporidiosis	99,800		8,372,000
	Amoebiasis	55,500		2,237,000
	Other diarrhoeal diseases	458,300		21,916,000
Diarrhoea and gastrointestinal diseases	Total	1,445,800	7	89,514,000
	Malaria	1,169,500	11	82,685,000
	Leishmaniasis	51,600		3,317,000
	Chagas disease	10,300		546,000
	African trypanosomiasis	9100		560,000
Selected parasitic/ tropical diseases	Total	1,240,500		87,108,000
	Preterm birth complications	859,700	15	76,982,000
	Neonatal sepsis and other infections	513,700	25	44,236,000
Selected neonatal disorders	Total	1,373,400		121,218,000
	Acute hepatitis A	102,800		4,351,000
	Acute hepatitis B	132,200		4,674,000
	Acute hepatitis C	16,000		518,000
	Acute hepatitis E	56,600		3,715,000
Hepatitis	Total	307,700		13,258,000

Adapted from Lozano et al. (2012) and Murray et al. (2012).

provide a performance similar to tests performed by skilled staff in controlled laboratory conditions (Abou Tayoun, Burchard, Malik, Scherer, & Tsongalis, 2014; Caliendo et al., 2013; Mohd Hanafiah, Garcia, & Anderson, 2013; Rudolf & Lewandrowski, 2015). Whatever the configuration of the POC test, if it can provide sufficiently precise information to the medical staff, the information will facilitate the control of disease transmission or dissemination if contact isolation is required, facilitate timely and appropriate antimicrobial treatment if needed, and reduce the risk of selecting for antimicrobial-resistant microbes which may engender (outbreaks of) HAIs. It may also reduce the length-of-stay of patients, reduce direct and indirect healthcare costs, and above all, reduce morbidity and mortality due to infection (Beavers Vandiver, 2015; Loubiere & Moatti, 2010; Rudolf & Lewandrowski, 2015). Undoubtedly, rapid antigen-based and molecular microbiology POC tests have the potential to induce "a change in culture without (microbial) culture in healthcare" (Bergeron, 2008). However, we and others consider that POC should a determinant of the overall diagnostic strategy or programme and not only of the properties (user-friendliness and cost) and performance of the test. In this line of thinking and to refocus medicine on the patient, the decentralisation of healthcare becomes critical to the success of POC diagnostics, as decreasing the distance from the healthcare user to the analysis site will directly influence the TAT of the test and the efficiency of test result transmission (Bissonnette et al., 2015; Huckle, 2008; Pai, Vadnais, Denkinger, Engel, & Pai, 2012; Rudolf & Lewandrowski, 2015; Schito et al., 2012; Zwerling & Dowdy, 2013).

To a certain extent, the development and implementation of low-cost diagnostic tests for improving global health in developing countries may provide insight on how to develop POC tests for industrialised countries. The World Health Organization (WHO) has proposed a series of criteria to orient the development of diagnostic tests deployable in resource-limited settings, the "ASSURED" (Affordable, Sensitive, Specific, User-friendly, Rapid, Equipment-free, Delivered to those in need) criteria (Mabey, Peeling, Ustianowski, & Perkins, 2004; Urdea et al., 2006). Accessible tests developed according to the ASSURED criteria and responding to performance requirements in the order of 85–95% sensitivity and 85–95% specificity could dramatically reduce to impact of infectious diseases in developing countries (Abou Tayoun et al., 2014; Urdea et al., 2006). Many simple rapid diagnostic tests on the market already conform to the ASSURED criteria, but the adoption of ASSURED nucleic acid, microfluidics, and/or nanotechnology-based testing technologies is hampered by issues related to cost, user-friendliness, and reliance to high-tech equipment. In recent years however, the WHO has been involved in promoting and benchmarking in developing countries (relatively) rapid molecular diagnostic tests and novel technologies to better diagnose and control infectious diseases such as HIV/AIDS, sexually transmitted infections (STI), and tuberculosis. The experience gained in these settings plus the advice and opinions from experts in the field, may provide evidence supporting or facilitating the development, regulatory approval, and implementation of tests and technologies (at the POC) in industrialised countries, although at a higher cost (Lewandrowski, 2008; Peeling & McNerney, 2014). In this chapter, we provide an overview of commercially available tests that are used at POC and describe some of the anticipated developments in the field of POC molecular diagnostics.

2 THE CURRENT UTILISATION OF INFECTIOUS DISEASES POC TESTING

In 1988, the United States of America (USA) Congress has adopted the Clinical Laboratory Improvement Amendments (CLIA), a set of regulations and requirements establishing the quality standards for laboratory testing to ensure the accuracy, reliability, and timeliness of patient test results, regardless of where the test is performed

(CMS, 2014a; Table 2). Under CLIA, devices and/or tests are generally categorised as waived, of moderate or high complexity, and the settings in which POC tests are performed must comply with standards related to personnel requirements, proficiency testing, quality control, quality assurance, and test management (Ammirati, 2015; Shawar & Weissfeld, 2011). POC tests generally fall under the "waived" category where they are defined as "simple laboratory examinations and procedures that have an insignificant risk of an erroneous result".

Table 3 shows a representative but not exhaustive list of (families of) tests currently waived for infectious disease diagnostics at POC, a market essentially dominated by immunology-based (lateral flow immunochromatographic, LFI) tests for the rapid detection of microbial antigens and by a limited number of enzymatic tests for the assessment of protein biomarkers. It is well known that many rapid POC tests are developed outside of clinical laboratories but often lack sensitivity and/or

Table 2 Recommendations and Requirements of the United States Food and Drug Administration for CLIA-Waived Tests

Test Platform/ Instrument	Specimen	Test and Reagent(s)	User
– Unitised or self-contained test, or fully automated instrument – Produces a direct readout or results that require no calibration, interpretation, or calculation – No operator intervention during analysis – No electronic or mechanical maintenance	– Uses unprocessed specimens (only whole blood- and urine-based assays may be waived) – Basic, non-technique-dependent specimen manipulation – Provides instructions and materials for obtaining/ shipping specimens for confirmatory testing, if clinically advisable	– Quick reference instruction sheet written at no higher than 7th grade reading level – Test performance comparable to traceable reference method – Basic, non-technique-dependent reagent manipulation – Produces a direct readout or results that require no calibration, interpretation, or calculation – Produces results that are clear to read (positive/negative), a direct readout of numerical values, the clear presence/ absence of a line, or obvious colour gradation(s)	– No operator intervention during analysis – No technical or specialised training related to troubleshooting or interpretation of multiple or complex error codes

Adapted from Ammirati (2015).

Table 3 Examples of Commercially Available Tests for Infectious Diseases Granted Waived Status Under CLIA (as of July 1, 2014)

Microorganism(s) or Infection	Representative Commercially Available Test	Manufacturer (URL)	Description/Recommended Use(s)
General infection			
Urinary tract infection	Foresight U120 Urine Analyzer	Acon Laboratories, Inc. (www.aconlabs.com)	Strip/dipstick-based system for screening of urine to monitor/diagnose various diseases/conditions including urinary tract infections
Urinary tract infection	CLINITEK Status+ Analyzer	Siemens Healthcare Diagnostics (www.healthcare.siemens.com)	Strip/dipstick-based system for screening of urine to monitor/diagnose various diseases/conditions including urinary tract infections and the semi-quantitative measurement of albumin and creatinine
Viruses			
Adenoviral conjunctivitis	AdenoPlus	Rapid Pathogen Screening, Inc. (www.rpsdetectors.com)	Immunochromatographic test for the qualitative detection of adenoviral antigens from eye fluid
Epstein–Barr virus (mononucleosis)	Acceava® Mono II Test	Alere (www.alere.com)	Rapid chromatographic immunoassay for the qualitative detection of infectious mononucleosis heterophile antibodies in whole blood, serum, and plasma to aid in the diagnosis of infectious mononucleosis
Hepatitis C virus (HCV)	OraQuick HCV Rapid Antibody Test	OraSure Technologies, Inc. (www.orasure.com)	Qualitative immunoassay to detect antibodies to hepatitis C virus in fingerstick whole blood and venipuncture whole blood specimens
Human immunodeficiency virus type 1 (HIV-1)	OraQuick Advance Rapid HIV-1/2 Antibody Test	OraSure Technologies, Inc. (www.orasure.com)	Qualitative immunoassay to detect antibodies to human immunodeficiency virus type 1 (HIV-1) and type 2 (HIV-2) in fingerstick whole blood, venipuncture whole blood, and/or oral fluid specimens
Human immunodeficiency virus type 1 (HIV-1)	OraQuick In-Home HIV Test (Oral Fluid)	OraSure Technologies, Inc. (www.orasure.com)	Qualitative immunoassay to detect antibodies to human immunodeficiency virus type 1 (HIV-1) and type 2 (HIV-2) in oral fluid specimens
Influenza virus (flu)	QuickVue® Influenza Test	Quidel Corporation (www.quidel.com)	Qualitative detection of influenza type A and type B antigens from nasal swab, nasal wash, or nasal aspirate specimens *that does not differentiate* between influenza types A and B
Influenza A/B virus (flu)	BinaxNOW® Influenza A & B Test in nasopharyngeal (NP) swab and nasal wash/aspirate specimens	Alere (www.alere.com)	*In vitro* immunochromatographic card assay for the qualitative detection of influenza A and B nucleoprotein antigens in nasopharyngeal (NP) swab, nasal swab samples, and nasal wash/aspirate specimens, for the rapid differential diagnosis of influenza A and B viral infections

Influenza A/B virus (flu)	Alere™ i Influenza A & B	Alere (www.alere.com)	*First NAT to receive a CLIA waiver.* Alere™ i Influenza A & B utilises isothermal nucleic acid amplification technology for the differential and qualitative detection of influenza A and influenza B viral nucleic acids. Detection from nasal swabs and viral transport media. The reaction tubes contain the reagents required for the amplification of target nucleic acids and an internal control. The test utilises a pair of templates (similar to primers) for the specific amplification of RNA from influenza A and B. Amplified RNA targets are specifically detected using a fluorescently labelled molecular beacon
Respiratory syncytial virus (RSV)	Alere BinaxNOW® RSV Card	Alere (www.alere.com)	*In vitro* rapid immunochromatographic card assay used to detect respiratory syncytial virus (RSV) fusion protein antigen in nasal wash and nasopharyngeal (NP) swab specimens from symptomatic patients
Bacteria			
Helicobacter pylori (peptic ulcer disease and some gastric cancers)	FlexSure® HP Whole Blood	Beckman Coulter, Inc. (www.beckmancoulter.com)	Non-invasive, visually read, qualitative, and immunochromatographic method for rapid detection of antibodies to *H. pylori*; results within 4 min
Borrelia burgdorferi (Lyme disease)	PreVue™ *B. burgdorferi* Antibody Detection Assay	Wampole Laboratories, Inc. (www.wampolelabs.com)	Enzyme-linked immunosorbent assay for the presumptive detection of IgG antibodies to *B. burgdorferi* in whole blood or serum. This ELISA should only be used for patients with signs and symptoms that are consistent with Lyme disease
Streptococcus pyogenes (Group A Streptococci; GAS)	ACON Strep A Twist Test	Acon Laboratories, Inc. (www.aconlabs.com)	Rapid detection of GAS antigen from throat swabs; used as an aid in the diagnosis of GAS infection, which typically causes strep throat, tonsillitis, and scarlet fever
Bacterial vaginosis	Gryphus Diagnostics BVBlue	Gryphus Diagnostics, LLC (http://bvblue.com)	Enzyme activity test for the detection of sialidase activity in vaginal fluid specimens, an enzyme produced by bacterial pathogens such as *Gardnerella vaginalis*, *Bacteroides* spp., *Prevotella* spp., and *Mobiluncus* spp.
Protozoa/parasites			
Trichomonas vaginalis	OSOM *Trichomonas* Rapid Test	Genzyme Diagnostics (www.genzymediagnostics.com)	Immunoassay for the qualitative detection of *Trichomonas vaginalis* antigens from vaginal swabs or from the saline solution prepared when making wet mounts from vaginal swabs

Adapted from Centers for Medicare and Medicaid Services (2014b), United States Food and Drug Administration (2014a), and from information made publicly available by manufacturers.

specificity. In many instances however, they may provide sufficiently good diagnostic information to facilitate more timely changes to patient management and improve adherence to treatment, as long as a physician is available to act upon the test result (Clerc & Greub, 2010; Gaydos & Hardick, 2014; Nichols, 2007, 2008; Niemz, Ferguson, & Boyle, 2011; Overturf, 2008; Rajan & Glorikian, 2009; Trevino & Weissfeld, 2007; Tucker, Bien, & Peeling, 2013; Watchirs Smith et al., 2012).

2.1 VIRAL RESPIRATORY INFECTIONS

In 2010, the global mortality from respiratory tract infections was estimated to more than 5.6 million (Table 1). Each year in the USA, almost 70 million medical visits are attributable to viral respiratory tract infections, while influenza is responsible for more than 200,000 hospitalisations and 3000–49,000 deaths (Chartrand, Leeflang, Minion, Brewer, & Pai, 2012). Rapid antigen-based CLIA-waived tests for the detection of the influenza virus, essentially influenza A and/or B virus, dominate this market but, at a cost (15–20$ per test) similar to a molecular amplification assay, their performance is somewhat variable and hence, there is ample room for improvement (Chartrand et al., 2012; He & Hoe, 2015; IDSA, 2011; Tayo, Ellis, Phillips, Simpson, & Ward, 2012). The genetic variability of the influenza virus, as well as the limited specificity and sensitivity of rapid antigen tests against H1N1 virus for example might limit their clinical utility (Landry, 2011). As humans, infants, and the elderly are increasingly exposed and susceptible to a very large variety of respiratory viruses, the clinical value of rapid immunology-based tests might be decreasing in favour of more adaptable molecular multiplex assays, many of which already integrated in automated systems (He & Hoe, 2015; IDSA, 2011; Tayo et al., 2012).

2.2 HIV/AIDS, TUBERCULOSIS, SEXUALLY TRANSMITTED INFECTIONS

The global public health burden of STI, in terms of morbidity and mortality, is mainly attributable to the human immunodeficiency virus (HIV/AIDS; Table 1). Each year, 448 million new cases worldwide (20 million in the USA alone) of STI caused by *Treponema pallidum*, *Chlamydia trachomatis*, *Neisseria gonorrhoeae*, or *Trichomonas vaginalis* incur significant costs related to patient management, treatment, and loss of productivity (Gaydos & Hardick, 2014; Tucker et al., 2013). Rapid oral tests for home-based testing of HIV can contribute to increase the coverage of populations at risk and are seen as a way of better controlling epidemics, especially in high-prevalence settings where the test positive predictive value with oral specimens is estimated to 98.7%. In industrialised countries, not only accurate, rapid, and inexpensive POC tests are needed to control the STI epidemics, but rapid POC testing done in outpatient clinics can increase compliance to treatment while impacting the transmission rate within the population, if the result can be obtained while the patient is still in the clinic (Gaydos & Hardick, 2014; Hsieh et al., 2010; Khabbaz et al., 2014; Loubiere & Moatti, 2010; Rompalo et al., 2013; Tucker et al., 2013; Turner et al., 2013; Watchirs Smith et al., 2012).

In developing countries, the availability of low-cost POC tests for malaria and HIV has transformed the management of these diseases, but the situation is much more complex in the context of synergistic epidemics of HIV and multidrug and extensively drug-resistant tuberculosis where molecular diagnostics performed at POC would have a great impact on global public health (Chan et al., 2013; Dheda, Ruhwald, Theron, Peter, & Yam, 2013; Ellner, 2009; Günther, 2014; Pai & Pai, 2012; Weyer, Carai, & Nunn, 2011; Young, Perkins, Duncan, & Barry, 2008). For diagnosing (resistant forms of) TB, the smear microscopy method only has a limited sensitivity as it cannot distinguish between drug-sensitive and drug-resistant *M. tuberculosis* and other mycobacteria of the *M. tuberculosis* complex. Moreover, culture-based methods for determining antimicrobial resistance necessitate weeks to months to produce results, thereby increasing the risk of household and community disease transmission. The Xpert® MTB/RIF test (Cepheid, USA), introduced in TB clinics and peripheral laboratories of developing countries after the line probe assay, is an opportunity of demonstrating the impact of molecular diagnostics at POC for controlling this re-emerging infectious disease (Günther, 2014; Moore, 2013; Pai & Pai, 2012; Pai et al., 2012; Van Rie, Page-Shipp, Scott, Sanne, & Stevens, 2010; Weyer et al., 2011). This should also provide a model for the strategic development of similar platforms for tackling other infectious diseases.

2.3 DIARRHOEA AND GASTROINTESTINAL DISEASES

Each year, the consumption of contaminated water or food results in hundreds of millions of cases (reported and unreported) of diarrhoea and gastroenteritis, associated with significant morbidity and mortality worldwide (Table 1). Gastrointestinal pathogens include a wide variety of viruses, bacteria, and protozoan parasites (Mead et al., 1999; Straub & Chandler, 2003) which inherently complicates their detection unless a persistent or life-threatening infection or outbreak is encountered. Instead, the risk of waterborne gastrointestinal diseases is generally determined by culture-based methods for the detection of faecal contamination indicators, *Escherichia coli* or enterococci for example.

Significant gastrointestinal pathogens such as norovirus (*Caliciviridae*) and the protozoan parasites *Cryptosporidium* and *Giardia* cannot be easily cultivated and, even if rapid antigen-based tests have been developed for stool screening, their detection by multiplex molecular assays would be preferable. In resource-limited settings however, the reference method for detecting *Cryptosporidium* and *Giardia* the microscopic evaluation of stained stool smears has limited diagnostic utility due to time constraints and lack of trained microscopists; therefore, rapid LFI tests are considered viable alternatives although they have a reduced analytical performance when compared to conventional methods (Minak et al., 2012; Shimelis & Tadesse, 2014).

2.4 SURVEILLANCE, PREVENTION, AND DIAGNOSIS OF HEALTHCARE-ACQUIRED INFECTIONS

Catalysed by the evolution and dissemination of antimicrobial resistance genes, microorganisms causing HAIs have been threatening patient safety and the economics of healthcare systems worldwide. Each year, it is estimated that 440,000–722,000 HAIs occur among USA adult inpatients, costing more than $9.8 billion; importantly, 50–75% of HAIs are considered preventable (Magill et al., 2014; Zimlichman et al., 2013). Although, this is a controversial issue both strategically and economically, the preventative war against antibiotic-resistant pathogens and HAIs associated to methicillin-resistant *S. aureus* (MRSA), vancomycin-resistant enterococci (VRE), carbapenemase-producing enterobacteria (CPE), or *Clostridium difficile* should ideally start upon admission in a healthcare facility such that patients suspected of being infected and/or potential carriers are quickly identified and subjected to contact isolation measures (Akova, Daikos, Tzouvelekis, & Carmeli, 2012; Barbut et al., 2014; Beavers Vandiver, 2015; Brenwald, Baker, & Oppenheim, 2010; Cunningham et al., 2007; Hardy et al., 2010; Koo et al., 2014; Laxminarayan et al., 2013; Parcell & Phillips, 2014; Peterson, 2010; Robicsek et al., 2008; van der Zee, Hendriks, Roorda, Ossewaarde, & Buitenwerf, 2013). Due to the complexity of coupling microbial identification of these important pathogens, especially CPE, with the determination of their genotypic potential for resistance to antimicrobial agents, rapid multiplex nucleic acid-based platforms would be more appropriate than rapid and specific molecular POC tests unless the local epidemiology or occurrence of an outbreak would indicate otherwise (Bissonnette & Bergeron, 2012b). Within 5–10 years, the availability of new POC platforms offering a reasonable multiplexing capacity should resolve this difficulty (Bissonnette et al., 2015).

2.5 ANTICIPATED POC TESTING ADVANCES THROUGH BIOMARKERS OR IMPROVED TECHNOLOGIES

2.5.1 Sepsis

In some countries, procalcitonin is a biomarker assessed in clinical practice to indicate the presence and severity of bacterial infections such as community-acquired pneumonia and sepsis (Hattori et al., 2014; Kasamatsu et al., 2012; Kibe, Adams, & Barlow, 2011; Lee & Cho, 2015; Schuetz, Albrich, & Mueller, 2011). In the case of sepsis, the suggestion has been made to use procalcitonin serum levels as an antimicrobial stewardship tool (Kibe et al., 2011; Schuetz et al., 2011). A rapid semi-quantitative solid-phase immunoassay (BRAHMS PCT-Q) and a POC assay (BRAHMS PCT™ direct) are commercially available (www.procalcitonin.com), but not yet CLIA waived. Kemmler and colleagues have described a POC immunofluorescence-based device for the measurement of four protein biomarkers (C-reactive protein, interleukin-6, procalcitonin, and neopterin), from plasma or serum, indicating the presence of sepsis (Kemmler, Sauer, Schleicher,

Preininger, & Brandenburg, 2014). A POC device for quantifying procalcitonin directly from blood was recently described (Rascher et al., 2014).

Recently, Jones and colleagues have demonstrated that a spontaneous migration phenotype of neutrophils in a microfluidic platform is highly correlated with the occurrence of sepsis in patients admitted with burns covering more than 20% of their body (Jones et al., 2014). The TAT of this experimental procedure is estimated to be 4 h.

2.5.2 Technological advances in rapid immunochromatographic platforms

To increase the utilisation of rapid diagnostic tests requires an increase in their analytical performance by the incorporation or integration of technological features, increasing assay sensitivity, or augmenting the capacity for multiplexed analysis. Improvements in microfluidic flow control, pumping and valving, advances in microfabrication and stability of ligands, (bio)sensors and detection mechanisms, portability, and communication of results are considered fields of opportunity for researchers and biotechnology companies (Bier & Schumacher, 2013; Gervais, de Rooij, & Delamarche, 2011; Gubala, Harris, Ricco, Tan, & Williams, 2012; Han, Li, & Seong, 2013; Sin, Mach, Wong, & Liao, 2014; Truong, Abraham, Kost, & Louie, 2015; Wang, Inci, De Libero, Singhal, & Demirci, 2013).

2.5.3 Paper-based diagnostics

Rapid LFI tests are essentially fabricated with nitrocellulose and plastic materials. In 2007, scientific reports describing that a paper matrix could be functionalised with hydrophobic polymers, to create hydrophilic and hydrophobic regions enabling to perform the rapid and economical detection of analytes opened a new field of investigation. The incorporation of more complex capillary-based fluidics on this material will certainly bring about novel solutions for low-cost POC diagnostics (Byrnes, Thiessen, & Fu, 2013; Chan et al., 2013; Hu et al., 2014; Phillips & Lewis, 2014).

2.6 PATIENT EMPOWERMENT IS KEY FOR ANY POC TEST

Glucose and pregnancy testing are classic examples of home-based tests, however with regard to infectious diseases, we predict that POC testing will be initially performed within a decentralised healthcare facility by minimally skilled staff, such that the result can be analysed by a qualified healthcare professional (physician, nurse, laboratory technician, or pharmacist) because the test result might lead to a therapeutic prescription (Kaman, Andrinopoulou, & Hays, 2013). However, home-based infectious disease POC testing will expand when tests like the OraQuick In-Home HIV Test gain a greater market share.

3 THE ROAD TO POC MOLECULAR DIAGNOSTICS IS PAVED WITH GOOD INTENTIONS AND TECHNOLOGIES

There are very few POC tests for detecting important infections such as HAIs caused by *S. aureus*, *C. difficile*, enterococci or multi-resistant enterobacteria, and syndromic infections such as or sepsis or those of the respiratory tract (Table 2). More than 2 million people in the USA each year develop an infection caused by antibiotic-resistant bacteria; according to the Centers for Disease Control and Prevention (CDC), an excess of 23,000 people will die from those infections (Laxminarayan et al., 2013). As syndromic infections are seldom polymicrobial by nature and anti-microbial resistance biomarkers are not easily detected by immunology-based tests, the diagnostic challenge(s) would be more appropriately addressed by multiplex NATs or post-culture approaches like mass spectrometry (Bissonnette & Bergeron, 2012b; He & Hoe, 2015; Mahmoudzadeh, Tran, Louie, Curtis, & Kost, 2015).

At the beginning of the century, an international panel of experts determined that "modified molecular technologies for affordable, simple diagnosis of infectious diseases" was the top priority for improving health in developing countries (Daar et al., 2002). Since then, we have witnessed a significant expansion of the spectrum of NATs that have been approved by the United States Food and Drug Administration (USFDA) for the diagnosis of infectious diseases (USFDA, 2014b), but market development is still complicated by financial, clinical, and acceptability challenges (Caliendo et al., 2013; Gibbs, 2011; Huckle, 2008, 2010; Palamountain et al., 2012). A new Frost & Sullivan report predicted that the global infectious disease diagnostics market, estimated to $7.92 billion in 2013 will grow at a compound annual growth rate of 10%, to reach $12.78 billion by 2018, essentially driven by molecular testing (Parmar, 2014). The Infectious Diseases Society of America (IDSA) recently published two documents calling for the development and implementation of molecular technologies to advance clinical microbiology testing and provide tools to prevent healthcare-acquired *C. difficile* infection and better control antibiotic resistance (Caliendo et al., 2013; IDSA, 2011). Despite the advantages of molecular technologies in terms of speed, adaptability, and analytical performance, no test had yet met the requirements for operating at POC under CLIA at the end of 2014. This is thought to be due to complex sample preparation methods, limited test menu, qualification issues, and/or insufficient phenotypic information about the detected pathogens provided by technologies (Table 2; Ammirati, 2015; Luppa, Müller, Schlichtiger, & Schlebusch, 2011; Mohd Hanafiah et al., 2013; Moore, 2013; Overturf, 2008; Tantra & van Heeren, 2013; Tran, Godwin, Steele, & Howes, 2015). To operate at POC and therefore compete with current antigen-based POC tests, a nucleic acid-based diagnostic platform or test should comply with quite stringent requirements: (1) 24/7 availability for *stat* testing or minimal batching, (2) specific and sensitive results obtained within a TAT of a few hours (ideally less than one) to effectively impact on clinical management, outcome, or public health, (3) minimal technical skills, (4) stable reagents with extended shelf life (at least 6 months), (5) cost effectiveness, and (6) compatibility with a wide range of clinical samples. We and others consider that POC should qualify the overall diagnostic strategy and not only the properties and performance of the test itself. In this line of thinking, the decentralisation of healthcare becomes highly critical to the success of POC diagnostics, as the distance from the healthcare user to the site of bioanalysis would directly influence the TAT of the test i.e., the length of the diagnostic cycle (Bissonnette et al., 2015; Pai et al., 2012; Schito et al., 2012; Zwerling & Dowdy, 2013). We also propose that public health studies demonstrating the feasibility of performing nucleic acid-based testing at POC should be undertaken to assess the acceptability of novel technology platforms by healthcare workers including physicians, nurses, hospital and clinic administrators, and government authorities. If the cost- and time-effectiveness of decentralised testing are confirmed, this should induce a strategic reorganisation of diagnostic practices including near-patient POC molecular testing done in dedicated infrastructures (near-POC laboratories; Cohen-Bacrie et al., 2011; Nougairede et al., 2010), outpatient clinics, dispensaries in remote areas or developing countries, relatively mobile or adaptable settings, and ultimately at home. Several studies have already shown that rapid molecular testing done in central laboratories impacted positively on the health of patients and healthcare costs, as well as on antimicrobial stewardship and preventing the dissemination of infections, especially HAIs (Akova et al., 2012; Barbut et al., 2014; Cunningham et al., 2007; Hardy et al., 2010; Koo et al., 2014; Perez et al., 2013; Peterson, 2010; Robicsek et al., 2008; van der Zee et al., 2013). In addition, other reports suggest that having molecular diagnostics performed at POC can enhance the benefits for patients, by accelerating disease management and by saving more lives and costs (Beavers Vandiver, 2015; Brenwald et al., 2010; Parcell & Phillips, 2014).

3.1 NEAR-POC TESTING

Before introducing POC testing, clinical molecular microbiology labs should step outside the central laboratory model towards satellite (near-POC) laboratories. These should be mandated, staffed, and equipped with instrumentation to shorten the time and physical length of the diagnostic process. Operated and staffed by trained technologists, such laboratories should offer a comprehensive menu of rapid molecular tests, according to an organisational model that is in operation in Marseille (France) (Cohen-Bacrie et al., 2011; Nougairede et al., 2010). In addition to its potential to impact on infectious disease management, near-POC testing may provide useful insights on how to implement novel platforms and tests in healthcare, and determine how these tools could be improved to be delivered closer to the patient.

3.2 ISOTHERMAL AMPLIFICATION TECHNOLOGIES

To a certain extent, the development of molecular POC tests must be envisaged with the ASSURED criteria in mind, as a test that works efficiently in tropical conditions would perform as well in our hospitals and clinics. Electrical power and cold storage requirements are arguments seldom raised against the implementation of PCR-based technologies in resource-limited settings and this has provided a window of opportunity for the development of isothermal amplification technologies. Isothermal methods, many of which relying on polyenzymatic cocktails, such as helicase-dependent amplification (HDA), loop-mediated amplification (LAMP), nucleic acid sequence-based amplification (NASBA), recombinase polymerase amplification (RPA), strand displacement amplification (SDA), and enzyme-assisted target recycling (EATR) have been used to develop (FDA approved) nucleic acid detection assays (de Paz, Brotons, & Muñoz-Almagro, 2014; Gerasimova & Kolpashchikov, 2014; Niemz et al., 2011; Yan et al., 2014), using lateral flow-like or luminescence detection. These have enabled the detection of a broader range of microbial targets compared to rapid LFI tests but less than what is possible using multiplex PCR amplification. Since isothermal amplification-based tests conceptually resemble rapid LFI tests, it is anticipated that the first CLIA-waived NATs will arise from this sector of molecular diagnostic manufacturing. Early in 2015, the Alere™ i Influenza A & B test became the first NAT to receive a CLIA waiver by the USFDA. Alere™ i is a platform, composed of a sample receiver containing an elution buffer, a test base comprising one or two sealed tubes with lyophilised reagents, a cartridge for transferring the eluted sample to the base and the Alere™ i instrument, which relies on the proprietary "Molecular In Minutes™ (MIM)" isothermal nucleic acid amplification technology for the qualitative detection of microbial pathogens (http://www.alere.com/ww/en/product-details/alere-i-influenza-ab.html). Alere™ i tests for the (POC) detection of respiratory syncytial virus (RSV), *Streptococcus pyogenes* (Strep A), *C. difficile*, and chlamydia (presumably gonorrhoea also) are also under development.

3.3 INTEGRATED MOLECULAR DIAGNOSTICS PLATFORMS

The demonstration of the rapid detection of a neonatal pathogen, *Streptococcus agalactiae* (Group B streptococci or GBS), by PCR and real-time PCR (rtPCR) directly from a clinical sample constitutes a cornerstone event in the history of molecular diagnostics of infectious diseases (Bergeron et al., 2000). Soon after this achievement, the IDI-StrepB™ assay, operated on the SmartCycler® instrument (Cepheid), became the first rtPCR test approved by the FDA for the detection of a microbial pathogen.

These first generations of molecular diagnostic tests have demonstrated that a streamlined workflow, encompassing elution of microbial cells or particles from a swab, nucleic acid extraction and rtPCR amplification, facilitates the rapid detection of pathogens in 1–2 h. Therefore their integration into closed microfluidic devices could provide a level of simplification making such workflows amenable to the POC test format (Bier & Schumacher, 2013; Chin, Linder, & Sia, 2012; Park, Sabour, Son, Lee, & Lee, 2014; Robinson & Dittrich, 2013; Schumacher, Lüdecke, Ehrentreich-Förster, & Bier, 2013; Yeh, Nisic, Yu, Xia, & Zheng, 2014; Table 4).

Table 4 Integrated Platforms or Systems Anticipated to Enable Nucleic Acid-Based Diagnostics of Infectious Diseases at the POC

Platform (Manufacturer; URL)	Description/Features/Reference
Alere™ q (Alere; www.alere.com/ww/en/product-details/alere-q-analyser.html)	– Fully automated nucleic acid testing platform enabling (multiplex) rtPCR – Built in battery enabling operation even in case of power outage – Alere™ q HIV-1/2 detect assay provides virological HIV-1 and HIV-2 results at POC within 50 min
FilmArray™ (BioFire Diagnostics, now a bioMérieux company; www.idahotech.com)	– Microfluidic cartridge for sample preparation, nucleic acid extraction, and multiplex PCR – Platform is FDA cleared – FDA-cleared applications include Blood Culture Identification (BCID) Panel, Respiratory (RP) Panel, and Gastrointestinal (GI) Panel
GenePOC (GenePOC Inc.; www.genepoc-diagnostics.com)	– Centripetal microfluidic platform enabling rapid lysis and nucleic acid extraction coupled to rtPCR amplification, in approximately 1 h – Each device can detect up to 12 genetic targets – Can perform 1–8 tests simultaneously – Bissonnette et al. (2015)
GeneXpert® (Cepheid; www.cepheid.com/us/cepheid-solutions/systems/genexpert-systems/genexpert-i/63?view=products)	– Random access system based on a fluidic cartridge enabling rapid lysis, nucleic acid extraction, and rtPCR/RT-rtPCR amplification – Platform is FDA cleared and CLIA moderate complexity – Available configuration can handle 1, 2, 4, 16, or 48 cartridges – Applications include diagnosis of healthcare-acquired infections, respiratory diseases (tuberculosis, flu, respiratory syncytial virus), and detection of *Streptococcus agalactiae* (GBS), *Chlamydia trachomatis* (CT), and *Neisseria gonorrhoeae* (NG)
IsoAmp® (BioHelix Corporation; www.biohelix.com)	– Isothermal amplification by helicase-dependent amplification (HDA) to amplify an HSV-specific target sequence – Amplification products are detected visually by a lateral flow strip enclosed in a disposable cassette – IsoAmp® HSV Assay is cleared by the FDA for the direct, qualitative detection of herpes simplex virus (HSV-1 and HSV-2) DNA in male and female genital and oral lesions within 1.5 h
Laboratory-in-a-tube (Liat) (Roche Molecular Diagnostics; http://molecular.roche.com/instruments/Pages/cobasLIATsystem.aspx)	– Developed by IQuum – Liat Analyzer and Liat Influenza A/B Assay are CE marked and FDA cleared
VerePLEX™ Biosystem (Veredus Laboratories; www.vereduslabs.com)	– Versatile multiplexing platform capable of performing qualitative analysis of nucleic acids from various sample types – Up to 50 tests in a single reaction performed in a lab-on-a-chip (VereChip™) – VereChip™ is a microfluidic device integrating multiplexed nucleic acid amplification and detection by molecular hybridisation – Applications include diagnostics or detection of seasonal and pandemic influenza, tuberculosis and resistance markers, tropical diseases, biothreat agents, and foodborne pathogens

Adapted from Niemz et al. (2011) and Peeling and McNerney (2014).

3.3.1 The GeneXpert® platform

To demonstrate the feasibility of bringing molecular diagnostics closer to the bedside of patients, Cepheid (Sunnyvale, CA; www.cepheid.com) implemented a sample-to-answer test for the rapid detection of *S. agalactiae* on the GeneXpert® integrated fluidic system. In 2006, the Xpert® GBS test was approved by the USFDA and categorised as "moderate complexity" under CLIA. Since then, the GeneXpert® molecular diagnostics platform has attracted considerable attention in industrialised countries, but also in developing countries where the Xpert® TB test has been implemented to diagnose tuberculosis (Pai & Pai, 2012; Schito et al., 2012; Van Rie et al., 2010; Weyer et al., 2011; Zwerling & Dowdy, 2013). The GeneXpert® platform is a random access instrument which utilises fluidic cartridges that automate nucleic acid extraction and rtPCR or reverse transcription rtPCR (RT-rtPCR) amplification to detect a range of pathogens genetic targets and human genetic markers. In a recent review focusing on the Xpert® Flu assay, the authors concluded that the properties and performance of this test should make it amenable to POC testing of respiratory diseases associated with the influenza A and B viruses, as it could easily compensate for the lack of sensitivity or speed of rapid influenza diagnostic tests (Salez et al., 2014).

The regulatory acceptance of the GeneXpert® platform constitutes another key event in the history of *in vitro* diagnostics, as it led to the development of a number of microfluidic platforms in academia and industry for incorporating genomics, transcriptomics, or proteomics-based bioanalysis in POC tests (Bissonnette & Bergeron, 2012b; Chin et al., 2012; He & Hoe, 2015; Kim, 2013; Mahmoudzadeh et al., 2015; Niemz et al., 2011; Peeling & McNerney, 2014; Table 4).

3.3.2 The GenePOC platform

For several years now, we and others have been advocating for the implementation of rapid molecular diagnostics at POC. Our academic-industrial research programme has led to the development of a unique centripetal microfluidics platform, that of GenePOC Inc. of Québec City, Canada (www.genepoc-diagnostics.com; Bissonnette et al., 2015). GenePOC is planning to commercialise a simple fully automated nucleic acid-based testing platform for the rapid (less than 1 h) detection of infectious microorganisms, in a configuration amenable to operation at POC. With less than 1 min of hands-on time, the GenePOC generic consumable (pie cartridge) has the capacity to process a wide range of clinical samples and detect up to twelve genetic targets by rtPCR. Furthermore, the instrument is designed to perform between 1 and 8 independent molecular tests during a single run. In 2013, GenePOC Inc. was named North American Molecular Diagnostics Entrepreneurial Company of the Year by Frost and Sullivan (2013).

CONCLUSIONS

Contrary to expectations, improvements in hygiene, sanitation, and public health have not sufficed to limit the morbidity and mortality of infectious diseases. The emergence or re-emergence of microbial pathogens and the evolution of antimicrobial resistance have complicated the management and treatment of infectious diseases. Although the phrase "*An ounce of prevention is worth a pound of cure*" (Benjamin Franklin, circa 1735) was raised as a directive in fire safety, it is highly pertinent to human health. Indeed, immunisation certainly is the most efficient measure against vaccine-preventable diseases (Duclos, Okwo-Bele, Gacic-Dobo, & Cherian, 2009). However, considering the range and genetic variability of human pathogens, diagnostic tests must be used to determine the cause of infection and appropriate treatment options. What is true for many vaccine-preventable diseases can be exemplified by tuberculosis where the availability of a rapid molecular diagnostic test is desired to curb epidemics of disease, by alleviating the need for a more effective vaccine, the time delay and cumbersomeness of culture-based methods for microbial identification and determination of antimicrobial resistance, and the wish for a shorter antibiotic regimen (Ellner, 2009; Nossal, 2013; Young et al., 2008).

In this chapter, we have provided an overview of the rapid POC tests that have been CLIA waived for the diagnosis of a limited number of infectious diseases and also provided a vision of biomarkers and technologies anticipated to enter the POC testing arena in the years to come. Hopefully, the spectrum of antigen-based tests will be quickly complemented by more specific, sensitive, and rapid NATs that will confirm the feasibility of doing molecular testing at POC and be simple and robust enough to obtain regulatory clearance. From 2013 to 2018, it is estimated that a greater demand for more decentralised (POC) molecular testing will be an important driver of the growth of the global infectious diseases diagnostics market (Parmar, 2014), thereby reversing a tendency we have witnessed in healthcare in the last decade. The simpler format of isothermal amplification technologies may enhance their acceptability and the development of integrated microfluidic nucleic acid-based platforms facilitating multiplex detection of microbial pathogens and antimicrobial resistance markers. The gain in time realised by deploying rapid molecular diagnostics at POC will be accompanied by a wide array of mostly positive direct and indirect effects, impacting the patient and the community, healthcare workers, as well as the organisation and economics of healthcare. The very recent CLIA waiver of the Alere™ i Influenza A & B test will certainly provide an impetus and a window of opportunity for developers of POC molecular diagnostics; there is an exciting road ahead of us.

ACKNOWLEDGEMENTS

This work was supported in part by a grant from the *Fonds de partenariat pour un Québec innovant et en santé* (*Ministère de l'Économie, de l'Innovation et des Exportations, Gouvernement du Québec*, Canada).

Conflict of Interest Declaration: L.B. declares having no conflict of interest, while M.G.B. is the founder of GenePOC Inc. of Québec City (Québec), Canada.

REFERENCES

Abou Tayoun, A. N., Burchard, P. R., Malik, I., Scherer, A., & Tsongalis, G. J. (2014). Democratizing molecular diagnostics for the developing world. *American Journal of Clinical Pathology, 141*, 17–24.

Akova, M., Daikos, G. L., Tzouvelekis, L., & Carmeli, Y. (2012). Interventional strategies and current clinical experience with carbapenemase-producing Gram-negative bacteria. *Clinical Microbiology and Infection, 18*, 439–448.

Ammirati, E. B. (2015). Bringing novel technologies to market: Addressing Food and Drug Administration, Clinical Laboratory Amendment, and international regulatory challenges for *in vitro* diagnostics. In G. J. Kost & C. M. Corbin (Eds.), *Global point of care—Strategies for disasters, emergencies, and public health resilience* (pp. 585–593). Washington: AACC Press.

Barbut, F., Surgers, L., Eckert, C., Visseaux, B., Cuingnet, M., Mesquita, C., et al. (2014). Does a rapid diagnosis of *Clostridium difficile* infection impact on quality of patient management? *Clinical Microbiology and Infection, 20*, 136–144.

Beavers Vandiver, T. K. (2015). Collaborative medicine: Weaving microbiology into point-of-care clinical responsiveness. In G. J. Kost & C. M. Corbin (Eds.), *Global point of care—Strategies for disasters, emergencies, and public health resilience* (pp. 277–289). Washington: AACC Press.

Bergeron, M. G. (2008). Revolutionizing the practice of medicine through rapid (<1h) DNA-based diagnostics. *Clinical and Investigative Medicine, 31*, E265–E271.

Bergeron, M. G., Ke, D., Ménard, C., Picard, F. J., Gagnon, M., Bernier, M., et al. (2000). Rapid detection of group B streptococci in pregnant women at delivery. *New England Journal of Medicine, 343*, 175–179.

Bier, F. F., & Schumacher, S. (2013). Integration in bioanalysis: Technologies for point-of-care testing. *Advances in Biochemical Engineering/Biotechnology, 133*, 1–14.

Bissonnette, L., & Bergeron, M. G. (2010). Diagnosing infections—Current and anticipated technologies for point-of-care diagnostics and home-based testing. *Clinical Microbiology and Infection, 16*, 1044–1053.

Bissonnette, L., & Bergeron, M. G. (2012a). Infectious disease management through point-of-care personalized medicine molecular diagnostic technologies. *Journal of Personalized Medicine, 2*, 50–70.

Bissonnette, L., & Bergeron, M. G. (2012b). Multiparametric technologies for the diagnosis of syndromic infections. *Clinical Microbiology Newsletter, 34*, 159–168.

Bissonnette, L., Chapdelaine, S., Peytavi, R., Huletsky, A., Stewart, G., Boissinot, M., et al. (2015). A revolutionary microfluidic stand-alone platform (GenePOC) for nucleic-acid based point-of-care diagnostics. In G. J. Kost & C. M. Corbin (Eds.), *Global point of care—Strategies for disasters, emergencies, and public health resilience* (pp. 235–247). Washington: AACC Press.

Brenwald, N. P., Baker, N., & Oppenheim, B. (2010). Feasibility study of a real-time PCR test for methicillin-resistant *Staphylococcus aureus* in a point of care setting. *Journal of Hospital Infection, 74*, 245–249.

Byrnes, S., Thiessen, G., & Fu, E. (2013). Progress in the development of paper-based diagnostics for low-resource point-of-care settings. *Bioanalysis, 5*, 2821–2836.

Caliendo, A. M., Gilbert, D. N., Ginocchio, C. C., Hanson, K. E., May, L., Quinn, T. C., et al. (2013). Better tests, better care: Improved diagnostics for infectious diseases. *Clinical Infectious Diseases, 57*(S3), S139–S170.

Centers for Medicare and Medicaid Services (U.S. Department of Health and Human Services). (2014a). *Clinical Laboratory Improvement Amendments (CLIA).* http://www.cms.gov/Regulations-and-Guidance/Legislation/CLIA. (Accessed December 2014).

Centers for Medicare and Medicaid Services (U.S. Department of Health and Human Services). (2014b). *Tests Granted Waived Status Under CLIA.* http://www.cms.gov/Regulations-and-Guidance/Guidance/Transmittals/downloads/R2919CP.pdf and http://www.cms.hhs.gov/CLIA/downloads/waivetbl.pdf (Accessed December 2014).

Chan, C. P. Y., Mak, W. C., Cheng, K. Y., Sin, K. K., Yu, C. M., Rainer, T. H., et al. (2013). Evidence-based point-of-care diagnostics: Current status and emerging technologies. *Annual Review of Analytical Chemistry, 6,* 191–211.

Chartrand, C., Leeflang, M. M. G., Minion, J., Brewer, T., & Pai, M. (2012). Accuracy of rapid influenza diagnostic tests—A meta-analysis. *Annals of Internal Medicine, 156,* 500–511.

Chin, C. D., Linder, V., & Sia, S. K. (2012). Commercialization of microfluidic point-of-care devices. *Lab on a Chip, 12,* 2118–2134.

Clerc, O., & Greub, G. (2010). Routine use of point-of-care tests: Usefulness and application in clinical microbiology. *Clinical Microbiology and Infection, 16,* 1054–1061.

Cohen-Bacrie, S., Ninove, L., Nougairède, A., Charrel, R., Richet, H., Minodier, P., et al. (2011). Revolutionizing clinical microbiology laboratory organization in hospitals with *in situ* point-of-care. *PLoS One, 6,* e22403.

Cunningham, R., Jenks, P., Northwood, J., Wallis, M., Ferguson, S., & Hunt, S. (2007). Effect on MRSA transmission of rapid PCR testing of patients admitted to critical care. *Journal of Hospital Infection, 65,* 24–28.

Daar, A. S., Thorsteinsdóttir, H., Martin, D. K., Smith, A. C., Nast, S., & Singer, P. A. (2002). Top ten biotechnologies for improving health in developing countries. *Nature Genetics, 32,* 229–232.

de Paz, H. D., Brotons, P., & Muñoz-Almagro, C. (2014). Molecular isothermal techniques for combating infectious diseases: Towards low-cost point-of-care diagnostics. *Expert Review of Molecular Diagnostics, 14,* 827–843.

Dheda, K., Ruhwald, M., Theron, G., Peter, J., & Yam, W. C. (2013). Point-of-care diagnosis of tuberculosis: Past, present and future. *Respirology, 18,* 217–232.

Duclos, P., Okwo-Bele, J.-M., Gacic-Dobo, M., & Cherian, T. (2009). Global immunization: Status, progress, challenges and future. *BMC International Health and Human Rights, 9*(Suppl. 1), S2.

Ellner, J. J. (2009). The emergence of extensively drug-resistant tuberculosis: A global health crisis requiring new interventions: Part II: Scientific advances that may provide solutions. *Clinical and Translational Science, 2,* 80–84.

Frost & Sullivan (2013). *North American molecular diagnostics entrepreneurial company of the year award.* http://www.genepoc-diagnostics.com/News/Frost–Sullivan-honors-GenePOC-with-the-2013-Best-Practices-Award.shtml, (Last accessed December 2014).

Gaydos, C., & Hardick, J. (2014). Point of care diagnostics for sexually transmitted infections: Perspectives and advances. *Expert Review of Anti-Infective Therapy, 12,* 657–672.

Gerasimova, Y. V., & Kolpashchikov, D. M. (2014). Enzyme-assisted target recycling (EATR) for nucleic acid detection. *Chemical Society Reviews, 43,* 6405–6438.

Gervais, L., de Rooij, N., & Delamarche, E. (2011). Microfluidic chips for point-of-care immunodiagnostics. *Advances in Materials, 23,* H151–H176.

Gibbs, J. N. (2011). Regulating molecular diagnostic assays: Developing a new regulatory structure for a new technology. *Expert Review of Molecular Diagnostics, 8,* 367–381.

Gubala, V., Harris, L. F., Ricco, A. J., Tan, M. X., & Williams, D. E. (2012). Point of care diagnostics: Status and future. *Analytical Chemistry, 84*, 487–515.

Günther, G. (2014). Multidrug-resistant and extensively drug-resistant tuberculosis: A review of current concepts and future challenges. *Clinical Medicine, 14*, 279–285.

Han, K. N., Li, C. A., & Seong, G. H. (2013). Microfluidic chips for immunoassays. *Annual Review of Analytical Chemistry, 6*, 119–141.

Hardy, K., Price, C., Szczepura, A., Gossain, S., Davies, R., Stallard, N., et al. (2010). Reduction in the rate of methicillin-resistant *Staphylococcus aureus* acquisition in surgical wards by rapid screening for colonization: A prospective, cross-over study. *Clinical Microbiology and Infection, 16*, 333–339.

Hattori, T., Nishiyama, H., Kato, H., Ikegami, S., Nagayama, M., Asami, S., et al. (2014). Clinical value of procalcitonin for patients with suspected bloodstream infection. *American Journal of Clinical Pathology, 141*, 43–51.

He, J., & Hoe, J. (2015). Rapid molecular diagnostics for major respiratory viruses. In G. J. Kost & C. M. Corbin (Eds.), *Global point of care—Strategies for disasters, emergencies, and public health resilience* (pp. 221–234). Washington: AACC Press.

Hsieh, Y. H., Hogan, M. T., Barnes, M., Jett-Goheen, M., Huppert, J., Rompalo, A. M., et al. (2010). Perceptions of an ideal point-of-care test for sexually transmitted infections—A qualitative study of focus group discussions with medical providers. *PLoS One, 5*, e14144.

Hu, J., Wang, S., Wang, L., Li, F., Pingguan-Murphy, B., Lu, T. J., et al. (2014). Advances in paper-based point-of-care diagnostics. *Biosensors and Bioelectronics, 54*, 585–597.

Huckle, D. (2008). Point-of-care diagnostics: An advancing sector with nontechnical issues. *Expert Review of Molecular Diagnostics, 8*, 679–688.

Huckle, D. (2010). Point-of-care diagnostics—Is this driven by supply or demand? *Expert Opinion on Medical Diagnostics, 4*, 189–200.

Infectious Diseases Society of America. (2011). An unmet medical need: Rapid molecular diagnostics tests for respiratory tract infections. *Clinical Infectious Diseases, 52*(S4), S384–S395.

Jones, C. N., Moore, M., Dimisko, L., Alexander, A., Ibrahim, A., Hassell, B. A., et al. (2014). Spontaneous neutrophil migration patterns during sepsis after major burns. *PLoS One, 9*, e114509.

Kaman, W. E., Andrinopoulou, E.-R., & Hays, J. P. (2013). Perceptions of point-of-care infectious disease testing among European medical personnel, point-of-care test kit manufacturers, and the general public. *Patient Preference and Adherence, 7*, 559–577.

Kasamatsu, Y., Yamaguchi, T., Kawaguchi, T., Tanaka, N., Oka, H., Nakamura, T., et al. (2012). Usefulness of a semi-quantitative procalcitonin test and the A-DROP Japanese prognostic scale for predicting mortality among adults hospitalized with community-acquired pneumonia. *Respirology, 17*, 330–336.

Kemmler, M., Sauer, U., Schleicher, E., Preininger, C., & Brandenburg, A. (2014). Biochip point-of-care device for sepsis diagnostics. *Sensors and Actuators B, 192*, 205–215.

Khabbaz, R. F., Moseley, R. R., Steiner, R. J., Levitt, A. M., & Bell, B. P. (2014). Challenges of infectious diseases in the USA. *Lancet, 384*, 53–63.

Kibe, S., Adams, K., & Barlow, G. (2011). Diagnostic and prognostic biomarkers of sepsis in critical care. *Journal of Antimicrobial Chemotherapy, 66*(suppl. 2), ii33–ii40.

Kim, L. (2013). Overview of the microfluidic diagnostics commercial landscape. *Methods in Molecular Biology, 949*, 65–83.

Koo, H. L., Van, J. N., Zhao, M., Ye, X., Revell, P. A., Jiang, Z.-D., et al. (2014). Real-time polymerase chain reaction detection of asymptomatic *Clostridium difficile* colonization and rising *C. difficile*-associated disease rates. *Infection Control and Hospital Epidemiology, 35*, 667–673.

Landry, M. L. (2011). Diagnostic tests for influenza infection. *Current Opinion in Pediatrics, 23*, 91–97.

Laxminarayan, R., Duse, A., Wattal, C., Zaidi, A. K. M., Wertheim, H. F. L., Sumpradit, N., et al. (2013). Antibiotic resistance—The need for global solutions. *Lancet Infectious Diseases, 13*, 1057–1098.

Lee, H. J., & Cho, S. Y. (2015). Utiliy of procalcitonin in the clinical laboratory. In G. J. Kost & C. M. Corbin (Eds.), *Global point of care—Strategies for disasters, emergencies, and public health resilience* (pp. 161–164). Washington: AACC Press.

Lewandrowski, K. (2008). Three wishes for POCT testing: A compendium of unmet needs from the perspective of practitioners in the field. *Point of Care, 7*, 86–88.

Loubiere, S., & Moatti, J.-P. (2010). Economic evaluation of point-of-care diagnostic technologies for infectious diseases. *Clinical Microbiology and Infection, 16*, 1070–1076.

Lozano, R., Naghavi, M., Foreman, K., Lim, S., Shibuya, K., Aboyans, V., et al. (2012). Global and regional mortality from 235 causes of death for 20 age groups in 1990 and 2010: A systematic analysis for the Global Burden of Disease Study 2010. *Lancet, 380*, 2095–2128.

Luppa, P. B., Müller, C., Schlichtiger, A., & Schlebusch, H. (2011). Point-of-care testing (POCT): Current techniques and future perspectives. *Trends in Analytical Chemistry, 30*, 887–898.

Mabey, D., Peeling, R. W., Ustianowski, A., & Perkins, M. D. (2004). Diagnostics for the real world. *Nature Reviews Microbiology, 2*, 231–240.

Magill, S. S., Edwards, J. R., Bamberg, W., Beldavs, Z. G., Dumyati, G., Kainer, M. A., et al. (2014). Multistate point-prevalence survey of health care-associated infections. *New England Journal of Medicine, 370*, 1198–1208.

Mahmoudzadeh, S., Tran, N. K., Louie, R., Curtis, C. M., & Kost, G. J. (2015). Rapid molecular diagnosis of sepsis in critical, emergency, and disaster care. In G. J. Kost & C. M. Corbin (Eds.), *Global point of care—Strategies for disasters, emergencies, and public health resilience* (pp. 191–220). Washington: AACC Press.

Mead, P. S., Slutsker, L., Dietz, V., McCaig, L. F., Bresee, J. S., Shapiro, C., et al. (1999). Food-related illness and death in the United States. *Emerging Infectious Diseases, 5*, 607–625.

Minak, J., Kabir, M., Mahmud, I., Liu, Y., Haque, R., & Petri, W. A., Jr. (2012). Evaluation of rapid antigen point-of-care tests for detection of *Giardia* and *Cryptosporidium* in human fecal specimens. *Journal of Clinical Microbiology, 50*, 154–156.

Mohd Hanafiah, K., Garcia, M., & Anderson, D. (2013). Point-of-care testing and the control of infectious diseases. *Biomarkers in Medicine, 7*, 333–347.

Moore, C. (2013). Point-of-care tests for infection control: Should rapid testing be in the laboratory or at the front line? *Journal of Hospital Infection, 85*, 1–7.

Munoz-Price, L. S., Poirel, L., Bonomo, R. A., Schwaber, M. J., Daikos, G. L., & Cormican, M. (2013). Clinical epidemiology of the global expansion of *Klebsiella pneumoniae* carbapenemases. *Lancet Infectious Diseases, 13*, 785–796.

Murray, C. J. L., Vos, T., Lozano, R., Naghavi, M., Flaxman, A. D., & Michaud, C. (2012). Disability-adjusted life years (DALYs) for 291 diseases and injuries in 21 regions,

1990–2010: A systematic analysis for the Global Burden of Disease Study 2010. *Lancet*, *380*, 2197–2223.

Nichols, J. H. (2007). Point of care testing. *Clinics in Laboratory Medicine*, *27*, 893–908.

Nichols, J. H. (2008). The future of point of care testing. *Point of Care*, *7*, 271–273.

Niemz, A., Ferguson, T. M., & Boyle, D. S. (2011). Point-of-care nucleic acid testing for infectious diseases. *Trends in Biotechnology*, *29*, 240–250.

Nossal, G. J. V. (2013). Immunology and world health: Key contributions from the global community. *Annals of the New York Academy of Sciences*, *1283*, 1–7.

Nougairede, A., Ninove, L., Zandotti, C., de Lamballerie, X., Gazin, C., Drancourt, M., et al. (2010). Point of care strategy for rapid diagnosis of novel A/H1N1 influenza virus. *PLoS One*, *5*, e9215.

Overturf, G. D. (2008). CLIA waived testing in infectious diseases. *Pediatric Infectious Disease Journal*, *27*, 1009–1011.

Pai, N. P., & Pai, M. (2012). Point-of-care diagnostics for HIV and tuberculosis: Landscape, pipeline and unmet needs. *Discovery Medicine*, *13*, 35–45.

Pai, N. P., Vadnais, C., Denkinger, C., Engel, N., & Pai, M. (2012). Point-of-care testing for infectious diseases: Diversity, complexity, and barriers in low- and middle-income countries. *PLoS Medicine*, *9*, e1001306.

Palamountain, K. M., Baker, J., Cowan, E. P., Essajee, S., Mazzola, L. T., Metzler, M., et al. (2012). Perspectives on introduction and implementation of new point-of-care diagnostic tests. *Journal of Infectious Diseases*, *205*(Suppl. 2), S181–S190.

Parcell, B. J., & Phillips, G. (2014). Use of Xpert™ MRSA PCR point-of-care testing beyond the laboratory. *Journal of Hospital Infection*, *87*, 119–121.

Park, S.-m, Sabour, A. F., Son, J. H., Lee, S. H., & Lee, L. P. (2014). Toward integrated molecular diagnostic system (iMDx): Principles and applications. *IEEE Transactions on Biomedical Engineering*, *61*, 1506–1521.

Parmar, A. (2014). Molecular testing to drive infectious diseases diagnostics market. *MDDI Medical Device and Diagnostic Industry*, (Accessed December 2014). www.mddionline.com/print/12115.

Peeling, R. W., & McNerney, R. (2014). Emerging technologies in point-of-care molecular diagnostics for resource-limited settings. *Expert Review of Molecular Diagnostics*, *14*, 525–534.

Perez, K. K., Olsen, R. J., Musick, W. L., Cernoch, P. L., Davis, J. R., & Land, G. A. (2013). Integrating rapid pathogen identification and antimicrobial stewardship significantly decreases hospital costs. *Archives of Pathology & Laboratory Medicine*, *137*, 1247–1254.

Peterson, L. R. (2010). To screen or not to screen for methicillin-resistant *Staphylococcus aureus*. *Journal of Clinical Microbiology*, *48*, 683–689.

Phillips, S. T., & Lewis, G. G. (2014). The expanding role of paper in point-of-care diagnostics. *Expert Review of Molecular Diagnostics*, *14*, 123–125.

Rajan, A., & Glorikian, H. (2009). Point-of-care diagnostics: Market trends and growth drivers. *Expert Opinion on Medical Diagnostics*, *3*, 1–4.

Rascher, D., Geerlof, A., Kremmer, E., Krämer, P., Michael, S., Hartmann, A., et al. (2014). Total internal reflection (TIRF)-based quantification of procalcitonin for sepsis diagnosis—A point-of-care testing application. *Biosensors and Bioelectronics*, *59*, 251–258.

Robicsek, A., Beaumont, J. L., Paule, S. M., Hacek, D. M., Thomson, R. B., Jr., Kaul, K. L., et al. (2008). Universal surveillance for methicillin-resistant *Staphylococcus aureus* in 3 affiliated hospitals. *Annals of Internal Medicine*, *148*, 409–418.

Robinson, T., & Dittrich, P. S. (2013). Microfluidic technology for molecular diagnostic. *Advances in Biochemical Engineering/Biotechnology, 133*, 89–114.

Rompalo, A. M., Hsieh, Y.-H., Hogan, T., Barnes, M., Jett-Goheen, M., Huppert, J.-S., et al. (2013). Point-of-care tests for sexually transmissible infections: What do "end users" want? *Sexual Health, 10*, 541–545.

Rudolf, J., & Lewandrowski, K. (2015). Point-of-care testing in the emergency department: Impact on operations and outcomes. In G. J. Kost & C. M. Corbin (Eds.), *Global point of care—Strategies for disasters, emergencies, and public health resilience* (pp. 69–82). Washington: AACC Press.

Salez, N., Nougairede, A., Ninove, L., Zandotti, C., de Lambellerie, X., & Charrel, R. N. (2014). Xpert Flu for point-of-care diagnosis of human influenza in industrialized countries. *Expert Review of Molecular Diagnostics, 14*, 411–418.

Schito, M., Peter, T. F., Cavanaugh, S., Piatek, A. S., Young, G. J., Alexander, H., et al. (2012). Opportunities and challenges for cost-efficient implementation of new point-of-care diagnostics for HIV and tuberculosis. *Journal of Infectious Diseases, 205*, S169–S180.

Schuetz, P., Albrich, W., & Mueller, B. (2011). Procalcitonin for diagnosis of infection and guide to antibiotic decisions: Past, present and future. *BMC Medicine, 9*, 107.

Schumacher, S., Lüdecke, C., Ehrentreich-Förster, E., & Bier, F. F. (2013). Platform technologies for molecular diagnostics near the patient's bedside. *Advances in Biochemical Engineering/Biotechnology, 133*, 75–87.

Shawar, R., & Weissfeld, A. S. (2011). FDA regulation of clinical microbiology diagnostic devices. *Journal of Clinical Microbiology, 49*, S80–S84.

Shimelis, T., & Tadesse, E. (2014). Performance evaluation of point-of-care test for detection of *Cryptosporidium* stool antigen in children and HIV infected adults. *Parasites & Vectors, 7*, 227.

Sin, M. L. Y., Mach, K. E., Wong, P. K., & Liao, J. C. (2014). Advances and challenges in biosensor-based diagnosis of infectious diseases. *Expert Review of Molecular Diagnostics, 14*, 225–244.

Straub, T. M., & Chandler, D. P. (2003). Toward a unified system for detecting waterborne pathogens. *Journal of Microbiological Methods, 53*, 185–197.

Tantra, R., & van Heeren, H. (2013). Product qualification: A barrier to point-of-care microfluidic-based diagnostics. *Lab on a Chip, 13*, 2199–2201.

Tayo, A., Ellis, J., Phillips, L. L., Simpson, S., & Ward, D. J. (2012). Emerging point of care tests for influenza: Innovation or status quo. *Influenza and Other Respiratory Viruses, 6*, 291–298.

Tran, N. K., Godwin, Z., Steele, A. N., & Howes, M. H. (2015). The disaster-critical care interface: Closing the gaps with value propositions and harmonization. In G. J. Kost & C. M. Corbin (Eds.), *Global point of care—Strategies for disasters, emergencies, and public health resilience* (pp. 83–93). Washington: AACC Press.

Trevino, E. A., & Weissfeld, A. S. (2007). The case for point-of-care testing in infectious-disease diagnosis. *Clinical Microbiology Newsletter, 29*, 177–179.

Truong, A.-T., Abraham, A., Kost, G. J., & Louie, R. F. (2015). Smartphone-enabled point of care. In G. J. Kost & C. M. Corbin (Eds.), *Global point of care—Strategies for disasters, emergencies, and public health resilience* (pp. 177–187). Washington: AACC Press.

Tucker, J. D., Bien, C. H., & Peeling, R. W. (2013). Point-of-care testing for sexually transmitted infections: Recent advances and implications for disease control. *Current Opinion in Infectious Diseases, 26*, 73–79.

Turner, S. D., Anderson, K., Slater, M., Quigley, L., Dyck, M., & Guiang, C. B. (2013). Rapid point-of-care HIV testing in youth: A systematic review. *Journal of Adolescent Health, 53*, 683–691.

United States Food and Drug Administration (2014a). *CLIA—Tests waived by FDA from January 2000 to present.* http://www.accessdata.fda.gov/scripts/cdrh/cfdocs/cfclia/testswaived.cfm (Accessed December 2014).

United States Food and Drug Administration (2014b). *Nucleic acid based tests.* http://www.fda.gov/MedicalDevices/ProductsandMedicalProcedures/InVitroDiagnostics/ucm330711.htm (Accessed December 2014).

Urdea, M., Penny, L. A., Olmsted, S. S., Giovanni, M. Y., Kaspar, P., Shepherd, A., et al. (2006). Requirements for high impact diagnostics in the developing world. *Nature, 444*(Suppl. 1), 73–79.

van der Zee, A., Hendriks, Willem D., Roorda, L., Ossewaarde, J. M., & Buitenwerf, J. (2013). Review of a major epidemic of methicillin-resistant *Staphylococcus aureus*: The costs of screening and consequences of outbreak management. *American Journal of Infection Control, 41*, 204–209.

Van Rie, A., Page-Shipp, L., Scott, L., Sanne, I., & Stevens, W. (2010). Xpert® MTB/RIF for point-of-care diagnosis of TB in high-HIV burden, resource-limited countries: Hype or hope? *Expert Reviews of Molecular Diagnostics, 10*, 937–946.

Wang, S., Inci, F., De Libero, G., Singhal, A., & Demirci, U. (2013). Point-of-care assays for tuberculosis: Role of nanotechnology/microfluidics. *Biotechnology Advances, 31*, 438–449.

Watchirs Smith, L. A., Hillman, R., Ward, J., Whiley, D. M., Causer, L., Skov, S., et al. (2012). Point-of-care tests for the diagnosis of *Neisseria gonorrhoeae* infection: A systematic review of operational and performance characteristics. *Sexually Transmitted Infections, 89*, 320–326.

Weyer, K., Carai, S., & Nunn, P. (2011). Viewpoint TB diagnostics: What does the world really need? *Journal of Infectious Diseases, 204*, S1196–S1202.

Yan, L., Zhou, J., Zheng, Y., Gamson, A. S., Roembke, B. T., Nakayama, S., et al. (2014). Isothermal amplified detection of DNA and RNA. *Molecular BioSystems, 10*, 970–1003.

Yeh, Y. _T., Nisic, M., Yu, X., Xia, Y., & Zheng, S.-Y. (2014). Point-of-care microdevices for blood plasma analysis in viral infectious diseases. *Annals of Biomedical Engineering, 42*, 2333–2343.

Young, D. B., Perkins, M. D., Duncan, K., & Barry, C. E., III (2008). Confronting the scientific obstacles to global control of tuberculosis. *Journal of Clinical Investigation, 118*, 1255–1265.

Zimlichman, E., Henderson, D., Tamir, O., Franz, C., Song, P., Yamin, C. K., et al. (2013). Health care-associated infections—A meta-analysis of costs and financial impact on the US health care system. *JAMA Internal Medicine, 173*, 2039–2046.

Zwerling, A., & Dowdy, D. (2013). Economic evaluations of point of care testing strategies for active tuberculosis. *Expert Review of Pharmacoeconomics and Outcomes Research, 13*, 313–325.

Molecular Diagnostics in the Diagnosis of Parasitic Infection

4

Bobbi S. Pritt[1]

Mayo Clinic, Rochester, Minnesota, USA

[1]Corresponding author: e-mail address: pritt.bobbi@mayo.edu

1 INTRODUCTION

Parasites are responsible for a significant burden of disease worldwide. Malaria alone was responsible for an estimated 198 million infections and 584,000 deaths in 2013 (World Health Organisation, 2014a), while millions more were infected with a diverse range of other protozoa and helminths. The significant global disease burden has led to increased recognition of parasitic infections. In 2010, the World Health Organisation (WHO) identified 17 neglected tropical diseases (NTDs), of which 11 are parasitic in nature (World Health Organisation, 2010), and in 2014, the Centers for Disease Control and Prevention (CDC) identified five parasitic diseases with significant public health implications based on disease burden, severity of illness and availability of preventative measures and therapies (Centers for Disease Prevention and Control, 2014).

Parasitic diagnostics have traditionally centred on morphologic techniques using light microscopy. Although microscopy is a useful method for identifying a wide range of parasites in clinical specimens, it is subjective and requires considerable skill, experience and expertise for accurate interpretation. Today, microscopic methods are increasingly supplemented or replaced by other laboratory techniques such as antigen detection, serology and molecular diagnostics. Over the past three decades, nucleic acid amplification tests (NAATs), primarily in the form of the polymerase chain reaction (PCR), have been described for nearly all human parasites, with many published assays existing for parasites of significant public health importance. Advances in post-amplification analysis methods such as high-throughput sequencing, Luminex xMAP bead-based detection, microarray analysis and mass spectrometry have also been described. Inclusion of an internal process control (recovery template) may not be necessary for all assays and specimen types but may be important for instances where a high frequency of inhibition has been observed (e.g. PCR using formalin-fixed paraffin-embedded (FFPE) tissue) (Buckwalter et al., 2014). The most commonly used formats for molecular parasitic testing have been recently described (Ndao, 2009; Verweij & Stensvold, 2014).

Methods in Microbiology, Volume 42, ISSN 0580-9517, http://dx.doi.org/10.1016/bs.mim.2015.05.001

Table 1 Advantages and Disadvantages of Molecular Diagnostics Compared to Conventional Microscopy-Based Methods[a]

Potential advantages of molecular tests	Potential advantages of conventional tests
1. Increased sensitivity	1. Time-honoured methods
2. Increased specificity, particularly for organisms with similar morphologies	2. Stains and reagents readily available
3. Ability to differentiate species and strains beyond what is possible by morphology alone	3. Simple equipment requirements
4. Not negatively impacted by poor or altered morphology	4. Single method may detect multiple parasites without prior knowledge of expected parasite
	5. Relatively rapid
Potential disadvantages of molecular tests	**Potential disadvantages of conventional tests**
1. Non-targeted organisms are not detected	1. Subjective interpretation, usually based on subtle morphologic features
2. Paucity of commercial and CE-marked/ FDA-approved assays	2. Considerable expertise and experience required for accurate interpretation
3. Lack of standardisation amongst laboratory-developed methods	3. Difficult to maintain expertise in non-endemic settings due to rarity of positives
4. High complexity testing; not widely available	4. Dwindling expertise in many non-endemic settings
5. Requires expensive and sophisticated equipment	5. Lower sensitivity; concentration techniques and/or examination of multiple specimens may be required to increase sensitivity
6. Potential for nucleic acid contamination	
7. Possible amplification inhibitors	6. Morphology may be negatively impacted by multiple factors (e.g. previous drug therapy, prolonged transport time, delayed placement in fixative)
8. Does not work well with all fixatives (e.g. formalin)	
9. Presence of DNA does not necessarily indicate active or symptomatic infection	
10. May not be performed rapidly	

[a]Performance of individual molecular assays may vary greatly.

Molecular methods offer a number of advantages over conventional microscopy-based techniques (Table 1). Importantly, they are generally more sensitive and may be particularly useful when the parasite load is low. Molecular methods may also provide a level of parasite identification beyond what is possible with morphology alone, such as differentiation between morphologically identical species (e.g. *Entamoeba histolytica* vs. *Entamoeba dispar*) and molecular strain typing. Finally, they may play an important role in identifying and differentiating parasites when their morphologic features are ambiguous or altered. Morphology can be compromised when the patient has recently received anti-parasitic therapy or if there is a significant delay in transporting the specimen to the laboratory. Incorporation of PCR into the clinical microbiology lab has been facilitated by use of automated nucleic acid extraction platforms which allow for high-throughput testing, while the introduction of real-time PCR has greatly decreased the risk of amplified nucleic acid carryover to subsequent reactions causing false-positive results. Multiple targets can be easily combined in a single real-time PCR, thus allowing parasite testing to be grouped by specimen type of clinical syndrome.

Unfortunately, molecular methods have several disadvantages compared to conventional techniques (Table 1). For example, NAATs are not well suited to use in resource-poor settings where many parasites are endemic, since they require expensive and sophisticated equipment, highly trained personnel, molecular grade reagents, reliable electricity, humidity and temperature controls, and appropriate transportation and storage conditions for specimens and reagents. Introduction of cartridge-based tests which automate nucleic acid processing, extraction and amplification, as well as introduction of relatively simple isothermal techniques such as nucleic acid sequence-based amplification (NASBA) and loop-mediated isothermal amplification (LAMP), have made NAATs more accessible in some settings. However, deficiencies in infrastructure and staffing continue to impede the widespread implementation of NAATs worldwide.

The dearth of commercially available and CE-labelled/US Food and Drug Administration (FDA)-cleared or approved assays is another potential limitation to NAAT adoption. Most NAATs used for clinical diagnosis are based on non-standardised laboratory-developed methods. Components of a laboratory-developed test (LDT) that may vary across tests for a given analyte include the type of specimen tested, methods for specimen processing, nucleic acid extraction and amplification, the choice of nucleic acid target(s), instrumentation and equipment used and reporting methods. This amount of variability can lead to significant differences in performance characteristics (e.g. accuracy, lower limit of detection) amongst assays, thus making it challenging to compare results between LDTs. In addition, LDTs require considerable expertise to validate, implement and maintain, and are therefore used primarily in reference laboratories, specialised research facilities and large public health laboratories.

Finally, important questions remain regarding the incorporation of molecular testing into routine patient care algorithms. Molecular methods, specifically NAATs, are generally more sensitive than conventional methods and may detect nucleic acid at levels below those that are considered clinically significant. Furthermore, detection of DNA does not necessarily correlate with active disease as DNA may be present for days to weeks after successful treatment. Depending on the individual testing platform, NAATs may also not be performed in a timely manner, particularly if they are only performed during certain shifts or in specimen batches.

Despite these potential limitations and challenges, molecular diagnostics have been slowly introduced to the market and incorporated into practise guidelines. The past few years have seen the introduction of several FDA-cleared/CE-labelled tests including multiplex NAATs for gastrointestinal parasites and fully automated testing options for *Trichomonas vaginalis*, while PCR for certain parasites (e.g. *Toxoplasma gondii*, *Leishmania* species) is now considered an important component of laboratory testing. Continued standardisation and technological advances should allow for broader use of molecular testing in both resource rich and limited settings. This review will focus on the use of molecular assays for common and important parasites.

2 BLOOD PARASITES

2.1 *PLASMODIUM* SPECIES (MALARIA)

2.1.1 Background

Malaria is caused by intraerythrocytic protozoan parasites in the genus *Plasmodium*. Human infection occurs throughout much of the tropics and subtropics and is caused primarily by five *Plasmodium* species: *Plasmodium falciparum*, *Plasmodium vivax*, *Plasmodium ovale*, *Plasmodium malariae* and *Plasmodium knowlesi*. *P. falciparum* is responsible for the bulk of malaria-related deaths, but infection with the other species is associated with significant morbidity and occasional mortality (World Health Organisation, 2014a). Infection is most commonly transmitted through the bite of an infected *Anopheles* mosquito. Other important means of infection are through blood transfusion and across the placenta during pregnancy or delivery.

Malaria is arguably one of the most important infectious diseases worldwide with among the highest associated morbidity and mortality of any human parasite (World Health Organisation, 2014a). Despite worldwide malaria prevention and treatment initiatives, there were still an estimated 584,000 deaths in 2013, with the vast majority (90%) occurring in Africa children (World Health Organisation, 2014a). In non-endemic countries such as the United States, Canada and Europe, malaria is almost exclusively an imported disease from individuals travelling from regions with ongoing transmission (Cullen & Arguin, 2014).

Plasmodium species descriptions were originally based on morphologic characteristics observed by light microscopy such as the size of the infected red blood cell, cytoplasmic inclusions (e.g. Schüffner's dots) and features of the parasite stages. Unsurprisingly, advances in molecular techniques have revealed complexity in *Plasmodium* phylogeny that could not be appreciated by morphology alone. In 2010, multilocus sequence analysis revealed the existence of two genetically distinct forms of *P. ovale* (Sutherland et al., 2010), which were later named *P. ovale curtisi* and *P. ovale wallikeri* (Fuehrer & Noedl, 2014), while *P. knowlesi* was recognised as an important and unexpected cause of zoonotic malaria in regions of Southeast Asia through the use of conventional PCR (Singh et al., 2004).

2.1.2 Diagnostic modalities

Malaria is a potentially life-threatening infection and testing should be performed on an immediate or 'STAT' basis. The conventional method for laboratory detection of malaria parasites is microscopic examination of Giemsa-stained thick and thin blood films. The thick film is typically used for parasite screening since it allows for examination of the largest volume of blood (20–30 layers of red blood cells), while the thin films is ideal for differentiating the *Plasmodium* species based on morphologic characteristics. Using the thick film, an experienced and skilled microscopist can detect as few as 10–50 parasites per microlitre of blood, which correlates with an approximate 0.0002–0.001% parasitemia (assuming an erythrocyte count of $5 \times 10^{6}/\mu L$ of blood) (Garcia, 2007). Unfortunately, the estimated sensitivity can be significantly lower in field conditions, particularly where there is a lack of well-trained technologists and robust laboratory quality controls (Ochola, Vounatsou, Smith, Mabaso, & Newton, 2006).

Recently, rapid immunochromatographic tests for *Plasmodium* antigens have been used to supplement or replace microscopy in many endemic and non-endemic settings due to their widespread availability and ease of use. These rapid diagnostic tests (RDTs) generally perform well for detection of *P. falciparum* at moderate or high levels of parasitemia, but may fail to detect lower levels of parasites and infections with non-*falciparum* species, thus resulting in false-negative results (Wilson, 2012). False-positive results may also be observed (Luchavez et al., 2011; Wilson, 2012), and it is, therefore, widely recommended that all results be followed up by traditional blood film examination (Wilson, 2012).

Molecular tests can overcome some of the limitations of other malaria diagnostic techniques and provide a viable diagnostic alternative in certain settings. There are currently no FDA-cleared or approved options in the United States, but several kits have been CE marked for *in vitro* diagnostic use (IVD), such as the Geno-Sen's® Malaria PCR (Genome Diagnostics Pvt. Ltd., New Delhi, India) and RealStar® Malaria PCR (Altona Diagnostics, Hamburg, Germany) kits. Unfortunately, there is a distinct lack of published studies using commercially available kits, with most studies employing laboratory-developed methods.

2.1.3 Design and performance of molecular tests

The majority of described molecular assays are PCR-based including conventional (Mixson-Hayden, Lucchi, & Udhayakumar, 2010; Singh et al., 1999) and real-time (Babady, Sloan, Rosenblatt, & Pritt, 2009; Cnops, Jacobs, & Van Esbroeck, 2011; Kamau et al., 2011; Perandin et al., 2004; Shokoples, Ndao, Kowalewska-Grochowska, & Yanow, 2009) formats. *Plasmodium* species are most commonly differentiated by PCR using species-specific primers and/or probes, but can also be differentiated using melting curve analysis (Figure 1). Isothermal formats including NASBA (Mens, Schoone, Kager, & Schallig, 2006) and LAMP (Polley et al., 2010; Poon et al., 2006; Sirichaisinthop et al., 2011; Tao et al., 2011) have also been described. Nearly all NAATs target the 18S small subunit ribosomal RNA gene (*ssu rDNA*) (Table 2), which is present in approximately five copies per *Plasmodium* spp. genome. The reticulocyte-binding protein 2 (*rbp2*) gene and *P. o. curtisi* and *P. o. wallikeri* tryptophan-rich antigen (*poctra* and *powtra*) genes have been used to differentiate *P. o. curtisi* and *P. o. wallikeri* (Oguike et al., 2011). Targeting mitochondrial DNA (e.g. *cox1*) which is present in 30–150 copies/parasite has also been reported to provide increased sensitivity over *18S rDNA* using LAMP-based assays (Polley et al., 2010; Poon et al., 2006; Tao et al., 2011). The performance characteristics vary widely between published assays. However, NAATs generally have equal or superior sensitivity to conventional blood film examination, with lower limits of detection (LOD) less than 10 parasites/μL of blood.

In addition to detection and species differentiation, molecular tests can be used to detect gene mutations or amplifications associated with resistance to commonly used

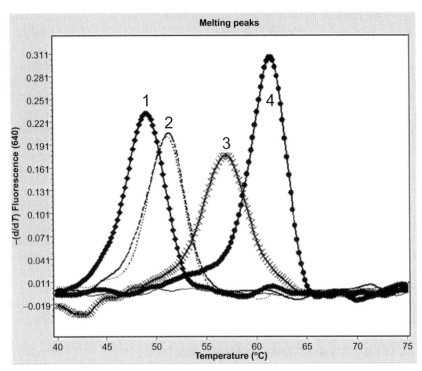

FIGURE 1

An example of a real-time PCR assay using a single *Plasmodium*-specific primer and probe set to detect all human-infecting species (Babady et al., 2009). *P. ovale* (1), *P. vivax* (2), *P. malariae* (3) and *P. falciparum* (4) are differentiated by their melting temperature peaks.

antimalarials. Well-described markers of resistance have been recognised for chloroquine (*P. falciparum* chloroquine-resistance transporter gene, *Pfcrt*), sulfadoxine (dihydropteroate synthase gene, *dhps*), pyrimethamine (dihydrofolate reductase gene, *dhft*) and mefloquine (*P. falciparum* multi-drug-resistance gene 1, *Pfmdr-1*), while mutations in *pfATPase6* been implicated in decreased parasite sensitivity to artemisinin-based compounds (Vestergaard & Ringwald, 2007). Mutation testing is simpler than standard *in vivo* and *in vitro* sensitivity testing and can be performed through use of single PCR and microarray-based formats (Alifrangis et al., 2005; Vestergaard & Ringwald, 2007). However, the full relationship between molecular markers and therapeutic efficacy is not fully understood and it is likely that some instances of treatment failure are multifactorial (Picot et al., 2009; Vestergaard & Ringwald, 2007). Creation of a global database for molecular markers of drug resistance has been recognised as a priority for guiding malaria treatment and prevention efforts and facilitating identification of new resistance markers (Plowe et al., 2007).

Table 2 Commonly Used Nucleic Acid Targets for Select Parasites

Parasite	Associated Disease	Nucleic Acid Target
Ascaris lumbricoides	Ascariasis	ITS1, ITS2, Cyt *b*
Babesia species	Babesiosis	18S rDNA, 16S-like rRNA gene, ITS, thiamine pyrophosphokinase gene
Brugia malayi	Lymphatic filariasis	*Hha I* repeat region
Cryptosporidium species	Cryptosporidiosis	18S rDNA, *Cryptosporidium parvum*-specific 452-bp fragment, *Cryptosporidium* oocyte wall protein (COWP) gene, DnaJ-like protein gene
Dientamoeba fragilis	*D. fragilis* infection	SSU rDNA, 5.8S rDNA
Entamoeba histolytica	Amebiasis	SSU rDNA
Free-living amoebae	Amoebic keratitis, granulomatous amoebic encephalitis, primary amoebic meningoencephalitis	18S rDNA
Giardia duodenalis	Giardiasis	SSU rDNA, β-Giardin gene, glutamate dehydrogenase gene, elongation factor 1(EF1)-α gene, triosephosphate isomerase gene
Hookworms (*Necator americanus*, *Ancylostoma duodenale*)	Hookworm infection	ITS2
Leishmania species	Leishmaniasis	kDNA, 16S rDNA, Glucose phosphate isomerase gene, glucose-6-phosphate dehydrogenase gene, ITS1, ITS2, GPI gene
Loa loa	Loaiasis	Expressed sequence tags (LLMF72 and LLMF269), 15r3 repeat region
Microsporidia species	Microsporidiosis	16S rDNA, ITS
Onchocerca volvulus	Onchocerciasis	O-150 repeat sequence
Plasmodium genus and species	Malaria	18S rDNA, Mitochondrial DNA, reticulocyte-binding protein 2 (rbp2) gene, *P. ovale curtisi* and *P. o. wallikeri* tryptophan-rich antigen (*poctra* and *powtra*) genes
Schistosoma species	Schistosomiasis	Highly repeated short 0.64-kb DNA sequence, cytochrome c oxidase gene
Strongyloides stercoralis	Strongyloidiasis	SSU rDNA, 28S

Continued

Table 2 Commonly Used Nucleic Acid Targets for Select Parasites—cont'd

Parasite	Associated Disease	Nucleic Acid Target
Toxoplasma gondii	Toxoplasmosis	B1 gene, RE gene (REP-52)
Trichomonas vaginalis	Trichomoniasis	16S rRNA, RNA probe
Trichuris trichiura	Trichuriasis	SSU rDNA
Trypanosoma cruzi	Chagas disease	18S rDNA, kDNA, nuclear DNA (nDNA/minisatellite TCZ region)
Trypanosoma brucei	Human African Trypanosomiasis (HAT)	177-bp satellite repeat, ITS1 rDNA, 18S rDNA, SRA gene, expression-site-associated genes 6 and 7 (ESAG6/7)
Wuchereria bancrofti	Lymphatic filariasis	188 bp DNA sequence *Ssp-I*

Abbreviations: bp, base pair; cyt, cytochrome; ITS, internal spacer region; kDNA, kinetoplastid DNA; PCR, polymerase chain reaction; rDNA, ribosomal DNA; SSU, small subunit.

2.1.4 *Clinical implementation of molecular tests*

Molecular tests offer certain advantages over standard microscopy and malaria RDTs which make them desirable for use in clinical care (Table 1). In general, they are highly specific and offer improved sensitivity over other methods. The increased sensitivity may be particularly useful for detecting sub-patent infections in endemic settings where the goal is to eliminate malaria. Depending on the assay design, molecular tests may also be superior to standard microscopy for detecting mixed species infections (Mixson-Hayden et al., 2010). Since they do not rely on subjective morphologic features, molecular tests can particularly useful for confirming blood smear results when parasite morphology is altered due to prolonged exposure to EDTA or prior exposure to antimalarial therapy. They can also be useful for differentiating *P. falciparum* from the similar-appearing *Babesia* parasites.

Unfortunately, molecular tests cannot completely replace blood films at this time. Few diagnostic laboratories offer molecular testing for malaria, and those that do rarely perform testing on a STAT basis. Furthermore, blood film examination is still required to calculate the number or percent of parasitised cells in a given sample, which is in turn used to predict patient prognosis and guide clinical care. Ideally, quantitative PCR could be used in place of subjective manual cell counts that suffer from poor intra-observer variability. However, PCR values do not correlate well with manual counts since the former counts each copy of target nucleic acid, while the latter counts only the number of infected cells. Thus, PCR would detect and count each parasite within an infected cell, including each merozoite in a schizont. Extracellular forms, gametocytes and free DNA would also be detected by PCR, but would not be included in a standard-infected cell counts (Garcia, 2007). Further studies are therefore needed to correlate quantitative PCR values with patient outcomes and response to therapy. Use of the WHO quantitated *P. falciparum* standard (Shokoples et al., 2009) should allow for standardisation and comparison of quantitative measurements among different laboratory-developed and commercial assays.

2.2 *BABESIA* SPECIES (BABESIOSIS)

2.2.1 *Background*

Babesiosis is caused by intracellular protozoan parasites in the genus *Babesia*. The parasite is found primarily in temperate climates, including parts of Europe and North America and is most commonly transmitted through the bite of an infected ixodid tick (Vannier & Krause, 2012). Blood transfusion and congenital transmission are also important means of infection (Leiby, 2011). Most cases in the United States are caused by *Babesia microti*, with *Babesia duncani* and *Babesia divergens/Babesia divergens*-like parasites causing a smaller number of infections (Conrad et al., 2006; Herwaldt et al., 1996, 2004). In Europe, *B. divergens* is the primary human pathogen, with a smaller number of infections being caused by *B. microti* and *Babesia venatorum* (Criado-Fornelio, Martinez-Marcos, Buling-Sarana, & Barba-Carretero, 2003; Herwaldt et al., 2003; Hildebrandt et al., 2007).

Babesiosis is an increasingly important tickborne disease in the United States. It is found primarily in the northeastern and upper midwestern states and caused a reported 1124 case in 2011 (Centers for Disease Prevention and Control, 2012). The primary tick vector, *Ixodes scapularis*, is an important vector for *Borrelia burgdorferi* and *Anaplasma phagocytophilum*, the agents of Lyme disease and Anaplasmosis, respectively, and co-infections with these agents are not uncommon. In Europe, babesiosis is an uncommon human disease, but has a substantial economic impact due to the large number of bovine infections with *B. divergens* (Zintl, Mulcahy, Skerrett, Taylor, & Gray, 2003). Most infections with *Babesia* spp. are thought to be asymptomatic but symptoms may range from mild to life-threatening (Gray, Zintl, Hildebrandt, Hunfeld, & Weiss, 2010; Vannier & Krause, 2012).

2.2.2 *Diagnostic modalities*

Like malaria, the traditional method for laboratory diagnosis of babesiosis is microscopic examination of thick and thin blood films (Garcia, 2009). *Babesia* parasites share many morphologic similarities with *Plasmodium* species, particularly *P. falciparum*, but can usually be differentiated by an experienced microscopist. Correlation with travel history is also essential. For cases with low parasitemia or non-characteristic morphology, PCR may be useful alternative diagnostic method. There are currently no FDA-approved/cleared or CE-marked options at this time, but several real-time and conventional PCR LDTs for *B. microti* (Chan, Marras, & Parveen, 2013; Krause et al., 1996; Liu, 2013; Persing et al., 1992; Teal, Habura, Ennis, Keithly, & Madison-Antenucci, 2012) and *B. divergens* (Liu, 2013) have been described. Serology plays no role in diagnosis of acute infection but may be useful for blood donor screening and epidemiologic studies, alone or in combination with PCR.

2.2.3 Design and performance of molecular tests

Published PCR assays target portions of the 18S ssu rRNA gene (Bloch et al., 2013; Teal et al., 2012), 16S-like small subunit rRNA gene (Persing et al., 1992), internal transcribed spacers (ITS) gene region of nuclear rRNA (Wilson et al., 2014) and the thiamine pyrophosphokinase gene (Chan et al., 2013). Assays may be species specific (Chan et al., 2013; Teal et al., 2012; Wilson et al., 2014) or use subsequent sequence analysis or electrospray ionisation MS for differentiation to the species level (Eshoo et al., 2014; Herwaldt et al., 2003; Liu, 2013; Persing et al., 1992). The described assays have similar or improved sensitivity compared to conventional thick film examination. Based on calculations of *Plasmodium* parasites, the LOD of the thick blood film under ideal conditions for detection of *Babesia* spp. parasites is 10–50 parasites/μL or 0.0002–0.001% parasitemia. In comparison, a recently described real-time *18S rDNA* PCR assay for *B. microti* reported an LOD of significantly less than 1 parasite/μL of whole blood (0.006 parasites/μL) (Bloch et al., 2013), while another *18S rDNA* PCR reported an estimated LOD of 5–10 parasites/μL (Teal et al., 2012). The number of copies of 18S rDNA in *B. microti* is unknown, but is expected to be similar to that seen with *Plasmodium* (approximately 5 copies/parasite) (Garcia, 2007).

2.2.4 Clinical implementation of molecular tests

Babesia PCR may be performed as an individual test or incorporated into a panel of PCR tests for other *Ixodes*-transmitted agents such as *B. burgdorferi*, *Borrelia miyamotoi*, *A. phagocytophilum* and Deer Tick virus (Chan et al., 2013). PCR can be used for primary diagnosis of babesiosis if it is available during all shifts and performed on a rapid basis. Unfortunately, limitations in PCR availability usually require the lab to maintain expertise in blood smear examination for routine use or coverage during times that PCR is not performed. More commonly, PCR is used for secondary/confirmatory testing to aid in differentiating *Babesia* from *Plasmodium* parasites and non-parasite artefacts such as Howell–Jolly and Pappenheimer bodies. It is also useful when morphology is suboptimal due to specimen age, prolonged exposure to EDTA and patient receipt of anti-babesial drugs. Finally, PCR may be useful for screening cases where the parasite load is low, such as for blood donor screening in endemic settings (Moritz et al., 2014; Young et al., 2012). The American Red Cross recently completed an investigational study using both an indirect immunofluorescence assay and PCR for screening blood donor units obtained in highly endemic states for babesiosis and found this approach to be feasible and cost-effective (Susan Stramer, personal communication, American Red Cross). It is important to remember that detection of *Babesia* DNA may occur for some time after successful treatment and this may be especially true in asplenic individuals due to delayed of infected RBC clearance.

2.3 *TRYPANOSOMA BRUCEI* (HUMAN AFRICAN TRYPANOSOMIASIS)

2.3.1 Background

Human African trypanosomiasis (HAT), also known as 'African Sleeping Sickness', is caused by the protozoan haemoflagellate protozoan, *Trypanosoma brucei rhodesiense* (east sub-Saharan African) and *Trypanosoma brucei gambiense* (west/central sub-Saharan Africa) (Garcia, 2007). Transmission to humans is via the bite of an infected tsetse fly. Acute infection is characterised by a chancre at the inoculation site, followed by fever and lymphadenopathy during the haemolymphatic phase of disease. The chronic, meningoencephalic stage of illness occurs with invasion of the central nervous system (CNS) and results in altered mental status, sleep disruption and eventually coma and death. The East African form of disease is more severe and rapidly progressive than the West African form, but both forms are fatal if untreated (Garcia, 2007).

2.3.2 Diagnostic modalities

The traditional 'gold standard' method for laboratory diagnosis of HAT is microscopic identification of extracellular flagellated trypomastigotes in peripheral blood, CSF and lymph node aspirates. There is often low parasitemia in these specimens and, therefore, concentration methods such as the mini anion exchange column technique and microhaematocrit centrifugation technique are widely employed. Unfortunately, microscopy is still estimated to miss up to 30% of cases (Robays, Bilengue, Van der Stuyft, & Boelaert, 2004). Furthermore, subspecies identification, which is important for selecting the appropriate treatment, cannot be performed using morphology alone. Usually, the geographic exposure history of the patient can be used to presumptively determine the infecting subspecies but patients with a broad travel history or specifically travel to Uganda where both subspecies circulate may present a diagnostic dilemma (Picozzi et al., 2005).

The Card Agglutination Test for Trypanosomiasis (CATT) is another commonly used method for diagnosis of HAT. This simple and rapid test can be used to detect antibodies reacting with the variant surface glycoprotein LiTat 1.3 of *T. b. gambiense* in blood, serum or plasma, with reported sensitivities of 87–98% (Chappuis, Loutan, Simarro, Lejon, & Buscher, 2005; Garcia, 2007). This test is commonly used for field-testing in endemic regions, in conjunction with physical examination for signs and symptoms of sleeping sickness (e.g. enlarged lymph nodes). Unfortunately, false CATT negatives can occur when the parasite gene that encodes for LiTat 1.3 is absent (seen in several foci in West Africa), while false-positives may occur due to cross-reactivity with non-pathogenic trypanosomes and other parasites (Chappuis et al., 2005). A similar test does not exist for east African trypanosomiasis.

Given the limitations of microscopy and serologic methods, molecular testing provides an attractive alternative for both screening and confirmatory testing. Several NAATs have been described, although none are FDA-approved/cleared or CE-marked at this time.

2.3.3 Design and performance of molecular tests

Early published NAATs used conventional PCR (Kabiri et al., 1999; Kyambadde, Enyaru, Matovu, Odiit, & Carasco, 2000; Moser et al., 1989; Penchenier et al., 2000) and later, real-time PCR (Becker et al., 2004) formats for detection of trypanosomes in clinical specimens. PCR targets include the 177 base pair satellite gene (Moser et al., 1989), and expression-site-associated genes 6 and 7 (Kabiri et al., 1999) and *ITS1 rDNA* (Njiru et al., 2005) *18S rDNA*, which allow for detection to the *Trypanozoon* subgenus level (including *T. brucei* and non-human trypanosomes). These multi-copy targets allow for sensitive detection of trypanosomes, but identification to the subspecies level requires the use of different targets. The genes encoding the serum-resistance-associated protein (*SRA*) and *T. b. gambiense*-specific glycoprotein (*TgsGP*) genes can be used to specifically identify *T. b. rhodesiense* and *T. b. gambiense*, respectively (Deborggraeve & Buscher, 2010) and assays have been developed for this purpose (Mathieu-Daude, Bicart-See, Bosseno, Breniere, & Tibayrenc, 1994; Radwanska, Chamekh, et al., 2002; Radwanska, Claes, et al., 2002; Welburn et al., 2001).

Although PCR assays have high sensitivity and specificity, the need for complex post-amplification processing and expensive equipment make them poorly suited for field use in endemic settings. One approach for simplifying testing has been to incorporate an oligochromatographic dipstick device for amplicon detection after conventional PCR (Deborggraeve et al., 2006). However, amplification can be further simplified through use of isothermal amplification methods such as LAMP (Kuboki et al., 2003; Njiru, Mikosza, Armstrong, et al., 2008; Njiru, Mikosza, Matovu, et al., 2008; Njiru, Traub, Ouma, Enyaru, & Matovu, 2011) and NASBA (Matovu et al., 2010; Mugasa et al., 2009). These isothermal techniques have demonstrated high analytical sensitivity and specificity, comparable or superior to PCR. A LAMP assay targeting 5.8S rRNA *ITS2* (Thekisoe et al., 2007) for detection of species-specific detection of *T. b. gambiense* and an assay targeting random insertion mobile elements for detection of the *Trypanozoon* subgenus each demonstrated a lower detection limit of 0.001 trypanosomes per reaction (Mugasa et al., 2009). Other LAMP targets include the SRA gene for specific detection of *T. b. rhodesiense* (Njiru, Mikosza, Armstrong, et al., 2008) and the paraflagella rod protein A gene for detection of *Trypanozoon* subgenus (Kuboki et al., 2003). A NASBA assay targeting *18S rDNA* of *T. brucei* has demonstrated an LOD of 10 parasites/mL blood (Mugasa et al., 2009, 2008). This assay was further simplified by using membrane oligochromatography (OC) for rapid detection of RNA following amplification (Mugasa et al., 2009). A recent comparison of a LAMP assay and the NASBA-OC assay showed the latter to have a higher level of positive and negative agreement with microscopy for detection of *T. b. gambiense* (Mugasa et al., 2014). The NASBA-OC assay used in this study was also compared to direct microscopy for detection of *T. brucei* in clinical specimens and showed a significantly higher sensitivity (73% vs. 57% for microscopy) compared to a composite gold standard (Mugasa et al., 2009).

Finally, methods that can directly detect parasite nucleic acid may offer a simpler, albeit less sensitive, alternative to amplification-based techniques. A fluorescence *in situ* hybridisation assay with peptide nucleic acid probes targeting rRNA has been described (Radwanska, Magez, et al., 2002) and may possibly be used with the recently introduced battery-powered light-emitting diode-based fluorescence microscopes (Jones, Nyalwidhe, Tetley, & Barrett, 2007). Use of RNA aptamers (small, serum-stable RNAs) to target parasite variant surface glycoproteins has also been described (Lorger, Engstler, Homann, & Goringer, 2003). Further studies are needed to fully explore the use of these techniques in different laboratory settings.

2.3.4 Clinical implementation of molecular tests

Given the high sensitivity and specificity of NAATs, particularly NASBA-OC, these methods can be used for both primary and confirmatory diagnosis of HAT. Some assays can also be used for subspecies identification. NASBA-OC and LAMP-based assays are particularly attractive because they are relatively rapid and simple to perform compared to PCR-based methods. The WHO and the Foundation for Innovative New Diagnostics (http://www.finddiagnostics.org) are collaborating to develop diagnostic tools that are affordable, easy to use, require minimal training and use reagents and equipment that are stable at ambient temperatures. Unfortunately, many challenges remain for widespread implementation of molecular diagnostics in clinical settings and national control programmes (Deborggraeve & Buscher, 2010).

2.4 *TRYPANOSOMA CRUZI* (CHAGAS DISEASE)

2.4.1 Background

Chagas disease, also known as American Trypanosomiasis, is caused by the heamoflagellate protozoan, *T. cruzi* (Garcia, 2007). The parasite is found primarily in Latin America, where an estimated 11 million individuals are infected, although rare cases of autochthonous (local) transmission have also been reported in the southern United States (Beard et al., 2003; Dorn et al., 2007; Garcia, 2007). Transmission usually occurs after the reduviid bug vector takes a blood meal from the human host and deposits faeces containing infectious metacyclic trypomastigotes near the bite site. Subsequent accidental inoculation of the faeces into the bite site or mucous membrane introduces the parasite into the host. Other important means of transmission are through ingestion of food or beverages contaminated with infected reduviid bug faeces, receipt of blood or organs from an infected donor and transplacental spread to the developing foetus. In particular, transmission through blood products is a growing concern in non-endemic settings such as the United States that receive a large number of migrants from Latin America. Initial infection is usually asymptomatic, but may be associated with mild symptoms including localised swelling at the inoculation site (chagoma, Romaña's sign), fever, malaise and lymphadenopathy (Garcia, 2007). If the patient is not treated, then infection is thought to be life-long, with the host experiencing intermittent, low-level parasitemia. Most chronically infected individuals are asymptomatic, but approximately 30% will go on to have chronic symptomatic disease with cardiac or gastrointestinal manifestations, including arrhythmias, cardiomyopathy, sudden cardiac death, megaesophagus and megacolon (Garcia, 2007).

There is a significant amount of genetic variability amongst *T. cruzi* parasites, resulting in recent recognition of six major *T. cruzi* typing units, named TcI–TcVI (Zingales et al., 2009). Genotypes are classified according to sequences of several gene loci including the intergenic region of the miniexon gene, the 24Sα rRNA gene, the 18S rRNA gene, ITS regions, miniexon gene and several housekeeping genes (Lewis et al., 2009; Westenberger, Sturm, & Campbell, 2006). Like *T. brucei*, *T. cruzi* parasites also contain a specialised structure called a kinetoplast which houses mitochondrial DNA (Rodrigues, Godinho, & de Souza, 2014). This structure is found only in trypanosomes and *Leishmania* parasites and can therefore be a useful target for molecular detection.

2.4.2 Diagnostic modalities

The preferred diagnostic modality for Chagas disease varies depending on the immune status of the host and the phase of infection (Bern, Kjos, Yabsley, & Montgomery, 2011). In the acute phase, definitive diagnosis is by microscopic identification of extracellular trypomastigotes in peripheral blood. The parasites have a similar appearance to the trypomastigotes of *T. brucei*, but have a significantly larger kinetoplast and often assume a 'C-shaped' configuration (Garcia, 2007). Concentration procedures such as the microhaematocrit method increase the sensitivity of microscopic detection, allowing for detection of as few as 40 parasites per millilitre of specimen (Torrico et al., 2005). Less commonly, haemoculture using standard parasite media or xenodiagnosis are used. Parasitemia subsequently decreases within 90 days of infection, so that diagnosis of chronic infection is usually by detection of IgG antibodies to *T. cruzi* (Bern et al., 2011). Serologic testing is also the preferred method for blood donor screening (Centers for Disease Control and Prevention, 2007). The most commonly used methods are the immunofluorescent-antibody assay and enzyme-linked immunosorbent assay (ELISA), although the radioimmunoprecipitation assay (RIPA) and trypomastigote excreted-secreted antigen immunoblot are considered to be reference tests. The sensitivity of these serologic assays likely varies by host response that the strain of the infecting parasite, and there is no true gold standard test for this stage of disease (Bern et al., 2011). Detection of non-motile amastigotes within affected tissues such as the heart using histopathology may also be used. Detection of congenital infection also depends on the stage of disease, with acute infection early in life being detected by microscopic examination of neonatal or cord blood. Later in life, conventional serologic testing is used for diagnosis (Bern et al., 2011).

Compared to conventional techniques, molecular tools such as PCR offer improved sensitivity for detection of acute and early congenital disease and are now considered the test of choice in these settings (Bern et al., 2011). Unfortunately, there are no FDA-approved/cleared or CE-marked assays at this time. For patients in the United States, PCR testing is available through the CDC (Qvarnstrom et al., 2012).

2.4.3 Design and performance of molecular tests

PCR is the most popular format for *T. cruzi* molecular tests, with multiple published conventional and real-time assays (Schijman et al., 2011). Most assays target nuclear satellite DNA (nDNA or satellite DNA) or kinetoplastic/minicircle DNA (kDNA). Both targets generally provide high sensitivity since they are present in multiple (approximately 100,000) copies per parasite, although there may be significant variability in performance characteristics amongst assays. Other less common PCR targets include 18S rRNA, 24Sα rRNA, miniexon and CO II subunit genes. A recent analysis of 48 different PCR assays from 26 laboratories was performed using three standardised specimen sets including parasite culture dilutions and positive whole blood specimens (Schijman et al., 2011). Of the assays tested, four provided superior performance for all three sample sets, with detection limits of at least 0.5 parasites/mL; three of these assays targeted satellite DNA and one targeted kDNA (Schijman et al., 2011). In comparison, assays targeting miniexon, ribosomal or CO II subunit genes were not able to reliably detect 10 fg/μL or less of purified parasite DNA, which the authors considered to be necessary for clinical use. Testing buffy coat specimens rather than whole blood may increase PCR sensitivity. Qvarnstrom and colleagues were able to diagnose reactivated Chagas disease with buffy coat specimens as much as 2 weeks earlier than with whole blood specimens (Qvarnstrom et al., 2012).

2.4.4 Clinical implementation of molecular tests

Given the improved sensitivity over conventional blood films, PCR is the preferred test for detection of acute-phase disease and early congenital disease (Bern et al., 2011). Parasite DNA may be detected in blood shortly after birth in infants that are ultimately determined not to have *T. cruzi* infection; therefore, it is prudent to take another specimen, 1–3 months after birth, prior to making a diagnosis of congenital Chagas disease (Bern et al., 2009; Oliveira, Torrico, Munoz, & Gascon, 2010). Another potential use for PCR is to monitor for infection after accidental laboratory exposure or receipt of an infected blood product/organ, since the PCR test becomes positive days to weeks before parasites are detectable on standard blood smears (Bern et al., 2011; Chin-Hong et al., 2011; Schijman et al., 2000). For immunosuppressed patients with chronic asymptomatic *T. cruzi* infection, serial testing of blood specimens using quantitative PCR may help monitor for reactivation (Diez et al., 2007; Duffy et al., 2009). PCR may be also useful for monitoring parasitologic response to treatment, even when parasitologic cure is not achieved (Britto, 2009; Duffy et al., 2009; Galvao et al., 2003; Schijman et al., 2003; Solari et al., 2001; Zulantay et al., 2004). Recent efforts have been directed to international standardisation of quantitative PCR for monitoring patient response to benznidazole therapy (Moreira et al., 2013). Finally, PCR can be performed on tissue specimens when histopathologic findings are not definitive for Chagas disease (Lages-Silva et al., 2001).

Unfortunately, PCR is significantly less useful for detecting chronic Chagas disease due to low, sporadic levels of circulating parasite DNA. Conventional IgG serology is the preferred test in this setting.

2.5 FILARIASIS

2.5.1 Background

Eight nematodes (roundworms) cause human filariasis, with *Wuchereria bancrofti*, *Brugia malayi*, *Loa loa* and *Onchocerca volvulus* being responsible for the majority of infections (Garcia, 2007). Disease manifestations vary by organism. *W. bancrofti* and *B. malayi* cause lymphatic filariasis, while *O. volvulus* causes river blindness. Infection with *L. loa* (loaiasis) is associated with transient subcutaneous (Calabar) swellings caused by migration of the adult worm; occasionally the worm is detected as it migrates across the surface of the eye. All filarial worms are transmitted through the bite of specific insect vectors.

Filariasis is mainly a disease of the tropics and subtropics, infecting millions of individuals worldwide (Garcia, 2007). Lymphatic filariasis presents the greatest burden of disease, with more than 120 million infected individuals (World Health Organisation, 2014b).

2.5.2 Diagnostic modalities

The preferred diagnostic modality varies by the type of filarial infection. Conventional testing entails microscopic examination of thick and thin blood films (*W. bancrofti*, *Brugia* species, *L. loa*, *Mansonella perstans* and *Mansonella ozzardi*) or skin snips (*O. volvulus*, *Mansonella streptocerca*) (Garcia, 2007). Concentration methods such as centrifugation of a blood specimen lysed in 2% formalin (Knott concentration) or filtration through a Nuclepore® membrane (Whatman Inc., Florham Park, NJ) increases the sensitivity of blood specimen examination. Drawbacks to conventional microscopy include the high degree of technical expertise required to identify the infecting filaria to the genus and species level and the need for timed blood draws to match the periodicity of certain species (Garcia, 2007). This can be particularly challenging in non-endemic settings where these techniques are not commonly practised (Fink, Fahle, Fischer, Fedorko, & Nutman, 2011). Sensitive and specific rapid immunochromatographic tests are available for *W. bancrofti* and obviate the need for microscopy and timed blood draws; unfortunately, there are not FDA-cleared/approved tests at this time and their use is limited primarily to endemic settings and control programmes. Filarial serology, particularly for detection of IgG4, may be useful for detecting infections caused by the major filarial parasites, but suffers from poor sensitivity and specificity and cannot distinguish between current and previous infection. Finally, multiple molecular assays have been described for the various filariae but are not commonly used for routine diagnosis. There are no FDA-cleared/approved molecular diagnostic assays for the filariae at this time, but some commercially available CE-marked kits exist for *W. bancrofti* (e.g. Geno-Sen's Filaria Real-Time PCR Kit by Genome Diagnostics, India).

2.5.3 Design and performance of molecular tests

Molecular assays, including conventional and real-time PCR formats and LAMP-based methods, have been described for the four major filarial pathogens (Drame, Fink, Kamgno, Herrick, & Nutman, 2014; Fischer, Supali, Wibowo, Bonow, & Williams, 2000; Hassan et al., 2005; Kluber, Supali, Williams, Liebau, & Fischer, 2001; Lizotte, Supali, Partono, & Williams, 1994; McCarthy et al., 1996; Ramzy, 2002; Rao, Atkinson, et al., 2006; Rao, Weil, Fischer, Supali, & Fischer, 2006; Takagi et al., 2011; Toe et al., 1998; Toure et al., 1997, 1998; Zhong et al., 1996; Zimmerman et al., 1994). Common targets are the 188 bp DNA non-coding DNA repeat sequence *Ssp-I* for *W. bancrofti* (Ramzy, 2002; Rao, Atkinson, et al., 2006; Takagi et al., 2011; Zhong et al., 1996), *Hha I* repeat region for *B. malayi* (Fischer et al., 2000; Lizotte et al., 1994; Rao, Weil et al., 2006), *LLMF72* or a repeat region (15r3) for *L. loa* (Drame et al., 2014; Fink, Kamgno, & Nutman, 2011; Toure et al., 1998), and *O-150* repeat sequence for *O. volvulus* (Toe et al., 1998; Zimmerman et al., 1994).

A recent study performed at the National Institutes of Health compared PCR and conventional parasitologic techniques on samples from 200 patients with suspected filarial infection (Fink, Fahle, et al., 2011). They found that real-time PCR was at least as sensitive for detection of *W. bancrofti* and *L. loa* in blood specimens, and significantly more sensitive for detecting *O. volvulus* in skin snips. Furthermore, their PCR methods required smaller specimen volumes than conventional tests (Fink, Fahle, et al., 2011). Similarly, Rao and colleagues found *Hha I* PCR using a minor groove binder probe had comparable sensitivity to blood membrane filtration (Rao, Weil et al., 2006), while Fink and colleagues detected 5 of 16 cases of low-level *L. loa* parasitemia in specimens that were negative by microscopy (Fink, Kamgno, et al., 2011).

2.5.4 Clinical implementation of molecular tests

NAATs are not widely used clinically, but may be particularly useful in settings where filarial infections are not commonly seen and expertise diagnosing infections with conventional microscopy is lacking (Fink, Fahle, et al., 2011). They may also play an important role in epidemiologic studies and large-scale filarial disease control programmes, since they can be used human as well as vector samples and thus be used for monitoring vector infection rates and asymptomatic disease in humans (Dissanayake, Min, & Piessens, 1991; Farid et al., 2001; Rao, Atkinson, et al., 2006; Takagi et al., 2011). NAATs for filarial infections may also be useful components of multiplex assays for non-specific febrile syndromes or for use in endemic regions where patients are commonly infected with multiple pathogens. For example, a multiplex PCR assays targeting the four main *Plasmodium* species and *W. bancrofti* has been described (Mehlotra et al., 2010). Similarly, assays that detect *L. loa* and *O. volvulus* co-infection may be useful since fatal reactions may be observed when patients with heavy infections receive ivermectin.

Unfortunately, it appears that the timing of blood draws may influence the sensitivity of NAATs for filaria with periodic release of microfilariae into the blood.

One of the major drawbacks of conventional microscopic techniques for *W. bancrofti* and *B. malayi* is that blood must be drawn at night for maximum sensitivity, and an assay that obviates the need for nocturnal blood draw would be highly desirable. However, Rao and colleagues found that PCR targeting *Hhal I* was less sensitive when tested on samples drawn during the day compared to conventional membrane filtration using nocturnally drawn specimens (Rao, Atkinson, et al., 2006; Rao, Weil et al., 2006). In comparison, antigen-based assays for *W. bancrofti* can be performed on both diurnal and nocturnal specimens with good sensitivity. There are few studies comparing NAATs to antigen-based assays, but a recent multicentre study found antigen positivity rates to be substantially higher than both PCR and traditional microscopy (Gass et al., 2012). Regardless, antigen tests for *W. bancrofti* are not widely available in the United States and Europe, and few labs possess the skills and knowledge to design and implement their own antigen detection assays. In comparison, a potentially larger number of labs are capable of designing and implementing laboratory-developed PCR assays, thus making this a more attractive option.

3 TISSUE PARASITES

3.1 *LEISHMANIA* SPECIES (LEISHMANIASIS)

3.1.1 Background

Human leishmaniasis is caused by 21 species in the genus *Leishmania* (Garcia, 2007). Infection is transmitted through the bite of an infected sandfly in the tropics and subtropics worldwide. The two main forms of disease are visceral leishmaniasis (VL) and cutaneous leishmaniasis (CL), and these are caused by different *Leishmania* species. VL is most commonly caused by *Leishmania infantum*, *Leishmania chagasi* (southwest and central Asia, the Mediterranean and South American) and by *Leishmania donovani* (Indian subcontinent, Asia and Africa), while CL is caused by *Leishmania major*, *Leishmania tropica*, *Leishmania aethiopica* (Old World leishmaniasis), *L. infantum* (Caspian sea and the Mediterranean basin), *Leishmania amazonensis*, *Leishmania braziliensis*, *Leishmania mexicana*, *Leishmania* (*Viannia*) *panamensis*, *Leishmania* (*Viannia*) *peruviana* and *Leishmania* (*Viannia*) *guyanensis* (New World leishmaniasis) (Garcia, 2007; Murray, Berman, Davies, & Saravia, 2005). Importantly, CL caused by members of the *Viannia* subgenus are capable of disseminating to the mucous membranes of the mouth and nose and causing a destructive form of disease called mucocutaneous leishmaniasis (MCL). This form of disease requires aggressive systemic treatment, and therefore, CL caused by species that can cause MCL should also be treated aggressively (Murray et al., 2005).

3.1.2 Diagnostic modalities

Diagnosis is most commonly achieved using microscopic- and culture-based methods. Non-motile amastigote forms within macrophages can be identified in a variety of microscopy preparations including touch/impression smears and histopathologic sections from skin (CL), spleen (VL) and bone marrow (VL)

(Garcia, 2007). These morphology-based tests are often a good first diagnostic step, but do not allow for species identification, which is important when there is a concern for MCL. Furthermore, the sensitivity of these methods varies tremendously with the sampling technique, age of the skin lesions (CL) and/or skill and knowledge of the microscopist. Culture of the sampled material using conventional parasitology media such as NNN media can improve the sensitivity of detection and provides amplified material for subsequent species identification using isoenzyme analysis or PCR. It is not widely available, but is performed at no charge for patients in the United States by the CDC. Finally, a variety of serodiagnostic methods (e.g. ELISA, rapid rK39 strip dipstick test) can be used to diagnose VL, but there are no useful serologic methods for CL (Garcia, 2007). A final serologic method worth mentioning is the Montenegro test (a.k.a. leishmanin skin test (LST)), in which leishmanin antigen is injected intradermally and examined for induration. This test is not available in the United States but is widely used in endemic settings and may be useful for both VL and CL diagnosis (Garcia, 2007).

NAATs are an attractive option for sensitive *Leishmania* detection and species identification using clinical specimens or culture isolates. In addition to commercially available CE-marked kits (e.g. STAT-NAT® *Leishmania* spp. real-time PCR kit, Sentinel Diagnostics, Italy), the FDA has granted 510 (k) clearance for the SMART Leish real-time PCR assay for detection of *Leishmania* species and identification of *L. major* in skin scrapings or biopsies (Food and Drug Administration, 2011). Unfortunately, use of this assay is largely restricted to the US Department of Defense.

3.1.3 Design and performance of molecular tests

Both PCR- and LAMP-based assays have been described for *Leishmania* species (Boggild et al., 2010; Castilho, Camargo, McMahon-Pratt, Shaw, & Floeter-Winter, 2008; Disch et al., 2005; Food and Drug Administration, 2011; Kumar, Bumb, Ansari, Mehta, & Salotra, 2007; Takagi et al., 2009; Wortmann et al., 2005). Genus level identification can be performed by targeting *16S rDNA* (Food and Drug Administration, 2011) or *ITS2* (Centers for Disease Prevention and Control, 2013), while the *Viannia* subgenus can be specifically detected by targeting kDNA (Disch et al., 2005). In general it is difficult to create an assay to identify and differentiate all *Leishmania* species, although the glucose-6-phosphate dehydrogenase gene, as well as *GPI* and *ITS1* have been used for specific detection of targeted species (Castilho et al., 2008; Food and Drug Administration, 2011; Kumar et al., 2007). An assay for detection of the *Leishmania* complexes (*L. Viannia*, *L. mexicana*, *L. donovani/infantum* and *L. major*) targeting the glucosephosphate isomerase gene has also been described.

In general, molecular assays have greater sensitivity than conventional microscopy, culture and serology-based methods. A recent meta-analysis found that PCR had a superior diagnostic odds ratio to serology for detection of VL in HIV-infected patients in Europe (Cota, de Sousa, Demarqui, & Rabello, 2012). Similarly, Boggild and colleagues found that conventional PCR targeting kDNA had superior sensitivity (96%) to culture (57.8%), microculture (78.3%), smear microscopy

(71.4%) and the LST (78.2%) when using a composite gold standard (Boggild et al., 2010). In this study, PCR also detected an additional 14 lesions that were not positive by any other method. Using sub-analysis of patients with chronic disease, the authors found that the LST performed as well as PCR (sensitivity 88%), while culture and smear performance was poor (24% and 44% sensitivity, respectively).

3.1.4 Clinical implementation of molecular tests

Given their superior sensitivity over conventional tests and the ability of some assays to provide identification to the species level, *Leishmania* NAATs are increasingly used for primary diagnosis at specialised reference centres, including the CDC and Walter Reed National Military Medical Center in the United States (Centers for Disease Prevention and Control, 2013; Federal Drug Administration, 2013). Because of the heterogeneity between *Leishmania* species, the CDC has adopted an approach where initial PCR targeting the ITS2 gene is used to broadly amplify *Leishmania* DNA; species level identification is then performed using sequencing analysis (Centers for Disease Prevention and Control, 2013). The higher degree of sensitivity makes NAATs an attractive test for detecting cases of MCL in which the parasite load is typically low and the sensitivity of microscopy- and culture-based assays are less than 50% (Disch et al., 2005). Finally, quantitative real-time PCR has been used to monitor the response to treatment in patients with VL (Aoun et al., 2009; Sudarshan, Weirather, Wilson, & Sundar, 2011).

3.2 *TOXOPLASMA GONDII* (TOXOPLASMOSIS)

3.2.1 Background

Toxoplasmosis is caused by the intracellular protozoan parasite *T. gondii* (Garcia, 2007). Humans become infected when ingesting cysts containing bradyzoites in undercooked meat or oocysts that are shed in cat faeces and have contaminated food or water. Transplacental transmission and through blood transfusion/organ transplantation are also important means of transmission (Garcia, 2007). In immunocompetent hosts, primary infection is often asymptomatic or may cause a self-limiting mononucleosis-like illness. Primary infection during pregnancy is also problematic since it can be transmitted across the placenta and cause significant morbidity and mortality in the developing foetus (Garcia, 2007). Once infected, parasites are thought to remain alive for the life of the host and infection can reactivate in states of profound immune compromise such as HIV. In these immunocompromised patients, infection may cause ocular disease and severe life-threatening brain involvement. Toxoplasmosis is found throughout the world and is most common in countries where raw/undercooked meat is commonly consumed or cat faeces widely contaminate the environment (Robert-Gangneux & Darde, 2012).

Multiple *T. gondii* genotypes have been identified using PCR-RFLP and multilocus enzyme electrophoresis (Su, Shwab, Zhou, Zhu, & Dubey, 2010). Interestingly, the prevalence of different genotypes varies geographically, with only a few dominant genotypes in the northern hemisphere and hundreds of coexisting genotypes in the southern hemisphere (Shwab et al., 2014). The clinical importance of these genotypes is unknown at this time (Higa, Garcia, Su, Rossini, & Falavigna-Guilherme, 2014; Shwab et al., 2014).

3.2.2 Diagnostic modalities

Serology and NAAT are the primary diagnostic modalities for toxoplasmosis. Serologic testing is used primarily for immunocompetent patients, whereas NAATs play a primary role in testing immunocompromised patients and detecting congenital disease. It is important to know that quantitative serology results may vary between methods, thus making it difficult to compare results performed at different laboratories. Furthermore, positive serology results cannot differentiate previous exposure to active disease. When positive IgM and IgG results are detected, repeat testing, preferably at a reference laboratory, is recommended. IgG-Avidity testing can also be useful to determine how recently the infection was acquired (McAuley, Jones, & Singh, 2011).

PCR assays are widely used in the United States and Europe for diagnosis of human toxoplasmosis. There are a few CE-marked kits available (e.g. Bio-Evolution *T. gondii* detection kit, France), but unfortunately no FDA-cleared/approved assays at this time. Instead, laboratory-developed assays are available through diagnostic reference laboratories for use with a variety of specimen types including blood, CSF, amniotic fluid and tissue.

3.2.3 Design and performance of molecular tests

Several multi-copy targets have been described for *T. gondii* NAATs including the repeated element (*RE*) sequence (GenBank accession number AF146527), and the ITS1, 18S rRNA and B1 genes (Burg, Grover, Pouletty, & Boothroyd, 1989; Homan, Vercammen, De Braekeleer, & Verschueren, 2000). The *RE* gene is present in higher copy numbers than *B1* and therefore is generally considered to be a more sensitive target. A recent study comparing LAMP and nested-PCR for both targets found the highest amount of positive results using the *RE* LAMP assay, followed by the *B1* LAMP assay, *RE* PCR and *B1* PCR (listed in order of decreasing sensitivities) (Fallahi, Seyyed Tabaei, Pournia, Zebardast, & Kazemi, 2014). A multicentre comparison of the CE-marked Bio-Evolution *T. gondii* PCR assay and eight laboratory-developed assays, all targeting *RE* (also referred to as *rep529*) demonstrated excellent (>99%) concordance between assays and uniform 86% sensitivity and 100% specificity for detection of *T. gondii* DNA in amniotic fluid (Filisetti et al., 2015). A lyophilised *T. gondii* standard was recently validated by members of the French National Reference Center for Toxoplasmosis and should facilitate standardisation between different *T. gondii* NAATs (Varlet-Marie et al., 2014).

3.2.4 Clinical implementation of molecular tests

The use of *T. gondii* NAATs fall into three main areas: (1) diagnosis of congenital toxoplasmosis, (2) detection of reactivated infection in immunocompromised patients and (3) identification of ocular toxoplasmosis (McAuley et al., 2011). These uses are described in further detail later.

PCR has become the test of choice for diagnosing congenital toxoplasmosis and has been used in France since the 1990s when it replaced culture as the gold standard diagnostic method (Filisetti et al., 2015). Initial testing usually begins with maternal serologic screening for at-risk mothers. When serology is positive and indicative of recent infection (i.e. IgM positive and IgG seroconversion, rising IgG titres, low IgG avidity), amniotic fluid should be tested by *T. gondii* PCR to determine foetal involvement. PCR on amniotic fluid is reported to have close to 100% positive and negative predictive value for maternal infection in the first or second trimester of pregnancy (Robert-Gangneux & Darde, 2012). Unfortunately, negative amniotic fluid PCR does not fully exclude congenital toxoplasmosis due to low levels of parasites (<5–10 parasites/mL) and further testing, including serologic studies and PCR of cord blood or placenta should be performed postnatally (Bastien, Jumas-Bilak, Varlet-Marie, & Marty, 2007; Robert-Gangneux & Darde, 2012; Yera et al., 2009).

PCR also plays an important role in testing immunocompromised patients, particularly those with AIDS and seropositive allogeneic stem cell transplant recipients who seem to be at the highest risk of reactivation disease (Edvinsson, Lundquist, Ljungman, Ringden, & Evengard, 2008; McAuley et al., 2011). Testing can be performed on a variety of sample types including blood, CSF and tissue. Older studies using conventional PCR reported sensitivities as low as 33% and 16% on CSF and blood, respectively, when testing patients with cerebral toxoplasmosis (Robert-Gangneux & Darde, 2012). Fortunately, recent work has demonstrated sensitivities of up to >90% when testing blood or CSF from patients with AIDS and cerebral toxoplasmosis (Colombo et al., 2005; Mesquita, Ziegler, Hiramoto, Vidal, & Pereira-Chioccola, 2010). It is important to note these results were generated using a variety of different assays. Additional studies using standardised and calibrated assays would allow for better appreciation of the sensitivity and specificity of PCR on blood and CSF in immunosuppressed patients.

The last clinical category that deserves mention is diagnosis of *T. gondii* retinochoroiditis. Although this is traditionally an ophthalmologic diagnosis based on identification of characteristic retinal lesions, testing of aqueous or vitreous humour can be useful for supporting or confirming the clinical diagnosis, particularly when the findings are atypical or the patient does not have the expected response to treatment (McAuley et al., 2011). Sensitivity is modest at 16–55% (Robert-Gangneux & Darde, 2012).

3.3 FREE-LIVING AMOEBAE: *NAEGLERIA FOWLERI*, *ACANTHAMOEBA* SPECIES AND *BALAMUTHIA MANDRILLARIS*

3.3.1 Background

Acanthamoeba species, *Balamuthia mandrillaris* and *Naegleria fowleri* are free-living amoebae (FLA) they may rarely become parasites of humans and cause significant, even fatal, disease (Garcia, 2007). They are found in multiple environmental sources worldwide including soil, sewage, water (lakes, tap water, swimming pools), heating and air-conditioning units and even dust. *N. fowleri* is the most virulent of the three amoebae and causes primary amoebic meningoencephalitis (PAM), an acute and rapidly fatal infection found predominantly in children and young adults. *Acanthamoeba* spp. and *B. mandrillaris* can a more chronic but equally fatal infection called granulomatous amoebic encephalitis (GAE). Rapid diagnosis is important in selecting and initiating treatment since early therapy, particularly for GAE, has resulted in recovery in some cases. *Acanthamoeba* spp. can also cause amoebic keratitis (AK), primarily in contact lens wearers. This infection is not fatal, but can be recalcitrant to treatment and lead to significant morbidity including blindness. Finally, *Sappinia pedata* (previously thought to be *Sappinia diploidea*) and *Paravahlkampfia francinae* have been recognised as rare causes of amoebic CNS infection (Qvarnstrom, da Silva, Schuster, Gelman, & Visvesvara, 2009; Visvesvara et al., 2009).

Unlike *B. mandrillaris* and *N. fowleri*, which are the only species in their genus to cause human disease, multiple *Acanthamoeba* species have been identified in cases of human ocular or CNS infection (*Acanthamoeba astronyxis*, *Acanthamoeba castellanii*, *Acanthamoeba culbertsoni*, *Acanthamoeba divionensis*, *Acanthamoeba hatchetti*, *Acanthamoeba healyi*, *Acanthamoeba polyphaga* and *Acanthamoeba rhysodes*) (Visvesvara, 2011). In nature, as many as 24 species have been identified and divided into 15 sequence types (T1–T15 genotypes) using analysis of nuclear *ssu* rDNA. T4 has been the predominant genotype detected from humans in both ocular and brain specimens, although several other genotypes have been less commonly identified (Visvesvara, Moura, & Schuster, 2007).

3.3.2 Diagnostic modalities

Diagnosis of infection with the FLA is traditionally through microscopic examination of amoebic cysts and/or trophozoites in clinical specimens such as corneal scrapings (AK), contact lenses (AK), CSF (PAM) and brain biopsies (PAM, GAE). The utility of this method is potentially hampered by scarcity of diagnostic forms, and therefore culture can be used for improved sensitivity. *N. fowleri* and *Acanthamoeba* spp. can be cultured on non-nutrient agar overlaid with bacteria for a food source, while *B. mandrillaris* can only be cultured in specialised cell lines. Both culture techniques are time consuming, can take several days to become positive, and are not widely available. Molecular methods are thus a welcome tool for diagnosis of FLA infection and can help expedite diagnosis and treatment. Unfortunately, there are currently no FDA-approved/cleared or CE-marked kits at this time.

3.3.3 Design and performance of molecular tests

Many conventional and real-time singleplex and multiplex PCR assays have been described for detection of FLA in human and environmental samples (Booton, Visvesvara, Byers, Kelly, & Fuerst, 2005; Kao et al., 2013; Qvarnstrom, Visvesvara, Sriram, & da Silva, 2006; Visvesvara et al., 2007; Yagi, Booton, Visvesvara, & Schuster, 2005). LAMP-based assays have also been described (Lek-Uthai, Passara, Roongruangchai, Buddhirakkul, & Thammapalerd, 2009; Yang et al., 2013). The multiplex Taqman probe-based real-time assay described by CDC investigators (Qvarnstrom et al., 2006) targets the nuclear small subunit 18S rRNA gene and allows for the simultaneous detection and differentiation of *Acanthamoeba* species, *B. mandrillaris* and *N. fowleri* in a single reaction. The assay was initially validated for use with CSF and brain tissue and has a reported lower LOD of 1 amoeba per processed specimen. This assay has also been successfully adapted for use with FFPE tissue with inclusion of a recovery template for identification of PCR inhibition (Norgan, Sloan, Giannini, & Pritt, 2014). Khairnar, Tamber, Ralevski, and Pillai (2011) recently compared the Qvarnstrom assay with another real-time PCR assay targeting *18S rDNA* (Riviere, Szczebara, Berjeaud, Frere, & Hechard, 2006) and two conventional PCR assays (targeting *rDNA*) for detection of *Acanthamoeba* from ocular specimens, using direct microscopy and culture as the gold standard. These authors found that the real-time assays had superior sensitivity (82.1–89.3%) to conventional PCR (50–53.6%). A separate study comparing the same two real-time PCR assays evaluated in the previously mentioned study found the Qvarnstrom 18S rDNA assay to have greater sensitivity for detection of *Acanthamoeba* spp. in water and biofilms from cooling towers (Chang, Wu, & Ming, 2010). Of note, the Qvarnstrom assay can detect genotypes T1, T4, T7 and T10, while the Rivera assay only detects T1 and T4, and not the T7 and T10 genotypes (Visvesvara et al., 2007). The T7 and T10 genotypes are known to cause GAE.

A few special considerations exist for the design, validation and implementation of FLA NAATs. Given the widespread presence of FLA in water sources, it is important test molecular reagents for contamination with FLA DNA during assay design and verification studies. Similarly, if non-human specimens such as contact lenses, contact cases and contact solution will be accepted for NAAT *Acanthamoeba* testing, then the laboratory should first establish the prevalence in asymptomatic individuals, since these sources could potentially be contaminated with *Acanthamoeba* spp. DNA. Goldschmidt and colleagues found that extraction of DNA from *Acanthamoeba* spp. cysts was facilitated by use of proteinase K and commercial extraction kits (e.g. Roche MagNA Pure) (Goldschmidt et al., 2008). Finally, Thompson, Kowalski, Shanks, and Gordon (2008) found the commonly used topical ophthalmic drug, polyhexamethylene biguanide, has inhibitory effects on PCR and therefore they recommended that ocular specimens be obtained for PCR prior to its application. These authors did not observe inhibition with several other topical ophthalmic drugs including prednisolone and antimicrobial agents.

3.3.4 Clinical implementation of molecular tests

Nucleic acid amplification assays for FLA offer significant potential advantages over conventional microscopy and culture-based assays. Specifically, their rapid and highly sensitive nature makes them ideal diagnostic tools for potentially fatal FLA infections involving the CNS. When testing for *Acanthamoeba* spp., it is important for the laboratory and physicians to known which genotypes are detected by the assay being employed.

4 INTESTINAL PROTOZOA AND MICROSPORIDIA

4.1 BACKGROUND

The intestinal protozoa and microsporidia are important causes of gastroenteritis worldwide. *Giardia duodenalis* (also known as *Giardia lamblia*, *Giardia intestinalis*), *Cryptosporidium* spp. and *E. histolytica* contribute significantly to the morbidity and mortality of children worldwide, particularly in endemic regions such as sub-Saharan Africa (Boschi-Pinto, Lanata, Mendoza, & Habte, 2006). In developed countries, *Cryptosporidium* spp. and *G. duodenalis* protozoa can cause periodic outbreaks as well as life-threatening infection in immunocompromised individuals such as those with AIDS. *E. histolytica* infection is less common in developed countries but may be seen in travellers and immigrants. Other protozoa associated with diarrhoea and gastrointestinal symptoms worldwide are *Dientamoeba fragilis*, *Cyclospora cayetanensis*, *Cystoisospora belli* and *Balantidium coli*. *Blastocystis hominis* may also cause gastrointestinal symptoms in some settings but the pathogenicity of this organism is not fully understood (Garcia, 2007). The microsporidia were originally thought to be protozoa, but are now known to be highly specialised fungi in the phylum Microspora. Since testing for the microsporidia is commonly performed in the parasitology lab using stool (and less commonly urine and other sources), they are included in this section. Transmission of intestinal protozoa and microsporidia is primarily thought to occur through the faecal-oral route with ingestion of contaminated food or water.

4.2 DIAGNOSTIC MODALITIES

Laboratory diagnosis of intestinal protozoa and microsporidia is traditionally based on microscopic examination of stained and unstained stool preparations (i.e. the 'ova and parasite' exam or 'O&P'). Unfortunately, the sensitivity of detection is compromised by intermittent parasite shedding (requiring examination of multiple stool specimens) and variability in technical knowledge and expertise. Pathogenic parasites must be differentiated from a wide variety of faecal materials as well as from non-pathogenic parasites. *E. histolytica* specifically shares many morphologic similarities with the other non-pathogenic amoebae commonly found in stool such as *Entamoeba hartmanni*, *Entamoeba coli*, *Entamoeba gingivalis*, *Endolimax nana* and *Iodamoeba buetschlii*. Furthermore, it is morphologically indistinguishable from *E. dispar*, *Entamoeba moshkovskii* and *Entamoeba bangladeshi* in stool specimens, unless erythrocytes are seen within the trophozoite forms, which is pathognomonic for *E. histolytica* (Clark & Diamond, 1991; Kebede, Verweij, Petros, & Polderman, 2004; Royer et al., 2012; World Health Organisation, 1997). Of these four protozoa, *E. histolytica* is considered to be the only definitive pathogen and should be differentiated from other amoebae to allow for appropriate treatment (World Health Organisation, 1997). Given these challenges, it is not surprising that a recent study of several European reference laboratories found only poor to moderate agreement for detection of protozoa (Utzinger et al., 2010).

Antigen detection methods are widely available for the major protozoan pathogens and are the non-molecular test of choice for *G. duodenalis*, *Cryptosporidium* spp. and *E. histolytica*. Some methods are capable of differentiating between *E. histolytica* and *E. dispar* infection and can be used for primary screening or confirmatory testing following microscopy. Despite the advantages of antigen methods, the sensitivity and specificity vary widely between tests and multiple exams may still be required for optimal sensitivity (Hanson & Cartwright, 2001).

Molecular diagnostics offer an attractive alternative since they do not depend on subjective morphologic features and can often provide a level of identification beyond what is possible with light microscopy. Although there are some CE-marked kits for individual protozoa, the most exciting advances in intestinal protozoa come with the introduction of commercial multiplex kits for protozoal, bacterial and viral causes of gastroenteritis. A variety of FDA-cleared/approved and CE-marked options now exists; these are discussed in further detail later.

4.3 DESIGN AND PERFORMANCE OF MOLECULAR TESTS

Many singleplex and multiplex NAATs have been described for detection of the major intestinal protozoa. Some of the commonly used targets for laboratory-developed assays are listed in Table 2. Targets for *E. histolytica*/*E. dispar* include *ssu-rDNA* (Haque, Ali, Akther, & Petri, 1998; Liu et al., 2013; Stark et al., 2008; Verweij et al., 2004), p30 (gene encoding the 30 kDa molecule) (Tachibana, Kobayashi, Okuzawa, & Masuda, 1992) and the hemolysin gene (HLY6) (Zindrou, Orozco, Linder, Tellez, & Bjorkman, 2001). Some assays are designed for detection *E. histolytica* and do not detect *E. dispar* (Zindrou et al., 2001). Gene targets for detection and/or genotyping of *G. duodenalis* include *ssu 18S rDNA* (Hopkins et al., 1997; Liu et al., 2013; Verweij et al., 2003), elongation factor 1-α (Monis, Andrews, Mayrhofer, & Ey, 1999), glutamate dehydrogenase (Monis et al., 1999; Read, Monis, & Thompson, 2004), β-giardin (Caccio, De Giacomo, & Pozio, 2002; Guy, Payment, Krull, & Horgen, 2003) and triosephosphate isomerase genes (Sulaiman et al., 2003), while targets for *Cryptosporidium* spp. include *18S rDNA* (Liu et al., 2013), the *Cryptosporidium parvum*-specific 452-bp fragment (Fontaine & Guillot, 2002; Laxer, Timblin, & Patel, 1991; Verweij et al., 2004) and the *Cryptosporidium* oocyte wall protein gene (Guy et al., 2003). Multiplex options for *E. histolytica*, *Cryptosporidium* spp. and *G. duodenalis* include a triplex real-time PCR assay (Haque et al., 2007), a heptaplex PCR-Luminex assay that includes four common intestinal nematodes (Taniuchi et al., 2011) and an ambitious multiplex Taqman Array for 19 enteropathogens including bacterial and viral pathogens (Liu et al., 2013).

Multiple PCR-based assays have also been described for the common microsporidia found in stool, *Enterocytozoon bieneusi* and *Encephalitozoon intestinalis* (Notermans et al., 2005; Verweij, Ten Hove, Brienen, & van Lieshout, 2007; Wolk, Schneider, Wengenack, Sloan, & Rosenblatt, 2002). Compared to examination of stool for the small (1–5 μm) spores of microsporidia, PCR offers potentially improved sensitivity and specificity (Notermans et al., 2005; Wolk et al., 2002).

The laboratory-developed assays for intestinal protozoa and microsporidia have significantly different performance characteristics, but in general show similar or greater sensitivity compared to conventional microscopy and antigen detection. It is beyond the scope of this chapter to fully discuss and compare the different LDTs. The reader is referred to the thorough review of molecular testing for intestinal parasites by Verweij and Stensvold (2014).

A promising development for molecular diagnosis of protozoan parasites is the recent introduction of several commercially available multiplex assays. Many assays detect the protozoan parasites, *G. duodenalis*, *E. histolytica* and *Cryptosporidium* spp. as well as bacterial and viral targets, and some include other protozoa such *C. cayetanensis* and *D. fragilis*. Among the more commonly used assays are the Luminex® xTAG Gastrointestinal Pathogen Panel (xTAG GPP) (Luminex Corporation, Austin, Texas, USA), FilmArray® GI panel (BioFire® Diagnostics, Salt Lake City, Utah, USA) and the RIDA® GENE Parasitic Stool Panel (R-Biopharm AG, Darmstadt, Germany). Reddington and colleagues recently reviewed several of these assays (Reddington, Tuite, Minogue, & Barry, 2014), and clinical evaluations are available for some (Buss et al., 2015; Khare et al., 2014; Perry, Corden, & Howe, 2014; Stark, Roberts, Ellis, Marriott, & Harkness, 2014).

At this time, only the xTAG GPP and FilmArray assays are FDA-cleared for IVD use. The xTAG GPP was CE-marked for IVD used in 2011 and received FDA clearance in 2013. A single multiplex real-time PCR assay amplifies DNA from extracted DNA from bacterial, viral and protozoan parasites, followed by qualitative detection using the xTAG bead-based tag sorting system. It is important to note that the FDA-cleared version only detects *Cryptosporidium* spp. and *G. duodenalis*, while the CE-marked assay also detects *E. histolytica* (Luminex, 2015). A run of 24 specimens can be completed in approximately 5 h (Luminex, 2015). When compared to conventional microscopy, the xTAG GPP showed excellent accuracy for detection of *G. duodenalis* (100% sensitivity, 96.9% specificity), *Cryptosporidium* (91.7% sensitivity, 100% specificity) and *E. histolytica* (sensitivity not calculated, 99% specificity) (Claas, Burnham, Mazzulli, Templeton, & Topin, 2013). Performance was also excellent compared to a multiplex real-time PCR assay, with sensitivities from 91.3% (*Cryptosporidium*) to 100% (*G. duodenalis*, *E. histolytica*) (Verweij et al., 2004). In comparison, the FilmArray GI panel is FDA-cleared/CE-marked for detection of *G. lamblia* (*G. duodenalis*), *Cryptosporidium* spp., *E. histolytica* and *C. cayetanensis*. This nested-PCR assay includes DNA extraction and amplification in a pouch-based format and produces results for one patient in 1 h (BioFire, 2014). A multicentre study comparing the FilmArray to PCR comparator assays demonstrated sensitivities and specificities of >99% (Buss et al., 2015).

4.4 CLINICAL IMPLEMENTATION OF MOLECULAR TESTS

At this time, laboratory-developed singleplex and multiplex NAATs are used primarily in specialised reference centres and research settings. They have not been widely implemented into diagnostic laboratories in the United States or Canada, but are more commonly found in larger European diagnostic laboratories. The reluctance to incorporate molecular tests may be due to the widespread availability of reliable, affordable and relatively accurate antigens tests for *G. duodenalis*, *Cryptosporidium* spp. and *E. histolytica* and lack of commercial testing options that are FDA-cleared/approved. The introduction of commercial multiplex PCR-based assays should, therefore, make NAATs more accessible for routine diagnosis of some protozoan parasites. The commercial multiplex NAATs are relatively expensive, but their use may be cost-effective when replacing a number of culture, microscopy, antigen and PCR-based tests for various analytes. Additionally, some include a number of viral and bacterial targets for which testing is not routinely performed (e.g. astrovirus, sapovirus, enteroaggregative *Escherichia coli*). Commercial NAATs may be especially useful for patients at risk of severe disease, but would not likely be cost-effective for patients with acute, self-limited viral gastroenteritis. Furthermore, there will likely continue to demand for conventional pathogen-specific tests when an individual pathogen is suspected (e.g. outbreak situations). When considering the implementation of a commercial multiplex assays, laboratories must decide which of the various options best meet their needs for number and type of analytes detected, specimen capacity, complexity and turn-around-time.

5 INTESTINAL HELMINTHS
5.1 BACKGROUND

Helminths (worms) that inhabit the human intestinal tract are a genetically diverse group of parasites that include several different nematodes (round worms), cestodes (tapeworms) and trematodes (flukes). Many of these helminths are transmitted by ingestion of parasite eggs in contaminated food or water (i.e. the faecal-oral route of infection). These infections are most common in regions of the world that lack proper sanitation and water treatment facilities. The WHO recognised several intestinal worms (specifically the soil-transmitted helminths) as one of the 17 NTDs and estimates that approximately 890 million children worldwide are in need of annual treatment and preventive therapy for these worms (World Health Organisation, 2010, 2013). The most common intestinal roundworms worldwide are *Ascaris lumbricoides*, hookworms (*Necator americanus* and *Ancylostoma duodenale*) and whipworm (*Trichuris trichiura*), while *Taenia solium*, *Taenia saginata* and *Hymenolepis nana* are common cestodes in many parts of the world. The prevalence of intestinal trematodes varies tremendously by geographic region, with *Fasciolopsis buski* being the most common, particularly in parts of Southeastern Asia (Garcia, 2007; World Health Organisation, 2010).

Morbidity is usually proportional to the burden of intestinal worms. Heavy infections are commonly associated with nutritional deficits and poor school performance and may also cause complications that necessitate surgical therapy such as rectal prolapse and intestinal obstruction. Some infections such as disseminated strongyloidiasis are life-threatening. Annual administration of antihelminths such as ivermectin or mebendazole significantly decreases the disease burden and associated morbidity among at-risk populations.

5.2 DIAGNOSTIC MODALITIES

As with the intestinal protozoa, intestinal helminths are traditionally detected by microscopic examination of stool. This remains the most commonly used method worldwide since it allows for simultaneous detection of multiple parasites without the need for expensive instruments and reagents. However, stool microscopy is subjective and requires examination of three or more specimens for optimal sensitivity (Thomson, Haas, & Thompson, 1984). For detection of the medically important nematode, *Strongyloides stercoralis*, six or more specimens are needed (Siddiqui & Berk, 2001). The requirement for examination of multiple specimens is burdensome for both the laboratory and the patient and may hinder diagnosis. Furthermore, reliable microscopic detection and differentiation of helminth eggs requires a considerable degree of expertise that may not be available in non-endemic settings.

Serologic testing may also be used for detection of select intestinal helminths. Unfortunately, helminth-specific tests are not widely available, often suffer from poor sensitivity and exhibit a high degree of cross-reactivity with other helminths (Garcia, 2007). Given the limitations of traditional methods, there is a need for affordable and sensitive multiplex assays for the simultaneous detection of numerous intestinal helminths. At this time, there are no CE-marked or FDA-cleared/approved singleplex or multiplex assays for intestinal helminths.

5.3 DESIGN AND PERFORMANCE OF MOLECULAR TESTS

A variety of laboratory-developed assays have been described for all of the common intestinal nematodes, including *A. lumbricoides* (Basuni et al., 2011; Leles, Araujo, Vicente, & Iniguez, 2009; Liu et al., 2013; Mejia et al., 2013), *A. duodenale* (Basuni et al., 2011; Gasser, Cantacessi, & Campbell, 2009; Mejia et al., 2013), *N. americanus* (Basuni et al., 2011; Gasser et al., 2009; Mejia et al., 2013), *S. stercoralis* (Kramme et al., 2011; Mejia et al., 2013; ten Hove et al., 2009; Verweij et al., 2009) and *T. trichiura* (Andersen, Roser, Nejsum, Nielsen, & Stensvold, 2013; Basuni et al., 2011; Liu et al., 2013; Mejia et al., 2013). The molecular targets for the more common pathogens are listed in Table 2. In general, these assays have been shown to have superior sensitivity over conventional microscopy. A study by Basuni and colleagues showed that a pentaplex real-time PCR was significantly more sensitive than direct wet prep and Kato-Katz microscopy for detection of *A. lumbricoides*, *N. americanus*, *S. stercoralis* and *Ancylostoma* species, with parasites detected in 48 of 77 specimens (62.3%) compared to microscopy (7.8%). Similarly, Mejia and colleagues compared the performance of a quantitative multi-parallel real-time PCR assay for detection of five common soil-transmitted nematodes and found higher detection rates for all parasites compared to direct microscopy (Mejia et al., 2013). Unfortunately, these two studies used direct stool preparations rather than more sensitive concentrated preparations for comparison. Finally, ten Hove and colleagues found increased detection of *S. stercoralis* over conventional microscopy using a multiplex real-time PCR assay for *S. stercoralis* and three pathogenic protozoa, with detection of 21 cases by PCR compared to only three using microscopy (formalin-ether concentration) and Baermann technique (ten Hove et al., 2009).

Molecular assays have also been described for food-borne trematodes, cestodes and schistosomes. The reader is referred to the in-depth review by Verweij and Stensvold (2014) for further information on these assays.

5.4 CLINICAL IMPLEMENTATION OF MOLECULAR TESTS

Despite their superior sensitivity, NAATs for intestinal helminths are used primarily in the research setting and for epidemiologic studies rather than for clinical diagnosis. This is likely due to the general lack of intestinal helminths in industrialised countries where advanced laboratory testing is widely available. Another likely contributor is the lack of commercial assays for common intestinal helminths. Finally, a major barrier to implementation of molecular assays is the extremely large diversity of helminths that can be found in the human intestinal tract, and thus the large number of analytes that would need to be included in a comprehensive stool parasite panel.

Despite these limitations, there are some instances where NAATs can be selectively used for clinical diagnosis. For example, PCR for *S. stercoralis* would obviate the need for laborious stool concentration techniques such as coproculture and the Baermann method and decrease laboratory personnel exposure to stool containing infectious *S. stercoralis* larvae. Similarly, PCR could replace urine concentration methods for detection of *Schistosoma haematobium* eggs. Finally, NAATs could be used to provide a higher degree of differentiation than what is possible by morphology alone, such as identification of the infecting *Taenia* species (i.e. *T. solium* vs. *T. saginata*) or differentiating similar-appearing eggs (e.g. hookworm, *Trichostrongylus*, *Oesophagostomum* spp.).

6 *TRICHOMONAS VAGINALIS*
6.1 BACKGROUND

T. vaginalis is a flagellate protozoan parasite responsible for trichomoniasis, the most common curable sexually transmitted disease (STD) worldwide. It was recently ranked as one of the five neglected parasitic infections in the United States based on the large number of individuals infected, the potential severity of illness and the availability of good preventative and treatment modalities (Centers for Disease Prevention and Control., 2014; Secor, Meites, Starr, & Workowski, 2014; Workowski & Berman, 2010). Approximately 3.7 million individuals in the United

States are thought to be infected, though most do not have signs or symptoms of infection. Populations with particularly high prevalences of infection include incarcerated women (9–32%), incarcerated men (2–9%) and patients attending STD clinics (26% of symptomatic women) (Workowski & Bolan, 2014). Most infected men and women (70–85%) are asymptomatic. When present, symptoms include malodorous vaginal discharge, dysuria and dyspareunia in women and urethral discharge and dysuria in men (Secor et al., 2014; Workowski & Bolan, 2014). Complications may include pelvic inflammatory disease in women, adverse pregnancy outcomes and increased risk of HIV transmission and acquisition (Secor et al., 2014; Workowski & Berman, 2010; Workowski & Bolan, 2014).

6.2 DIAGNOSTIC MODALITIES

Diagnosis is traditionally performed by examination of unfixed or fixed and stained vaginal secretions, urine, urethral swabs or semen for identification of characteristic trophozoites. However, microscopy has a limited sensitivity of only 60–70% when used on vaginal secretions and is significantly less sensitive for use with urine and specimens from men (Workowski & Berman, 2010). Furthermore, sensitivity may decrease by 20% within the first hour after collection, necessitating the need for immediate microscopic examination (Secor et al., 2014). Culture methods provide improved sensitivity (75–96%) compared to microscopy and were considered the gold standard method for *T. vaginalis* detection until molecular diagnostic methods became available (Nye, Schwebke, & Body, 2009). Antigen test methods such as the FDA-approved OSOM Trichomonas Rapid Test (Sekisui Diagnostics, Framingham, Massachusetts, USA) may also be useful, particularly as a point-of-care test since it is CLIA waived and provides results in 10 min. Sensitivities of 82–95% have been reported (Campbell, Woods, Lloyd, Elsayed, & Church, 2008; Huppert et al., 2007).

Nucleic acid detection methods offer a significant advance in laboratory diagnosis of trichomoniasis. Although non-amplification techniques such as the Becton Dickinson Affirm VP III (Sparks, Maryland, USA) are commercially available, NAATs offer the higher sensitivity and specificity and are now considered the preferred test for detection of *T. vaginalis* where resources allow their use (Andrea & Chapin, 2011; Domeika et al., 2010; Sherrard et al., 2014; Workowski & Bolan, 2014). The only FDA-cleared/approved NAAT at this time is the Aptima® *Trichomonas vaginalis* assay (Gen-Probe®, Hologic™, Bedford, Massachusetts, USA). This assay was cleared by the FDA in 2011 for use on the TIGRIS instrument and, in 2013 also received clearance for use on the fully automated PANTHER system. The test is cleared for testing symptomatic and asymptomatic women using urine, vaginal swabs and specimens collected in PreservCyt solution, but is not cleared for use with male specimens (GEN-PROBE, 2013). CE-marked *T. vaginalis* PCR kits from several manufacturers (e.g. Cepheid®, Diagenode Diagnostics, AmpliSens®) are also available.

6.3 DESIGN AND PERFORMANCE OF MOLECULAR TESTS

A variety of laboratory-developed conventional and real-time PCR assays have been described that offer improved sensitivity over microscopy and culture (Crucitti et al., 2003; Jordan, Lowery, & Trucco, 2001; Kaydos-Daniels et al., 2003; Lawing, Hedges, & Schwebke, 2000; Pillay, Radebe, Fehler, Htun, & Ballard, 2007; Radonjic et al., 2006; Schirm, Bos, Roozeboom-Roelfsema, Luijt, & Moller, 2007; Schwebke & Lawing, 2002; Simpson, Higgins, Qiao, Waddell, & Kok, 2007; Smith et al., 2005; Van den Eede et al., 2009; Wendel, Erbelding, Gaydos, & Rompalo, 2003). Common PCR targets include *18S rDNA,* β-tubulin gene and 261 bp repeat sequence. PCR provides sensitivities of 88–97%, depending on the specimen type, population and reference standard (Sherrard et al., 2014). A LAMP assay targeting a 2-kbp repeated sequence has also been described and demonstrated 10–1000 times higher sensitivity than the comparator *18S rDNA* PCR (Reyes, Solon, & Rivera, 2014), with an LoD of 1 trichomonad per reaction. These NAATs can be used to test a variety of specimen types including vaginal and urethral swabs, urine and semen.

The FDA-cleared Aptima *T. vaginalis* assay has equal or superior sensitivity to LDTs and may be more easily incorporated into diagnostic laboratories (Hardick, Hardick, Wood, & Gaydos, 2006; Nye et al., 2009; Schwebke & Lawing, 2002). This assay uses target capture, TMA and chemiluminescent probe hybridisation for detection of *18S rDNA*. In a large prospective US clinical trial, Schwebke and colleagues found the Aptima test to have a clinical sensitivity and specificity of 100% and ≥99%, respectively for detection of *T. vaginalis* in vaginal swabs, endocervical swabs and ThinPrep samples from asymptomatic and symptomatic women (Schwebke & Lawing, 2002). Slightly lower sensitivity and specificity (95.2% and 98.9%) were reported for female urine specimens. Importantly, the sensitivity was not significantly different between symptomatic and asymptomatic patients. This is in contrast to the performance of the DNA-probe-based Affirm assay, in which sensitivity may be higher among symptomatic women (Andrea & Chapin, 2011; Brown, Fuller, Jasper, Davis, & Wright, 2004). In a similar study, Nye and colleagues compared the Aptima assay to β-tubulin PCR, conventional wet mount microscopy and culture for detection of *T. vaginalis* in adult men and women who attended an STD clinic in Alabama, USA and found the Aptima assay to have superior sensitivity when testing both men and women (Nye et al., 2009). In women, the use of vaginal swabs for APTIMA testing provided the greatest sensitivity (96.6%), compared to endocervical swabs (89.8%) and urine (87.5%), while in men, urethral swabs provided the greatest sensitivity (95.2%) compared to urine (73.8%). Finally, when compared to the Affirm assay, the Aptima *T. vaginalis* assay demonstrated significantly greater sensitivity (100% compared to 63.4%), detecting 36.6% more positive patients (Andrea & Chapin, 2011).

Although the Aptima *T. vaginalis* NAAT is not cleared for use with male specimens, the reagents can be purchased as analyte-specific reagents and validated for per CLIA specifications for this population. Dize and colleagues have shown self-collected penile-meatal swabs to provide greater sensitivity than urine (80% vs. 39%, respectively) with the Aptima *T. vaginalis* assay (Dize et al., 2013). Semen may also be tested (Workowski & Bolan, 2014).

6.4 CLINICAL IMPLEMENTATION OF MOLECULAR TESTS

The high sensitivity and specificity of NAATs make them ideal for screening patients at risk of trichomoniasis. In comparison, the widely used microscopic exam has insufficient sensitivity to serve as a stand-alone test for diagnosis of infection and should be supplemented or replaced by molecular diagnostics when possible. The advent of easy-to-use commercial assays and improved sensitivity over some laboratory-developed PCRs position them to become the test of choice in resource rich settings.

At this time, NAATs screening is recommended for women who are considered to be at high risk for infection, such as those with new or multiple sexual partners, use of injection drugs, history of previous STDs and women who perform sex for payment (Workowski & Berman, 2010). Testing is also recommended for women seeking medical evaluation for vaginal discharge and for women with HIV infection (Secor et al., 2014; Workowski & Berman, 2010). In low prevalence settings such as the general UK population, NAAT screening of all patients may not be cost-effective (Hathorn et al., 2015), and the CDC does not recommend routine screening during pregnancy at this time (Workowski & Berman, 2010).

For women with a positive *T. vaginalis* result, the CDC recommends retesting at 3 months, regardless of whether or not the sexual partner was also treated, due to the high risk of reinfection (Workowski & Bolan, 2014). Testing for other STDs is also an important consideration for at-risk populations. The same specimen may be tested for *Chlamydia trachomatis* and *Neisseria meningitidis*. Testing and treating sexual partners for *T. vaginalis* and other STDs is an important component of patient management (Workowski & Bolan, 2014).

7 SUMMARY

Molecular methods are a welcome addition to diagnostic techniques for parasitic diseases (see Tables 1 and 2). While many assays find their primarily utility in the research lab, others have become an important component of the diagnostic parasitology test menu and have been added to international clinical algorithms and guidelines. The continued introduction of commercially available assays will no doubt increase the utility of molecular tests into both resource rich and resource limited settings.

REFERENCES

Alifrangis, M., Enosse, S., Pearce, R., Drakeley, C., Roper, C., Khalil, I. F., et al. (2005). A simple, high-throughput method to detect *Plasmodium falciparum* single nucleotide polymorphisms in the dihydrofolate reductase, dihydropteroate synthase, and P. falciparum chloroquine resistance transporter genes using polymerase chain reaction- and enzyme-linked immunosorbent assay-based technology. *The American Journal of Tropical Medicine and Hygiene*, 72(2), 155–162.

Andersen, L. O., Roser, D., Nejsum, P., Nielsen, H. V., & Stensvold, C. R. (2013). Is supplementary bead beating for DNA extraction from nematode eggs by use of the NucliSENS easyMag protocol necessary? *Journal of Clinical Microbiology*, 51(4), 1345–1347.

Andrea, S. B., & Chapin, K. C. (2011). Comparison of Aptima *Trichomonas vaginalis* transcription-mediated amplification assay and BD affirm VPIII for detection of *T. vaginalis* in symptomatic women: Performance parameters and epidemiological implications. *Journal of Clinical Microbiology*, 49(3), 866–869.

Aoun, K., Chouihi, E., Amri, F., Ben Alaya, N., Raies, A., Mary, C., et al. (2009). Short report: Contribution of quantitative real-time polymerase chain reaction to follow-up of visceral leishmaniasis patients treated with meglumine antimoniate. *The American Journal of Tropical Medicine and Hygiene*, 81(6), 1004–1006.

Babady, N. E., Sloan, L. M., Rosenblatt, J. E., & Pritt, B. S. (2009). Detection of *Plasmodium knowlesi* by real-time polymerase chain reaction. *The American Journal of Tropical Medicine and Hygiene*, 81(3), 516–518.

Bastien, P., Jumas-Bilak, E., Varlet-Marie, E., & Marty, P. (2007). Three years of multilaboratory external quality control for the molecular detection of *Toxoplasma gondii* in amniotic fluid in France. *Clinical Microbiology and Infection*, 13(4), 430–433.

Basuni, M., Muhi, J., Othman, N., Verweij, J. J., Ahmad, M., Miswan, N., et al. (2011). A pentaplex real-time polymerase chain reaction assay for detection of four species of soil-transmitted helminths. *The American Journal of Tropical Medicine and Hygiene*, 84(2), 338–343.

Beard, C. B., Pye, G., Steurer, F. J., Rodriguez, R., Campman, R., Peterson, A. T., et al. (2003). Chagas disease in a domestic transmission cycle, southern Texas, USA. *Emerging Infectious Diseases*, 9(1), 103–105.

Becker, S., Franco, J. R., Simarro, P. P., Stich, A., Abel, P. M., & Steverding, D. (2004). Real-time PCR for detection of *Trypanosoma brucei* in human blood samples. *Diagnostic Microbiology and Infectious Disease*, 50(3), 193–199.

Bern, C., Kjos, S., Yabsley, M. J., & Montgomery, S. P. (2011). *Trypanosoma cruzi* and Chagas' disease in the United States. *Clinical Microbiology Reviews*, 24(4), 655–681.

Bern, C., Verastegui, M., Gilman, R. H., Lafuente, C., Galdos-Cardenas, G., Calderon, M., et al. (2009). Congenital *Trypanosoma cruzi* transmission in Santa Cruz, Bolivia. *Clinical Infectious Diseases*, 49(11), 1667–1674.

BioFire. (2014). *FilmArray gastrointestinal panel*. Retrieved January 17, 2015, from, http://www.biofiredx.com/media/InfoSheet-FilmArray-GI-Panel-0234.pdf.

Bloch, E. M., Lee, T. H., Krause, P. J., Telford, S. R., 3rd, Montalvo, L., Chafets, D., et al. (2013). Development of a real-time polymerase chain reaction assay for sensitive detection and quantitation of *Babesia microti* infection. *Transfusion*, 53(10), 2299–2306.

Boggild, A. K., Ramos, A. P., Espinosa, D., Valencia, B. M., Veland, N., Miranda-Verastegui, C., et al. (2010). Clinical and demographic stratification of test performance: A pooled analysis of five laboratory diagnostic methods for American cutaneous leishmaniasis. *The American Journal of Tropical Medicine and Hygiene*, *83*(2), 345–350.

Booton, G. C., Visvesvara, G. S., Byers, T. J., Kelly, D. J., & Fuerst, P. A. (2005). Identification and distribution of *Acanthamoeba* species genotypes associated with nonkeratitis infections. *Journal of Clinical Microbiology*, *43*(4), 1689–1693.

Boschi-Pinto, C., Lanata, C. F., Mendoza, W., & Habte, D. (2006). Diarrheal diseases. In D. T. Jamison, R. G. Feachem, M. W. Makgoba, E. R. Bos, F. K. Baingana, & K. J. Hofman, et al. (Eds.), *Disease and mortality in Sub-Saharan Africa* (2nd ed.). Washington, DC: The World Bank.

Britto, C. C. (2009). Usefulness of PCR-based assays to assess drug efficacy in Chagas disease chemotherapy: Value and limitations. *Memórias do Instituto Oswaldo Cruz*, *104*(Suppl. 1), 122–135.

Brown, H. L., Fuller, D. D., Jasper, L. T., Davis, T. E., & Wright, J. D. (2004). Clinical evaluation of affirm VPIII in the detection and identification of *Trichomonas vaginalis*, *Gardnerella vaginalis*, and *Candida species* in vaginitis/vaginosis. *Infectious Diseases in Obstetrics and Gynecology*, *12*(1), 17–21.

Buckwalter, S. P., Sloan, L. M., Cunningham, S. A., Espy, M. J., Uhl, J. R., Jones, M. F., et al. (2014). Inhibition controls for qualitative real-time PCR assays: Are they necessary for all specimen matrices? *Journal of Clinical Microbiology*, *52*(6), 2139–2143.

Burg, J. L., Grover, C. M., Pouletty, P., & Boothroyd, J. C. (1989). Direct and sensitive detection of a pathogenic protozoan, *Toxoplasma gondii*, by polymerase chain reaction. *Journal of Clinical Microbiology*, *27*(8), 1787–1792.

Buss, S. N., Leber, A., Chapin, K., Fey, P. D., Bankowski, M. J., Jones, M. K., et al. (2015). Multicenter evaluation of the BioFire FilmArray gastrointestinal panel for the etiologic diagnosis of infectious gastroenteritis. *Journal of Clinical Microbiology*, *53*, 915–925.

Caccio, S. M., De Giacomo, M., & Pozio, E. (2002). Sequence analysis of the beta-giardin gene and development of a polymerase chain reaction-restriction fragment length polymorphism assay to genotype *Giardia duodenalis* cysts from human faecal samples. *International Journal for Parasitology*, *32*(8), 1023–1030.

Campbell, L., Woods, V., Lloyd, T., Elsayed, S., & Church, D. L. (2008). Evaluation of the OSOM *Trichomonas* rapid test versus wet preparation examination for detection of *Trichomonas vaginalis* vaginitis in specimens from women with a low prevalence of infection. *Journal of Clinical Microbiology*, *46*(10), 3467–3469.

Castilho, T. M., Camargo, L. M., McMahon-Pratt, D., Shaw, J. J., & Floeter-Winter, L. M. (2008). A real-time polymerase chain reaction assay for the identification and quantification of American *Leishmania* species on the basis of glucose-6-phosphate dehydrogenase. *The American Journal of Tropical Medicine and Hygiene*, *78*(1), 122–132.

Centers for Disease Control and Prevention. (2007). Blood donor screening for chagas disease—United States, 2006–2007. *Morbidity and Mortality Weekly Report*, *56*(7), 141–143.

Centers for Disease Prevention and Control. (2012). Babesiosis surveillance—18 States, 2011. *Morbidity and Mortality Weekly Report*, *61*(27), 505–509.

Centers for Disease Prevention and Control. (2013). *Leishmaniasis*. Retrieved January 11, 2015, from, http://www.cdc.gov/dpdx/leishmaniasis/dx.html.

Centers for Disease Prevention and Control. (2014). *Neglected Parasitic Infections (NPIs) in the United States*. Retrieved January 17, 2015, from, http://www.cdc.gov/parasites/npi/.

Chan, K., Marras, S. A., & Parveen, N. (2013). Sensitive multiplex PCR assay to differentiate Lyme spirochetes and emerging pathogens *Anaplasma phagocytophilum* and *Babesia microti*. *BMC Microbiology*, *13*, 295.

Chang, C. W., Wu, Y. C., & Ming, K. W. (2010). Evaluation of real-time PCR methods for quantification of *Acanthamoeba* in anthropogenic water and biofilms. *Journal of Applied Microbiology*, *109*(3), 799–807.

Chappuis, F., Loutan, L., Simarro, P., Lejon, V., & Buscher, P. (2005). Options for field diagnosis of human African trypanosomiasis. *Clinical Microbiology Reviews*, *18*(1), 133–146.

Chin-Hong, P. V., Schwartz, B. S., Bern, C., Montgomery, S. P., Kontak, S., Kubak, B., et al. (2011). Screening and treatment of chagas disease in organ transplant recipients in the United States: Recommendations from the chagas in transplant working group. *American Journal of Transplantation*, *11*(4), 672–680.

Claas, E. C., Burnham, C. A., Mazzulli, T., Templeton, K., & Topin, F. (2013). Performance of the xTAG(R) gastrointestinal pathogen panel, a multiplex molecular assay for simultaneous detection of bacterial, viral, and parasitic causes of infectious gastroenteritis. *Journal of Microbiology and Biotechnology*, *23*(7), 1041–1045.

Clark, C. G., & Diamond, L. S. (1991). The Laredo strain and other 'Entamoeba histolytica-like' amoebae are *Entamoeba moshkovskii*. *Molecular and Biochemical Parasitology*, *46*(1), 11–18.

Cnops, L., Jacobs, J., & Van Esbroeck, M. (2011). Validation of a four-primer real-time PCR as a diagnostic tool for single and mixed *Plasmodium* infections. *Clinical Microbiology and Infection*, *17*(7), 1101–1107.

Colombo, F. A., Vidal, J. E., Penalva de Oliveira, A. C., Hernandez, A. V., Bonasser-Filho, F., Nogueira, R. S., et al. (2005). Diagnosis of cerebral toxoplasmosis in AIDS patients in Brazil: Importance of molecular and immunological methods using peripheral blood samples. *Journal of Clinical Microbiology*, *43*(10), 5044–5047.

Conrad, P. A., Kjemtrup, A. M., Carreno, R. A., Thomford, J., Wainwright, K., Eberhard, M., et al. (2006). Description of *Babesia duncani* n.sp. (Apicomplexa: Babesiidae) from humans and its differentiation from other piroplasms. *International Journal for Parasitology*, *36*(7), 779–789.

Cota, G. F., de Sousa, M. R., Demarqui, F. N., & Rabello, A. (2012). The diagnostic accuracy of serologic and molecular methods for detecting visceral leishmaniasis in HIV infected patients: Meta-analysis. *PLoS Neglected Tropical Diseases*, *6*(5), e1665.

Criado-Fornelio, A., Martinez-Marcos, A., Buling-Sarana, A., & Barba-Carretero, J. C. (2003). Molecular studies on *Babesia*, *Theileria* and *Hepatozoon* in southern Europe. Part II. Phylogenetic analysis and evolutionary history. *Veterinary Parasitology*, *114*(3), 173–194.

Crucitti, T., Van Dyck, E., Tehe, A., Abdellati, S., Vuylsteke, B., Buve, A., et al. (2003). Comparison of culture and different PCR assays for detection of *Trichomonas vaginalis* in self collected vaginal swab specimens. *Sexually Transmitted Infections*, *79*(5), 393–398.

Cullen, K. A., & Arguin, P. M. (2014). Malaria surveillance—United States, 2012. *Morbidity and Mortality Weekly Report. Surveillance Summaries*, *63*(12), 1–22.

Deborggraeve, S., & Buscher, P. (2010). Molecular diagnostics for sleeping sickness: What is the benefit for the patient? *The Lancet Infectious Diseases*, *10*(6), 433–439.

Deborggraeve, S., Claes, F., Laurent, T., Mertens, P., Leclipteux, T., Dujardin, J. C., et al. (2006). Molecular dipstick test for diagnosis of sleeping sickness. *Journal of Clinical Microbiology*, *44*(8), 2884–2889.

Diez, M., Favaloro, L., Bertolotti, A., Burgos, J. M., Vigliano, C., Lastra, M. P., et al. (2007). Usefulness of PCR strategies for early diagnosis of Chagas' disease reactivation and treatment follow-up in heart transplantation. *American Journal of Transplantation*, *7*(6), 1633–1640.

Disch, J., Pedras, M. J., Orsini, M., Pirmez, C., de Oliveira, M. C., Castro, M., et al. (2005). *Leishmania* (*Viannia*) subgenus kDNA amplification for the diagnosis of mucosal leishmaniasis. *Diagnostic Microbiology and Infectious Disease*, *51*(3), 185–190.

Dissanayake, S., Min, X., & Piessens, W. F. (1991). Detection of amplified *Wuchereria bancrofti* DNA in mosquitoes with a nonradioactive probe. *Molecular and Biochemical Parasitology*, *45*(1), 49–56.

Dize, L., Agreda, P., Quinn, N., Barnes, M. R., Hsieh, Y. H., & Gaydos, C. A. (2013). Comparison of self-obtained penile-meatal swabs to urine for the detection of *C. trachomatis*, *N. gonorrhoeae* and *T. vaginalis*. *Sexually Transmitted Infections*, *89*(4), 305–307.

Domeika, M., Zhurauskaya, L., Savicheva, A., Frigo, N., Sokolovskiy, E., Hallen, A., et al. (2010). Guidelines for the laboratory diagnosis of trichomoniasis in East European countries. *Journal of the European Academy of Dermatology and Venereology*, *24*(10), 1125–1134.

Dorn, P. L., Perniciaro, L., Yabsley, M. J., Roellig, D. M., Balsamo, G., Diaz, J., et al. (2007). Autochthonous transmission of *Trypanosoma cruzi*, Louisiana. *Emerging Infectious Diseases*, *13*(4), 605–607.

Drame, P. M., Fink, D. L., Kamgno, J., Herrick, J. A., & Nutman, T. B. (2014). Loop-mediated isothermal amplification for rapid and semiquantitative detection of *Loa loa* infection. *Journal of Clinical Microbiology*, *52*(6), 2071–2077.

Duffy, T., Bisio, M., Altcheh, J., Burgos, J. M., Diez, M., Levin, M. J., et al. (2009). Accurate real-time PCR strategy for monitoring bloodstream parasitic loads in chagas disease patients. *PLoS Neglected Tropical Diseases*, *3*(4), e419.

Edvinsson, B., Lundquist, J., Ljungman, P., Ringden, O., & Evengard, B. (2008). A prospective study of diagnosis of *Toxoplasma gondii* infection after bone marrow transplantation. *Acta Pathologica, Microbiologica et Immunologica Scandinavica*, *116*(5), 345–351.

Eshoo, M. W., Crowder, C. D., Carolan, H. E., Rounds, M. A., Ecker, D. J., Haag, H., et al. (2014). Broad-range survey of tick-borne pathogens in Southern Germany reveals a high prevalence of *Babesia microti* and a diversity of other tick-borne pathogens. *Vector Borne and Zoonotic Diseases*, *14*(8), 584–591.

Fallahi, S., Seyyed Tabaei, S. J., Pournia, Y., Zebardast, N., & Kazemi, B. (2014). Comparison of loop-mediated isothermal amplification (LAMP) and nested-PCR assay targeting the RE and B1 gene for detection of *Toxoplasma gondii* in blood samples of children with leukaemia. *Diagnostic Microbiology and Infectious Disease*, *79*(3), 347–354.

Farid, H. A., Hammad, R. E., Hassan, M. M., Morsy, Z. S., Kamal, I. H., Weil, G. J., et al. (2001). Detection of *Wuchereria bancrofti* in mosquitoes by the polymerase chain reaction: A potentially useful tool for large-scale control programmes. *Transactions of the Royal Society of Tropical Medicine and Hygiene*, *95*(1), 29–32.

Federal Drug Administration. (2013). *FDA Drug Safety Communication: FDA approves label changes for antimalarial drug mefloquine hydrochloride due to risk of serious psychiatric and nerve side effects*. Retrieved November 25, 2013, from, http://www.fda.gov/drugs/drugsafety/ucm362227.htm.

Filisetti, D., Sterkers, Y., Brenier-Pinchart, M. P., Cassaing, S., Dalle, F., Delhaes, L., et al. (2015). Multicentric comparative assessment of the bio-evolution *Toxoplasma gondii* detection kit with eight laboratory-developed PCR assays for molecular diagnosis of congenital toxoplasmosis. *Journal of Clinical Microbiology, 53*(1), 29–34.

Fink, D. L., Fahle, G. A., Fischer, S., Fedorko, D. F., & Nutman, T. B. (2011). Toward molecular parasitologic diagnosis: Enhanced diagnostic sensitivity for filarial infections in mobile populations. *Journal of Clinical Microbiology, 49*(1), 42–47.

Fink, D. L., Kamgno, J., & Nutman, T. B. (2011). Rapid molecular assays for specific detection and quantitation of *Loa loa* microfilaremia. *PLoS Neglected Tropical Diseases, 5*(8), e1299.

Fischer, P., Supali, T., Wibowo, H., Bonow, I., & Williams, S. A. (2000). Detection of DNA of nocturnally periodic *Brugia malayi* in night and day blood samples by a polymerase chain reaction-ELISA-based method using an internal control DNA. *The American Journal of Tropical Medicine and Hygiene, 62*(2), 291–296.

Fontaine, M., & Guillot, E. (2002). Development of a TaqMan quantitative PCR assay specific for *Cryptosporidium parvum*. *FEMS Microbiology Letters, 214*(1), 13–17.

Food and Drug Administration. (2011). *510 (k) Summary—k081868 (Smart Leish)*. Retrieved January 11, 2015, from, http://www.accessdata.fda.gov/cdrh_docs/pdf8/K081868.pdf.

Fuehrer, H. P., & Noedl, H. (2014). Recent advances in detection of *Plasmodium ovale*: Implications of separation into the two species *Plasmodium ovale wallikeri* and *Plasmodium ovale curtisi*. *Journal of Clinical Microbiology, 52*(2), 387–391.

Galvao, L. M., Chiari, E., Macedo, A. M., Luquetti, A. O., Silva, S. A., & Andrade, A. L. (2003). PCR assay for monitoring *Trypanosoma cruzi* parasitemia in childhood after specific chemotherapy. *Journal of Clinical Microbiology, 41*(11), 5066–5070.

Garcia, L. S. (2007). *Diagnostic medical parasitology* (5th ed.). Washington, DC: ASM Press; Oxford: Blackwell [distributor].

Garcia, L. S. (2009). *Practical guide to diagnostic parasitology* (2nd ed.). Washington, DC: ASM Press.

Gass, K., Beau de Rochars, M. V., Boakye, D., Bradley, M., Fischer, P. U., Gyapong, J., et al. (2012). A multicenter evaluation of diagnostic tools to define endpoints for programs to eliminate bancroftian filariasis. *PLoS Neglected Tropical Diseases, 6*(1), e1479.

Gasser, R. B., Cantacessi, C., & Campbell, B. E. (2009). Improved molecular diagnostic tools for human hookworms. *Expert Review of Molecular Diagnostics, 9*(1), 17–21.

GEN-PROBE. (2013). *APTIMA Trichomonas vaginalis Assay (PANTHER System)*. Retrieved January 17, 2015, from, http://www.hologic.com/sites/default/files/package%20inserts/503684-EN-RevA.pdf.

Goldschmidt, P., Degorge, S., Saint-Jean, C., Yera, H., Zekhnini, F., Batellier, L., et al. (2008). Resistance of *Acanthamoeba* to classic DNA extraction methods used for the diagnosis of corneal infections. *The British Journal of Ophthalmology, 92*(1), 112–115.

Gray, J., Zintl, A., Hildebrandt, A., Hunfeld, K. P., & Weiss, L. (2010). Zoonotic babesiosis: Overview of the disease and novel aspects of pathogen identity. *Ticks and Tick-borne Diseases, 1*(1), 3–10.

Guy, R. A., Payment, P., Krull, U. J., & Horgen, P. A. (2003). Real-time PCR for quantification of *Giardia* and *Cryptosporidium* in environmental water samples and sewage. *Applied and Environmental Microbiology, 69*(9), 5178–5185.

Hanson, K. L., & Cartwright, C. P. (2001). Use of an enzyme immunoassay does not eliminate the need to analyze multiple stool specimens for sensitive detection of *Giardia lamblia*. *Journal of Clinical Microbiology, 39*(2), 474–477.

Haque, R., Ali, I. K., Akther, S., & Petri, W. A., Jr. (1998). Comparison of PCR, isoenzyme analysis, and antigen detection for diagnosis of *Entamoeba histolytica* infection. *Journal of Clinical Microbiology*, *36*(2), 449–452.

Haque, R., Roy, S., Siddique, A., Mondal, U., Rahman, S. M., Mondal, D., et al. (2007). Multiplex real-time PCR assay for detection of *Entamoeba histolytica*, *Giardia intestinalis*, and *Cryptosporidium* spp. *The American Journal of Tropical Medicine and Hygiene*, *76*(4), 713–717.

Hardick, A., Hardick, J., Wood, B. J., & Gaydos, C. (2006). Comparison between the Gen-Probe transcription-mediated amplification *Trichomonas vaginalis* research assay and real-time PCR for *Trichomonas vaginalis* detection using a Roche LightCycler instrument with female self-obtained vaginal swab samples and male urine samples. *Journal of Clinical Microbiology*, *44*(11), 4197–4199.

Hassan, M., Sanad, M. M., el-Karamany, I., Abdel-Tawab, M., Shalaby, M., el-Dairouty, A., et al. (2005). Detection of DNA of *W. bancrofti* in blood samples by QC-PCR-ELISA-based. *Journal of the Egyptian Society of Parasitology*, *35*(3), 963–970.

Hathorn, E., Ng, A., Page, M., Hodson, J., Gaydos, C., & Ross, J. D. (2015). A service evaluation of the Gen-Probe APTIMA nucleic acid amplification test for Trichomonas vaginalis: Should it change whom we screen for infection? *Sexually Transmitted Infections*, *91*, 81–86.

Herwaldt, B. L., Caccio, S., Gherlinzoni, F., Aspock, H., Slemenda, S. B., Piccaluga, P., et al. (2003). Molecular characterization of a non-*Babesia divergens* organism causing zoonotic babesiosis in Europe. *Emerging Infectious Diseases*, *9*(8), 942–948.

Herwaldt, B. L., de Bruyn, G., Pieniazek, N. J., Homer, M., Lofy, K. H., Slemenda, S. B., et al. (2004). *Babesia divergens*-like infection, Washington State. *Emerging Infectious Diseases*, *10*(4), 622–629.

Herwaldt, B., Persing, D. H., Precigout, E. A., Goff, W. L., Mathiesen, D. A., Taylor, P. W., et al. (1996). A fatal case of babesiosis in Missouri: Identification of another piroplasm that infects humans. *Annals of Internal Medicine*, *124*(7), 643–650.

Higa, L. T., Garcia, J. L., Su, C., Rossini, R. C., & Falavigna-Guilherme, A. L. (2014). *Toxoplasma gondii* genotypes isolated from pregnant women with follow-up of infected children in southern Brazil. *Transactions of the Royal Society of Tropical Medicine and Hygiene*, *108*(4), 244–246.

Hildebrandt, A., Hunfeld, K. P., Baier, M., Krumbholz, A., Sachse, S., Lorenzen, T., et al. (2007). First confirmed autochthonous case of human *Babesia microti* infection in Europe. *European Journal of Clinical Microbiology & Infectious Diseases*, *26*(8), 595–601.

Homan, W. L., Vercammen, M., De Braekeleer, J., & Verschueren, H. (2000). Identification of a 200- to 300-fold repetitive 529 bp DNA fragment in *Toxoplasma gondii*, and its use for diagnostic and quantitative PCR. *International Journal for Parasitology*, *30*(1), 69–75.

Hopkins, R. M., Meloni, B. P., Groth, D. M., Wetherall, J. D., Reynoldson, J. A., & Thompson, R. C. (1997). Ribosomal RNA sequencing reveals differences between the genotypes of *Giardia* isolates recovered from humans and dogs living in the same locality. *The Journal of Parasitology*, *83*(1), 44–51.

Huppert, J. S., Mortensen, J. E., Reed, J. L., Kahn, J. A., Rich, K. D., Miller, W. C., et al. (2007). Rapid antigen testing compares favorably with transcription-mediated amplification assay for the detection of *Trichomonas vaginalis* in young women. *Clinical Infectious Diseases*, *45*(2), 194–198.

Jones, D., Nyalwidhe, J., Tetley, L., & Barrett, M. P. (2007). McArthur revisited: Fluorescence microscopes for field diagnostics. *Trends in Parasitology*, *23*(10), 468–469.

Jordan, J. A., Lowery, D., & Trucco, M. (2001). TaqMan-based detection of *Trichomonas vaginalis* DNA from female genital specimens. *Journal of Clinical Microbiology, 39*(11), 3819–3822.

Kabiri, M., Franco, J. R., Simarro, P. P., Ruiz, J. A., Sarsa, M., & Steverding, D. (1999). Detection of *Trypanosoma brucei gambiense* in sleeping sickness suspects by PCR amplification of expression-site-associated genes 6 and 7. *Tropical Medicine & International Health, 4*(10), 658–661.

Kamau, E., Tolbert, L. S., Kortepeter, L., Pratt, M., Nyakoe, N., Muringo, L., et al. (2011). Development of a highly sensitive genus-specific quantitative reverse transcriptase real-time PCR assay for detection and quantitation of plasmodium by amplifying RNA and DNA of the 18S rRNA genes. *Journal of Clinical Microbiology, 49*(8), 2946–2953.

Kao, P. M., Tung, M. C., Hsu, B. M., Tsai, H. L., She, C. Y., Shen, S. M., et al. (2013). Real-time PCR method for the detection and quantification of *Acanthamoeba* species in various types of water samples. *Parasitology Research, 112*(3), 1131–1136.

Kaydos-Daniels, S. C., Miller, W. C., Hoffman, I., Banda, T., Dzinyemba, W., Martinson, F., et al. (2003). Validation of a urine-based PCR-enzyme-linked immunosorbent assay for use in clinical research settings to detect *Trichomonas vaginalis* in men. *Journal of Clinical Microbiology, 41*(1), 318–323.

Kebede, A., Verweij, J. J., Petros, B., & Polderman, A. M. (2004). Short communication: Misleading microscopy in amoebiasis. *Tropical Medicine & International Health, 9*(5), 651–652.

Khairnar, K., Tamber, G. S., Ralevski, F., & Pillai, D. R. (2011). Comparison of molecular diagnostic methods for the detection of *Acanthamoeba* spp. from clinical specimens submitted for keratitis. *Diagnostic Microbiology and Infectious Disease, 70*(4), 499–506.

Khare, R., Espy, M. J., Cebelinski, E., Boxrud, D., Sloan, L. M., Cunningham, S. A., et al. (2014). Comparative evaluation of two commercial multiplex panels for detection of gastrointestinal pathogens by use of clinical stool specimens. *Journal of Clinical Microbiology, 52*(10), 3667–3673.

Kluber, S., Supali, T., Williams, S. A., Liebau, E., & Fischer, P. (2001). Rapid PCR-based detection of *Brugia malayi* DNA from blood spots by DNA Detection Test Strips. *Transactions of the Royal Society of Tropical Medicine and Hygiene, 95*(2), 169–170.

Kramme, S., Nissen, N., Soblik, H., Erttmann, K., Tannich, E., Fleischer, B., et al. (2011). Novel real-time PCR for the universal detection of *Strongyloides* species. *Journal of Medical Microbiology, 60*(Pt 4), 454–458.

Krause, P. J., Telford, S., 3rd, Spielman, A., Ryan, R., Magera, J., Rajan, T. V., et al. (1996). Comparison of PCR with blood smear and inoculation of small animals for diagnosis of Babesia microti parasitemia. *Journal of Clinical Microbiology, 34*(11), 2791–2794.

Kuboki, N., Inoue, N., Sakurai, T., Di Cello, F., Grab, D. J., Suzuki, H., et al. (2003). Loop-mediated isothermal amplification for detection of African trypanosomes. *Journal of Clinical Microbiology, 41*(12), 5517–5524.

Kumar, R., Bumb, R. A., Ansari, N. A., Mehta, R. D., & Salotra, P. (2007). Cutaneous leishmaniasis caused by *Leishmania tropica* in Bikaner, India: Parasite identification and characterization using molecular and immunologic tools. *The American Journal of Tropical Medicine and Hygiene, 76*(5), 896–901.

Kyambadde, J. W., Enyaru, J. C., Matovu, E., Odiit, M., & Carasco, J. F. (2000). Detection of trypanosomes in suspected sleeping sickness patients in Uganda using the polymerase chain reaction. *Bulletin of the World Health Organisation, 78*(1), 119–124.

Lages-Silva, E., Crema, E., Ramirez, L. E., Macedo, A. M., Pena, S. D., & Chiari, E. (2001). Relationship between *Trypanosoma cruzi* and human chagasic megaesophagus: Blood and tissue parasitism. *The American Journal of Tropical Medicine and Hygiene*, *65*(5), 435–441.

Lawing, L. F., Hedges, S. R., & Schwebke, J. R. (2000). Detection of trichomonosis in vaginal and urine specimens from women by culture and PCR. *Journal of Clinical Microbiology*, *38*(10), 3585–3588.

Laxer, M. A., Timblin, B. K., & Patel, R. J. (1991). DNA sequences for the specific detection of *Cryptosporidium parvum* by the polymerase chain reaction. *The American Journal of Tropical Medicine and Hygiene*, *45*(6), 688–694.

Leiby, D. A. (2011). Transfusion-transmitted *Babesia* spp.: Bull's-eye on *Babesia microti*. *Clinical Microbiology Reviews*, *24*(1), 14–28.

Lek-Uthai, U., Passara, R., Roongruangchai, K., Buddhirakkul, P., & Thammapalerd, N. (2009). Rapid identification of *Acanthamoeba* from contact lens case using loop-mediated isothermal amplification method. *Experimental Parasitology*, *121*(4), 342–345.

Leles, D., Araujo, A., Vicente, A. C., & Iniguez, A. M. (2009). Molecular diagnosis of ascariasis from human feces and description of a new *Ascaris* sp. genotype in Brazil. *Veterinary Parasitology*, *163*(1–2), 167–170.

Lewis, M. D., Ma, J., Yeo, M., Carrasco, H. J., Llewellyn, M. S., & Miles, M. A. (2009). Genotyping of *Trypanosoma cruzi*: Systematic selection of assays allowing rapid and accurate discrimination of all known lineages. *The American Journal of Tropical Medicine and Hygiene*, *81*(6), 1041–1049.

Liu, D. (2013). *Molecular detection of human parasitic pathogens*. Retrieved on February 2, 2015 from, http://marc.crcnetbase.com/isbn/9781439812433.

Liu, J., Gratz, J., Amour, C., Kibiki, G., Becker, S., Janaki, L., et al. (2013). A laboratory-developed TaqMan Array Card for simultaneous detection of 19 enteropathogens. *Journal of Clinical Microbiology*, *51*(2), 472–480.

Lizotte, M. R., Supali, T., Partono, F., & Williams, S. A. (1994). A polymerase chain reaction assay for the detection of *Brugia malayi* in blood. *The American Journal of Tropical Medicine and Hygiene*, *51*(3), 314–321.

Lorger, M., Engstler, M., Homann, M., & Goringer, H. U. (2003). Targeting the variable surface of African trypanosomes with variant surface glycoprotein-specific, serum-stable RNA aptamers. *Eukaryotic Cell*, *2*(1), 84–94.

Luchavez, J., Baker, J., Alcantara, S., Belizario, V., Jr., Cheng, Q., McCarthy, J. S., et al. (2011). Laboratory demonstration of a prozone-like effect in HRP2-detecting malaria rapid diagnostic tests: Implications for clinical management. *Malaria Journal*, *10*, 286.

Luminex. (2015). *xTAG Gastrointestinal Pathogen Panel (GPP)*. Retrieved on January 17, 2015, from, http://www.luminexcorp.com/Products/Assays/ClinicalDiagnostics/xTAGGPP/.

Mathieu-Daude, F., Bicart-See, A., Bosseno, M. F., Breniere, S. F., & Tibayrenc, M. (1994). Identification of *Trypanosoma brucei gambiense* group I by a specific kinetoplast DNA probe. *The American Journal of Tropical Medicine and Hygiene*, *50*(1), 13–19.

Matovu, E., Mugasa, C. M., Ekangu, R. A., Deborggraeve, S., Lubega, G. W., Laurent, T., et al. (2010). Phase II evaluation of sensitivity and specificity of PCR and NASBA followed by oligochromatography for diagnosis of human African trypanosomiasis in clinical samples from D.R. Congo and Uganda. *PLoS Neglected Tropical Diseases*, *4*(7), e737.

McAuley, J. B., Jones, J. L., & Singh, K. (2011). Toxoplasma. In J. Versalovic, K. C. Carroll, G. Funke, J. H. Jorgensen, M. L. Landry, & D. W. Warnock (Eds.), *Manual of clinical microbiology: Vol. 2* (10th ed., pp. 2127–2138). Washington, DC: ASM Press.

McCarthy, J. S., Zhong, M., Gopinath, R., Ottesen, E. A., Williams, S. A., & Nutman, T. B. (1996). Evaluation of a polymerase chain reaction-based assay for diagnosis of *Wuchereria bancrofti* infection. *The Journal of Infectious Diseases, 173*(6), 1510–1514.

Mehlotra, R. K., Gray, L. R., Blood-Zikursh, M. J., Kloos, Z., Henry-Halldin, C. N., Tisch, D. J., et al. (2010). Molecular-based assay for simultaneous detection of four *Plasmodium* spp. and *Wuchereria bancrofti* infections. *The American Journal of Tropical Medicine and Hygiene, 82*(6), 1030–1033.

Mejia, R., Vicuna, Y., Broncano, N., Sandoval, C., Vaca, M., Chico, M., et al. (2013). A novel, multi-parallel, real-time polymerase chain reaction approach for eight gastrointestinal parasites provides improved diagnostic capabilities to resource-limited at-risk populations. *The American Journal of Tropical Medicine and Hygiene, 88*(6), 1041–1047.

Mens, P. F., Schoone, G. J., Kager, P. A., & Schallig, H. D. (2006). Detection and identification of human *Plasmodium* species with real-time quantitative nucleic acid sequence-based amplification. *Malaria Journal, 5*, 80.

Mesquita, R. T., Ziegler, A. P., Hiramoto, R. M., Vidal, J. E., & Pereira-Chioccola, V. L. (2010). Real-time quantitative PCR in cerebral toxoplasmosis diagnosis of Brazilian human immunodeficiency virus-infected patients. *Journal of Medical Microbiology, 59*(Pt 6), 641–647.

Mixson-Hayden, T., Lucchi, N. W., & Udhayakumar, V. (2010). Evaluation of three PCR-based diagnostic assays for detecting mixed *Plasmodium* infection. *BMC Research Notes, 3*, 88.

Monis, P. T., Andrews, R. H., Mayrhofer, G., & Ey, P. L. (1999). Molecular systematics of the parasitic protozoan *Giardia intestinalis*. *Molecular Biology and Evolution, 16*(9), 1135–1144.

Moreira, O. C., Ramirez, J. D., Velazquez, E., Melo, M. F., Lima-Ferreira, C., Guhl, F., et al. (2013). Towards the establishment of a consensus real-time qPCR to monitor *Trypanosoma cruzi* parasitemia in patients with chronic Chagas disease cardiomyopathy: A substudy from the BENEFIT trial. *Acta Tropica, 125*(1), 23–31.

Moritz, E. D., Winton, C. S., Johnson, S. T., Krysztof, D. E., Townsend, R. L., Foster, G. A., et al. (2014). Investigational screening for *Babesia microti* in a large repository of blood donor samples from nonendemic and endemic areas of the United States. *Transfusion, 54*(9), 2226–2236.

Moser, D. R., Cook, G. A., Ochs, D. E., Bailey, C. P., McKane, M. R., & Donelson, J. E. (1989). Detection of *Trypanosoma congolense* and *Trypanosoma brucei* subspecies by DNA amplification using the polymerase chain reaction. *Parasitology, 99*(Pt 1), 57–66.

Mugasa, C. M., Katiti, D., Boobo, A., Lubega, G. W., Schallig, H. D., & Matovu, E. (2014). Comparison of nucleic acid sequence-based amplification and loop-mediated isothermal amplification for diagnosis of human African trypanosomiasis. *Diagnostic Microbiology and Infectious Disease, 78*(2), 144–148.

Mugasa, C. M., Laurent, T., Schoone, G. J., Kager, P. A., Lubega, G. W., & Schallig, H. D. (2009). Nucleic acid sequence-based amplification with oligochromatography for detection of *Trypanosoma brucei* in clinical samples. *Journal of Clinical Microbiology, 47*(3), 630–635.

Mugasa, C. M., Schoone, G. J., Ekangu, R. A., Lubega, G. W., Kager, P. A., & Schallig, H. D. (2008). Detection of *Trypanosoma brucei* parasites in blood samples using real-time nucleic acid sequence-based amplification. *Diagnostic Microbiology and Infectious Disease*, *61*(4), 440–445.

Murray, H. W., Berman, J. D., Davies, C. R., & Saravia, N. G. (2005). Advances in leishmaniasis. *Lancet*, *366*(9496), 1561–1577.

Ndao, M. (2009). Diagnosis of parasitic diseases: Old and new approaches. *Interdisciplinary Perspectives on Infectious Diseases*, *2009*, 278246.

Njiru, Z. K., Constantine, C. C., Guya, S., Crowther, J., Kiragu, J. M., Thompson, R. C., et al. (2005). The use of ITS1 rDNA PCR in detecting pathogenic African trypanosomes. *Parasitology Research*, *95*(3), 186–192.

Njiru, Z. K., Mikosza, A. S., Armstrong, T., Enyaru, J. C., Ndung'u, J. M., & Thompson, A. R. (2008). Loop-mediated isothermal amplification (LAMP) method for rapid detection of *Trypanosoma brucei rhodesiense*. *PLoS Neglected Tropical Diseases*, *2*(1), e147.

Njiru, Z. K., Mikosza, A. S., Matovu, E., Enyaru, J. C., Ouma, J. O., Kibona, S. N., et al. (2008). African trypanosomiasis: Sensitive and rapid detection of the sub-genus *Trypanozoon* by loop-mediated isothermal amplification (LAMP) of parasite DNA. *International Journal for Parasitology*, *38*(5), 589–599.

Njiru, Z. K., Traub, R., Ouma, J. O., Enyaru, J. C., & Matovu, E. (2011). Detection of Group 1 *Trypanosoma brucei gambiense* by loop-mediated isothermal amplification. *Journal of Clinical Microbiology*, *49*(4), 1530–1536.

Norgan, A. P., Sloan, L., Giannini, C., & Pritt, B. S. (2014). *Detection of Naegleria fowleri, Acanthamoeba spp. and Balamuthia mandrillaris in formalin-fixed paraffin-embedded tissues by real-time multiplex PCR*. In *Abstract presented at the American Society of Tropical Medicine and Hygiene 63rd Annual Meeting*.

Notermans, D. W., Peek, R., de Jong, M. D., Wentink-Bonnema, E. M., Boom, R., & van Gool, T. (2005). Detection and identification of *Enterocytozoon bieneusi* and *Encephalitozoon* species in stool and urine specimens by PCR and differential hybridization. *Journal of Clinical Microbiology*, *43*(2), 610–614.

Nye, M. B., Schwebke, J. R., & Body, B. A. (2009). Comparison of APTIMA *Trichomonas vaginalis* transcription-mediated amplification to wet mount microscopy, culture, and polymerase chain reaction for diagnosis of trichomoniasis in men and women. *American Journal of Obstetrics and Gynecology*, *200*(2)188 e181–188 e187.

Ochola, L. B., Vounatsou, P., Smith, T., Mabaso, M. L., & Newton, C. R. (2006). The reliability of diagnostic techniques in the diagnosis and management of malaria in the absence of a gold standard. *The Lancet Infectious Diseases*, *6*(9), 582–588.

Oguike, M. C., Betson, M., Burke, M., Nolder, D., Stothard, J. R., Kleinschmidt, I., et al. (2011). *Plasmodium ovale curtisi* and *Plasmodium ovale wallikeri* circulate simultaneously in African communities. *International Journal for Parasitology*, *41*(6), 677–683.

Oliveira, I., Torrico, F., Munoz, J., & Gascon, J. (2010). Congenital transmission of Chagas disease: A clinical approach. *Expert Review of Anti-Infective Therapy*, *8*(8), 945–956.

Penchenier, L., Simo, G., Grebaut, P., Nkinin, S., Laveissiere, C., & Herder, S. (2000). Diagnosis of human trypanosomiasis, due to *Trypanosoma brucei gambiense* in central Africa, by the polymerase chain reaction. *Transactions of the Royal Society of Tropical Medicine and Hygiene*, *94*(4), 392–394.

Perandin, F., Manca, N., Calderaro, A., Piccolo, G., Galati, L., Ricci, L., et al. (2004). Development of a real-time PCR assay for detection of *Plasmodium falciparum*, *Plasmodium vivax*, and *Plasmodium ovale* for routine clinical diagnosis. *Journal of Clinical Microbiology*, *42*(3), 1214–1219.

Perry, M. D., Corden, S. A., & Howe, R. A. (2014). Evaluation of the Luminex xTAG Gastrointestinal Pathogen Panel and the Savyon Diagnostics Gastrointestinal Infection Panel for the detection of enteric pathogens in clinical samples. *Journal of Medical Microbiology*, *63*(Pt 11), 1419–1426.

Persing, D. H., Mathiesen, D., Marshall, W. F., Telford, S. R., Spielman, A., Thomford, J. W., et al. (1992). Detection of *Babesia microti* by polymerase chain reaction. *Journal of Clinical Microbiology*, *30*(8), 2097–2103.

Picot, S., Olliaro, P., de Monbrison, F., Bienvenu, A. L., Price, R. N., & Ringwald, P. (2009). A systematic review and meta-analysis of evidence for correlation between molecular markers of parasite resistance and treatment outcome in falciparum malaria. *Malaria Journal*, *8*, 89.

Picozzi, K., Fevre, E. M., Odiit, M., Carrington, M., Eisler, M. C., Maudlin, I., et al. (2005). Sleeping sickness in Uganda: A thin line between two fatal diseases. *BMJ*, *331*(7527), 1238–1241.

Pillay, A., Radebe, F., Fehler, G., Htun, Y., & Ballard, R. C. (2007). Comparison of a TaqMan-based real-time polymerase chain reaction with conventional tests for the detection of *Trichomonas vaginalis*. *Sexually Transmitted Infections*, *83*(2), 126–129.

Plowe, C. V., Roper, C., Barnwell, J. W., Happi, C. T., Joshi, H. H., Mbacham, W., et al. (2007). World Antimalarial Resistance Network (WARN) III: Molecular markers for drug resistant malaria. *Malaria Journal*, *6*, 121.

Polley, S. D., Mori, Y., Watson, J., Perkins, M. D., Gonzalez, I. J., Notomi, T., et al. (2010). Mitochondrial DNA targets increase sensitivity of malaria detection using loop-mediated isothermal amplification. *Journal of Clinical Microbiology*, *48*(8), 2866–2871.

Poon, L. L., Wong, B. W., Ma, E. H., Chan, K. H., Chow, L. M., Abeyewickreme, W., et al. (2006). Sensitive and inexpensive molecular test for *falciparum* malaria: Detecting *Plasmodium falciparum* DNA directly from heat-treated blood by loop-mediated isothermal amplification. *Clinical Chemistry*, *52*(2), 303–306.

Qvarnstrom, Y., da Silva, A. J., Schuster, F. L., Gelman, B. B., & Visvesvara, G. S. (2009). Molecular confirmation of *Sappinia pedata* as a causative agent of amoebic encephalitis. *The Journal of Infectious Diseases*, *199*(8), 1139–1142.

Qvarnstrom, Y., Schijman, A. G., Veron, V., Aznar, C., Steurer, F., & da Silva, A. J. (2012). Sensitive and specific detection of *Trypanosoma cruzi* DNA in clinical specimens using a multi-target real-time PCR approach. *PLoS Neglected Tropical Diseases*, *6*(7), e1689.

Qvarnstrom, Y., Visvesvara, G. S., Sriram, R., & da Silva, A. J. (2006). Multiplex real-time PCR assay for simultaneous detection of *Acanthamoeba* spp., *Balamuthia mandrillaris*, and *Naegleria fowleri*. *Journal of Clinical Microbiology*, *44*(10), 3589–3595.

Radonjic, I. V., Dzamic, A. M., Mitrovic, S. M., Arsic Arsenijevic, V. S., Popadic, D. M., & Kranjcic Zec, I. F. (2006). Diagnosis of *Trichomonas vaginalis* infection: The sensitivities and specificities of microscopy, culture and PCR assay. *European Journal of Obstetrics, Gynecology, and Reproductive Biology*, *126*(1), 116–120.

Radwanska, M., Chamekh, M., Vanhamme, L., Claes, F., Magez, S., Magnus, E., et al. (2002). The serum resistance-associated gene as a diagnostic tool for the detection of *Trypanosoma brucei rhodesiense*. *The American Journal of Tropical Medicine and Hygiene*, *67*(6), 684–690.

Radwanska, M., Claes, F., Magez, S., Magnus, E., Perez-Morga, D., Pays, E., et al. (2002). Novel primer sequences for polymerase chain reaction-based detection of *Trypanosoma brucei gambiense*. *The American Journal of Tropical Medicine and Hygiene*, *67*(3), 289–295.

Radwanska, M., Magez, S., Perry-O'Keefe, H., Stender, H., Coull, J., Sternberg, J. M., et al. (2002). Direct detection and identification of African trypanosomes by fluorescence in situ hybridization with peptide nucleic acid probes. *Journal of Clinical Microbiology*, *40*(11), 4295–4297.

Ramzy, R. M. (2002). Recent advances in molecular diagnostic techniques for human lymphatic filariasis and their use in epidemiological research. *Transactions of the Royal Society of Tropical Medicine and Hygiene*, *96*(Suppl. 1), S225–S229.

Rao, R. U., Atkinson, L. J., Ramzy, R. M., Helmy, H., Farid, H. A., Bockarie, M. J., et al. (2006). A real-time PCR-based assay for detection of *Wuchereria bancrofti* DNA in blood and mosquitoes. *The American Journal of Tropical Medicine and Hygiene*, *74*(5), 826–832.

Rao, R. U., Weil, G. J., Fischer, K., Supali, T., & Fischer, P. (2006). Detection of *Brugia* parasite DNA in human blood by real-time PCR. *Journal of Clinical Microbiology*, *44*(11), 3887–3893.

Read, C. M., Monis, P. T., & Thompson, R. C. (2004). Discrimination of all genotypes of *Giardia duodenalis* at the glutamate dehydrogenase locus using PCR-RFLP. *Infection, Genetics and Evolution*, *4*(2), 125–130.

Reddington, K., Tuite, N., Minogue, E., & Barry, T. (2014). A current overview of commercially available nucleic acid diagnostics approaches to detect and identify human gastroenteritis pathogens. *Biomolecular Detection and Quantification, 1* (3–7).

Reyes, J. C., Solon, J. A., & Rivera, W. L. (2014). Development of a loop-mediated isothermal amplification assay for detection of *Trichomonas vaginalis*. *Diagnostic Microbiology and Infectious Disease*, *79*(3), 337–341.

Riviere, D., Szczebara, F. M., Berjeaud, J. M., Frere, J., & Hechard, Y. (2006). Development of a real-time PCR assay for quantification of *Acanthamoeba* trophozoites and cysts. *Journal of Microbiological Methods*, *64*(1), 78–83.

Robays, J., Bilengue, M. M., Van der Stuyft, P., & Boelaert, M. (2004). The effectiveness of active population screening and treatment for sleeping sickness control in the Democratic Republic of Congo. *Tropical Medicine & International Health*, *9*(5), 542–550.

Robert-Gangneux, F., & Darde, M. L. (2012). Epidemiology of and diagnostic strategies for toxoplasmosis. *Clinical Microbiology Reviews*, *25*(2), 264–296.

Rodrigues, J. C., Godinho, J. L., & de Souza, W. (2014). Biology of human pathogenic trypanosomatids: Epidemiology, lifecycle and ultrastructure. *Sub-Cellular Biochemistry*, *74*, 1–42.

Royer, T. L., Gilchrist, C., Kabir, M., Arju, T., Ralston, K. S., Haque, R., et al. (2012). *Entamoeba bangladeshi* nov. sp., Bangladesh. *Emerging Infectious Diseases*, *18*(9), 1543–1545.

Schijman, A. G., Altcheh, J., Burgos, J. M., Biancardi, M., Bisio, M., Levin, M. J., et al. (2003). Aetiological treatment of congenital Chagas' disease diagnosed and monitored by the polymerase chain reaction. *The Journal of Antimicrobial Chemotherapy*, *52*(3), 441–449.

Schijman, A. G., Bisio, M., Orellana, L., Sued, M., Duffy, T., Mejia Jaramillo, A. M., et al. (2011). International study to evaluate PCR methods for detection of *Trypanosoma cruzi* DNA in blood samples from Chagas disease patients. *PLoS Neglected Tropical Diseases*, *5*(1), e931.

Schijman, A. G., Vigliano, C., Burgos, J., Favaloro, R., Perrone, S., Laguens, R., et al. (2000). Early diagnosis of recurrence of *Trypanosoma cruzi* infection by polymerase chain reaction after heart transplantation of a chronic Chagas' heart disease patient. *The Journal of Heart and Lung Transplantation*, *19*(11), 1114–1117.

Schirm, J., Bos, P. A., Roozeboom-Roelfsema, I. K., Luijt, D. S., & Moller, L. V. (2007). *Trichomonas vaginalis* detection using real-time TaqMan PCR. *Journal of Microbiological Methods*, *68*(2), 243–247.

Schwebke, J. R., & Lawing, L. F. (2002). Improved detection by DNA amplification of *Trichomonas vaginalis* in males. *Journal of Clinical Microbiology*, *40*(10), 3681–3683.

Secor, W. E., Meites, E., Starr, M. C., & Workowski, K. A. (2014). Neglected parasitic infections in the United States: Trichomoniasis. *The American Journal of Tropical Medicine and Hygiene*, *90*(5), 800–804.

Sherrard, J., Ison, C., Moody, J., Wainwright, E., Wilson, J., & Sullivan, A. (2014). United Kingdom National Guideline on the Management of *Trichomonas vaginalis* 2014. *International Journal of STD & AIDS*, *25*(8), 541–549.

Shokoples, S. E., Ndao, M., Kowalewska-Grochowska, K., & Yanow, S. K. (2009). Multiplexed real-time PCR assay for discrimination of *Plasmodium* species with improved sensitivity for mixed infections. *Journal of Clinical Microbiology*, *47*(4), 975–980.

Shwab, E. K., Zhu, X. Q., Majumdar, D., Pena, H. F., Gennari, S. M., Dubey, J. P., et al. (2014). Geographical patterns of *Toxoplasma gondii* genetic diversity revealed by multilocus PCR-RFLP genotyping. *Parasitology*, *141*(4), 453–461.

Siddiqui, A. A., & Berk, S. L. (2001). Diagnosis of *Strongyloides stercoralis* infection. *Clinical Infectious Diseases*, *33*(7), 1040–1047.

Simpson, P., Higgins, G., Qiao, M., Waddell, R., & Kok, T. (2007). Real-time PCRs for detection of *Trichomonas vaginalis* beta-tubulin and 18S rRNA genes in female genital specimens. *Journal of Medical Microbiology*, *56*(Pt 6), 772–777.

Singh, B., Bobogare, A., Cox-Singh, J., Snounou, G., Abdullah, M. S., & Rahman, H. A. (1999). A genus- and species-specific nested polymerase chain reaction malaria detection assay for epidemiologic studies. *The American Journal of Tropical Medicine and Hygiene*, *60*(4), 687–692.

Singh, B., Kim Sung, L., Matusop, A., Radhakrishnan, A., Shamsul, S. S., Cox-Singh, J., et al. (2004). A large focus of naturally acquired *Plasmodium knowlesi* infections in human beings. *Lancet*, *363*(9414), 1017–1024.

Sirichaisinthop, J., Buates, S., Watanabe, R., Han, E. T., Suktawonjaroenpon, W., Krasaesub, S., et al. (2011). Evaluation of loop-mediated isothermal amplification (LAMP) for malaria diagnosis in a field setting. *The American Journal of Tropical Medicine and Hygiene*, *85*(4), 594–596.

Smith, K. S., Tabrizi, S. N., Fethers, K. A., Knox, J. B., Pearce, C., & Garland, S. M. (2005). Comparison of conventional testing to polymerase chain reaction in detection of *Trichomonas vaginalis* in indigenous women living in remote areas. *International Journal of STD & AIDS*, *16*(12), 811–815.

Solari, A., Ortiz, S., Soto, A., Arancibia, C., Campillay, R., Contreras, M., et al. (2001). Treatment of *Trypanosoma cruzi*-infected children with nifurtimox: A 3 year follow-up by PCR. *The Journal of Antimicrobial Chemotherapy*, *48*(4), 515–519.

Stark, D., Roberts, T., Ellis, J. T., Marriott, D., & Harkness, J. (2014). Evaluation of the EasyScreen enteric parasite detection kit for the detection of *Blastocystis* spp., *Cryptosporidium* spp., *Dientamoeba fragilis*, *Entamoeba complex*, and *Giardia intestinalis* from clinical stool samples. *Diagnostic Microbiology and Infectious Disease*, *78*(2), 149–152.

Stark, D., van Hal, S., Fotedar, R., Butcher, A., Marriott, D., Ellis, J., et al. (2008). Comparison of stool antigen detection kits to PCR for diagnosis of amebiasis. *Journal of Clinical Microbiology, 46*(5), 1678–1681.

Su, C., Shwab, E. K., Zhou, P., Zhu, X. Q., & Dubey, J. P. (2010). Moving towards an integrated approach to molecular detection and identification of *Toxoplasma gondii*. *Parasitology, 137*(1), 1–11.

Sudarshan, M., Weirather, J. L., Wilson, M. E., & Sundar, S. (2011). Study of parasite kinetics with antileishmanial drugs using real-time quantitative PCR in Indian visceral leishmaniasis. *The Journal of Antimicrobial Chemotherapy, 66*(8), 1751–1755.

Sulaiman, I. M., Fayer, R., Bern, C., Gilman, R. H., Trout, J. M., Schantz, P. M., et al. (2003). Triosephosphate isomerase gene characterization and potential zoonotic transmission of *Giardia duodenalis*. *Emerging Infectious Diseases, 9*(11), 1444–1452.

Sutherland, C. J., Tanomsing, N., Nolder, D., Oguike, M., Jennison, C., Pukrittayakamee, S., et al. (2010). Two nonrecombining sympatric forms of the human malaria parasite *Plasmodium ovale* occur globally. *The Journal of Infectious Diseases, 201*(10), 1544–1550.

Tachibana, H., Kobayashi, S., Okuzawa, E., & Masuda, G. (1992). Detection of pathogenic *Entamoeba histolytica* DNA in liver abscess fluid by polymerase chain reaction. *International Journal for Parasitology, 22*(8), 1193–1196.

Takagi, H., Itoh, M., Islam, M. Z., Razzaque, A., Ekram, A. R., Hashighuchi, Y., et al. (2009). Sensitive, specific, and rapid detection of *Leishmania donovani* DNA by loop-mediated isothermal amplification. *The American Journal of Tropical Medicine and Hygiene, 81*(4), 578–582.

Takagi, H., Itoh, M., Kasai, S., Yahathugoda, T. C., Weerasooriya, M. V., & Kimura, E. (2011). Development of loop-mediated isothermal amplification method for detecting *Wuchereria bancrofti* DNA in human blood and vector mosquitoes. *Parasitology International, 60*(4), 493–497.

Taniuchi, M., Verweij, J. J., Noor, Z., Sobuz, S. U., Lieshout, L., Petri, W. A., Jr., et al. (2011). High throughput multiplex PCR and probe-based detection with Luminex beads for seven intestinal parasites. *The American Journal of Tropical Medicine and Hygiene, 84*(2), 332–337.

Tao, Z. Y., Zhou, H. Y., Xia, H., Xu, S., Zhu, H. W., Culleton, R. L., et al. (2011). Adaptation of a visualized loop-mediated isothermal amplification technique for field detection of *Plasmodium vivax* infection. *Parasites & Vectors, 4*, 115.

Teal, A. E., Habura, A., Ennis, J., Keithly, J. S., & Madison-Antenucci, S. (2012). A new real-time PCR assay for improved detection of the parasite *Babesia microti*. *Journal of Clinical Microbiology, 50*(3), 903–908.

ten Hove, R. J., van Esbroeck, M., Vervoort, T., van den Ende, J., van Lieshout, L., & Verweij, J. J. (2009). Molecular diagnostics of intestinal parasites in returning travellers. *European Journal of Clinical Microbiology & Infectious Diseases, 28*(9), 1045–1053.

Thekisoe, O. M., Kuboki, N., Nambota, A., Fujisaki, K., Sugimoto, C., Igarashi, I., et al. (2007). Species-specific loop-mediated isothermal amplification (LAMP) for diagnosis of trypanosomosis. *Acta Tropica, 102*(3), 182–189.

Thompson, P. P., Kowalski, R. P., Shanks, R. M., & Gordon, Y. J. (2008). Validation of real-time PCR for laboratory diagnosis of *Acanthamoeba* keratitis. *Journal of Clinical Microbiology, 46*(10), 3232–3236.

Thomson, R. B., Jr., Haas, R. A., & Thompson, J. H., Jr. (1984). Intestinal parasites: The necessity of examining multiple stool specimens. *Mayo Clinic Proceedings, 59*(9), 641–642.

Toe, L., Boatin, B. A., Adjami, A., Back, C., Merriweather, A., & Unnasch, T. R. (1998). Detection of *Onchocerca volvulus* infection by O-150 polymerase chain reaction analysis of skin scratches. *The Journal of Infectious Diseases, 178*(1), 282–285.

Torrico, M. C., Solano, M., Guzman, J. M., Parrado, R., Suarez, E., Alonzo-Vega, C., et al. (2005). [Estimation of the parasitemia in *Trypanosoma cruzi* human infection: High parasitemias are associated with severe and fatal congenital Chagas disease]. *Revista da Sociedade Brasileira de Medicina Tropical, 38*(Suppl. 2), 58–61.

Toure, F. S., Bain, O., Nerrienet, E., Millet, P., Wahl, G., Toure, Y., et al. (1997). Detection of *Loa loa*-specific DNA in blood from occult-infected individuals. *Experimental Parasitology, 86*(3), 163–170.

Toure, F. S., Kassambara, L., Williams, T., Millet, P., Bain, O., Georges, A. J., et al. (1998). Human occult loiasis: Improvement in diagnostic sensitivity by the use of a nested polymerase chain reaction. *The American Journal of Tropical Medicine and Hygiene, 59*(1), 144–149.

Utzinger, J., Botero-Kleiven, S., Castelli, F., Chiodini, P. L., Edwards, H., Kohler, N., et al. (2010). Microscopic diagnosis of sodium acetate-acetic acid-formalin-fixed stool samples for helminths and intestinal protozoa: A comparison among European reference laboratories. *Clinical Microbiology and Infection, 16*(3), 267–273.

Van den Eede, P., Van, H. N., Van Overmeir, C., Vythilingam, I., Duc, T. N., Hung le, X., et al. (2009). Human *Plasmodium knowlesi* infections in young children in central Vietnam. *Malaria Journal, 8*, 249.

Vannier, E., & Krause, P. J. (2012). Human babesiosis. *The New England Journal of Medicine, 366*(25), 2397–2407.

Varlet-Marie, E., Sterkers, Y., Brenier-Pinchart, M. P., Cassaing, S., Dalle, F., Delhaes, L., et al. (2014). Characterization and multicentric validation of a common standard for *Toxoplasma gondii* detection using nucleic acid amplification assays. *Journal of Clinical Microbiology, 52*(11), 3952–3959.

Verweij, J. J., Blange, R. A., Templeton, K., Schinkel, J., Brienen, E. A., van Rooyen, M. A., et al. (2004). Simultaneous detection of *Entamoeba histolytica*, *Giardia lamblia*, and *Cryptosporidium parvum* in fecal samples by using multiplex real-time PCR. *Journal of Clinical Microbiology, 42*(3), 1220–1223.

Verweij, J. J., Canales, M., Polman, K., Ziem, J., Brienen, E. A., Polderman, A. M., et al. (2009). Molecular diagnosis of *Strongyloides stercoralis* in faecal samples using real-time PCR. *Transactions of the Royal Society of Tropical Medicine and Hygiene, 103*(4), 342–346.

Verweij, J. J., Schinkel, J., Laeijendecker, D., van Rooyen, M. A., van Lieshout, L., & Polderman, A. M. (2003). Real-time PCR for the detection of *Giardia lamblia*. *Molecular and Cellular Probes, 17*(5), 223–225.

Verweij, J. J., & Stensvold, C. R. (2014). Molecular testing for clinical diagnosis and epidemiological investigations of intestinal parasitic infections. *Clinical Microbiology Reviews, 27*(2), 371–418.

Verweij, J. J., Ten Hove, R., Brienen, E. A., & van Lieshout, L. (2007). Multiplex detection of *Enterocytozoon bieneusi* and *Encephalitozoon* spp. in fecal samples using real-time PCR. *Diagnostic Microbiology and Infectious Disease, 57*(2), 163–167.

Vestergaard, L. S., & Ringwald, P. (2007). Responding to the challenge of antimalarial drug resistance by routine monitoring to update national malaria treatment policies. *The American Journal of Tropical Medicine and Hygiene, 77*(6 Suppl.), 153–159.

Visvesvara, G. S. (2011). Pathogenic and opportunistic free-living amebae. In J. Versalovic, K. C. Carroll, G. Funke, J. H. Jorgensen, M. L. Landry, & D. W. Warnock (Eds.), *Manual of clinical microbiology: Vol. 2* (10th ed., pp. 2139–2148). Washington, DC: ASM Press.

Visvesvara, G. S., Moura, H., & Schuster, F. L. (2007). Pathogenic and opportunistic free-living amoebae: *Acanthamoeba* spp., *Balamuthia mandrillaris*, *Naegleria fowleri*, and *Sappinia diploidea*. *FEMS Immunology and Medical Microbiology*, *50*(1), 1–26.

Visvesvara, G. S., Sriram, R., Qvarnstrom, Y., Bandyopadhyay, K., Da Silva, A. J., Pieniazek, N. J., et al. (2009). *Paravahlkampfia francinae* n. sp. masquerading as an agent of primary amoebic meningoencephalitis. *The Journal of Eukaryotic Microbiology*, *56*(4), 357–366.

Welburn, S. C., Picozzi, K., Fevre, E. M., Coleman, P. G., Odiit, M., Carrington, M., et al. (2001). Identification of human-infective trypanosomes in animal reservoir of sleeping sickness in Uganda by means of serum-resistance-associated (SRA) gene. *Lancet*, *358*(9298), 2017–2019.

Wendel, K. A., Erbelding, E. J., Gaydos, C. A., & Rompalo, A. M. (2003). Use of urine polymerase chain reaction to define the prevalence and clinical presentation of *Trichomonas vaginalis* in men attending an STD clinic. *Sexually Transmitted Infections*, *79*(2), 151–153.

Westenberger, S. J., Sturm, N. R., & Campbell, D. A. (2006). *Trypanosoma cruzi* 5S rRNA arrays define five groups and indicate the geographic origins of an ancestor of the heterozygous hybrids. *International Journal for Parasitology*, *36*(3), 337–346.

Wilson, M. L. (2012). Malaria rapid diagnostic tests. *Clinical Infectious Diseases*, *54*(11), 1637–1641.

Wilson, M., Glaser, K. C., Adams-Fish, D., Boley, M., Mayda, M., & Molestina, R. E. (2014). Development of droplet digital PCR for the detection of *Babesia microti* and *Babesia duncani*. *Experimental Parasitology*, *149C*, 24–31.

Wolk, D. M., Schneider, S. K., Wengenack, N. L., Sloan, L. M., & Rosenblatt, J. E. (2002). Real-time PCR method for detection of *Encephalitozoon intestinalis* from stool specimens. *Journal of Clinical Microbiology*, *40*(11), 3922–3928.

Workowski, K. A., & Berman, S. (2010). Sexually transmitted diseases treatment guidelines, 2010. *Morbidity and Mortality Weekly Report*, *59*(RR-12), 1–110.

Workowski, K. A., & Bolan, G. (2014). *Sexually Transmitted Diseases Treatment Guidelines, 2014 (Draft for Public Comment)*. Retrieved on January 17, 2015 from, http://www.cdc.gov/std/treatment/2014/2014-std-guidelines-peer-reviewers-08-20-2014.pdf.

World Health Organisation. (1997). Amoebiasis. *Weekly Epidemiological Record*, *72*, 97–100.

World Health Organisation. (2010). *Working to overcome the global impact of neglected tropical diseases: First WHO report on neglected tropical diseases*. Geneva: World Health Organisation.

World Health Organisation. (2013). *Sustaining the drive to overcome the global impact of neglected tropical diseases: Second WHO report on neglected tropical diseases*. Geneva: World Health Organisation.

World Health Organisation. (2014a). *World malaria report 2014*. Geneva: World Health Organisation.

World Health Organisation. (2014b). *Lymphatic Filarisis*. Retrieved on January 6, 2015, from, http://www.who.int/mediacentre/factsheets/fs102/en/.

Wortmann, G., Hochberg, L., Houng, H. H., Sweeney, C., Zapor, M., Aronson, N., et al. (2005). Rapid identification of *Leishmania* complexes by a real-time PCR assay. *The American Journal of Tropical Medicine and Hygiene*, *73*(6), 999–1004.

Yagi, S., Booton, G. C., Visvesvara, G. S., & Schuster, F. L. (2005). Detection of *Balamuthia* mitochondrial 16S rRNA gene DNA in clinical specimens by PCR. *Journal of Clinical Microbiology*, *43*(7), 3192–3197.

Yang, H. W., Lee, Y. R., Inoue, N., Jha, B. K., Danne, D. B., Kim, H. K., et al. (2013). Loop-mediated isothermal amplification targeting 18S ribosomal DNA for rapid detection of *Acanthamoeba*. *The Korean Journal of Parasitology*, *51*(3), 269–277.

Yera, H., Filisetti, D., Bastien, P., Ancelle, T., Thulliez, P., & Delhaes, L. (2009). Multicenter comparative evaluation of five commercial methods for toxoplasma DNA extraction from amniotic fluid. *Journal of Clinical Microbiology*, *47*(12), 3881–3886.

Young, C., Chawla, A., Berardi, V., Padbury, J., Skowron, G., & Krause, P. J. (2012). Preventing transfusion-transmitted babesiosis: Preliminary experience of the first laboratory-based blood donor screening program. *Transfusion*, *52*(7), 1523–1529.

Zhong, M., McCarthy, J., Bierwert, L., Lizotte-Waniewski, M., Chanteau, S., Nutman, T. B., et al. (1996). A polymerase chain reaction assay for detection of the parasite *Wuchereria bancrofti* in human blood samples. *The American Journal of Tropical Medicine and Hygiene*, *54*(4), 357–363.

Zimmerman, P. A., Guderian, R. H., Aruajo, E., Elson, L., Phadke, P., Kubofcik, J., et al. (1994). Polymerase chain reaction-based diagnosis of *Onchocerca volvulus* infection: Improved detection of patients with onchocerciasis. *The Journal of Infectious Diseases*, *169*(3), 686–689.

Zindrou, S., Orozco, E., Linder, E., Tellez, A., & Bjorkman, A. (2001). Specific detection of *Entamoeba histolytica* DNA by hemolysin gene targeted PCR. *Acta Tropica*, *78*(2), 117–125.

Zingales, B., Andrade, S. G., Briones, M. R., Campbell, D. A., Chiari, E., Fernandes, O., et al. (2009). A new consensus for *Trypanosoma cruzi* intraspecific nomenclature: Second revision meeting recommends TcI to TcVI. *Memórias do Instituto Oswaldo Cruz*, *104*(7), 1051–1054.

Zintl, A., Mulcahy, G., Skerrett, H. E., Taylor, S. M., & Gray, J. S. (2003). *Babesia divergens*, a bovine blood parasite of veterinary and zoonotic importance. *Clinical Microbiology Reviews*, *16*(4), 622–636.

Zulantay, I., Honores, P., Solari, A., Apt, W., Ortiz, S., Osuna, A., et al. (2004). Use of polymerase chain reaction (PCR) and hybridization assays to detect *Trypanosoma cruzi* in chronic chagasic patients treated with itraconazole or allopurinol. *Diagnostic Microbiology and Infectious Disease*, *48*(4), 253–257.

Clinical Applications of Quantitative Real-Time PCR in Virology

Julia Engstrom-Melnyk[1], Pedro L. Rodriguez, Olivier Peraud, Raymond C. Hein
Medical and Scientific Affairs, Roche Diagnostic Corporation, Indianapolis, Indiana, USA
[1]*Corresponding author: e-mail address: julia.engstrom-melnyk@roche.com*

1 INTRODUCTION

Over the last several decades, the polymerase chain reaction (PCR) (Mullis & Faloona, 1987; Saiki et al., 1985) has become synonymous with molecular diagnostics and is a staple in most clinical laboratories. The technique, relying on thermo-stable *Taq* polymerase (Saiki et al., 1988) and sequential three-step temperature cycling to denature and amplify specific nucleic acid sequences, is a quick, easy way to create unlimited copies of DNA from just one original strand (Figure 1). The copied DNA can then be used in a broad range of applications from screening, diagnosing, or monitoring diseases to evaluating medical and therapeutic decision points and assessing cure rates, all of which lend to its appeal in the clinical laboratory. Through its dependence on target amplification, PCR is able to be highly sensitive, detecting the presence of just a few copies of the target. In addition, by targeting and amplifying highly unique sequences of nucleic acid, PCR can be designed to be extremely specific, detecting only the intended target in a complex sample.

Unlike traditional PCR, which relies on target detection and analysis at the completion of the thermal cycling, real-time PCR (Holland, Abramson, Watson, & Gelfand, 1991) simultaneously amplifies and detects DNA in real time and, thus, yields faster results. The improved speed offered by real-time PCR is due largely to removal of post-PCR detection procedures and the use of fluorescent labels and sensitive methods of detecting their emissions. It can be multiplexed for detection of up to six targets, and because it analyses the PCR product within the exponential phase of amplification (at which point the target DNA concentration doubles at each three-step temperature cycle), it is able to provide not only reliable and reproducible qualitative results (target detected/not detected) but also quantitative results for exact measures of the starting template copy number with accuracy and high sensitivity over a wide dynamic range. In addition to being a very accurate and sensitive methodology, real-time PCR allows for high-throughput, automated

Methods in Microbiology, Volume 42, ISSN 0580-9517, http://dx.doi.org/10.1016/bs.mim.2015.04.005

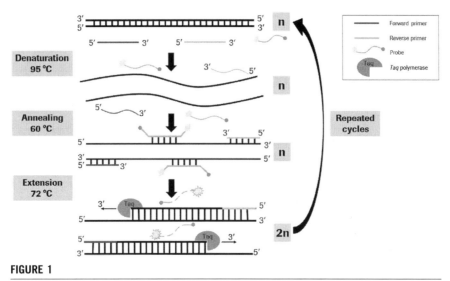

FIGURE 1

Example of amplification and detection of target nucleic acid by real-time PCR.

processes, small sample volumes and can be utilised in a wide variety of applications, making it the method of choice in today's molecular laboratories.

1.1 QUANTITATIVE REAL-TIME PCR

Through the aid of fluorescent signalling probes to measure amplification of DNA at each PCR cycle, at the point of exponential DNA accumulation, real-time PCR is able to provide broader linear dynamic ranges and increased assay performance as determined by sensitivity, specificity, precision, and reproducibility. Due to the consistency in signal intensity changes during the exponential growth phase of PCR, it is also easily adaptable for quantitative reporting. However, there are three properties that are uniquely associated with quantitative real-time PCR: quantification, standardisation, and lower limit resulting.

The accumulation of fluorescence signal is measured at each PCR cycle of the reaction and the cycle at which this signal exceeds a predetermined background fluorescence threshold during the logarithmic phase of amplification is referred to as the cycle threshold (C_T). The C_T value is inversely proportional to the viral copy number in the specimen, and through comparisons of this value to an external calibration curve or an internal quantitation standard, the initial nucleic acid target concentration can be calculated (Heid, Stevens, Livak, & Williams, 1996; Livak & Schmittgen, 2001). However, accurate quantitation within each sample is hindered when relying solely on an external standard as amplification efficiencies for each individual sample may be variable and inconsistent. By utilising a standard internal reference template, with the rationale that any variable influencing amplification efficiency should

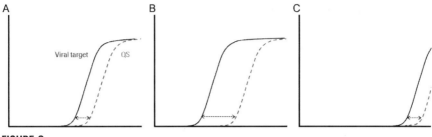

FIGURE 2

Quantitation of viral target using competitive quantitation standard (QS). The QS compensates for effects of inhibition and controls the preparation and amplification processes, allowing a more accurate quantitation viral target in each specimen. The competitive QS contains sequences with identical primer binding sites as the viral target to ensure equivalent amplification efficiency and a unique probe binding region that distinguishes the two amplicons. The competitive QS is added to each specimen at a known copy number and is carried through the subsequent steps of specimen preparation, reverse transcription (when applicable), simultaneous PCR amplification, and detection. Viral target concentration in the test specimens is calculated by comparing the viral target signal (solid line) to the QS signal (dashed line) for each specimen and control (A, B). In the presence of inhibitors, both QS and viral target are equally suppressed and yield accurate viral load calculations (C).

affect both template and target similarly, inhibition and amplification effects are compensated for which allows for more accurate quantitation (Figure 2). This control can be further enhanced when incorporating an internal reference that utilises the same primer sequence as the target since any potential additional effects on PCR efficiency for each of the two targets is eliminated. Thus, the competitive real-time PCR strategy is the most reliable approach for nucleic acid quantitation (Diviacco et al., 1992; Gilliland, Perrin, & Bunn, 1990; Stieger, Demolliere, Ahlborn-Laake, & Mous, 1991; Wang, Doyle, & Mark, 1989; Zentilin & Giacca, 2007) and is the basis for the majority of present-day virology assays.

It is equally important to utilise appropriate quantitation standards, when available, to ensure accurate quantitative results, inter-laboratory correlation, and overall standardisation. Standardisation of reported viral loads ensures not only inter-laboratory consistency but also high clinical utility of viral load monitoring, sets the foundation for establishing clinical correlations and critical thresholds leading to better management of infections and treatments, and are critical for the development of clinical guidelines (Miller et al., 2011). With the wide availability of assay methods, viral targets, specimen type, and lack of standard reference material (Hayden et al., 2012), viral load variability across laboratories can range significantly, as high as 4.3 log copies/mL (Pang et al., 2009). Specifically, results from proficiency testing/external quality assessment programmes as well as inter-laboratory specimen exchange studies have demonstrated that there is significant variability in quantitative results for assays that lack appropriate standards

(Hayden et al., 2008; Pang et al., 2009; Preiksaitis et al., 2009; Wolff, Heaney, Neuwald, Stelrecht, & Press, 2009). Findings such as these reinforce the fact that with this high degree of variability and discrepancy, clinicians are unable to compare test results between two different laboratories and, further, clinically relevant cut-offs set by one test would not apply to results of another (Caliendo et al., 2009). Without standardisation, the quality of patient care is dramatically impacted, preventing meaningful inter-laboratory comparison of patient results and influencing disease prevention and management programmes (Kraft, Armstrong, & Caliendo, 2012). This is especially critical for transplant patients, who may be initially monitored at one institution and then transferred to another for longer-term follow-up receiving results that no longer correlate. Therefore, whenever possible, viral load monitoring tests must report results in IU/mL and be fully traceable to the higher-order first WHO International standard. They must generate highly accurate and reliable results based on a robust calibration methodology (Caliendo et al., 2009) and have excellent reproducibility across the dynamic range of the test with demonstrated co-linearity to the WHO standard.

Lastly, there exist two distinct end-points with quantitative real-time PCR, which should be of consideration for result interpretation and reporting: the lower limit of detection (LLOD/LOD) and the lower limit of quantitation (LLOQ/LOQ). These two limits are assessed differently and are not equivalent in either definition or, in some cases, their assigned values. The LOD (also referred to as analytical sensitivity) represents the lowest viral load level at which $\geq 95\%$ of tested samples are detected (CLSI EP17-A, 2004); theoretically, viral levels at or below the LOD are not detected $\geq 5\%$ of the time. It differentiates between 'detectable' and 'undetectable' results. The LLOQ, on the other hand, is the lowest viral level that is within the linear and analytically acceptable range of the assay (CLSI EP17-A, 2004). In other words, the LLOQ is the lowest point at which an accurate viral load can be assigned and determines which 'detectable' sample will have a reported viral load. A common misconception is that the LOD of the assay is the minimum viral level for a 'detected' result but 'undetectable' and 'detectable' viral levels are never differentiated by a single theoretical viral threshold as viral levels less than the LOD may still have a high probability of being detected. This probability spans a broad range in which the lower the viral titre, the more likely the 'undetectable' result. Ultimately, the statistical probability will favour the 'undetectable' result (Figure 3). And because the LLOQ can be equal or greater than the LOD on some viral load assays, it is not unusual for 'detectable but below the LOQ' (detectable/BLOQ) result reporting (Cobb et al., 2011). Further, the 'detectable/BLOQ' results should not be inferred that the actual viral concentration of the sample is between the LOD and LOQ.

1.2 CO-EVOLUTION OF REAL-TIME PCR AND DIAGNOSTICS

The clinical demand has driven and shaped the evolution of PCR and continues to do so as we gain a greater understanding of the infections we monitor and treat. Through the study of the natural history and disease progression attributed to specific viral

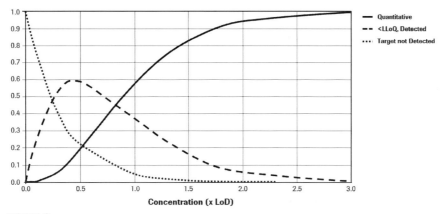

FIGURE 3

Likelihoods of different test results given different viral concentration. When the viral concentration tends to 0, the proportion of 'Target not Detected' increases to 1 (dotted line), increasing the likelihood of 'Not Detected' results. As the concentration tends to LLOQ (dashed line), the likelihood of 'Detected but <LLOQ' results peaks. When the concentration tends to infinity, the proportion of quantitative results tends to 1 (solid line), resulting in a continual increase in the likelihood of 'Detected: Quantitative' results. At any concentration, the sum of the three types of reported results is always 100% and throughout the concentration continuum, variations in result reporting exist. As the concentration of viral levels approaches the LLOQ, near equal likelihood of 'Detected: Quantitative' and 'Detected but <LLOQ' results are possible, as are 'Not detected' results. Decreasing concentrations will further shift the likelihoods and increase the chance of 'Detected but <LLOQ' or 'Not Detected' results. Diagram assumes LLOQ=LLOD. (See the color plate.)

Reprinted from Cobb, Vaks, Do, and Vilchez (2011): Copyright (2011), with permission from Elsevier.

infections, the need for sensitive, accurate, precise, reproducible, and reliable quantitative measurements of viral levels has become a necessity.

With the deeper understanding of the natural history of human immunodeficiency virus (HIV) infections, it is now well understood that progressive immunosuppression and the onset and development of clinical disease are strictly associated with increasing viral burden (Furtado, Kingsley, & Wolinsky, 1995; Ho, Moudgil, & Alam, 1989; Mathez et al., 1990; Nicholson et al., 1989; Schnittman et al., 1990). Thus, quantitative real-time PCR is critical for monitoring patients infected with HIV (Hufert et al., 1991; Mellors et al., 1995) and those undergoing antiretroviral therapy (ART) to ensure viral replication is sufficiently and effectively suppressed and to monitor potential for viral resistance to the medication (DHHS HIV, 2014). This monitoring and maintained viral suppression is absolutely necessary not only to maintain progression-free survival of HIV-infected patients but also to reduce subsequent HIV transmission (Cohen et al., 2011; Diffenbach, 2012). Due to the significance of viral load monitoring and maintaining viral suppression, the demand for increasingly more sensitive assays for HIV has driven the innovation of diagnostic tests and continues to push the limits of the LOD and LOQ ever lower (Glaubitz et al., 2011; Sizmann et al., 2010).

Additionally, the amount of viral DNA in samples from either hepatitis B virus (HBV) or cytomegalovirus (CMV) chronic carriers is indicative of active viral replication in liver cells and is correlated with liver disease progression (Chen, Lin, et al., 2006; Chen, Yang, et al., 2006; Emery et al., 2000; Humar et al., 1999; Humar, Kumar, Boivin, & Caliendo, 2002; Iloeje et al., 2006; Wursthorn, Manns, & Wedemeyer, 2008), highlighting the need for quantitative viral levels in chronic disease monitoring. Similarly, hepatitis C virus (HCV) levels have been linked to prognosis and treatment outcomes (Trepo, 2000), treatment response (Pearlman & Ehleben, 2014; Zeuzem et al., 2009), as well as assessing sustained virologic response (SVR), or 'cure', following treatment (Pearlman & Traub, 2011). More recently, as new antiviral therapies for HCV treatment rely on targeting specific biological steps of the viral replication cycle, reliable quantitative monitoring of the viral burden at specific time-points to ensure treatment compliance and resistance emergence during treatment is recommended (Au & Pockros, 2014; AASLD/IDSA/IAS-USA, 2014).

Alongside the changing clinical needs, several instrument and manufacturing innovations have been introduced to meet these newer requirements. The first PCR-based diagnostic test and the first automated system were both introduced in the early 1990s, allowing for improved standardised technique, increased efficiency, and reduction in error and contamination. In 1996, to meet the clinical needs of monitoring patients infected with HIV, the first 'personalised healthcare' diagnostic test was FDA approved, that monitored whether ART was working and whether a patient was on optimal treatment. Entire systems were introduced in the early 2000s that automated the up-front sample preparation step, leading to further reduction of hands-on time, increased reliability and reproducibility. With the introduction of real-time PCR techniques to further enhance assay performance, the first PCR tests that allowed for simultaneous amplification and detection were introduced in 2003. Throughout the decade, improvements in assays and instrumentation began pushing the boundary for HIV-1 detection, achieving greater and greater sensitivity, a feat that has proven to be incredibly relevant in understanding risks of viral load rebound and virologic failure (Doyle et al., 2012; Estevez et al., 2013; Pascual-Pareja et al., 2010). More recently, shortly after the release of two new higher-order standards for CMV (first WHO International Standard and NIST CMV Standard), fully automated real-time quantitative PCR assays were FDA approved to meet laboratory and clinical demands for standardisation. Not only does the instrumentation reduce assay design variability and inconsistency across laboratories, but also the assays are standardised to the WHO and report in international units, providing accuracy across the entire dynamic range that is only achieved by calibration and traceability to these higher-order standards. But it is not until these FDA-approved tests gain widespread use will inter-laboratory agreement for CMV viral load results truly improve (Kraft et al., 2012).

The objective of test innovation is to evolve and adapt to clinical and laboratory needs. As patient outcome gaps are identified, and as we gain greater understanding and insight into the clinical progression of disease and disease management, technological achievements facilitate advancements in quality of diagnostics and patient outcomes. Life-threatening illnesses convert to chronic/manageable disease with the aid of viral load monitoring. And pressure exists to develop more precise, accurate, sensitive assays, which, in turn, drives the development of more efficacious drugs. This synergy demonstrates the value of diagnostics: it is an integral part of the patient care continuum.

2 APPLICATIONS IN THE CONTINUUM OF CARE

Applications of quantitative real-time PCR for virology are extensive. It is especially necessary when antibody seroconversion is delayed after an acute infection and early diagnosis are essential (DiBiasi & Tyler, 2004; Thomson et al., 2009), in immunocompromised patients that may not have an optimal antibody response (Kadmon et al., 2013), and for the diagnosis of congenital or perinatally acquired viral infections (Park, Streicher, & Rothberg, 1987; Young, Nelson, & Good, 1990). Additionally, it is critical in maintaining a high level of patient care at each stage of the disease and infection (Figure 4).

2.1 SCREENING

Screening tests are designed with one key parameter in mind: exceptional sensitivity (Herman, Gill, Eng, & Fajardo, 2002). They must ensure that disease is not missed in a population that is generally free from risk to allow for appropriate early intervention, thereby effectively reducing mortality and morbidity. Although traditional

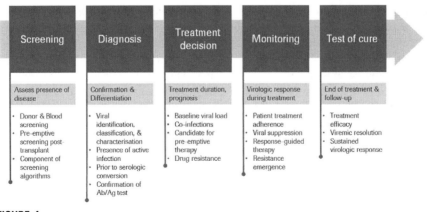

FIGURE 4

Applications of real-time PCR in the continuum of care and patient management.

population-based screening programmes and recommendations do not often utilise nucleic acid testing (NAT) for virology targets, relying more on immunoassays and antigen testing, NATs and real-time PCR assays are still integral components.

Donor-eligibility determination ensures that a donor is eligible to donate cells or tissues to be used as human cells, tissues, and cellular and tissue-based products (HCT/Ps), which can include haematopoietic stem/progenitor cell, organ, semen, and other types of donations. In part, living and cadaveric HCT/P donor eligibility is granted if screening shows that the donor is free from risk factors for, and clinical evidence of, infection due to relevant communicable disease agents and diseases such as HIV type 1 and 2, HBV, HCV, and West Nile Virus (WNV) (FDA Testing, 2014). The FDA testing recommendations stipulate that HCT/Ps tests for HIV and HCV may include FDA-licensed NAT blood donor screening and that, specifically, an FDA-licensed NAT should be used to assess infection with WNV. Despite the fact that these NAT tests provide qualitative as opposed to quantitative results, the molecular technology utilised is often real-time PCR due to its enhanced accurate and reliable resulting (FDA Assays, 2014).

Additionally, pre-emptive virology screening post-transplant may reduce subsequent complications and provide a more cost-effective management strategy (Evers, 2013). Pre-emptive therapy utilises routine viral screening to initiate therapy at the first indications of viraemia, prior to clinical manifestation of disease. This strategy reduces the overall morbidity and mortality post-transplant compared to strategies in which treatment is initiated at the onset of clinical disease (Schönberger et al., 2010). The pre-emptive treatment model utilising routine screening by quantitative real-time PCR of asymptomatic post-transplant patients has been shown to be especially effective for paediatric patients undergoing haematopoietic stem cell transplant, who are uniquely at a high risk of CMV and Epstein–Barr virus (EBV) infections (Evers, 2013). This strategy and prospective screening utilising a quantitative real-time PCR assay for BK polyoma virus (BKV) is recommended as part of routine post-transplant follow-up of all kidney transplants since early identification and management of BKV infection may prevent future incidence of polyoma virus-associated nephropathy (Hirsch et al., 2005; KDIGO, 2009).

NATs and real-time PCR are also utilised as vital parts of certain screening algorithms. The Center for Disease Control (CDC) currently recommends that HCV RNA testing be utilised for anyone who may have been exposed to HCV within the preceding 6 months. In addition, NATs would identify active HCV infection among persons who have tested anti-HCV positive or those with an indeterminate antibody test indicating need for referral for further medical evaluation and care (CDC HCV, 2013).

2.2 DIAGNOSIS

Viral diagnostic tests are used to determine presence or absence of current or previous infection. Because many viral infections present with similar symptoms, accurate diagnosis is critical as each requires unique and vastly different interventions and/or management strategies. Historically, diagnosis of viral infections has relied on viral growth in cell culture, immunoassays, antigen assays (ELISA), haemagglutination testing, and electron microscopy (Krishna & Cunnion, 2012). The introduction of molecular methods including NATs and real-time PCR assays has vastly improved viral diagnosis with their superior sensitivity, specificity, and rapid result reporting (Emmadi et al., 2011). NAT-based infectious disease testing, providing rapid results, aids in outbreak detection (Ebola), genotype identification (HCV), and identification of possible drug resistance (HIV-1), which can lead to rapid clinical therapeutic decisions and early infection control to prevent spread of disease (Espy et al., 2006).

Viral culture was traditionally considered to be the 'gold standard' for viral diagnosis because of increased sensitivity compared to rapid antigen-testing methods. However, a significant limitation of viral culture was a long time to result, reaching 14 days for CMV (Gleaves, Smith, Shuster, & Pearson, 1985). Improvements in culture included the introduction of shell viral culture, which reduced the result turnaround-time (TAT) for CMV to 24 h. Although considered to be a vast improvement, certain viruses require immediate treatment intervention to prevent life-threatening infection and therefore, a 24-h TAT is simply much too long. Further, reliance on viral culture for diagnostics testing introduces other pronounced drawbacks, the most noteworthy being that not all routine viruses grow in culture and that virus viability, and thus its culture ability, could be impacted by sample collection, transport conditions, or prior patient treatment. With these limitations in mind, NAT testing has tremendous advantages and has replaced culture for most viruses due to its greatly reduced TAT (typically less than 8 h) while still retaining high sensitivity.

The applications of NAT testing, and more specifically, real-time quantitative PCR technology, in viral diagnostics and confirmatory testing continue to expand. The CDC-updated recommendations for HIV testing state that specimens that are reactive on the initial antigen/antibody combination immunoassay and non-reactive or indeterminate on the HIV-1/HIV-2 antibody differentiation immunoassay should be tested with an FDA-approved HIV-1 NAT test (Branson, 2010; Branson et al., 2014). Additionally, NATs have demonstrated utility in high-risk populations, in which antibody testing alone might miss a considerable percentage of HIV infections that are otherwise detectable by NAT virologic tests (Priddy et al. 2007; Stekler et al., 2009). Particularly, immunoassays for HIV diagnosis are limited by the marked delay between infection and seroconversion, a time when HIV viral levels are at their peak (Figure 5). For this reason, NATs are the recommended method for diagnosis of HIV during the acute phase of infection (10–50 days post-infection). HIV viral RNA is the first marker to manifest itself approximately 10 days post-infection at initiation of the acute phase of infection (Lindback et al., 2000). Not until 4–10 days after initial detection of HIV RNA do the HIVp24 antigen levels rise to detectable levels using fourth-generation immunoassays. This marked delay and the need for rapid diagnosis by the identification of HIV RNA is especially critical when an early diagnosis is medically warranted. During the acute phase

patient predisposition, prior treatment experience, viral genotype, viral resistance profile, and host genetic profile, among others. And depending on the viral infection, a quantitative baseline viral load may also serve an important role in helping to guide treatment decision (Table 1).

Over the past 25 years, pharmaceutical development and clinical trials investigating cutting-edge antiviral treatments have relied heavily on the data generated from PCR-based—and eventually quantitative real-time PCR-based—technology to determine safe and effective drug use (Cobb et al., 2011; MacKay, Arden, & Nitsche, 2002). Practice guidelines continue to reference registrational and non-registrational studies utilising quantitative real-time PCR to guide both clinicians and laboratories in the proper implementation of treatment and testing in order to deliver the most effective personalised care to patients (AASLD/IDSA/IAS-USA,

Table 1 Utility of Quantitative Real-Time PCR in Antiviral Treatment Decisions

Viral Infection	Established Baseline VL Cut-off	Other Baseline Considerations	Rx Eligibility
HCV	<6,000,000 IU/mL	• Genotype 1, *and* • Rx naïve, *and* • Non-cirrhotic	Shorter Rx eligible (8 weeks) with combined regimen sofosbuvir/ledipasvir
HIV	<100,000 cp/mL	• Rx naïve, *and* • CD4 >200 cells/mm^3, *or* • HLA-B*5701 negative	Eligible for Rx with specific NNRTI- and PI-based regimens
HBV	>20,000 IU/mL	• HBeAg positive/negative, *and* • ALT >2× normal, *and/or* • Moderate/severe inflammation • Significant fibrosis	Eligible for Rx initiation with appropriate regimens.
CMV	*Unknown*	• *Unknown*	*Eligible for pre-emptive Rx and for specific duration*

Established viral load (VL) cut-offs—as per referenced prescribing information and treatment guidelines—are shown for each viral infection. In addition to other baseline factors, a patient's treatment initiation, regimen, and/or duration can be impacted by quantitative viral load. Quantitative accuracy and precision are required at each respective cut-off by the real-time PCR assay in order to appropriately manage patients to a given treatment strategy. No baseline viral load threshold has been widely established for pre-emptive treatment of CMV post-transplant; however, broader standardisation to the WHO CMV international standard may allow for determination of viraemic threshold for pre-emptive treatment initiation as well as the appropriate duration of antiviral treatment. This table only summarises antiviral treatment recommendations. For more detailed and complete recommendations, see referenced prescribing information or treatment guideline. NNRTI=Non-Nucleoside Reverse Transcriptase Inhibitor; PI=Protease Inhibitor.
HARVONI (2014), DHHS (2014), Lok and McMahon (2009) and Kotton et al. (2013).

2014; DHHS HIV, 2014; Kotton et al., 2013). Because of this extensive co-utilisation of real-time PCR, it is widely accepted as the gold-standard technology for measuring a patient's quantitative viral load before, during, and after the course of treatment.

Once the decision to initiate treatment for chronic HCV infection is made, several baseline factors are routinely considered (AASLD/IDSA/IAS-USA, 2014). Among these are complications like liver fibrosis stage, co-infection with HIV, hepatocellular carcinoma, end-stage liver disease, genetic variations like HCV genotype, subtype and resistance markers, and prior treatment experience and outcome. The association of baseline viral load, measured by quantitative real-time PCR, to chronic HCV treatment outcome has been well documented with earlier therapies (pegylated-interferon plus ribavirin) but, given the lack of alternative therapeutic options, has not been recommended as a therapeutic decision factor (Jensen et al., 2006; Pawlotsky, 2012). Historically, practice guidelines have recommended the measurement of baseline viral load to serve only as an initial time-point required for effective monitoring of treatment without any prognostic indication (Yee, Currie, Darling, & Wright, 2006). Currently, tremendous advancements in the treatment of chronic HCV infection that employ direct-acting antivirals (DAAs) reported cure rates (as determined by SVR) of >90% for even the once most difficult to treat HCV genotype-1 patients, the most predominant in the United States. Because of this high potency of these drugs across patient populations and the greater importance of numerous other factors, including HCV genotype and prior treatment experience, in determining the appropriate course of treatment, the most recent AASLD/IDSA practice guidelines still do not recommend a baseline quantitative viral load as a therapeutic decision factor.

However, in the rapidly evolving field of HCV treatment, the recent FDA approval of a fixed-dose combination drug consisting of two DAAs (sofosbuvir and ledipasvir) for the treatment of HCV genotype-1, the manufacturer's drug label now includes a new indication for quantitative real-time PCR. It is indicated that treatment naïve and non-cirrhotic patients with a specific baseline viral load are eligible for shortened therapy, an indication with tremendous implications. According to the prescribing information, patients with a baseline viral load below 6 million IU/mL are eligible to have shorter therapy duration of 8 weeks, much shorter than the 12- or 24-week duration for other patient populations (HARVONI, 2014). This therapeutic decision practice is the first of its kind in treatment of chronic HCV infection and is likely to be a recurring theme as DAA manufacturers strive to develop high efficacy regimens requiring shorter treatment durations. Additionally, shorter treatment durations are more favourable to patients and payers when considering the cost of achieving SVR with DAAs and may improve patient drug adherence and completion of therapy (Hep C Online, 2014). As much as quantitative real-time PCR helped to develop this claim for this particular regimen, this technology will also be employed by numerous laboratories to aid in this part of therapeutic decision.

In contrast to chronic infection, treatment of patients presenting in the acute phase of HCV infection, within the first 6 months after exposure, is not recommended by AASLD/IDSA for patients in whom HCV infection spontaneously clears (AASLD/IDSA/IAS-USA, 2014). Therefore, careful monitoring of HCV RNA by a sensitive nucleic acid test is required in order to confirm spontaneous clearance, defined as HCV RNA negative at two specific measurements. Quantitative and qualitative real-time PCR assays are both widely used for this purpose, given their comparable sensitivity.

Factors influencing ART decision for HIV-infected patients include determination of pregnancy, AIDS-defining conditions, acute opportunistic infections, low CD4 counts, HIV-associated nephropathy, potential drug interactions, co-infection with HCV or HBV, HIV resistance testing, and prior treatment experience (DHHS HIV, 2014). Plasma HIV RNA viral load, performed widely by quantitative real-time PCR, is also recommended as a pre-ART decision factor specifically for treatment naïve patients. The Department of Health and Human Services (DHHS HIV) recommends that only ART-naïve patients with a plasma HIV viral load below 100,000 cp/mL can be prescribed various regimen options, which they otherwise should be restricted from taking with higher viral load. This is primarily due to inferior virologic responses in patients with higher viral loads observed in clinical studies (Sax et al., 2009). These clinical trial studies employed quantitative real-time PCR in order to help determine this cut-off and many labs have utilised the same technology to help guide HIV-treating clinicians in this decision.

In the case of chronic HBV infection, several studies have shown that Hepatitis B 'e' antigen (HBeAg) and high levels of HBV DNA are independent risk factors for the subsequent development of cirrhosis and hepatocellular carcinoma (Chen, Lin, et al., 2006; Chen, Yang, et al., 2006; Iloeje et al., 2006). However, due to the fluctuating nature of chronic HBV infection, the prognostic utility of one high HBV DNA level at a single time-point is limited. Thus, HBV baseline DNA viral load, along with HBeAg, alanine aminotransferase (ALT) levels, and fibrosis, collectively aids in the decision to treat with antiviral agents as well as which HBV antiviral regimen to choose and duration of treatment (Lok & McMahon, 2009). Typically, patients with an HBV DNA viral load >20,000 IU/mL, signs of liver disease (i.e. high ALT levels and/or significant fibrosis), and loss of HBeAg are considered for immediate treatment with antivirals, whereas patients <2000 IU/mL are closely monitored for viral load changes prior to treatment. Patients who fall in between this range are monitored for persistent viraemia and signs of liver disease before deciding to treat. Quantitative real-time PCR, therefore, plays a crucial role in the care of chronic HBV patients who, if not treated at the appropriate time with the appropriate regimen and duration, are at greater risk of liver complications.

Unlike treatment guidelines for HCV, HIV, and HBV, management of CMV after solid organ transplant is not associated with specific quantitative CMV viral load cut-offs in order to make therapeutic decisions (Kotton et al., 2013). This is partly due to the historical lack of an international standard and varying assay designs, which has led to poor inter-institutional correlation of quantitative NATs. In addition, the widespread practice of universal prophylaxis, where CMV antiviral medication is administered to patients early in the post-transplant period and continued for a finite period of time, has diminished the clinical utility of baseline viral loads for making therapeutic decisions. However, with the recent availability of the WHO CMV International Reference Standard, the establishment of viral load cut-offs that can be applied to pre-emptive monitoring of patients prior to treatment initiation may soon become more widely accepted (Pang et al., 2009). Until then, institutions are required to determine their own test performance characteristics and clinical cut-offs.

Several studies have shown that a low CMV virologic threshold (e.g. detectable viraemia) using quantitative real-time PCR should be used for starting pre-emptive therapy especially in high-risk cases where the organ donor screens positive and the receptor screens negative for CMV serology (Atabani et al., 2012; Couzi et al., 2012; Sun, Cacciarelli, Wagener, & Singh, 2010). Among a variety of baseline risk factors that may indicate longer CMV treatment duration, significant predictive value has been demonstrated with higher baseline viral loads where longer treatment duration may prevent CMV disease relapse (Kotton et al., 2013; Sia et al., 2000). Clinical trial studies supporting the recent FDA approval of a quantitative real-time PCR CMV test calibrated to the WHO International Standard also demonstrated clinical value for baseline testing of patients with CMV disease who are undergoing treatment with the anti-CMV drugs ganciclovir or valganciclovir (Razonable et al., 2013). Data from this study suggested that patients with a baseline CMV viral load <18,200 IU/mL are likely to resolve CMV disease more rapidly than those who have a higher baseline viral load. Further studies are needed to determine universal thresholds for pre-emptive therapy initiation and predictive value for CMV baseline viral load in defining optimal treatment duration.

There exists a clear application for quantitative real-time PCR technology in baseline determination of patients with significant viral infections, and in fact, quantitative viral load determination plays a critical role in therapeutic decision for many other viral infections. High baseline viral load has been shown to correlate with advanced disease during infection with numerous viruses such as BKV, HSV-1, EBV, and Adenovirus and may potentiate the need for longer duration therapies in certain scenarios (Cincinnati Children's Hospital Medical Center, 2012; Domingues, Lakeman, Mayo, & Whitley, 1998; Gustafson et al., 2008; Randhawa et al., 2004).

After the patient's baseline assessment or pre-emptive monitoring suggests if treatment is available, which treatment regimen to choose and perhaps the duration of therapy, the patient can move on to therapeutic administration. Quantitative real-time PCR has helped and continues to set the stage for decisions that potentially saves lives, reduces complications, decreases morbidity, and lessens the economic burden to both the patient and the healthcare system.

2.4 MONITORING

Serial measures of viral load serve as an individualised map of a viral infection through the estimation of the amount of virus found within an infected person. Tracking viral load in the continuum of care is a vital tool used predominantly to monitor treatment response and its effectiveness, early signs of resistance emergence during therapy of chronic viral infections, and viral activation or reactivation in immuno-compromised patients following bone marrow or solid organ transplantation.

While the goal of treatment for chronic HCV infection is SVR, patients may fail therapy due to non-response, on-treatment breakthrough, or post-treatment relapse (Figure 6). The early change in quantitative viral load over time may be predictive of treatment efficacy and a shorten therapy for patients who respond rapidly to treatment (Yee et al., 2006). This 'response-guided therapy' (RGT) is best exemplified during treatment of chronic HCV patients. Specifically, the sooner a patient becomes HCV RNA undetectable during treatment, the lower the relapse rate when treatment is shortened. Conversely, the longer it takes for a patient to become HCV RNA undetectable, the longer they need to remain on treatment to limit relapse. However, given the poorer efficacy of earlier regimens, not all patients who received therapy achieved SVR. For this reason, 'futility rules' or 'stopping rules' were also developed, which required that failure of a patient to respond (target not detected or viral load cut-off) by a given time-point indicated the need to immediately discontinue therapy.

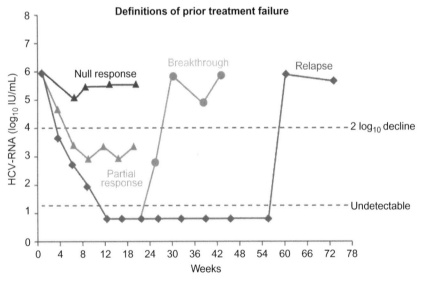

FIGURE 6

Monitoring HCV viral loads during treatment. Despite advances in treatment for HCV patients, failure to achieve SVR is still a reality. Patients who do not achieve SVR fall into four categories: (1) null responders (black line) achieve less than 2-log decrease in hepatitis C viral load upon treatment; (2) partial responders (red line; light grey in the print version) experiences at least a 2-log decrease in hepatitis C viral load during HCV treatment but fail to proceed to an undetectable viral load level; (3) breakthrough patients (orange line; light grey in the print version) have an undetectable HCV viral load, but the virus rebounded during treatment; (4) relapsers (blue line; dark grey in the print version) have had an undetectable HCV viral load, but the virus rebounded after they completed HCV treatment. (See the color plate.)

Although these RGT notions were originally developed from observations made during treatment with the older therapies, peg-IFN and ribavirin, RGT was also required during treatment with the much more potent first-generation DAAs, telaprevir and boceprevir, and stopping rules were put in place during treatment with the second-generation DAA, simeprevir (AASLD/IDSA/IAS-USA, 2014; Ghany et al., 2009; Yee et al., 2006). Newer IFN-free DAA regimens targeting HCV, which are better tolerated by patients and by virtue of the targets they inhibit, have a higher barrier to resistance, yield more rapidly declining viral kinetics, and, thus, do not contain treatment indications for RGT in their prescribing information (HARVONI, 2014; OLYSIO, 2014; SOVALDI, 2013; VIEKIRA, 2014). While RGT was a major driver for regular viral load monitoring during antiviral therapy, it is not the only reason to monitor HCV viral load. In the interval between baseline measurement and assessment of SVR, the 2014 AASLD/IDSA guidelines also include recommendations for monitoring initial response (week 4 on treatment with a repeat at week 6 if detectable) and end of treatment in order to provide an assessment of drug compliance/early efficacy and predict treatment outcomes, respectively (AASLD/IDSA/IAS-USA, 2014). In the most recent revision to these Web-based guidelines, it is recommended that an HCV viral load increase of greater than 10-fold on repeat testing at week 6 (or thereafter) should prompt a discontinuation of HCV treatment. Many clinicians also closely monitor and report the declining viral loads to their patients in order to demonstrate treatment efficacy, motivating patients to continue treatment and remain adherent to the drug regimen until the next follow-up appointment (Fusfeld et al., 2013). Regardless of monitoring during HCV treatment for RGT, adherence/compliance, patient motivation, early treatment efficacy, etc., quantitative real-time PCR is widely used by laboratories due to its sensitivity, accuracy, and reproducibility of each consecutive viral load test.

For patients infected with chronic viral infections, such as HIV, the lifelong regimen of highly active ART aims to suppress HIV viral levels to near undetectable levels, ensuring progression-free survival (delay or all together prevention of the progression to AIDS) and reducing potential transmission. Alongside monitoring immune function and immunologic efficacy through CD4 T-cell count, HIV viral levels are critical in the clinical evaluation and assessment of HIV-infected patients undergoing ART. Determining a patient's HIV viral load is indicated prior to entry into care, at the initiation of ART, at 2–8 weeks after ART initiation, and then typically every 3–4 months while on treatment: (1) to establish a baseline level of HIV viral load; (2) to establish viral response to the therapy to assess the virologic efficacy of ART; and (3) to monitor for abnormalities that may be associated with antiretroviral drugs (DHHS HIV, 2014).

The baseline HIV viral load is not only linked to treatment options (Sax et al., 2009) but also helps to establish the magnitude of viral load decline after initiation of ART and provides prognostic information about the probability of progression to AIDS or death (Marschner et al., 1998; Murray, Elashoff, Iacono-Connors, Cvetkovich, & Struble, 1999; Thiebaut et al., 2000). Once treatment is initiated, the goal is to reach and maintain suppressed HIV replication as determined by undetected viral levels utilising highly sensitive NAT tests, which is generally achieved within 8–24 weeks after ART initiation. The need for sensitive assays to effectively assess viral suppression hinges on the need to suppress HIV replication to the extent that viral evolution and drug resistance mutations do not emerge, which typically do not occur in patients whose HIV RNA levels are maintained below the LLOD of current real-time quantitative PCR assays (Kieffer et al., 2004).

Due to the introduction of more sensitive real-time PCR assays, which can detect as few as 20 viral copies/mL, natural variability in HIV viral levels over time, even in patients with effective suppression, is much more evident (Lima, Harrigan, & Montaner, 2009; Gatanaga et al., 2009; Willig et al., 2010). Although controversy exists between the clinical significance of viral loads between LLOD and <200 copies/mL, there are reports suggesting that this low-level viraemia is predictive of virologic rebound (Doyle et al., 2012; Eron et al., 2013; Laprise, de Pokomandy, Baril, Dufresne, & Trottier, 2013), virologic failure (Estevez et al., 2013), and indication of drug resistance (Taiwo et al., 2010), signifying the need for highly sensitive assays. Viraemic blips, a single detectable HIV viral load (<500 copies/mL) in an otherwise seemingly suppressed patient (Figure 7), however, do not indicate subsequent virologic failure or development of resistance mutations (Castro et al., 2013; Lee, Kieffer, Siliciano, & Nettles, 2006; Nettles et al., 2005). Blips are not unusual (Havlir et al., 2001) and appear to be more common in winter, suggesting that host-related and seasonal factors are associated with the occurrence of viraemia (van Sighem et al., 2008). On the other hand, persistent HIV RNA levels ≥200 copies/mL are often evidence of viral evolution and accumulation of drug resistance

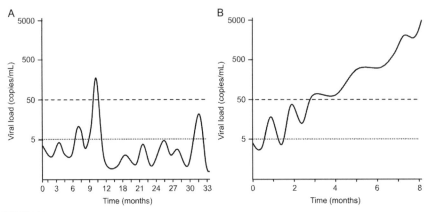

FIGURE 7

On-treatment HIV patient monitoring. (A) HIV viral loads will fluctuate as patients are on treatment, and, in most instances, will remain 'undetectable' (at or below dotted line); viral 'blips' are not uncommon and will result in transient 'detectable' and even quantifiable results (above the dashed line). (B) Virologic failure will lead to a sustained high-level viral titre that, without intervention, will increase with time.

mutations (Aleman, Soderbarg, Visco-Comandini, Sitbon, & Sonnerborg, 2002; Karlsson et al., 2004). Once treatment failure is confirmed, immediate intervention is recommended to avoid progressive accumulation of resistance mutations and effective response of new regimen (DHHS HIV, 2014), which is benefited by low HIV RNA levels and/or higher CD4 cell counts (Eron et al., 2013), and even a brief interruption in therapy may lead to a rapid increase in HIV RNA and a decrease in CD4 cell count and increases the risk of clinical progression (Deeks et al., 2001; Lawrence et al., 2003). With the development and administration of newer drugs that target specific biological processes of HIV, routine and clinical monitoring of viral loads using a real-time quantitative PCR assay continues to be critical to predict treatment failure and early emergence of drug resistance mutations, within a timeframe that would increase subsequent treatment success.

Viral load monitoring is also essential when the recipient of a solid organ transplant is CMV seropositive and the decision is made to initiate treatment only once the CMV levels predictive of disease are reached. This strategy, known as pre-emptive therapy, utilises intensive monitoring for CMV activity by sensitive real-time quantitative PCR methods and short-term antiviral treatment is given only to those with significant viral counts before symptoms occur. CMV is one of the most common opportunistic pathogens that infect solid organ transplant recipients (Fishman, 2007) and is associated with increased morbidity and mortality (Sagedal et al., 2004; Schnitzler et al., 2003). Following primary infection, the virus establishes a lifelong latent infection in several sites of the body and may reactivate in the presence of immunosuppression, such as in transplant recipients. Once reactivated, CMV is able to modulate the immune system and is known to be a potent upregulator of alloantigens (Razonable, 2008), increasing the risk of chronic allograft dysfunction (Reischig, 2010; Sagedal et al., 2002; Smith et al., 2010) and acute rejection (Sagedal et al., 2004). Pre-emptive therapy reduces the incidence of CMV disease (Khoury et al., 2006; Reischig et al., 2008), which has been documented as a serious problem in randomised trials upon completion of universal antiviral prophylaxis therapy (Kalil, Levitsky, Lyden, Stoner, & Freifeld, 2005; Lowance et al., 1999; Paya et al., 2004). Long-term studies have demonstrated that patients receiving pre-emptive therapy, when compared to prophylaxis therapy, were less likely to develop moderate-to-severe kidney scaring and atrophy and significantly better survival of the transplanted organ (Reischig et al., 2012). However, challenges still exist around defining appropriate thresholds to initiate pre-emptive therapy (Humar & Snydman, 2009). But with new standardised real-time PCR assays, widespread adoption, and utilisation of these tests, pre-emptive therapy relying in intensive viral load monitoring may become the standard for certain at-risk patients.

2.5 TEST OF CURE

Test of cure, or end of treatment response, is assessed following a given therapeutic regimen for signs of treatment efficacy. In few cases, a quantitative viral load measurement serves as a way to establish a cure rate, but, in others, may only be used as a confirmation of virologic suppression as clinical cure may not yet be possible with current therapies or technical limitations by real-time PCR that limits the overall sensitivity of viral detection. Regardless of the clinical utility for measuring a virologic suppression, quantitative real-time PCRs with their current limits of detection and limits of quantitation are valuable tools in measuring low-level viraemia and establishing undetectable viral loads.

Utilisation of quantitative real-time PCR to assess virologic cure is perhaps best exemplified by treatment of patients with chronic HCV. According to the AASLD/IDSA guidelines, patients who have 'undetectable' HCV RNA in the serum, when assessed by a sensitive PCR assay, 12 or more weeks after completing treatment, are deemed to have achieved a sustained virologic response (SVR-12). Achieving an SVR is considered a virologic cure of HCV infection since, in these patients, hepatitis C-related liver injury stops and recurrence of infection is marginal, detected in <1% of patients after 5 years post-treatment (AASLD/IDSA/IAS-USA, 2014; Manns et al., 2013). In agreement with these guidelines, the FDA recommendation to pharmaceutical DAA manufacturers also stipulates that viral RNA clearance at SVR-12 be measured in clinical trials using an FDA-approved sensitive and specific quantitative HCV RNA assay (FDA HCV, 2013). According to prescribing information accompanying the current DAAs, the threshold of SVR-12 is defined as a quantitative threshold of HCV RNA <25 IU/mL at 12 weeks after the end of treatment (Feld et al., 2014; Kowdley et al., 2014; Lawitz et al., 2013). This is somewhat dissimilar to the AASLD/IDSA guidelines as 'undetected' viral levels are not equivalent to 'detected but below the limit of quantitation' (Figure 3). But, with the benefit of high sensitivity and reproducibility, quantitative real-time PCR has a clear established role in assessment of HCV virologic cure in both clinical trials and clinical practice and is able to meet the needs for assessing SVR.

Quantitative real-time PCR may also play a critical role in the assessment of CMV disease resolution. The consensus guidelines recommend that two consecutive negative samples be obtained with a minimum treatment course of 2 weeks before treatment is discontinued, which is thought to minimise the risk for development of resistance and disease recurrence (Asberg et al., 2009; Chou, 2001; Sia et al., 2000). Still, some transplant centres may extend treatment (secondary prophylaxis) in patients with compartmentalised disease for as long as necessary to reduce the likelihood of recurrent CMV infection (Kotton et al., 2013). Resolving CMV disease has the long-term benefits of reducing mortality, potential allograft rejection, and the risk of bacterial, fungal, or viral opportunistic infections, among many other transplant- and non-transplant-specific effects (Arthurs et al., 2008; Fishman, 2007).

Although there is currently no cure for HIV infection, highly sensitive quantitative and qualitative real-time PCR tests targeting total HIV DNA and RNA have been used in clinical studies for both sterilisation (elimination of HIV-infected cells) and functional (controlled HIV in the absence of ART) cures (Kibirige, 2013; Lewin & Rouzioux, 2011). Improvements in real-time PCR technology may lead to profound increases in assay sensitivity and the ability to achieve single-copy detection (1 cp/mL) may lead us to a better understanding of HIV virology and what may be needed therapeutically to achieve a cure (Alidjinou, Bocket, & Hober, 2014). If therapeutic strategies are one day able to achieve an HIV cure, these highly sensitive tests will no doubt play a key role in the continuum of care for patients and, most importantly, in the confirmation of cure.

3　EXPANDING APPLICATIONS

Clinical laboratories have undergone changes to become more efficient and flexible while delivering the same high-quality results. When choosing to implement new testing, even beyond viral targets, laboratories have to consider first and foremost the performance and medical value of the test and then factors such as TAT, ease of use, and cost. Real-time PCR with its wide dynamic range, high specificity, and high sensitivity is considered the gold standard for the quantification and identification of a variety of targets including bacteria, fungi, viruses, or oncological mutations (Klein, 2002). Furthermore, the multiplexing capability of real-time PCR increases the number of targets and information gathered from the same test, further improving laboratory workflow, TAT, and costs (Deshpande & White, 2012). While novel technologies have entered clinical laboratories including mass spectrometry and next-generation sequencing, real-time PCR remains a staple and an attractive option for clinical laboratories aiming to create molecular laboratory-developed tests (LDTs). In addition, PCR can quickly be adapted to provide a robust test for the identification of emerging disease and molecular testing is now able to reach beyond the clinical laboratory and further enhance healthcare (Farrar & Wittwer, 2015; Foudeh, Didar, Veres, & Tabrizian, 2012).

3.1　PCR-BASED LABORATORY-DEVELOPED TESTS

Most molecular tests used in clinical laboratories are FDA-approved and commercially available. There are instances, however, when a test may not be available for a specific virus or the sample type and/or clinical indication used by the laboratory differs from those of the FDA-approved assay, typically leading a laboratory to design its own PCR-based test or modify existing assays. FDA defines an LDT as 'a type of *in vitro* diagnostic test that is designed, manufactured, and used within a single laboratory' and recognises that 'LDTs are important to the continued development of personalised medicine' (FDA LDT, 2014). Laboratory developed tests can be grouped into three categories, FDA-cleared or approved test that have been modified, tests that are not subject to FDA clearance or approval, and tests for which no performance specifications have been provided by the manufacturer (e.g. analyte-specific reagents or ASRs) (Burd, 2010; Code of Federal Regulations, 2009).

With alternative sample types or applications, FDA-approved tests are often modified to fit the testing needs of laboratories, including alternative collection media and sample types or expanded clinical applications. As an example, a recent gap was created in the HCV-screening algorithm for the confirmation of a positive enzyme immunoassay result following the discontinuation of the only FDA-approved confirmatory test (Alter, Kuhnert, & Finelli, 2003). In response, the CDC published recommendation for the use of FDA-approved tests detecting HCV viraemia (CDC HCV, 2013), despite the fact that most of these assays did not have specific claims for confirmatory testing; as a result, several laboratories chose to validate these assays as LDTs to meet the screening needs for HCV. Additionally, LDTs are the only option for the identification of the aetiologic agents of viral infections that can occur in transplant patients, such as EBV, adenovirus, VZV, and BK virus, that often present with non-specific clinical manifestations (Razonable, 2011) and for which FDA-approved assay options are lacking.

LDTs are an integral part of molecular laboratory testing. Whether created from the ground-up or modified from FDA-approved assays, LDTs are answering the clinicians' needs for information as an aid for diagnosis or treatment of patients. As with any clinical tests, LDTs have to meet the minimum standards set forth by CLIA prior to report patient results (Code of Federal Regulations, 2009). In July 2014, FDA informed Congress of the agency's LDT regulatory oversight framework (FDA LDT, 2014). FDA aims to address concerns over high-risk LDTs with inadequately supported claims, lack of appropriate controls, and falsification of data that may lead to inadequate treatment, possible harm to patients, and unnecessary healthcare cost. Presently, there is still a high degree of uncertainty as to what the final regulation scope will be and the possible impact on molecular laboratories will have to be seen.

3.2 EMERGING/RE-EMERGING INFECTIOUS DISEASES

Palaeopathology confirmed the truism that humanity, since its inception, has been exposed to genetic and infectious diseases with early documentation of trachoma (8000 B.C.E.), tuberculosis (7000 B.C.E.), and pneumonia (ca. 1150 B.C.E.) (Aufderheide & Rodreguez-martin, 1998; Hershkovitz et al., 2008; Roberts & Manchester, 1995; Webb, 1990). Even today, emerging infectious diseases (EIDs) continue to appear unpredictably driven by changes in human demographic, land use, and population behaviour (Lederberg, Hamburg, & Smolinski, 2003; Sehgal, 2010; Taylor, Latham, & Woolhouse, 2001). These infections can be classified as either newly emerging/a previously unknown disease or re-emerging infectious diseases/a previously known disease, that reappears after a significant reduction in incidence or elimination (Morens & Fauci, 2013).

EIDs are a threat not only to human health but also to global stability and economy. Efforts to monitor these EIDs are in place both at the global level spearheaded by the WHO and at the national level. In the United States, governmental agencies (Department of Health and Human Services, United States Agency for International development, Department of Defense) are supporting activities to detect, assess, and respond to potential outbreaks. Specifically, PCR and real-time PCR are easily adaptable to detect nucleic acid targets that are unique to each given pathogen, and as such, they play essential roles in the identification and detection of infectious pathogens and have been routinely used by health organisation agencies during epidemic outbreaks such as severe acute respiratory syndrome (SARS), H5N1, H1N1, and Ebola (Shuaib et al., 2014; WHO Influenza, 2011).

The SARS epidemic appeared in November 2002, in the Chinese province of Guangdong before reaching the adjacent Hong Kong in 2003 (WHO SARS, 2003). This SARS eventually spread to 26 countries and resulted in more than 8000 cases. In response, the CDC triggered its emergency operations centre and issued a draft genome in April 2003, 33 days after the initial WHO global alert (CDC SARS, 2013). Soon after, real-time quantitative PCR assays were described and put in use for the diagnosis of SARS (Drosten et al., 2003; Peiris et al., 2003). A host of measures were taken in order to contain this epidemic, and the molecular identification and diagnosis of the infectious agent by PCR played a key role in providing critical information to address the situation and contributed to the care of the patients infected. Additionally, the re-emerging 2014 Ebola epidemic (CDC Ebola, 2014; WHO Ebola, 2014) started in Guinea in March of 2014 before spreading to nearby West African countries and eventually reaching the United States and Europe (WHO Ebola, 2014). At the height of the epidemic, FDA issued an Emergency Use Authorisation (EUA) for the use of the first real-time RT-PCR assay (FDA EUA, 2014) and less than 4 months later, five additional real-time PCR tests were authorised under an EUA (FDA EUA, 2014) to provide an early diagnosis of the Ebola viral disease (CDC Ebola, 2014).

EIDs remain a constant and unpredictable threat to human health. The flexibility of real-time PCR technology continues to show how promptly it can be used for the detection of infectious agents. By providing a rapid diagnostic, real-time PCR can help in starting the appropriate treatment right away and maximise the chances of a positive outcome.

3.3 REAL-TIME PCR AND POINT-OF-CARE TESTING

The goal of point-of-care testing (POCT) is to quickly obtain a test result that will be used to implement the appropriate treatment for an improved clinical outcome. By definition, POCT is laboratory testing that takes place at or near the site of patients (CAP POC, 2013). The advantages of POCT are an improved TAT and result availability regardless of normal core laboratory hours, access to care in remote areas, and greater patient involvement.

The fight against AIDS largely contributed to the development of POCT devices with viral load capabilities (Hong, Studer, Hang, Anderson, & Quake, 2004; Lee et al., 2010; Marcus, Anderson, & Quake, 2006; Tanriverdi, Chen, & Chen, 2010; UNITAID, 2014; Vulto et al., 2009). Originally developed to meet the difficult conditions associated with remote places, far from any core laboratory facility often found in the developing world, the design and convenience of a portable POCT device with fast turnaround and accurate results extends the reach of healthcare. With this in mind, these POCT systems could easily be used in developed nations at hospitals, within clinics a physicians' offices, pharmacies, correction facilities, or mobile health units, to target pathogens that benefit from immediate actionable results, for which not only accurate but also quick results are critical (Kiechle & Holland, 2009). Ultimately, test menu available on these platforms will drive its implementation as a complement for the clinical laboratory core testing.

The ideal molecular POCT system that includes medical value, simplicity, fast TAT, and ruggedness remains an ongoing engineering challenge. However, the latest advances in microfluidics are a great example of the potential of these devices and brings the real-time PCR lab-on-a-chip closer to mainstream diagnostic use. This is an exciting time for molecular POCT and the upcoming years should bring new systems and perhaps a paradigm change in the world of healthcare.

4 DISCUSSION

As the needs of the clinicians, laboratory, and patients continue to evolve, so do the applications of molecular diagnostics and PCR. Over the past decade, quantitative real-time PCR technology has been increasingly phased into clinical practice and all of the potential present-day applications of real-time PCR-based methods are enumerable. They serve to advance experimental approaches within biological fields, pushing the boundaries of what we know and what we can learn, as well as to diminish empiric medical identification and management of viral diseases.

The high sensitivity of the technology has reduced risks of the most commonly transmitted transfusion illnesses and has become an integral part of managing a variety of viral infections by providing pretreatment prognostic information, therapeutic effectiveness through monitoring, and end of treatment response assessment. Quantitative real-time PCR complements serologic testing by detecting infections within the pre-seroconversion window period and infections with immunovariant viruses and are able to predict therapeutic failures sooner than traditional methods, allowing for a more timely management response. Real-time PCR assays can be rapidly developed in cases of emerging epidemic crises involving new pathogens that may result in significant health threats. The next few years are likely to see an even further increase in the expansion of the clinical applications of nucleic acid quantification, particularly following bone marrow and solid organ transplantation for which the newest standardised assays may provide an avenue for the development of consensus management guidelines for initiating pre-emptive anti-CMV treatment. Further, with the drive towards HIV eradication and complete elimination of the virus from within cells of infected patients, innovations in quantitative real-time PCR assay design will continue to push the boundaries of detection and introduce assays with progressively lower limits of detection. Thus, quantitative real-time PCR has and will facilitate advancements in the quality of diagnostics and of what we can achieve in research, medicine, and patient outcomes.

REFERENCES

AASLD/IDSA/IAS–USA. (2014). *Recommendations for testing, managing, and treating hepatitis C.* http://www.hcvguidelines.org (Accessed 10/28/14).

Aleman, S., Soderbarg, K., Visco-Comandini, U., Sitbon, G., & Sonnerborg, A. (2002). Drug resistance at low viraemia in HIV-1-infected patients with antiretroviral combination therapy. *AIDS, 16*(7), 1039–1044.

Alidjinou, E. K., Bocket, L., & Hober, D. (2014). Quantification of viral DNA during HIV-1 infection: A review of relevant clinical uses and laboratory methods. *Pathologie et Biologie, S0369–8114*(14), 00122–00129. http://dx.doi.org/10.1016/j.patbio.2014.07.007.

Alter, M. J., Kuhnert, W. L., & Finelli, L. (2003). Guidelines for laboratory testing and result reporting of antibody to hepatitis C virus. *CDC MMWR, 52*(RR03), 1–16.

Arthurs, S. K., Eid, A. J., Pedersen, R. A., Kremers, W. K., Cosio, F. G., Patel, R., et al. (2008). Delayed-onset primary cytomegalovirus disease and the risk of allograft failure and mortality after kidney transplantation. *Clinical Infectious Diseases, 46*, 840–846. http://dx.doi.org/10.1086/528718.

Asberg, A., Humar, A., Jardine, A. G., Rollag, H., Pescovitz, M. D., Mouas, H., et al. (2009). Long-term outcomes of CMV disease treatment with valganciclovir versus IV ganciclovir in solid organ transplant recipients. *American Journal of Transplantation, 9*, 1205. http://dx.doi.org/10.1111/j.1600-6143.2009.02617.x.

Atabani, S. F., Smith, C., Atkinson, C., Aldridge, R. W., Rodriguez-Perálvarez, M., Rolando, N., et al. (2012). Cytomegalovirus replication kinetics in solid organ transplant recipients managed by preemptive therapy. *American Journal of Transplantation, 12*, 2457–2464. http://dx.doi.org/10.1111/j.1600-6143.2012.04087.x.

Au, J. S., & Pockros, P. J. (2014). Novel therapeutic approaches for hepatitis C. *Clinical Pharmacology and Therapeutics, 95*(1), 78–88.

Aufderheide, A. C., & Rodreguez-martin, C. (1998). *The Cambridge encyclopedia of human paleopathology.* Cambridge University Press.

Branson, B. M. (2010). The future of HIV testing. *Journal of Acquired Immune Deficiency Syndromes, 55*(Suppl. 2), 102–105. http://dx.doi.org/10.1097/QAI.0b013e3181fbca44.

Branson, B., Owen, S., Wesolowski, L., Bennett, B., Werner, B., Wroblewski, K., et al. (2014). Centers for Disease Control and Prevention and Association of Public Health Laboratories. *Laboratory testing for the diagnosis of HIV infection: Updated recommendations.* Available at http://stacks.cdc.gov/view/cdc/23447. Published June 27, 2014. Accessed 22.10.14.

Burd, E. M. (2010). Validation of laboratory-developed molecular assays for infectious diseases. *Clinical Microbiology Reviews, 23*, 550–576. http://dx.doi.org/10.1128/CMR.00074-09.

Caliendo, A. M., Shahbazianm, M. D., Schaper, C., Ingersoll, J., Abdul-Ali, D., Boonyaratanakornkit, J., et al. (2009). A commutable cytomegalovirus calibrator is required to improve the agreement of viral load values between laboratories. *Clinical Chemistry, 55*(9), 1701–1710. http://dx.doi.org/10.1373/clinchem.2009.124743.

CAP POC. (2013). *Point of care testing.* http://www.cap.org/apps//cap.portal?_nfpb=true&cntvwrPtlt_actionOverride=%2Fportlets%2FcontentViewer%2Fshow&_windowLabel=cntvwrPtlt&cntvwrPtlt%7BactionForm.contentReference%7D=policies%2Fpolicy_appII.html&_state=maximized&_pageLabel=cntvwr (Accessed 11/12/14).

Castro, P., Plana, M., González, R., López, A., Vilella, A., Nicolas, J. M., et al. (2013). Influence of episodes of intermittent viremia ("blips") on immune responses and viral load rebound in successfully treated HIV-infected patients. *AIDS Research and Human Retroviruses*, *29*(1), 68–76.

CDC Ebola. (2014). *Centers for disease control and prevention. Ebola virus disease diagnosis.* http://www.cdc.gov/vhf/ebola/diagnosis/index.html(Accessed 11/12/14).

CDC HCV. (2012). Centers for disease control and prevention. Recommendations for the identification of chronic hepatitis C virus infection among persons born during 1945–1965. *MMWR: Morbidity and Mortality Weekly Report*, *61*(4), 1–32.

CDC HCV. (2013). Centers for disease control and prevention. Testing for HCV infection: An update of guidance for clinicians and laboratorians. *MMWR: Morbidity and Mortality Weekly Report*, *62*, 1–4.

CDC Influenza. (2014). *Centers for disease control and prevention. Guidance for clinicians on the use of RT-PCR and other molecular assays for diagnosis of influenza virus infection.* http://www.cdc.gov/flu/professionals/diagnosis/molecular-assays.htm(Accessed 11/3/2014).

CDC SARS. (2013). *Centers for disease control and prevention. SARS response timeline.* http://www.cdc.gov/about/history/sars/timeline.htm(Accessed 11/12/14).

Chen, G., Lin, W., Shen, F., Iloeje, U. H., London, W. T., & Evans, A. A. (2006). Past HBV viral load as predictor of mortality and morbidity from HCC and chronic liver disease in a prospective study. *The American Journal of Gastroenterology*, *101*, 1797–1803. http://dx.doi.org/10.1111/j.1572-0241.2006.00647.x.

Chen, C. J., Yang, H. I., Su, J., Jen, C. L., You, S. L., Lu, S. N., et al. (2006). Risk of hepatocellular carcinoma across a biological gradient of serum hepatitis B virus DNA level. *JAMA*, *295*, 65–73.

Chou, S. W. (2001). Cytomegalovirus drug resistance and clinical implications. *Transplant Infectious Disease*, *3*(2), 20–24. http://dx.doi.org/10.1034/j.1399-3062.2001.00004.x.

Chu, C., & Selwyn, P. A. (2010). Diagnosis and initial management of acute HIV infection. *American Family Physician*, *81*(10), 1239–1244.

Cincinnati Children's Hospital Medical Center. (2012). *Evidence based clinical practice guideline for management of EBV-associated post-transplant lymphoproliferative disease (PTLD) in solid organ transplant.* http://www.guideline.gov/content.aspx?id=38418 (Accessed 10/10/2014).

CLSI EP17-A. (2004). *Protocols for determination of limits of detection and limits of quantitation: Approved Guideline.* Pennsylvania: Clinical and Laboratory Standards Institute1-56238-551-8.

Cobb, B. R., Vaks, J. E., Do, T., & Vilchez, R. A. (2011). Evolution in the sensitivity of quantitative HIV-1 viral load tests. *Journal of Clinical Virology*, *52*(1), 77–82. http://dx.doi.org/10.1016/j.jcv.2011.09.015.

Code of Federal Regulations. (2009). *Title 42. Public health, vol. 4, chapter V. Health care financing administration, department of health and human services, part 493. Laboratory requirements, section 493.1253. Standard: Establishment and verification of performance specifications.* Washington, DC: U.S. Government Printing Office.

Cohen, M. S., Chen, Y. Q., McCauley, M., Gamble, T., Hosseinipour, M. C., Kumarasamy, N., et al. (2011). Prevention of HIV-1 infection with early antiretroviral therapy. *The New England Journal of Medicine*, *365*, 493–505. http://dx.doi.org/10.1056/NEJMoa1105243.

Couzi, L., Helou, S., Bachelet, T., Moreau, K., Martin, S., Morel, D., et al. (2012). High incidence of anticytomegalovirus drug resistance among D+R- kidney transplant recipients receiving preemptive therapy. *American Journal of Transplantation*, *1*, 202–209. http://dx.doi.org/10.1111/j.1600-6143.2011.03766.x.

Deeks, S. G., Wrin, T., Liegler, T., Hoh, R., Hayden, M., Barbour, J. D., et al. (2001). Virologic and immunologic consequences of discontinuing combination antiretroviral-drug therapy in HIV-infected patients with detectable viremia. *The New England Journal of Medicine*, *344*(7), 472–480.

Deshpande, A., & White, P. S. (2012). Multiplexed nucleic acid-based assay for molecular diagnostics of human disease. *Expert Review of Molecular Diagnostics*, *12*(6), 645–659. http://dx.doi.org/10.1586/erm.12.60.

DHHS HIV. (2014). *Panel on antiretroviral guidelines for adults and adolescents. Guidelines for the use of antiretroviral agents in HIV-1-infected adults and adolescents.* Department of Health and Human Services. http://aidsinfo.nih.gov/contentfiles/lvguidelines/AdultandAdolescentGL.pdf Accessed 28.10.14.

DiBiasi, R. L., & Tyler, K. L. (2004). Molecular methods for diagnosis of viral encephalitis. *Clinical Microbiology Reviews*, *17*(4), 903–925.

Diffenbach, C. W. (2012). Preventing HIV transmission through antiretroviral treatment-mediated virologic suppression: Aspects of an emerging scientific agenda. *Current Opinion in HIV and AIDS*, *7*(2), 106–110. http://dx.doi.org/10.1097/COH.0b013e32834f3f13.

Diviacco, S., Norio, P., Zentilin, L., Menzo, S., Clementi, M., Biamonti, G., et al. (1992). A novel procedure for quantitative polymerase chain reaction by coamplification of competitive templates. *Gene*, *122*, 3013–3020.

Domingues, R. B., Lakeman, F. D., Mayo, M. S., & Whitley, R. J. (1998). Application of competitive PCR to cerebrospinal fluid samples from patients with herpes simplex encephalitis. *Journal of Clinical Microbiology*, *36*, 2229–2234.

Doyle, T., Smith, C., Vitiello, P., Cambiano, V., Johnson, M., Owen, A., et al. (2012). Plasma HIV-1 RNA detection below 50 copies/ml and risk of virologic rebound in patients receiving highly active antiretroviral therapy. *Clinical Infectious Diseases*, *54*(5), 724–732. http://dx.doi.org/10.1093/cid/cir936.

Drosten, C., Günther, S., Preiser, W., van der Werf, S., Brodt, H. R., Becker, S., et al. (2003). Identification of a novel coronavirus in patients with severe acute respiratory syndrome. *The New England Journal of Medicine*, *348*(20), 1967–1976.

Emery, V. C., Sabin, C. A., Cope, A. V., Gor, D., Hassan-Walker, A. F., & Griffiths, P. D. (2000). Application of viral-load kinetics to identify patients who develop cytomegalovirus disease after transplantation. *The Lancet*, *355*(9220), 2032–2036.

Emmadi, R., Boonyarantanakornkit, J. B., Selvarangan, R., Venkatakrishna, S., Zimmer, B. L., Williams, L., et al. (2011). Molecular methods and platforms for infectious disease testing: A review of FDA-approved and cleared assays. *The Journal of Molecular Diagnostics*, *13*, 583–604. http://dx.doi.org/10.1016/j.jmoldx.2011.05.011.

Eron, J. J., Cooper, D. A., Steigbigel, R. T., Clotet, B., Gatell, J. M., Kumar, P. N., et al. (2013). Efficacy and safety of raltegravir for treatment of HIV for 5 years in the BENCHMRK studies: Final results of two randomised, placebo-controlled trials. *The Lancet Infectious Diseases*, *13*(7), 587–596. http://dx.doi.org/10.1016/S1473-3099(13)70093-8.

Espy, M., Uhl, J., Sloan, L., Buckwalter, S., Jones, M., Vetter, E., et al. (2006). Real-time PCR in clinical microbiology: Applications for routine laboratory testing. *Clinical Microbiology Reviews*, *19*, 165–256.

Estevez, M. A., Porcuna, N. C., Suay, V. G., Monge, A. P., Garcia, F. G., Medina, L. M., et al. (2013). Quantification of viral loads lower than 50 copies per milliliter by use of the Cobas AmpliPrep/Cobas TaqMan HIV-1 test, version 2.0, can predict the likelihood of subsequent virological rebound to >50 copies per milliliter. *Journal of Clinical Microbiology*, *51*(5), 1555–1557. http://dx.doi.org/10.1128/JCM.00100-13.

Evers, P. D. (2013). Pre-emptive virology screening in the pediatric hematopoietic stem cell transplant population: A cost effective analysis. *Hematology/Oncology and Stem Cell Therapy*, *6*(3–4), 81–88.

Farrar, J. S., & Wittwer, C. T. (2015). Extreme PCR: Efficient and specific DNA amplification in 15–60 seconds. *Clinical Chemistry*, *61*(1), 145–153. http://dx.doi.org/10.1373/clinchem.2014.228304.

FDA Assays. (2014). *Complete list of donor screening assays for infectious agents and HIV diagnostic assays.* http://www.fda.gov/BiologicsBloodVaccines/BloodBloodProducts/ApprovedProducts/LicensedProductsBLAs/BloodDonorScreening/InfectiousDisease/UCM080466(Accessed 10/08/2014).

FDA EUA. (2014). *Emergency use authorizations.* http://www.fda.gov/medicaldevices/safety/emergencysituations/ucm161496.htm(Accessed 11/12/14).

FDA HCV. (2013). *Guidance for industry chronic hepatitis C virus infection: Developing direct-acting antiviral drugs for treatment, draft guidance.* October Revision 1. http://www.fda.gov/downloads/Drugs/GuidanceComplianceRegulatoryInformation/Guidances/UCM225333.pdf(Accessed 10/20/2014).

FDA LDT. (2014). *FDA laboratory developed tests.* http://www.fda.gov/MedicalDevices/ProductsandMedicalProcedures/InVitroDiagnostics/ucm407296.htm(Accessed 11/11/2014).

FDA Testing. (2014). *FDA testing HCT/P donors for relevant communicable disease agents and diseases.* http://www.fda.gov/BiologicsBloodVaccines/SafetyAvailability/TissueSafety/ucm095440.htm (Accessed 10/08/2014).

Feld, J. J., Kowdley, K. V., Coakley, E., Sigal, S., Nelson, D. R., Crawford, D., et al. (2014). Treatment of HCV with ABT-450/r-ombitasvir and dasabuvir with ribavirin. *The New England Journal of Medicine*, *370*, 1594–1603. http://dx.doi.org/10.1056/NEJMoa1315722.

Fishman, J. A. (2007). Infection in solid-organ transplant recipients. *The New England Journal of Medicine*, *357*(25), 2601–2614. http://dx.doi.org/10.1056/NEJMra064928.

Foudeh, A. M., Didar, T. F., Veres, T., & Tabrizian, M. (2012). Microfluidic designs and techniques using lab-on-a-chip devices for pathogen detection for point-of-care diagnostics. *Lab on a Chip*, *12*, 3249–3266. http://dx.doi.org/10.1039/c2lc40630f.

Furtado, M. R., Kingsley, L. A., & Wolinsky, S. M. (1995). Changes in the viral mRNA expression pattern correlate with a rapid rate of CD4+ T-cell number decline in human immunodeficiency virus type 1-infected individuals. *Journal of Virology*, *69*(4), 2092–2100.

Fusfeld, L., Aggarwal, J., Dougher, C., Vera-Llonch, M., Bubb, S., Donepudi, M., et al. (2013). Assessment of motivating factors associated with the initiation and completion of treatment for chronic hepatitis C virus (HCV) infection. *BMC Infectious Diseases*, *13*, 234. http://dx.doi.org/10.1186/1471-2334-13-234.

Gatanaga, H., Tsukada, K., Honda, H., Tanuma, J., Yazaki, H., Watanabe, T., et al. (2009). Detection of HIV type 1 load by the Roche Cobas TaqMan assay in patients with viral loads previously undetectable by the Roche Cobas Amplicor Monitor. *Clinical Infectious Diseases*, *48*(2), 260–262. http://dx.doi.org/10.1086/595707.

Ghany, M., Strader, D., Thomas, D., & Seeff, L. (2009). AASLD practice guidelines, diagnosis, management, and treatment of hepatitis C: An update. *Hepatology*, *49*(4), 1335–1374. http://dx.doi.org/10.1002/hep.22759.

Gilliland, G., Perrin, S., & Bunn, H. F. (1990). Competitive PCR for quantitation of mRNA. In M. A. Innis, D. H. Gelfand, J. J. Sninsky, & T. J. White (Eds.), *PCR protocols. A guide to methods and applications* (pp. 60–69). San Diego: Academic Press.

Glaubitz, J., Sizmann, D., Simon, C. O., Hoffmann, K. S., Drogan, D., Hesse, M., et al. (2011). Accuracy to 2nd International HIV-1 RNA WHO Standard: Assessment of three generations of quantitative HIV-1 RNA nucleic acid amplification tests. *Journal of Clinical Virology*, *50*, 119–124. http://dx.doi.org/10.1016/j.jcv.2010.10.017.

Gleaves, C. A., Smith, T. F., Shuster, E. A., & Pearson, G. R. (1985). Comparison of standard tube and shell vial cell culture techniques for the detection of cytomegalovirus in culture specimens. *Journal of Clinical Microbiology*, *21*(2), 217–221.

Gustafson, I., Lindblom, A., Yun, Z., Omar, H., Engstrom, L., Lewensohn-Fuchs, I., et al. (2008). Quantification of adenovirus DNA in unrelated donor hematopoietic stem cell transplant recipients. *Journal of Clinical Virology*, *43*(1), 79–85. http://dx.doi.org/10.1016/j.jcv.2008.04.014.

HARVONI® (sofosbuvir and ledipasvir) [Full Prescribing Information]. Gilead Sciences, Foster City, CA. Revised October 2014.

Havlir, D. V., Bassett, R., Levitan, D., Gilbert, P., Tebas, P., Collier, A. C., et al. (2001). Prevalence and predictive value of intermittent viremia with combination HIV therapy. *JAMA*, *286*(2), 171–179.

Hayden, R. T., Hokanson, K. M., Pounds, S. B., Bankowski, M. J., Belzer, S. W., Carr, L., et al. (2008). Multicenter comparison of different real-time PCR assays for quantitative detection of Epstein–Barr virus. *Journal of Clinical Microbiology*, *46*(1), 157–163.

Hayden, R. T., Yan, X., Rodriguez, A. B., Xiong, X., Ginocchio, C. C., Mitchell, M. J., et al. (2012). Factors contributing to variability of quantitative viral PCR results in proficiency testing samples: A multivariate analysis. *Journal of Clinical Microbiology*, *50*(2), 337–345. http://dx.doi.org/10.1128/JCM.01287-11.

Heid, C., Stevens, J., Livak, K., & Williams, P. (1996). Real time quantitative PCR. *Genome Research*, *6*, 986–994.

Hep C Online. (2014). Medications to treat HCV. University of Washington, In D. H. Spach & N. N. Kim (Eds.), Hepatitis C Online (Accessed 11/2014). http://www.hepatitisc.uw.edu/page/treatment/drugs.

Herman, C. R., Gill, H. K., Eng, J., & Fajardo, L. L. (2002). Screening for preclinical disease: Test and disease characteristics. *American Journal of Roentgenology*, *179*(4), 825–831.

Hershkovitz, I., Donoghue, H. D., Minnikin, D. E., Besra, G. S., Lee, O. Y., Gernaey, A. M., et al. (2008). Detection and molecular characterization of 9,000-year-old Mycobacterium tuberculosis from a Neolithic settlement in the Eastern Mediterranean. *PLoS One*, *3*(10), e3426. http://dx.doi.org/10.1371/journal.pone.0003426, 1–6.

Hirsch, H. H., Brennan, D. C., Drachenberg, C. B., Ginevri, F., Gordon, J., Limaye, A. P., et al. (2005). Polyomavirus-associated nephropathy in renal transplantation: Interdisciplinary analyses and recommendations. *Transplantation*, *79*, 1277–1286.

Ho, D. D., Moudgil, T., & Alam, M. (1989). Quantitation of human immunodeficiency virus type 1 in the blood of infected persons. *The New England Journal of Medicine*, *321*(24), 1621–1625.

Holland, P. M., Abramson, R. D., Watson, R., & Gelfand, D. H. (1991). Detection of specific polymerase chain reaction product by utilizing the $5'–3'$ exonuclease activity of Thermus aquaticus. *Proceedings of the National Academy of Sciences of the United States of America*, *88*(16), 7276–7280.

Hong, J. W., Studer, V., Hang, G., Anderson, W. F., & Quake, S. R. (2004). A nanoliter-scale nucleic acid processor with parallel architecture. *Nature Biotechnology*, *22*(4), 435–439. http://dx.doi.org/10.1038/nbt951.

Hufert, F. T., von Laer, D., Fenner, T. E., Schwander, S., Kern, P., & Schmitz, H. (1991). Progression of HIV-1 infection. Monitoring of HIV-1 DNA in peripheral blood mononuclear cells by PCR. *Archives of Virology*, *120*(3–4), 233–240.

Humar, A., Gregson, D., Caliendo, A. M., McGeer, A., Malkan, G., Krajden, M., et al. (1999). Clinical utility of quantitative cytomegalovirus viral load determination for predicting cytomegalovirus disease in liver transplant recipients. *Transplantation*, *68*(9), 1305–1311.

Humar, A., Kumar, D., Boivin, G., & Caliendo, A. M. (2002). Cytomegalovirus (CMV) virus load kinetics to predict recurrent disease in solid-organ transplant patients with CMV disease. *The Journal of Infectious Diseases*, *186*(6), 829–833.

Humar, A., & Snydman, D. (2009). AST infectious diseases community of practice. Cytomegalovirus in solid organ transplant recipients. *American Journal of Transplantation*, *9*(4), S78–S86.

Iloeje, U. H., Yang, H. I., Su, J., Jen, C. L., You, S. L., & Chen, C. J. (2006). Predicting cirrhosis risk based on the level of circulating hepatitis B viral load. *Gastroenterology*, *130*(3), 678–686.

Jensen, D. M., Morgan, T. R., Marcellin, P., Pockros, P. J., Reddy, K. R., Hadziyannis, S. J., et al. (2006). Early identification of HCV genotype 1 patients responding to 24 weeks peginterferon alpha-2a (40 kd)/ribavirin therapy. *Hepatology*, *43*(5), 954–960.

Kadmon, G., Levy, I., Mandelboim, M., Nahum, E., Stein, J., Dovrat, S., et al. (2013). Polymerase-chain-reaction-based diagnosis of viral pulmonary infections in immunocompromised children. *Acta Paediatrica*, *102*(6), e263–e268. http://dx.doi.org/10.1111/apa.12207.

Kalil, A. C., Levitsky, J., Lyden, E., Stoner, J., & Freifeld, A. G. (2005). Meta-analysis: The efficacy of strategies to prevent organ disease by cytomegalovirus in solid organ transplant recipients. *Annals of Internal Medicine*, *143*(12), 870–880.

Karlsson, A. C., Younger, S. R., Martin, J. N., Grossman, Z., Sinclair, E., Hunt, P. W., et al. (2004). Immunologic and virologic evolution during periods of intermittent and persistent low-level viremia. *AIDS*, *18*(7), 981–989.

KDIGO. (2009). Kidney Disease: Improving Global Outcomes Transplant Work Group. KDIGO clinical practice guideline for the care of kidney transplant. *American Journal of Transplantation*, *9*(Suppl. 3), S1–S155. http://dx.doi.org/10.1111/j.1600-6143.2009.02834.x.

Khoury, J. A., Storch, G. A., Bohl, D. L., Schuessler, R. M., Torrence, S. M., Lockwood, M., et al. (2006). Prophylactic versus preemptive oral valganciclovir for the management of cytomegalovirus infection in adult renal transplant recipients. *American Journal of Transplantation*, *6*(9), 2134–2143.

Kibirige, C. (2013). The use of ultra-sensitive molecular assays in HIV cure-related research. *Journal of AIDS & Clinical Research*, (Suppl. 6), 1–9. http://dx.doi.org/10.4172/2155-6113.S6-002.

Kiechle, F. L., & Holland, C. A. (2009). Point-of-care testing and molecular diagnostics: Miniaturization required. *Clinics in Laboratory Medicine*, *29*(3), 555–560. http://dx.doi.org/10.1016/j.cll.2009.06.013.

Kieffer, T. L., Finucane, M. M., Nettles, R. E., Quinn, T. C., Broman, K. W., Ray, S. C., et al. (2004). Genotypic analysis of HIV-1 drug resistance at the limit of detection: Virus production without evolution in treated adults with undetectable HIV loads. *The Journal of Infectious Diseases*, *189*(8), 1452–1465.

Klein, D. (2002). Quantification using real-time PCR technology: Applications and limitations. *Trends in Molecular Medicine, 8*(6), 257–260.

Kotton, C. N., Kumar, D., Caliendo, A. M., Asberg, A., Chou, S., Danzinger-Isakov, L., et al. (2013). Updated international consensus guidelines on the management of cytomegalovirus in solid-organ transplantation. *Transplantation, 96*, 333–360. http://dx.doi.org/10.1097/TP.0b013e31829df29d.

Kowdley, K. V., Gordon, S. C., Reddy, K. R., Rossaro, L., Bernstein, D. E., Lawitz, E., et al. (2014). Ledipasvir and sofosbuvir for 8 or 12 weeks for chronic HCV without cirrhosis. *The New England Journal of Medicine, 370*(20), 1879–1888. http://dx.doi.org/10.1056/NEJMoa1402355.

Kraft, C. S., Armstrong, W. S., & Caliendo, A. M. (2012). Interpreting quantitative cytomegalovirus DNA testing: Understanding the laboratory perspective. *Clinical Infectious Diseases, 54*(12), 1793–1797. http://dx.doi.org/10.1093/cid/cis212.

Krishna, N. K., & Cunnion, K. J. (2012). Role of molecular diagnostics in the management of infectious disease emergencies. *The Medical Clinics of North America, 96*(6), 1067–1078. http://dx.doi.org/10.1016/j.mcna.2012.08.005.

Laprise, C., de Pokomandy, A., Baril, J. G., Dufresne, S., & Trottier, H. (2013). Virologic failure following persistent low-level viremia in a cohort of HIV-positive patients: Results from 12 years of observation. *Clinical Infectious Diseases, 57*(10), 1489–1496. http://dx.doi.org/10.1093/cid/cit529.

Lawitz, E., Mangia, A., Wyles, D., Rodriguez-Torres, M., Hassanein, T., Gordon, S. C., et al. (2013). Sofosbuvir for previously untreated chronic hepatitis C infection. *The New England Journal of Medicine, 368*, 1878–1887. http://dx.doi.org/10.1056/NEJMoa1214853.

Lawrence, J., Mayers, D. L., Hullsiek, K. H., Collins, G., Abrams, D. I., Reisler, R. B., et al. (2003). Structured treatment interruption in patients with multidrug-resistant human immunodeficiency virus. *The New England Journal of Medicine, 349*(9), 837–846. http://dx.doi.org/10.1056/NEJMoa035103.

Lederberg, J., Hamburg, M. A., & Smolinski, M. S. (Eds.), (2003). *Microbial threats to health: Emergence, detection, and response.* Washington, DC: National Academies Press.

Lee, H. H., Dineva, M. A., Chua, Y. L., Ritchie, A., Ushiro-Lumb, I., & Wisniewski, C. A. (2010). Simple amplification-based assay: A nucleic acid-based point-of-care platform for HIV-1 testing. *Journal of Infectious Diseases, 201*(Suppl. 1), S65–S71. http://dx.doi.org/10.1086/650385.

Lee, P. K., Kieffer, T. L., Siliciano, R. F., & Nettles, R. E. (2006). HIV-1 viral load blips are of limited clinical significance. *The Journal of Antimicrobial Chemotherapy, 57*(5), 803–805.

Lewin, S. R., & Rouzioux, C. (2011). HIV cure and eradication: How will we get from the laboratory to effective clinical trials? *AIDS, 25*(7), 885–897. http://dx.doi.org/10.1097/QAD.0b013e3283467041.

Lima, V., Harrigan, R., & Montaner, J. S. (2009). Increased reporting of detectable plasma HIV-1 RNA levels at the critical threshold of 50 copies per milliliter with the Taqman assay in comparison to the Amplicor assay. *Journal of Acquired Immune Deficiency Syndromes, 51*(1), 3–6.

Lindback, S., Karlsson, A. C., Mittler, J., Blaxhult, A., Carlsson, M., Briheim, G., et al. (2000). Viral dynamics in primary HIV-1 infection. Karolinska institutet primary HIV infection study group. *AIDS, 14*(15), 2283–2291.

Livak, K., & Schmittgen, T. (2001). Analysis of relative gene expression data using real-time quantitative PCR and the $2^{-\Delta\Delta CT}$ method. *Methods, 25*(4), 402–408.

Lok, A. S. F., & McMahon, B. J. (2009). Chronic hepatitis B: Update 2009. *Hepatology, 50*(3), 661–662. http://dx.doi.org/10.1002/hep.23190.

Lowance, D., Neumayer, H. H., Legendre, C. M., Squifflet, J. P., Kovarik, J., Brennan, P. J., et al. (1999). International valacyclovir cytomegalovirus prophylaxis transplantation study group: Valacyclovir for the prevention of cytomegalovirus disease after renal transplantation. *The New England Journal of Medicine, 340*, 1462–1470.

MacKay, I. M., Arden, K. E., & Nitsche, A. (2002). Survey and summary: Real-time PCR in virology. *Nucleic Acids Research, 30*(6), 1292–1305.

Manns, M. P., Pockros, P. J., Norkrans, G., Smith, C. I., Morgan, T. R., Häussinger, D., et al. (2013). Long-term clearance of hepatitis C virus following interferon alpha-2b or peginterferon alpha-2b, alone or in combination with ribavirin. *Journal of Viral Hepatitis, 20*(8), 524–529. http://dx.doi.org/10.1111/jvh.12074.

Marcus, J. S., Anderson, W. F., & Quake, S. R. (2006). Parallel picoliter RT-PCR assays using microfluidics. *Analytical Chemistry, 78*(3), 956–958.

Marschner, I. C., Collier, A. C., Coombs, R. W., D'Aquila, R. T., DeGruttola, V., Fischl, M. A., et al. (1998). Use of changes in plasma levels of human immunodeficiency virus type 1 RNA to assess the clinical benefit of antiretroviral therapy. *The Journal of Infectious Diseases, 177*(1), 40–47.

Mathez, D., Paul, D., de Belilovsky, C., Sultan, Y., Deleuze, J., Gorin, I., et al. (1990). Productive human immunodeficiency virus infection levels correlate with AIDS-related manifestations in the patient. *Proceedings of the National Academy of Sciences of the United States of America, 87*(19), 7438–7442. http://dx.doi.org/10.1073/pnas.87.19.7438.

Mellors, J. W., Kingsley, L. A., Rinaldo, C. R., Todd, J. A., Hoo, B. S., Kokka, R. P., et al. (1995). Quantitation of HIV-1 RNA in plasma predicts outcome after seroconversion. *Annals of Internal Medicine, 122*(8), 573–579.

Miller, W. G., Myers, G. L., Gantzer, M. L., Kahn, S. E., Schonbrunner, E. R., Theinpont, L. M., et al. (2011). Roadmap for harmonization of clinical laboratory measurement procedures. *Clinical Chemistry, 57*(8), 1108–1117. http://dx.doi.org/10.1373/clinchem.2011.164012.

Morens, D. M., & Fauci, A. S. (2013). Emerging infectious diseases: Threats to human health and global stability. *PLoS Pathogens, 9*(7), e1003467. http://dx.doi.org/10.1371/journal.ppat.1003467.

Mullis, K. B., & Faloona, F. A. (1987). Specific synthesis of DNA in vitro via a polymerase-catalyzed chain reaction. *Methods in Enzymology, 155*, 335–350.

Murray, J. S., Elashoff, M. R., Iacono-Connors, L. C., Cvetkovich, T. A., & Struble, K. A. (1999). The use of plasma HIV RNA as a study endpoint in efficacy trials of antiretroviral drugs. *AIDS, 13*(7), 797–804.

Nettles, R. E., Kieffer, T. L., Kwon, P., Monie, D., Han, Y., Parsons, T., et al. (2005). Intermittent HIV-1 viremia (Blips) and drug resistance in patients receiving HAART. *JAMA, 293*(7), 817–829.

Nicholson, J. K., Spira, T. J., Aloisio, C. H., Jones, B. M., Kennedy, M. S., Holman, R. C., et al. (1989). Serial determinations of HIV-1 titers in HIV-infected homosexual men: Association of rising titers with CD4 T cell depletion and progression to AIDS. *AIDS Research and Human Retroviruses, 5*(2), 205–215.

OLYSIO® (simeprevir) [Full Prescribing Information]. Janssen Therapeutics, Titusville, NJ. Revised November 2014.

Pang, X. L., Fox, J. D., Fenton, J. M., Miller, G. G., Caliendo, A. M., & Preiksaaitis, J. K. (2009). Interlaboratory comparison of cytomegalovirus viral load assays. *American Journal of Transplantation*, *9*(2), 256–268. http://dx.doi.org/10.1111/j.1600-6143.2008.02513.x.

Park, C. L., Streicher, H., & Rothberg, R. (1987). Transmission of human immunodeficiency virus from parents to only one dizygotic twin. *Journal of Clinical Microbiology*, *25*(6), 1119–1121.

Pascual-Pareja, J. F., Martínez-Prats, L., Luczkowiak, J., Fiorante, S., Rubio, R., Pulido, F., et al. (2010). Detection of HIV-1 at between 20 and 49 copies per milliliter by the Cobas TaqMan HIV-1 v2.0 assay is associated with higher pretherapy viral load and less time on antiretroviral therapy. *Journal of Clinical Microbiology*, *48*(5), 1911–1912. http://dx.doi.org/10.1128/JCM.02388-09.

Pawlotsky, J.-M. (2012). Measuring HCV RNA and assessing virologic response. In M. L. Shiffman (Ed.), *Chronic hepatitis C virus: Advances in treatment, promise for the future* (pp. 149–157). New York: Springer.

Paya, C., Humar, A., Dominguez, E., Washburn, K., Blumberg, E., Alexander, B., et al. (2004). Valganciclovir solid organ transplant study group: Efficacy and safety of valganciclovir vs. oral ganciclovir for prevention of cytomegalovirus disease in solid organ transplant recipients. *American Journal of Transplantation*, *4*(4), 611–620.

Pearlman, B. L., & Ehleben, C. (2014). Hepatitis C genotype 1 virus with low viral load and rapid virologic response to peginterferon/ribavirin obviates a protease inhibitor. *Hepatology*, *59*(1), 71–77. http://dx.doi.org/10.1002/hep.26624.

Pearlman, B. L., & Traub, N. (2011). Sustained virologic response to antiviral therapy for chronic hepatitis C virus infection: A cure and so much more. *Clinical Infectious Diseases*, *52*(7), 889–900. http://dx.doi.org/10.1093/cid/cir076.

Peiris, J. S., Lai, S. T., Poon, L. L., Guan, Y., Yam, L. Y., Lim, W., et al. (2003). Coronavirus as a possible cause of severe acute respiratory syndrome. *The Lancet*, *361*(9366), 1319–1325. http://dx.doi.org/10.1016/S0140-6736(03)13077-2.

Preiksaitis, J. K., Pang, X. L., Fox, J. D., Fenton, J. M., Caliendo, A. M., Miller, G. G., et al. (2009). Interlaboratory comparison of Epstein–Barr virus viral load assays. *American Journal of Transplantation*, *9*(2), 269–279. http://dx.doi.org/10.1111/j.1600-6143.2008.02514.x.

Priddy, F. H., Pilcher, C. D., Moore, R. H., Tambe, P., Park, M. N., Fiscus, S. A., et al. (2007). Detection of acute HIV infections in an urban HIV counseling and testing population in the United States. *Journal of Acquired Immune Deficiency Syndromes*, *44*(2), 196–202.

Prober, C. G., Sullender, W. M., Yasukawa, L. L., Au, D. S., Yeager, A. S., & Arvin, A. M. (1987). Low risk of herpes simplex virus infections in neonates exposed to the virus at the time of vaginal delivery to mothers with recurrent genital herpes simplex virus infections. *The New England Journal of Medicine*, *316*(5), 240–244.

Randhawa, P., Ho, A., Shapiro, R., Vats, A., Swalsky, P., Finkelstein, S., et al. (2004). Correlates of quantitative measurement of BK polyomavirus (BKV) DNA with clinical course of BKV infection in renal transplant patients. *Journal of Clinical Microbiology*, *42*(3), 1176–1180. http://dx.doi.org/10.1128/JCM.42.3.1176-1180.2004.

Razonable, R. R. (2008). Cytomeglavirus infection after liver transplantation: Current concepts and challenges. *World Journal of Gastroenterology*, *14*(31), 4849–4860.

Razonable, R. R. (2011). Management of viral infections in solid organ transplant recipients. *Expert Review of Anti-Infective Therapy*, *9*(6), 685–700. http://dx.doi.org/10.1586/eri.11.43.

Razonable, R. R., Åsberg, A., Rollag, H., Duncan, J., Biosvert, D., Yao, J. D., et al. (2013). Virologic suppression measured by a cytomegalovirus (CMV) DNA test calibrated to the World Health Organization international standard is predictive of CMV disease resolution in transplant recipients. *Clinical Infectious Diseases*, *56*(11), 1546–1553. http://dx.doi.org/10.1093/cid/cit096.

Reischig, T. (2010). Cytomegalovirus-associated renal allograft rejection: New challenges for antiviral preventive strategies. *Expert Review of Anti-Infective Therapy*, *8*(1), 903–910. http://dx.doi.org/10.111/j.1600-6143.2007.02031.x.

Reischig, T., Hribova, P., Jindra, P., Hes, O., Bouda, M., Treska, V., et al. (2012). Long-term outcomes of pre-emptive valganciclovir compared with valacyclovir prophylaxis for prevention of cytomegalovirus in renal transplantation. *Journal of the American Society of Nephrology*, *23*(9), 1588–1597. http://dx.doi.org/10.1681/ASN.2012010100.

Reischig, T., Jindra, P., Hes, O., Svecová, M., Klaboch, J., & Treska, V. (2008). Valacyclovir prophylaxis versus preemptive valganciclovir therapy to prevent cytomegalovirus disease after renal transplantation. *American Journal of Transplantation*, *8*, 69–77.

Roberts, C. A., & Manchester, K. (1995). *The archaeology of disease*. Ithaca, Gloucester: Cornell University Press, Sutton Publishing.

Sagedal, S., Hartmann, A., Nordal, K. P., Osnes, K., Leivestad, T., Foss, A., et al. (2004). Impact of early cytomegalovirus infection and disease on long-term recipient and kidney graft survival. *Kidney International*, *66*(1), 329–337.

Sagedal, S., Nordal, K. P., Hartmann, A., Sund, S., Scott, H., Degré, M., et al. (2002). The impact of cytomegalovirus infection and disease on rejection episodes in renal allograft recipients. *American Journal of Transplantation*, *2*(9), 850–856. http://dx.doi.org/10.1034/j.1600-6143/2002.20907.x.

Saiki, R. K., Gelfand, D. H., Stoffel, S., Scharf, S. J., Higuchi, R., Horn, G. T., et al. (1988). Primer-directed enzymatic amplification of DNA with a thermostable DNA polymerase. *Science*, *239*(4839), 487–491.

Saiki, R. K., Scharf, S., Faloona, F., Mullis, K. B., Horn, G. T., Erlich, H. A., et al. (1985). Enzymatic amplification of beta-globin genomic sequences and restriction site analysis for diagnosis of sickle cell anemia. *Science*, *230*(4732), 1350–1354.

Sax, P. E., Tierney, C., Collier, A. C., Fischl, M. A., Mollan, K., Peeples, L., et al. (2009). S AIDS clinical trials group study A5202 team. Abacavir-lamivudine versus tenofovir-emtricitabine for initial HIV-1 therapy. *The New England Journal of Medicine*, *361*, 2230–2240. http://dx.doi.org/10.1056/NEJMoa0906768.

Schnittman, S. M., Greenhouse, J. J., Psallidopoulos, M. C., Baseler, M., Salzman, N. P., Fauci, A. S., et al. (1990). Increasing viral burden in CD4+ T cells from patients with human immunodeficiency virus infection reflects rapidly progressive immunosuppression and clinical disease. *Annals of Internal Medicine*, *113*(6), 438–443.

Schnitzler, M. A., Lowell, J. A., Hardinger, K. L., Boxerman, S. B., Bailey, T. C., & Brennan, D. C. (2003). The association of cytomegalovirus sero-pairing with outcomes and costs following cadaveric renal transplantation prior to the introduction of oral ganciclovir CMV prophylaxis. *American Journal of Transplantation*, *3*(4), 445–451. http://dx.doi.org/10.1034/j.1600-1643.2003.00069.x.

Schönberger, S., Meisel, R., Adams, O., Pufal, Y., Laws, H. J., Enczmann, J., et al. (2010). Prospective, comprehensive, and effective viral monitoring in children undergoing allogeneic hematopoietic stem cell transplantation. *Biology of Blood and Marrow Transplantation*, *16*(10), 1428–1435. http://dx.doi.org/10.1016/j.bbmt.2010.04.008.

Sehgal, R. N. M. (2010). Deforestation and avian infectious diseases. *The Journal of Experimental Biology*, *213*(6), 955–960. http://dx.doi.org/10.1242/jeb.037663.

Shuaib, F., Gunnala, R., Musa, E. O., Mahoney, F. J., Oguntimehin, O., Nguku, P. M., et al. (2014). Ebola virus disease outbreak—Nigeria, July–September 2014. *MMWR: Morbidity and Mortality Weekly Report*, *63*(39), 867–872.

Sia, I. G., Wilson, J. A., Groettum, C. M., Espy, M. J., Smith, T. F., & Paya, C. V. (2000). Cytomegalovirus (CMV) DNA load predicts relapsing CMV infection after solid organ transplantation. *The Journal of Infectious Diseases*, *181*(2), 717–720.

Sizmann, D., Glaubitz, J., Simon, C. O., Goedel, S., Buergisser, P., Drogan, D., et al. (2010). Improved HIV-1 RNA quantitation by COBAS AmpliPrep/COBAS TaqMan HIV-1 Test, v2.0 using a novel dual-target approach. *Journal of Clinical Virology*, *49*(1), 41–46. http://dx.doi.org/10.1016/j.jcv.2010.06.004.

Smith, J. M., Corey, L., Bittner, R., Finn, L. S., Healey, P. J., Davis, C. L., et al. (2010). Subclinical viremia increases risk for chronic allograft injury in pediatric renal transplantation. *Journal of the American Society of Nephrology*, *21*(9), 1579–1586. http://dx.doi.org/10.1681/ASN.2009111188.

SOVALDI® (sofosbuvir) [Full Prescribing Information]. Gilead Sciences, Foster City, CA. Revised December 2013.

Stekler, J., Swenson, P., Coombs, R., Dragavon, J., Thomas, K., Brennan, C., et al. (2009). HIV testing in a high-incidence population: Is antibody testing alone good enough? *Clinical Infectious Diseases*, *49*(3), 444–453.

Stieger, M., Demolliere, C., Ahlborn-Laake, J., & Mous, J. (1991). Competitive polymerase chain reaction assay for quantitation of HIV-1 DNA and RNA. *Journal of Virological Methods*, *34*(2), 149–160.

Sun, H. Y., Cacciarelli, T. V., Wagener, M. M., & Singh, N. (2010). Preemptive therapy for cytomegalovirus based on real-time measurement of viral load in liver transplant recipients. *Transplant Immunology*, *23*(4), 166–169. http://dx.doi.org/10.1016/j.trim.2010.06.013.

Taiwo, B., Gallien, S., Aga, S., Ribaudo, H., Haubrich, R., Kurzkes, D., et al. (2010). HIV drug resistance evolution during persistent near-target viral suppression. *Antiviral Therapy*, *15*, A38.

Tang, Y.-W., & Ou, C.-Y. (2012). Past, present and future molecular diagnosis and characterization of human immunodeficiency virus infections. *Emerging Microbes & Infections*, *1*, e19. http://dx.doi.org/10.1038/emi.2012.15.

Tanriverdi, S., Chen, L., & Chen, S. (2010). A rapid and automated sample-to-result HIV load test for near-patient application. *Journal of Infectious Diseases*, *201*(Suppl. 1), S52–S58. http://dx.doi.org/10.1086/650387.

Taubenberger, J. K., & Morens, D. M. (2006). 1918 influenza: The mother of all pandemics. *Emerging Infectious Diseases*, *12*(1), 15–22. http://wwwnc.cdc.gov/eid/article/12/1/05-0979_article/ (Accessed 11.10.14).

Taylor, L. H., Latham, S. M., & Woolhouse, M. E. (2001). Risk factors for human disease emergence. *Philosophical Transactions of the Royal Society of London. Series B: Biological Sciences*, *356*(1411), 983–989. http://dx.doi.org/10.1098/rstb.2001.0888.

Thiebaut, R., Morlat, P., Jacqmin-Gadda, H., Neau, D., Mercié, P., Dabis, F., et al. (2000). Clinical progression of HIV-1 infection according to the viral response during the first year of antiretroviral treatment. Groupe d'Epidemiologie du SIDA en Aquitaine (GECSA). *AIDS*, *14*(8), 971–978.

Thomson, E. C., Nastouli, E., Main, J., Karayiannis, P., Eliahoo, J., Muir, D., et al. (2009). Delayed anti-HCV antibody response in HIV-positive men acutely infected with HCV. *AIDS*, *23*(1), 89–93. http://dx.doi.org/10.1097/QAD.0b013e32831940a3.

Trepo, C. (2000). Genotype and viral load as prognostic indicators in the treatment of hepatitis C. *Journal of Viral Hepatitis*, *7*(4), 250–257.

UNITAID. (2014). *HIV/AIDS diagnostics technology landscape* (4th ed., pp. 1–147). Geneva, Switzerland: UNITAID: World Health Organization.

van Rossum, A. M., Fraaij, P. L., & de Groot, R. (2002). Efficacy of highly active antiretroviral therapy in HIV-1 infected children. *The Lancet Infectious Diseases*, *2*(2), 93–102.

van Sighem, A., Zhang, S., Reiss, P., Gras, L., van der Ende, M., Kroon, F., et al. (2008). Immunologic, virologic, and clinical consequences of episodes of transient viremia during suppressive combination antiretroviral therapy. *Journal of Acquired Immune Deficiency Syndromes*, *48*(1), 104–108. http://dx.doi.org/10.1097/QAI.0b013e31816a1d4f.

VIEKIRA PAK™ (ombitasvir, paritaprevir, and ritonavir tablets; dasabuvir tablets) [Full Prescribing Information]. Abbvie Inc., North Chicago, IL. Revised December 2014.

Vulto, P., Hermann, C., Zahn, P., Maier, U., Dame, G., & Urban, G. A. (2009). A microchip for automated extraction of RNA from gram-positive bacteria. *Transducers*, 108–111. http://dx.doi.org/10.1109/SENSOR.2009.5285553.

Wang, A. M., Doyle, M. V., & Mark, D. F. (1989). Quantitation of mRNA by the polymerase chain reaction. *Proceedings of the National Academy of Sciences of the United States of America*, *86*(24), 9717–9721.

Webb, S. G. (1990). Prehistoric eye disease (trachoma?) in Australian aborigines. *American Journal of Physical Anthropology*, *81*(1), 91–100.

WHO Ebola. (2014). *World health organization. Global alert and response: Ebola outbreak.* http://www.who.int/csr/disease/ebola/en/ (Accessed 11/12/14).

WHO Influenza. (2011). World health organization. The use of PCR in the surveillance and diagnosis of influenza. In *Report of the 4th meeting of the WHO working group on polymerase chain reaction protocols for detecting subtype influenza A viruses. Geneva, Switzerland 14–15 June 2011.* http://www.who.int/influenza/gisrs_laboratory/final_who_pcr__meeting_report_aug_2011_en.pdf (Accessed 11/3/2014).

WHO SARS. (2003). *World health organization. Severe acute respiratory syndrome.* http://www.who.int/ith/diseases/sars/en/ (Accessed 11/12/14).

Willig, J. H., Nevin, C. R., Raper, J. L., Saag, M. S., Mugavero, M. J., Willig, A. L., et al. (2010). Cost ramifications of increased reporting of detectable plasma HIV-1 RNA levels by the Roche COBAS AmpliPrep/COBAS TaqMan HIV-1 version 1.0 viral load test. *Journal of Acquired Immune Deficiency Syndromes*, *54*(4), 442–444. http://dx.doi.org/10.1097/QAI.0b013e3181d01d1d.

Wolff, D. J., Heaney, D. L., Neuwald, P. D., Stelrecht, K. A., & Press, R. D. (2009). Multi-site PCR-based CMV viral load assessment-assays demonstrate linearity and precision, but lack numeric standardization: A report of the Association for Molecular Pathology. *The Journal of Molecular Diagnostics*, *11*(2), 87–92. http://dx.doi.org/10.2353/jmoldx.2009.080097.

Wursthorn, K., Manns, M. P., & Wedemeyer, H. (2008). Natural history: The importance of viral load, liver damage and HCC. *Best Practice & Research. Clinical Gastroenterology*, *22*(6), 1063–1079. http://dx.doi.org/10.1016/j.bpg.2008.11.006.

Yee, H. S., Currie, S. L., Darling, J. M., & Wright, T. L. (2006). Management and treatment of hepatitis C Viral Infection: Recommendations from the Department of Veterans Affairs

Hepatitis C Resource Center Program and the National Hepatitis C Program office. *The American Journal of Gastroenterology, 101*(10), 2360–2378.

Young, K. Y., Nelson, R. P., & Good, R. A. (1990). Discordant human immunodeficiency virus infection in dizygotic twins detected by polymerase chain reaction. *The Pediatric Infectious Disease Journal, 9*(6), 454–456.

Zentilin, L., & Giacca, M. (2007). Competitive PCR for precise nucleic acid quantification. *Nature Protocols, 2*(9), 2092–2104.

Zeuzem, S., Berg, T., Moeller, B., Hinrichsen, H., Mauss, S., Wedemeyer, H., et al. (2009). Expert opinion on the treatment of patients with chronic hepatitis C. *Journal of Viral Hepatitis, 16*(2), 75–90. http://dx.doi.org/10.1111/j.1365-2893.2008.01012.x.

Low-Density TaqMan® Array Cards for the Detection of Pathogens

Jude Heaney*,†, Kathryn Rolfe‡, Nicholas S. Gleadall‡,
Jane S. Greatorex‡, Martin D. Curran‡,1

*Department of Medicine, University of Cambridge, Cambridge, United Kingdom
†Institute of Hepatology, Foundation for Liver Research, London, United Kingdom
‡Public Health England, Cambridge University Hospitals NHS, Cambridge, United Kingdom
1Corresponding author: e-mail address: martin.curran@addenbrookes.nhs.uk

1 INTRODUCTION

Clinicians have a complex task when attempting to identify infectious disease aetiologies, particularly in critically ill patients. Current diagnostic practices typically detect and report single pathogen analysis from individual patient samples, often being curtailed when a pathogen is detected. Adopting a syndromic approach for the rapid identification of infection(s) would not only change current clinical diagnostic practices but also provide a rapid diagnostic tool to inexpensively supply evidence of specific infections and super-infections, particularly in paediatric, critically ill and immunosuppressed patients. Test selection may be based on limited information and therefore screening for many relevant pathogens may not occur, simply because they were not requested by the referring clinician, consequently infectious aetiologies can remain unidentified. Delays may also occur if reference laboratories are required to run tests for less common pathogens, greatly increasing the turn-around time for results. The patient may remain for extended periods on broad-spectrum therapy pending the results of different tests, with the associated increased drug toxicity, unwarranted selective pressure for antibiotic resistance, cost and length of hospital stay. There is, therefore, a great need to simplify complex workflows for pathogen detection and identification to allow the rapid initiation of tailored treatment regimens.

Over the past 10 years, the application of both qualitative and quantitative nucleic acid detection techniques has dramatically altered clinical diagnostic practices. PCR and, more recently, real-time PCR have revolutionised diagnostic practices in clinical microbiology laboratories (Bankowski & Anderson, 2004; Cockerill, 2003; Mackay, 2004), allowing rapid pathogen identification. The combination of excellent sensitivity and specificity, low contamination risk, speed, reduced hands on time and ease of use has made real-time PCR technology an appealing alternative to

Methods in Microbiology, Volume 42, ISSN 0580-9517, http://dx.doi.org/10.1016/bs.mim.2015.06.002

conventional culture-based or immunoassay-based testing methods, which are often time consuming and labour intensive. As a direct result of the introduction of real-time PCR into clinical microbiology, turn-around times have decreased considerably with obvious benefits at the bedside. The use of molecular diagnostics has proven advantageous resulting in more accurate results with overall better patient care and outcomes.

When molecular diagnostics were first introduced into clinical microbiology, there were few commercial assays available; therefore, in-house PCR and real-time PCR assays were developed for the majority of targets. In more recent years, an expanded portfolio of commercial assays has become available, enabling laboratories for whom extensive assay development and validation is not practicable to still utilise molecular-based methods. All in-house assays should now be developed and validated according to the Conformité Européene—*in vitro* diagnostic (CE-IVD) recommendations and should be considered for CE marking or ISO 15189 accreditation. All PCR-based assays, qualitative and quantitative, should meet Minimum Information for Publication of Quantitative Real-Time PCR Experiments (MIQE) requirements as detailed in the literature (Bustin et al., 2010, 2013; Johnson, Nour, Nolan, Huggett, & Bustin, 2014; Taylor, Wakem, Dijkman, Alsarraj, & Nguyen, 2010) and comply with the guidelines outlined for the development and validation of diagnostic tests that depend on nucleic acid amplification and detection (Saunders et al., 2013). There is also a requirement for the use of standardised materials and participation in external quality control programmes.

2 TaqMan® ARRAY CARDS

PCR and real-time PCR technology have evolved from monoplex to multiplex assays with mastermixes comprising several primer/probe sets for multiple pathogen detection. Array technology has revisited monoplex assays but utilising them as 'simultaneous singleplex' to detect many more pathogens from the full spectrum of organisms, thus introducing a syndromic rather than pathogen-focused approach to diagnosis. The TaqMan® Array Card (TAC) (Life Technologies, Carlsbad, CA) formally known as TaqMan® Low-density Array (TLDA) is a microfluidic card which utilises primers and probes pre-loaded and lyophilised onto wells. The TAC platform offers a 384-well, single-plate real-time PCR in an 8 sample by 48-well format with a single target tested in each of 48 available wells/pods (including controls). Microfluidic technology distributes the sample into individual PCRs and real-time PCR and detection takes place within an analyser (ViiA7 or QuantStudio 7) with results generated in less than 1 h. The process and individual steps involved in setting up a card are schematically outlined in Figure 1.

To our knowledge, TACs are the only available real-time PCR platform with the inherent ability to utilise in-house-validated PCR assays onto a single array and detect and quantify up to 48 targets simultaneously on a single specimen in such a simple, robust and easily transferable application. Nucleic acid extract (20–75 µl) is added to

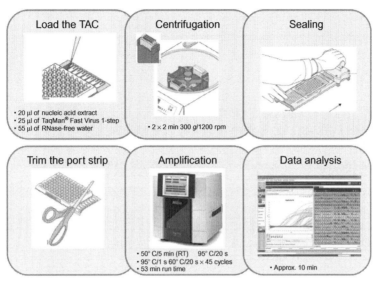

FIGURE 1

Processing steps involved in running a TAC. Once the nucleic acid extract/RT-PCR mastermix (100 μl) is added to the port/reservoir for each specimen, centrifugation is used to channel the mix into the 48 wells/pods. These contain the pairs of lyophilised primers and probes for the specific targets to be amplified by real-time PCR on each of the eight clinical specimens applied to the plate.

a mastermix (containing enzyme, buffer and nucleotides). The $4 \times$ TaqMan® Fast Virus 1-Step mastermix (Life Technologies) is recommended for use with TAC but others are widely available. The reaction volume is adjusted to 100 μl with nuclease-free water, mixed and transferred to a single reservoir/port (Figure 1). Once eight specimens have been added into the 8 reservoirs/ports, the plate is centrifuged twice for 2 min at 1200 rpm/$300 \times g$, sealed, the ports trimmed off with scissors and loaded into the real-time PCR ViiA 7 instrument (Life Technologies): the run time is approximately 53 min (Figure 1).

In practical terms, the advantages of TAC are numerous. Handling time and training required are both minimal, substantially improving the real cost of clinical diagnosis. Data interpretation is straightforward as data output is a single, user-friendly file. The lyophilisation of the primers and probes onto the card ensures a shelf-life of up to 2 years with refrigeration, allowing less commonly required cards to be stored but readily available, with no loss of sensitivity. TACs can also be shipped at room temperature. An added advantage of the TAC is the ability to include confirmatory assays for many of the pathogens with assays being designed to detect multiple targets per pathogen, enhancing confidence in diagnosis (Figure 2).

The TAC format has some disadvantages, and some groups have reported a reduced sensitivity when compared to monoplex assays (Kodani et al., 2011). This is

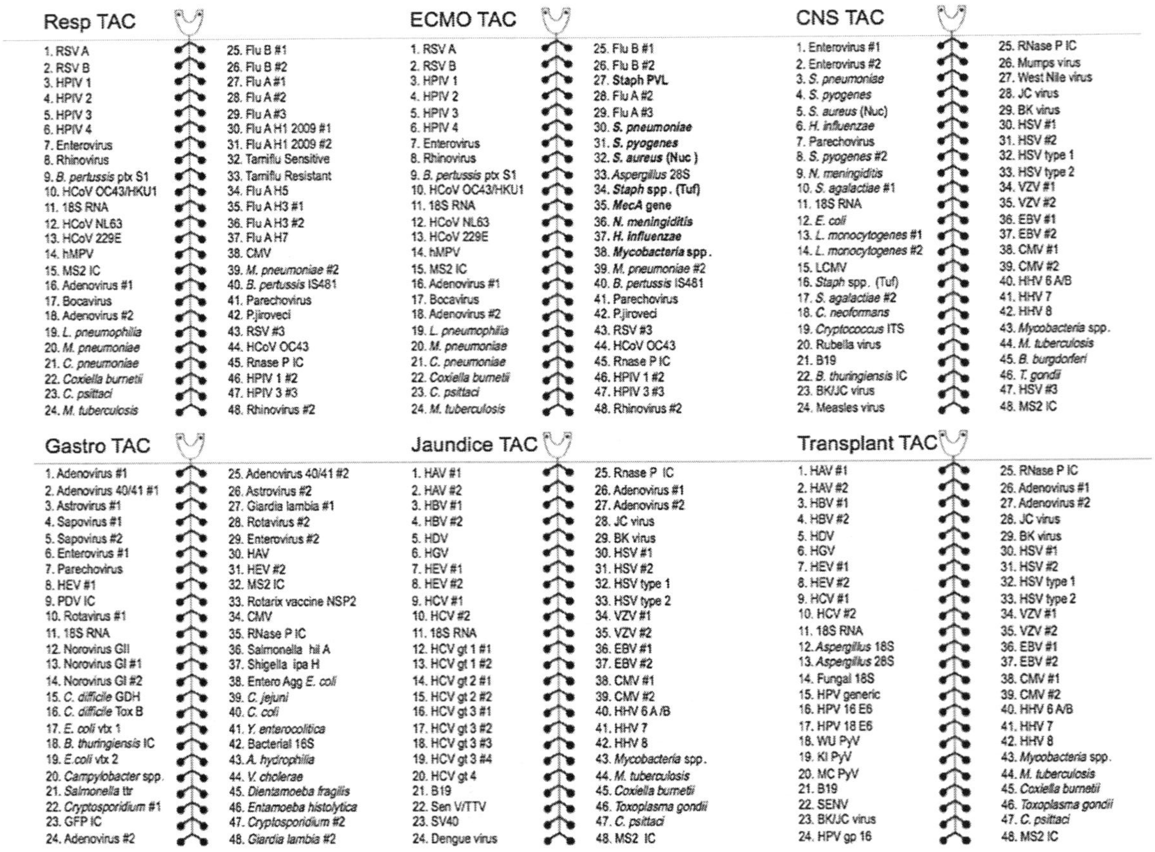

Resp TAC

1. RSV A
2. RSv B
3. HPV 1
4. HPV 2
5. HPV 3
6. HPV 4
7. Enterovirus
8. Rhinovirus
9. B. pertussis ptx S1
10. HCoV OC43/HKU1
11. 18S RNA
12. HCoV NL63
13. HCoV 229E
14. hMPV
15. MS2 IC
16. Adenovirus #1
17. Bocavirus
18. Adenovirus #2
19. L. pneumophila
20. M. pneumoniae
21. C. pneumoniae
22. Coxiella burnetii
23. C. psittaci
24. M. tuberculosis
25. Flu B #1
26. Flu B #2
27. Flu A #1
28. Flu A #2
29. Flu A #3
30. Flu A H1 2009 #1
31. Flu A H1 2009 #2
32. Tamiflu Sensitive
33. Tamiflu Resistant
34. Flu A H5
35. Flu A H3 #1
36. Flu A H3 #2
37. Flu A H7
38. CMV
39. M. pneumoniae #2
40. B. pertussis IS481
41. Parechovirus
42. P.jiroveci
43. RSV #3
44. HCoV OC43
45. Rnase P IC
46. HPV 1 #2
47. HPV 3 #3
48. Rhinovirus #2

ECMO TAC

1. RSV A
2. RSV B
3. HPV 1
4. HPV 2
5. HPV 3
6. HPV 4
7. Enterovirus
8. Rhinovirus
9. B. pertussis ptx S1
10. HCoV OC43/HKU1
11. 18S RNA
12. HCoV NL63
13. HCoV 229E
14. hMPV
15. MS2 IC
16. Adenovirus #1
17. Bocavirus
18. Adenovirus #2
19. L. pneumophila
20. M. pneumoniae
21. C. pneumoniae
22. Coxiella burnetii
23. C. psittaci
24. M. tuberculosis
25. Flu B #1
26. Flu B #2
27. Staph PVL
28. Flu A #2
29. Flu A #3
30. S. pneumoniae
31. S. pyogenes
32. S. aureus (Nuc)
33. Aspergillus 28S
34. Staph spp. (Tuf)
35. MecA gene
36. N. meningitidis
37. H. influenzae
38. Mycobacteria spp.
39. M. pneumoniae #2
40. B. pertussis IS481
41. Parechovirus
42. P.jiroveci
43. HCoV OC43
44. HCoV OC43
45. Rnase P IC
46. HPV 1 #2
47. HPV 3 #3
48. Rhinovirus #2

CNS TAC

1. Enterovirus #1
2. Enterovirus #2
3. S. pneumoniae
4. S. pyogenes
5. S. aureus (Nuc)
6. H. influenzae
7. Parechovirus
8. S. pyogenes #2
9. N. meningitidis
10. S. agalactiae #1
11. 18S RNA
12. E. coli
13. L. monocytogenes #1
14. L. monocytogenes #2
15. LCMV
16. Staph spp. (Tuf)
17. S. agalactiae #2
18. C. neoformans
19. Cryptococcus ITS
20. B. thuringiensis IC
21. B19
22. B. thuringiensis IC
23. BK/JC virus
24. Measles virus
25. RNase P IC
26. Mumps virus
27. West Nile virus
28. JC virus
29. BK virus
30. HSV #1
31. HSV #2
32. HSV type 1
33. HSV type 2
34. VZV #1
35. VZV #2
36. EBV #1
37. EBV #2
38. CMV #1
39. CMV #2
40. HHV 6 A/B
41. HHV 7
42. HHV 8
43. Mycobacteria spp.
44. M. tuberculosis
45. B. burgdorferi
46. T. gondii
47. HSV #3
48. MS2 IC

Gastro TAC

1. Adenovirus #1
2. Adenovirus 40/41 #1
3. Astrovirus #1
4. Sapovirus #1
5. Sapovirus #2
6. Enterovirus #1
7. Parechovirus
8. HEV #1
9. PDV IC
10. Rotavirus #1
11. 18S RNA
12. Norovirus GII
13. Norovirus GI #1
14. Norovirus GI #2
15. C. difficile GDH
16. C. difficile Tox B
17. E. coli vtx 1
18. B. thuringiensis IC
19. E. coli vtx 2
20. Campylobacter spp.
21. Salmonella tir
22. Cryptosporidium #1
23. GFP IC
24. Adenovirus #2
25. Adenovirus 40/41 #2
26. Astrovirus #2
27. Giardia lambia #1
28. Rotavirus #2
29. Enterovirus #2
30. HAV
31. HEV #2
32. MS2 IC
33. Rotarix vaccine NSP2
34. CMV
35. RNase P IC
36. Salmonella hil A
37. Shigella ipa H
38. Entero Agg E. coli
39. C. jejuni
40. C. coli
41. Y. enterocolitica
42. Bacterial 16S
43. A. hydrophilia
44. V. cholerae
45. Dientamoeba fragilis
46. Entamoeba histolytica
47. Cryptosporidium #2
48. Giardia lambia #2

Jaundice TAC

1. HAV #1
2. HAV #2
3. HBV #1
4. HBV #2
5. HDV
6. HGV
7. HEV #1
8. HEV #2
9. HCV #1
10. HCV #2
11. 18S RNA
12. HCV gt 1 #1
13. HCV gt 1 #2
14. HCV gt 2 #1
15. HCV gt 2 #2
16. HCV gt 3 #1
17. HPV gt 3 #2
18. HCV gt 3 #3
19. HCV gt 3 #4
20. HCV gt 4
21. B19
22. Sen V/TTV
23. SV40
24. Dengue virus
25. Rnase P IC
26. Adenovirus #1
27. Adenovirus #2
28. JC virus
29. BK virus
30. HSV #1
31. HSV #2
32. HSV type 1
33. HSV type 2
34. VZV #1
35. VZV #2
36. EBV #1
37. EBV #2
38. CMV #1
39. CMV #2
40. HHV 6 A/B
41. HHV 7
42. HHV 8
43. Mycobacteria spp.
44. M. tuberculosis
45. Coxiella burnetii
46. Toxoplasma gondii
47. C. psittaci
48. M. tuberculosis

Transplant TAC

1. HAV #1
2. HAV #2
3. HBV #1
4. HBV #2
5. HDV
6. HGV
7. HEV #1
8. HEV #2
9. HCV #1
10. HCV #2
11. 18S RNA
12. Aspergillus 18S
13. Aspergillus 28S
14. Fungal 18S
15. HPV generic
16. HPV 16 E6
17. HPV 18 E6
18. WU PyV
19. KI PyV
20. MC PyV
21. B19
22. SENV
23. BK/JC virus
24. HPV gp 16
25. RNase P IC
26. Adenovirus #1
27. Adenovirus #2
28. JC virus
29. BK virus
30. HSV #1
31. HSV #2
32. HSV type 1
33. HSV type 2
34. VZV #1
35. VZV #2
36. EBV #1
37. EBV #2
38. CMV #1
39. CMV #2
40. HHV 6 A/B
41. HHV 7
42. HHV 8
43. Mycobacteria spp.
44. M. tuberculosis
45. Coxiella burnetii
46. Toxoplasma gondii
47. C. psittaci
48. MS2 IC

FIGURE 2

TaqMan array card layouts with all the pathogens included for each sample. For each sample, one internal positive control with bacteriophage MS2 (MS2 IC) and two human DNA/RNA controls (RNase P/18S RNA) are included. Some cards include additional internal positive controls, i.e., phocine distemper virus (PDV), Bacillus thuringiensis and Escherichia coli green fluorescent protein (GFP IC). For several pathogens, more than one genetic target is included (indicated with #number).

usually restricted to a one $\log_{10}$ lower limit of detection (LOD) when compared to monoplex assays (Kodani, Mixson-Hayden, Drobeniuc, & Kamili, 2014; Kodani et al., 2011; Rachwal et al., 2012). However, careful optimisation of TAC such as testing multiple targets for some pathogens within the card (see Figure 2), using efficient nucleic acid extraction methods (particularly for samples such as blood which may have a low yield of organism) and increasing the extraction volume (as well as the nucleic acid input volume), may all help to increase the sensitivity to that suitable for use in routine diagnostics laboratories (Diaz et al., 2013). Additional drawbacks of TAC technology include the necessity to screen eight samples in parallel to avoid waste plus the limited number of samples per plate (maximum of eight), lack of automation and inaccessibility of the cards for downstream analysis of PCR products.

The array requires a single annealing temperature for all assays on a card; therefore, individual assays must be optimised to work efficiently at this temperature. This requires a careful validation process so that all assays perform optimally under the universal conditions of the array card. However, once this validation has been undertaken, addition of other assays to the panel or replacement of existing assays with new, improved versions can be carried out without the need for extensive revalidation, allowing for a flexible approach. The reported drop in sensitivity compared to monoplex PCRs (pertaining to the small volume analysed) (Kodani et al., 2011) does not have a negative impact because pathogens at clinically relevant levels are detected. LODs of single copy number are achievable, and this has been demonstrated for our in-house-developed respiratory TAC (Figure 3) where three of our independent influenza B real-time array assays were demonstrated to detect down to 3 copies per reaction using a synthetic puc57 plasmid control (www.genscript.com) containing all three target sequences.

At present, quantitation is not performed on TACs; however, when attention is focused upon what organism is responsible for infection, a detected or not detected answer is usually sufficient. Adaptation of TAC to perform quantitation is possible if there is clinical need for this function. Multiple organisms may be detected from a single specimen, requiring careful interpretation by clinicians to determine which of the identified organisms are responsible for disease and which may be considered commensals. This relies on clinical judgement; however, the cycle threshold (Ct) value at which a particular organism is detected on the TAC (and which represents the relative amount of that organism within the sample) can aid in this interpretation. A good example to illustrate this point is the multiple infections found in a specimen recently processed on our Gastro TAC which is currently undergoing validation in our network of Public Health England (PHE) laboratories (Figure 4).

One of the major attractions of array technology is the ability to design cards with specific syndromes in mind. Examples of these, some of which are described within this chapter, include infectious respiratory disease cards, hepatitis/jaundice cards, infectious gastrointestinal disease cards, central nervous system (CNS) infection cards and sexually transmitted infection (STI) cards (see Figure 2 for the layouts of the cards developed to date in our laboratory). There is the possibility to include assays for antibiotic and anti-viral resistance relevant to a particular organism/syndrome,

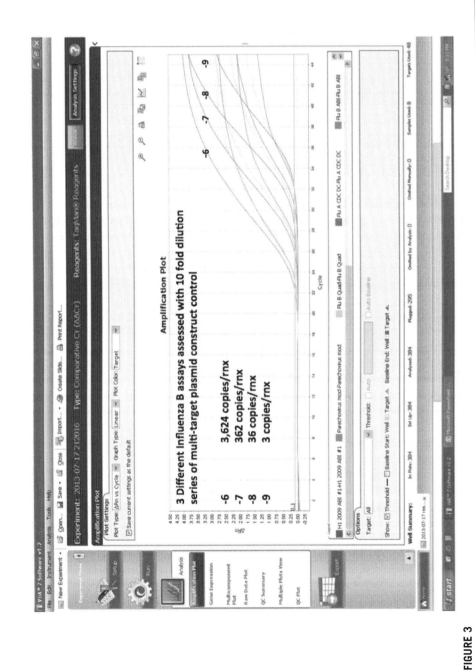

FIGURE 3

Assessment of analytical sensitivity of three influenza B real-time PCR assays on the TAC.

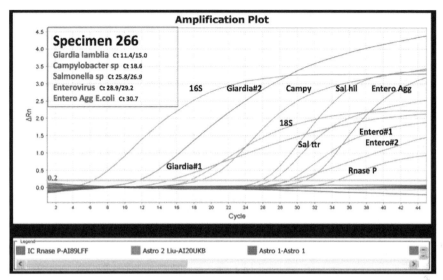

FIGURE 4

Gastro TAC specimen result displaying the numerous pathogens that were detected, with a broad range of Ct values.

for example, carbapenamases or oseltamivir resistance. This technique can also be used in surveillance for pathogens which are not sought routinely, for example, toroviruses, kobuviruses and parvoviruses in gastrointestinal disease. Array cards can be used in outbreak response, for example, with respiratory outbreaks whereby outbreak specimens can be rapidly tested for multiple respiratory pathogens allowing for rapid initiation of appropriate public health response, implementation of infection control precautions and prophylaxis/treatment where necessary.

Although this chapter focuses on TAC technology, it is important to recognise that there are other types of array technology which are currently in development or diagnostic use. FilmArray technology (BioFire Diagnostics Inc., Salt Lake City, UT) is a 'pouch-format' array which incorporates nucleic acid extraction, multiplex PCR followed by nested PCR in wells of 1 μl volume each to produce a result within 1 h (Pierce, Elkan, Leet, McGowan, & Hodinka, 2012; Popowitch, O'Neill, & Miller, 2013). The single-sample format makes this technology inadequate for high-throughput testing but does make it ideal for point-of-care testing which could be deployed in critical situations such as field-based needs for the armed forces.

This chapter explores how TACs are being developed to aid diagnosis of infection and to individualise the process according to the presenting syndrome based predominantly on our experiences. In the fight against infectious disease where emerging and re-emerging pathogens add to the complexity, it is important to apply contemporary knowledge and novel technologies to aid diagnosis and to simplify the complex workflows that are required for pathogen identification.

3 DEVELOPING A TAC BASED ON IN-HOUSE REAL-TIME PCR ASSAYS

Both MIQE and IVD guidelines (Saunders et al., 2013) were followed when designing and optimising assays on TACs, taking into consideration: experimental design, sample properties, nucleic acid extraction and quality assessment, reverse transcription (RT), target information, primer and probe design, real-time PCR protocol optimisation and validation details, and data analysis.

Existing, previously validated, in-house assays were taken from a variety of formats including monoplex and multiplex real-time PCR with differing fluorescent probes and quenchers. The initial step was to transform all assays to a single monoplex format, with high specificity and sensitivity while using identical reaction conditions, chemistry, probes with the same reporter fluorophore and quencher. Although each assay had previously been extensively validated, the changes required for transformation to TAC assays necessitated re-optimisation and validation. This opportunity provided a chance to re-evaluate each assay individually and perform BLAST analysis (www.ncbi.nlm.gov/Blast) of primers and probes to ensure specificity (utilising recently deposited sequences in the GenBank database); to examine primer and probe melting temperatures (T_m) and whether any adverse primer/probe dimer and heterodimer interactions existed using OligoAnalyzer 3.1 (www.idtdna.com) that required modification. In some instances, the primers needed to be modified at the 5' end (adding target-matched extensions) to raise their T_m to approximately 55–60 °C to ensure uniform amplification across 48 different assays at the single defined temperature (60 °C). Ideally, the amplicon length should be between 60 and 150 nucleotides but we have had acceptable results with amplicons up to 200 nucleotides in length. A huge benefit is that once validated on the TAC format, an assay can be incorporated into any syndromic TAC.

Although alternatives may be considered, for ease of use and interpretation, probes with a single fluorescent 5' dye (FAM) and minor groove binder (MGB) incorporated into a non-fluorescent quencher at the 3' end are favoured for the development of TACs. These probes have proven to generate better precision in quantitation due to a lower background signal. Further, the MGB moiety stabilises the hybridised probe, effectively raising the melting temperature (T_m), and increasing specificity. MGB probes can therefore be shorter than traditional dual-labelled probes, making them better suited for applications such as allelic discrimination, or when designing probes in regions of high AT content, for genotyping or when designing assays to detect mutations conferring drug resistance. The majority of assays using traditional dual-labelled quenchers transfer to the MGB format with minimal re-optimisation required; indeed, in many cases the shorter MGB probes increase the sensitivity and efficiency of the reaction, as well as the amplitude of the fluorescence signal giving high rising sigmoidal amplification curves. In our experience, designing the T_m of the MGB probe to be between 48 and 55 °C works effectively with probe lengths ranging from 14 to 24 nucleotides.

The TaqMan® Fast Virus 1-Step mastermix (Life Technologies) is our preferred chemistry for TAC assays; therefore, initial development work must ensure all existing primer and probe sets perform adequately using this chemistry and fast ramping and cycling times. If existing primer and probe sets do not perform adequately, modified or new assays must be designed. The Fast Virus 1-Step mastermix amplifies both RNA and DNA with high sensitivity allowing for mix-and-match RT-PCR and PCR on any TAC. It has an added benefit of retaining high sensitivity even in the presence of RT-PCR inhibitors often found in blood, stool and other typically inhibitory specimens. Although not essential, for ease of use, we chose to have one cut-off threshold value for all assays on each developed TAC (Figure 2). Again, this requires careful assay design ensuring all true amplification curves (with high rising amplitudes) cross this threshold effectively for each assay.

In the initial developmental stages, assay optimisation should utilise the real-time platform routinely used for in-house assays to allow for ease of comparison. Assay sensitivity, specificity, efficiency and fluorescence intensity using both known standards and patient samples should be determined. A range of sample types including plasma, stool, nasal swabs, bronchoalveolar lavage, naso-pharyngeal aspirates (NPAs), CNS, fresh biopsy tissue and formalin-fixed, paraffin-embedded specimens may be included in assay optimisation. Extraction volumes can also be optimised and increasing the volume can be beneficial when low levels of pathogen are suspected, e.g., very early or late in the infection. An internal control such as bacteriophage MS2 (or *Bacillus thuringiensis* cells) should be included in all extraction protocols to serve as a process control (Rolfe et al., 2007).

The next step is to transfer the assays to a 384-well PCR plate format using fast reaction conditions (see Figure 1) on the ViiA7 and determine sensitivity, specificity and LODs for each assay, again using known standards and patient samples. Although a larger volume (20 µl) is used in the 384-well PCR plate format on the ViiA7, it does provide a good indication of how assays will perform on a TAC (1 µl reaction volume) due to the proximity to the fluorescence detectors within the ViiA7 instrument.

When a TAC has been spotted with primer and probe sets lyophilised into individual pods, validation procedures must be undertaken. Known standards (such as viruses and diluted nucleic acid extracts), United Kingdom National External Quality Assessment (NEQAS) and Quality Control for Molecular Diagnostics (QCMD) (www.qcmd.org), external quality assurance panels, patient specimens derived from different sample types and engineered synthetic plasmid controls can be used for validation (Kodani & Winchell, 2012). Employing such a combination allows for optimal assay scrutiny, providing information on LOD, sensitivity, specificity and any reaction inhibitors from different sample types. NEQAS and QCMD panels provide a good indication if a generated TAC is of sufficient sensitivity to be used in an accredited diagnostic service. A panel of engineered plasmid controls can serve multiple purposes: initially to determine specificity of each assay on a TAC and their LODs, and subsequently to determine any batch variation. For our TAC development, we used a commercial company, GenScript (www.genscript.com), to generate a panel of synthetic control plasmids containing all our target sequences (with 20 nucleotides each side of the primer target sites also included) combined together. These plasmid panels also serve to quality-check each new batch of TAC plates, in a checker board fashion, to ensure all the primers and probes have been spotted correctly into their assigned pods. Extraction and internal controls (MS2, RNase P gene and 18S) can be used to determine sample quality, the reaction failing if any of the control reactions are outside of pre-determined ranges. Specific positioning of the control assays throughout the TAC also enables the user to determine whether sufficient centrifugation has taken place and the mastermix has reached all pods on the TAC.

4 SYNDROMIC TAC

As emphasised in Section 1, adopting a syndromic approach to infectious disease diagnoses has many advantages. However, it also requires careful planning and input from both clinicians and diagnostic laboratory teams. Designing syndromic TACs involves an initial decision as to which pathogens are considered important in the targeted syndromes: those identified most frequently and/or those with serious complications and adverse consequences for the patient. Syndromic diagnosis allows not only for rapid turn-around times in routine diagnosis but also rapid response times in outbreak situations.

The syndromic approach to infectious disease diagnosis is not new and both in-house and commercial systems have been described using techniques ranging from multiplex PCR and microarrays to MALDI-TOF and sequencing or combinations of these approaches (Elnifro, Ashishi, Cooper, & Klapper, 2000; Gray & Coupland, 2014; Platts-Mills, Operario, & Houpt, 2012; Wolk, Kaleta, & Wysocki, 2012).

4.1 RESPIRATORY TACs

The first syndromic TAC for infectious diseases to be described was a respiratory array card developed by Kodani et al. (2011). This was capable of detecting 21 targets (13 viruses and 8 bacteria) plus control, aimed at diagnosing the cause of acute respiratory infection. Others, including ourselves, have expanded the range of organisms to include those causing atypical pneumonias, a wider range of specific organisms (e.g. influenza A subtypes) and those causing infections in specific patient groups, e.g., those receiving extracorporeal membrane oxygenation (ECMO) therapy for severe respiratory failure due to a suspected infectious aetiology (see Figure 2 for respiratory/ECMO card layouts). One of the inherent advantages of the TAC development process is the ability to make changes and improvements to individual assays when ordering the next batch of plates (minimum order is 50 plates/400 specimens), which take 6–10 weeks to manufacture. Our current respiratory TAC is now in its eighth version and has dramatically improved during this evolutionary process. Its performance with a commercially available respiratory ZeptoMetrix verification panel (www.zeptometrix.com) compared to Cepheid GeneXpert Flu/RSV test, Bio-fire FilmArray respiratory panel and our routine multiplex real-time respiratory assays is clearly demonstrated in Table 1. Moreover, performance with all the available QCMD respiratory pathogen panels (www.qcmd.org) over the last few years has been excellent and is equal to and in most cases now superior to our current routine molecular real-time respiratory assays. Improvements observed in TAC performance during development of the plate have correlated with increased analytical sensitivity of the individual assays and reflect the fact that all assays on the eighth version of the respiratory card can detect down to approximately 3 copies per reaction using our panel of synthetic plasmid controls. A recent comparison with our routine multiplex assays on 417 consecutive respiratory specimen demonstrated the TAC performance (Table 2), with discrepant results seen only in specimens with very late Ct values (low viral load) and highlighting that the card now outperforms the gold standard test, i.e., more sensitive for RSV and adenovirus.

A modification of our respiratory TAC plate tailored to our ECMO service (adjusted assays on the ECMO card are highlighted in bold font in Figure 2) has been in evaluation in our laboratory since November 2013. To date (April 2015), 55 patients (151 specimens) have been processed on the ECMO card in parallel with routine investigations. In addition to confirming all routine investigative findings, the ECMO array card has had a significant beneficial impact, directly influencing clinical management in some patients. Notable infections identified include *Mycoplasma pneumoniae* (two cases), *Aspergillus fumigatus* (one case), *Streptococcus pyogenes* (two cases, both dual infections with influenza A), *Mycobacterium tuberculosis* (one case) and *Streptococcus pneumoniae* (six cases).

4.2 GASTROINTESTINAL TACs

It has long been recognised that mass spectator events represent huge challenges to public health—bringing the possibility of imported infections, mass transmissions and food-borne infection outbreaks. Transmission of gastrointestinal infections can cause major issues at mass spectator events and identifying causative pathogens can be problematic. Repeated studies have shown that the overall positivity of routine microscopy and culture of stool samples from symptomatic individuals are poor compared to conventional PCR. Selected references in the reference list demonstrate this finding (Morgan, Paillart, & Thompson, 1998; Santos & Rivera, 2013; Stensvold & Nielsen, 2012). The need for comprehensive screening assays in both outbreak settings and routine clinical investigation has therefore been recognised and the latter developed and used in a number of settings (Liu et al., 2013, 2014; Pholwat et al., 2015). To date, a single screening assay has not been available for mass event settings but such assays are under development.

The TAC recently developed by Liu et al. (2013), which simultaneously detects 19 enteropathogens, constitutes a significant advance in diagnostics for gastrointestinal pathogens. Not only do the cards allow fast, accurate and quantitative detection of a broad spectrum of enteropathogens (bacteria, viruses and parasites), the authors

Table 1 ZeptoMetrix NATtrol™ Flu Verification Panel (NATFVP-C) Challenge

Panel Member	Strain	Multiplex RT-PCR[a] Routine Results Ct	Xpert Flu/RSV Assay Results Ct	BioFire FilmArray Results	TLDA Array Card Results Ct
Influenza A H1 (1:100)[b]	A/New Caledonia/20/99	Flu A 29.1	Flu A 30.6/31.3	Flu A Pos H1 pos	Flu A 33.6/32.9 H1 seasonal 31/32.3
Influenza A H1 (1:100)	A/Brisbane/59/07	Flu A 29.6	Flu A 33.6/33.4	Flu A Pos equivocal	Flu A 35.2/34.5 H1 seasonal 36.2/34.1
Influenza A H3 (1:100)	A/Brisbane/10/07	Flu A 28.8 H3 28.6	Flu A 27.0/28.9	Flu A pos H3 Pos	Flu A 33.3/33.5 H3 seasonal 31.7/31.1
Influenza A H3 (1:100)	A/Wisconsin/67/05	Flu A 29.3 H3 28.8	Flu A 30.5/32.8	Flu A Pos equivocal	Flu A 34.0/33.0 H3 seasonal 31.9/31.3
Influenza A 2009 H1N1 (1:100)	Canada/6294/09	Flu A 26.7 H1 2009 25.6	Flu A 32.3/32.7	Flu A Pos H1 2009 Pos	Flu A 32.3/30.3 H12009 30.9/28.7 Tamiflu Sens 32.5
Influenza A 2009 H1N1 (1:100)	NY/02/09	Flu A 29.5 H1 2009 27.8	Flu A 31.4/32.6	Flu A Pos equivocal	Flu A 35.0/32.6 H12009 33.8/29.9 Tamiflu Sens 35.2
Influenza B (1:100)	B/Florida/02/06	Flu B 27.1	Flu B 29.4	Flu B Pos	Flu B 31.4/31.1
Influenza B (1:100)	B/Malaysia/2506/04	Flu B 25.4	Flu B 28.8	Flu B Pos	Flu B 30.6/30.0
Respiratory Syncytial Virus A	NA	RSV 14.8	RSV 25.2	RSV A Pos	RSV A 19.5 RSV 19.7
Respiratory Syncytial Virus B	NA	RSV 13.9	RSV 22.4	RSV B Pos	RSV B 18.1 RSV 19.2
Rhinovirus 1A	CH93 (18)-18	Rhino 21.9	Flu A, Flu B & RSV negative	Rhinovirus Pos	Rhino 28.0/39.6
Parainfluenza virus type 1	NA	HPIV1 22.9	Flu A, Flu B & RSV negative	HPIV 1 Pos	HPIV 1 27.9/26.7
Echovirus type 30	NA	Enterovirus 20.4	Flu A, Flu B & RSV negative	Enterovirus Pos	Enterovirus 24.8
Coxsackievirus type A9	NA	Enterovirus 18.6	Flu A, Flu B & RSV negative	Enterovirus Pos	Enterovirus 24.9
M. pneumoniae	M129	Negative (NT)	Flu A, Flu B & RSV negative	M. pneumoniae Pos	M. pneumoniae 29.3
N. meningitidis serotype A	NA	Negative (NT)	Flu A, Flu B & RSV negative	Negative (NT)	N. meningitidis 25.0

[a] In-house routine multiplex real-time respiratory PCR assays (Clark et al., 2014).
[b] Influenza samples diluted 1:100 in virus transport medium and then processed in all assays to increase the challenge.

Table 2 Comparison of the TAC Results to Gold Standard Routine Real-Time Multiplex PCR Testing (417 Consecutive Specimens Dec 2014/Jan 2015)

Pathogen	True Positive	True Negative	Positive Array	False Positive	False Negative	Sensitivity (%)	Specificity (%)
Flu A	15	402	14	0	1	93.75	100.0
RSV	92	325	103	11	0	100.0	96.73
HPIV 1–4	12	405	12	1	1	92.3	99.75
Adenovirus	9	408	16	8	1	90.0	98.08
Rhinovirus	84	333	85	6	5	94.38	98.23
HMPV	10	407	10	1	1	90.91	99.75
Coronaviruses (Gp 1 and 2)	18	399	19	2	1	94.74	99.5
Overall	205	212	213	27	10	95.35	88.7

Table 4 QCMD 2014 *Clostridium difficile* DNA EQA Programme

Sample	Matrix[a]	Sample Contents	Ct Value[b]	Sample Status	Sample Type	Real-Time In-House PCR[c] Ct Value	Gastro TaqMan Array Card Results Ct Value
CD14-01	BHI broth	C. difficile 027 (Toxin: A+/B+)	27.7	Detected	Educational	22.9/25.3	C. difficile GDH 26.8 Tox B 28.5
CD14-02	BHI broth	C. difficile 017 (Toxin: A−/B+)	26.6	Detected	Educational	22.5/25.0	C. difficile GDH 26.9 Tox B 28.0
CD14-03	BHI broth	C. difficile 027 (Toxin: A+/B+)	24.2	Frequently detected	Core	20.0/21.8	C. difficile GDH 24.3 Tox B 25.6
CD14-04	BHI broth	C. difficile 017 (Toxin: A−/B+)	19.8	Frequently detected	Core	15.7/17.6	C. difficile GDH 20.0 Tox B 21.4
CD14-05	BHI broth	C. sordellii (Toxin: A−/B−)	–	Negative	Educational	–	Negative
CD14-06	BHI broth	C. difficile 027 (Toxin: A+/B+)	20.9	Frequently detected	Core	16.6/18.8	C. difficile GDH 20.7 Tox B 22.0
CD14-07	BHI broth	Clostridium Negative	–	Negative	Core	–	Negative
CD14-08	BHI broth	C. difficile 017 (Toxin: A−/B+)	23.0	Frequently detected	Core	19.2/21.7	C. difficile GDH 23.1 Tox B 25.0
CD14-09	BHI broth	C. difficile 027 (Toxin: A+/B+)	24.2	Frequently detected	Core	19.9/22.3	C. difficile GDH 23.6 Tox B 25.6
CD14-10	BHI broth	C. difficile 027 (Toxin: A+/B+)	30.9	Infrequently detected	Educational	27.3/29.4	C. difficile GDH 30.4 Tox B 31.9

[a]BHI broth: Brain Heart Infusion broth.
[b]The values provided are specific to the QCMD reference target and methodology used for the qualification of panel members.
[c]In-house routine real-time Clostridium difficile (GDH/Tox B) PCR assay (McElgunn et al., 2014).

Table 4 QCMD 2014 *Clostridium difficile* DNA EQA Programme

Sample	Matrix[a]	Sample Contents	Ct Value[b]	Sample Status	Sample Type	Real-Time In-House PCR[c] Ct Value	Gastro TaqMan Array Card Results Ct Value
CD14-01	BHI broth	*C. difficile* 027 (Toxin: A+/B+)	27.7	Detected	Educational	22.9/25.3	*C. difficile* GDH 26.8 Tox B 28.5
CD14-02	BHI broth	*C. difficile* 017 (Toxin: A−/B+)	26.6	Detected	Educational	22.5/25.0	*C. difficile* GDH 26.9 Tox B 28.0
CD14-03	BHI broth	*C. difficile* 027 (Toxin: A+/B+)	24.2	Frequently detected	Core	20.0/21.8	*C. difficile* GDH 24.3 Tox B 25.6
CD14-04	BHI broth	*C. difficile* 017 (Toxin: A−/B+)	19.8	Frequently detected	Core	15.7/17.6	*C. difficile* GDH 20.0 Tox B 21.4
CD14-05	BHI broth	*C. sordellii* (Toxin: A−/B−)	–	Negative	Educational	–	Negative
CD14-06	BHI broth	*C. difficile* 027 (Toxin: A+/B+)	20.9	Frequently detected	Core	16.6/18.8	*C. difficile* GDH 20.7 Tox B 22.0
CD14-07	BHI broth	*Clostridium* Negative	–	Negative	Core	–	Negative
CD14-08	BHI broth	*C. difficile* 017 (Toxin: A−/B+)	23.0	Frequently detected	Core	19.2/21.7	*C. difficile* GDH 23.1 Tox B 25.0
CD14-09	BHI broth	*C. difficile* 027 (Toxin: A+/B+)	24.2	Frequently detected	Core	19.9/22.3	*C. difficile* GDH 23.6 Tox B 25.6
CD14-10	BHI broth	*C. difficile* 027 (Toxin: A+/B+)	30.9	Infrequently detected	Educational	27.3/29.4	*C. difficile* GDH 30.4 Tox B 31.9

[a]BHI broth: Brain Heart Infusion broth.
[b]The values provided are specific to the QCMD reference target and methodology used for the qualification of panel members.
[c]In-house routine real-time Clostridium difficile (GDH/Tox B) PCR assay (McElgunn et al., 2014).

Table 3 QCMD 2014 Norovirus RNA EQA Programme

Sample	Matrix[a]	Sample Contents	Ct Value[b]	Sample Status	Sample Type	Real-Time In-House PCR[c] Ct Value	Gastro TaqMan Array Card Results Ct Value
NV14-01	TM	Norovirus GII.4	27.3	Detected	Educational	25.7	GII 29.0
NV14-02	TM	Norovirus GI.3	26.8	Detected	Educational	28.5	GI#1 32.8 GI#2 34.4
NV14-03	TM	Norovirus GII.4	25.8	Frequently detected	Core	20.1	GII 27.5
NV14-04	TM	NV Negative	–	Negative	Core	–	Negative
NV14-05	TM	Norovirus GII.4	23.8	Frequently detected	Core	20.5	GII 27.7
NV14-06	TM	Norovirus GI.3	23.8	Detected	Core	22.0	GI#1 27.7 GI#2 29.0
NV14-07	TM	Norovirus GII.4	31.6	Detected	Educational	29.3	GII 37.5
NV14-08	Buffer	Norovirus I.7 RNA	14.9	Detected	Educational	13.3	GI#1 21.2 GI#2 18.7
NV14-09	Buffer	NV Negative	–	Negative	Core	–	Negative
NV14-10	Buffer	Norovirus II.b RNA	15.8	Frequently detected	Core	15.7	GII 21.7
NV14-11	Buffer	Norovirus I.8 RNA	18.5	Frequently detected	Core	14.9	GI#1 24.0 GI#2 22.6
NV14-12	Buffer	Norovirus II.4 RNA	13.7	Frequently detected	Core	12.9	GII 21.0

[a]TM: transport medium. Buffer: 10 mM Tris–Cl, 1 mM EDTA.
[b]The values provided are specific to the QCMD reference target and methodology used for the qualification of panel members.
[c]In-house routine real-time Norovirus PCR assay (Rolfe et al., 2007).

also concluded that they were well suited for surveillance or clinical purposes. In a follow-up seminal study (Liu et al., 2014) assessing the performance of their TAC alongside two other molecular platforms (PCR Luminex and multiplex real-time PCR; both in-house) against comparator methods (bacterial culture, ELISA and PCR) using over 1500 specimens, a molecular quantitative approach was clearly superior 'Of the laboratories participating in this study, the TaqMan® array card platform was viewed the most favourable for a complete syndromic screen because implementation and procurement was simple, risk of contamination scant, and quantification robust'. We too have developed a gastrointestinal TAC incorporating a few of the assays described by Liu et al. (2013) but expanding it with additional assays (e.g. enterovirus, Norovirus group I, parechovirus and Hepatitis A & E viruses), outlined in Figure 2. While validation of this plate is currently underway in our network of regional laboratories within PHE, the data generated (>500 samples processed) so far are extremely encouraging, and this is clearly illustrated in the performance obtained for the recent Norovirus QCMD (Table 3) and *Clostridium difficile* QCMD (Table 4) panels.

4.3 OTHER SYNDROMIC TACs

Rachwal and colleagues published a study in which they described an array for the detection of biothreat organisms (Rachwal et al., 2012). These included *Bacillus anthracis*, *Francisella tularensis*, *Yersinia pestis*, *Burkholderia mallei* and *Burkholderia pseudomallei*. As in the initial paper from Kodani et al. (2011), these authors pointed out that the array system is around 10-fold less sensitive than singleplex RT-PCR assays performed alongside. This is a factor that has to be borne in mind when considering the clinical application of these assays but since the relevance of very low Ct value results is always a point of contention, this issue is not new to molecular diagnostics. However, as outlined above for our respiratory TAC, striving to improve assays during card development and ensuring only sublime assays (with exquisite analytical sensitivity) migrate to the finalised validated diagnostic card should mitigate and address this point.

Additional TACs have been developed within our laboratory. Like the respiratory and Gastro TACs shown in Figure 2, a jaundice TAC was designed incorporating assays considered important in causing or complicating jaundice in patients. Assays on the jaundice TAC currently include hepatitis viruses A–G, HCV genotyping (1–4), erythrovirus B19, SEN virus (SEN-V), SV40, Dengue virus, pan-adenovirus, polyomaviruses BK and JC, herpes viruses 1–8, *Toxoplasma gondii*, *M. tuberculosis* complex, *Chlamydophila psittaci* and internal controls (Figure 2). A similar TAC has also been developed at the Centers for Disease Control and Prevention (CDC, Atlanta) but is limited to the hepatitis viruses (Kodani et al., 2014). Including additional pathogens allows for a more rapid and comprehensive diagnosis without the requirement for multiple blood or biopsy samples to be sought from a patient. It also removes or limits the possibility of missing a complicating infection once an initial diagnosis is made.

Table 2 Comparison of the TAC Results to Gold Standard Routine Real-Time Multiplex PCR Testing (417 Consecutive Specimens Dec 2014/Jan 2015)

Pathogen	True Positive	True Negative	Positive Array	False Positive	False Negative	Sensitivity (%)	Specificity (%)
Flu A	15	402	14	0	1	93.75	100.0
RSV	92	325	103	11	0	100.0	96.73
HPIV 1–4	12	405	12	1	1	92.3	99.75
Adenovirus	9	408	16	8	1	90.0	98.08
Rhinovirus	84	333	85	6	5	94.38	98.23
HMPV	10	407	10	1	1	90.91	99.75
Coronaviruses (Gp 1 and 2)	18	399	19	2	1	94.74	99.5
Overall	205	212	213	27	10	95.35	88.7

Another area in which TACs are being applied is in the diagnosis of infections in immunocompromised patients. The number of transplants per year continues to rise, placing increased demands on the clinical care teams and therefore also on diagnostic laboratories. Despite technical innovations, morbidity and mortality rates have not improved in the past 20 years, mainly due to infection, cardiovascular disease and malignancy. Significant advances in early diagnosis of infections will have a beneficial impact on morbidity and mortality rates and therefore will continue to be the focus of current diagnostic algorithm improvements.

A modification of the jaundice TAC, replacing the HCV genotyping assays with 12 additional assays, carefully selected to include those considered problematic in transplant and immunocompromised patients has generated the first comprehensive diagnostic algorithm for infections in this setting (Figure 2). Assays on this 'transplant' TAC include herpes viruses 1–8, human polyomaviruses (BK/JC/WU/KI/MC), human papillomaviruses (pan HPV, E6 mRNA genotype 16 and 18), erythrovirus B19, pan-adenovirus, hepatitis viruses A–G, Sen V, *T. gondii*, *M. tuberculosis* complex and fungal species (*Aspergillus* and *Candida* species).

There are many other TACs in development, e.g., for CNS infections (Figure 2) and STIs, both in the United Kingdom and elsewhere. The versatility of this approach and its adaptability highlights the fact that it can be readily recruited or used in a 'mix-and-match' style depending on local or outbreak requirements. Furthermore, due to their ease of use, they can be considered a parachute technology, fast tracked when need arises or in resource poor settings.

CONCLUSIONS

This chapter has outlined the development and use of TACs in the clinical diagnostic laboratory. The number of different TACs available continues to expand and their versatility has been explained. As a clinical intervention tool, they offer many possibilities. Locally, the use of TAC assays has provided rapid diagnosis in seriously ill patients requiring ECMO. It has also been instrumental in identifying B19 as the possible causative agent of recurring rhabdomyolysis in a young child, HSV type 1 pneumonia in an immunocompetent patient and a case of BK virus-induced pneumonia in an immunocompromised patient, all of which would have otherwise remained undiagnosed. In addition, the real potential for the use of TAC in outbreak situations has been recognised in its ability to identify *M. pneumoniae* as the causative agent in an outbreak among university students (Waller et al., 2014). In that outbreak, the TAC exhibited 100% sensitivity and specificity when compared to multiplex real-time PCR and allowed the outbreak to be quickly recognised and appropriately managed. The syndromic approach offers many advantages over a monoplex targeted one. The assays are not without their limitations, and the concommitant loss of sensitivity when using relatively small amounts of nucleic acid extract has to be considered. That aside, the TAC heralds a step forward in the development of versatile molecular assays for diagnosis of infectious disease.

ACKNOWLEDGEMENTS

The authors wish to thank Dr. Marijke Reydners and Dr. Patrick Descheemaeker (Az Sint-Jan Hospital, Bruges, Belgium) for their collaborative contribution on the development of the respiratory TAC. We would also like to thank all our collaborators within PHE Network for their efforts in developing and validating the Gastro TAC, namely, Dr. Andrew Sails, Dr. John Magee, Mr. Gary Eltringham, Dr. Malcolm Guiver, Dr. Barry Vipond, Dr. Derren Ready, Dr. David Allen, Dr. Kathie Grant, Dr. Claire Jenkins and Dr. Peter Marsh.

In addition, we are appreciative of Dr. Emma Hutley, Mr. Andrew Dixon (Royal Centre for Defence Medicine, ICT Centre, Birmingham Research Park) and Dr. Simon Weller (Defence Science and Technology Laboratory, Porton Down) for their collaborative work on the Biofire™ FilmArray platform.

We also would like to thank Dr. Graeme Alexander (Department of Medicine, University of Cambridge) for his invaluable input with the Jaundice and Transplant TAC. Lastly, we thank Dr. Richard Stevens and Mr. Surendra Parmar for their continued contributions to TAC development and all the staff of the Clinical Microbiology & Public Health laboratory for all their technical help and expertise.

REFERENCES

Bankowski, M. J., & Anderson, S. M. (2004). Real-time nucleic acid amplification in clinical microbiology. *Clinical Microbiology Newsletter*, *26*, 9–15.

Bustin, S. A., Beaulieu, J. F., Huggett, J., Jaggi, R., Kibenge, F. S., Olsvik, P. A., et al. (2010). MIQE precis: Practical implementation of minimum standard guidelines for fluorescence-based quantitative real-time PCR experiments. *BMC Molecular Biology*, *11*, 74.

Bustin, S. A., Benes, V., Garson, J., Hellemans, J., Huggett, J., Kubista, M., et al. (2013). The need for transparency and good practices in the qPCR literature. *Nature Methods*, *10*(11), 1063–1067.

Clark, T. W., Medina, M., Batham, S., Curran, M. D., Parmar, S., & Nicholson, K. G. (2014). C-reactive protein level and microbial aetiology in patients hospitalised with acute exacerbation of COPD. *The European Respiratory Journal*, *45*(1), 76–86.

Cockerill, F. R., 3rd (2003). Application of rapid-cycle real-time polymerase chain reaction for diagnostic testing in the clinical microbiology laboratory. *Archives of Pathology & Laboratory Medicine*, *127*(9), 1112–1120.

Diaz, M. H., Waller, J. L., Napoliello, R. A., Islam, M. S., Wolff, B. J., Burken, D. J., et al. (2013). Optimization of multiple pathogen detection using the TaqMan Array Card: Application for a population-based study of neonatal infection. *PLoS One*, *8*(6), e66183.

Elnifro, E., Ashishi, A., Cooper, R., & Klapper, P. (2000). Multiplex PCR: Optimization and application in diagnostic virology. *Clinical Microbiology Reviews*, *13*(4), 559–570.

Gray, J., & Coupland, L. J. (2014). The increasing application of multiplex nucleic acid detection tests to the diagnosis of syndromic infections. *Epidemiology and Infection*, *142*(1), 1–11.

Johnson, G., Nour, A. A., Nolan, T., Huggett, J., & Bustin, S. (2014). Minimum information necessary for quantitative real-time PCR experiments. *Methods in Molecular Biology*, *1160*, 5–17.

Kodani, M., Mixson-Hayden, T., Drobeniuc, J., & Kamili, S. (2014). Rapid and sensitive approach to simultaneous detection of genomes of hepatitis A, B, C, D and E viruses. *Journal of Clinical Virology*, *61*(2), 260–264.

Kodani, M., & Winchell, J. M. (2012). Engineered combined-positive-control template for real-time reverse transcription-PCR in multiple-pathogen-detection assays. *Journal of Clinical Microbiology*, *50*(3), 1057–1060.

Kodani, M., Yang, G., Conklin, L. M., Travis, T. C., Whitney, C. G., Anderson, L. J., et al. (2011). Application of TaqMan low-density arrays for simultaneous detection of multiple respiratory pathogens. *Journal of Clinical Microbiology*, *49*(6), 2175–2182.

Liu, J., Gratz, J., Amour, C., Kibiki, G., Becker, S., Janaki, L., et al. (2013). A laboratory-developed TaqMan Array Card for simultaneous detection of 19 enteropathogens. *Journal of Clinical Microbiology*, *51*(2), 472–480.

Liu, J., Kabir, F., Manneh, J., Lertsethtakarn, P., Begum, S., Gratz, J., et al. (2014). Development and assessment of molecular diagnostic tests for 15 enteropathogens causing childhood diarrhoea: A multicentre study. *The Lancet Infectious Diseases*, *14*(8), 716–724.

Mackay, I. M. (2004). Real-time PCR in the microbiology laboratory. *Clinical Microbiology and Infection*, *10*(3), 190–212.

McElgunn, C. J., Pereira, C. R., Parham, N. J., Smythe, J. E., Wigglesworth, M. J., Smielewska, A., et al. (2014). A low complexity rapid molecular method for detection of *Clostridium difficile* in stool. *PLoS One*, *9*(1), e83808.

Morgan, U. M., Paillart, L., & Thompson, R. C. A. (1998). Comparison of PCR and microscopy for detection of *Cryptosporidium parvum* in human fecal specimens in a clinical trial. *Journal of Clinical Microbiology*, *36*(4), 995–998.

Pholwat, S., Liu, J., Stroup, S., Gratz, J., Banu, S., Rahman, S., et al. (2015). Integrated microfluidic card with Taqman probes and high-resolution melt analysis to detect tuberculosis in 10 genes. *mBio*, *6*, 2273–2287.

Pierce, V. M., Elkan, M., Leet, M., McGowan, K. L., & Hodinka, R. L. (2012). Comparison of the Idaho Technology FilmArray system to real-time PCR for detection of respiratory pathogens in children. *Journal of Clinical Microbiology*, *50*(2), 364–371.

Platts-Mills, J. A., Operario, D. J., & Houpt, E. R. (2012). Molecular diagnosis of diarrhea: Current status and future potential. *Current Infectious Disease Reports*, *14*(1), 41–46.

Popowitch, E. B., O'Neill, S. S., & Miller, M. B. (2013). Comparison of the Biofire FilmArray RP, Genmark eSensor RVP, Luminex xTAG RVPv1, and Luminex xTAG RVP fast multiplex assays for detection of respiratory viruses. *Journal of Clinical Microbiology*, *51*(5), 1528–1533.

Rachwal, P. A., Rose, H. L., Cox, V., Lukaszewski, R. A., Murch, A. L., & Weller, S. A. (2012). The potential of TaqMan Array Cards for detection of multiple biological agents by real-time PCR. *PLoS One*, *7*(4), e35971.

Rolfe, K. J., Parmar, S., Mururi, D., Wreghitt, T. G., Jalal, H., Zhang, H., et al. (2007). An internally controlled, one-step, real-time RT-PCR assay for norovirus detection and genogrouping. *Journal of Clinical Virology*, *39*(4), 318–321.

Santos, H. J., & Rivera, W. L. (2013). Comparison of direct fecal smear microscopy, culture, and polymerase chain reaction for the detection of *Blastocystis* sp. in human stool samples. *Asian Pacific Journal of Tropical Medicine*, *6*(10), 780–784.

Saunders, N., Zambon, M., Sharp, I., Siddiqui, R., Bermingham, A., Ellis, J., et al. (2013). Guidance on the development and validation of diagnostic tests that depend on nucleic acid amplification and detection. *Journal of Clinical Virology*, *56*(3), 260–270.

Stensvold, C. R., & Nielsen, H. V. (2012). Comparison of microscopy and PCR for detection of intestinal parasites in Danish patients supports an incentive for molecular screening platforms. *Journal of Clinical Microbiology, 50*(2), 540–541.

Taylor, S., Wakem, M., Dijkman, G., Alsarraj, M., & Nguyen, M. (2010). A practical approach to RT-qPCR-Publishing data that conform to the MIQE guidelines. *Methods, 50*(4), S1–S5.

Waller, J. L., Diaz, M. H., Petrone, B. L., Benitez, A. J., Wolff, B. J., Edison, L., et al. (2014). Detection and characterization of *Mycoplasma pneumoniae* during an outbreak of respiratory illness at a university. *Journal of Clinical Microbiology, 52*(3), 849–853.

Wolk, D. M., Kaleta, E. J., & Wysocki, V. H. (2012). PCR-electrospray ionization mass spectrometry: The potential to change infectious disease diagnostics in clinical and public health laboratories. *The Journal of Molecular Diagnostics, 14*(4), 295–304.

Invasive Fungal Infections and Approaches to Their Diagnosis

7

Michael A. Pfaller[*,†,1]

T2Biosystems, Lexington, Massachusetts, USA
†*University of Iowa College of Medicine and College of Public Health, Iowa City, Iowa, USA*
¹*Corresponding author: e-mail address: michael-pfaller@uiowa.edu*

1 INTRODUCTION

Historically, the diagnosis of invasive fungal infection (IFI) has been limited to the correlation of clinical signs and symptoms of disease with recovery of the infecting organism from or histopathologic detection of the organism in clinical specimens (Alexander & Pfaller, 2006; Arvanitis, Anagnostou, Fuchs, Caliendo, & Mylonakis, 2014; Guarner & Brandt, 2011; Marcos & Pincus, 2013; Pfaller, Pappas, & Wingard, 2006; Pfaller & McGinnis, 2009). The frequency of fungal disease, particularly that caused by systemic and opportunistic pathogens, has increased substantially during the past several decades (Brown et al., 2012; Diekema & Pfaller, 2004; Fridkin, 2005; Hajjeh et al., 2004; McNeil et al., 2001; Nucci & Marr, 2005; Pfaller & Diekema, 2004a, 2004b, 2007; Rees, Pinner, Hajjeh, Brandt, & Reingold, 1998; Trick et al., 2002; Walsh et al., 2004). This increase is primarily due to expanding patient populations at high risk for the development of opportunistic life-threatening fungal infections, which includes persons with AIDS, neoplastic disease, extremes of age, immunosuppressive therapy, and those undergoing organ transplantation (both haematologic and solid organ) and aggressive surgery (Apisarnthanarak & Powderly, 2005; Baddley, Stroud, Salzman, & Pappas, 2001; Blumberg et al., 2001; Bow, 2005; Foster & Chanock, 2005; Grazziatti, Anaissie, & Wingard, 2005; Kauffman, 2001; Kaufman & Fairchild, 2004; Marr, 2005; Marr, Carter, Crippa, Wald, & Corey, 2002; Marty, Baden, & Rubin, 2005; McNeil et al., 2001; Patterson, Kirkpatrick, & White, 2000; Pfaller & Diekema, 2007; Rees et al., 1998; Roilides et al., 2004; Saiman et al., 2000; Segal et al., 2006; Singh, 2003; Singh & Paterson, 2005; Walsh & Groll, 1999; Wisplinghoff et al., 2004; Zaoutis et al., 2005). Infections in these populations are clearly important causes of morbidity and mortality. Serious infections are being reported with an ever-increasing array of pathogens (Table 1) including the well-known pathogenic fungi such as *Candida albicans*, *Cryptococcus neoformans*, and *Aspergillus fumigatus* (Mirza et al., 2003; Patterson et al., 2000; Pfaller, 2001; Pfaller & Diekema, 2007). Yeasts such as *Trichosporon*, *Rhodotorula*,

Methods in Microbiology, Volume 42, ISSN 0580-9517, http://dx.doi.org/10.1016/bs.mim.2015.05.002

Table 1 Spectrum of Opportunistic Fungal Pathogens[a]

Organism Group	Examples of Specific Pathogens	
Candida	C. albicans	C. krusei
	C. glabrata	C. lusitaniae
	C. parapsilosis	C. guilliermondii
	C. tropicalis	C. rugosa
Other yeasts	Cryptococcus neoformans/ gattii	Saccharomyces species
		Rhodotorula species
	Trichosporon species	Malassezia species
	Blastoschizomyces capitatus	
Aspergillus	A. fumigatus	A. versicolor
	A. flavus	A. terreus
	A. niger	A. calidoustus
Mucormycetes	Rhizopus species	Apophysomyces species
	Rhizomucor species	Cunninghamella bertholletiae
	Mucor species	Saksenaea species
	Lichtheimia (Absidia) species	
Other hyaline moulds	Fusarium species	Trichoderma species
	Sarocladium (Acremonium) species	Purpureocilium (Paecilomyces) lilacinus
	Scedosporium species	Chrysosporium species
Dematiaceous moulds	Alternaria species	Cladophialophora species
	Bipolaris species	Phialophora species
	Exophiala species	Dactylaria species
	Ramichloridium species	Wangiella species
Dimorphic moulds	Histoplasma capsulatum	Sporothrix schenckii
	Coccidioides immitis/ posadasii	Talaromyces (Penicillium) marneffei
	Blastomyces dermatitidis	
	Paracoccidioides brasiliensis	
Other	Pneumocystis jirovecii	
	Microsporidia species	

[a]List not all inclusive.

and *Blastoschizomyces* spp., species of *Candida* other than *C. albicans*, hyaline hyphomycetes including *Fusarium* and *Scedosporium* spp., and a wide variety of dematiaceous fungi are increasing in importance (Arendrup et al., 2014; Asada et al., 2006; Fleming & Anaissie, 2005; Fridkin, 2005; Jensen, Gahrn-Hansen, Arendrup, & Bruun, 2004; Lamaris et al., 2006; Lo Cascio, Ligozzi, Maccacaro, & Fontana, 2004; Matsue, Uryu, Koseki, Asada, & Takeuchi, 2006; Pendle et al., 2004; Pfaller & Diekema, 2004a, 2004b, 2007, 2010; Pfaller, Diekema, Colombo, et al., 2006; Pfaller, Diekema, Rex, et al., 2006; Pfaller, Diekema, & Sheehan, 2006; Pfaller, Pappas, & Wingard, 2006; Tucker, Beresford, Sigler, & Rogers, 2004). Modern medical mycology has become the study of infections caused by a variety of taxonomically diverse opportunistic fungi (Fleming & Anaissie, 2005; Nucci & Marr, 2005; Schell, 1995; Walsh & Groll, 1999; Walsh et al., 2004). It is now clear that there are no non-pathogenic fungi; virtually any fungus can cause a lethal mycosis in an immunocompromised host.

Owing to the complexity of the various patient populations subject to infection and the increasing variety of fungal pathogens, opportunistic mycoses pose a significant diagnostic challenge to clinicians and microbiologists alike (Nucci & Marr, 2005; Pfaller & Diekema, 2004a, 2004b; Pfaller, Richter, & Diekema, 2005). It is absolutely essential that institutions caring for high-risk immunocompromised patients place a high priority on maximising their diagnostic capabilities for the early detection of opportunistic fungal infections (Fridkin, 2005; Nucci & Marr, 2005; Pfaller, Richter, et al., 2005). Successful diagnosis and management of such infections in the compromised patient is highly dependent on a team approach involving clinicians, microbiologists, pharmacists, and pathologists (Ananda-Rajuah, Slavin, & Thursky, 2012; Chandler & Watts, 1987; Koneman & Roberts, 2002; O'Shaughnessy, Shea, & Witebsky, 2003; Schell, 1995). Due to the recognition that delays in administering timely and appropriate antifungal therapy is directly related to patient mortality (Chamilos, Lewis, & Kontoyiannis, 2008; Garey et al., 2006; Morrel, Fraser, & Kollef, 2005), increasingly emphasis has been placed on achieving earlier and less invasive means of diagnosis, in hopes of improving the mortality rates associated with these infections (Alexander & Pfaller, 2006; Arvanitis et al., 2014; Griffin & Hanson, 2014; Zhang, 2013). This chapter provides a review of both the conventional diagnostic approaches (i.e. direct microscopy, culture, and histopathologic methods) and the newer culture-independent methods (i.e. immunologic, biochemical, proteomic, and molecular methods) for both identification and detection of fungal pathogens (Table 2).

2 SPECTRUM OF OPPORTUNISTIC PATHOGENS

As noted previously, the spectrum of possible opportunistic fungal pathogens is very broad (Table 1). Nevertheless, *C. albicans* remains the most common fungal pathogen (Azie et al., 2012; Pfaller & Diekema, 2007; Pfaller, Chaturvedi, et al., 2012; Pfaller, Neofytos, et al., 2012; Pfaller, Woosley, Messer, Jones, & Castanheira, 2012; Pfaller, Andes, et al., 2014; Pfaller, Diekema, Procop, & Wiederhold, 2014; Pfaller, Jones, & Castanheira, 2014) and *A. fumigatus* is the commonest mould pathogen (Azie et al., 2012; Morgan et al., 2005; Pfaller, Pappas, & Wingard, 2006). Likewise, *Candida* spp., *Aspergillus* spp., and *C. neoformans* account for more than 80% of all fungal infections in blood and marrow transplant (BMT), solid-organ transplant, and other immunocompromised patient populations (Azie et al., 2012; Bow, 2005; Marcos & Pincus, 2013; Marr, 2005; Marty et al.,

Table 2 Laboratory Diagnosis of Invasive Fungal Infections

A Conventional Microbiologic
    1 Direct microscopy (Gram, Giemsa, and Calcofluor stains)
    2 Culture
    3 Identification
    4 Susceptibility testing
B Histopathologic
    1 Conventional microscopy
        a Routine stains (H&E)
        b Special stains (GMS, Mucicarmine, PAS)
    2 Direct immunofluorescence
    3 *In situ* hybridisation
C Immunologic
    1 *Aspergillus* lateral-flow device (LFD) antigen test
    2 *Blastomyces* antigen test
    3 Cryptococcal antigen tests (LA, EIA, LFD)
    4 *Histoplasma* antigen test
    5 Galactomannan test
    6 Mannan/anti-mannan test
    7 Antibody detection
D Molecular and proteomic methods
    1 Direct detection
    2 Identification
    3 Strain typing
    4 Biochemical
E Biochemical
    1 Cell wall components (β-D-glucan)

Abbreviations: H&E, haematoxylin and eosin; GMS, Gomori's methenamine silver; PAS, periodic acid–Schiff.

2005; Pfaller & Diekema, 2010; Pfaller, Andes, et al., 2014; Pfaller, Diekema, et al., 2014; Pfaller, Jones, et al., 2014) (Table 3).

The frequency of fungal infections, predilection for specific fungal pathogens, and the time of onset of various fungal infections differs for different patient groups (Azie et al., 2012; Dictar, Maioo, Alexander, Jacob, & Veron, 2000; Pfaller & McGinnis, 2009; Pfaller, Boyken, Hollis, et al., 2005; Pfaller, Boyken, Messer, et al., 2005; Pfaller, Richter, et al., 2005) (Table 3). Unique risk factors, exposures, and the degree and duration of immunosuppression likely account for this variability (Fleming & Anaissie, 2005). Thus, invasive aspergillosis (IA) exerts a major impact on patients following allogeneic BMT, lung and liver transplant, whereas infections due to *Candida* sp. are most problematic among neonates, post-surgical intensive care unit patients, and in those patients post liver, pancreas, and small bowel transplantation (Azie et al., 2012; Collins et al., 1994; Pfaller & Diekema, 2007; Pfaller, Diekema, Colombo, et al., 2006; Pfaller, Diekema, Rex, et al., 2006; Pfaller, Diekema, & Sheehan, 2006; Pfaller, Pappas, & Wingard, 2006) (Table 3). Aside from *Candida* and *Aspergillus* spp., the less common and "emerging" fungal pathogens are often found in the environment and tend to cause infection in those

Table 3 Relative Frequency of Opportunistic Mycoses Among Different Patient Groups[a]

Patient Group[b]	Mycosis[c]											
	Asp.	Can.	Cryp.	Tri.	PCP.	Hyal.	Phae.	Blas.[d]	Hist.[d]	Cocci.[d]	Tmar.[d]	Mucor
Transplant												
Allo. BMT	++++	++	++	++	+++	++	+	(+)	(+)	(+)	(+)	++
Liver	+++	++++	+++	+	+	+	+	(+)	(+)	(+)	(+)	+
Lung	++++	+++	++	+	+	+	+	(+)	(+)	(+)	(+)	+
Kidney	++	+++	++	+	+	+	+	(+)	(+)	(+)	(+)	+
Heart	++++	+++	++	+	+	+	+	(+)	(+)	(+)	(+)	+
Pancreas	++	++++	+	+	+	+	+	(+)	(+)	(+)	(+)	+
Sm. Bowel	++	++++	+	+	+	+	+	(+)	(+)	(+)	(+)	+
Malignancy												
Haeme	+++	++++	++	+	++	+	+	(+)	(+)	(+)	(+)	+
Solid	++	++++	++	+	++	+	+	(+)	(+)	(+)	(+)	+
HIV/AIDS	++	++	+++		++++	+	+	(++)	(++++)	(++++)	(++++)	+
Critical care												
Adult	+	++++		+	+	+						+
Neonate		++++		+	+							+

[a]Relative frequency of mycoses within each patient group indicated as ++++ (most frequent) to + (least frequent).
[b]Patient group abbreviations: Allo BMT, allogeneic blood and marrow transplant; Sm. bowel, small bowel; Haeme, haematologic malignancy; Solid, solid tumor malignancy.
[c]Mycosis abbreviations: Asp., Aspergillosis; Can., Candidiasis; Tric., Trichosporonosis; Cryp. Cryptococcosis; PCP., Pneumocystis jirovecii pneumonia; Hyal., Hyalohyphomycosis; Phae., Phaeohyphomycosis; Blas., Blastomycosis; Hist., Histoplasmosis; Cocci., Coccidioidomycosis; Tmar., Talaromyces (Penicillium) marneffei; Mucor, Mucormycosis. See Table 1 for specific examples within each group.
[d]Frequency of endemic mycoses indicated by (+) to (++++) within endemic regions only.
Adapted from Azie et al. (2012).

individuals with long-term immunosuppression (Fleming & Anaissie, 2005; Pfaller & Diekema, 2004a, 2004b; Pfaller & Diekema, 2010; Walsh & Groll, 1999). The greater the degree and duration of immunosuppression, the broader the spectrum of potential opportunistic fungal pathogens (Fleming & Anaissie, 2005; Nucci & Marr, 2005; Segal et al., 2006). The diversity of potential fungal pathogens places a great deal of pressure on the clinical laboratory to make a rapid and accurate aetiologic diagnosis. This is not a simple task and often requires the use of several different approaches and the collaboration of clinicians, mycologists, and pathologists.

3 CLINICAL RECOGNITION OF FUNGAL INFECTION

The increased frequency of invasive mycoses requires that clinicians caring for immunocompromised individuals have an enhanced index of clinical suspicion and a greater appreciation and recognition of the major risk factors that predispose patients to fungal infections (Garey et al., 2006; McLintock et al., 2004; Ostrosky-Zeichner, 2003; Pfaller, Diekema, Colombo, et al., 2006; Pfaller, Diekema, Rex, et al., 2006; Pfaller, Diekema, & Sheehan, 2006; Pfaller, Pappas, & Wingard, 2006; Prentice, Kibbler, & Prentice, 2000). Clinical suspicion, thorough history, and physical examination, including evaluation for skin and mucosal lesions, inspection of all intravascular devices, a careful ophthalmologic examination, diagnostic imaging of appropriate organ systems, and finally procurement of appropriate specimens for laboratory diagnosis, are critical steps that must be taken. Specific fungal infections may be associated with well-known clinical scenarios such as endophthalmitis and macronodular skin lesions (candidiasis), onychomycosis in a neutropenic patient (fusariosis), sinus infection in diabetic ketoacidosis (mucormycosis), or myalgias and myositis (candidaemia due to *C. tropicalis*) (Pfaller, Boyken, Hollis, et al., 2005; Pfaller, Boyken, Messer, et al., 2005; Pfaller, Richter, et al. 2005). Other clinical signs and symptoms that may be associated with fungal infections include suppurative thrombophlebitis (*Candida*), hepatitis (hepatosplenic candidiasis), purpura fulminans and bullous dermatitis (*Candida*, *Aspergillus*, and *Fusarium* spp.), and osteomyelitis (*Candida*) (Fleming, Walsh, & Anaissie, 2002; Pfaller, 1992). Although these are useful clinical clues, they are not entirely specific and often are observed rather late in the course of the infection. Thus, diagnosis usually depends upon three basic laboratory approaches (Table 2): (i) microbiologic, (ii) immunologic, and (iii) histopathologic (Alexander & Pfaller, 2006; Chandler & Watts, 1987; Pfaller & McGinnis, 2009; Pfaller, Boyken, Hollis, et al., 2005; Pfaller, Boyken, Messer, et al., 2005; Pfaller, Richter, et al., 2005).

The newer molecular, proteomic, and biochemical diagnostic tests are increasingly useful as rapid approaches for diagnosis of fungal infections (Alexander & Pfaller, 2006; Arvanitis et al., 2014; Chen, Holliday, & Meyer, 2002; Iwen, Hinrichs, & Rupp, 2002; Lopez-Ribot & Patterson, 2005; Marcos & Pincus, 2013; McLintock & Jones, 2004; Odds, 2003; Yeo & Wong, 2002; Zhang, 2013). Unfortunately, the clinician frequently cannot wait for definitive laboratory results and may have to decide to treat solely on the basis of the available clinical information and the clinician's subjective assessment (Alexander & Pfaller, 2006; Pfaller & Diekema, 2007; Pfaller & McGinnis, 2009; Ostrosky-Zeichner, 2003).

4 LABORATORY DIAGNOSIS
4.1 STAINS AND DIRECT EXAMINATION

Direct microscopic examination of clinical material is perhaps the most rapid, useful, and cost-effective means of diagnosing fungal infections (Chandler & Watts, 1987; Connor, Chandler, Schwartz, Manz, & Lack, 1997; Guarner, 2014; Guarner & Brandt, 2011; Koneman & Roberts, 2002; Shea, 2007). Detection of fungal elements microscopically may provide a diagnosis in less than an hour, whereas culture results are often not available for days or even weeks. In contrast with histopathologic examination, direct microscopy as performed in the microbiology laboratory does not utilise fixed tissue and relies instead on rapid examination of wet mounts, Gram-, Giemsa-, or Calcofluor-stained material to visualise fungal elements in clinical specimens (Table 4).

Often infections are caused by organisms that can be specifically identified by direct microscopy because they possess a distinctive morphology. For example, if typical yeast cells, spherules, or other structures are observed microscopically, an aetiologic diagnosis can be made for infections caused by *Histoplasma capsulatum*, *Blastomyces dermatitidis*, *C. neoformans/gattii*, *Coccidioides immitis/posadasii*, and *Pneumocystis jirovecii*. In other infections, such as aspergillosis, candidiasis, and mucormycosis, the morphologic appearance may lead to a diagnosis of the type of infection but not the actual species identification of the aetiologic agent (Table 5).

Detection of fungi in tissue and clinical material by direct microscopic examination is often helpful in determining the significance of culture results. This is especially true when the fungi isolated in culture are known components of the normal human flora or normally occur in the environment. Finally, detection of specific fungal elements by microscopy can assist the laboratory in selecting the most appropriate means by which to culture the clinical specimen. For example, the presence of hyphae of a mucormycetous organism should prompt the use of malt agar or even sterile bread without preservatives for its isolation.

Although direct examination may be quite valuable in diagnosing fungal infections, one must keep in mind that both false-negative and false-positive results may occur. As in other areas of microbiology, direct examinations are less sensitive than culture, and a negative microscopic examination does not rule out a fungal infection.

A number of different stains and techniques may be used to help demonstrate the presence of fungi by direct microscopic examination (Chandler & Watts, 1987; Guarner & Brandt, 2011; Koneman & Roberts, 2002; Shea, 2007) (Table 4). Most commonly, microscopy as performed in the clinical laboratory consists of examination of clinical material placed in 10–20% potassium hydroxide (KOH) containing

Table 4 Methods and Stains Available for Direct Microscopic Detection of Fungal Elements

Method/Stain	Use	Comments
Alcian blue stain	Detection of *C. neoformans* in CSF	Rapid (2 min); insensitive and not commonly used
Calcofluor white stain	Detection of all fungi including *P. jirovecii*	Rapid (1–2 min); detects fungal cell wall chitin by bright fluorescence. Used in combination with KOH. Requires fluorescent microscope and proper filters. Background fluorescence may make examination of some specimens difficult
Fluorescent monoclonal antibody treatment	Examination of respiratory specimens for *P. jirovecii*	Sensitive and specific method for detecting the cysts of *P. jirovecii*. Does not stain the extracystic (trophozoite) forms
Fontana-Masson stain	Melanin stain for histologic sections	Confirms the presence of melanin in lightly pigmented cells of dematiaceous fungi when present in tissue sections. Useful for distinguishing *C. neoformans/ gattii* (positive) from most other yeasts (e.g. *Candida* spp. are negative for melanin)
Giemsa stain	Examination of bone marrow, peripheral smears, touch preparations, and respiratory specimens	Detects intracellular *H. capsulatum* and both intracystic and extracystic (trophozoite) forms of *P. jirovecii*. Does not stain cysts of *P. jirovecii*. Does stain organisms other than *H. capsulatum* and *P. jirovecii*
Gram stain	Detection of bacteria and fungi	Rapid (2–3 min); commonly performed on clinical specimens. Will stain most yeasts and hyphal elements. Most fungi stain Gram-positive but some, such as *C. neoformans/gattii*, exhibit stippling or appear Gram-negative
Haematoxylin and eosin(H&E) stain	General purpose histologic stain	Best stain to demonstrate host reaction in infected tissue. Stains most fungi but small numbers of organisms may be difficult to differentiate from background. Useful in demonstrating natural pigment in dematiaceous fungi

Table 4 Methods and Stains Available for Direct Microscopic Detection of Fungal Elements—cont'd

Method/Stain	Use	Comments
India ink	Detection of encapsulated yeasts	Rapid (1 min); insensitive (40%) means of detecting *C. neoformans/gattii* in CSF
KOH treatment	Clearing specimens of cellular debris to make fungi more visible	Rapid (5 min); some specimens may be difficult to clear and require an additional 5–10 min. May produce confusing artefacts. Most useful when combined with Calcofluor white
Methylene blue treatment	Detection of fungi in skin scrapings	Rapid (2 min); may be used in combination with KOH. Largely replaced by Calcofluor white (improved sensitivity and specificity)
Methenamine silver stain (GMS)	Detection of fungi in histologic sections and *P. jirovecii* cysts in respiratory specimens	Staining of tissue may take up to 1 h. Respiratory specimens more rapid (5–10 min). Best stain for detection of all fungi. Usually performed in Cytopathology laboratory
Mucicarmine stain	Histopathologic stain for mucin	Useful for demonstrating capsular material of *C. neoformans/gattii*. May also stain the cell walls of *B. dermatitidis* and *Rhinosporidium sieberi*
Papanicolaou stain (PAP)	Cytologic stain used primarily to detect malignant cells	Stains most fungal elements; yeasts >hyphae. Allows cytologist to detect fungal elements
Periodic acid–Schiff (PAS) stain	Histologic stain for detection of fungi	Stains both yeasts and hyphae in tissue. *B. dermatitidis* may appear pleomorphic. PAS-positive artefacts may resemble yeast cells
Toluidine blue stain	Examination of respiratory specimens for *P. jirovecii*	Stains *P. jirovecii* cysts a purple colour. Does stain other fungi. Largely displaced by fluorescent antibody and Calcofluor white treatments
Wright stain	Examination of bone marrow, peripheral smears, and touch preparations	Similar to Giemsa stain. Detects intracellular *H. capsulatum*

Adapted from Pfaller, Boyken, Hollis, et al. (2005), Pfaller, Boyken, Messer, et al. (2005), Pfaller, Richter, et al. (2005) and Shea (2007).

Table 5 Characteristic Microscopic Features of Opportunistic and Pathogenic Fungi in Clinical Specimens

Fungi	Microscopic Morphologic Features in Clinical Specimens
Candida species	Oval budding yeasts 2–6 µm in diameter. Pseudohyphae and true hyphae may be present
Cryptococcus neoformans/gattii	Spherical budding yeasts of variable size, 2–15 µm in diameter. Capsule may be present or absent. No hyphae or pseudohyphae
Trichosporon species	Hyaline arthroconidia, blastoconidia, and pseudohyphae, 2–4 by 8 µm in size
Malassezia species	Small oval budding yeasts. "Bowling pin" appearance with collarette. Both hyphal and yeast forms seen on skin scrapings
Aspergillus species	Hyaline, septate, dichotomously branched hyphae of uniform width (3–6 µm). Conidial heads may be seen in cavitary lesions
Other hyaline Hyphomycetes[a]	Hyaline, septate, dichotomously branching hyphae. Angioinvasion is common. Adventitious conidiation may be present. May be indistinguishable from *Aspergillus* species
Mucormycetes	Broad, thin-walled, pausi-septate hyphae, 6–25 µm wide, with non-parallel sides and random branches. May stain poorly with GMS
Dematiaceous fungi	Pigmented (brown, tan or black) hyphae, 2–6 µm wide. May be branched or unbranched. Often constricted at the point of septation
Histoplasma capsulatum	Small (2–4 µm in diameter), intracellular, budding yeasts
Coccidioides immitis/ posadasii	Spherical, thick-walled spherules, 20–30 µm in diameter. Mature spherules contain small (2–5 µm in diameter) endospores. Released endospores may be mistaken for yeast. Arthroconidia and hyphae may form in cavitary lesions
Blastomyces dermatitidis	Large (8–15 µm in diameter), thick-walled yeast cells. The junction between mother and daughter cells is typically broad-based. Cells may appear multinucleate
Sporothrix schenckii	Elongated or "cigar-shaped" yeast cells of varying size (rare). Tissue reaction forms asteroid bodies
Talaromyces (Penicillium) marneffei	Oval, intracellular yeast cells bisected with a septum (fission yeast)
Pneumocystis jirovecii	Cysts are round, collapsed, or crescent shaped. Trophozoites seen on staining with Giemsa or immunofluorescent stains

[a]*Includes* Fusarium, Scedosporium, Sarocladium, *and* Purpureocilium (Penicillium) *species.*
Adapted from Alexander and Pfaller (2006).

the fluorescent reagent Calcofluor white, or staining of individual smears or touch preparations by either Gram, Giemsa, periodic acid–Schiff (PAS), or any combination of these stains. The Calcofluor white binds to the chitin in the fungal cell wall and fluoresces blue-white or green, providing a rapid and sensitive means of detecting fungi in clinical material (Shea, 2007). This technique has some limitations if the fungal structures are heavily melanised. The Gram stain is useful for the detection of *Candida* and *Cryptococcus* spp. and also stains the hyphal elements of moulds such as *Aspergillus*, the mucormycetes, and *Fusarium* spp. Fungi are typically Gram-positive but may appear speckled or Gram-negative. The capsular material of *C. neoformans* often appears as an orange-red precipitate around the cells. Many fungi will stain blue with the Giemsa stain but this stain is especially useful in detecting *H. capsulatum* intracellularly in bone marrow, peripheral blood, bronchoalveolar lavage (BAL) specimens, or touch preparations of lymph nodes or other tissues.

The morphologic characteristics of fungi seen on direct microscopic examination include budding yeasts, hyphae, and pseudohyphae. The combination of budding yeast cells and pseudohyphae is characteristic of *Candida*; however, these structures may also be seen with *Trichosporon* and *Geotrichum* (Table 5).

Among the moulds, *Aspergillus* spp. typically shows hyaline, dichotomous, acute angle branching, septate hyphae; however, this appearance is also typical of other hyaline moulds such as *Fusarium*, *Sarocladium* (*Acremonium*), *Purpureocilium* (*Paecilomyces*), and *Trichoderma* (Table 5). In contrast, mucormycetes (e.g. *Rhizopus*, *Mucor*) characteristically show broad, ribbon-like, aseptate, or sparsely septate hyphae. Finally, the dematiaceous fungi often present as darkly pigmented yeast-like and hyphal forms that may be visualised on unstained material and further characterised by the Fontana-Masson stain for melanin.

The laboratory diagnosis of a *P. jirovecii* infection is commonly made in the clinical laboratory by the direct examination of induced sputum and specimens collected by bronchoscopy. In addition to the more general stains such as Gomori's methenamine silver stain (GMS), Giemsa, and toluidine blue (Table 4), the commercial availability of fluorescent monoclonal antibody-based conjugates have enhanced the detection of this organism and provide a sensitive and highly specific diagnosis (Cushion, 2011).

Cytologic and histologic stains, including the Papanicolaou, haematoxylin and eosin (H&E), GMS, and PAS stains, are used for the detection of fungi in cytologic preparations, fine-needle aspirates, tissues, body fluids, and exudates (Table 4). The Papanicolaou stain is usually performed in the cytopathology laboratory. These stains can detect fungi such as *Candida* spp., *B. dermatitidis*, *C. neoformans*, *C. immitis/posadasii*, *H. capsulatum*, and the hyphae of mucormycetes, as well as *Aspergillus* spp. and other fungi. When present in sufficient numbers, most fungi can be seen in tissue stained with H&E; however, *Candida* and *Aspergillus* spp. may be missed in H&E-stained sections. Special stains such as GMS and PAS are essential for detecting small numbers of organisms and for clearly seeing their morphology. Although not widely available, specific immunofluorescent stains may be extremely helpful in confirming a presumptive histologic identification of some fungi such as *Candida*, *Aspergillus*, *Fusarium*, the dimorphic fungi, and others (Hayden et al., 2003; Hayden, Qian, Procop, Roberts, & Lloyd, 2002; Kaufman, Standard, Jalberl, & Kraft, 1997). Histologic examination of fixed tissue provides the opportunity to determine whether the fungus is in viable tissue, information that is useful in distinguishing between active infection and colonisation (Guarner, 2014; Guarner & Brandt, 2011). The microscopic morphologic features of several of the more common fungal pathogens are presented in Table 5.

4.2 CULTURE

The most sensitive means of diagnosing a fungal infection is generally considered to be the isolation of the infecting agent on culture media. Having said this, false-negative cultures are well documented in the face of disseminated fungal infection and even when positive, the results may be delayed or difficult to interpret (Avni, Leibovici, & Paul, 2011; Chen et al., 2002; Garey et al., 2006; Morrel et al., 2005; Yeo & Wong, 2002). In most instances, culture is necessary to specifically identify the aetiologic agent and, if indicated, to determine the *in vitro* susceptibility to various antifungal agents.

Multiple factors must be considered in order to optimise the recovery of fungi from clinical material. In addition to adequate specimen procurement, a variety of culture media should be employed to ensure the isolation of a broad range of medically important fungi. Generally, both selective and non-selective media are used for the primary recovery of fungi from clinical specimens. Nonselective media such as brain heart infusion (BHI) agar or SABHI (Sabouraud glucose plus BHI) agar will permit recovery of both rapidly growing moulds and yeasts as well as the more slowly growing, or fastidious, fungi. Sabouraud glucose agar is generally considered inferior to these media for primary isolation and should not be used for this purpose. Routine bacteriologic media, such as sheep blood agar and chocolate agar, will support the growth of most fungi; however, growth may be slow and not detectable in the short time period (3–5 days) allowed for the incubation of most bacterial cultures. A blood-containing medium such as BHI with 5% or 10% sheep blood may be used, in addition to the non-selective primary isolation media, as an aid in recovering fastidious dimorphic fungi such as *H. capsulatum*, *B. dermatitidis*, or *C. immitis/posadasii*. Although cycloheximide is often added to BHI-blood agar to inhibit the growth of rapidly growing yeasts and moulds that may "contaminate" the culture, it is important to understand that this agent will inhibit the growth of many opportunistic fungi that may also serve as the aetiologic agent of the infection. For this reason, media that has been supplemented with cycloheximide must always be complemented with the same media without the inhibitory agent. Specimens that may be contaminated with bacteria should also be cultured on a selective medium such as inhibitory mould agar, SABHI, or BHI plus antibacterial agents (e.g., gentamicin, chloramphenicol, or ciprofloxacin plus streptomycin). CHROMagar (Hardy Diagnostics; Becton-Dickinson) is a chromogenic medium that is both selective for fungi and differential for certain *Candida* species (Murray, Zinchuk, & Larone, 2005; Odds & Bernaerts, 1994; Pfaller, Houston, & Coffman, 1996; Tan & Peterson, 2005). This medium inhibits bacterial growth and because different species of *Candida* appear as different coloured colonies it is useful in detecting mixed cultures of *Candida* and other fungi (Odds & Bernaerts, 1994; Pfaller et al., 1996). In certain situations, it is necessary to use specialised media for recovery of specific fungi. For example, the use of malt agar, or sterile bread without preservatives, will enhance the recovery of the mucormycetes whereas medium supplemented with olive oil is necessary for recovery of the lipophillic yeast, *Malassezia furfur* (McGowan, 2011).

Although not all serious fungal infections are marked by haematogenous dissemination and fungaemia, detection of fungaemia is useful in diagnosing opportunistic infection due to *Candida* spp., *C. neoformans*, *Trichosporon* spp., *Malassezia* spp., *Fusarium* spp., and occasionally, *Sarocladium* (*Acremonium*) spp., *Purpureocilium* (*Paecilomyces*) spp., *Scedosporium* spp., and *Aspergillus terreus* (Arendrup et al., 2011, 2014; O'Shaughnessy et al., 2003). Blood cultures may be negative in the face of disseminated disease (Avni et al., 2011; Clancy & Nguyen, 2013); however, advances in blood culture technology have markedly improved the ability of laboratories to detect fungaemia (Magadia & Weinstein, 2001; Reimer, Wilson, & Weinstein, 1997; Riedel & Carroll, 2010). The lysis centrifugation method (Isolator, Wampole) and the continuous monitoring automated blood culture systems are all widely employed methods for the detection of fungaemia due to *Candida* spp. (Magadia & Weinstein, 2001; Riedel & Carroll, 2010). The development of specialised broth media containing lytic agents, resins, charcoal, or diatomaceous earth coupled with continuous agitation has contributed to the improved performance of the broth-based systems (Arendrup et al., 2011; Horvath, George, Murray, Harrison, & Hospenthal, 2004; Magadia & Weinstein, 2001; Meyer, Letscher-Bru, Jaulhac, Waller, & Candolfi, 2004; Reimer et al., 1997; Riedel & Carroll, 2010; Sandven et al., 2006). However, recovery of *C. glabrata*, *C. neoformans*, *H. capsulatum*, *M. furfur*, and *Fusarium* spp. may be inferior with broth-based systems compared with the lysis centrifugation method (Arendrup et al., 2011; Magadia & Weinstein, 2001). On the other hand, culture contamination occurs more frequently with the more labour-intensive lysis centrifugation method.

Among the four automated, continuous-monitoring blood culture systems that have been developed, the Bactec (BD Diagnostics) and the BactT/Alert (bioMerieux) systems are superior in their capability to recover yeasts from blood. Studies have demonstrated that these systems match the performance of the lysis centrifugation method for the detection of *Candida* spp. and *C. neoformans* (Geha & Roberts, 1994; Reimer et al., 1997). Likewise, the Bactec MycoF-lytic bottle (Becton-Dickinson) also matches the lysis centrifugation method for the recovery of dimorphic fungi (Magadia & Weinstein, 2001; Pfaller, 1992; Reimer et al., 1997; Riedel & Carroll, 2010). It has been proposed that optimal detection of fungaemia requires the collection of adequate volumes of blood (20–30 ml per BC set) and the use of both a broth (vented, agitated) and an agar-based (lysis centrifugation) blood culture method (Pfaller, 1992). It should be noted that the Bactec aerobic blood culture bottle and medium are notorious for poor recovery of *C. glabrata* and requires the use of a fungal medium (MycoF-lytic) for optimal recovery of this important species (Arendrup et al., 2011; Beyda, Alam, & Garey, 2013; Horvath et al., 2004; Meyer et al., 2004; Riedel, Eisinger, Dam, Stamper, & Carroll, 2011).

Blood cultures remain an established approach to the diagnosis of invasive candidiasis (IC; candidaemia as well as infections of normally sterile sites, tissues, and abscesses) (Cornely et al., 2012; Cuenca-Estrella et al., 2012; Pappas et al., 2009; Ullmann et al., 2012). This is despite the fact that blood cultures have always been considered to be too slow and insensitive to serve as an early diagnostic method (Arvanitis et al., 2014; Clancy & Nguyen, 2013; Meyer et al., 2004; Zhang, 2013). It is increasingly apparent that as many as one-third of patients with IC never have a positive blood culture and even when positive, species-level results may not be available for 48–72 h or longer (Clancy & Nguyen, 2013; Nguyen et al., 2012; Taur, Cohen, Dubnow, Paskovaty, & Seo, 2010). Notably, in a review of 415 autopsy-proven cases of IC, Clancy and Nguyen (2013) demonstrated that blood cultures were positive in only 38% of cases. This low yield of blood cultures was confirmed by Avni et al. (2011) who reported a pooled blood culture positivity rate of 38% in a meta-analysis of 10 studies for the diagnosis of IC. Cultures of blood and other clinical specimens may be rendered falsely negative by the use of prophylactic or empiric antifungal therapy further confounding efforts to establish a firm diagnosis (Neely et al., 2013; Riedel et al., 2011). Despite these negative features, blood cultures remain at the heart of care guidelines for IC in hospitalised individuals (Cornely et al., 2012; Pappas et al., 2009). It is now understood that the information required to limit mortality and the emergence of antifungal resistance is time-critical and as such rapid culture-independent diagnostic tests are required to complement blood cultures in the management of high-risk patients (Arvanitis et al., 2014; Clancy & Nguyen, 2013; Zhang, 2013).

Interpretation of the results of fungal cultures may be difficult due to the frequent colonisation of certain body sites (e.g. respiratory, gastrointestinal (GI), and genitourinary (GU) tracts) and contamination of specimens or cultures by environmental organisms, many of which can also serve as aetiologic agents of opportunistic mycoses. Whereas most isolates of *Candida* spp., *C. neoformans*, *H. capsulatum*, and *Fusarium* spp. obtained from blood cultures are clinically significant (Magadia & Weinstein, 2001; Pfaller, 1992; Pien et al., 2010; Reimer et al., 1997), others such as *Aspergillus* spp. (with the exception of *A. terreus*) and *Penicillium* spp. (with the exception of *Talaromyces* [*Penicillium*] *marneffei*) most probably represent pseudofungaemia or contamination (Duthie & Denning, 1995; Kontoyiannis et al., 2000; Lionakis, Bodey, Tarrand, Raad, & Kontoyannis, 2004; Pien et al., 2010).

Cultures of any clinical specimens that are positive for any of the endemic dimorphic pathogens (*H. capsulatum*, *B. dermatitidis*, *C. immitis/posadasii*) are virtually always considered to be clinically significant. Isolation of *Aspergillus* spp. from cultures of respiratory tract specimens is especially problematic, because this organism is common in the environment, and it may colonise the airways of an individual without causing overt disease. The clinical significance of isolation of *Aspergillus* spp. from respiratory tract cultures may be confirmed upon direct microscopic visualisation of the organism in viable tissue (Guarner & Brandt, 2011); however, this is often not possible due to the inability of the patient to tolerate an invasive procedure.

Table 6 Risk of Invasive Aspergillosis for Patients with Respiratory Tract Cultures Positive for *Aspergillus* species, by Study and Risk Characteristics

Risk Category	Study, Percentage Risk (No. of Patients with Positive Culture/ Total No. of Patients)		
	Yu et al. (1986)	Horvath and Dummer (1996)	Perfect et al. (2001)
High[a]	100 (17/17)	72 (34/47)	57 (117/206)
Intermediate[b]	37 (20/54)	58 (14/24)	15 (228/1510)
Low[c]	0 (0/9)	14 (1/7)	<1 (1/155)

[a]*Includes allogeneic BMT recipients, patients with neutropenia, and patients with haematologic cancer.*
[b]*Includes autologous BMT and solid-organ transplant recipients, patients receiving therapy with corticosteroids, HIV-infected patients, and patients with malnutrition, diabetes, underlying pulmonary disease, or solid-organ cancer.*
[c]*Includes HIV-infected patients, patients with cystic fibrosis or connective tissue disease, and other nonimmunosuppressed patients.*

There is now considerable evidence indicating that the interpretation of respiratory tract cultures (e.g. expectorated sputum, BAL) yielding *Aspergillus* spp. may be aided by considering the risk group of the patient (Horvath & Dummer, 1996; Perfect, Cox, Lee, et al., 2001; Yu, Muder, & Poorsattar, 1986) (Table 6). Among patients considered to be at high risk for IA (e.g. allogeneic BMT recipients, patients with haematologic malignancies, and patients with neutropenia), a positive culture that yields *Aspergillus* spp. is often associated with invasive disease. The positive predictive value (PPV) of a culture positive for *Aspergillus* spp. is lessened for autologous BMT recipients, solid-organ transplant recipients, and HIV-infected patients (Horvath & Dummer, 1996; Perfect et al., 2001; Yu et al., 1986). In addition, the specific identification of the fungus isolated from respiratory culture specimens can also help in determining clinical significance; *Aspergillus niger* is rarely a pathogen, whereas *A. terreus* and *A. flavus* have been shown to be statistically associated with IA when isolated from cultures of respiratory tract specimens (Perfect et al., 2001).

Although most fungal cultures are performed for the express purpose of making an aetiologic diagnosis of a fungal infection in order to optimise specific antifungal therapy, fungal surveillance cultures have been studied as potential predictors of invasive mycoses in select patient populations (LaRocco & Burgert, 1997; Muñoz, Burillo, & Bouza, 2001; Pfaller, Cabezudo, Koontz, Bale, & Gingrich, 1987; Sandford, Merz, Wingard, Charache, & Saral, 1980; Snydman, 2001). Although active surveillance of high-risk patients may enhance case detection of IC and IA, the data is quite variable (Blumberg et al., 2001; Kusne et al., 1992; Nalesnk, Myerowitz, Jenkins, Lenskey, & Herbert, 1980; Pfaller et al., 1987; Riley, Pavia, Beatty, Denton, & Carroll, 1995; Rogers, Visscher, Bartlett, & Smith, 1985; Saiman et al., 2002; Sandford et al., 1980; Tollemar, Ericzon, Holmberg, & Andersson, 1990; Tollemar, Holmberg, Ringden, & Lonquist, 1989; Yu et al., 1986). In light of this problem, guidelines for the development of microbiologic surveillance protocols have been published (LaRocco & Burgert, 1997; Snydman, 2001; Walker, 1991) and include consideration of (i) the probability and severity of a specific infection in the patient population; (ii) the time period in which the risk of infection exists relative to the onset of immunosuppression; (iii) the accuracy, timeliness, and the cost of the tests used for surveillance; (iv) the sensitivity, specificity, PPV and the negative predictive value (NPV) of the test result in an asymptomatic patient; (v) the efficacy of prophylactic therapy as well as the risk of development of resistance; and (vi) the expected impact on clinical outcome.

It is widely acknowledged that the primary reservoir for haematogenously disseminated candidiasis is the GI tract and to a lesser extent the GU system (Eggimann, Garbino, & Pittet, 2003; Garbino et al., 2002; Solomkin, 2005). As such, one might expect that those immunocompromised patients with demonstrated colonisation of several anatomic sites by *Candida* spp. to be at increased risk of IC compared to similar patients without colonisation (Garbino et al., 2002; Pfaller & Diekema, 2007; Pittet, Monod, Suter, Frenck, & Auckenthaler, 1994; Solomkin, 2005).

Although several studies have shown this to be the case, it is clear that colonisation without infection is very common, with the result that the PPV of a positive surveillance culture for *Candida* is quite low (Blumberg et al., 2001; Pfaller et al., 1987; Riley et al., 1995; Snydman, 2001; Solomkin, 2005). Conversely, given the high NPV of surveillance cultures, persistently negative cultures may be useful in identifying those patients at low risk for IC (Blumberg et al., 2001; Pfaller et al., 1987; Saiman et al., 2002; Sandford et al., 1980; Solomkin, 2005).

The PPV of surveillance cultures for *Candida* may be improved when species such as *C. tropicalis*, *C. glabrata*, or *C. krusei* are found colonising high-risk patients (Pfaller et al., 1987; Sandford et al., 1980; Wingard et al., 1991). Studies of surveillance cultures in BMT patients and those with haematologic malignancies found that as many as 60–80% of neutropenic patients who were colonised with *C. tropicalis* eventually developed IC with this species (Kontoyiannis et al., 2001; Pfaller & Diekema, 2007; Pfaller et al., 1987; Sandford et al., 1980). Detection of *C. glabrata* and *C. krusei* as colonising species may be of additional importance given that these species are typically less susceptible to azoles and other systemically active antifungal agents than *C. albicans* (Pfaller & Diekema, 2004a, 2004b, 2007; Pfaller et al., 2003; Wingard et al., 1991), thus prompting a change in empirical antifungal coverage for patients colonised with these species (Pappas et al., 2009; Spellberg, Filler, & Edwards, 2006).

The poor sensitivity of routine nasal cultures for *Aspergillus* spp. limits the usefulness of surveillance cultures for early detection of IA (LaRocco & Burgert, 1997; Perfect et al., 2001). The isolation of *Aspergillus* spp. from respiratory specimens is often considered to represent colonisation or contamination (LaRocco & Burgert, 1997); however, a positive respiratory culture for *Aspergillus* in a host at high risk for IA (e.g. neutropenic, BMT) has been shown to be predictive of IA and is an indication for more aggressive attempts at diagnosis and consideration of early empiric antifungal therapy (Kusne et al., 1992; Nalesnk et al., 1980; Perfect et al., 2001; Rogers et al., 1985; Yu et al., 1986) (Table 6).

Thus, the performance of surveillance cultures for *Candida* and *Aspergillus* spp. may have some value in highly selected patient groups (Pittet et al., 1994; Snydman, 2001; Solomkin, 2005). However, the use of routine surveillance cultures of asymptomatic immunocompromised patients remains of questionable value and generally should be discouraged (Dykewicz, 2001; Riley et al., 1995; Snydman, 2001).

4.3 IDENTIFYING CHARACTERISTICS OF FUNGI

Identification of fungi to genus and species is increasingly important as the spectrum of opportunistic pathogens continues to expand (Fridkin, 2005; Nucci & Marr, 2005; Pfaller & Diekema, 2004a, 2004b) (Table 1). Although the clinical presentation of many fungal infections may be indistinguishable, specific identification of the aetiologic agent may have a direct bearing on the management of the infectious process. It is increasingly apparent that one cannot rely on a single therapeutic approach (e.g. administration of amphotericin B) for the management of all, or even most, fungal infections (Groll et al., 2003; Ostrosky-Zeichner, Casdevall, Galgiani, Odds, & Rex, 2010; Roilides, Lyman, Panagopoulu, & Chanock, 2003; Spanakis, Aparis, & Mylonakis, 2006). Furthermore, the identification of fungal pathogens may have additional diagnostic and epidemiologic implications. In the case of the more unusual mycoses, specific aetiologic identification may provide access to the literature and the experience of others regarding the probable course of infection and response to therapy.

4.3.1 Identification of yeasts

Yeasts are usually characterised morphologically as single cells that reproduce by simple budding; however, under certain conditions some yeasts may form true hyphae, pseudohyphae, capsules, arthroconidia, and other reproductive structures (Larone, 2014) (Table 5). Colonies form on most agar media within a few (2–5) days and are usually round, opaque, moist, or mucoid, and white or cream coloured. Although the appearance of small hyphal projections or "feet" from the edge of a colony has been cited as characteristic of *C. albicans* (Calvin et al., 1998; Nagashi & Baron, 1997; National Committee for Clinical Laboratory Standards, 2002), further investigation has shown this to be unreliable as both *C. tropicalis* and *C. krusei* are capable of producing this phenotype (Buschelman, Jones, Pfaller, Koontz, & Doern, 1999).

Since *C. albicans* constitutes the vast majority of yeasts recovered from clinical specimens, several rapid and simple tests have been devised to distinguish it from other yeasts (Alexander, Dodds-Ashley, Reller, & Reed, 2006; Baumgartner, Freydiere, & Gille, 1996; Forrest et al., 2006; Howell & Hazen, 2011; Marcos & Pincus, 2013; NCCLS, 2002; Odds & Bernaerts, 1994; Oliveira, Haase, Kurtzman, Hyldig-Nielsen, & Stender, 2001; Pfaller et al., 1996; Rigby et al., 2002; Tan & Peterson, 2005; Wilson et al., 2005). The most widely used test for identification of *C. albicans* is the germ tube test (Howell & Hazen, 2011; Larone, 2014). *C. albicans* forms germ tubes within 3 h when incubated in serum or plasma at 35 °C. Other species of *Candida* are capable of germ tube formation but require extended incubation. *C. dubliniensis* and *C. stellatoidea* are capable of forming germ tubes within 3 h and may be difficult or impossible to differentiate from *C. albicans* without performing additional physiologic, immunologic, or nucleic acid-based testing (Gales et al., 1999; Marcos & Pincus, 2013; Oliveira et al., 2001; Pinjon, Sullivan, Salkin, Shanley, & Coleman, 1998; Rigby et al., 2002).

Chromogenic media, such as CHROMagar, and rapid (<24 h) colorimetric tests based on the detection of *C. albicans*-specific enzymes (L-proline aminopeptidase and β-galactose-aminidase), have proven useful in the rapid presumptive identification of *C. albicans* (Howell & Hazen, 2011).

Although a single presumptive identification test is not sufficient for identifying most yeasts, a positive germ tube or colorimetric test or characteristic green colony on CHROMagar medium is generally considered to be acceptable for the identification of *C. albicans* (Howell & Hazen, 2011; NCCLS, 2002).

Among the more than 200 species of *Candida* that have been identified, five-*C. albicans*, *C. glabrata*, *C. parapsilosis*, *C. tropicalis*, and *C. krusei*—account for 95–98% of cases of IC (Hajjeh et al., 2004; Pfaller & Diekema, 2004a, 2004b, 2007). Recent reports indicate that shifts have occurred in the distribution of non-*albicans* species with the emergence of *C. glabrata*, *C. krusei*, *C. lusitaniae*, and other less common species (Fridkin, 2005; Nucci & Marr, 2005; Pfaller, 2001; Pfaller & Diekema, 2007; Pfaller, Andes, et al., 2014; Pfaller, Diekema, et al., 2014; Pfaller, Jones, et al., 2014). Infections with these various species may require different therapeutic considerations (Arendrup et al., 2011, 2014; Dykewicz, 2001; Groll et al., 2003; Ostrosky-Zeichner et al., 2010; Pappas et al., 2009; Roilides et al., 2003; Spanakis et al., 2006; Spellberg et al., 2006) and so further identification of all germ tube-negative or colorimetric test-negative yeasts is mandatory for isolates obtained from blood and other normally sterile body fluids (Pappas et al., 2009; Pfaller, Boyken, Hollis, et al., 2005; Pfaller, Boyken, Messer, et al., 2005; Pfaller, Richter, et al., 2005; Rex & Pfaller, 2002; Spellberg et al., 2006). Due to the pathogenic potential of *C. neoformans/gattii*, all encapsulated yeasts from any body site should also be identified. There are several rapid screening tests that may be used for the presumptive identification of *C. neoformans* including the urease test (positive), nitrate test (negative), and production of phenol oxidase (positive) (Howell & Hazen, 2011). Other important non-*Candida* yeasts include *Blastoschizomyces capitatus*, *Rhodotorula* spp., *Saccharomyces* spp., and *Trichosporon* spp. (Arendrup et al., 2014; Nucci & Marr, 2005; Pfaller & Diekema, 2004a, 2004b; Spanakis et al., 2006; Walsh et al., 2004). These organisms are notable due to their broad antifungal resistance profiles (Pfaller & Diekema, 2004a, 2004b; Pfaller, Boyken, Hollis, et al., 2005; Pfaller, Boyken, Messer, et al., 2005; Pfaller, Richter, et al., 2005; Spanakis et al., 2006; Walsh et al., 2004), and require the use of biochemical, physiologic, and morphologic characterisation for accurate identification (Howell & Hazen, 2011). More recently, application of molecular and proteomic methods have greatly expanded the abilities of the clinical laboratory to identify *Candida* and other yeasts to species level (Arvanitis et al., 2014; Marcos & Pincus, 2013).

Further identification of germ tube-negative yeasts to species requires the determination of biochemical and physiologic profiles as well as an assessment of their morphology when grown on a medium such as cornmeal agar or yeast morphology agar (Howell & Hazen, 2011; Larone, 2014; Marcos & Pincus, 2013). In addition to the identification of *C. albicans*, colony morphology on CHROMagar allows the presumptive identification of *C. tropicalis* and *C. krusei* (Odds & Bernaerts, 1994; Pfaller et al., 1996). Likewise, *C. glabrata* may be identified by a rapid trehalose test (Howell & Hazen, 2011) or differential growth on blood agar (no growth or slow growth) versus eosin methylene blue agar (rapid growth) (Bale, Yang, & Pfaller, 1997). Given the fact that *C. glabrata* does not produce hyphae or pseudohyphae under most laboratory conditions (Howell & Hazen, 2011; Larone, 2014), the presence of "feet" along the edge of a colony may be sufficient to exclude *C. glabrata* from the list of possible species. Carbohydrate assimilation tests provide definitive identification for most *Candida* and non-*Candida* yeasts and may be performed by using one of several commercial identification systems (Howell & Hazen, 2011; Marcos & Pincus, 2013). Differentiation of yeasts with similar biochemical profiles can usually be accomplished by observing their microscopic characteristics on cornmeal agar (Howell & Hazen, 2011; Larone, 2014) (Table 5).

One of the most important advances in the post-culture identification of fungi is that of matrix-assisted laser desorption ionisation-time of flight (MALDI-TOF) mass spectrometry (MS) (Buchan & Ledeboer, 2014; Clark, Kaleta, Arora, & Wolk, 2013). MALDI-TOF MS uses species-specific patterns of peptides and protein masses to identify microorganisms. It has been shown to be highly accurate in identifying a broad array of bacteria and recently has been shown to provide a rapid and reliable method for the identification of yeasts, yeast-like fungi, and some moulds (Clark et al., 2013). The technique involves the use of whole-cell preparations or an extract of proteins from the fungal cells, spotting of the specimen on a grid, and overlaying the spot with a matrix. The proteins are ionised by a laser and migrate through a charged field in a vacuum tube towards a detector. The spectrum is generated rapidly (approximately 10 min per specimen) and is compared to a reference database. Presently, there are four commercial systems based on this method that are able to identify yeast and mould species. Studies evaluating their performances are promising, showing that this method is able to accurately and rapidly identify *Candida* spp. and *Aspergillus* spp. from positive cultures, with a high concordance (>90%) in comparison to both conventional and molecular methods (Arvanitis et al., 2014; Clark et al., 2013). In some instances, MALDI-TOF MS has been shown to be superior to conventional methods (Clark et al., 2013).

4.3.2 Identification of moulds

In contrast to yeasts, the identification of moulds is largely based upon morphologic features such as gross colony appearance and microscopic morphology (Larone, 2014) (Table 5). Visible growth on agar media may be obtained within 1–5 days for the Mucormycetes, most hyaline (light coloured hyphae and conidia) hyphomycetes, and some, but not all, dematiaceous (dark pigmented hyphae and conidia) fungi. In contrast, the dimorphic fungi (*H. capsulatum*, *B. dermatitidis*, *C. immitis/posadasii*, *Sporothrix schenckii*, *Paracoccidioides brasiliensis*) grow much more slowly and may require 2–4 weeks of incubation on solid media. Furthermore, the dimorphic fungi are not inhibited by cycloheximide, an agent that inhibits the growth of most of the more rapidly growing moulds that may represent either clinically unimportant "contaminants" or opportunistic pathogens causing infections in immunocompromised hosts.

The macroscopic appearance of many filamentous fungi when grown on solid media may provide clues as to the identification of the fungus. Variations in colonial morphology that may be medium- or strain-dependent precludes the use of this feature as the sole criterion for identification. Surface texture, topography, colour, reverse pigmentation, growth at 37 °C, and requirements for specific vitamins are all useful characteristics (Larone, 2014; Verweij & Brandt, 2007). Potato glucose agar and cornmeal agar are considered two of the more reliable media for assessment of gross colonial morphology. Colour development may be dependent on exposure to light.

Definitive identification of most moulds is dependent upon visualisation of the microscopic morphology of the fungus. Key morphologic features include the size, shape, method of production, and arrangement of the conidia or spores, as well as the size and appearance of the hyphal structures (Larone, 2014; Verweij & Brandt, 2007). Material must be prepared for microscopic examination in such a way as to minimise any disruption of the relationship of the conidia or spores to their respective reproductive structures. This is usually best accomplished by the use of slide cultures, whereby a coverslip is placed on the agar in such a way that the fungus spreads over the glass surface. The coverslip is then removed and placed on a slide for examination under the microscope. Determination of cell wall melanin and temperature-regulated dimorphism are also important characteristics. The dimorphic pathogens may also be characterised by immunologic- or nucleic acid probe-based methods in addition to morphology and thermal dimorphism (Larone, 2014). The typical features of selected filamentous and dimorphic pathogens are listed in Table 5.

Whereas MALDI-TOF MS has been shown to be a highly accurate and rapid method for the identification of *Candida* species from positive cultures, it remains limited for the identification of filamentous fungi (Clark et al., 2013). This is largely due to deficiencies in the available spectral libraries. MALDI-TOF MS has been shown to accurately identify several different species of *Aspergillus* (Alanio et al., 2011) and supplementation of the spectral library with an "in-house" database has allowed species-level identification of other filamentous fungi (Lau, Drake, Calhoun, Henderson, & Zelazny, 2013). Further expansion of the databases of the various MALDI-TOF MS systems using sequence-based identification of reference panels of moulds is ongoing.

4.3.3 Molecular methods for identification of yeasts and moulds

The use of both direct nucleic acid probes and amplification-based molecular approaches provide more rapid and objective identification of yeasts and moulds compared with traditional phenotypic methods (Arvanitis et al., 2014; Borman, Szekely, Palmer, & Johnson, 2012; Brandt et al., 2005; Buzina, Lang-Loidot, Braun, Freudenschuss, & Stammberger, 2001; Hall, Wohlfiel, & Roberts, 2003, 2004; Hinrickson, Hurst, Lott, Warnock, & Morrison, 2005; Hsiao et al., 2005; Iwen et al., 2002; Padhye, Smith, McLaughlin, Standard, & Kaufman, 1992; Padhye, Smith, Standard, McLaughlin, & Kaufman, 1994; Pfaller, Woosley, et al., 2012; Pounder et al., 2007; Rakeman et al., 2005; Schwarz et al., 2006). The development of direct nucleic acid probes has advanced the identification of dimorphic fungal pathogens once they are recovered in culture. These chemiluminescent-labelled DNA probes (AccuProbe, Hologic) are specific for target fungal rRNA and are commercially available for use in clinical laboratories. When applied to a lysate of the organism, the probes have a sensitivity similar to that of the more labour-intensive exoantigen test but demonstrate slightly less specificity, depending on the fungus tested (Brandt et al., 2005; Padhye et al., 1994). The probes demonstrate 100% specificity for *C. immitis* (Padhye et al., 1994); however, specificity is slightly less (99–99.7%) for both *B. dermatitidis* and *H. capsulatum* due to demonstrated cross-reactivity with *P. brasiliensis* and *Chrysosporium* species, respectively (Brandt et al., 2005; Padhye et al., 1992, 1994).

Among the newer rapid, post-culture methods for identification of *Candida* are the techniques of peptide nucleic acid (PNA)—fluorescence *in situ* hybridisation (FISH) (Arvanitis et al., 2014). The PNA-FISH tests (AdvanDx, Woburn, MA) are based on a fluorescein-labelled PNA probe that specifically detects *C. albicans*, *C. tropicalis*, or *C. glabrata* as individual species or detects a yeast species group (e.g. *C. albicans* and *C. parapsilosis* fluoresce green, *C. tropicalis* fluoresces yellow, and *C. glabrata* and *C. krusei* fluoresce red with the Yeast Traffic Light™ PNA FISH kit) in blood cultures by targeting species-specific rRNA sequences. The probes are added to smears made directly from the contents of the blood culture bottle and are hybridised for 90 min. Recent modifications to the probes and reagents have resulted in a second-generation test (*Quick*FISH™) that shortens the assay time to 30 min. Smears are subsequently examined by fluorescent microscopy. The test has been shown to have excellent sensitivity (99%), specificity (100%), PPV (100%), and NPV (99.3%) (Hall, LeFebre, Demi, Wohlfiel, & Wengenach, 2012). This approach may provide a time savings of 24–48 h, compared with conventional laboratory methods used for identification. It allows physicians to be notified of the yeasts identity along with positive blood culture results. Rapid, accurate identification of *C. albicans*, *C. tropicalis*, *C. parapsilosis*, *C. glabrata*, and *C. krusei* should promote optimal antifungal therapy with the most cost-effective antifungal agents, resulting in improved outcomes and significant antifungal savings for hospitals. A cost minimisation model was developed by Alexander et al. (2006) to assess cost savings associated with implementation of the PNA-FISH test in a hospital with a rate of 40% for *C. albicans* candidaemias and in which caspofungin was used over fluconazole for empiric treatment of IC. In their study, they predicted that the use of the PNA-FISH test would result in a cost savings of approximately $1800 per patient from reduced caspofungin use (switch to fluconazole in patients infected with *C. albicans*), despite the fact that laboratory costs for doing the *C. albicans* PNA-FISH test ($82.72 per test) exceeded those for the *C. albicans* screen test ($2.83) or the germ tube test ($4.42). A subsequent study at the University of Maryland (Forrest et al., 2006) used clinical data to show the effect of PNA-FISH testing for *C. albicans* and validated the decision model of Alexander et al. (2006). The Maryland investigators found that 43% of candidaemias in their hospital were due to *C. albicans* and that the most pronounced effects of the PNA-FISH test was on caspofungin usage in patients with *C. albicans* fungaemia. In this group, there was significant reduction in the usage of caspofungin (shift to fluconazole) with a corresponding decrease in antifungal costs of $1978 per patient (Forrest et al., 2006). The overall cost savings in reducing caspofungin usage surpassed the cost of the PNA-FISH test (net savings of $1729 per patient) and led to the development of straightforward hospital-specific algorithms (Forrest et al., 2006).

Amplification-based methods are increasingly being applied to the post-culture identification of yeasts and moulds (Brandt et al., 2005; Buchan & Ledeboer, 2014; Buzina et al., 2001; Clark et al., 2013; Hall et al., 2003, 2004; Griffin & Hanson, 2014; Hinrickson et al., 2005; Hsiao et al., 2005; Iwen et al., 2002; Kothari, Morgan, & Haake, 2014; Pfaller, Chaturvedi, et al., 2012; Pfaller, Neofytos, et al., 2012; Pfaller, Woosley, et al., 2012; Pounder et al., 2007; Rakeman et al., 2005; Schwarz et al., 2006). These methods are especially useful in identifying nonsporulating moulds that are unable to be identified by conventional methods (Rakeman et al., 2005). Both ribosomal targets and internal transcribed spacer regions have proven useful for the molecular identification of a wide variety of fungi. A major limitation of this approach is the variable quality and accuracy of the existing sequence databases (Hall et al., 2004; Hinrickson et al., 2005). Presently, with the availability of improved sequencing techniques, broader and more reliable databases, and more readily available kits and software, this technology has become a competitive alternative to the classic mycological identification methods used for clinically important fungi (Alexander & Pfaller, 2006; Bhally et al., 2006; Borman et al., 2012; Marcos & Pincus, 2013; Page, Shields, Merz, & Kurtzman, 2006; Pfaller, Chaturvedi, et al., 2012; Pfaller, Neofytos, et al., 2012; Pfaller, Woosley, et al., 2012; Pounder et al., 2007).

Multiplex polymerase chain reaction (PCR) platforms have been developed for the identification of *Candida* species in positive blood culture samples (Griffin & Hanson, 2014). The FilmArray™ blood culture identification panel (BioFire, Salt Lake City, UT) and the xTAG™ fungal analyte-specific reagent assay (Luminex Molecular Diagnostics, Toronto, Canada) have demonstrated sensitivities and specificities of 100% and 99%, respectively, for the five most common species of *Candida* (Arvanitis et al., 2014; Griffin & Hanson, 2014). The FilmArray™ approach provides fully automated nucleic acid extraction, amplification, and detection and has been FDA cleared for post-culture identification of bacteria and *Candida* species. Thus, molecular methods for fungal identification from positive blood cultures improve the time to identification when compared to conventional methods of identification and may be useful in antimicrobial stewardship interventions (Ananda-Rajuah et al., 2012; Caliendo et al., 2013). The disadvantages of these post-culture approaches are that they are still dependent on a positive blood culture, which may be negative in >50% of IC cases (Avni et al., 2011; Clancy & Nguyen, 2013). Pan-fungal nucleic acid amplification test platforms are emerging and offer promise for an expanded menu of fungal targets (Arvanitis et al., 2014).

4.4 IMMUNOLOGIC, BIOCHEMICAL, AND NUCLEIC ACID-BASED METHODS OF DIAGNOSIS

Although culture and histopathology remain the primary means of diagnosing fungal infections, there continues to be a need for more rapid, non-culture methods for diagnosis (Alexander & Pfaller, 2006; Arvanitis et al., 2014; Chen et al., 2002; Christie, 2003; Donnelly, 2006; Garey et al., 2006; Griffin & Hanson, 2014; Kawazu et al., 2004; Lopez-Ribot & Patterson, 2005; Maertens et al., 2005; Marcos & Pincus, 2013; McLintock & Jones, 2004; Mennink-Kersten, Donnelly, & Verweij, 2004; Morrel et al., 2005; Odds, 2003; Pfeiffer, Fin, & Safdar, 2006; Pham et al., 2003; Quindos, 2006; Rex, 2006; White, Archer, & Barnes, 2005; White, Linton, Perry, Johnson, & Barnes, 2006; Wingard, 2002; Yeo et al., 2006; Yoo et al., 2005; Zhang, 2013). Tests for detection of antibodies, rapid detection of specific fungal antigens and cell wall components, and fungal species-specific RNA or DNA sequences have the potential to yield rapid diagnostic information that can guide the early and appropriate use of antifungal therapy (Alexander & Pfaller, 2006; Arvanitis et al., 2014; Griffin & Hanson, 2014; Liesenfeld, Lehman, Hunfeld, & Kost, 2014; Lopez-Ribot & Patterson, 2005; Maertens et al., 2005; Quindos, 2006; Rex, 2006; White et al., 2005; Wingard, 2002; Yeo & Wong, 2002). Although a great deal of progress has been made in these areas, the true impact on the diagnosis and outcome of IFI is yet to be realised (Arvanitis et al., 2014; Christie, 2003; Donnelly, 2006; Griffin & Hanson, 2014; Odds, 2003; Rex, 2006; Zhang, 2013).

4.4.1 Antibody detection

Serologic tests can provide a rapid means of diagnosing fungal infections, as well as a means to monitor the progression of the infection and the patient's response to therapy by comparing serial determinations of antibody or antigen titers (Alexander & Pfaller, 2006; Lopez-Ribot & Patterson, 2005; Persat et al., 2002; Philip et al., 2005; Saubolle, McKellar, & Sussland, 2007; Sendid et al., 2003; Wheat et al., 1983). Most conventional serologic tests are based on detection of antibodies against specific fungal antigens. Often, this serodiagnostic approach is ineffective, because many patients who are at risk for IFI are not capable of mounting a specific antibody response to infection. In addition, determination of the presence of an acute infection typically requires a comparison of the type and quantity of antibody present in both acute-phase and convalescent-phase serum samples, an exercise that is not helpful during the acute presentation, when therapeutic interventions are being decided (Alexander & Pfaller, 2006; Lopez-Ribot & Patterson, 2005).

Among the most reliable and widely used conventional serodiagnostic tests in mycology are the antibody tests for histoplasmosis and coccidioidomycosis (Brandt, Gomez, & Warnock, 2011; Saubolle et al., 2007; Wheat et al., 1983). Both the complement fixation (CF) and the immunodiffusion (ID) tests have been found useful for diagnosis of these infections. CF titers of >1:32 may be diagnostically significant, whereas lower titers may represent early infection, a cross-reaction, or residual antibodies from a previous infection (Saubolle et al., 2007; Wheat et al., 1983). ID tests are generally less sensitive than CF tests but may be useful in identifying cross-reactions. The results of the ID test for histoplasmosis can give false-positive results if a histoplasmosis skin test has been administered to the patient more than 5–7 days before obtaining sera. Importantly, the CF and ID test detect different antibodies, and both should be performed for maximum diagnostic sensitivity. In contrast to serologic tests for other fungal diseases, these tests use well-standardised commercially available reagents (Brandt et al., 2011). More recently, novel enzyme immunoassay (EIA) antibody tests for blastomycosis and coccidioidomycosis have been developed and show improved sensitivities (87.8% for blastomycosis and 95.5% for coccidioidomycosis) and specificities (94% for blastomycosis and 98.5% for coccidioidomycosis) (Ampel, 2010; Richer et al., 2014).

Antibody tests for *Candida* and *Aspergillus* may be performed; however, these tests are frequently unable to distinguish between active and past infection on the one hand and colonisation on the other (Kappe, Schulze-Berge, & Sonntag, 1996; Persat et al., 2002; Philip et al., 2005). Furthermore, a negative serologic test does not rule out infection because immunocompromised patients and some individuals with disseminated infection may not mount an antibody response to the infecting organism.

Several commercial enzyme-linked immunosorbent assay (ELISA) techniques are now available which detect anti-*Candida* antibodies in an effort to improve the diagnosis of IC (Persat et al., 2002; Philip et al., 2005; Sendid et al., 1999, 2002). These kits have shown sensitivities ranging from 50% to 90% and specificities of ~15% to 65% (Persat et al., 2002). The evaluation of an ELISA-based kit (Syscan 3; Rockeby, Biomed Ltd.) for detection of anti-*Candida* antibodies (anti-enolase and intracytoplasmic antigens) demonstrated a sensitivity, specificity, PPV, and NPV for IC of 74%, 75%, 62%, and 84% in a group of immunocompetent patients and 15%, 60%, 1.7%, and 93% in an immunocompromised group (Philip et al., 2005). Despite a moderately high NPV, it is difficult to see how such testing would be of much value especially among high-risk patients.

Other groups have reconsidered the value of detection of anti-mannan antibodies in the diagnosis of IC. The Platelia *Candida* antibody test (BioRad, Redmond, WA) uses an ELISA format to capture circulating anti-mannan antibodies in sera from patients, with reported specificity and sensitivity values of 94% and 53%, respectively (Sendid et al., 1999). When performed simultaneously in combination with a mannan

antigen detection test, the method gave a sensitivity of 80% and a specificity of 93% (Sendid et al., 1999, 2002). Other authors showed a sensitivity and specificity of 59% and 63%, respectively, for the Platelia anti-mannan test with an improved sensitivity of 95% and a lower specificity of 53%, when combined with a test for mannanaemia (Persat et al., 2002). It appears from these results that the diagnosis of IC cannot be made using a single test for antibodies alone. Rather, a strategy based on detection of mannanaemia and anti-mannan antibodies may prove to be the most useful (Persat et al., 2002; Sendid et al., 2003). In a recent meta-analysis Mikulska, Calandra, Sanguinetti, Puolain, and Viscoli (2010) reported a combined mannan/anti-mannan sensitivity and specificity for candidiasis diagnosis of 83.0% and 86.0%, respectively (Table 7): separate sensitivities and specificities of 58% and 93%, respectively, for mannan antigen alone and 59% and 83%, respectively, for anti-mannan antibodies alone. Furthermore, it appears that regular (at least twice weekly) serum sampling is critical to achieving an early diagnosis of IC (Mikulska et al., 2010; Persat et al., 2002; Sendid et al., 1999, 2002; Year, Sendid, Francois, Camus, & Poulain, 2001).

Similar difficulties are apparent when one considers using antibody detection for the serodiagnosis of IA. Kappe et al. (1996) evaluated eight different anti-*Aspergillus* antibody detection assays and found that sensitivity ranged from 14% to 36% and specificity from 72% to 99%. Clearly, the commercially available antibody detection assays are inadequate for the serodiagnosis of IA.

Table 7 Culture-Independent Testing Methods for the Diagnosis of Invasive Candidiasis and Invasive Aspergillosis

		No.	%	
Organism	**Diagnostic Test**	**Patients**	**Sens**	**Spec**
Candida spp.	β-D-glucan (meta-analysis)	2979	75.3	85.0
	Mannan/anti-mannan (meta-analysis)	1220	83.0	86.0
	SeptiFast (meta-analysis)	6012	61.0	99.0
	PCR (meta-analysis)	4694	95.0	92.0
	T2Candida	1801	91.1	99.4
Aspergillus spp.	Galactomannan (meta-analysis)	1585	71.0/90.0 (serum/BAL)	89.0/94.0 (serum/BAL)
	β-D-glucan (meta-analysis)	2979	77.0	85.0
	Lateral-flow device	103	81.8	98.0
	PCR (meta-analysis)	1618/1585 (serum/BAL)	88.0/90.2 (serum/BAL)	75.0/96.4 (serum/BAL)

Data compiled from Arvanitis et al. (2014), Avni et al. (2011, 2012), Chang et al. (2013), Karageorgopoulos et al. (2011), Mengoli, Cruciani, Barnes, Loeffler, and Donnelly (2009), Mikulska et al. (2010), Mylonakis et al. (2015), Pfeiffer et al. (2006), White, Parr, Thornton, and Barnes (2013).

4.4.2 *Immunologic and biochemical antigen detection methods*

Tests to detect fungal antigens in serum or other body fluids represent the most direct means of providing a serodiagnosis of IFI (Alexander & Pfaller, 2006; Antinori et al., 2005; Arvanitis et al., 2014; Busca et al., 2006; Fujita, Takamura, Nagahara, & Hashimoto, 2006; Fortun et al., 2001; Kawazu et al., 2004; Klont, Mennink-Kersten, & Verweij, 2004; Kwak et al., 2004; Maertens et al., 2005; Marr et al., 2004; Mennink-Kersten et al., 2004; Muñoz, Guinea, & Bouza, 2006; Musher et al., 2004; Odabasi et al., 2004; Pfeiffer et al., 2006; Pickering, Sant, Bowles, Roberts, & Woods, 2005; Quindos, 2006; Rex, 2006; Sendid et al., 1999, 2002, 2003, 2004; Upton et al., 2005; Weisser et al., 2005; Wheat, 2003; Wheat, Garringer, Brizendine, & Connolly, 2002; Yeo et al., 2006; Zedek & Miller, 2006). Significant advances have been made in recent years (Alexander & Pfaller, 2006; Arvanitis et al., 2014; Griffin & Hanson, 2014; Lopez-Ribot & Patterson, 2005; Marcos & Pincus, 2013; Quindos, 2006); however, for most fungal infections a widely acceptable method is not available. Although several tests for the detection of fungal antigens have been standardised and are now available commercially, issues still remain concerning the sensitivity and specificity of the various tests in certain patient populations, which populations should be monitored, how often should testing be performed, how the test behaves over time in relation to disease progression or improvement, what testing strategies are the most practical and cost-effective, and what is the true impact of such testing on patient outcome (de Pauw, 2005; McLintock et al., 2004; Rex, 2006; Zedek & Miller, 2006)?

Presently, the most established and widely used fungal antigen tests are the latex agglutination (LA) and EIA tests for the detection of the capsular polysaccharide antigen of *C. neoformans* in cerebrospinal fluid (CSF) and serum. The commercially available tests for cryptococcal antigen detect >95% of cryptococcal meningitis and approximately 67% of disseminated cryptococcal infections (Antinori et al., 2005). These antigen tests are well standardised, widely available and supplant India Ink (sensitivity <40%) for the diagnosis of cryptococcal meningitis (Antinori et al., 2005). A newly developed method to detect cryptococcal antigen employs a lateral-flow device (LFD) immunoassay and has been shown to provide results that are comparable to those of EIA and LA methods in both serum and CSF (Arvanitis et al., 2014; McMullan et al., 2012). The low cost, ease of use, and high degree of accuracy of the LFD make it very promising as a point-of-care diagnostic method for use in low-resource settings (Lawn, 2012).

Another useful antigen test available from a reference laboratory (MiraVista Diagnostics, Indianapolis, IN) is the test for *Histoplasma* antigen (Wheat et al., 1992, 2002; Wheat, Kohler, & Tewari, 1986). The *Histoplasma* antigen test has been shown to be rapid (<24 h), sensitive (55–99%), specific (>98%), and reproducible (Wheat et al., 2002). The test uses an EIA format and detects a *Histoplasma*-specific polysaccharide antigen present in body fluids. Urine and serum are the most common specimens tested; however, the antigen may be detected in the spinal fluid of 42–67% of patients with *Histoplasma* meningitis and in the BAL fluid of 70% of patients with AIDS and severe pulmonary histoplasmosis (Wheat et al., 2002).

Tests for the detection of galactomannan antigens from *C. immitis/posadasii* and *B. dermatitidis* are available from MiraVista Diagnostics (Indianapolis, IN). These test both show cross-reaction in patients with histoplasmosis, but may be useful in the right epidemiological context.

Mannan is the major circulating antigen in patients with IC. Detection of mannan is complicated by rapid clearance from the patient's sera and binding by anti-mannan antibody. Although circulating mannan may be detected by several methods, a dissociation of antigen-antibody complexes is required for optimal sensitivity (Yeo & Wong, 2002). Sensitivities of 25–100% and specificities of 92–100% have been reported with EIA assays for mannan detection (Fujita et al., 2006; Kohno et al., 1993; Mitsutake et al., 1996; Pfaller et al., 1993; Sendid et al., 1999, 2002, 2003, 2004; White et al., 2005; Yeo & Wong, 2002). An early commercial system to detect mannan used a LA format (Pastorex *Candida* test, Sanofi Diagnostics Pasteur, Mames-la-Coquette, France) and demonstrated poor sensitivity (0–25%) due to the rapid clearance of mannan from patients' sera and the insensitive LA format (Gutierrez, Maroto, Piedrola, Martin, & Perez, 1993). More recently, the Platelia *Candida* antigen test (BioRad) uses a monoclonal antibody-based double sandwich EIA format with a resultant increase in sensitivity and a limit of detection of 0.1 ng of mannan per ml of serum (Sendid et al., 1999, 2002, 2003). Clinical evaluations of the Platelia *Candida* antigen test report sensitivities ranging from 40% to 86% and specificities from 79% to 98% (Fujita et al., 2006; Kohno et al., 1993; Mikulska et al., 2010; Sendid et al., 1999, 2002; White et al., 2005). Virtually every study evaluating the detection of the mannanaemia has shown that multiple serial samples are required to overcome the rapid clearance of mannan from patient's sera and to optimise diagnostic sensitivity

Sendid et al. (2004) took advantage of differences in clearance of α-mannan (rapid) and β-mannan (slower) to demonstrate that simultaneous detection of both forms of mannan improves the sensitivity of a test for mannanaemia from 69% when either α-mannan or β-mannan was tested for alone to 85% when both forms were tested for in the diagnosis of IC. The specificity of the single tests was 98% (α-mannan) and 95% (β-mannan) and for the combined test was 95%. They also demonstrated that mannanaemia preceded early clinical symptoms and isolation of *Candida* in culture by an average of 4.7 days. Although the current Platelia *Candida* antigen test is specific for α-mannan only, these results suggest that joint detection of both epitopes is a rational approach that contributes to increases in the sensitivity and timeliness of diagnosis.

Although a test for β-mannan is not yet available, Sendid and colleagues (Sendid et al., 1999, 2002, 2003) have applied a similar rationale to the use of the Platelia *Candida* anti-mannan antibody test in combination with the Platelia *Candida* antigen test to maximise the ability to diagnose IC. As discussed previously, simultaneous testing for mannanaemia and anti-mannan antibodies resulted in an improved sensitivity from 40% to 53% with the single tests to 80% with combined testing (Mikulska et al., 2010; Sendid et al., 1999, 2002). Specificity remained high (93%) with the combined testing strategy. Whereas the Platelia *Candida* antigen test is specific for α-mannan, the Platelia antibody test detects antibodies against the whole mannan oligomannose repertoire containing both α- and β-mannan epitopes (Sendid et al., 1999, 2004). These and other investigators emphasise the importance of regular (twice weekly) serial monitoring of at-risk patients in maximising the sensitivity of serodiagnostic testing for IC (Persat et al., 2002; Sendid et al., 1999).

A simple and commercially available test (CAND-TEC *Candida* Detection System, Ramco Laboratories, Houston, Texas) relies on the detection of a structurally uncharacterised 56 °C heat-labile antigen of *C. albicans* using an LA format (Gentry, Wilkinson, Lea, & Price, 1983). Although easy to perform, the test suffers from low sensitivity (as low as 0–16%) and specificity, and its usefulness for the reliable diagnosis of IC is limited (Fujita et al., 2006; Lopez-Ribot & Patterson, 2005; Mitsutake et al., 1996; Phillips et al., 1990).

Several different protein antigens of *Candida* have been explored for their diagnostic potential (Lopez-Ribot & Patterson, 2005; Yeo & Wong, 2002). These antigens include secreted aspartyl proteinases, *C. albicans* heat-shock protein (hsp) 90, and enolase (Matthews & Burnie, 1988; Mitsutake et al., 1996; Ruchel, Boning-Stutzer, & Mari, 1988; Walsh et al., 1991). One confounding factor that all of these protein antigens share is the fact that they form immune complexes, which negatively impacts the sensitivity of antigen detection assays.

Detection of other compounds released by fungal cells during infections has also been explored for the diagnosis of IC (Arvanitis et al., 2014; Karageorgopoulos et al., 2011; Lopez-Ribot & Patterson, 2005). The most promising among these various targets is detection of serum β-D-glucan (BDG; for a non-specific fungal diagnosis including IC). BDG is an important component of the cell wall of *Candida*, *Aspergillus*, *Pneumocystis*, and many other pathogenic fungi. Although BDG is not immunogenic, the fact that it can be found circulating in the bloodstream of patients with IFI has been exploited for use diagnostically and as a surrogate marker of infection (Kawazu et al., 2004; Kohno et al., 1993; Mitsutake et al., 1996; Miyazaki et al., 1995; Obayashi et al., 1995; Odabasi et al., 2004; Pazos, Ponton, & Del Palacio, 2005; Pickering et al., 2005). The Fungitell BDG assay (Associates of Cape Cod, Inc., Falmouth, MA) is an FDA-cleared commercially available colorimetric assay that can indirectly determine the concentration of BDG in the serum. The detection system is based on the activation of a BDG-sensitive proteolytic coagulation cascade, the components of which are purified from the horseshoe crab (Tamura, Tanaka, Ikeda, Obayashi, & Hashimoto, 1997). The assay can measure picogram amounts of BDG and has been used to demonstrate the presence of the polysaccharide in the serum of patients with IC and IA, but not cryptococcosis or mucormycosis (organisms lack BDG) (Karageorgopoulos et al., 2011; Kawazu et al., 2004; Kohno et al., 1993; Mitsutake et al., 1996; Miyazaki et al., 1995; Obayashi et al., 1995; Odabasi et al., 2004; Pazos et al., 2005). Several studies have demonstrated a modest degree of sensitivity and specificity in the diagnosis of IC (78–97% sensitivity and 88–100% specificity) and IA (50–87.5% sensitivity and 81–89% specificity) (Kami et al., 2001; Karageorgopoulos et al., 2011; Kawazu et al., 2004; Kohno et al., 1993; Mitsutake et al., 1996; Odabasi et al., 2004; Ostrosky-Zeichner et al., 2005; Pazos et al., 2005; Pickering et al., 2005).

BDG is also detectable in patients with infections caused by species of *Pneumocystis*, *Fusarium*, *Trichosporon*, *Saccharomyces*, and *Sarocladium* (*Acremonium*) (Karageorgopoulos et al., 2011; Yoshida et al., 1997).

The Fungitell BDG assay has been evaluated for the early diagnosis of IFI in patients with haematologic malignancies (Odabasi et al., 2004; Pazos et al., 2005) and in a multicentre study of patients with IFIs and healthy control subjects (Ostrosky-Zeichner et al., 2005). In the latter, study sera was obtained from 170 fungal infection-negative control subjects and from 163 patients with proven or probable IFI diagnosed at one of 6 participating medical centres. Overall, the sensitivity and specificity of the assay were 69.9% and 87.1%, respectively, with a PPV of 83.8% and a NPV of 75.1%. The sensitivity of the BDG test was 81.3% among the 107 patients with proven IC and 80% in the 10 patients with IA. Measurement of serum BDG was considered to be a useful diagnostic adjunct for IFI. Another study of patients with acute myelogenous leukaemia suggested that the sensitivity, specificity, and PPV of the assay increased significantly if sera were obtained twice weekly (Odabasi et al., 2004). Obtaining multiple samples increased the sensitivity, PPV, and NPV of the BDG assay to >98% for subjects with leukaemia who were receiving antifungal prophylaxis. A recent meta-analysis of 16 studies measuring serum or plasma BDG for the diagnosis of IC reported pooled sensitivity and specificity values of 75.3% and 85.0%, respectively (Karageorgopoulos et al., 2011) (Table 7). Although the BDG test may be performed in a hospital-based laboratory, it is a send-out test for most institutions resulting in a turnaround time for results measured in days rather than hours.

In contrast to these rather favourable studies, other investigators have pointed out considerable problems with the specificity of the BDG assay for IFI (Digby et al., 2003; Sendid et al., 2004). Pickering et al. (2005) tested sera from healthy blood donors and patients with candidaemia and found a sensitivity and specificity of 92.9% and 100%, respectively. When bacteraemic patients were included in their assessment of the performance of the BDG assay, the specificity and PPV fell to 77.2% and 51.9%, respectively, due to a high number of false-positive results, especially in samples from patients with Gram-positive bacteraemia. These investigators demonstrated that excess manipulation of a sample can result in contamination by BDG (Pickering et al., 2005). They also found that haemolysis would cause false-positive BDG test results and that high concentrations of bilirubin and triglycerides were inhibitory and would cause false-negative results. These confounding factors can be added to a list of other causes of false-positive BDG test results including haemodialysis with cellulose membranes, patients treated with intravenous immunoglobulins, albumin, coagulation factors, or plasma protein factor, or patients exposed to gauze or other materials that contain glucans (Pickering et al., 2005). Thus, a negative BDG test result may be useful for ruling out an IFI due to most fungal pathogens with the exception of cryptococci or the Mucormycetes; however, a single positive result should be confirmed by testing another specimen (two consecutive positive tests) and a thorough review for potential sources of false positivity should be conducted.

Recently, it has been shown that the BDG test may have utility in the diagnosis of pneumonia due to *P. jirovecii* (Onishi et al., 2012). In a meta-analysis, Karageorgopoulos et al. (2013) demonstrated that the BDG test performed on serum had a sensitivity and specificity of 94.8% and 86.3%, respectively, for the diagnosis of *Pneumocystis* pneumonia. Likewise, a large retrospective cohort analysis showed that a positive BDG result correlated well with BAL fluid fungal loads (Costa et al., 2012). Thus, the BDG test may serve as an excellent screening tool to rule out *Pneumocystis* pneumonia in an at-risk population (Onishi et al., 2012), although additional confirmatory tests may be necessary due to the high rate of false-positive results (Costa et al., 2012).

Often other tests, in addition to BDG, are required both to confirm a positive result and to identify the fungal pathogen (Alexander & Pfaller, 2006; Arvanitis et al., 2014). In a study designed to assess the benefit of monitoring patients for the presence of both BDG and galactomannan (GM), Pazos et al. (2005) found that the combination of the two tests improved both specificity (to 100%) and PPV (to 100%) for the diagnosis of IA, without affecting sensitivity or NPV. Although both tests were useful for early diagnosis of IA, the BDG test was positive earlier than the GM assay. Thus, the BDG assay may be a useful screening test for IFI. Testing should be performed with a minimal amount of sample manipulation and sequential positive results should be required for a "true-positive" test result (Odabasi et al., 2004). A positive BDG result should prompt further diagnostic work up that may include testing for GM or the use of the PCR to define and identify the IFI.

GM is an important component of the cell wall of *Aspergillus* spp. Similar to mannan in IC, GM has been detected in biological fluids (serum, urine, bronchoalveolar lavage [BAL] fluid) obtained from patients with IA (Mennink-Kersten et al., 2004) (Table 7). Detection of GM for diagnosis of IA has been facilitated by the development of monoclonal antibody-based EIA methods which can detect as little as 0.5–1 ng of GM per ml of serum (Lopez-Ribot & Patterson, 2005; Mennink-Kersten et al., 2004). In 2003, the Platelia *Aspergillus* EIA test (Bio-Rad) was approved by the US FDA for use in the diagnosis of IA in BMT recipients and in patients with leukaemia.

GM EIA results are reported as a ratio between the optical density of the patient's serum sample and that of a control with a low, but detectable, amount of GM and data are expressed as the serum GM index (GMI). Most published studies use a cut-off GMI of <1.0 as a negative value, a value greater than 1.5 as positive, and those between 1.0 and 1.5 as indeterminate (Pfeiffer et al., 2006). In addition, the manufacturer recommends that two or more samples be tested positive for the GM results to be considered as indicative of IA. The data submitted to the FDA used a cut-off of 0.5 as positive and required that the test be positive on two aliquots of the same sample- rather than multiple samples testing positive (U.S. Food and Drug Administration, 2003). These criteria resulted in a sensitivity of 80.7% and specificity of 89.2% for the diagnosis of IA in a multicentre study conducted on serially collected serum samples from 179 BMT recipients and patients with leukaemia (31 with IA) (FDA, 2003).

Table 8 Sensitivity and Specificity of the Galactomannan (GM) Assay for Diagnosis of Proven or Probable Invasive Aspergillosis

	No.	%	
Patient Group	**Patients**	**Sens**	**Spec**
All studies[a]	3467	61	93
Haematologic malignancy	2761	58	95
BMT recipients	75	65	65
SOT recipients	269	41	85
Positive GM cut-off			
Value			
0.5	658	79	86
1.0	1401	65	94
1.5	1408	48	95

[a]*A total of 27 studies.*
Data compiled from Pfeiffer et al. (2006).

In contrast to the data submitted to the FDA, several studies have used GMI values of 1.0 as a positive cut-off and have required that two consecutive samples must test positive to declare a positive assay (Pfeiffer et al., 2006) (Table 8). It is now clear that the reported results with this test can be influenced by the extent of IA at the time of diagnosis, the prevalence of aspergillosis among the patients studied, exposure of the patient to mould-active antifungals, the cut-off ratio used, and whether multiple consecutive positive tests were or were not required for significance (Lopez-Ribot & Patterson, 2005; Mennink-Kersten et al., 2004; Pfeiffer et al., 2006). Thus, it is not surprising that reported sensitivities of GM testing for the diagnosis of proven IA range from 17% to 100% and specificities from 38% to 98% (Pfeiffer et al., 2006).

In a meta-analysis of 27 studies in which GM EIA was used to diagnose proven/probable IA, Pfeiffer et al. (2006) found a pooled sensitivity and specificity of 61% and 93%, respectively, for the diagnosis of proven IA (Tables 7 and 8). The high NPV (98%) and low PPV (26%) suggest that the GM assay is good for ruling out disease but is less useful for confirming the diagnosis of IA (Rex, 2006). The test was found to be most useful in patients with haematologic malignancy or who have undergone BMT than in solid-organ transplant recipients (Pfeiffer et al., 2006) (Table 8). Although the use of a lower GMI threshold may be important in patients receiving mould-active antifungal agents (Marr et al., 2004) and may provide an earlier diagnosis, the accuracy of the test clearly improves with a high GMI threshold (Pfeiffer et al., 2006) (Table 8). Irrespective of the GMI threshold employed, it is apparent that in approximately two-thirds of patients with IA, circulating GM can be detected at a mean of 8 days before diagnosis by another means (Mennink-Kersten et al., 2004). When performed in serial fashion, a gradual increase in GMI in consecutive samples is a very strong indication of infection and should be considered when interpreting the results. Likewise, the course of the antigen seems to correlate well with outcome and could be important in monitoring therapeutic response (Wheat, 2003).

In addition to variable sensitivity, false-positive GM results have been a significant issue, occurring in approximately 10% of patients in most series (Pfeiffer et al., 2006). False-positive GM assay results have been reported in children (Letscher-Bru et al., 1998), in patients receiving piperacillin–tazobactam (Adam et al., 2004; Sulahian, Touratier, & Ribaud, 2003; Viscoli et al., 2004) and amoxicillin–clavulanate (Maertens, Theunissen, Verhoef, & van Eldere, 2004; Mattei et al., 2004), in BAL fluid containing Plasmalyte (Hage, Reynolds, Durkin, Wheat, & Knox, 2007) and possibly in patients consuming GM-rich foods (Mennink-Kersten et al., 2004).

In addition to methodological factors, such as GMI cut-off and definition of a positive test, other unresolved issues may contribute towards variability in test performance including issues affecting the release of the antigen from the hypha (e.g. fungal strain and stage of growth), leakage of the antigen from the site of infection into the bloodstream, binding of the antigen by blood substances, clearance of GM from the circulation, and host factors such as location and extent of fungal disease, antifungal therapy, and patient age (Mennink-Kersten et al., 2004). Another issue that has a direct bearing on assay performance is the frequency of GM testing and the use of tests for GM in combination with other diagnostic tests (Rex, 2006). Testing approximately twice weekly seems practical and should overcome many of the confounding variables, listed above and enable reasonably prompt diagnoses. Unfortunately, it appears that the strategy of test utilisation in practice is quite variable. In a 12-month study of the utilisation of GM EIA in the setting of haematologic malignancies, Zedek and Miller (2006) found that 33% of 86 patients had multiple serial tests and 67% had only one test. Furthermore, of the nine patients with biopsy-proven IA, six patients did not receive a GM test, and one patient was GM positive. These data underscore less than ideal test utilisation, a practice that is likely to impact negatively on the clinical utility of the GM assay.

Studies of combining GM with other diagnostic modalities such as CT scans of the chest suggest that this approach may be useful in establishing a likely diagnosis of IA (Becker et al., 2003; Busca et al., 2006; Kawazu et al., 2004; Maertens et al., 2005; Musher et al., 2004; Severens et al., 1997). Busca et al. (2006) demonstrated that sequential (twice weekly) GM detection combined with early radiologic evaluation (chest CT) were useful tools to detect minimal changes of IA and initiation of antifungal therapy. Maertens et al. (2005) assessed the feasibility of daily GM monitoring and clinical evaluation coupled with high-resolution thoracic CT and bronchoscopy with lavage in identifying patients to receive pre-emptive anti-mould antifungal therapy. This approach was successful in reducing the number of patients exposed to toxic antifungal therapy (liposomal amphotericin B) by 78% and lead to the early initiation of antifungal therapy in 10 episodes that were clinically not suspected of being IFI. No undetected cases of IA were identified, although one case of mucormycosis was missed. Pazos et al. (2005) has shown that using GM in combination with BDG testing provided an increase in specificity for the diagnosis of IA compared to either test alone. Musher et al. (2004) found that GM EIA and quantitative PCR added to the sensitivity of BAL for diagnosing IA in high-risk patients and Becker et al. (2003) found that the sensitivity of serial serum GM determinations increased from 47% to 85% when combined with CT-directed BAL fluid GM testing.

Although not approved for non-serum samples, measurement of GM in other fluids such as BAL has been proposed (Arvanitis et al., 2014; Becker et al., 2003; Klont et al., 2004; Musher et al., 2004). Sensitivities of the assay in BAL fluid range from 61% to 76% and positive and negative predictive values were 100% when BAL testing was combined with high-resolution CT scanning (Becker et al., 2003; Musher et al., 2004). GM detection in CSF has also been used to diagnose *Aspergillus* meningitis and to follow the course of infection in a limited number of patients (Verweij, Brinkman, Kremer, Kullberg, & Meis, 1999).

The recent introduction of an LFD for the detection of an *Aspergillus* glucoprotein antigen in serum and BAL fluid has shown great promise as a simple and rapid (results in 15 min) means of diagnosing IA (Thornton, 2008) (Table 7). The antigen is secreted during the active growth of *A. fumigatus* and when detected has shown improved sensitivity and specificity compared to that of the BDG and Platelia GM assays (Held, Schmidt, Thornton, Kotter, & Bertz, 2013; Wiederhold et al., 2009). Due to the subjective nature of the interpretation of LFD results, it has been recommended that they be used for confirmation or exclusion of IA in combination with other tests such as PCR (Held et al., 2013; White et al., 2013).

4.4.3 Nucleic acid detection

As in other areas of microbiology, the application of molecular biology, specifically the PCR, offers great promise for the rapid diagnosis of fungal inflections (Alexander et al., 2006; Arvanitis et al., 2014; Avni et al., 2011, 2012; Baron, 2006; Chang et al., 2013; Chen et al., 2002; Christie, 2003; Iwen et al., 2002; Liesenfeld et al., 2014; Lu et al., 2011; Marcos & Pincus, 2013; McLintock & Jones, 2004; Mengoli et al., 2009; Mylonakis et al., 2015; Neely et al., 2013; Quindos, 2006; Sun et al., 2011; White et al., 2013; Yeo & Wong, 2002; Zhang, 2013). At present, most of the research has been focused on the diagnosis of IC (Avni et al., 2011; Chang et al., 2013; Chen et al., 2002; Lopez-Ribot & Patterson, 2005; Moreira-Oliveira et al., 2005; Mylonakis et al., 2015; Neely et al., 2013; Yeo & Wong, 2002; Yeo et al., 2006) and IA (Avni et al., 2012; Buchheidt et al., 2001, 2004; Florent et al., 2006; Hebart, Löffler, Meisner, et al., 2000; Hebart, Löffler, Reitze, et al., 2000; Hope, Walsh, & Denning, 2005; Kami et al., 2001; Kawazu et al., 2004; Lass-Flörl et al., 2001; Mengoli et al., 2009; Muñoz et al., 2006; Musher et al., 2004; Skladney et al., 1999; Sun et al., 2011; White et al., 2006, 2013; Yeo et al., 2006) (Table 7); however, PCR has also been applied to the diagnosis of other IFI (Arvanitis et al., 2014; Chen et al., 2002; Christie, 2003; Einsele et al., 1997; Hebart, Löffler, Meisner, et al., 2000; Hebart, Löffler, Reitze, et al., 2000; Iwen et al., 2002; Jordanides et al., 2005; Lin, Lu, & Chen, 2001; Lu et al., 2011; Marcos & Pincus, 2013; McLintock & Jones, 2004; Pham et al., 2003; Quindos, 2006; Rappelli et al., 1998; Thompson et al., 2013). It should be noted, however, that despite a great deal of interest in molecular approaches to the diagnosis of infectious diseases, molecular methods are used in only 5% of laboratories providing diagnostic services in medical mycology (Baselski, Buchanan, Carey, Clarridge, & Weissfeld, 2005). Given the limitations of current fungal diagnostics, the use of PCR as a diagnostic adjunct for the diagnosis of IFI is especially promising; however, considerable additional research, standardisation of testing protocols, and prospective assessment is necessary before one can conclude that molecular biology has achieved a tangible clinical pay-off (Alexander & Pfaller, 2006; Arvanitis et al., 2014; Christie, 2003; Donnelly, 2006; Griffin & Hanson, 2014; McLintock & Jones, 2004; Zhang, 2013).

PCR-based methods for the diagnosis of IFI have been applied to a variety of specimen types including whole blood, serum, tissue, BAL fluid, and CSF (Arvanitis et al., 2014; Avni et al., 2012; Chen et al., 2002; Iwen et al., 2002; Yeo & Wong, 2002; Zhang, 2013). In addition to the use of whole blood or serum as the optimal specimen type, important procedural considerations need to be taken for removal of contaminating non-fungal DNA, breaking fungal cells for DNA extraction, and prevention of introduction of contaminating fungal DNA as well as minimising the destruction of target DNA (Arvanitis et al., 2014; Avni et al., 2011, 2012; Lopez-Ribot & Patterson, 2005; Neely et al., 2013; Yeo & Wong, 2002). Target sequences vary widely but include genus- and species-specific variable regions as well as highly conserved regions of the fungal genome (Arvanitis et al., 2014; Avni et al., 2011; Chen et al., 2002; Iwen et al., 2002; Yeo & Wong, 2002; Zhang, 2013). Both single (e.g. *hsp*90, lanosterol demethylase, chitin synthase, actin) and multicopy (e.g. ribosomal, intergenic transcribed spacer (ITS) regions, mitochondrial) gene targets have been studied, although molecular diagnostic methods targeting multicopy genes generally have better sensitivity than those targeting single copy genes (Arvanitis et al., 2014; Avni et al., 2011, 2012; Neely et al., 2013; Yeo & Wong, 2002). The use of multicopy ribosomal (18S rRNA, 28S rRNA, 5.8S rRNA) and ITS targets offer the potential for sensitive panfungal markers for detection of IFI, followed by identification at the genus or species level (Arvanitis et al., 2014; Avni et al., 2011; Chen et al., 2002; Iwen et al., 2002; Neely et al., 2013; Yeo & Wong, 2002).

PCR amplicon detection methods vary widely but most laboratory developed tests (LDT) employ capture probes in an ELISA format. Recent developments such as real-time PCR, gene chip technology, and the coupling of nanotechnology with T2 magnetic resonance (T2MR) detection will facilitate the broad use of this technology (Arvanitis et al., 2014; Griffin & Hanson, 2014; Liesenfeld et al., 2014; Mylonakis et al., 2015; Neely et al., 2013; Nguyen et al., 2012).

Irrespective of the technology used, most reports in the literature indicate that the sensitivity of PCR-based diagnosis is equal to or better than other currently used diagnostic techniques (Arvanitis et al., 2014; Avni et al., 2011; Chen et al., 2002; Lopez-Ribot & Patterson, 2005; Neely et al., 2013; Yeo & Wong, 2002; Zhang, 2013). The major impediment to the application of PCR to the diagnosis of IFI has been the lack of standardisation and the need for nucleic acid extraction and purification from clinical samples (Liesenfeld et al., 2014; Zhang, 2013). The studies reported in the literature must be viewed with the caveat that the vast majority are derived from in-house LDT protocols developed by different groups of investigators using different samples (e.g. whole blood vs. serum), different protocols for sample preparation, different molecular targets, and different PCR detection platforms (Arvanitis et al., 2014; Avni et al., 2011, 2012; Chen et al., 2002; Donnelly, 2006; Lopez-Ribot & Patterson, 2005; Mengoli et al., 2009; Yeo & Wong, 2002; Zhang, 2013). Thus, when considering the use of PCR for the diagnosis of either IC or IA, one encounters a range of sensitivities and specificities of 77–100% and 66–100%, respectively, for IC and of 43–100% and 64–100% , respectively, for IA (Arvanitis et al., 2014; Avni et al., 2011, 2012). Among several reasons for the reported low levels of sensitivity include requirements for nucleic acid extraction and the use of optical methods of detection resulting in excessively high limit of detection (LOD) (Arvanitis et al., 2014; Griffin & Hanson, 2014; Neely et al., 2013; Zhang, 2013).

The lack of standardisation makes any comparison among studies very difficult and further hinders efforts to perform prospective, multicentre clinical trials (Avni et al., 2011, 2012; Christie, 2003; Donnelly, 2006; Lopez-Ribot & Patterson, 2005; Mengoli et al., 2009; Zhang, 2013). Thus, despite the availability of promising new approaches for the rapid diagnosis of IFI, we have little data from prospective, multicentre clinical trials of sufficient size to determine the performance characteristics of the diagnostic methods or the most cost-effective manner in which to use them (Arvanitis et al., 2014; Zhang, 2013).

PCR-amplified *Candida*-specific DNA has been recovered from blood and other body fluids obtained from infected patients (Arvanitis et al., 2014; Avni et al., 2011; Burgener-Kainuz et al., 1994; Chen et al., 2002; Flahaut, Sanglard, Monod, Bille, & Rossier, 1998; Griffin & Hanson, 2014; Jordan, 1994; Kan, 1993; Morace et al., 1999; Mylonakis et al., 2015; Neely et al., 2013; Nguyen et al., 2012; Yeo & Wong, 2002). The most frequently employed targets for the diagnosis of IC are the multicopy broad-range panfungal genes such as the 18S, 5.8S, and 28S ribosomal RNA genes (rRNA), and the ITS regions within the rRNA gene cluster (Arvanitis et al., 2014; Avni et al., 2011; Chen et al., 2002; Iwen et al., 2002; Neely et al., 2013; Nguyen et al., 2012; Yeo & Wong, 2002). In a systematic meta-analysis evaluating PCR assays for the diagnosis of IC, pooled analysis of 54 studies and almost 5000 patients found that the optimal conditions for the detection of *Candida* in blood using PCR were (i) the use of whole blood, (ii) a multicopy target (rRNA or P450 genes), and (iii) a limit of detection (LOD) of ≤10 CFU per ml of blood (Avni et al., 2011). Under these conditions, the pooled sensitivity and specificity of PCR in diagnosing IC was 95% and 92%, respectively (Avni et al., 2011) (Table 7). Importantly, PCR-based tests for *Candida* DNA in blood are negative in most subjects with GI colonisation with *Candida* species and the specificity of these tests is quite high (Avni et al., 2011; Baron, 2006; Burnie, Golband, & Matthews, 1997; Chang et al., 2013; Chryssanthou et al., 1999; Mylonakis et al., 2015; White et al., 2005) (Table 7).

Aside from using a consensus process to develop a standardised LDT molecular assay for the diagnosis of IC, standardisation may also be achieved by either a centralised model whereby a molecular test is offered and performed by a publically available reference laboratory or by commercialisation of a molecular test to be performed on-site (Liesenfeld et al., 2014; Mylonakis et al., 2015; Neely et al., 2013; Nguyen et al., 2012; Zhang, 2013). Commercialisation of molecular tests not only standardises the method but also facilitates large-scale "real-world" clinical validation, leading to the implementation of the molecular test for clinical use (Zhang, 2013).

Presently, there are only three commercially available amplification-based molecular tests that have been evaluated for use in the diagnosis of IC directly from the clinical specimen in either the United States (US; Viracor *Candida* Real-Time PCR Panel [Viracor-IBT] and the T2Candida Panel [T2Biosystems]) or Europe (LightCycler SeptiFast Test [Roche Diagnostics] and theT2Candida Panel [T2Biosystems]) (Chang et al., 2013; Liesenfeld et al., 2014; Mylonakis et al., 2015; Neely et al., 2013; Nguyen ct al., 2012).

The Viracor (www.viracoribt.com) reference laboratory offers a real-time PCR panel for the detection of the five major species of *Candida* (*C. albicans*, *C. tropicalis*, *C. parapsilosis*, *C. glabrata*, and *C. krusei*) in serum or plasma. This is an LDT test and has not been cleared by the FDA. The Viracor *Candida* Real-Time PCR Panel requires nucleic acid extraction and purification and reports an LOD of ≤1 CFU/ml in serum or plasma (Nguyen et al., 2012). In a single-centre clinical evaluation, the sensitivity and specificity of the Viracor Panel for the diagnosis of IC was 80% and 70%, respectively (Nguyen et al., 2012). Although the in-laboratory turnaround time (TAT) for the Viracor Panel is reported to be within 6–8 h, the send-out nature of the test means that the actual TAT for reporting results to the clinician is measured in days rather than hours.

Another real-time PCR test for IC is offered in the United States by Quest Diagnostics reference laboratories (www.questdiagnostics.com). Similar to the Viracor *Candida* Real-Time PCR Panel, the Quest *Candida* DNA, Qualitative, Real-Time PCR test is not FDA cleared, requires nucleic acid extraction and purification from serum, and detects the five most common species of *Candida*. The LOD for the Quest *Candida* test is reported to be ≤ 1–350 CFU/ml depending on the species. Thus far the clinical sensitivity and specificity of this test has not been reported in the literature.

The LightCycler SeptiFast Test uses a real-time PCR platform and has been extensively evaluated for use in Europe (CE-marked) but has not been cleared by the FDA for use in the United States (Chang et al., 2013; Liesenfeld et al., 2014). SeptiFast detects nucleic acids extracted from whole blood for five species of *Candida* as well as 19 species of bacteria and *A. fumigatus*. The LOD for *Candida* using SeptiFast ranges from 30 CFU/ml (*C. albicans*, *C*, *parapsilosis*, *C. tropicalis*, and *C. krusei*) to 100 CFU/ml (*C. glabrata*) (Chang et al., 2013). When performed in the laboratory the TAT for a single test is approximately 6 h. In a meta-analysis, Chang et al. (2013) reported a pooled sensitivity of 61% and a specificity of 99% for the detection of *Candida* in whole blood following extraction (Table 7).

The T2Candida Panel is an FDA-cleared rapid diagnostic approach that employs a proprietary formulation enabling inhibition-free target amplification of the multicopy ITS2 (internal transcribed spacer region 2) region of the *Candida* genome coupled with detection nanoparticles coated with oligonucleotide capture probes to enable sensitive and specific detection of the amplified product directly in whole blood without the need for culture or nucleic acid extraction steps (Neely et al., 2013). The T2Candida Panel uses T2MR relaxometry to measure the magnetic properties of the water molecules in the specimen (providing signal amplification and non-optical detection) and not just the amplified target to achieve excellent sensitivity in complex clinical samples (Neely et al., 2013). Results from a formal LOD study of the T2Candida Panel showed an LOD of 1–3 CFU/ml of blood (1 CFU/ml for *C. tropicalis* and *C. krusei*, 2 CFU/ml for *C. albicans* and *C. glabrata*, and 3 CFU/ml for *C. parapsilosis*) with results available as fast as 3 h compared to >48 h for blood culture (Beyda et al., 2013; Neely et al., 2013). In a recently completed 1801 subject clinical trial, the T2Candida Panel showed a sensitivity and specificity for the detection of candidaemia of 91.1% and 99.4%, respectively (Mylonakis et al., 2015) (Table 7). Importantly the median time for species identification in this trial was 4.4 h for T2Candida and 129 h for blood culture. Likewise the median time for a negative result was 4.2 h for T2Candida and >120 h for blood culture. The T2Candida Panel allows for the identification of the five major species of *Candida* (*C. albicans*, *C. tropicalis*, *C. parapsilosis*, *C. glabrata*, and *C. krusei*). The test is completely automated and requires <5 min of hands-on time.

PCR has also been used successfully for early detection of *Aspergillus* DNA in peripheral blood and in BAL fluid (Avni et al., 2012; Chen et al., 2002; Donnelly, 2006; Hope et al., 2005; Kami et al., 2001; Mengoli et al., 2009; Muñoz et al., 2006; Musher et al., 2004; Skladney et al., 1999; White et al., 2006; Yeo & Wong, 2002; Yoo et al., 2005). Sensitivities and specificities range from 43% to 100% and 64% to 100%, respectively, in patients with proven/probable IA (Arvanitis et al., 2014). In a meta-analysis of 16 studies (1618 patients) in the literature evaluating the diagnostic accuracy of several in-house PCR LDTs on whole blood, serum, or plasma from patients with IA, Mengoli et al. (2009) reported a pooled sensitivity of 88% and a specificity of 75% (Table 7). Likewise, Avni et al. (2012) performed a systematic review of the literature concerning the use of PCR on BAL fluid and reported summary sensitivity and specificity values for diagnosis of IA of 90.2% and 96.4%, respectively (Table 7). False-positive results may be seen when BAL fluid is tested, most likely due to the transient presence of conidia in the respiratory tract. The use of whole blood, serum, or plasma may be preferable to the use of respiratory tract specimens, because contamination with conidia is much less likely (Alexander & Pfaller, 2006). The combination of GM antigen testing and PCR using serum or BAL fluid has proven to be useful in regular screening for IA in patients with haematologic disorders (Avni et al., 2012; Florent et al., 2006; Kawazu et al., 2004; Musher et al., 2004; Yoo et al., 2005; White et al., 2013). The combined use of PCR and GM assays increased the sensitivity and NPVs of each individual test to 83.3% and 97.6%, respectively, for the early diagnosis of IA in patients with haematologic malignancies

(Florent et al., 2006). Recently White et al. (2013) demonstrated that the combination of PCR and antigen detection in serum using an LFD resulted in a 100% sensitivity and NPV, allowing IA to be confidently excluded if both tests were negative. Musher et al. (2004) found that the combination of quantitative PCR and GM testing of BAL fluid resulted in improved sensitivity and specificity compared to either test alone, in the diagnosis of invasive pulmonary aspergillosis. It is now apparent that PCR monitoring of high-risk patients allows early diagnosis of IA with good sensitivity and specificity, both in BMT recipients and in patients with haematologic malignancies, and that in patients with a negative PCR result the probability of IFI is extremely low (Arvanitis et al., 2014; Avni et al., 2012; Buchheidt et al., 2004; Donnelly, 2006; Florent et al., 2006; Hebart, Löffler, Meisner, et al., 2000; Hebart, Löffler, Reitze, et al., 2000; Jordanides et al., 2005; Kawazu et al., 2004; Lass-Flörl et al., 2001; McLintock & Jones, 2004; Mengoli et al., 2009; White et al., 2006, 2013).

The changing epidemiology of IFI and the increasing recognition of the pathogenic potential of a bewildering array of fungal species in the high-risk patient population (Table 1), underscores the need for diagnostic testing approaches to detect a wide range of medically important fungi (Alexander & Pfaller, 2006; Arvanitis et al., 2014; Lopez-Ribot & Patterson, 2005; Pfaller & Diekema, 2004a, 2004b; Pfaller, Diekema, Colombo, et al., 2006; Pfaller, Diekema, Rex, et al., 2006; Pfaller, Diekema, & Sheehan, 2006; Pfaller, Pappas, & Wingard, 2006). Thus, the application of broad-range (panfungal) PCR to amplify a target (e.g. 18S rRNA) from all or most common opportunistic fungi is an important strategy (Espy et al., 2006; Hebart, Löffler, Meisner, et al., 2000; Hebart, Löffler, Reitze, et al., 2000; Iwen et al., 2002; Jordanides et al., 2005; Lin et al., 2001; McLintock & Jones, 2004; Quindos, 2006). This approach combined with hybridisation using specific or panfungal oligonucleotide probes enables the detection of pathogenic fungi with an analytical sensitivity of 1–10 fg of DNA per ml (Iwen et al., 2002). Universal primer pairs can specifically amplify the 18S rRNA gene segment from several medically important fungal pathogens including *Candida* species, *Aspergillus* species, *Trichosporon* species, *Fusarium* species, and *Cryptococcus* species, but not from bacteria or human samples. Real-time PCR techniques combined with automated DNA extraction methods (Löeffler et al., 2000) and panfungal primers enable the detection of several different pathogens in a single reaction (Espy et al., 2006). This approach both minimises contamination and provides a degree of cost-effectiveness and standardisation that should make it ideal for rapid screening of high-risk haemato-oncology patients (Iwen et al., 2002; Jordanides et al., 2005). Indeed, sensitivity approaching 100% has been documented in prospective clinical studies, with two consecutive positive results being strongly associated with the development of IFI. Sequential PCR positivity has preceded both clinical signs and symptoms and clinical diagnosis by several days (Iwen et al., 2002; Hebart, Löffler, Meisner, et al., 2000; Hebart, Löffler, Reitze, et al., 2000; Jordanides et al., 2005; Lass-Flörl et al., 2001; Lin et al., 2001).

The ability to detect the presence of fungal pathogens directly in clinical specimens offers exciting new possibilities for the early diagnosis and management of IFIs in high-risk patients (Arvanitis et al., 2014; Mylonakis et al., 2015; Neely et al., 2013;

Zhang, 2013). A potential strategy for the use of non-culture-based methods for the diagnosis of IFI is to stratify patients according to risk and conduct prospective screening using the rapid test coupled with other diagnostic tests (e.g. cultures and imaging studies) (Barnes et al., 2009; Clancy & Nguyen, 2014; McLintock et al., 2004; Prentice et al., 2000; Wenzel & Gennings, 2005). This strategy should facilitate earlier species-level diagnosis thereby allowing targeted pre-emptive antifungal therapy. A highly sensitive and specific test should also help to reduce empirical therapy with concomitant reduction in selection pressure for resistance as well as cost savings for the hospital (Ananda-Rajuah et al., 2012; Clancy & Nguyen, 2014; Diekema & Pfaller, 2013; Hull, Purdy, & Moody, 2014; Mylonakis et al., 2015). With a rapid and simple test, the opportunity for monitoring the response to antifungal therapy may also be possible.

5 ANTIFUNGAL SUSCEPTIBILITY TESTING

Antifungal susceptibility testing *in vitro* is playing an increasing role in antifungal drug selection, as an aid in drug development studies, and as a means of tracking the development of antifungal resistance in epidemiological studies (Alexander & Pfaller, 2006; Alexander et al., 2013; Arendrup, 2013; Castanheira, Messer, Jones, Farrell & Pfaller, 2014; Cleveland et al., 2012; Pfaller, 2005; Pfaller, Messer, Rhomberg, Jones, & Castanheira, 2013; Rex & Pfaller, 2002; Spellberg et al., 2006). The Clinical and Laboratory Standards Institute (CLSI) Subcommittee for Antifungal Testing has developed standardised broth microdilution (BMD) (CLSI, 2008a, 2008b, 2012) and disk diffusion methods (CLSI, 2009, 2010) for *in vitro* susceptibility testing of yeasts and moulds. These methods are reproducible, accurate, and provide clinically useful information that is comparable to that of antibacterial testing (Pfaller, Andes, Diekema, Espinel-Ingroff, & Sheehan, 2010; Pfaller & Jones, 2006; Rex & Pfaller, 2002; Pfaller, Diekema, & Sheehan, 2006; Pfaller, Boyken, Hollis, et al., 2004; Pfaller, Boyken, Messer, et al., 2004; Pfaller, Espinel-Ingroff, & Jones, 2004; Pfaller, Hazen, et al., 2004; Pfaller, Boyken, Hollis, et al., 2005; Pfaller, Boyken, Messer, et al., 2005; Pfaller, Richter, et al., 2005; Pfaller, Diekema, Colombo, et al., 2006; Pfaller, Diekema, Rex, et al., 2006; Pfaller, Diekema, & Sheehan, 2006; Pfaller, Pappas, & Wingard, 2006; Pfaller, Diekema, Procop, & Rinaldi, 2007a, 2007b; Pfaller, Messer, et al., 2007; Pfaller, Andes, et al., 2011; Pfaller, Diekema, et al., 2011).

Species-specific clinical interpretive breakpoints (CBPs) for the most common species of *Candida* (*C. albicans*, *C. tropicalis*, *C. parapsilosis*, *C. glabrata*, and *C. krusei*) and five systemically active antifungal agents (fluconazole, voriconazole, anidulafungin, caspofungin, and micafungin) have been developed by considering data relating the MICs to known resistance mechanisms, the MIC (and zone diameter) distribution profiles, pharmacokinetic and pharmacodynamic parameters, and the relationship between *in vitro* activity (MIC or zone diameter) and clinical outcomes, as determined by the available clinical efficacy studies (CLSI, 2012;

Pfaller & Diekema, 2012; Pfaller, Diekema, Colombo, et al., 2006; Pfaller, Diekema, Rex, et al., 2006; Pfaller, Diekema, & Sheehan, 2006; Pfaller, Pappas, & Wingard, 2006; Pfaller et al., 2010; Pfaller, Andes, et al., 2011; Pfaller, Diekema, et al., 2011; Rex & Pfaller, 2002; Rex et al., 1997, 2001) (Table 9). Although CBPs have not been established for any antifungal agent and the uncommon species of *Candida*, the non-*Candida* yeasts, *Aspergillus* spp., or the non-*Aspergillus* moulds, the development of epidemiological cut-off values (ECVs) for amphotericin B, flucytosine, the azoles,

Table 9 Antifungal Susceptibility Testing: Clinical Interpretive Breakpoints for *Candida* with Use of Clinical and Laboratory Standards Institute Broth Microdilution Methods

Antifungal agent	Species[a]	Category (μg/ml)[a]		
		S	SDD/I	R
Fluconazole	CA	≤2	4	≥8
	CT	≤2	4	≥8
	CP	≤2	4	≥8
	CG	–	≤32	≥64
	CK	NA[b]		
Voriconazole	CA	≤0.12	0.25–0.5	≥1
	CT	≤0.12	0.25–0.5	≥1
	CP	≤0.12	0.25–0.5	≥1
	CG	NA		
	CK	≤0.5	1	≥2
Anidulafungin	CA	≤0.25	0.5	≥1
	CT	≤0.25	0.5	≥1
	CP	≤2	4	≥8
	CG	≤0.12	0.25	≥0.5
	CK	≤0.25	0.5	≥1
Caspofungin	CA	≤0.25	0.5	≥1
	CT	≤0.25	0.5	≥1
	CP	≤2	4	≥8
	CG	≤0.12	0.25	≥0.5
	CK	≤0.25	0.5	≥1
Micafungin	CA	≤0.25	0.5	≥1
	CT	≤0.25	0.5	≥1
	CP	≤2	4	≥8
	CG	≤0.06	0.12	≥0.5
	CK	≤0.25	0.5	≥1

[a]*Abbreviations: CA, C. albicans; CT, C. tropicalis; CP, C. parapsilosis; CG, C. glabrata; CK, C. krusei.; S, susceptible; SDD, susceptible dose dependent (applies to azoles); I, intermediate (applies to echinocandins); R, resistant.*
[b]*NA, not applicable. There are no clinical breakpoints.*
Data compiled from Pfaller and Diekema (2012).

and the echinocandins has aided in the detection of non-wild-type strains (non-WT; strains that harbour intrinsic or acquired resistance mechanisms) for six species of *Candida* (*C. lusitaniae*, *C. guilliermondii*, *C. dubliniensis*, *C. kefyr*, *C. orthopsilosis*, and *C. pelliculosa*) (Pfaller & Diekema, 2012), *C. neoformans/gattii* (Espinel-Ingroff et al., 2012), and four species of *Aspergillus* (*A. fumigatus*, *A. flavus*, *A. terreus*, and *A. niger*) (Espinel-Ingroff et al., 2010). In the absence of clinical outcomes data for these organism-drug combinations, the ECVs should help to detect those strains of fungi that are likely to harbour antifungal resistance mechanisms and thus exhibit reduced susceptibility to the respective agents.

The CLSI BMD method for testing amphotericin B, 5FC, fluconazole, itraconazole, voriconazole, and the echinocandins has served as the reference standard for the development of both broth and agar-based procedures designed to provide simple, flexible, and commercially available alternative susceptibility testing methods for use in the clinical laboratory (Cuenca-Estrella, Gomez-Lopez, Mellado, & Rodriguez-Tudela, 2005; Espinel-Ingroff et al., 2004; Matar et al., 2003; Morace et al., 2002; Pfaller, 2005; Pfaller, Diekema, et al., 2007a, 2007b; Pfaller, Messer, et al., 2007; Pfaller et al., 2002; Pfaller, Boyken, Hollis, et al., 2004; Pfaller, Boyken, Messer, et al., 2004; Pfaller, Espinel-Ingroff, et al., 2004; Pfaller, Hazen, et al., 2004; Pfaller, Boyken, Hollis, et al., 2005; Pfaller, Boyken, Messer, et al., 2005; Pfaller, Richter, et al., 2005). Although a number of antifungal testing systems are now commercially available, the performance has been quite variable (Arendrup, Pfaller, & Danish Fungemia Study Group, 2012; Astvad, Perlin, Johansen, Jensen, & Arendrup, 2013; Cuenca-Estrella et al., 2005, 2010; Espinel-Ingroff et al., 1999, 2004; Matar et al., 2003; Morace et al., 2002; Pfaller & Jones, 2006; Pfaller, Diekema, Procop, & Rinaldi, 2013; Pfaller, Diekema, et al., 2014; Pfaller, Boyken, Hollis, et al., 2004; Pfaller, Boyken, Messer, et al., 2004; Pfaller, Espinel-Ingroff, et al., 2004; Pfaller, Hazen, et al., 2004; Pfaller, Chaturvedi, et al., 2012; Pfaller, Neofytos, et al., 2012; Pfaller, Woosley, et al., 2012; Posteraro et al., 2009). The Sensititre YeastOne System (TREK, Cleveland, OH) is a colorimetric BMD panel that has been approved by the U.S. Food and Drug Administration (FDA) for *in vitro* susceptibility testing of *Candida* against fluconazole, itraconazole, flucytosine, voriconazole, caspofungin, and micafungin (Pfaller, 2005; Pfaller & Jones, 2006; Pfaller, Chaturvedi, et al., 2012; Pfaller, Neofytos, et al., 2012; Pfaller, Woosley, et al., 2012). Both the agar-based Etest (AB BIODISK, Solna Sweden) and the VITEK 2 Yeast Susceptibility Test (bioMerieux, Hazelwood, MO) are approved by the FDA for testing fluconazole against *Candida* (Pfaller, 2005; Pfaller & Jones, 2006; Pfaller, Diekema, et al., 2007a, 2007b, 2013, 2014; Pfaller, Messer, et al., 2007, 2013; Pfaller, Andes, et al., 2014; Pfaller, Jones, et al., 2014). All three of these test methods also allow for testing of amphotericin B, 5FC, voriconazole, and the echinocandins

The VITEK 2 Yeast Susceptibility Test determines growth spectrophotometrically and allows fully automated antifungal susceptibility testing of *Candida* using the VITEK 2 microbiology system (Astvad et al., 2013; Cuenca-Estrella et al., 2010;

Pfaller, Diekema, et al., 2007a, 2007b, 2013, 2014; Pfaller, Messer, et al., 2007, 2013; Pfaller, Andes, et al., 2014; Pfaller, Jones, et al., 2014; Posteraro et al., 2009). This automated approach allows for the standardisation of all the critical parameters known for antifungal susceptibility testing: inoculum preparation, filling of the device, duration and temperature of incubation, and MIC endpoint determination. A multicentre evaluation of the VITEK 2 Yeast Susceptibility Test for fluconazole demonstrated a high level of reproducibility (100%), an essential agreement (MIC ± 1-$\log_2$ dilution) of 93.7–97.9%, and a categorical agreement of 88.3–97.2% when compared to the reference BMD results (Pfaller, Diekema, et al., 2007a,2007b, 2013; Pfaller, Messer, et al., 2007, 2013; Pfaller, Andes, et al., 2014; Pfaller, Diekema, Procop, & Wiederhold, 2014; Pfaller, Jones, et al., 2014). Notably, the VITEK 2 fluconazole MIC results were available after only 10–26 h of incubation (mean, 13 h). Similar performance was seen in an evaluation of this system for testing of 5FC, voriconazole, and amphotericin B against *Candida* (Pfaller, Diekema, et al., 2007a, 2007b; Pfaller, Messer, et al., 2007). The introduction of the VITEK 2 system in the clinical laboratory will be an important step towards the optimisation of antifungal therapy of candidiasis.

Establishing a clinical correlation between *in vitro* susceptibility tests and clinical outcome has been difficult; however, it is now clear that antifungal susceptibility testing can predict outcome in several situations including candidaemia and mucosal candidiasis (Pfaller et al., 2010; Rex & Pfaller, 2002; Pfaller, Diekema, Colombo, et al., 2006; Pfaller, Diekema, Rex, et al., 2006; Pfaller, Diekema, & Sheehan, 2006; Pfaller, Pappas, & Wingard, 2006; Pfaller, Andes, et al., 2011; Pfaller, Diekema, et al., 2011; Rex et al., 1997). Similar to antibacterial susceptibility testing, antifungal susceptibility testing can be said to predict the outcome of treatment with an accuracy that has been summarised as the "90–60" rule (Pfaller et al., 2010; Pfaller, Andes, et al., 2011; Pfaller, Diekema, et al., 2011; Rex & Pfaller, 2002). According to this rule, infections due to susceptible isolates respond to therapy approximately 90% of the time, whereas infections due to resistant isolates respond to therapy approximately 60% of the time. Thus, low MICs are not entirely predictive of clinical success, and high MICs help to predict which patients are less likely to have a favourable response to a given antifungal agent. The 90–60 rule reflects the fact that the *in vitro* susceptibility of an infecting organism to the antifungal (or antibacterial) agent is only one of several factors that may influence the likely success of therapy for an infection (Alexander & Pfaller, 2006; Rex & Pfaller, 2002).

As a result of the efforts of numerous investigators in North America and Europe, antifungal susceptibility testing is now widely accepted as a useful tool for informing decision making during the management of patients with IFIs, to the extent that recommendations for testing are now included in management guidelines (Cuenca-Estrella et al., 2012; Pappas et al., 2009). Guidelines for the use of laboratory studies, including antifungal susceptibility testing, have been developed (Pfaller & Diekema, 2012; Pfaller, Boyken, Hollis, et al., 2005; Pfaller, Boyken, Messer, et al., 2005; Pfaller, Richter, et al., 2005; Rex & Pfaller, 2002) (Table 10). Selective application of *in vitro* susceptibility testing, coupled with broader identification of fungi to the

Table 10 Recommendations for Studies of Fungal Isolates in the Clinical Laboratory

Clinical Setting	Recommendation
Routine	Species-level identification of all *Candida* isolates from deep sites Genus-level identification of moulds (species level preferred for *Aspergillus*) Routine antifungal testing of fluconazole and an echinocandin against *Candida glabrata* isolated from blood and normally sterile body fluids and tissue Routine testing of fluconazole and an echinocandin against other species of *Candida* possibly helpful but susceptibility usually predicted by species identification Use CBPs[a] or ECVs[b] to interpret results as appropriate (Table 9) Consider cross-resistance between fluconazole and all other azoles to be complete for *C. glabrata* Create an antifungogram
Oropharyngeal candidiasis	Determination of azole susceptibility may be helpful but not routinely necessary Susceptibility testing may be useful for patients unresponsive to azole therapy
Invasive disease with clinical failure of initial therapy	Consider susceptibility testing as an adjunct – *A. fumigatus* and itraconazole, isavuconazole, posaconazole and voriconazole – *Candida* species and amphotericin B, fluconazole, voriconazole, and an echinocandin – *C. neoformans* and fluconazole, flucytosine, or amphotericin B – *H. capsulatum* and fluconazole Consultation with an experienced microbiologist recommended
Infection with species with high rates of intrinsic or acquired resistance	Susceptibility testing not necessary when intrinsic resistance is known *Candida lusitaniae* and amphotericin B *Candida krusei* and fluconazole, flucytosine *Candida guilliermondii* and echinocandins With high rates of acquired resistance, monitor closely for signs of failure and perform susceptibility testing *Candida glabrata* and fluconazole, amphotericin B and echinocandins *Candida krusei* and amphotericin B

Continued

Table 10 Recommendations for Studies of Fungal Isolates in the Clinical Laboratory—cont'd

Clinical Setting	Recommendation
New treatment options (e.g. caspofungin, voriconazole, posaconazole, isavuconazole) or unusual organisms	*Candida guilliermondii* and amphotericin B *Candida rugosa* and amphotericin B, fluconazole, and echinocandins Susceptibility of *Candida* spp. to echinocandins may be assumed unless initial response is suboptimal Susceptibility testing is warranted if prior exposure to echinocandins or fluconazole Select therapy based on published consensus guidelines and review of survey data on the organism–drug combination in question Susceptibility testing may be helpful when patient is not responding to what should be effective therapy
Patients who respond to therapy despite being infected with an isolate later found to be resistant	Best approach not clear Take into account severity of infection, patient immune status, consequences of recurrence of infection, etc. Consider alternative therapy for infections with isolates that appear to be highly resistant to therapy selected
Mould infections	Susceptibility testing not recommended as a routine If azole-resistant *Aspergillus* spp. consider testing itraconazole, isavuconazole, posaconazole and voriconazole Clinical breakpoints have not been established but epidemiological cut-off values (ECVs) are published to aid in detection of non-wild-type strains
Selection of susceptibility testing method	Standardised methods CLSI broth-based methods Yeasts; M27-A3 Moulds; M38-A2 EUCAST EDef 7.1 Commercial BMD and Automated Methods Sensititre YeastOne Colorimetric VITEK 2 Yeast Susceptibility Test Agar-based methods Etest, numerous agents, yeasts and moulds Disk (fluconazole, voriconazole), CLSI M44-A2 method for yeasts

[a]CBPs, clinical breakpoints.
[b]ECVs, epidemiological cut-off values.
Adapted from Rex and Pfaller (2002) and Pfaller and Diekema (2012).

species level, should prove to be useful, especially in difficult to manage fungal infections (Cuenca-Estrella et al., 2012; Pappas et al., 2009; Spellberg et al., 2006). Future efforts will be directed towards further validation of interpretive breakpoints for established antifungal agents and developing them for newly introduced systemically active agents. In addition, procedures must be optimised for testing non-*Candida* yeasts (e.g. *C. neoformans*, *Trichosporon*) and moulds.

SUMMARY AND CONCLUSIONS

The infectious fungi now constitute one of the most important threats to the survival of immunocompromised hosts. There is little doubt that in addition to *C. albicans* and *A. fumigatus*, a vast array of fungi, previously considered to be non-pathogenic, may serve as significant human pathogens. Recognition of these emerging fungal pathogens has resulted in a better understanding of their clinical presentation and response to the available therapeutic measures. Conventional laboratory-based methods for diagnosis of fungal infection remain useful but are often slow and lack sensitivity. The newer rapid, sensitive, and specific culture-independent methods for the diagnosis of IC and IA offer great promise for improved diagnosis and management of these serious infections.

The continuing threat to public health posed by mycoses (Brown et al., 2012; Pfaller & Diekema, 2007; Rosner, Reiss, Warren, Shadomy, & Lipman, 2002), combined with the severity of IFIs and the difficulty encountered in their treatment (Spanakis et al., 2006), requires a skilled laboratory workforce with the capacity to rapidly identify causative agents in time to institute effective antifungal therapy. Key findings of a cross-sectional survey of U.S. laboratories revealed a lack of ongoing training in mycology and a need for more clinically appropriate and cost-effective laboratory practices (Rosner et al., 2002). The increasing complexity of medical mycology practice underscores the need for laboratory staff to have access to high-quality and clinically relevant continuing education (Rosner et al., 2002). Currently, training is needed in basic isolation procedures and in advanced topics such as identification of problematic moulds and yeasts, antifungal susceptibility testing, and in molecular and proteomic methods for detection and identification of fungal pathogens (Rosner et al., 2002).

Newer broad-spectrum antifungal agents may prove useful in the management of IFI but in turn may require more sensitive methods for diagnosing the infections, as well as for estimating the extent of disease, in order to significantly impact disease outcome. Broad application of both new and established antifungal agents may also select more resistant organisms from the vast pool of environmental opportunistic fungi. Such "emerging" fungal pathogens will pose yet another set of diagnostic and therapeutic challenges and will require that they are both visualised in tissue and identified in culture to truly define their pathogenesis and response to treatment.

Although progress in the diagnosis of mycotic infections has been made in the past two decades-including the development of GM- and BDG-based antigen assays, the PNA-FISH and MALDI-TOF MS approaches for rapid identification of *Candida* from blood cultures, the introduction of rapid, sensitive, and specific approaches, such as T2MR, for direct detection of *Candida* from whole blood, and the standardisation of antifungal susceptibility testing—a great deal of work remains to be done. Future efforts must be directed towards standardising and validating additional molecular assays for the diagnosis of fungal disease, expanding the availability of the new FDA-approved and CLSI-recommended tests, and establishing and validating the interpretive CBPs for all medically important fungi to all licenced antifungal agents (Alexander & Pfaller, 2006). Finally, the lack of training in medical mycology is a growing crisis in clinical microbiology laboratories (Rosner et al., 2002; Wilson, 2002). It is impossible for laboratory personnel to provide the testing necessary to keep up with the changes in medical mycology without adequate training (Wilson, 2002).

REFERENCES

Adam, O., Auperin, A., Wilquin, F., Bourhis, J. H., Gachot, B., & Chachaty, E. (2004). Treatment with piperacillin-tazobactam and false-positive Aspergillus galactomannan antigen test results for patients with hematological malignancies. *Clinical Infectious Diseases, 38,* 917–920.

Alanio, A., Beretti, J. L., Dauphin, B., Mellado, E., Quesne, G., Lacroix, C., et al. (2011). Matrix-assisted laser desorption ionization time-of-flight mass spectrometry for fast and accurate identification of clinically relevant Aspergillus species. *Clinical Microbiology and Infection, 17,* 750–755.

Alexander, B. D., Dodds-Ashley, E. S., Reller, L. B., & Reed, S. D. (2006). Cost savings with implementation of PNA FISH testing for identification of Candida albicans in blood cultures. *Diagnostic Microbiology and Infectious Disease, 54,* 277–282.

Alexander, B. D., Johnson, M. D., Pfeiffer, C. D., Jiminez-Ortigosa, C., Catania, J., Booker, R., et al. (2013). Increasing echinocandin resistance in Candida glabrata: Clinical failure correlates with presence of FKS mutations and elevated minimum inhibitory concentrations. *Clinical Infectious Diseases, 56,* 1724–1732.

Alexander, B. D., & Pfaller, M. A. (2006). Contemporary tools for the diagnosis and management of invasive mycoses. *Clinical Infectious Diseases, 43*(Suppl. 1), S15–S27.

Ampel, N. M. (2010). The diagnosis of coccidioidomycosis. *F1000 Medicine Reports, 2,* 2.

Ananda-Rajuah, M. R., Slavin, M. A., & Thursky, K. T. (2012). The case for antifungal stewardship. *Current Opinion in Infectious Diseases, 25,* 107–115.

Antinori, S., Radice, A., Galimberti, L., Magni, C., Fasan, M., & Parravicini, C. (2005). The role of cryptococcal antigen assay in diagnosis and monitoring of cryptococcal meningitis. *Journal of Clinical Microbiology, 43,* 5828–5829.

Apisarnthanarak, A., & Powderly, W. G. (2005). Epidemiology of fungal infection in patients with human immunodeficiency virus. In J. R. Wingard & E. J. Anaissie (Eds.), *Fungal infections in the immunocompromised patient* (pp. 129–148). Boca Raton: Taylor & Francis.

Arendrup, M. C. (2013). Candida and candidaemia: Susceptibility and epidemiology. *Danish Medical Journal, 60,* B4698.

Arendrup, M. C., Boekhout, T., Akova, M., Meis, J. F., Cornely, O. A., & Lortholary, O. (2014). ESCMID and ECMM joint clinical guidelines for the diagnosis and management of rare invasive yeast infections. *Clinical Microbiology and Infection*, 20(Suppl. 3), S76–S78.

Arendrup, M. C., Pfaller, M. A., & Danish Fungemia Study Group. (2012). Caspofungin Etest susceptibility testing of Candida species: Risk of misclassifaction of susceptible isolates of Candida glabrata and Candida krusei when adopting the revised CLSI caspofungin breakpoints. *Antimicrobial Agents and Chemotherapy*, 56, 3965–3968.

Arendrup, M. C., Sulim, S., Holm, A., Nielsen, L., Nielsen, S. D., Knudsen, J. D., et al. (2011). Diagnostic issues, clinical characteristics, and outcomes for patients with fungemia. *Journal of Clinical Microbiology*, 49, 3300–3308.

Arvanitis, M., Anagnostou, T., Fuchs, B. B., Caliendo, A. M., & Mylonakis, E. (2014). Molecular and nonmolecular diagnostic methods for invasive fungal infections. *Clinical Microbiology Reviews*, 27, 490–526.

Asada, N., Uryu, H., Koseki, M., Takeuchi, M., Komatsu, M., & Matsue, K. (2006). Successful treatment of breakthrough Trichosporon asahii fungemia with voriconazole in a patient with acute myeloid leukemia. *Clinical Infectious Diseases*, 43, e39–e41.

Astvad, K. M., Perlin, D. S., Johansen, H. K., Jensen, R. H., & Arendrup, M. C. (2013). Evaluation of caspofungin susceptibility testing by the new Vitek 2 AST-YS06 Yeast Card using a unique collection of FKS wild-type and hot spot mutant isolates, including the five most common Candida species. *Antimicrobial Agents and Chemotherapy*, 57, 177–182.

Avni, T., Leibovici, L., & Paul, M. (2011). PCR diagnosis of invasive candidiasis: Systematic review and meta-analysis. *Journal of Clinical Microbiology*, 49, 665–670.

Avni, T., Levy, I., Sprecher, H., Yahav, D., Leibovici, L., & Paul, M. (2012). Diagnostic accuracy of PCR alone compared to galactomannan in bronchoalveolar lavage fluid for diagnosis of invasive pulmonary aspergillosis: A systematic review. *Journal of Clinical Microbiology*, 50, 3652–3658.

Azie, N., Neofytos, D., Pfaller, M., Meier-Kriesche, H. U., Quan, S. P., & Horn, D. (2012). The PATH (Prospective Antifungal Therapy) Alliance registry and invasive fungal infections: Update 2012. *Diagnostic Microbiology and Infectious Disease*, 73, 293–300.

Baddley, J. W., Stroud, T. P., Salzman, D., & Pappas, P. G. (2001). Invasive mold infections in allogeneic bone marrow transplant recipients. *Clinical Infectious Diseases*, 32, 1319–1324.

Bale, M. J., Yang, C., & Pfaller, M. A. (1997). Evaluation of growth characteristics on blood agar and eosin methylene blue agar for the identification of Candida (Torulopsis) glabrata. *Diagnostic Microbiology and Infectious Disease*, 29, 65–67.

Barnes, R. A., White, P. L., Bygrave, C., Evans, N., Healy, B., & Kell, J. (2009). Clinical impact of enhanced diagnosis of invasive fungal disease in high-risk haematology and stem cell treansplant patients. *Journal of Clinical Pathology*, 62, 64–69.

Baron, E. J. (2006). Implications of new technology for infectious disease practice. *Clinical Infectious Diseases*, 43, 1318–1323.

Baselski, V., Buchanan, K., Carey, R., Clarridge, J., & Weissfeld, A. (2005). *Survey of clinical microbiology laboratory workloads, productivity rates, and staffing vacancies.* Washington, DC: American Society for Microbiology Office of Public and Scientific Affairs.

Baumgartner, C., Freydiere, A. M., & Gille, Y. (1996). Direct identification and recognition of yeast species from clinical material by using Albicans ID and CHROMagar Candida plates. *Journal of Clinical Microbiology*, 34, 454–456.

Beyda, N. D., Alam, M. J., & Garey, K. W. (2013). Comparison of the T2DX instrument with T2Candida assay and automated blood culture in the detection of Candida species using seeded blood samples. *Diagnostic Microbiology and Infectious Disease*, 77, 324–326.

Becker, M. J., Lugtenburg, E. J., Cornelissen, J. J., Van Der Schee, C., Hoogsteden, H. C., & De Marie, S. (2003). Galactomannan detection in computerized tomography-based bronchoalveolar lavage fluid and serum in haematological patients at risk for invasive pulmonary aspergillosis. *British Journal of Haematology*, *121*, 448–451.

Bhally, H. S., Jain, S., Shields, C., Halsey, N., Cristofalo, E., & Merz, W. G. (2006). Infections in a neonate caused by Pichia fabianii: Importance of molecular identification. *Medical Mycology*, *44*, 185–187.

Blumberg, H. M., Jarvis, W. R., Soucie, J. M., Edwards, J. E., Patterson, J. E., Pfaller, M. A., et al. (2001). Risk factors for candidal bloodstream infections in surgical intensive care unit patients: The NEMIS prospective multicenter study. *Clinical Infectious Diseases*, *33*, 177–186.

Borman, A. M., Szekely, A., Palmer, M. D., & Johnson, E. M. (2012). Assessment of accuracy of identification of pathogenic yeasts in microbiology laboratories in the United Kingdom. *Journal of Clinical Microbiology*, *50*, 2639–2644.

Bow, E. J. (2005). Assessment of the risk for invasive fungal infection among oncology patients. In J. R. Wingard & E. J. Anaissie (Eds.), *Fungal infections in the immunocompromised patient* (pp. 97–128). Boca Raton: Taylor & Francis.

Brandt, M. E., Gaunt, D., Iqbal, N., McClinton, S., Hambleton, S., & Sigler, L. (2005). False-positive Histoplasma capsulatum Gen-Probe chemiluminescent test result caused by a Chrysosporium species. *Journal of Clinical Microbiology*, *43*, 1456–1458.

Brandt, M. E., Gomez, B. L., & Warnock, D. W. (2011). *Histoplasma, blastomyces, coccidioides*, and other dimorphic fungi causing systemic mycoses. In J. Versalovic, K. L. Carroll, G. Funke, J. H. Jorgensen, M. L. Landty, & D. W. Warnock (Eds.), *Manual of clinical microbiology* (10th ed., pp. 1902–1918). Washington DC: ASM Press.

Brown, G. D., Denning, D. W., Gow, N. A., Levitz, S. M., Netea, M. G., & White, T. C. (2012). Hidden killers: Human fungal infections. *Science Translational Medicine*, *4*, 165rv13.

Buchan, B. W., & Ledeboer, N. A. (2014). Emerging technologies for the clinical microbiology laboratory. *Clinical Microbiology Reviews*, *27*, 783–822.

Buchheidt, D., Baust, C., Skladny, H., Ritter, J., Suedhoff, T., Baldus, M., et al. (2001). Detection of Aspergillus species in blood and bronchoalveolar lavage samples from immunocompromised patients by means of 2-step polymerase chain reaction: Clinical results. *Clinical Infectious Diseases*, *33*, 428–435.

Buchheidt, D., Hummel, M., Schleiermacher, D., Spiess, V., Schwerdtferger, R., Cornely, O. A., et al. (2004). Prospective clinical evaluation of a LightCycler™–mediated polymerase chain reaction assay, a nested-PCR assay and a galactomannan enzyme-linked immunosorbent assay for detection of invasive aspergillosis in neutropenic cancer patients and haematological stem cell transplant patients. *British Journal of Haematology*, *125*, 196–202.

Burgener-Kainuz, P., Zuber, J. P., Jaunin, P., Buchman, T. G., Bille, J., & Rossier, M. (1994). Rapid detection and identification of Candida albicans and Torulopsis (Candida) glabrata in clinical specimens by species-specific nested PCR amplification of a cytochrome P-450 lanosterol-alpha-demethylase (L1A1) gene fragment. *Journal of Clinical Microbiology*, *32*, 1902–1907.

Burnie, J. P., Golband, N., & Matthews, R. C. (1997). Semiquantitative polymerase chain reaction enzyme immunoassay for diagnosis of disseminated candidiasis. *European Journal of Clinical Microbiology & Infectious Diseases*, *16*, 346–350.

Busca, A., Locatelli, F., Barbul, A., Limerutti, G., Serra, R., Libertucci, D., et al. (2006). Usefulness of sequential Aspergillus galactomannan antigen detection combined with early radiologic evaluation for diagnosis of invasive pulmonary aspergillosis in patients undergoing allogeneic stem cell transplantation. *Transplantation Proceedings*, *38*, 1610–1613.

Buschelman, B., Jones, R. N., Pfaller, M. A., Koontz, F. P., & Doern, G. V. (1999). Colony morphology of Candida spp. as a guide to species identification. *Diagnostic Microbiology and Infectious Disease, 35*, 89–91.

Buzina, W., Lang-Loidot, D., Braun, H., Freudenschuss, K., & Stammberger, H. (2001). Development of molecular methods for identification of Schizophyllum commune from clinical samples. *Journal of Clinical Microbiology, 39*, 2391–2396.

Caliendo, A. M., Gilbert, D. N., Ginocchio, C. C., Hanson, K. E., May, L., et al. (2013). Better tests, better care: Improved diagnostics for infectious diseases. *Clinical Infectious Diseases, 57*(Suppl. 3), S139–S170.

Calvin, C., Freeman, C., Masterson, R., Miles, D., Straedey, V., & Vineyard, M. (1998). Identification of Candida albicans by colony morphology. *Clinical Microbiology Newsletter, 20*, 16.

Castanheira, M., Messer, S. A., Jones, R. N., Farrell, D. J., & Pfaller, M. A. (2014). Activity of echinocandins and triazoles against a contemporary (2012) worldwide collection of yeast and moulds collected from invasive infections. *International Journal of Antimicrobial Agents, 44*, 320–326.

Chamilos, G., Lewis, R. E., & Kontoyiannis, D. P. (2008). Delaying amphotericin B-based frontline therapy significantly increases mortality among patients with hematologic malignancy who have zygomycosis. *Clinical Infectious Diseases, 47*, 503–509.

Chandler, F. W., & Watts, J. C. (Eds.). (1987). *Pathologic diagnosis of fungal infections.* Chicago: ASCP Press.

Chang, S. S., Hsieh, W. H., Liu, T. S., Wang, C. H., Chou, H. C., et al. (2013). Multiplex PCR system for rapid detection of pathogens in apteients with presumed sepsis-a systematic review and meta-analysis. *PLoS One, 8*, e62323. http://dx.doi.org/10.1371/journal.pone.0062323.

Chen, S. C. A., Holliday, C. L., & Meyer, W. (2002). A review of nucleic acid-based diagnostic tests for systemic mycoses with an emphasis on polymerase chain reaction-based assays. *Medical Mycology, 40*, 333–357.

Christie, J. D. (2003). Diagnosis of invasive mold infection: Is PCR the answer? *American Journal of Clinical Pathology, 119*, 11–13.

Chryssanthou, E., Klingspor, L., Tollemar, J., Petrini, B., Larsson, L., Christensson, B., et al. (1999). PCR and other non-culture methods for diagnosis of invasive Candida infections in allogeneic bone marrow and solid organ transplant recipients. *Mycoses, 42*, 239–247.

Clancy, C. J., & Nguyen, M. H. (2013). Finding the "missing 50%" of invasive candidiasis: How nonculture diagnostics will improve the understanding of disease spectrum and transform patient care. *Clinical Infectious Diseases, 56*, 1284–1292.

Clancy, C. J., & Nguyen, M. H. (2014). Undiagnosed invasive candidiasis: Incorporating nonculture diagnostics into rational prophylactic and preemptive antifungal strategies. *Expert Review of Anti-Infective Therapy, 12*, 731–734.

Clark, A. E., Kaleta, E. J., Arora, A., & Wolk, D. M. (2013). Matrix-assisted laser desorption ionization-time of flight mass spectrometry: A fundamental shift in the routine practice of clinical microbiology. *Clinical Microbiology Reviews, 26*, 547–603.

Cleveland, A. A., Farley, M. M., Harrison, L. H., Stein, B., Hollick, R., Lockhart, S. R., et al. (2012). Changes in incidence and antifungal drug resistance in candidemia: Results from population-based laboratory surveillance in Atlanta and Baltimore, 2008–2011. *Clinical Infectious Diseases, 55*, 1352–1361.

Clinical and Laboratory Standards Institute. (2008a). *Reference method for broth dilution antifungal susceptibility testing of filamentous fungi: Approved standard.* M38-A2 document (2nd ed.). Wayne, PA: Clinical and Laboratory Standards Institute.

Clinical and Laboratory Standards Institute. (2008b). *Reference method for broth dilution antifungal susceptibility testing of yeasts: Approved standard.* M27-A3 document (3rd ed.). Wayne, PA: Clinical and Laboratory Standards Institute.

Clinical and Laboratory Standard Institute. (2009). *Method for antifungal disk diffusion susceptibility testing of yeasts: Approved guideline.* M44-A2 document (2nd ed.). Wayne, PA: Clinical and Laboratory Standards Institute.

Clinical and Laboratory Standards Institute. (2010). *Method for antifungal disk diffusion susceptibility testing of non-dermatophyte filamentous fungi, Approved standard (M51-A1 document).* Wayne, PA: Clinical and Laboratory Standards Institute.

Clinical and Laboratory Standards Institute (2012). *Reference method for broth dilution antifungal susceptibility testing of yeasts; 4th informational supplement (M27-S4 document).* Wayne, PA: Clinical and Laboratory Standards Institute.

Collins, L. A., Samore, M. H., Roberts, M. S., Luzzati, R., Jenkins, R. L., Lewis, W. D., et al. (1994). Risk factors for invasive fungal infections complicating orthotopic liver transplantation. *The Journal of Infectious Diseases, 170,* 644–652.

Connor, D. H., Chandler, F. W., Schwartz, D. A., Manz, H. J., & Lack, E. E. (Eds.). (1997). *Pathology of infectious diseases.* Stamford, Connecticut: Appleton & Lange.

Cornely, O. A., Bassetti, M., Calandra, T., Garbino, J., Kullberg, B. J., Lortholary, O., et al. (2012). ESCMID guideline for the diagnosis and management of Candida diseases 2012: Non-neutropenic adult patients. *Clinical Microbiology and Infection, 18*(Suppl. 7), S19–S37.

Costa, J. M., Botterel, F., Cabaret, O., Foulet, F., Cotdonnier, C., & Bretagne, S. (2012). Association between circulating DNA, serum (1 –> 3)-beta-D-glucan, and pulmonary fungal burden in Pneumocystis pneumonia. *Clinical Infectious Diseases, 55,* e5–e8.

Cuenca-Estrella, M., Gomez-Lopez, A., Alastruey-Izquierdo, A., Bernal-Martinez, L., Cuesta, I., Buitrago, M. J., et al. (2010). Comparison of the Vitek 2 antifungal susceptibility system with the clinical and laboratory standards institute (CLSI) and European Committee on Antimicrobial Susceptibility Testing (EUCAST) broth microdilution reference methods and with the Sensititre YeastOne and Etest techniques for in vitro detection of antifungal resistance in yeast isolates. *Journal of Clinical Microbiology, 48,* 1782–1786.

Cuenca-Estrella, M., Gomez-Lopez, A., Mellado, E., & Rodriguez-Tudela, J. L. (2005). Correlation between the procedure for antifungal susceptibility testing for Candida spp. of the European Committee on Antibiotic Susceptibility Testing (EUCAST) and four commercial techniques. *Clinical Microbiology and Infection, 11,* 486–492.

Cuenca-Estrella, M., Verweij, P. E., Arendrup, M. C., Arikan-Akdagli, S., Bille, J., Donnelly, J. P., et al. (2012). ESCMID guidelines for the diagnosis and management of Candida diseases 2012: Diagnostic procedures. *Clinical Microbiology and Infection, 18*(Suppl. 7), S9–S18. http://dx.doi.org/10.1111/1469-0691.12038.

Cushion, M. T. (2011). Pneumocystis. In J. Versalovic, K. L. Carroll, G. Funke, J. H. Jorgensen, M. L. Landry, & D. W. Warnock (Eds.), *Manual of clinical microbiology* (10th ed., pp. 1822–1835). Washington DC: American Society for Microbiology.

de Pauw, B. E. (2005). Between over- and under-treatment of invasive fungal disease. *Clinical Infectious Diseases, 41,* 1251–1253.

Dictar, M. O., Maioo, E., Alexander, B., Jacob, N., & Veron, M. T. (2000). Mycoses in the transplanted patient. *Medical Mycology, 38*(Suppl. 1), 251–258.

Diekema, D. J., & Pfaller, M. A. (2004). Nosocomial candidemia: An ounce of prevention is better than a pound of cure. *Infection Control and Hospital Epidemiology, 25,* 624–626.

Diekema, D. J., & Pfaller, M. A. (2013). Rapid detection of antibiotic-resistant organism carriage for infection prevention. *Clinical Infectious Diseases, 56,* 1614–1620.

Digby, J., Kalbfleisch, J., Glenn, A., Larsen, A., Browder, W., & Williams, D. (2003). Serum glucan levels are not specific for presence of fungal infections in intensive care units. *Clinical and Diagnostic Laboratory Immunology, 10,* 882–885.

Donnelly, J. P. (2006). Polymerase chain reaction for diagnosing invasive aspergillosis: Getting closer but still a ways to go. *Clinical Infectious Diseases, 42,* 487–489.

Duthie, R., & Denning, D. W. (1995). Aspergillus fungemia: Report of 2 cases and review. *Clinical Infectious Diseases, 20,* 598–605.

Dykewicz, C. A. (2001). Hospital infection control in hematopoietic stem cell transplant recipients. *Emerging Infectious Diseases, 7,* 263–267.

Eggimann, P., Garbino, J., & Pittet, D. (2003). Epidemiology of *Candida* species infections in critically ill non-immunosuppressed patients. *The Lancet Infectious Diseases, 3,* 685–702.

Einsele, H., Hebart, H., Roller, G., Löffler, J., Rothenhöfer, I., Müller, C. A., et al. (1997). Detection and identification of fungal pathogens in blood by using molecular probes. *Journal of Clinical Microbiology, 35,* 1353–1360.

Espinel-Ingroff, A., Aller, A. I., Canton, E., Castanon-Olivares, L. R., Chowdhary, A., Cordoba, S., et al. (2012). Cryptococcus neoformans-Cryptococcus gattii species complex: An international study of wild-type susceptibility endpoint distributions and epidemiological cutoff values for fluconazole, itraconazole, posaconazole and voriconazole. *Antimicrobial Agents and Chemotherapy, 56,* 5898–5906.

Espinel-Ingroff, A., Diekema, D. J., Fothergill, A., Johnson, E., Pelaez, T., Pfaller, M. A., et al. (2010). Wild-type distributions and epidemiological cutoff values for the triazoles and six Aspergillus spp. for the CLSI broth microdilution method (M38-A2 document). *Journal of Clinical Microbiology, 48,* 3251–3257.

Espinel-Ingroff, A., Pfaller, M., Messer, S. A., Knapp, C. C., Holliday, N., & Killian, S. B. (2004). Multicenter comparison of the Sensititre YeastOne colorimetric antifungal panel with the NCCLS M27-A2 reference method for testing new antifungal agents against clinical isolates of Candida spp. *Journal of Clinical Microbiology, 42,* 718–721.

Espinel-Ingroff, A., Pfaller, M., Messer, S. A., Knapp, C. C., Killian, S., Norris, H. A., et al. (1999). Multicenter comparison of the Sensititre YeastOne Colorimetric Antifungal Panel with the National Committee for Clinical Laboratory Standards M27-A reference method for testing clinical isolates of common and emerging Candida spp., Cryptococcus spp., and other yeasts and yeast-like organisms. *Journal of Clinical Microbiology, 37,* 591–595.

Espy, M. J., Uhl, J. R., Sloan, L. M., Buckwalter, S. P., Jones, M. F., Vetter, E. A., et al. (2006). Real-time PCR in clinical microbiology: Applications for routine laboratory testing. *Clinical Microbiology Reviews, 19,* 165–256.

Flahaut, M., Sanglard, D., Monod, M., Bille, J., & Rossier, M. (1998). Rapid detection of *Candida albicans* in clinical samples by DNA amplification of common regions from C. albicans-secreted aspartic proteinase genes. *Journal of Clinical Microbiology, 36,* 395–401.

Fleming, R., & Anaissie, E. J. (2005). Emerging fungal infections. In J. R. Wingard & E. J. Anaissie (Eds.), *Fungal infections in the immunocompromised patient* (pp. 311–340). Boca Raton: Taylor & Francis.

Fleming, R. V., Walsh, T. J., & Anaissie, E. J. (2002). Emerging and less common fungal pathogens. *Infectious Disease Clinics of North America, 16,* 915–934.

Florent, M., Katsahian, S., Vekhoff, A., Levy, V., Rio, B., Marie, J. P., et al. (2006). Prospective evaluation of a polymerase chain reaction—ELISA targeted to Aspergillus fumigatus and Aspergillus flavus for the early diagnosis of invasive aspergillosis in patients with hematological malignancies. *The Journal of Infectious Diseases, 193,* 741–747.

Forrest, G. N., Mankes, K., Jabra-Rizk, M. A., Weekes, E., Johnson, J. K., Lincalis, D. P., et al. (2006). Peptide nucleic acid fluorescence in situ hybridization-based identification of Candida albicans and its impact on mortality and antifungal therapy costs. *Journal of Clinical Microbiology, 44*, 3381–3383.

Fortun, J., Martin-Davila, P., Alvarez, M. E., Sanchez-Sousa, A., Quereda, C., Navas, E., et al. (2001). Aspergillus antigenemia sandwich-enzyme immunoassay test as a serodiagnostic method for invasive aspergillosis in liver transplant recipients. *Transplantation, 71*, 145–149.

Foster, C. B., & Chanock, S. J. (2005). Fungal infections in immunocompromised hosts: Host defenses, risks and epidemiology in special patient groups—Pediatrics. In J. R. Wingard & E. J. Anaissie (Eds.), *Fungal infections in the immunocompromised patient* (pp. 185–197). Boca Raton: Taylor & Francis.

Fridkin, S. K. (2005). The changing face of fungal infections in health care settings. *Clinical Infectious Diseases, 41*, 1455–1460.

Fujita, S., Takamura, T., Nagahara, M., & Hashimoto, T. (2006). Evaluation of a newly developed down-flow immunoassay for detection of serum mannan antigens in patients with candidaemia. *Journal of Medical Microbiology, 55*, 537–543.

Gales, A. C., Pfaller, M. A., Houston, A. K., Joly, S., Sullivan, D. J., Coleman, D. C., et al. (1999). Identification of Candida dubliniensis based on temperature and utilization of xylose and α-methyl-D-glucoside as determined with the API 20C AUX and Vitek YBC systems. *Journal of Clinical Microbiology, 37*, 3804–3808.

Garbino, J., Lew, D. P., Romand, J. A., Hugnonnet, S., Auckenthaler, R., & Pittet, D. (2002). Prevention of severe Candida infections in non-neutropenic, high-risk, critically ill patients: A randomized, double-blind, placebo-controlled trial in patients treated by selective digestive-decontamination. *Intensive Care Medicine, 28*, 1708–1717.

Garey, K. W., Rege, M., Pai, M. P., Mingo, D. E., Sude, K. J., Turpin, R. S., et al. (2006). Time to initiation of fluconazole therapy impacts mortality in patients with candidemia: A multi-institutional study. *Clinical Infectious Diseases, 43*, 25–31.

Geha, D. J., & Roberts, G. D. (1994). Laboratory detection of fungemia. *Clinics in Laboratory Medicine, 14*, 83–97.

Gentry, L. O., Wilkinson, I. D., Lea, A. S., & Price, M. F. (1983). Latex agglutination test for detection of Candida antigen in patients with disseminated disease. *European Journal of Clinical Microbiology, 2*, 122–128.

Grazziatti, M., Anaissie, E. J., & Wingard, J. R. (2005). Overview of host defenses: Innate and acquired immunity. In J. R. Wingard & E. J. Anaissie (Eds.), *Fungal infections in the immunocompromised patient* (pp. 1–44). Boca Raton: Taylor & Francis.

Griffin, A. T., & Hanson, K. E. (2014). Update on fungal diagnostics. *Current Infectious Disease Reports, 16*, 415–425.

Groll, A. H., Gea-Banacloche, J. C., Glasmacher, A., Just-Nüebblings, G., Maschmeyer, G., & Walsh, T. J. (2003). Clinical pharmacology of antifungal compounds. *Infectious Disease Clinics of North America, 17*, 159–191.

Guarner, J. (2014). Incorporating pathology in the practice of infectious disease: Myths and reality. *Clinical Infectious Diseases, 59*, 1133–1141.

Guarner, J., & Brandt, M. E. (2011). Histopathologic diagnosis of fungal infections in the 21st century. *Clinical Microbiology Reviews, 24*, 247–280.

Gutierrez, J., Maroto, C., Piedrola, G., Martin, E., & Perez, J. A. (1993). Circulating Candida antigens and antibodies: Useful markers of candidemia. *Journal of Clinical Microbiology, 31*, 2550–2552.

Hage, C. A., Reynolds, J. M., Durkin, M., Wheat, L. J., & Knox, K. S. (2007). Plasmalyte as a cause of false-positive results for Aspergillus galactomannan in bronchoalveolar lavage fluid. *Journal of Clinical Microbiology*, *45*, 676–677.

Hajjeh, R. A., Sofair, A. N., Harrison, I. H., Lyon, G., Arthington-Skaggs, B. A., Mirza, S. A., et al. (2004). Incidence of bloodstream infections due to *Candida* species and in vitro susceptibilities of isolates collected from 1998 to 2000 in a population-based active surveillance program. *Journal of Clinical Microbiology*, *432*, 159–1527.

Hall, L., LeFebre, K. M., Demi, S. M., Wohlfiel, S. L., & Wengenach, N. L. (2012). Evaluation of the Yeast Traffic Light™ PNA FISH; probes for the identification of Candida species from positive blood cultures. *Journal of Clinical Microbiology*, *50*, 1446–1448.

Hall, L., Wohlfiel, S., & Roberts, G. D. (2003). Experience with the MicroSeq D2 large-subunit ribosomal DNA sequencing kit for identification of commonly encountered, clinically important yeast species. *Journal of Clinical Microbiology*, *41*, 5099–5102.

Hall, L., Wohlfiel, S., & Roberts, G. D. (2004). Experience with the MicroSeq D2 large-subunit ribosomal DNA sequencing kit for identification of filamentous fungi encountered in the clinical laboratory. *Journal of Clinical Microbiology*, *42*, 622–626.

Hayden, R. T., Isotalo, P. A., Parrett, T., Wolk, D. M., Qian, X., Roberts, G. D., et al. (2003). In situ hybridization for the differentiation of Aspergillus, Fusarium, and Pseudallescheria species in tissue sections. *Diagnostic Molecular Pathology*, *12*, 21–26.

Hayden, R. T., Qian, X., Procop, G. W., Roberts, G. D., & Lloyd, R. V. (2002). In situ hybridization for the identification of filamentous fungi in tissue sections. *Diagnostic Molecular Pathology*, *11*, 119–126.

Hebart, H., Löffler, J., Meisner, C., Serey, F., Schmidt, D., Böhme, A., et al. (2000). Early detection of Aspergillus infection after allogeneic stem cell transplantation by polymerase chain reaction screening. *The Journal of Infectious Diseases*, *181*, 1713–1719.

Hebart, H., Löffler, J., Reitze, H., Engel, A., Schumacher, U., Klingebiel, T., et al. (2000). Prospective screening by a panfungal polymerase chain reaction assay in patients at risk for fungal infections: Implications for the management of febrile neutropenia. *British Journal of Haematology*, *111*, 635–640.

Held, J., Schmidt, T., Thornton, C. R., Kotter, E., & Bertz, H. (2013). Comparison of a novel Aspergillus lateral-flow device and Platelia™ galactomannan assay for the diagnosis of invasive aspergillosis following haematopoietic stem cell transplantation. *Infection*, *41*, 1163–1169.

Hinrickson, H. P., Hurst, S. F., Lott, T. J., Warnock, D. W., & Morrison, C. J. (2005). Assessment of ribosomal large-subunit D1-D2, internal transcribed spacer 1, and internal transcribed spacer 2 regions as targets for molecular identification of medically important Aspergillus species. *Journal of Clinical Microbiology*, *43*, 2092–2103.

Hope, W. W., Walsh, T. J., & Denning, D. W. (2005). Laboratory diagnosis of invasive aspergillosis. *The Lancet Infectious Diseases*, *5*, 609–622.

Horvath, J. A., & Dummer, S. (1996). The use of respiratory-tract cultures in the diagnosis of invasive pulmonary aspergillosis. *The American Journal of Medicine*, *100*, 171–178.

Horvath, L. L., George, B. J., Murray, C. K., Harrison, L. S., & Hospenthal, D. R. (2004). Direct comparison of the BACTEC 9240 and BacT/ALERT 3D automated blood culture systems for Candida growth detection. *Journal of Clinical Microbiology*, *42*, 115–118.

Howell, S. A., & Hazen, K. C. (2011). Candida, Cryptococcus and other yeasts of medical importance. In J. Versalovic, K. C. Carroll, G. Funke, J. H. Jorgensen, M. L. Landry, & D. W. Warnock (Eds.), *Manual of clinical microbiology* (10th ed., pp. 1793–1821). Washington DC: ASM Press.

Hsiao, C. R., Huang, L., Bouchara, J. P., Barton, R., Li, H. C., & Chang, T. C. (2005). Identification of medically important molds by an oligonucleotide array. *Journal of Clinical Microbiology, 43*, 3760–3768.

Hull, C. M., Purdy, N. J., & Moody, S. C. (2014). Mitigation of human-pathogenic fungi that exhibit resistance to medical agents: Can antifungal stewardship help? *Future Microbiology, 9*, 307–325.

Iwen, P. C., Hinrichs, S. H., & Rupp, M. E. (2002). Utilization of the internal transcribed spacer regions as molecular targets to detect and identify human fungal pathogens. *Medical Mycology, 40*, 87–109.

Jensen, T. G., Gahrn-Hansen, B., Arendrup, M., & Bruun, B. (2004). Fusarium fungaemia in immunocompromised patients. *Clinical Microbiology and Infection, 10*, 499–501.

Jordanides, N. E., Allan, E. K., McLintock, L. A., Copland, M., Devaney, M., Stewart, K., et al. (2005). A prospective study of real-time panfungal PCR for the early diagnosis of invasive fungal infections in haemato-oncology patients. *Bone Marrow Transplantation, 35*, 389–395.

Jordan, J. A. (1994). PCR identification of four medically important Candida species by using a single primer pair. *Journal of Clinical Microbiology, 32*, 2962–2967.

Kami, M., Fukui, T., Ogawa, S., Kazuyama, Y., Machida, U., Tanaka, Y., et al. (2001). Use of real-time PCR on blood samples for diagnosis of invasive aspergillosis. *Clinical Infectious Diseases, 33*, 1504–1512.

Kan, V. L. (1993). Polymerase chain reaction for the diagnosis of candidemia. *The Journal of Infectious Diseases, 168*, 779–783.

Kappe, R., Schulze-Berge, A., & Sonntag, H. G. (1996). Evaluation of eight antibody tests and one antigen test for the diagnosis of invasive aspergillosis. *Mycoses, 39*, 13–23.

Karageorgopoulos, D. E., Qu, J. M., Korbila, I. P., Zhu, Y. G., Vasileiou, V. A., & Falagas, M. E. (2013). Accuracy of beta-D-glucan for the diagnosis of Pneumocystis jirovecii pneumonia: A meta-analysis. *Clinical Microbiology and Infection, 19*, 39–49.

Karageorgopoulos, D. E., Vouloumanou, E. K., Ntziora, F., Michalopoulos, A., Rafailidis, P. I., & Falagas, M. E. (2011). Beta-D-glucan assay for the diagnosis of invasive fungal infections: A meta-analysis. *Clinical Infectious Diseases, 52*, 750–770.

Kauffman, C. A. (2001). Fungal infections in older adults. *Clinical Infectious Diseases, 33*, 550–555.

Kaufman, D., & Fairchild, K. K. (2004). Clinical microbiology of bacterial and fungal sepsis in very-low-birth-weight infants. *Clinical Microbiology Reviews, 17*, 638–680.

Kaufman, L., Standard, P. G., Jalberl, M., & Kraft, D. E. (1997). Immunohistologic identification of Aspergillus spp. and other hyaline fungi by using polyclonal fluorescent antibodies. *Journal of Clinical Microbiology, 35*, 2206–2209.

Kawazu, M., Kanda, Y., Nannya, Y., Aoki, K., Kurokawa, M., Chiba, S., et al. (2004). Prospective comparison of the diagnostic potential of real-time PCR, double-sandwich enzyme-linked immunosorbent assay for galactomannan, and a $(1 \rightarrow 3)$-β-D-glucan test in weekly screening for invasive aspergillosis in patients with hematologic disorders. *Journal of Clinical Microbiology, 42*, 2733–2741.

Klont, R. R., Mennink-Kersten, M. A. S. H., & Verweij, P. D. (2004). Utility of Aspergillus antigen detection in specimens other than serum specimens. *Clinical Infectious Diseases, 39*, 1467–1474.

Kohno, S., Mitsutake, K., Masesaki, S., Yasuoka, A., Miyazaki, T., Kaku, M., et al. (1993). An evaluation of serodiagnostic tests in patients with candidemia: Beta-glucan, mannan, candida antigen by Cand-Tec and D-arabinitol. *Microbiology and Immunology, 37*, 207–212.

Koneman, E. W., & Roberts, G. D. (2002). The appearance of fungi in tissues. *Laboratory Medicine, 33*, 927–933.

Kontoyiannis, D. P., Sumoza, D., Tarrand, J., Bodey, G. P., Storey, R., & Raad, II. (2000). Significance of aspergillemia in patients with cancer: A 10-year study. *Clinical Infectious Diseases, 31*, 88–91.

Kontoyiannis, D. P., Vaziri, I., Hanna, H. A., Boktour, M., Thornby, J., Hachem, R., et al. (2001). Risk factors for Candida tropicalis fungemia in patients with cancer. *Clinical Infectious Diseases, 33*, 1676–1681.

Kothari, A., Morgan, M., & Haake, D. (2014). Emerging technologies for rapid identification of bloodstream pathogens. *Clinical Infectious Diseases, 59*, 272–278.

Kusne, S., Torre-Cisneros, J., Monez, R., Rish, W., Marten, M., Fung, J., et al. (1992). Factors associated with invasive lung aspergillosis and the significance of positive Aspergillus culture after liver transplantation. *The Journal of Infectious Diseases, 66*, 1379–1383.

Kwak, E. J., Husain, S., Obman, A., Meinke, L., Stout, J., Kusne, S., et al. (2004). Efficacy of galactomannan antigen in the Platelia Aspergillus enzyme immunoassay for diagnosis of invasive aspergillosis in liver transplant recipients. *Journal of Clinical Microbiology, 42*, 435–438.

Lamaris, G. A., Chamilos, G., Lewis, R. E., Safdar, A., Raad, I. I., & Kontoyiannis, D. P. (2006). Scedosporium infection in a tertiary care cancer center: A review of 25 cases from 1989–2006. *Clinical Infectious Diseases, 43*, 1580–1584.

LaRocco, M. T., & Burgert, S. J. (1997). Infection in the bone marrow transplant recipient and role of the microbiology laboratory in clinical transplantation. *Clinical Microbiology Reviews, 10*, 277–297.

Larone, D. H. (2014). *Medically important fungi: A guide to identification* (5th ed.). Washington DC: American Society for Microbiology.

Lass-Flörl, C., Aigner, J., Gunsilius, E., Petzer, A., Nachbaur, D., Gastl, G., et al. (2001). Screening for Aspergillus spp. using polymerase chain reaction of whole blood samples from patients with haematological malignancies. *British Journal of Haematology, 113*, 180–184.

Lau, A. F., Drake, S. K., Calhoun, L. B., Henderson, C. M., & Zelazny, A. M. (2013). Development of a clinically comprehensive database and a simple procedure for identification of molds from solid media by matrix-assisted laser desorption ionization-time of flight mass spectrometry. *Journal of Clinical Microbiology, 51*, 828–834.

Lawn, S. D., & Wood, R. (2012). Point-of-care urine antigen screening tests for tuberculosis and cryptococcosis: Potential for mortality reduction in antiretroviral treatment programs in Africa. *Clinical Infectious Diseases, 54*, 730–740.

Letscher-Bru, V., Cavalier, A., Pernot-Marino, E., Koenig, H., Eyer, D., Waller, J., et al. (1998). Aspergillus galactomannan antigen detection with Platelia (R) Aspergillus: Multiple positive antigenemia without Aspergillus infection. *Journal de Mycologie Médicale, 8*, 112–113.

Liesenfeld, O., Lehman, L., Hunfeld, K. P., & Kost, G. (2014). Molecular diagnosis of sepsis: New aspects and recent developments. *European Journal of Microbiology & Immunology, 4*, 1–25.

Lin, M. T., Lu, H. C., & Chen, W. L. (2001). Improving efficacy of antifungal therapy by polymerase chain reaction-based strategy among patients with neutropenia and cancer. *Clinical Infectious Diseases, 33*, 1621–1627.

Lionakis, M. S., Bodey, G. P., Tarrand, J. J., Raad, I. I., & Kontoyannis, D. P. (2004). The significance of blood cultures positive for emerging saprophytic moulds in cancer patients. *Clinical Microbiology and Infection, 10*, 922–925.

Lo Cascio, G., Ligozzi, M., Maccaro, L., & Fontana, R. (2004). Utility of molecular identification in opportunistic mycotic infections: A case of cutaneous Alternaria infectoria infection in a cardiac transplant recipient. *Journal of Clinical Microbiology, 42*, 5334–5336.

Löeffler, J., Henke, N., Hebart, H., Schmidt, D., Hagmeyer, L., Schumacher, U., et al. (2000). Quantification of fungal DNA by using fluorescence resonance energy transfer and the light cycler system. *Journal of Clinical Microbiology, 38*, 586–590.

Lopez-Ribot, J. L., & Patterson, T. F. (2005). Serological and molecular approaches to the diagnosis of invasive fungal infections in immunocompromised patients. In J. R. Wingard & E. J. Anaissie (Eds.), *Fungal infections in the immunocompromised patient* (pp. 383–406). Boca Raton: Taylor & Francis.

Lu, Y., Ling, G., Qiang, C., Ming, Q., Wu, C., Wang, K., et al. (2011). PCR diagnosis of Pneumocystis pneumonia: A bivariate meta-analysis. *Journal of Clinical Microbiology, 49*, 4361–4363.

Maertens, J., Theunissen, K., Verhoef, G., & van Eldere, J. (2004). False-positive Aspergillus galactomannan antigen test results. *Clinical Infectious Diseases, 39*, 289–290.

Maertens, J., Theunissen, K., Verhoef, G., Verschakelen, J., Lagrosu, K., Verbeken, E., et al. (2005). Galactomannan and computed tomotgraphy-based preemptive antifungal therapy in neutropenic patients at high risk for invasive fungal infection: A prospective feasibility study. *Clinical Infectious Diseases, 41*, 1242–1250.

Magadia, R., & Weinstein, M. P. (2001). Laboratory diagnosis of bacteremia and fungemia. *Infectious Disease Clinics of North America, 15*, 1009–1024.

Marcos, J. Y., & Pincus, D. H. (2013). Fungal diagnostics: Review of commercially available methods. *Methods in Molecular Biology, 968*, 24–45.

Marr, K. A. (2005). Fungal infections in blood and marrow transplant recipients. In J. R. Wingard & E. J. Anaissie (Eds.), *Fungal infections in the immunocompromised patient* (pp. 75–95). Boca Raton: Taylor & Francis.

Marr, K. A., Balajee, S. A., McLaughlin, L., Tabouret, M., Bentsen, C., & Walsh, T. J. (2004). Detection of galactomannan antigenemia by enzyme immunoassay for the diagnosis of invasive aspergillosis: Variables that affect performance. *The Journal of Infectious Diseases, 190*, 641–649.

Marr, K. A., Carter, R. A., Crippa, F., Wald, A., & Corey, L. (2002). Epidemiology and outcome of mould infections in hematopoietic stem cell transplant recipients. *Clinical Infectious Diseases, 34*, 909–917.

Marty, F. M., Baden, L. R., & Rubin, R. H. (2005). Impact of invasive fungal infection on patients undergoing solid organ transplantation. In J. R. Wingard & E. J. Anaissie (Eds.), *Fungal infections in the immunocompromised patient* (pp. 45–74). Boca Raton: Taylor & Francis.

Matar, M. J., Ostrosky-Zeichner, L., Paetznick, V. L., Rodriguez, J. R., Chan, E., & Rex, J. H. (2003). Correlation between E-test, disk diffusion, and microdilution methods for antifungal susceptibility testing of fluconazole and voriconazole. *Antimicrobial Agents and Chemotherapy, 47*, 1647–1651.

Matsue, K., Uryu, H., Koseki, M., Asada, N., & Takeuchi, M. (2006). Breakthrough trichosporonosis in patients with hematologic malignancies receiving micafungin. *Clinical Infectious Diseases, 42*, 753–757.

Mattei, D., Rapezzi, D., Mordini, N., Cuda, F., Lo Nigro, C., Musso, M., et al. (2004). False-positive Aspergillus galactomannan enzyme-linked immunosorbent assay results in vivo during amoxicillin-clavulanic acid treatment. *Journal of Clinical Microbiology, 42*, 5362–5363.

Matthews, R., & Burnie, J. (1988). Diagnosis of systemic candidiasis by an enzyme-linked dot immunobinding assay for a circulating immunodominant 47-kilodalton antigen. *Journal of Clinical Microbiology*, *26*, 459–463.

McGowan, K. L. (2011). Specimen collection, transport, and processing: Mycology. In J. Versalovic, K. L. Carroll, G. Funke, J. H. Jorgensen, M. L. Landry, & D. W. Warnock (Eds.), *Manual of clinical microbiology* (10th ed., pp. 1756–1766). Washington DC: American Society for Microbiology.

McLintock, L. A., & Jones, B. L. (2004). Advances in the molecular and serological diagnosis of invasive fungal infection in haemato-oncology patients. *British Journal of Haematology*, *126*, 289–297.

McLintock, L. A., Jordanides, N. E., Alla, E. K., Copland, M., Stewart, K., Parker, A., et al. (2004). The use of a risk group stratification in the management of invasive fungal infection: A prospective validation. *British Journal of Haematology*, *124*, 403–404.

McMullan, B. J., Halliday, C., Sorrell, T. C., Judd, D., Sleiman, S., Marriott, D., et al. (2012). Clinical utility of the cryptococcal antigen lateral flow assay in a diagnostic mycology laboratory. *PLoS One*, *7*, e49541. http://dx.doi.org/10.1371/journal.pone.0049541.

McNeil, M. M., Nash, S. L., Hajjeh, R. A., Phelan, M. A., Conn, L. A., Plikaytis, B. D., et al. (2001). Trends in mortality due to invasive mycotic diseases in the United States, 1988–1997. *Clinical Infectious Diseases*, *33*, 641–647.

Mennink-Kersten, M. A. S. H., Donnelly, J. P., & Verweij, P. E. (2004). Detection of circulating galactomannan for the diagnosis and management of invasive aspergillosis. *The Lancet Infectious Diseases*, *4*, 349–357.

Mengoli, C., Cruciani, M., Barnes, R. A., Loeffler, J., & Donnelly, J. P. (2009). Use of PCr for diagnosis of invasive aspergillosis: Systematic review and meta-analysis. *The Lancet Infectious Diseases*, *9*, 89–96.

Meyer, M. H., Letscher-Bru, V., Jaulhac, B., Waller, J., & Candolfi, E. (2004). Comparison of Mycosis IC/F and Plus Aerobic/F media for diagnosis of fungemia by the Bactec 9240 system. *Journal of Clinical Microbiology*, *42*, 773–777.

Mikulska, M., Calandra, T., Sanguinetti, M., Puolain, D., & Viscoli, C. (2010). The use of mannan antigen and anti-mannan antibodies in the diagnosis of candidiasis: Recommendations from the Third European Conference on Infections in Leukemia. *Critical Care*, *14*, R222.

Mirza, S. A., Phelan, M., Rimland, D., Graviss, E., Hamill, R., Brandt, M. E., et al. (2003). The changing epidemiology of cryptococcosis: An update from population-based active surveillance in 2 large metropolitan areas, 1992–2000. *Clinical Infectious Diseases*, *36*, 789–794.

Mitsutake, K., Miyazaki, T., Tashiro, T., Yamamoto, Y., Kakeya, H., Otsubo, T., et al. (1996). Enolase antigen, mannan antigen, Cand-Tec antigen and beta-glucan in patients with candidemia. *Journal of Clinical Microbiology*, *34*, 1918–1921.

Miyazaki, T., Kohno, S., Mitsutake, K., Maesaki, S., Tanaka, K. I., Ishikawa, N., et al. (1995). Plasma (1–3)-β-D-glucan and fungal antigenemia in patients with candidemia, aspergillosis, and cryptococcosis. *Journal of Clinical Microbiology*, *33*, 3115–3118.

Morace, G., Amato, G., Bistoni, F., Fadda, G., Marone, P., Montagna, M. T., et al. (2002). Multicenter comparative evaluation of six commercial systems and the national committee for clinical laboratory standards M27-a broth microdilution method for fluconazole susceptibility testing of Candida species. *Journal of Clinical Microbiology*, *40*, 2953–2958.

Morace, G., Pagano, L., Sanguinetti, M., Posteraro, B., Mele, L., Equitani, F., et al. (1999). PCR-restriction enzyme analysis for detection of Candida DNA in blood from febrile patients with hematological malignancies. *Journal of Clinical Microbiology, 37,* 1871–1875.

Moreira-Oliveira, M. S., Mikami, Y., Miyaji, M., Imai, T., Schreiber, A. Z., & Moretti, M. L. (2005). Diagnosis of candidemia by polymerase chain reaction and blood culture: Prospective study in a high-risk population and identification of variables associated with development of candidemia. *European Journal of Clinical Microbiology & Infectious Diseases, 24,* 721–726.

Morgan, J., Wannemuchler, K. A., Marr, K. A., Hadley, S., Kontoyiannis, D. P., Walsh, T. J., et al. (2005). Incidence of invasive aspergillosis following hematopoietic stem cell and solid organ transplantation: Interim results of a prospective multicenter surveillance program. *Medical Mycology, 43*(Suppl. 1), S49–S58.

Morrel, M., Fraser, V. J., & Kollef, M. H. (2005). Delaying the empiric treatment of Candida bloodstream infection until positive blood culture results are obtained: A potential risk factor for hospital mortality. *Antimicrobial Agents and Chemotherapy, 49,* 3640–3645.

Muñoz, P., Burillo, A., & Bouza, E. (2001). Environmental surveillance and other control measures in the prevention of nosocomial fungal infections. *Clinical Microbiology and Infection, 7*(Suppl. 2), 38–45.

Muñoz, P., Guinea, J., & Bouza, E. (2006). Update on invasive aspergillosis: Clinical and diagnostic aspects. *Clinical Microbiology and Infection, 12*(Suppl. 7), 24–39.

Murray, M. P., Zinchuk, R., & Larone, D. H. (2005). CHROMagar Candida as the sole primary medium for isolation of yeasts and as a source medium for the rapid-assimilation-of-trehalose test. *Journal of Clinical Microbiology, 43,* 1210–1212.

Musher, B., Fredricks, D., Leisenring, W., Balajee, S. A., Smith, C., & Marr, K. A. (2004). Aspergillus galactomannan enzyme immunoassay and quantitative PCR for diagnosis of invasive aspergillosis with bronchoalveloar lavage fluid. *Journal of Clinical Microbiology, 42,* 5517–5522.

Mylonakis, E., Clancy, C. J., Ostrosky-Zeichner, L., Garey, K. W., Alangaden, G. J., Vazquez, J. A., et al. (2015). T2 magnetic resonance assay for the rapid diagnosis of candidemia in whole blood. *Clinical Infectious Diseases, 60,* 892–899.

Nagashi, K., & Baron, E. J. (1997). Identification of Candida albicans by colony morphology. *Clinical Microbiology Newsletter, 19,* 112.

Nalesnk, M. A., Myerowitz, R. L., Jenkins, R., Lenskey, J., & Herbert, D. (1980). Significance of Aspergillus species isolated from respiratory secretions in the diagnosis of invasive pulmonary aspergillosis. *Journal of Clinical Microbiology, 11,* 370–376.

National Committee for Clinical Laboratory Standards. (2002). *Abbreviated identification of bacteria and yeast; Approved Guideline NCCLS document M35-A.* Wayne, PA: NCCLS.

Neely, L. A., Audeh, M., Phung, N. A., Min, M., Suchocki, A., Plourde, D., et al. (2013). T2 magnetic resonance enables nanoparticle-mediated rapid detection of candidemia in whole blood. *Science Translational Medicine, 5,* 82ra54. http://dx.doi.org/10.1126/scitranslmed.3005377.

Nguyen, M. H., Wissel, M. C., Shields, R. K., Salomoni, M. A., Hao, B., Press, E. G., et al. (2012). Performance of Candida real-time polymerase chain reaction, beta-D-glucan assay, and blood cultures in the diagnosis of invasive candidiasis. *Clinical Infectious Diseases, 54,* 1240–1248.

Nucci, M., & Marr, K. A. (2005). Emerging fungal diseases. *Clinical Infectious Diseases, 4,* 521–526.

Obayashi, T., Yoshida, M., Mori, T., Goto, H., Yasuoka, A., Shimada, K., et al. (1995). Plasma $(1\rightarrow3)$-beta-D-glucan measurement in diagnosis of invasive deep mycosis and fungal febrile episodes. *Lancet, 345*, 17–20.

Odabasi, Z., Mattiuzzi, G., Estey, E., Kantarjian, H., Saeki, F., Ridge, R. J., et al. (2004). β-D-glucan as a diagnostic adjunct for invasive fungal infections: Validation, cutoff development, and performance in patients with acute myelogenous leukemia and myelodysplastic syndrome. *Clinical Infectious Diseases, 39*, 199–205.

Odds, F. C. (2003). Reflections on the question: What does molecular mycology have to do with the clinical treating of the patient? *Medical Mycology, 41*, 1–6.

Odds, F. C., & Bernaerts, R. (1994). CHROMagar Candida, a new differential isolation medium for presumptive identification of clinically important Candida species. *Journal of Clinical Microbiology, 32*, 1923–1929.

Oliveira, K., Haase, G., Kurtzman, C., Hyldig-Nielsen, J. J., & Stender, H. (2001). Differentiation of Candida albicans and Candida dubliniensis by fluorescent in situ hybridization with peptide nucleic acid probes. *Journal of Clinical Microbiology, 39*, 4138–4141.

Onishi, A., Sugiyama, D., Kogata, Y., Saegusa, J., Sugimoto, T., Kawano, S., et al. (2012). Diagnostic accuracy of serum 1,3-beta-D-glucan for Pneumocystis jirovecii pneumonia, invasive candidiasis, and invasive aspergillosis: Systematic review and meta-analysis. *Journal of Clinical Microbiology, 50*, 7–15.

O'Shaughnessy, E. M., Shea, Y. M., & Witebsky, F. G. (2003). Laboratory diagnosis of invasive mycoses. *Infectious Disease Clinics of North America, 17*, 135–158.

Ostrosky-Zeichner, L. (2003). New approaches to the risk of Candida in the intensive care unit. *Current Opinion in Infectious Diseases, 16*, 533–537.

Ostrosky-Zeichner, L., Alexander, B. D., Kett, D. H., Vazquez, J., Pappas, P. G., Saeki, F., et al. (2005). Multicenter clinical evaluation of the $(1\rightarrow3)$-β-D-glucan assay as an aid to diagnosis of fungal infections in humans. *Clinical Infectious Diseases, 41*, 654–659.

Ostrosky-Zeichner, L., Casdevall, A., Galgiani, J. N., Odds, F. C., & Rex, J. H. (2010). An insight into the antifungal pipeline: Selected new molecules and beyond. *Nature Reviews Drug Discovery, 9*, 719–727.

Padhye, A. A., Smith, G., McLaughlin, D., Standard, P. G., & Kaufman, L. (1992). Comparative evaluation of a chemiluminescent DNA probe and an exoantigen test for rapid identification of Histoplasma capsulatum. *Journal of Clinical Microbiology, 30*, 3108–3111.

Padhye, A. A., Smith, G., Standard, P. G., McLaughlin, D., & Kaufman, L. (1994). Comparative evaluation of chemiluminescent DNA probe assays and exoantigen tests for rapid identification of Blastomyces dermatitidis and Coccidiodies immitis. *Journal of Clinical Microbiology, 32*, 867–870.

Page, B. T., Shields, C. E., Merz, W. G., & Kurtzman, C. P. (2006). Rapid identification of ascomycetous yeasts from clinical specimens by a molecular method based on flow cytometry and comparison with identifications from phenotypic assays. *Journal of Clinical Microbiology, 44*, 3167–3171.

Pappas, P. G., Kauffman, C. A., Andes, D., Benjamen, D. K., Jr., Calandra, T. F., Edwards, J. E., Jr., et al. (2009). Clinical practice guidelines for the management of candidiasis: 2009 update by the Infectious Diseases Society of America. *Clinical Infectious Diseases, 48*, 503–535.

Patterson, T. F., Kirkpatrick, W. R., & White, M. (2000). Invasive aspergillosis: Disease spectrum, treatment practices, and outcomes. *Medicine, 79*, 250–260.

Pazos, C., Ponton, J., & Del Palacio, A. (2005). Contributions of $(1\rightarrow 3)$-β-D-glucan chromogenic assay to diagnosis and therapeutic monitoring of invasive aspergillosis in neutropenic adult patients: A comparison with serial screening for circulating galactomannan. *Journal of Clinical Microbiology, 43,* 299–305.

Pendle, S., Weeks, K., Priest, M., Gill, A., Hudson, B., Kotsiou, G., et al. (2004). Phaeohyphomycotic soft tissue infections caused by the coelomycetous fungus Microsphaeropsis arundinis. *Journal of Clinical Microbiology, 42,* 5315–5319.

Perfect, J. R., Cox, G. M., Lee, J. Y., Kauffman, C. A., deRepentigny, L., Chapman, S. W., et al. (2001). The impact of culture isolation of Aspergillus species: A hospital-based survey of aspergillosis. *Clinical Infectious Diseases, 33,* 1824–1833.

Persat, F., Topenot, R., Piens, M. A., Thiebaut, A., Dannaoui, E., & Picot, S. (2002). Evaluation of different commercial ELISA methods for the serodiagnosis of systemic candidosis. *Mycoses, 45,* 455–460.

Pfaller, M. A. (1992). Laboratory aids in the diagnosis of invasive candidiasis. *Mycopathologia, 120,* 65–72.

Pfaller, M. A. (2001). The fungal pathogen shift in the United States. *The Journal of Critical Illness, 16*(Suppl.), S35–S42.

Pfaller, M. A. (2005). Antifungal susceptibility testing methods. *Current Drug Targets, 6,* 929–943.

Pfaller, M. A., Andes, D., Arendrup, M. C., Diekema, D. J., Brown, S. D., Motyl, M., et al. (2011). Clinical breakpoints for voriconazole and Candida spp. revisited: Review of microbiologic, molecular, pharmacodynamics, and clinical data as they pertain to the development of species-specific interpretive criteria. *Diagnostic Microbiology and Infectious Disease, 70,* 330–343.

Pfaller, M. A., Andes, D., Diekema, D. J., Espinel-Ingroff, A., & Sheehan, D. (2010). Wildtype MIC distributions, epidemiological cutoff values and species-specific clinical breakpoints for fluconazole and Candida: Time for harmonization of CLSI and EUCAST broth microdilution methods. *Drug Resistance Updates, 13,* 180–195.

Pfaller, M. A., Andes, D. R., Diekema, D. J., Horn, D. L., Reboli, A. C., Rotstein, C., et al. (2014). Epidemiology and outcomes of invasive candidiasis due to non- albicans species of Candida in 2,496 patients: Data from the Prospective Antifungal Therapy (PATH) registry 2004–2008. *PLoS One, 9,* e101510. http://dx.doi.org/10.1371/journal.pone.0101510.

Pfaller, M. A., Boyken, L., Hollis, R. J., Messer, S. A., Tendolkar, S., & Diekema, D. J. (2004). Clinical evaluation of a dried commercially prepared microdilution panel for antifungal susceptibility testing of five antifungal agents against Candida spp. and Cryptococcus neoformans. *Diagnostic Microbiology and Infectious Disease, 50,* 113–117.

Pfaller, M. A., Boyken, L., Hollis, R. J., Messer, S. A., Tendolkar, S., & Diekema, D. J. (2005). In vitro susceptibilities of clinical isolates of Candida species, Cryptococcus neoformans, and Aspergillus species to itraconazole: Global survey of 9,359 isolates tested by clinical and laboratory standards institute broth microdilution methods. *Journal of Clinical Microbiology, 43,* 3807–3810.

Pfaller, M. A., Boyken, L., Messer, S. A., Tendolkar, S., Hollis, R. J., & Diekema, D. J. (2004). Evaluation of the Etest method using Mueller-Hinton agar with glucose and methylene blue for determining amphotericin B MICs for 4,936 clinical isolates of Candida species. *Journal of Clinical Microbiology, 42,* 4977–4979.

Pfaller, M. A., Boyken, L., Messer, S. A., Tendolkar, S., Hollis, R. J., & Diekema, D. J. (2005). Comparison of results of voriconazole disk diffusion testing for Candida species with results from a central reference laboratory in the ARTEMIS global antifungal surveillance program. *Journal of Clinical Microbiology, 43,* 5208–5213.

Pfaller, M. A., Cabezudo, I., Buschelman, B., Bale, M., Howe, T., Vitug, M., et al. (1993). Value of Hybritech ICON Candida assay in the diagnosis of invasive candidiasis in high risk patients. *Diagnostic Microbiology and Infectious Disease, 16*, 53–60.

Pfaller, M., Cabezudo, I., Koontz, F., Bale, M., & Gingrich, R. (1987). Predictive value of surveillance cultures for systemic infection due to Candida species. *European Journal of Clinical Microbiology, 6*, 628–633.

Pfaller, M. A., Chaturvedi, V., Diekema, D. J., Ghannoum, M. A., Holliday, N. M., Killian, S. B., et al. (2012). Comparison of the Senstitre YeastOne colorimetric antifungal panel with CLSI microdilution for antifungal susceptibility testing of the echinocandins against Candida spp., using new clinical breakpoints and epidemiological cutoff values. *Diagnostic Microbiology and Infectious Disease, 73*, 365–368.

Pfaller, M. A., & Diekema, D. J. (2004a). Rare and emerging opportunistic fungal pathogens: Concern for resistance beyond Candida albicans and Aspergillus fumigatus. *Journal of Clinical Microbiology, 42*, 4419–4431.

Pfaller, M. A., & Diekema, D. J. (2004b). Twelve years of fluconazole in clinical practice: Global trends in species distribution and fluconazole susceptibility of bloodstream isolates of Candida. *Clinical Microbiology and Infection, 10*(Suppl. 1), S11–S23.

Pfaller, M. A., & Diekema, D. J. (2007). The epidemiology of invasive candidiasis: A persistent public health problem. *Clinical Microbiology Reviews, 20*, 133–163.

Pfaller, M. A., & Diekema, D. J. (2010). Epidemiology of invasive mycoses in North America. *Critical Reviews in Microbiology, 36*, 1–53.

Pfaller, M. A., & Diekema, D. J. (2012). Progress in antifungal ausceptibility testing of Candida spp. by use of Clinical and Laboratory Standards Institute broth microdilution methods, 2010 to 2012. *Journal of Clinical Microbiology, 50*, 2846–2856.

Pfaller, M. A., Diekema, D. J., Andes, D., Arendrup, M. C., Brown, S. D., Lockhart, S. R., et al. (2011). Clinical breakpoints for the echinocandins and Candida revisited: Integration of molecular, clinical and microbiological data to arrive at species-specific interpretive criteria. *Drug Resistance Updates, 14*, 164–176.

Pfaller, M. A., Diekema, D. J., Colombo, A. L., Kibbler, C., Ng, K. P., Gibbs, D. L., et al. (2006). Candida rugosa, an emerging fungal pathogen with resistance to azoles: Geographic and temporal trends from the ARTEMIS DISK antifungal surveillance program. *Journal of Clinical Microbiology, 44*, 3578–3582.

Pfaller, M. A., Diekema, D. J., Procop, G. W., & Rinaldi, M. G. (2007a). Multicenter comparison of VITEK 2 yeast susceptibility test with the CLSI broth microdilution reference method for testing fluconazole against Candida spp. *Journal of Clinical Microbiology, 45*, 796–802.

Pfaller, M. A., Diekema, D. J., Procop, G. W., & Rinaldi, M. G. (2007b). Multicenter comparison of VITEK 2 yeast susceptibility test with the CLSI broth microdilution method for testing amphotericin B, flucytosine, and voriconazole against Candida spp. *Journal of Clinical Microbiology, 45*, 3522–3528.

Pfaller, M. A., Diekema, D. J., Procop, G. W., & Rinaldi, M. G. (2013). Comparison of the Vitek 2 yeast susceptibility system with CLSI microdilution for antifungal susceptibility testing of fluconazole and voriconazole against Candida spp., using new clinical breakpoints and epidemiological cutoff values. *Diagnostic Microbiology and Infectious Disease, 77*, 37–40.

Pfaller, M. A., Diekema, D. J., Procop, G. W., & Wiederhold, N. P. (2014). Multicenter evaluation of the new Vitek 2 yeast susceptibility test using new clinical breakpoints for fluconazole. *Journal of Clinical Microbiology, 52*, 2126–2130.

Pfaller, M. A., Diekema, D. J., Rex, J. H., Espinel-Ingroff, A., Johnson, E. M., Andes, D., et al. (2006). Correlation of MIC with outcome for Candida species tested against voriconazole:

Analysis and proposal for interpretive breakpoints. *Journal of Clinical Microbiology*, *44*, 819–826.

Pfaller, M. A., Diekema, D. J., & Sheehan, D. J. (2006). Interpretive breakpoints for fluconazole and Candida revisited: A blueprint for the future of antifungal susceptibility testing. *Clinical Microbiology Reviews*, *19*, 435–447.

Pfaller, M. A., Espinel-Ingroff, A., & Jones, R. N. (2004). Clinical evaluation of the Sensititre YeastOne colorimetric antifungal plate for antifungal susceptibility testing of the new triazoles voriconazole, posaconazole, and ravuconazole. *Journal of Clinical Microbiology*, *42*, 4577–4580.

Pfaller, M. A., Hazen, K. C., Messer, S. A., Boyken, L., Tendolkar, S., Hollis, R. J., et al. (2004). Comparison of results of fluconazole disk diffusion testing for Candida species with results from a central reference laboratory in the ARTEMIS global antifungal surveillance program. *Journal of Clinical Microbiology*, *42*, 3607–3612.

Pfaller, M. A., Houston, A., & Coffman, S. (1996). Application of CHROMagar Candida for rapid screening of clinical specimens for Candida albicans, Candida tropicalis, Candida krusei, and Candida (Torulopsis) glabrata. *Journal of Clinical Microbiology*, *34*, 58–61.

Pfaller, M. A., & Jones, R. N. (2006). Performance accuracy of antibacterial and antifungal susceptibility test methods: Report from the College of American Pathologists (CAP) Microbiology Surveys Program (2001–2003). *Archives of Pathology & Laboratory Medicine*, *130*, 767–778.

Pfaller, M. A., Jones, R. N., & Castanheira, M. (2014). Regional data analysis of Candida non-albicans strains collected in United States medical sites over a 6-year period, 2006–2011. *Mycoses*, *57*, 602–611.

Pfaller, M. A., & McGinnis, M. R. (2009). The laboratory and clinical mycology. In E. J. Anaissie, M. R. McGinnis, & M. A. Pfaller (Eds.), *Clinical mycology* (2nd ed., pp. 55–77). New York: Elsevier, Inc..

Pfaller, M. A., Messer, S. A., Boyken, L., Huyn, H., Hollis, R. J., & Diekema, D. J. (2002). In vitro activities of 5-fluorocytosine against 8,803 clinical isolates of Candida spp.: Global assessment of primary resistance using National Committee for Clinical Laboratory Standards susceptibility testing methods. *Antimicrobial Agents and Chemotherapy*, *46*, 3518–3521.

Pfaller, M. A., Messer, S. A., Boyken, L., Rice, C., Tendolkar, S., Hollis, R. J., et al. (2007). Use of fluconazole as a surrogate marker to predict susceptibility and resistance to voriconazole among 13,338 clinical isolates of Candida spp. tested by Clinical and Laboratory Standards Institute—Recommended broth microdilution methods. *Journal of Clinical Microbiology*, *45*, 70–75.

Pfaller, M. A., Messer, S. A., Boyken, L., Tendolkar, S., Hollis, R. J., & Diekema, D. J. (2003). Variation in susceptibility of blood-stream isolates of Candida glabrata to fluconazole according to patient age and geographic location. *Journal of Clinical Microbiology*, *41*, 2176–2179.

Pfaller, M. A., Messer, S. A., Rhomberg, P. R., Jones, R. N., & Castanheira, M. (2013). In vitro activities of isavuconazole and comparator antifungal agents against a global collection of opportunistic yeasts and molds. *Journal of Clinical Microbiology*, *51*, 2608–2616.

Pfaller, M. A., Neofytos, D., Diekema, D., Azie, N., Meier-Kriesche, H. U., Quan, S. P., et al. (2012). Epidemiology and outcomes of candidemia in 3648 patients: Data from The Prospective Antifungal Therapy (PATH Alliance[TM]) registry, 2004–2008. *Diagnostic Microbiology and Infectious Disease*, *74*, 323–331.

Pfaller, M. A., Pappas, P. G., & Wingard, J. R. (2006). Invasive fungal pathogens: Current epidemiological trends. *Clinical Infectious Diseases, 43*(Suppl.1), 3–14.

Pfaller, M. A., Richter, S. S., & Diekema, D. J. (2005). Conventional methods for the laboratory diagnosis of fungal infections in the immunocompromised host. In J. R. Wingard & E. J. Anaissie (Eds.), *Fungal infections in the immunocompromised patient* (pp. 341–381). Boca Raton: Taylor & Francis.

Pfaller, M. A., Woosley, L. N., Messer, S. A., Jones, R. N., & Castanheira, M. (2012). Significance of molecular identification and antifungal susceptibility of clinical yeasts and moulds in a global antifungal surveillance program. *Mycopathologia, 174,* 259–271.

Pfeiffer, C. D., Fin, J. P., & Safdar, N. (2006). Diagnosis of invasive aspergillosis using a galactomannan assay: A meta-analysis. *Clinical Infectious Diseases, 42,* 1417–1427.

Pham, A. S., Tarrand, J. J., May, G. S., Lee, M. S., Kontoyiannis, D. P., & Han, X. Y. (2003). Diagnosis of invasive mold infection by real-time quantitative PCR. *American Journal of Clinical Pathology, 119,* 38–44.

Philip, A., Odabasi, Z., Matiuzzi, G., Paetznick, V. L., Tan, S. W., Warmington, J., et al. (2005). Syscan 3, a kit for detection of anti-Candida antibodies for diagnosis of invasive candidiasis. *Journal of Clinical Microbiology, 43,* 4834–4835.

Phillips, P., Dowd, A., Jewesson, P., Radigan, G., Tweeddale, M. G., Clarke, A., et al. (1990). Nonvalue of antigen detection immunoassays for diagnosis of candidemia. *Journal of Clinical Microbiology, 28,* 2320–2326.

Pickering, J. W., Sant, H. W., Bowles, C. A. P., Roberts, W. L., & Woods, G. L. (2005). Evaluation of a $(1\rightarrow3)$-β-D-glucan assay for diagnosis of invasive fungal infections. *Journal of Clinical Microbiology, 43,* 5957–5962.

Pien, B. C., Sundaram, P., Raoof, N., Costa, S. F., Mirrett, S., Woods, C. W., et al. (2010). The clinical and prognostic importance of positive blood cultures in adults. *The American Journal of Medicine, 123,* 819–828.

Pinjon, E., Sullivan, D., Salkin, I., Shanley, D., & Coleman, D. (1998). Simple, inexpensive, reliable method for differentiation of Candida dubliniensis from Candida albicans. *Journal of Clinical Microbiology, 36,* 2093–2095.

Pittet, D., Monod, M., Suter, P. M., Frenck, E., & Auckenthaler, R. (1994). Candida colonization and subsequent infections in critically ill surgical patients. *Annals of Surgery, 220,* 751–758.

Posteraro, B., Martucci, R., La Sorda, M., Fiori, B., Sanglard, D., De Carolis, E., et al. (2009). Reliability of the Vitek 2 yeast susceptibility test for the detection of in vitro resistance to fluconazole and voriconazole in clinical isolates of Candida albicans and Candida glabrata. *Journal of Clinical Microbiology, 47,* 1927–1930.

Pounder, J. I., Simmon, K. E., Barton, C. A., Hohmann, S. L., Brandt, M. E., & Petti, C. A. (2007). Discovering potential pathogens among fungi identified as "nonsporulating molds" *Journal of Clinical Microbiology, 45,* 568–571.

Prentice, H. G., Kibbler, C. C., & Prentice, A. G. (2000). Towards a targeted, risk-based, antifungal strategy in neutropenic patients. *British Journal of Haematology, 110,* 273–284.

Quindos, G. (2006). New microbiological techniques for the diagnosis of invasive mycoses caused by filamentous fungi. *Clinical Microbiology and Infection, 12*(Suppl. 7), 40–52.

Rakeman, J. L., Bui, U., LaFe, K., Chen, Y. C., Honeycutt, R. J., & Cookson, B. T. (2005). Multilocus DNA sequence comparisons rapidly identify pathogenic molds. *Journal of Clinical Microbiology, 43,* 3324–3333.

Singh, N. (2003). Fungal infections in recipients of solid organ transplantation. *Infectious Disease Clinics of North America*, *17*, 113–134.

Singh, N., & Paterson, D. L. (2005). Aspergillus infections in transplant recipients. *Clinical Microbiology Reviews*, *18*, 44–69.

Skladney, H., Buchheidt, D., Baust, C., Krieg-Schneider, F., Seifarth, W., Leib-Mösch, C., et al. (1999). Specific detection of Aspergillus species in blood and bronchoalveolar lavage samples of immunocompromised patients by two-step PCR. *Journal of Clinical Microbiology*, *37*, 3865–3871.

Snydman, D. R. (2001). Posttransplant microbiological surveillance. *Clinical Infectious Diseases*, *3*(Suppl. 1), S2–S25.

Solomkin, J. S. (2005). *Candida* infections in the intensive care unit. In J. R. Wingard & E. J. Anaissie (Eds.), *Fungal infections in the immunocompromised patient* (pp. 149–183). Boca Raton: Taylor & Francis.

Spanakis, E. K., Aparis, G., & Mylonakis, E. (2006). New agents for the treatment of fungal infections: Clinical efficacy and gaps in coverage. *Clinical Infectious Diseases*, *42*, 1060–1068.

Spellberg, B. J., Filler, S. G., & Edwards, J. E., Jr. (2006). Current treatment strategies for disseminated candidiasis. *Clinical Infectious Diseases*, *42*, 244–251.

Sulahian, A., Touratier, S., & Ribaud, P. (2003). False positive test for aspergillus antigenemia related to concomitant administration of piperacillin and tazobactam. *The New England Journal of Medicine*, *349*, 2366–2367.

Sun, W., Wang, K., Gao, W., Su, X., Qian, Q., Lu, X., et al. (2011). Evaluation of PCR on bronchoalveolar lavage fluid for diagnosis of invasive aspergillosis: A bivariate metaanalysis and systematic review. *PLoS One*, *6*, e28467. http://dx.doi.org/10.1371/journal.pone.0028467.

Tamura, H., Tanaka, S., Ikeda, T., Obayashi, T., & Hashimoto, Y. (1997). Plasma ((1→3)-beta-D-glucan assay and immunohistochemical staining of (1→3)-beta-D-glucan in the fungal cell walls using a novel horseshoe crab protein (T-GBP) that specifically binds to (1→3)-beta-D-glucan. *Journal of Clinical Laboratory Analysis*, *11*, 104–109.

Tan, G. L., & Peterson, E. M. (2005). CHROMagar Candida medium for direct susceptibility testing of yeast from blood cultures. *Journal of Clinical Microbiology*, *43*, 1727–1731.

Taur, Y., Cohen, N., Dubnow, S., Paskovaty, A., & Seo, S. K. (2010). Effect of antifungal therapy on mortality in cancer patients with candidemia. *Antimicrobial Agents and Chemotherapy*, *54*, 184–190.

Thompson, G. R., Sharma, S., Bays, D. J., Pruitt, R., Engelthaler, D. M., Bowers, J., et al. (2013). Coccidioidomycosis: Adenosine deaminase levels, serologic parameters, culture results, and polymerase chain reaction testing in pleural fluid. *Chest*, *143*, 776–781.

Thornton, C. R. (2008). Development of an immunochromatographic lateral-flow device for rapid serodiagnosis of invasive aspergillosis. *Clinical and Vaccine Immunology*, *15*, 1095–1105.

Tollemar, J., Ericzon, B. G., Holmberg, K., & Andersson, J. (1990). The incidence and diagnosis of invasive fungal infections in liver transplant recipients. *Transplantation Proceedings*, *22*, 242–244.

Tollemar, J., Holmberg, K., Ringden, O., & Lonquist, B. (1989). Surveillance tests for the diagnosis of invasive fungal infections in bone marrow transplant recipients. *Scandinavian Journal of Infectious Diseases*, *21*, 205–212.

Trick, W. E., Fridkin, S. K., Edwards, J. R., Hajjeh, R. A., Gaynes, R. P., & The National Nosocomial Infections Surveillance System Hospitals. (2002). Secular trend of hospital-acquired candidemia among intensive care unit patients in the United States during 1989–1999. *Clinical Infectious Diseases, 35,* 622–630.

Tucker, D. L., Beresford, C. H., Sigler, L., & Rogers, K. (2004). Disseminated Beauveria bassiana infection in a patient with acute lymphoblastic leukemia. *Journal of Clinical Microbiology, 42,* 5412–5414.

Ullmann, A. J., Akova, M., Herbrecht, R., Viscoli, C., Arendrup, M. C., Arikan- Akdagli, S., et al. (2012). ESCMID guideline for the diagnosis and management of *Candida* diseases 2012: Adults with haematological malignancies and after haematopoietic stem cell transplantation (HCT). *Clinical Microbiology and Infection, 18*(Suppl. 7), S53–S67.

Upton, A., Guzel, A., Leisenring, W., Limaye, A., Alexander, B., Hayden, R., et al. (2005). Reproducibility of low galactomannan enzyme immunoassay index values tested in multiple laboratories. *Journal of Clinical Microbiology, 43,* 4796–4800.

U.S. Food and Drug Administration. (2003). *FDA clears rapid test for Aspergillus infection.* Retrieved from, http://www.fda.gov/bbs/topics/NEWS/2003/NEW00907.html.

Verweij, P. E., & Brandt, M. E. (2007). Aspergillus, Fusarium, and other opportunistic moniliaceous fungi. In P. R. Murray, E. J. Baron, J. H. Jorgensen, M. L. Landry, & M. A. Pfaller (Eds.), *Manual of clinical microbiology* (9th ed., pp. 1802–1838). Washington DC: ASM Press.

Verweij, P. E., Brinkman, K., Kremer, H. P. H., Kullberg, B. J., & Meis, J. (1999). Aspergillus meningitis: Diagnosis by non-culture-based microbiological methods management. *Journal of Clinical Microbiology, 37,* 1186–1189.

Viscoli, C., Machetti, M., Cappellano, P., Bucci, B., Bruzzi, P., Van Lint, M. T., et al. (2004). False-positive galactomannan platelia Aspergillus test results for patients receiving piperacillin-tazobactam. *Clinical Infectious Diseases, 38,* 913–916.

Walker, R. C. (1991). The role of the clinical microbiology laboratory in transplantation. *Archives of Pathology & Laboratory Medicine, 115,* 299–305.

Walsh, T. J., & Groll, A. H. (1999). Emerging fungal pathogens: Evolving challenges to immunocompromised patients for the twenty-first century. *Transplant Infectious Disease, 1,* 247–261.

Walsh, T. J., Groll, A., Hiemenz, J., Flemming, R., Roilides, E., & Anaissie, E. (2004). Infections due to emerging and uncommon medically important fungal pathogens. *Clinical Microbiology and Infection, 10*(Suppl. 1), 48–66.

Walsh, T. J., Hathorn, J. W., Sobel, J. D., Merz, W. G., Sanchez, V., Maret, S. M., et al. (1991). Detection of circulating Candida enolase by immunoassay in patients with cancer and invasive candidiasis. *The New England Journal of Medicine, 324,* 1026–1031.

Weisser, M., Rausch, C., Droll, A., Simcock, M., Sendi, P., Steffen, I., et al. (2005). Galactomannan does not precede major signs on a pulmonary computerized tomographic scan suggestive of invasive aspergillosis in patients with hematological malignancies. *Clinical Infectious Diseases, 41,* 1143–1149.

Wenzel, R. P., & Gennings, C. (2005). Bloodstream infections due to Candida species in the intensive care unit: Identifying especially high-risk patients to determine prevention strategies. *Clinical Infectious Diseases, 41*(Suppl. 6), S389–S393.

Wheat, L. J. (2003). Rapid diagnosis of invasive aspergillosis by antigen detection. *Transplant Infectious Disease, 5,* 158–166.

Wheat, L. J., Connolly-Stringfield, P., Williams, B., Connolly, K., Blair, R., Bartlett, M., et al. (1992). Diagnosis of histoplasmosis in patients with acquired immunodeficiency syndrome by detection of Histoplasma capsulaturm polysaccharide antigen in bronchoalveolar lavage fluid. *The American Review of Respiratory Disease, 145*, 1421–1424.

Wheat, L. J., Garringer, T., Brizendine, E., & Connolly, P. (2002). Diagnosis of histoplasmosis by antigen detection based upon experience at the histoplasmosis reference laboratory. *Diagnostic Microbiology and Infectious Disease, 43*, 29–37.

Wheat, L. J., Kohler, R. B., French, M. L., Garten, M., Kleiman, M., Zimmerman, S. E., et al. (1983). Immunoglobulin M and G histoplasmal antibody response in histoplasmosis. *The American Review of Respiratory Disease, 128*, 6570.

Wheat, L. J., Kohler, R. B., & Tewari, R. P. (1986). Diagnosis of disseminated histoplasmosis by detection of Histoplasma capsulatum antigen in serum and urine specimens. *The New England Journal of Medicine, 314*, 83–88.

White, P. L., Archer, A. E., & Barnes, R. A. (2005). Comparison of non-culture-based methods for detection of systemic fungal infections, with an emphasis on invasive Candida infections. *Journal of Clinical Microbiology, 43*, 2181–2187.

White, P. L., Linton, C. J., Perry, M. D., Johnson, E. M., & Barnes, R. A. (2006). The evolution and evaluation of a whole blood polymerase chain reaction assay for the detection of invasive aspergillosis in hematology patients in a routine clinical setting. *Clinical Infectious Diseases, 42*, 479–486.

White, P. L., Parr, C., Thornton, C., & Barnes, R. A. (2013). Evaluation of real-time PCR, galactomannan enzyme-linked immunosorbent assay (ELISA), and a novel lateral-flow device for diagnosis of invasive aspergillosis. *Journal of Clinical Microbiology, 51*, 1510–1516.

Wiederhold, N. P., Thornton, C. R., Najvar, L. K., Kirkpatrick, W. R., Bocanegra, R., & Patterson, T. F. (2009). Comparison of lateral-flow technology and galactomannan and $(1 \rightarrow 3)$-beta-D-glucan assays for detection of invasive pulmonary aspergillosis. *Clinical and Vaccine Immunology, 16*, 1844–1846.

Wilson, D. A., Joyce, M. J., Hall, L. S., Reller, L. B., Roberts, G. D., Hall, G. S., et al. (2005). Multicenter evaluation of a Candida albicans peptide nucleic acid fluorescent in situ hybridization probe for characterization of yeast isolates from blood cultures. *Journal of Clinical Microbiology, 43*, 2909–2912.

Wilson, M. L. (2002). Practice makes perfect. *American Journal of Clinical Pathology, 118*, 167–169.

Wingard, J. R. (2002). Antifungal chemoprophylaxis after blood and marrow transplantation. *Clinical Infectious Diseases, 34*, 1386–1390.

Wingard, J. R., Merz, W. G., Rinaldi, M. G., Johnson, T. R., Karp, J. E., & Saral, R. (1991). Increase in Candida krusei infection among patients with bone marrow transplantation and neturopenia treated prophylactically with fluconazole. *The New England Journal of Medicine, 325*, 1274–1277.

Wisplinghoff, H., Bischoff, T., Tallent, S. M., Sifert, H., Wenzel, R. P., & Edmond, M. B. (2004). Nosocomial bloodstream infections in US hospitals: Analysis of 24,179 cases from a prospective nationwide surveillance study. *Clinical Infectious Diseases, 39*, 309–317.

Year, H., Sendid, B., Francois, N., Camus, D., & Poulain, D. (2001). Contribution of serological tests and blood culture to the early diagnosis of systemic candidiasis. *European Journal of Clinical Microbiology & Infectious Diseases, 20*, 864–870.

Yeo, S. F., Huie, S., Sofair, A. N., Campbell, S., Durante, A., & Wong, B. (2006). Measurement of serum D -arabinitol/creatinine ratios for initial diagnosis and for predicting outcome in an unselected, population-based sample of patients with Candida fungemia. *Journal of Clinical Microbiology*, *44*, 3894–3899.

Yeo, S. F., & Wong, B. (2002). Current status of nonculture methods for diagnosis of invasive fungal infections. *Clinical Microbiology Reviews*, *15*, 465–484.

Yoo, U. H., Choi, S. M., Lee, D. G., Skin, W. H., Min, W. S., & Kim, C. C. (2005). Application of nucleic acid sequence-based amplification for diagnosis of and monitoring the clinical course of invasive aspergillosis in patients with hematologic diseases. *Clinical Infectious Diseases*, *40*, 392–398.

Yoshida, M., Obayashi, T., Iwama, A., Ito, M., Tsunoda, S., Suzuki, T., et al. (1997). Detection of plasma $(1{\rightarrow}3)$-β-D-glucan in patients with Fusarium, Trichosporon, Saccharomyces, and Acremonium fungaemias. *Journal of Medical and Veterinary Mycology*, *35*, 371–374.

Yu, V. L., Muder, R. R., & Poorsattar, A. (1986). Significance of isolation of Aspergillus from the respiratory tract in diagnosis of invasive pulmonary aspergillosis: Results from a three-year prospective study. *The American Journal of Medicine*, *81*, 249–254.

Zaoutis, T. E., Argon, J., Chu, J., Berlin, J. A., Walsh, T. J., & Feudtner, C. (2005). The epidemiology and attributable outcomes of candidemia in adults and children hospitalized in the United States: A propensity analysis. *Clinical Infectious Diseases*, *41*, 1232–1239.

Zedek, D. C., & Miller, M. B. (2006). Use of galactomannan enzyme immunoassay for diagnosis of invasive aspergillosis in a tertiary-care center over a 12-month period. *Journal of Clinical Microbiology*, *44*, 1601.

Zhang, S. X. (2013). Enhancing molecular approaches for diagnosis of fungal infections. *Future Microbiology*, *8*, 1599–1611.

Technical and Software Advances in Bacterial Pathogen Typing

8

Linda Chui*,†,1, Vincent Li*

*Provincial Laboratory for Public Health, Walter Mackenzie Health Sciences Centre, University of Alberta Hospital, Edmonton, Alberta, Canada
†Department of Laboratory Medicine and Pathology, University of Alberta, Edmonton, Alberta, Canada
1Corresponding author: e-mail address: linda.chui@albertahealthservices.ca

1 INTRODUCTION

Bacterial typing is an essential public health tool crucial for disease surveillance and outbreak detection, determining relatedness between bacterial strains, understanding pathogen transmission and evolution, and helping make informed decisions affecting public health. The fundamental concept of bacterial typing is the ability to differentiate bacteria based on their phenotypes and genotypes. Classical typing methods focused on phenotypic characteristics including biotypes, phage types, serotypes, and antibiograms. In the past 20 years, molecular typing methods based on bacterial genotypes have increased and improved, leading to enhanced analytical methods for strain classification, clonal relatedness, detection of virulence genes, and drug resistance markers, as well as for the study of genetic evolution over time across geographic regions. Concurrently, significant advances have been made in computer hardware and software that expanded the application of computers in all fields, including bacterial typing. In particular, current molecular typing methods benefit greatly from the advent of the internet, superior data storage and sharing, drastic increases in computing power, decreased cost and increased accessibility of personal computers, and improved data analysis and clustering tools. The combination of molecular and computational tools enhances public health surveillance, outbreak recognition, and response. This can aid in the implementation of control programs that determine the source and route of disease transmission and guide appropriate intervention initiatives.

The selection of appropriate bacterial typing methods depends on several criteria and the nature of the investigation. In epidemiological studies or surveillance, the typing technique used must have the capability to type all the isolates under investigation and also have enough discriminatory power to differentiate between sporadic and outbreak strains. This is an extremely important factor especially when a cluster of cases has been identified, as the typing results can help to pinpoint

Methods in Microbiology, Volume 42, ISSN 0580-9517, http://dx.doi.org/10.1016/bs.mim.2015.06.003

the source and prevent the spread of infection through food recall, disinfection procedures, education, and the implementation of infection control protocols and programs. Another important factor for selecting typing methods is the reproducibility of the assay, particularly if it involves surveillance data being collected over time and geographical regions. Other considerations include the speed and cost of the assay, the technical expertise required, and how easily results can be interpreted and shared.

Molecular typing data are often used in global settings due to international travel and food import and export. Therefore, standardisation of experimental protocols, nomenclature, and data interpretation are essential to ensure global inter-laboratory communication and comparison is possible. In recent years, the introduction of globally accessible online molecular typing databases that are curated and standardised by a central authority has played a major role in facilitating this process.

In the following sections, selected examples of typing techniques will be presented to illustrate the technical advancements of bacterial typing and concurrent improvements in associated software.

2 PULSED-FIELD GEL ELECTROPHORESIS

Pulsed-field gel electrophoresis (PFGE) has been the gold standard for bacterial typing for many years and has been used to characterise many different pathogens including *Escherichia coli* (*E. coli*), *Salmonella*, *Listeria*, *Staphylococcus aureus*, and *Shigella*. This is because PFGE has high discriminatory power and good concordance with epidemiological relatedness. As well, results are reproducible and can be shared between different laboratories when standardised protocols, equipment, and analysis tools are used. In PFGE, bacterial cells are embedded into an agarose gel to prevent shearing of chromosomal DNA, and DNA digestion is performed *in situ*. The bacterial chromosome is digested using a rare cutting enzyme that recognises specific DNA sequences ranging from 6 to 8 nucleotides, resulting in a limited number of DNA fragments of varying lengths. After DNA restriction, slices of the agarose blocks are inserted into wells of the agarose gel matrix. Restriction fragments that are megabases in size are impossible to separate by conventional agarose gel electrophoresis (Schwartz & Cantor, 1984; Schwartz & Koval, 1989; Schwartz et al., 1983). However, by applying an electric field that is periodically re-oriented, the different fragment sizes will zigzag through the pores in the agarose gel matrix, with smaller DNA molecules re-orienting in a shorter time than the larger ones. The gradual increase of the electrophoretic pulse time in different directions allows the separation of a variety of DNA fragment sizes ranging from kilobases to megabases. Different approaches based on a similar strategy, include orthogonal field gel electrophoresis (Carle & Olson, 1984), field inversion gel electrophoresis (Carle, Frank, & Olson, 1986), transverse alternating field electrophoresis (Gardiner, Laas, & Patterson, 1986), and contour-clamped homogeneous electric field electrophoresis (CHEF) (Chu, Vollrath, & Davis, 1986). CHEF is most commonly used in laboratories worldwide. This system consists of 24 electrodes in an octagon arrangement producing a constant electrophoresis gradient that switches from the primary to the secondary electrodes located at a 60° angle from the centre of the agarose gel. Consequently, this configuration causes the DNA molecules to re-orient in the agarose gel matrix over a 120° angle and allows the separation of different sized fragments over approximately 15–30 h of electrophoresis run time, depending on the bacteria, and generates a fingerprint profile that can be digitally imaged (Fig. 1).

There are many different computer programs available for analysis of the fingerprint image data such as BioNumerics (Applied Maths, Austin, TX, USA), Fingerprinting II (Bio-Rad, Hercules, CA, USA), GelQuest (SequentiX, Klein Raden, Germany), and PyElph (Pavel & Vasile, 2012). The aforementioned software and those like it have significantly improved laboratory capability for storing, analysing, and sharing PFGE data. For many reasons, the most commonly used software for processing PFGE data is BioNumerics. The BioNumerics software suite integrates the industry leading database engines (Oracle, MySQL, and Microsoft Access), allowing users to link epidemiological and laboratory testing data together in a single user-friendly interface. Furthermore, this software has many gel image preprocessing functions (image import and normalisation, automated lane and band detection, background reduction, and spot removal; Fig. 2), which help to maximise the quality of the PFGE data generated by the laboratory.

Cluster and statistical analysis tools available in BioNumerics further extend the software's suitability for PFGE analysis. For example, BioNumerics can determine the relatedness between PFGE fingerprint patterns using different clustering algorithms including un-weighted pair group method with arithmetic mean (UPGMA),

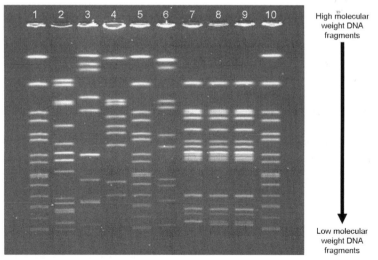

FIGURE 1

DNA fingerprint profiles generated by PFGE. Molecular weight ladders are located in gel lanes 1, 5, and 10.

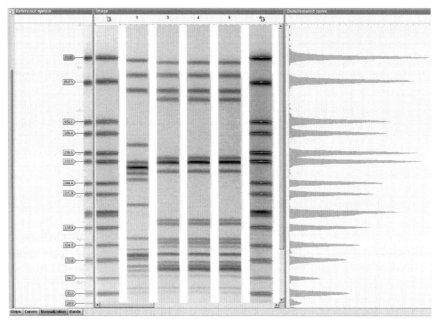

FIGURE 2

Processing a PFGE gel image using BioNumerics. Individual gel lanes are normalised using a user-defined reference system (shown in the reference system panel) based on PFGE banding patterns. The densitometric curve panel helps to guide user PFGE band assignment. (See the color plate.)

neighbour joining, and Ward's method (Fig. 3). The calculated similarities are represented in a dendrogram complete with error flags, cophenetic correlation values, and cluster cutoff values. If required, bootstrap analysis can be performed on the dendrogram to estimate the significance of each tree branch.

BioNumerics also allows the import and export of PFGE and epidemiological data between laboratories by directly connecting to external online BioNumerics databases and servers, making data transfer and comparisons seamless. Another feature of BioNumerics is the ability to accommodate typing data from a variety of different typing methods, consolidating all testing results from multiple sources for each specimen. The tools offered by BioNumerics and other similar software contribute to the enduring usage of PFGE for bacterial typing.

3 SEROTYPING FOR *SALMONELLA*

Salmonellosis is one of the leading foodborne infections worldwide. It causes intestinal infection accompanied by fever, abdominal cramps, and diarrhoea. *Salmonella* outbreaks have been associated with different types of foods (Bayer et al., 2014; Dechet et al., 2014; Deng et al., 2013; Garvey et al., 2013; Hawkey et al., 2013;

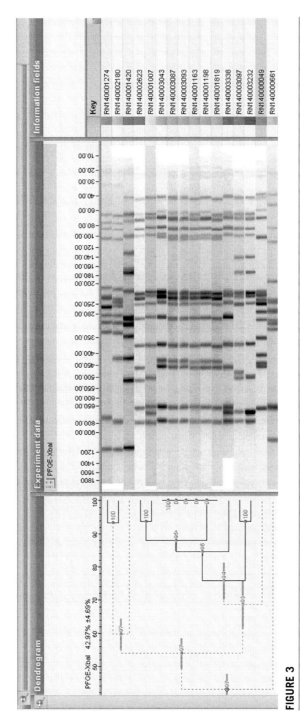

FIGURE 3

PFGE DNA fingerprint analysis using BioNumerics. Isolate information, associated PFGE DNA fingerprint patterns, and a dendrogram created with the UPGMA clustering algorithm are shown. (See the color plate.)

Kinross et al., 2014; Lettini et al., 2014; Paine et al., 2014; Painter et al., 2013; Raguenaud et al., 2012), exposure to pets (Bartholomew et al., 2014; Centers for Disease Control and Prevention (CDC), 2013; Mettee Zarecki et al., 2013; Sylvester et al., 2014) or farm/fair animals (Basler et al., 2014; Jay-Russell et al., 2014; Pabilonia et al., 2014), pet treats (Centers for Disease Control and Prevention (CDC), 2006; Clark et al., 2001; Imanishi et al., 2014; Pitout et al., 2003), and person-to-person transmission due to poor sanitation (Gicquelais et al., 2014; Smith et al., 2014). Onsite farm surveillance programs play a major role in controlling the disease in animals before it enters the food chain for human consumption. On the clinical side, *Salmonella* surveillance is important for identifying the infection source and to eradicate the problem using control programs (Arguello, Alvarez-Ordoñez, Carvajal, Rubio, & Prieto, 2013; Dewaele et al., 2012; Méroc et al., 2012; National Association of State Public Health Veterinarians, Inc. (NASPHV), & Centers for Disease Control and Prevention (CDC), 2011; Snary, Munday, Arnold, & Cook, 2010; Young et al., 2010).

Salmonella enterica isolates can be discriminated on the basis of O antigens (surface polysaccharide) and H antigens (flagella). There are over 150 different O and H antigens used to characterise more than 2500 *Salmonella* serovars (Bopp, Brenner, Fields, Wells, & Stockbine, 2003; Perch et al., 2003). The classical serotyping method is based on agglutination using specific antisera to determine the antigenic formulae by the Kaufmann–White scheme and updates are provided annually by the World Health Organization (Popoff, Bockemühl, & Gheesling, 2004). Although this standardised protocol is recognised worldwide, there are some drawbacks. It is time consuming and can take up to five days for full identification of some serotypes. As well, conventional serotyping can be technically challenging for those lacking experience and strains that lose their antigenic properties fail to be typed. The required antisera can be prepared in-house but it is very expensive to maintain an animal housing unit with personnel. Also, a quality control system for checking the antisera must be in place. Commercial antisera can be another alternative to in-house antisera but the price is high and accessibility is an issue. All these factors support the need for an alternative molecular assay for serotyping.

Different amplification assays such as conventional and real-time PCR (Kim et al., 2006; Park & Ricke, 2014), DNA typing (Shi, Singh, Ranieri, Wiedmann, & Moreno Switt, 2015), and DNA microarrays (Braun et al., 2012; Guo et al., 2013; McQuiston, Waters, Dinsmore, Mikoleit, & Fields, 2011; Peterson et al., 2010; Tankouo-Sandjong et al., 2008; Yoshida et al., 2014) have been published in recent years. Currently, a commercial kit known as the Check&Trace *Salmonella* (CTS) typing system (Check-Points, Wageningen, The Netherlands) is available for molecular serotyping (Wattiau, Weijers, et al., 2008). The methodology is based on multiplex ligation detection reactions (LDR) to generate circular DNA molecules that are subsequently amplified by PCR using a pair of primers, one of which is labelled with biotin. After amplification, the labelled strand will hybridise to the complimentary strand located on a low density array. There are different probes involved with the reaction. The *Salmonella* general LDR probes are species specific and serovar identification is based on a series of 14 LDR probes targeting DNA markers whose sequence vary within the different *Salmonella* serovars. Each LDR probe carries a ZIP code complimentary to a unique oligonucleotide immobilised on the microarray. The LDR probes bind to the *Salmonella* markers on the microarray, the hybridisation results are read using a single channel ATR03 reader (a charge-coupled device-based transmission detector; Clondiag, Jena, Germany), and the data is recorded on a connected computer. The combination of positive and negative hybridisation reactions results in a 14-digit binary code which is then converted into a unique identifier calculated using a specific formula as seen below (Wattiau, Van Hessche, Schlicker, Vander Veken, & Imberechts, 2008):

$$\sum_{x=1}^{14} 2^{x-1}k$$

where x is the spot number and k is a multiplication factor equal to 1 for positive hybridisation or 0 for negative hybridisation. Data is processed using the CTS software supplied by Check-Points, that automatically translates the microarray data and compares the corresponding pattern to the established library of previously observed *Salmonella* serovars provided in CTS software updates (Fig. 4). If no match is found in the database, a genovar number is provided and the user has the option to submit their data to Check-Points for inclusion in future CTS software updates.

Molecular serotyping of *Salmonella* with CTS is rapid with a reporting turn-around time of 10 h or less. It is also reproducible, easy to perform, and can currently identify over 300 serotypes. The assay is easily transferable to other laboratories,

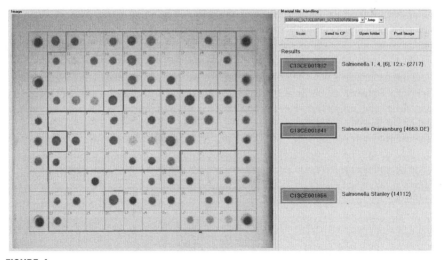

FIGURE 4

Graphical user interface of the CTS software showing microarray interpretation and genovar results. (See the color plate.)

inter-laboratory data comparisons are possible, and extensive staff training and expertise are not required. Currently, the CTS database of *Salmonella* serovars is not publicly accessible and is only limited to CTS users. As the database becomes more populated, a wider spectrum of *Salmonella* species deriving from agriculture, animals, and humans can be identified.

4 SEQUENCING-BASED MOLECULAR TYPING

4.1 *spa* TYPING

Methicillin-resistant *Staphylococcus aureus* (MRSA) is a pathogen with a great impact on global human health (Köck et al., 2010; Lee et al., 2013) and is associated with endocarditis (Levine, Fromm, & Reddy, 1991), bloodstream infections (Friedman et al., 2002), pneumonia (Francis et al., 2005; Lewis et al., 2014), skin and soft tissue infections (Hu et al., 2014; Seeleang, Manning, Saks, & Winstead, 2014), and necrotising fasciitis (de Carvalho Ferreira, Cisne Frota, Cavalcante, Abad, & Netto Dos Santos, 2012; Swain, Hatcher, Azadian, Soni, & De Souza, 2013). Historically, MRSA was primarily linked to healthcare settings and was the major cause of nosocomial infection (Popovich, Weinstein, & Hota, 2008). Presently, healthcare-associated MRSA strains persist but community-associated MRSA strains are also becoming increasingly prevalent in all settings (Braga et al., 2014; Gilbert et al., 2006; Landers, Harris, Wittum, & Stevenson, 2009; Lewis et al., 2014; Redziniak et al., 2009; Tanner et al., 2014; Tonn & Ryan, 2013), further increasing the burden of MRSA on healthcare systems. Surveillance is important for identifying trends in MRSA distribution, dissemination, evolution, and epidemiology in an effort to enhance patient outcome and reduce transmission (de Kraker et al., 2013; Iwamoto et al., 2013; Mera et al., 2011; Wilson et al., 2011).

In the past, PFGE was the gold standard for MRSA characterisation (Bannerman, Hancock, Tenover, & Miller, 1995). Some inherent disadvantages in PFGE (cost, assay time, and subjective data interpretation) led to the need for alternative typing methods. *spa* typing is an assay based on sequencing a variable number of direct repeats in the polymorphic X region of the *Staphylococcus* protein A (*spa*) gene (Shopsin et al., 1999). *spa* typing has near equivalent discriminatory power compared to PFGE and is less expensive and less labour intensive (Golding et al., 2008; Li et al., 2013). It is a rapid technique appropriate for high-throughput laboratories, it generates reproducible results, and comparison of *spa* data between different laboratories is seamless. Further, since MRSA is a global problem, the international recognition of *spa* typing data and *spa* types allows for easier epidemiological comparison and surveillance across different parts of the world.

The short direct repeats in the polymorphic X region of the *spa* gene are made up of 24 base pairs that vary in base composition and are amplified by targeted PCR.

Sequencing results from the PCR amplified product are imported and analysed using available software, and corresponding *spa* types are assigned based on the number, order, and sequences of repeats (Fig. 5).

BioNumerics has a module customised specifically for *spa* typing and is a great example of how software can assist bacterial typing. As previously discussed, BioNumerics integrates popular database engines that make storage and linkage of epidemiological and sequencing data simple. The BioNumerics *spa* module automatically detects and assigns repeats in the *spa* gene. BioNumerics users can then directly connect to the international online Ridom *spa* database (http://spa.ridom. de), developed by Ridom GmbH and curated by SeqNet.org (http://www.SeqNet. org), that currently contains information on over 14,300 *spa* types from 112 countries (Harmsen et al., 2003). By querying the Ridom *spa* database, BioNumerics automatically assigns Ridom *spa* types based on the repeats present. BioNumerics also has specialised *spa* typing analysis and clustering tools to determine genetic relatedness between different sequenced isolates (Fig. 6). This is an excellent example of technical and software improvements working together to enhance a typing method.

4.2 *emm* TYPING

Streptococcus pyogenes is a major pathogen that causes a range of human diseases including skin and soft tissue infections (Kohayagawa et al., 2014; Lin, Chang, Lai, Lin, & Chen, 2011), pneumonia (Montes et al., 2011), meningitis (de Almeida Torres, Fedalto, de Almeida Torres, Steer, & Smeesters, 2013), necrotising fasciitis (Carlin et al., 2014; Johansson, Thulin, Low, & Norrby-Teglund, 2010), and streptococcal toxic shock syndrome (Lin, Chang, Lai, Lin, & Chen, 2013; Montes et al., 2011). Global incidence of severe invasive disease caused by *S. pyogenes* have been increasing since the 1980s (Lamagni et al., 2005; Lamagni, Darenberg, et al., 2008; Lamagni, Neal, et al., 2008; O'Loughlin et al., 2007), highlighting the need for surveillance of this pathogen. *emm* typing is a sequence-based typing method for Group A beta haemolytic streptococci which targets the M protein gene (*emm*) (Beall, Facklam, & Thompson, 1996). This gene encodes for the cell surface antigen and virulence factor responsible for different serovars of *S. pyogenes* (Beall et al., 1996). Two conserved primers are used to amplify the 5′ junction of the *emm* gene which includes sequences of the membrane export signal and the M protein. The hypervariable sequence encoding M serospecificity is adjacent to one of the primers therefore direct sequencing for *emm* typing is possible.

Sequencing results are checked for quality and trimmed to a length of 160 base pairs. These sequences are subsequently queried against an online reference database of *emm* types that is hosted and curated by the CDC, using the online BLAST-*emm* tool (http://www2a.cdc.gov/ncidod/biotech/strepblast.asp). The corresponding *emm* type is assigned based on the returned E-values and bit scores for each submitted sequence with matches typically requiring similarity scores ≥95% over a 150 base pair span. In instances where a subtype assignment is not obtained, users are able to

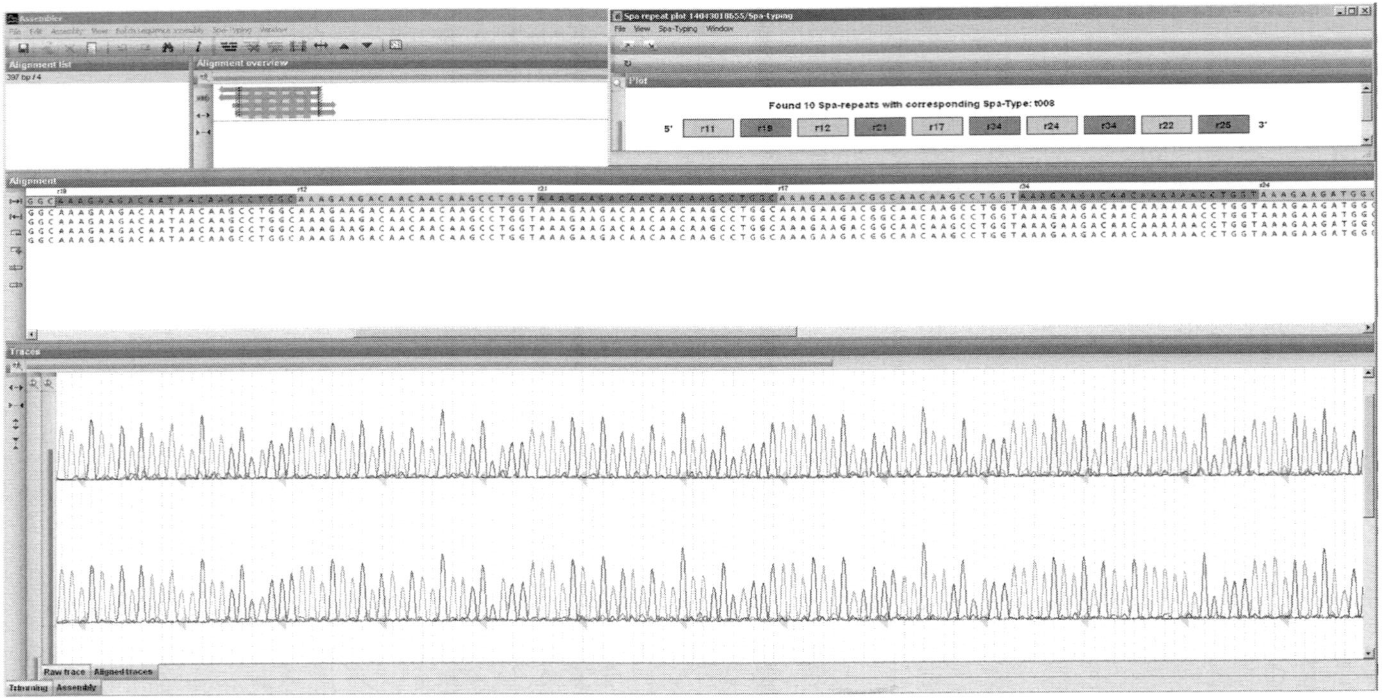

FIGURE 5

MRSA *spa* typing analysis in BioNumerics. Sequence alignments, base quality, *spa* repeat successions, and *spa* type are displayed in the BioNumerics *spa* typing analysis window. (See the color plate.)

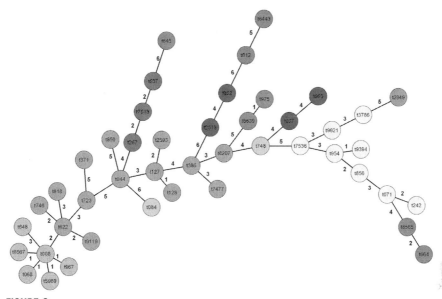

FIGURE 6

Minimum spanning tree (MST) of MRSA *spa* types. The MST was made using the BioNumerics *spa* module and was based on the duplication, substitutions, and indels model. Each circle in the tree represents a different *spa* type. Similarity values are shown on the connecting lines between *spa* types. *spa* type clusters are indicated by different colours or shades. (See the color plate.)

submit their novel sequences to the database curator. Upon verification, the novel sequence is added to the centralised online database and a new subtype is assigned. The CDC *emm* type database is available for user download and contains additional information about each subtype including the date and location of isolation, related literature, isolate source, clinical manifestation, and submitting agency.

4.3 SEROTYPING FOR SHIGA TOXIN-PRODUCING *E. COLI*

Shiga toxin-producing *E. coli* (STEC) have been associated with water and food-borne outbreaks in the developed world causing a spectrum of clinical presentations including non-specific diarrhoea, haemorrhagic colitis, and the potentially fatal hae-molytic uremic syndrome (HUS) (Chokoshvili et al., 2014; Pachkoria, Vashakidze, Megrelishvili, & Tevzadze, 2014; Paton & Paton, 1998). Although *E. coli* O157:H7 is the most prevalent serotype implicated in outbreaks and HUS cases, there is a growing concern about the risk of non-O157 STEC serotypes (Brooks et al., 2001, 2005; Buchholz et al., 2011; Friesema, van der Zwaluw, et al., 2014; Pérez et al., 2014), more than 200 of which have now been associated with human disease (Friesema, Schotsborg, Heck, & Van Pelt, 2014; Luna-Gierke et al., 2014; McCarthy et al., 2001; Miyajima et al., 2007; Wang et al., 2013). Since non-O157 STEC are a genetically diverse group, serotyping is essential to the identification of potential clusters and for epidemiological studies.

Agglutination is the conventional serotyping method and it requires antibody and technical experience. An alternative to the agglutination method is molecular typing via PCR amplification and sequencing of a region in the *gnd* gene (Gilmour, Olson, Andrysiak, Ng, & Chui, 2007) which encodes 6-phosphogluconate dehydrogenase. This method is able to serotype both O157 and non-O157 STEC and standardised PCR protocols and reference sequences for this method can be found online at the *E. coli* O-Typer homepage (https://www.corefacility.ca/ecoli_typer). Typically, sequencing produces a 712 base pair product that is trimmed to 643 base pairs to align with *E. coli* O-Typer input requirements. CLUSTALW2 (www.ebi.ac.uk/Tools/msa/) can be used to align the query sequence with provided reference sequences and identify the 643 base pair target for trimming. *E. coli* O-Typer will use the trimmed sequence as input, search through the *E. coli* O-Typer database, and output the matching serogroups. If no match is found, users are welcome to submit their sequence to the reference database and contribute to the *E. coli* serogroup coverage provided by the database.

4.4 MULTILOCUS SEQUENCE TYPING

Multilocus sequence typing (MLST) is a method that detects DNA sequence polymorphisms in a variable number of metabolic or housekeeping genes. The selected gene loci must be identified and validated with each bacterial species. MLST protocols involve PCR amplification and sequencing of the panel of targeted genes. Sequences of the gene loci are queried against an appropriate established database of known allelic sequences for the desired bacterial species via online submission. The first MLST scheme was developed for *Neisseria meningitidis* (Maiden et al., 1998) and then expanded to a wide spectrum of other bacteria species (http://pubmlst.org/databases.shtml). Currently, the major public MLST reference databases can be found online at http://www.mlst.net/databases/ and http://pubmlst.org/databases which are hosted by the University of Oxford, UK; the Imperial College, London, UK; the University of Warwick, UK; and the Pasteur Institute, Paris, France. Additional online bioinformatic tools are accessible from the websites for further sequence analysis and for determining genetic relatedness (i.e. BLAST, eBURST, genome comparisons, and codon usage). Furthermore, the MLST databases capture demographic data pertaining to each sequence type (ST), such as the submitting country and the date of submission. As with some of the other public online databases discussed in this chapter, users can submit novel allele sequences which are subsequently verified and added to the database. The standardisation of data and the portable nature of the STs allow this method to be used as a worldwide epidemiological tool to study the molecular evolution of pathogens over time in different geographic regions.

Another version of MLST has been recently developed and is known as multi-virulence-locus sequence typing (MVLST). This method is similar to MLST but targets virulence genes instead. MVLST has been applied for the subtyping of *Listeria monocytogenes* (Zhang, Jayarao, & Knabel, 2004), *Vibrio cholerae* (Teh, Chua, & Thong, 2011), *S. enterica* (Liu et al., 2011), and *S. aureus* (Verghese, Schwalm, Dudley, & Knabel, 2012). This assay has only recently been described and more data is required to correlate the typing and epidemiological data.

4.5 VARIABLE NUMBER OF TANDEM REPEAT TYPING

There are regions of the bacterial genome that contain repetitive sequences of DNA motifs ranging from a few base pairs to over a hundred base pairs in length. The repeats are arranged in tandem, they cluster together and orient in the same direction. The number of tandem repeats can vary and are prone to higher rates of mutation (van Belkum, 1999). Bacterial strains can acquire or lose the tandem repeats at a given locus creating a distinctive fingerprinting profile, even within a short time frame (Call, Orfe, Davis, Lafrentz, & Kang, 2008). Consequently, the fingerprinting profile can provide clonal information pertaining to strain relatedness.

Variable number of tandem repeat (VNTR) typing is based on PCR amplification of a specific selection of a number of tandem repeats located in the bacterial genome. The PCR primers are designed to bind to the non-repetitive sequences that flank the tandem repeat region. Amplified products are analysed using a gel matrix, capillary electrophoresis, or microfluidic technology (Sabat et al., 2012). PCR product sizes are determined using a size marker and is indicative of the number of repeats within that region. The final result is a numerical code corresponding to the repeat number in each VNTR locus. Standardised protocols pertaining to primer and target selection, the number of targets, and choice of matrices for fragment separation, allow VNTR data to be shared and compared between laboratories.

Multilocus VNTR analysis (MLVA) is an approach being used for differentiation of bacterial strains (Heilbronn, Munnoch, Butler, Merritt, & Durrheim, 2014; Holmes, Perry, Willshaw, Hanson, & Allison, 2014; Kobayashi et al., 2014; Saleh-Lakha et al., 2013). This assay can be used as a supplementary typing method for some enteric bacterial species in parallel with PFGE (Chenal-Francisque et al., 2013). The targeted loci, primers, and protocols are all standardised to ensure reproducible results and to facilitate data exchange (Holmes et al., 2014). MLVA is an excellent candidate for global epidemiological studies because MLVA data can be easily and objectively interpreted.

Unique VNTR sequences called Mycobacterial Interspersed Repetitive Units (MIRUs) are present in the genome of *Mycobacterium tuberculosis*. These genetic elements vary in size ranging from 50 to 100 base pairs, and they are located mainly in intergenic regions dispersed throughout the *M. tuberculosis* genome (Supply et al., 2001, 2006; Supply, Magdalena, Himpens, & Locht, 1997). Since there is strain-to-strain variation in the location and number of repeats, typing by MIRU can generate unique fingerprints. Currently, 24 loci are targeted in MIRU typing for

M. tuberculosis (Supply et al., 2006). MIRU-VNTRplus (http://www.miru-vntrplus. org) is a free online application developed by Weniger, Krawczyk, Supply, Niemann, and Harmsen (2010) that allows users to analyse their genotyping data independently or in comparison with the reference database of strains representing the major *M. tuberculosis* complex lineages. The web server also has the epidemiological information associated with each strain and clustering tools for dendrogram creation.

Repetitive sequence-based PCR (Rep-PCR) is another PCR typing method based on the repetitive genetic elements located throughout the bacterial genome. Primers are designed near the end of the repetitive sequences in an outward direction such that DNA sequences separating the repetitive elements are amplified. This generates DNA fragments of varying length since the location and number of the repetitive elements may vary among different strains. Fragment separation by size and charge can be used to generate a unique fingerprint profile. There are two Rep-PCRs being used to type enteric bacteria: a 38-bp repetitive extragenic palindromic element (REP) and a 126-bp enterobacterial repetitive intergenic consensus sequence (Versalovic, Koeuth, & Lupski, 1991). A BOX repetitive element which is highly conserved has been found in *Streptococcus pneumoniae*. BOX-PCR has been used as typing tool for *S. pneumoniae* and other bacterial species (Koeuth, Versalovic, & Lupski, 1995; van Belkum, Sluijuter, de Groot, Verbrugh, & Hermans, 1996).

Rep-PCR has been modified and adapted to a commercial automated format known as the DiversiLab system (bioMérieux Canada, Inc. QC, Canada) and has proved to be a valuable typing tool in clinical laboratories (Healy et al., 2005). The typing protocol for each organism is standardised using the DiversiLab system. DNA fragments generated from the amplification are separated with microfluidic chips (LabChip device; Caliper Technologies, Inc.) and placed in the 2100 bioanalyzer (Agilent Technologies, Inc., Palo Alto, CA, USA). The DiversiLab software uses standardised algorithms to determine distance matrices and create dendrograms (Fig. 7).

The data generated is encrypted for security and automatically transferred from the bioanalyzer to a password-protected customer website. Reports include the dendrogram, electropherograms, virtual gel images, scatter plots, and demographic fields (Fig. 7). Fingerprints from these isolates can be compared to each other or to those contained in secured and password-protected online databases limited to registered users only. This molecular typing process has a rapid turn-around time which is crucial, especially in outbreak situations.

5 COMPARATIVE GENOMIC FINGERPRINTING

Campylobacter jejuni is a leading cause of bacterial gastroenteritis affecting over 1.3 million persons worldwide annually (CDC, www.cdc.gov/nczved/divisions/dfbmd/ diseases/campylobacter). Campylobacteriosis is commonly caused by the ingestion of contaminated food or water. Clinically, campylobacteriosis is indistinguishable from infection by other gastrointestinal pathogens and symptoms include: diarrhoea,

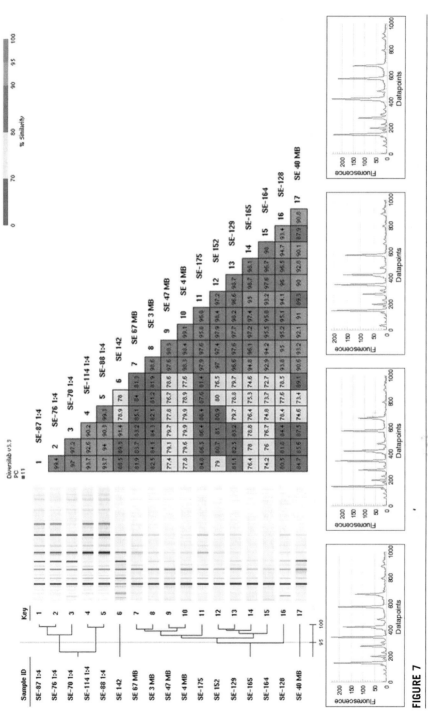

FIGURE 7

Fingerprint analysis using DiversiLab software. A dendrogram based on DNA fingerprint patterns was constructed using the UPGMA clustering algorithm. The corresponding similarity matrix and a subset of graphs highlighting visible fingerprint peaks are also shown. (See the color plate.)

fever, and abdominal pain (Allos, 2001). Diagnosis of campylobacteriosis is achieved by culture and identification of the organism from stool samples. Surveillance systems utilise numerous subtyping methods such as PFGE, restriction fragment length polymorphism analysis of the flagellin gene (flaA RFLP), and MLST (Fitzgerald, Sails, & Fields, 2005). PulseNet Canada, a national surveillance system used to identify and respond to foodborne disease outbreaks in Canada, currently utilises PFGE for surveillance of *C. jejuni*. However, this method has limited value for *C. jejuni* due to high genetic diversity (Michaud, Ménard, & Arbeit, 2005) and chromosomal rearrangements (Barton, Ng, Tyler, & Clark, 2007). An alternative typing method is comparative genomic fingerprinting (CGF), developed in the Laboratory for Foodborne Zoonoses by the Public Health Agency of Canada in Lethbridge, Alberta. CGF produces high-resolution allelic assessment of 40 loci widely distributed throughout the *C. jejuni* genome (Taboada et al., 2012). The basis of selection for the 40 genes assessed in this method is that they are highly variable accessory genes encoding a range of proteins and enzymes. Each gene was also required to meet five additional criteria (Taboada et al., 2012): (i) are expressed as clear presence or absence; (ii) are classified as unbiased genes showing adequate carriage across multiple strains; (iii) have a representative genomic distribution; (iv) have the ability to capture strain-to-strain relationships inferred from whole-genome comparative genomic analysis; and (v) are present in two or more of the publicly available draft *C. jejuni* genomes. Eight multiplex PCR assays consisting of five gene targets per assay were designed. DNA from *C. jejuni* isolates are amplified with each multiplex using conventional PCR. The 40 amplicons are separated based on size using the QIAxcel high-throughput capillary electrophoresis system, which outputs high-resolution gel images and electropherograms. Data is analysed and visualised using the BioCalculator v3.0 software (Qiagen, Mississauga, Canada). Each multiplex produces a single gel image and accompanying electropherograms for each of the 12 lanes (Fig. 8).

C. jejuni NCTC 11168 (ATCC 700819) acts as a positive control for the assay as it possesses all 40 CGF genes in its genome and each *C. jejuni* isolate is compared to this positive control to determine the presence or absence of genes. A gene is considered present if the peak generated in the electropherogram matches both in expected amplicon size and is above the minimal threshold expressed in relative fluorescence units to distinguish it from background. A gene is considered absent if the generated peak is not above the minimal threshold or if the amplicon size does not match the expected size. PCR data by the QIAxcel and BioCalculator software is converted to binary values in generation of a CGF report. For each of the 40 genes, a 1 denotes presence of the gene and a 0 denotes absence of the gene. Validation of data scoring is accomplished by both the computer software and user inspection for discrepancies. The compiled binary data comprises the genomic fingerprint of the *C. jejuni* isolates and can be clustered using appropriate algorithms to determine the genetic relatedness, which is especially valuable in outbreak investigations (Fig. 9). Further, high concordance is observed between results generated with CGF and those from PFGE and MLST (Taboada et al., 2012).

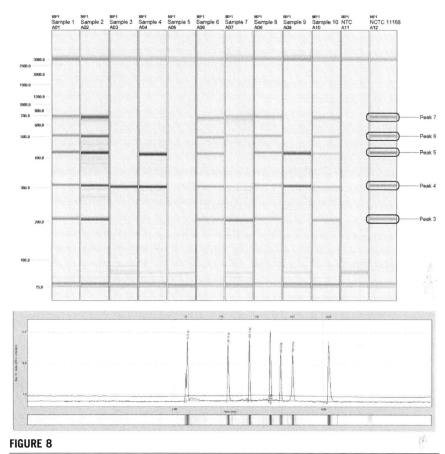

FIGURE 8

Gel image of CGF multiplex PCR reaction and corresponding electropherogram. Lanes A01-10 *C. jejuni* isolates, A11 no-template control (NTC), A12 positive control NCTC 11168. NCTC 11168 amplicons in gel image labelled to corresponding peaks generated in electropherogram.

6 TOOLS FOR PATHOGEN SURVEILLANCE AND REPORTING
6.1 PULSENET

As previously mentioned, standardised typing protocols for instrumentation, software, and analysis allow inter-laboratory comparison of bacterial fingerprints and form the foundation of international consensus standards for worldwide surveillance programs. PulseNet, established by the CDC (Atlanta) in 1996, is an example of this paradigm in practise. PulseNet is a network of public health and food regulatory agency laboratories that perform molecular subtyping of bacteria that cause foodborne diseases with established standardised laboratory protocols used by all

	MP1					MP2					MP3					MP4				
Isolate	cj0483 (612 bp)	cj0181 (486 bp)	cj0570 (405 bp)	cj0728 (296 bp)	cj0298c (198 bp)	cj1427c (613 bp)	cj0733 (441 bp)	cj1431c (307 bp)	cj0860 (282 bp)	cj0057 (175 bp)	cj1585 (630 bp)	cj0008 (486 bp)	cj0264 (406 bp)	cj1727c (369 bp)	cj0297c (300 bp)	cj0566 (558 bp)	cj1334 (462 bp)	cj0177 (399 bp)	cj1329 (307 bp)	cj1550c (188 bp)
1	0	0	1	0	1	0	0	0	1	0	0	0	1	0	1	0	0	0	0	0
2	0	0	1	0	1	0	0	0	1	0	0	0	1	0	1	0	0	0	0	0
3	0	0	1	0	1	0	0	0	1	0	0	0	1	0	1	0	0	0	0	0
4	0	0	1	0	1	0	0	0	1	0	0	0	1	0	1	0	0	0	0	0
5	0	0	1	0	1	0	0	0	1	0	0	0	0	0	1	0	0	0	0	0
6	0	0	1	1	1	0	1	0	1	0	0	0	1	0	1	0	1	0	1	0
7	1	1	1	1	1	0	1	0	1	1	0	0	0	0	1	1	0	1	0	0
8	0	1	1	1	1	1	1	0	0	1	1	1	1	1	1	1	1	1	1	0
9	0	1	1	1	1	1	1	0	0	1	1	1	1	1	1	1	1	1	1	0
10	1	1	1	1	1	0	1	0	0	1	1	0	0	1	1	0	1	1	1	0
11	1	1	1	1	1	0	1	0	0	1	1	0	0	1	1	0	1	1	1	0
NCTC 11168	1	1	1	1	1	1	1	1	1	1	1	1	1	1	1	1	1	1	1	1

	MP5					MP6					MP7					MP8					
Isolate	cj0625 (498 bp)	cj0569 (399 bp)	cj0486 (301 bp)	cj0033 (206 bp)	cj0421c (127 bp)	cj1136 (510 bp)	cj1141 (413 bp)	cj0967 (301 bp)	cj0736 (205 bp)	cj0755 (101 bp)	cj1679 (529 bp)	cj1721 (415 bp)	cj1439 (307 bp)	cj1552 (222 bp)	cj1134 (152 bp)	cj0035 (541 bp)	cj1324 (440 bp)	cj0307 (347 bp)	cj1551 (241 bp)	cj1294 (160 bp)	Cluster number 100%
1	1	0	0	0	1	0	0	1	0	0	1	0	0	0	0	1	1	1	0	1	1
2	1	0	0	0	1	0	0	1	0	0	1	0	0	0	0	1	1	1	0	1	1
3	1	0	0	0	1	0	0	1	0	0	1	0	0	0	0	1	1	1	0	1	1
4	1	0	0	0	1	0	0	1	0	0	1	0	0	0	0	1	1	1	0	1	1
5	1	0	0	0	1	0	0	1	0	0	1	0	0	0	0	1	0	1	0	1	2
6	1	0	0	0	1	0	0	1	1	0	0	0	0	0	1	1	1	1	0	1	3
7	1	1	1	0	1	0	0	1	1	1	0	0	0	0	0	1	1	1	0	1	4
8	1	1	0	0	1	0	0	1	1	0	1	0	0	0	1	1	1	1	0	1	5
9	1	1	0	0	1	0	0	1	1	0	1	0	0	0	1	1	1	1	0	1	5
10	1	1	1	0	1	0	0	1	1	1	0	0	0	0	0	1	0	1	0	1	6
11	1	1	1	0	1	0	0	1	1	1	0	0	0	0	0	1	0	1	0	1	6
NCTC 11168	1	1	1	1	1	1	1	1	1	1	1	1	1	1	1	1	1	1	1	1	7

FIGURE 9

Final CGF report. Presence of gene denoted by 1, absence of gene denoted by 0. *C. jejuni* isolates with 100% genetic similarity share identical genomic fingerprints and are clustered together. Cluster 1, isolates 1–4; cluster 5, isolates 8–9; cluster 6, isolates 10–11. NCTC 11168 serves as positive control. (See the color plate.)

member laboratories (Cooper et al., 2006; Graves & Swaminathan, 2001; Hyytiä-Trees, Smole, Fields, Swaminathan, & Ribot, 2006; Kam et al., 2008; Parsons et al., 2007; Ribot et al., 2006). PFGE is the current method used by PulseNet laboratories to type bacteria. The software used by PulseNet members for handling typing data is also standardised to BioNumerics, which helps data normalisation and sharing. The key role of this network is to improve surveillance and early detection of foodborne disease so outbreaks can be controlled in a timely manner (Swaminathan et al., 2006). At a national level, PulseNet maintains online discussion boards that track and manage enteric outbreaks across the country. In Canada, representatives from participating laboratories in each province create topics on the discussion board when a cluster of cases is observed. Details, such as the number of cases included in the cluster, the collection dates for the isolates, and images showing the PFGE fingerprint designations associated with the potential outbreak are shared amongst PulseNet members. Provincial designates will then respond to the board posting with information on whether regional laboratories in each respective province have observed isolates with similar or identical PFGE fingerprint patterns. This process helps to quickly determine whether an enteric outbreak is occurring at a regional or national level and is done in real-time. Further, all PulseNet members submit their PFGE data to a central PulseNet database, allowing PulseNet administrators to identify outbreaks spanning multiple provinces that would not have been detected by individual participating laboratories.

This network has expanded to international partners and PulseNet membership currently consists of 83 countries in six regions including Asia, Africa, Canada, Europe, Latin America, and the Caribbean and the Middle East (Fig. 10). Outbreak information and typing data on the online PulseNet Canada discussion board are also shared and regularly monitored by international PulseNet members from a number of different participating countries. This ensures that international outbreaks are recognised and promotes a rapid international response. The implementation of the PulseNet program in each of the laboratories requires a commitment of resources (human and materials) and a quality system that consists of training, certification, proficiency, and competency. As a result, PulseNet International has successfully tracked many outbreaks worldwide (Boxrud, Monson, Stiles, & Besser, 2010; Cartwright et al., 2013; Demczuk et al., 2010; Gaffga et al., 2012; Graves et al., 2005; Imanishi et al., 2014; Li et al., 2012; Terajima et al., 2006; Timbo, Keys, & Klontz, 2010) and response times have shortened tremendously throughout these years (Swaminathan et al., 2006).

6.2 EXPOSURE INVESTIGATION SYSTEM

Communication of data is a crucial component of pathogen surveillance and is particularly important for outbreak tracking and response. Software designed for this purpose must have timely and accurate reporting, be easily accessible, have appropriate data security features, and be able to convey the necessary information. An example of this is the online exposure investigation (EI) system developed by the

FIGURE 10

PulseNet International participant countries. (See the color plate.)

From www.cdc.gov/pulsenet

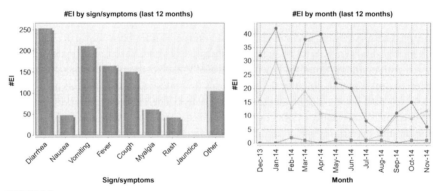

FIGURE 11

Real-time presentation of provincial exposure investigation trends in the EI system in Alberta, Canada. (See the color plate.)

Provincial Laboratory for Public Health and used in Alberta, Canada for tracking EIs that require laboratory testing and support (Figs. 11 and 12).

The EI system links pathogen typing results generated in frontline microbiology laboratories with outbreaks throughout the province. It is a comprehensive information hub that provides service to clients from multiple organisations across different geographic divisions. Users involved with each outbreak investigation, such as microbiologists and virologists, public health officials, nurses, laboratory personnel,

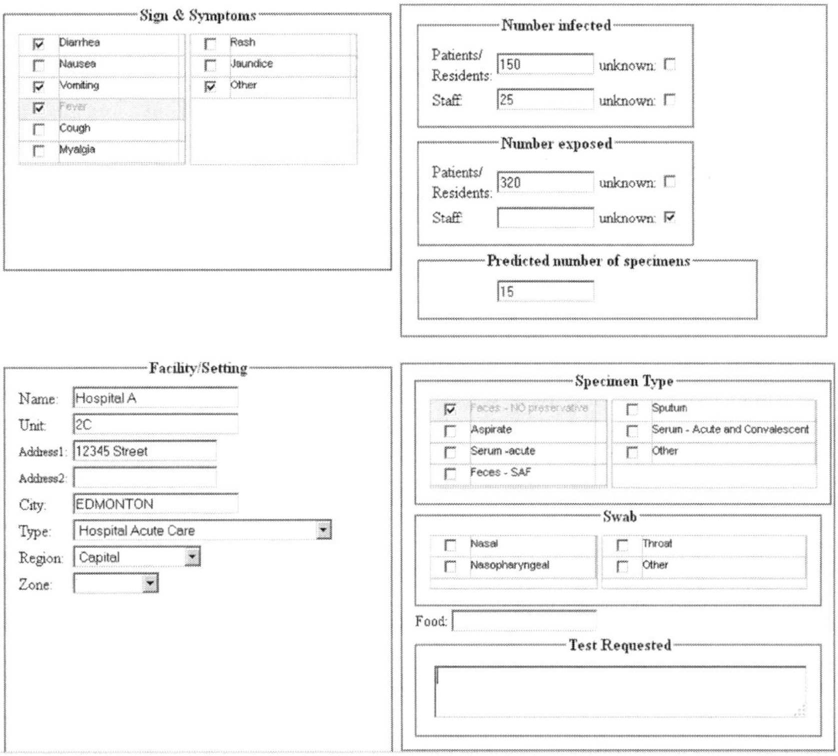

FIGURE 12

A subset of exposure investigation details stored in the EI system that is visible to users.

and epidemiologists, are sent real-time e-mail notifications as laboratory testing data becomes available. In addition to real-time typing results, the EI system allows users to view details associated with each outbreak including the onset date, type, and duration of the outbreak, the suspected and confirmed pathogen, the number of associated laboratory samples, the laboratory testing required, the city and facility linked to the outbreak, and the number of exposed and infected patients (Fig. 12).

The EI system is compatible with mobile devices and browsers, maximising accessibility and minimising response times. Further, the EI system has a refined access management system that restricts viewing permissions to appropriate users as well as extensive auditing functions. To date, the EI system has over 500 users and managed over 1700 outbreak investigations from 2012 to 2014.

6.3 DATA INTEGRATION FOR ALBERTA LABORATORIES TOOL

Rapid response to public health issues require accurate data management, prompt analysis and interpretation, and seamless information sharing to assist with public health decision making and emergency planning. Making sense of pathogen typing results is a challenge, particularly in high volume front line public health laboratories. The effectiveness of public health response and intervention is dependent on how efficiently trends are monitored and detected. This process often requires sophisticated methods for data storage, collation, extraction, and analysis that are adapted to the appropriate testing method(s) and pathogen. Further, the presentation and sharing of results and their interpretation is another important factor that must be considered. In recent years, different software has been designed to aid in handling these issues. One example is the web-based surveillance tool, data integration for Alberta laboratories (DIAL), created by the Provincial Laboratory for Public Health (ProvLab) in Alberta in partnership with the Canadian Network for Public Health Intelligence (CNPHI). DIAL is an internet-based, near real-time tool that allows rapid access of laboratory data, provides descriptive analysis and interpretation of relevant data to assist with emergency preparedness and response. DIAL can collate, extract, aggregate, and intuitively sort laboratory data. The simple point and click technology and user-friendly interface provides users with simplified data to examine pathogen trends and positivity rates in laboratory samples (Mukhi, May-Hadford, Plitt, Preiksaitis, & Lee, 2010). The system can report summarised laboratory data for public health intelligence, aiding public health policy decisions, pathogen surveillance, and outbreak detection in Alberta, Canada (Earn et al., 2012; Fathima et al., 2014; Fathima, Lee, May-Hadford, Mukhi, & Drews, 2012; Lee et al., 2011; Pabbaraju et al., 2011). Further, DIAL is able to work directly with other surveillance software systems such as the previously mentioned EI system. One of the biggest strengths of systems such as DIAL is the ability to adapt to technological advancements in laboratory testing algorithms and the resulting changes in laboratory data.

7 GEOGRAPHIC INFORMATIONS SYSTEMS

Geographic informations systems (GIS) include computer hardware and software, database engines, and data infrastructure that are used to store, process, visualise, and share spatial data. GIS can greatly enhance public health surveillance initiatives (Holt et al., 2013; Kelly-Hope et al., 2007; Uhlmann et al., 2009) and outbreak tracking and response (Bull, Hall, Leach, & Robesyn, 2012; Cromley, 2003; Schimmer et al., 2010). There are many commercial (ArcGIS (Environmental Systems Research Institute, Redlands, CA, USA)) and open source (QGIS (www.qgis.org)) GIS software available that are suitable for mapping epidemiological and laboratory typing data.

GIS analyses can integrate crucial geographic information with bacterial typing and epidemiological data to reveal the distribution, dissemination, and evolution of the pathogen (Fig. 13). In addition, these analyses can help identify risk factors associated with disease and assist in the determination of an appropriate course of action.

It is becoming increasingly common to use GIS for enhanced disease surveillance, outbreak tracking, and studies exploring geographic distribution of pathogens.

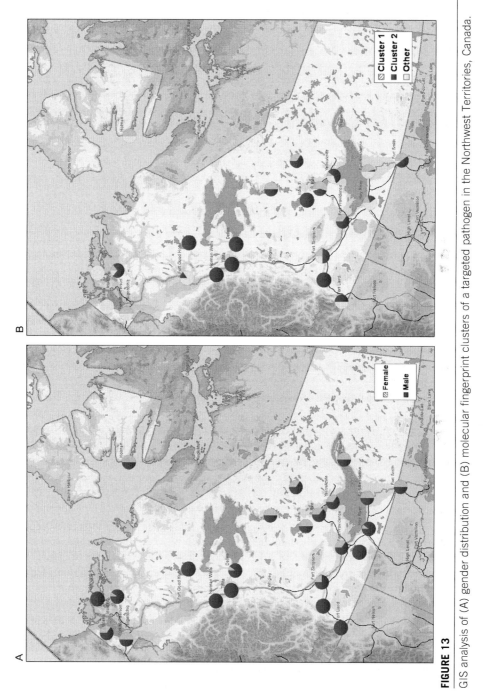

FIGURE 13

GIS analysis of (A) gender distribution and (B) molecular fingerprint clusters of a targeted pathogen in the Northwest Territories, Canada.

GIS has been used to map foodborne disease outbreaks (Gould et al., 2013), identify outbreak sources (Bedard, Kennedy, & Weimer, 2014), relate disease incidence with socioeconomic factors (Maciel et al., 2010), detail risk factors and possible interventions (Varga et al., 2013), find geographical associations between human and animal disease (Casey, Curriero, Cosgrove, Nachman, & Schwartz, 2013), and link climate change to infectious diseases (Semenza, Suk, Estevez, Ebi, & Lindgren, 2012). There are several web-based GIS initiatives designed for pathogen surveillance that currently exist including SRL-Maps (http://www.spa tialepidemiology.net/srl-maps/) for *S. aureus*, EpiScanGIS (http://www. episcangis.org) for meningococcal disease, and HealthMap (http://healthmap.org) for contagious diseases. These tools are publicly available and illustrate the utility of integrating GIS with bacterial typing.

8 WHOLE-GENOME SEQUENCING

In recent years, significant advancements have been made in sequencing technology, performance, and software. Whole-genome sequencing (WGS) using a high-throughput platform, is also known as next generation sequencing (NGS). This technology has emerged as one of the most valuable pathogen typing methods. The approach of NGS is different from the conventional "Sanger" sequencing method. In a single run, NGS can generate millions of contigs that vary in length. Many different sequencing platforms are available, each with differences in read length, sequence quality, error rates, yield, sensitivity, and associated cost and time. The information generated is tremendous and powerful, making WGS an attractive tool for epidemiological investigations (Croucher, Harris, Grad, & Hanage, 2013). The steady decrease in cost and turn-around-time associated with WGS has made the typing method even more appealing. There are many examples where this technology has been successfully applied in retrospective epidemiological investigations, especially in outbreaks (Ben Zakour, Venturini, Beatson, & Walker, 2012; Chin et al., 2011; Gilmour et al., 2010; Grad et al., 2012; Grad & Waldor, 2013; Köser, Ellington, et al., 2012; Köser, Holden, et al., 2012; Mellmann et al., 2011; Reimer et al., 2011; Török et al., 2013). WGS has the potential to replace current diagnostic and molecular epidemiological tools. This is an important feature for fastidious organisms, especially ones that are slow growing and hazardous and requires a high level of biosafety containment, such as *M. tuberculosis* (Köser, Ellington, et al., 2012). WGS data can also aid in bacterial identification (Köser, Fraser, et al., 2014) and characterisation, such as antibiotic susceptibilities (Grad et al., 2014; Köser et al., 2013), and to control antimicrobial resistance (Köser, Ellington, & Peacock, 2014). The availability of WGS data will provide diagnostic and reference laboratories with an unprecedented degree of flexibility and has great impact in both research and public health (Ehrlich & Post, 2013; Köser, Ellington, et al., 2012; Reuter et al., 2013).

WGS produces an incredible amount of sequencing data, making data analysis and interpretation a challenge. Recent advancements in WGS software has helped to alleviate this problem by automating much of the analysis. A general workflow includes: (i) evaluating the quality of a sequencing run; (ii) filtering poor reads; (iii) *de novo* or reference-guided genome assembly; (iv) genome annotation; (v) genome analysis (for example SNP analysis, phage finding, pangenome comparisons, gene prediction, antibiotic resistance predictions, and MLST analysis). Because there are many steps involved, a wide variety of specialised WGS software is available. Galaxy (galaxyproject.org; Blankenberg et al., 2010, 2011; Giardine et al., 2005; Goecks, Nekrutenko, Taylor, & The Galaxy Team, 2010) is an example of an open source and online platform that can process WGS data from start to finish. It is a collection of useful bioinformatic tools that work together to assemble and analyse genomic sequencing data from raw sequencing reads. One of the advantages of using Galaxy is that it is very user-friendly and does not require much computer programming experience, unlike many other tools currently available. Further, it is constantly updated, analyses can be easily reproduced and shared with other users, and it has an extensive documentation system. Examples of some common bioinformatics software in the Galaxy interface includes SPAdes (Bankevich et al., 2012) for genome assembly, PROKKA (Seemann, 2014) for automated gene prediction and annotation, BAMTools (Barnett, Garrison, Quinlan, Strömberg, & Marth, 2011) for variant calling and alignment management, BLAST (Altschul, Gish, Miller, Myers, & Lipman, 1990) and ClustalW (Thompson, Gibson, & Higgins, 2002) for clustering analysis and alignments, and Phyloviz (Francisco et al., 2012) for phylogenetic analysis, although many more are available. Since Galaxy is open source, new tools are constantly added and existing ones are frequently updated. Further, Galaxy is highly customisable allowing users to modify the platform to suit their needs. Without bioinformatics software advances to accompany the development of WGS, the processing and analysis of the high volume of data generated would be overwhelming for most laboratories and certainly could not be implemented in a routine laboratory setting.

9 DISCUSSION

This chapter has provided an overview of the molecular typing methods and software available for pathogen surveillance and outbreak detection. The selection of a molecular typing method depends on intra-laboratory and inter-laboratory reproducibility, discriminatory power, ease of use, turn-around time for reporting, data interpretation and portability, and costs associated with human resources, equipment, and materials. A high-throughput assay with good reproducibility and comparability is valuable, especially during national or international outbreaks. As typing methods and associated software continue to evolve, so does our understanding of bacterial pathogens. The shift towards molecular typing from classical methods has expanded our capacity for pathogen surveillance and outbreak detection. Typing techniques are becoming increasingly automated thus requiring less technical expertise. Also, there have been notable improvements in speed, cost, and volume of testing making pathogen typing accessible to laboratories of varying technical expertise and budget. The amount of laboratory data generated by current molecular typing assays is steadily increasing, giving an unprecedented level of detail but also creating new issues with storage analysis and sharing. This problem is mostly alleviated by improvements in computer hardware and software in recent years. With tests such as whole-genome sequencing, computer hardware becomes very important. Because of the sheer volume of data generated and because analyses are heavily automated, computer processing power, random access memory, storage space, and server availability can be bottlenecks that limit the utility of the assay. Without recent advances in computer hardware, some molecular typing methods would not be possible. As was explored throughout this chapter, the development of software to accommodate the specific types of laboratory data and analyses associated with the different typing methods was essential. Powerful software suites and platforms, such as BioNumerics and Galaxy, are able to integrate typing data from multiple sources, making interpretation of results easier and more robust. This also allows incorporation of historical data with data from recent advanced methods. As our understanding of pathogens grows, more and more molecular methods with better discriminatory power, speed, cost, reproducibility, and automation can be developed. As we move into the era of NGS, there will be paradigm shift for public health laboratories in terms of workflow, staff training, and laboratory setup. Standardisation of procedures, equipment, data interpretation, quality systems, and bioinformatics will play a major role in the "New World" of molecular typing.

ACKNOWLEDGEMENTS

The authors would like to thank the staff and students at the Provincial Laboratory for Public Health in Alberta for their contributions and their critical review of this book chapter. The authors would also like to thank the Northwest Territories Department of Health and Social Services for their GIS data contributions. Special thanks to Dr. Brendon Parsons for his final review of the manuscript.

REFERENCES

Allos, B. M. (2001). *Campylobacter jejuni* infections: Update on emerging issues and trends. *Clinical Infectious Diseases, 32*(8), 1201–1206.

Altschul, S. F., Gish, W., Miller, W., Myers, E. W., & Lipman, D. J. (1990). Basic local alignment search tool. *Journal of Molecular Biology, 215*(3), 403–410.

Arguello, H., Alvarez-Ordoñez, A., Carvajal, A., Rubio, P., & Prieto, M. (2013). Role of slaughtering in *Salmonella* spreading and control in pork production. *Journal of Food Protection, 76*(5), 899–911.

Bankevich, A., Nurk, S., Antipov, D., Gurevich, A. A., Dvorkin, M., Kulikov, A. S., et al. (2012). SPAdes: A new genome assembly algorithm and its applications to single-cell sequencing. *Journal of Computational Biology, 19*(5), 455–477.

Bannerman, T. L., Hancock, G. A., Tenover, F. C., & Miller, J. M. (1995). Pulsed-field gel electrophoresis as a replacement for bacteriophage typing of *Staphylococcus aureus*. *Journal of Clinical Microbiology, 33*(3), 551–555.

Barnett, D. W., Garrison, E. K., Quinlan, A. R., Strömberg, M. P., & Marth, G. T. (2011). BamTools: A C++ API and toolkit for analyzing and managing BAM files. *Bioinformatics*, *27*(12), 1691–1692.

Bartholomew, M. L., Heffernan, R. T., Wright, J. G., Klos, R. F., Monson, T., Khan, S., et al. (2014). Multistate outbreak of *Salmonella enterica* serotype enteritidis infection associated with pet guinea pigs. *Vector Borne and Zoonotic Diseases*, *14*(6), 414–421.

Barton, C., Ng, L.-K., Tyler, S. D., & Clark, C. G. (2007). Temperate bacteriophages affect pulsed-field gel electrophoresis patterns of *Campylobacter jejuni*. *Journal of Clinical Microbiology*, *45*(2), 386–391.

Basler, C., Forshey, T. M., Machesky, K., Erdman, M. C., Gomez, T. M., Nguyen, T. A., et al. (2014). Multistate outbreak of human *Salmonella* infections linked to live poultry from a mail-order hatchery in Ohio—March-September 2013. *Morbidity and Mortality Weekly Report*, *63*(10), 222.

Bayer, C., Bernard, H., Prager, R., Rabsch, W., Hiller, P., Malorny, B., et al. (2014). An outbreak of *Salmonella* Newport associated with mung bean sprouts in Germany and the Netherlands, October to November 2011. *Euro Surveillance*, *19*(1). Erratum in: Euro Surveill (2014) *19*(1). pii: 20665.

Beall, B., Facklam, R., & Thompson, T. (1996). Sequencing emm-specific PCR products for routine and accurate typing of group A streptococci. *Journal of Clinical Microbiology*, *34*(4), 953–958.

Bedard, B., Kennedy, B. S., & Weimer, A. C. (2014). Geographical information software and shopper card data, aided in the discovery of a *Salmonella* Enteritidis outbreak associated with Turkish pine nuts. *Epidemiology and Infection*, *142*(12), 2567–2571.

Ben Zakour, N. L., Venturini, C., Beatson, S. A., & Walker, M. J. (2012). Analysis of a *Streptococcus pyogenes* puerperal sepsis cluster by use of whole-genome sequencing. *Journal of Clinical Microbiology*, *50*(7), 2224–2228.

Blankenberg, D., Coraor, N., Von Kuster, G., Taylor, J., Nekrutenko, A., & Galaxy Team. (2011). Integrating diverse databases into an unified analysis framework: A galaxy approach. *Database: The Journal of Biological Databases and Curation*, *2011*, bar011. http://dx.doi.org/10.1093/database/bar011. Print 2011. PMID: 2153198.

Blankenberg, D., Von Kuster, G., Coraor, N., Ananda, G., Lazarus, R., Mangan, M., et al. (2010). Galaxy: A web-based genome analysis tool for experimentalists. *Current Protocols in Molecular Biology*. Chapter 19, Unit 19.10.1-21.

Bopp, C. A., Brenner, F. W., Fields, P. I., Wells, J. G., & Stockbine, N. A. (2003). *Escherichia*, *Shigella*, and *Salmonella*. In P. R. Murray, E. J. Baron, J. H. Jorgensen, M. A. Pfaller, & R. H. Yolken (Eds.), *Manual of clinical microbiology: Vol. 1.* (8th ed., pp. 654–671). Washington, DC: ASM Press.

Boxrud, D., Monson, T., Stiles, T., & Besser, J. (2010). The role, challenges, and support of pulsenet laboratories in detecting foodborne disease outbreaks. *Public Health Reports*, *125*(Suppl. 2), 57–62.

Braga, E. D., Aguiar-Alves, F., de Freitas Mde, F., de e Silva, M. O., Correa, T. V., Snyder, R. E., et al. (2014). High prevalence of *Staphylococcus aureus* and methicillin-resistant *S. aureus* colonization among healthy children attending public daycare centers in informal settlements in a large urban center in Brazil. *BMC Infectious Diseases*, *14*, 538.

Braun, S. D., Ziegler, A., Methner, U., Slickers, P., Keiling, S., Monecke, S., et al. (2012). Fast DNA serotyping and antimicrobial resistance gene determination of *Salmonella* enterica with an oligonucleotide microarray-based assay. *PLoS One*, *7*(10), e46489.

Brooks, H. J., Mollison, B. D., Bettelheim, K. A., Matejka, K., Paterson, K. A., & Ward, V. K. (2001). Occurrence and virulence factors of non-O157 Shiga toxin-producing *Escherichia coli* in retail meat in Dunedin, New Zealand. *Letters in Applied Microbiology, 32*(2), 118–122.

Brooks, J. T., Sowers, E. G., Wells, J. G., Greene, K. D., Griffin, P. M., Hoekstra, R. M., et al. (2005). Non-O157 Shiga toxin-producing *Escherichia coli* infections in the United States, 1983-2002. *The Journal of Infectious Diseases, 192*(8), 1422–1429.

Buchholz, U., Bernard, H., Werber, D., Böhmer, M. M., Remschmidt, C., Wilking, H., et al. (2011). German outbreak of *Escherichia coli* O104:H4 associated with sprouts. *The New England Journal of Medicine, 365*(19), 1763–1770.

Bull, M., Hall, I. M., Leach, S., & Robesyn, E. (2012). The application of geographic information systems and spatial data during Legionnaires disease outbreak responses. *Euro Surveillance, 17*(49). pii: 20331.

Call, D. R., Orfe, L., Davis, M. A., Lafrentz, S., & Kang, M. S. (2008). Impact of compounding error on strategies for subtyping pathogenic bacteria. *Foodborne Pathogens and Disease, 5*(4), 505–516.

Carle, G. F., Frank, M., & Olson, M. V. (1986). Electrophoretic separations of large DNA molecules by periodic inversion of the electric field. *Science, 232*(4746), 65–68.

Carle, G. F., & Olson, M. V. (1984). Separation of chromosomal DNA molecules from yeast by orthogonal-field-alternation gel electrophoresis. *Nucleic Acids Research, 12*(14), 5647–5664.

Carlin, P. S., Botella, A. G., Valladares, L. D., Sánchez-Pernaute, A., Santacruz, C. C., Granizo, R. M., et al. (2014). Necrotizing fasciitis caused by *Streptococcus pyogenes* in immunocompetent patient. *Surgical Infections, 15*(6), 859–860. PMID: 25402863.

Cartwright, E. J., Jackson, K. A., Johnson, S. D., Graves, L. M., Silk, B. J., & Mahon, B. E. (2013). Listeriosis outbreaks and associated food vehicles, United States, 1998-2008. *Emerging Infectious Diseases, 19*(1), 1–9. quiz 184.

Casey, J. A., Curriero, F. C., Cosgrove, S. E., Nachman, K. E., & Schwartz, B. S. (2013). High-density livestock operations, crop field application of manure, and risk of community-associated methicillin-resistant *Staphylococcus aureus* infection in Pennsylvania. *JAMA Internal Medicine, 173*(21), 1980–1990.

Centers for Disease Control and Prevention (CDC). (2006). Human salmonellosis associated with animal-derived pet treats—United States and Canada, 2005. *Morbidity and Mortality Weekly Report, 55*(25), 702–705.

Centers for Disease Control and Prevention (CDC). (2013). Notes from the field: Multistate outbreak of human *Salmonella* typhimurium infections linked to contact with pet hedgehogs—United States, 2011-2013. *Morbidity and Mortality Weekly Report, 62*(4), 73.

Centers for Disease Control and Prevention (CDC). http://www.cdc.gov/nczved/divisions/dfbmd/diseases/Campylobacter/.

Chenal-Francisque, V., Diancourt, L., Cantinelli, T., Passet, V., Tran-Hykes, C., Bracq-Dieye, H., et al. (2013). Optimized Multilocus variable-number tandem-repeat analysis assay and its complementarity with pulsed-field gel electrophoresis and multilocus sequence typing for *Listeria monocytogenes* clone identification and surveillance. *Journal of Clinical Microbiology, 51*(6), 1868–1880.

Chin, C. S., Sorenson, J., Harris, J. B., Robins, W. P., Charles, R. C., Jean-Charles, R. R., et al. (2011). The origin of the Haitian cholera outbreak strain. *The New England Journal of Medicine, 364*(1), 33–42.

Chokoshvili, O., Lomashvili, K., Malakmadze, N., Geleishvil, M., Brant, J., Imnadze, P., et al. (2014). Investigation of an outbreak of bloody diarrhea complicated with hemolytic uremic syndrome. *Journal of Epidemiology and Global Health*, *4*(4), 249–259.

Chu, G., Vollrath, D., & Davis, R. W. (1986). Separation of large DNA molecules by contour-clamped homogeneous electric fields. *Science*, *234*(4783), 1582–1585.

Clark, C., Cunningham, J., Ahmed, R., Woodward, D., Fonseca, K., Isaacs, S., et al. (2001). Characterization of *Salmonella* associated with pig ear dog treats in Canada. *Journal of Clinical Microbiology*, *39*(11), 3962–3968.

Cooper, K. L., Luey, C. K., Bird, M., Terajima, J., Nair, G. B., Kam, K. M., et al. (2006). Development and validation of a PulseNet standardized pulsed-field gel electrophoresis protocol for subtyping of *Vibrio cholerae*. *Foodborne Pathogens and Disease*, *3*(1), 51–58.

Cromley, E. K. (2003). GIS and disease. *Annual Review of Public Health*, *24*, 7–24.

Croucher, N. J., Harris, S. R., Grad, Y. H., & Hanage, W. P. (2013). Bacterial genomes in epidemiology—Present and future. *Philosophical Transactions of the Royal Society of London. Series B, Biological Sciences*, *368*(1614), 20120202.

de Almeida Torres, R. S., Fedalto, L. E., de Almeida Torres, R. F., Steer, A. C., & Smeesters, P. R. (2013). Group A streptococcus meningitis in children. *The Pediatric Infectious Disease Journal*, *32*(2), 110–114.

de Carvalho Ferreira, D., Cisne Frota, A. C., Cavalcante, F. S., Abad, E. D., & Netto Dos Santos, K. R. (2012). Necrotizing fasciitis secondary to community pneumonia by Panton-Valentine leukocidin-positive methicillin-resistant *Staphylococcus aureus*. *American Journal of Respiratory and Critical Care Medicine*, *186*(2), 202–203.

Dechet, A. M., Herman, K. M., Chen Parker, C., Taormina, P., Johanson, J., Tauxe, R. V., et al. (2014). Outbreaks caused by sprouts, United States, 1998-2010: Lessons learned and solutions needed. *Foodborne Pathogens and Disease*, *11*(8), 635–644.

de Kraker, M. E., Jarlier, V., Monen, J. C., Heuer, O. E., van de Sande, N., & Grundmann, H. (2013). The changing epidemiology of bacteraemias in Europe: Trends from the European antimicrobial resistance surveillance system. *Clinical Microbiology and Infection*, *19*(9), 860–868.

Demczuk, W. H., Finley, R., Nadon, C., Spencer, A., Gilmour, M., Ng, L. K., et al. (2010). Characterization of antimicrobial resistance, molecular and phage types of *Salmonella enterica* serovar Typhi isolations. *Epidemiology and Infection*, *138*(10), 1414–1426.

Deng, X., Salazar, J. K., Frezet, S., Maccannell, D., Ribot, E. M., Fields, P. I., et al. (2013). Genome sequence of *Salmonella* enterica serotype Tennessee strain CDC07-0191, implicated in the 2006-2007 multistate food-borne outbreak linked to peanut butter in the United States. *Genome Announcements*, *1*(3). e00260-13.

Dewaele, I., Rasschaert, G., Wildemauwe, C., Van Meirhaeghe, H., Vanrobaeys, M., De Graef, E., et al. (2012). Polyphasic characterization of *Salmonella* enteritidis isolates on persistently contaminated layer farms during the implementation of a national control program with obligatory vaccination: A longitudinal study. *Poultry Science*, *91*(11), 2727–2735.

Earn, D. J., He, D., Loeb, M. B., Fonseca, K., Lee, B. E., & Dushoff, J. (2012). Effects of school closure on incidence of pandemic influenza in Alberta, Canada. *Annals of Internal Medicine*, *156*(3), 173–181.

Ehrlich, G. D., & Post, J. C. (2013). The time is now for gene- and genome-based bacterial diagnostics: "You say you want a revolution". *JAMA Internal Medicine*, *173*(15), 1405–1406.

Fathima, S., Ferrato, C., Lee, B. E., Simmonds, K., Yan, L., Mukhi, S. N., et al. (2014). *Bordetella pertussis* in sporadic and outbreak settings in Alberta, Canada, July 2004-December 2012. *BMC Infectious Diseases, 14*, 48.

Fathima, S., Lee, B. E., May-Hadford, J., Mukhi, S., & Drews, S. J. (2012). Use of an innovative web-based laboratory surveillance platform to analyze mixed infections between human metapneumovirus (hMPV) and other respiratory viruses circulating in Alberta (AB), Canada (2009-2012). *Viruses, 4*(11), 2754–2765.

Fitzgerald, C., Sails, A., & Fields, P. (2005). *Campylobacter jejuni* strain variation. In J. Ketley & M. Konkel (Eds.), Campylobacter jejuni*: New perspectives in molecular and cellular biology.* Wymondham, UK: Horizon Bioscience.

Francis, J. S., Doherty, M. C., Lopatin, U., Johnston, C. P., Sinha, G., Ross, T., et al. (2005). Severe community-onset pneumonia in healthy adults caused by methicillin-resistant *Staphylococcus aureus* carrying the Panton-Valentine leukocidin genes. *Clinical Infectious Diseases, 40*(1), 100–107.

Francisco, A. P., Vaz, C., Monteiro, P. T., Melo-Cristino, J., Ramirez, M., & Carriço, J. A. (2012). PHYLOViZ: Phylogenetic inference and data visualization for sequence based typing methods. *BMC Bioinformatics, 13*, 87.

Friedman, N. D., Kaye, K. S., Stout, J. E., McGarry, S. A., Trivette, S. L., Briggs, J. P., et al. (2002). Health care—Associated bloodstream infections in adults: A reason to change the accepted definition of community-acquired infections. *Annals of Internal Medicine, 137*(10), 791–797.

Friesema, I. H., Schotsborg, M., Heck, M. E., & Van Pelt, W. (2014). Risk factors for sporadic Shiga toxin-producing *Escherichia coli* O157 and non-O157 illness in The Netherlands, 2008-2012, using periodically surveyed controls. *Epidemiology and Infection, 8*, 1–8.

Friesema, I., van der Zwaluw, K., Schuurman, T., Kooistra-Smid, M., Franz, E., van Duynhoven, Y., et al. (2014). Emergence of *Escherichia coli* encoding Shiga toxin 2f in human Shiga toxin-producing *E. coli* (STEC) infections in the Netherlands, January 2008 to December 2011. *Euro Surveillance, 19*(17), 26–32.

Gaffga, N. H., Barton Behravesh, C., Ettestad, P. J., Smelser, C. B., Rhorer, A. R., Cronquist, A. B., et al. (2012). Outbreak of salmonellosis linked to live poultry from a mail-order hatchery. *The New England Journal of Medicine, 366*(22), 2065–2073.

Gardiner, K., Laas, W., & Patterson, D. (1986). Fractionation of large mammalian DNA restriction fragments using vertical pulsed-field gradient gel electrophoresis. *Somatic Cell and Molecular Genetics, 12*(2), 185–195.

Garvey, P., McKeown, P., Kelly, P., Cormican, M., Anderson, W., Flack, A., et al. (2013). Investigation and management of an outbreak of *Salmonella* Typhimurium DT8 associated with duck eggs, Ireland 2009 to 2011. *Euro Surveillance, 18*(16), 20454.

Giardine, B., Riemer, C., Hardison, R. C., Burhans, R., Elnitski, L., Shah, P., et al. (2005). Galaxy: A platform for interactive large-scale genome analysis. *Genome Research, 15*(10), 1451–1455.

Gicquelais, R. E., Morris, J. F., Matthews, S., Gladden, L., Safi, H., Grayson, C., et al. (2014). Multiple-serotype *Salmonella* outbreaks in two state prisons—Arkansas, August 2012. *Morbidity and Mortality Weekly Report, 63*(8), 169–173.

Gilbert, M., MacDonald, J., Gregson, D., Siushansian, J., Zhang, K., Elsayed, S., et al. (2006). Outbreak in Alberta of community-acquired (USA300) methicillin-resistant *Staphylococcus aureus* in people with a history of drug use, homelessness or incarceration. *CMAJ, 175*(2), 149–154.

Gilmour, M. W., Graham, M., Van Domselaar, G., Tyler, S., Kent, H., Trout-Yakel, K. M., et al. (2010). High-throughput genome sequencing of two *Listeria monocytogenes* clinical isolates during a large foodborne outbreak. *BMC Genomics, 11*, 120.

Gilmour, M. W., Olson, A. B., Andrysiak, A. K., Ng, L. K., & Chui, L. (2007). Sequence-based typing of genetic targets encoded outside of the O-antigen gene cluster is indicative of Shiga toxin-producing *Escherichia coli* serogroup lineages. *Journal of Medical Microbiology, 56*(Pt. 5), 620–628.

Goecks, J., Nekrutenko, A., Taylor, J., & The Galaxy Team (2010). Galaxy: A comprehensive approach for supporting accessible, reproducible, and transparent computational research in the life sciences. *Genome Biology, 11*(8), R86.

Golding, G. R., Campbell, J. L., Spreitzer, D. J., Veyhl, J., Surynicz, K., Simor, A., et al. (2008). A preliminary guideline for the assignment of methicillin-resistant *Staphylococcus aureus* to a Canadian pulsed-field gel electrophoresis epidemic type using spa typing. *The Canadian Journal of Infectious Diseases & Medical Microbiology, 19*(4), 273–281.

Gould, L. H., Walsh, K. A., Vieira, A. R., Herman, K., Williams, I. T., Hall, A. J., et al. (2013). Surveillance for foodborne disease outbreaks—United States, 1998-2008. *MMWR Surveillance Summaries, 62*(2), 1–34.

Grad, Y. H., Kirkcaldy, R. D., Trees, D., Dordel, J., Harris, S. R., Goldstein, E., et al. (2014). Genomic epidemiology of *Neisseria gonorrhoeae* with reduced susceptibility to cefixime in the USA: A retrospective observational study. *The Lancet Infectious Diseases, 14*(3), 220–226.

Grad, Y. H., Lipsitch, M., Feldgarden, M., Arachchi, H. M., Cerqueira, G. C., Fitzgerald, M., et al. (2012). Genomic epidemiology of the *Escherichia coli* O104:H4 outbreaks in Europe, 2011. *Proceedings of the National Academy of Sciences of the United States of America, 109*(8), 3065–3070. Erratum in: Proceedings of the National Academy of Sciences USA (2012) *109*(14), 5547.

Grad, Y. H., & Waldor, M. K. (2013). Deciphering the origins and tracking the evolution of cholera epidemics with whole-genome-based molecular epidemiology. *mBio, 4*(5), e00670-13.

Graves, L. M., Hunter, S. B., Ong, A. R., Schoonmaker-Bopp, D., Hise, K., Kornstein, L., et al. (2005). Microbiological aspects of the investigation that traced the 1998 outbreak of listeriosis in the United States to contaminated hot dogs and establishment of molecular subtyping-based surveillance for *Listeria monocytogenes* in the PulseNet network. *Journal of Clinical Microbiology, 43*(5), 2350–2355.

Graves, L. M., & Swaminathan, B. (2001). PulseNet standardized protocol for subtyping *Listeria monocytogenes* by macrorestriction and pulsed-field gel electrophoresis. *International Journal of Food Microbiology, 65*(1-2), 55–62.

Guo, D., Liu, B., Liu, F., Cao, B., Chen, M., Hao, X., et al. (2013). Development of a DNA microarray for molecular identification of all 46 *Salmonella* O serogroups. *Applied and Environmental Microbiology, 79*(11), 3392–3399.

Harmsen, D., Claus, H., Witte, W., Rothgänger, J., Claus, H., Turnwald, D., et al. (2003). Typing of methicillin-resistant *Staphylococcus aureus* in a university hospital setting using a novel software for spa-repeat determination and database management. *Journal of Clinical Microbiology, 41*, 5442–5448.

Hawkey, J., Edwards, D. J., Dimovski, K., Hiley, L., Billman-Jacobe, H., Hogg, G., et al. (2013). Evidence of microevolution of *Salmonella* Typhimurium during a series of egg-associated outbreaks linked to a single chicken farm. *BMC Genomics, 14*, 800.

Healy, M., Huong, J., Bittner, T., Lising, M., Frye, S., Raza, S., et al. (2005). Microbial DNA typing by automated repetitive-sequence-based PCR. *Journal of Clinical Microbiology, 43*(1), 199–207.

Heilbronn, C., Munnoch, S., Butler, M. T., Merritt, T. D., & Durrheim, D. N. (2014). Timeliness of *Salmonella* Typhimurium notifications after the introduction of routine MLVA typing in NSW. *New South Wales Public Health Bulletin, 24*(4), 159–163.

Holmes, A., Perry, N., Willshaw, G., Hanson, M., & Allison, L. (2014). Inter-laboratory comparison of multi-locus variable-number tandem repeat analysis (MLVA) for verocytotoxin-producing *Escherichia coli* O157 to facilitate data sharing. *Epidemiology and Infection, 4*, 1–4.

Holt, K. E., Thieu Nga, T. V., Thanh, D. P., Vinh, H., Kim, D. W., Vu Tra, M. P., et al. (2013). Tracking the establishment of local endemic populations of an emergent enteric pathogen. *Proceedings of the National Academy of Sciences of the United States of America, 110*(43), 17522–17527.

Hu, Q., Cheng, H., Yuan, W., Zeng, F., Shang, W., Tang, D., et al. (2014). Panton-valentine leukocidin (PVL)-positive healthcare-associated methicillin-resistant *Staphylococcus aureus* isolates are associated with skin and soft tissue infections and colonized mainly by infective PVL-encoding bacteriophages. *Journal of Clinical Microbiology, 53*(1), 67–72. pii: JCM.01722-14. [Epub ahead of print] PMID: 25339405.

Hyytiä-Trees, E., Smole, S. C., Fields, P. A., Swaminathan, B., & Ribot, E. M. (2006). Second generation subtyping: A proposed PulseNet protocol for multiple-locus variable-number tandem repeat analysis of Shiga toxin-producing *Escherichia coli* O157 (STEC O157). *Foodborne Pathogens and Disease, 3*(1), 118–131.

Imanishi, M., Rotstein, D. S., Reimschuessel, R., Schwensohn, C. A., Woody, D. H., Jr., Davis, S. W., et al. (2014). Outbreak of *Salmonella enterica* serotype Infantis infection in humans linked to dry dog food in the United States and Canada, 2012. *Journal of the American Veterinary Medical Association, 244*(5), 545–553.

Iwamoto, M., Mu, Y., Lynfield, R., Bulens, S. N., Nadle, J., Aragon, D., et al. (2013). Trends in invasive methicillin-resistant *Staphylococcus aureus* infections. *Pediatrics, 132*(4), e817–e824.

Jay-Russell, M. T., Madigan, J. E., Bengson, Y., Madigan, S., Hake, A. F., Foley, J. E., et al. (2014). *Salmonella* Oranienburg isolated from horses, wild turkeys and an edible home garden fertilized with raw horse manure. *Zoonoses and Public Health, 61*(1), 64–71.

Johansson, L., Thulin, P., Low, D. E., & Norrby-Teglund, A. (2010). Getting under the skin: The immunopathogenesis of *Streptococcus pyogenes* deep tissue infections. *Clinical Infectious Diseases, 51*(1), 58–65.

Kam, K. M., Luey, C. K., Parsons, M. B., Cooper, K. L., Nair, G. B., Alam, M., et al. (2008). Evaluation and validation of a PulseNet standardized pulsed-field gel electrophoresis protocol for subtyping *Vibrio parahaemolyticus*: An international multicenter collaborative study. *Journal of Clinical Microbiology, 46*(8), 2766–2773.

Kelly-Hope, L. A., Alonso, W. J., Thiem, V. D., Anh, D. D., Canh do, G., Lee, H., et al. (2007). Geographical distribution and risk factors associated with enteric diseases in Vietnam. *The American Journal of Tropical Medicine and Hygiene, 76*(4), 706–712.

Kim, S., Frye, J. G., Hu, J., Fedorka-Cray, P. J., Gautom, R., & Boyle, D. S. (2006). Multiplex PCR-based method for identification of common clinical serotypes of *Salmonella* enterica subsp. enterica. *Journal of Clinical Microbiology, 44*(10), 3608–3615.

Kinross, P., van Alphen, L., Martinez Urtaza, J., Struelens, M., Takkinen, J., Coulombier, D., et al. (2014). Multidisciplinary investigation of a multicountry outbreak of *Salmonella* Stanley infections associated with turkey meat in the European Union, August 2011 to January 2013. *Euro Surveillance, 19*(19). pii: 20801.

Kobayashi, A., Takahashi, S., Ono, M., Tanaka, K., Kishima, M., Akiba, M., et al. (2014). Molecular typing of *Salmonella enterica* serovar enteritidis isolates from food-producing animals in Japan by multilocus variable-number tandem repeat analysis: Evidence of clonal dissemination and replacement. *Acta Veterinaria Scandinavica, 56*, 31.

Köck, R., Becker, K., Cookson, B., van Gemert-Pijnen, J. E., Harbarth, S., Kluytmans, J., et al. (2010). Methicillin-resistant *Staphylococcus aureus* (MRSA): Burden of disease and control challenges in Europe. *Euro Surveillance, 15*(41). Pii: 19688.

Koeuth, T., Versalovic, J., & Lupski, J. R. (1995). Differential subsequence conservation of interspersed repetitive *Streptococcus pneumoniae* BOX elements in diverse bacteria. *Genome Research, 5*(4), 408–418.

Kohayagawa, Y., Ishitobi, N., Yamamori, Y., Wakuri, M., Sano, C., Tominaga, K., et al. (2014). Streptococcal toxic shock syndrome from necrotizing soft-tissue infection of the breast caused by a mucoid type strain. *Journal of Infection and Chemotherapy, 21*(2), 144–147. http://dx.doi.org/10.1016/j.jiac.2014.08.031. PMID: 25260866, pii: S1341-321X(14)00314-6.

Köser, C. U., Bryant, J. M., Becq, J., Török, M. E., Ellington, M. J., Marti-Renom, M. A., et al. (2013). Whole-genome sequencing for rapid susceptibility testing of *M. tuberculosis*. *The New England Journal of Medicine, 369*(3), 290–292.

Köser, C. U., Ellington, M. J., Cartwright, E. J., Gillespie, S. H., Brown, N. M., Farrington, M., et al. (2012). Routine use of microbial whole genome sequencing in diagnostic and public health microbiology. *PLoS Pathogens, 8*(8), e1002824.

Köser, C. U., Ellington, M. J., & Peacock, S. J. (2014). Whole-genome sequencing to control antimicrobial resistance. *Trends in Genetics, 30*(9), 401–407.

Köser, C. U., Fraser, L. J., Ioannou, A., Becq, J., Ellington, M. J., Holden, M. T., et al. (2014). Rapid single-colony whole-genome sequencing of bacterial pathogens. *The Journal of Antimicrobial Chemotherapy, 69*(5), 1275–1281.

Köser, C. U., Holden, M. T., Ellington, M. J., Cartwright, E. J., Brown, N. M., Ogilvy-Stuart, A. L., et al. (2012). Rapid whole-genome sequencing for investigation of a neonatal MRSA outbreak. *The New England Journal of Medicine, 366*(24), 2267–2275.

Lamagni, T. L., Darenberg, J., Luca-Harari, B., Siljander, T., Efstratiou, A., Henriques-Normark, B., et al. (2008). Epidemiology of severe *Streptococcus pyogenes* disease in Europe. *Journal of Clinical Microbiology, 46*, 2359–2367.

Lamagni, T. L., Efstratiou, A., Vuopio-Varkila, J., Jasir, A., Schalén, C., & Strep-EURO. (2005). The epidemiology of severe *Streptococcus pyogenes* associated disease in Europe. *Euro Surveillance, 10*, 179–184.

Lamagni, T. L., Neal, S., Keshishian, C., Alhaddad, N., George, R., Duckworth, G., et al. (2008). Severe *Streptococcus pyogenes* infections, United Kingdom, 2003-2004. *Emerging Infectious Diseases, 14*(2), 202–209.

Landers, T. F., Harris, R. E., Wittum, T. E., & Stevenson, K. B. (2009). Colonization with *Staphylococcus aureus* and methicillin-resistant *S. aureus* among a sample of homeless individuals, Ohio. *Infection Control and Hospital Epidemiology, 30*(8), 801–803.

Lee, B. E., Mukhi, S. N., May-Hadford, J., Plitt, S., Louie, M., & Drews, S. J. (2011). Determination of the relative economic impact of different molecular-based laboratory algorithms for respiratory viral pathogen detection, including Pandemic (H1N1), using a secure web based platform. *Virology Journal, 8*, 277.

Lee, B. Y., Singh, A., David, M. Z., Bartsch, S. M., Slayton, R. B., Huang, S. S., et al. (2013). The economic burden of community-associated methicillin-resistant *Staphylococcus aureus* (CA-MRSA). *Clinical Microbiology and Infection, 19*(6), 528–536.

Lettini, A. A., Saccardin, C., Ramon, E., Longo, A., Cortini, E., Dalla Pozza, M. C., et al. (2014). Characterization of an unusual *Salmonella* phage type DT7a and report of a foodborne outbreak of salmonellosis. *International Journal of Food Microbiology, 189*, 11–17.

Levine, D. P., Fromm, B. S., & Reddy, B. R. (1991). Slow response to vancomycin or vancomycin plus rifampin in methicillin-resistant *Staphylococcus aureus* endocarditis. *Annals of Internal Medicine, 115*(9), 674–680.

Lewis, S. S., Walker, V. J., Lee, M. S., Chen, L., Moehring, R. W., Cox, C. E., et al. (2014). Epidemiology of methicillin-resistant *Staphylococcus aureus* pneumonia in community hospitals. *Infection Control and Hospital Epidemiology, 35*(12), 1452–1457.

Li, V., Chui, L., Louie, L., Simor, A., Golding, G. R., & Louie, M. (2013). Cost-effectiveness and efficacy of spa, SCCmec, and PVL genotyping of methicillin-resistant *Staphylococcus aureus* as compared to pulsed-field gel Electrophoresis. *PLoS One, 8*(11), e79149.

Li, W., Lu, S., Cui, Z., Cui, J., Zhou, H., Wang, Y., et al. (2012). PulseNet China, a model for future laboratory-based bacterial infectious disease surveillance in China. *Frontiers in Medicine, 6*(4), 366–375.

Lin, J. N., Chang, L. L., Lai, C. H., Lin, H. H., & Chen, Y. H. (2011). Clinical and molecular characteristics of invasive and noninvasive skin and soft tissue infections caused by group A *Streptococcus*. *Journal of Clinical Microbiology, 49*(10), 3632–3637.

Lin, J. N., Chang, L. L., Lai, C. H., Lin, H. H., & Chen, Y. H. (2013). Emergence of *Streptococcus pyogenes* emm102 causing toxic shock syndrome in Southern Taiwan during 2005-2012. *PLoS One, 8*(12), e81700.

Liu, F., Barrangou, R., Gerner-Smidt, P., Ribot, E. M., Knabel, S. J., & Dudley, E. G. (2011). Novel virulence gene and clustered regularly interspaced short palindromic repeat (CRISPR) multilocus sequence typing scheme for subtyping of the major serovars of *Salmonella enterica* subsp. *enterica*. *Applied and Environmental Microbiology, 77*(6), 1946–1956.

Luna-Gierke, R. E., Wymore, K., Sadlowski, J., Clogher, P., Gierke, R. W., Tobin-D'Angelo, M., et al. (2014). Multiple-aetiology enteric infections involving non-O157 Shiga toxin-producing *Escherichia coli*—FoodNet, 2001-2010. *Zoonoses and Public Health, 61*(7), 492–498.

Maciel, E. L., Pan, W., Dietze, R., Peres, R. L., Vinhas, S. A., Ribeiro, F. K., et al. (2010). Spatial patterns of pulmonary tuberculosis incidence and their relationship to socio-economic status in Vitoria, Brazil. *The International Journal of Tuberculosis and Lung Disease, 14*(11), 1395–1402.

Maiden, M. C., Bygraves, J. A., Feil, E., Morelli, G., Russell, J. E., Urwin, R., et al. (1998). Multilocus sequence typing: A portable approach to the identification of clones within populations of pathogenic microorganisms. *Proceedings of the National Academy of Sciences of the United States of America, 95*(6), 3140–3145.

McCarthy, T. A., Barrett, N. L., Hadler, J. L., Salsbury, B., Howard, R. T., Dingman, D. W., et al. (2001). Hemolytic-uremic syndrome and *Escherichia coli* O121 at a lake in Connecticut, 1999. *Pediatrics, 108*(4), E59.

McQuiston, J. R., Waters, R. J., Dinsmore, B. A., Mikoleit, M. L., & Fields, P. I. (2011). Molecular determination of H antigens of *Salmonella* by use of a microsphere-based liquid array. *Journal of Clinical Microbiology, 49*(2), 565–573.

Mellmann, A., Harmsen, D., Cummings, C. A., Zentz, E. B., Leopold, S. R., Rico, A., et al. (2011). Prospective genomic characterization of the German enterohemorrhagic *Escherichia coli* O104:H4 outbreak by rapid next generation sequencing technology. *PLoS One*, *6*(7), e22751.

Mera, R. M., Suaya, J. A., Amrine-Madsen, H., Hogea, C. S., Miller, L. A., Lu, E. P., et al. (2011). Increasing role of *Staphylococcus aureus* and community-acquired methicillin-resistant *Staphylococcus aureus* infections in the United States: A 10-year trend of replacement and expansion. *Microbial Drug Resistance*, *17*(2), 321–328.

Méroc, E., Strubbe, M., Vangroenweghe, F., Czaplicki, G., Vermeersch, K., Hooyberghs, J., et al. (2012). Evaluation of the *Salmonella* surveillance program in Belgian pig farms. *Preventive Veterinary Medicine*, *105*(4), 309–314.

Mettee Zarecki, S. L., Bennett, S. D., Hall, J., Yaeger, J., Lujan, K., Adams-Cameron, M., et al. (2013). US outbreak of human *Salmonella* infections associated with aquatic frogs, 2008-2011. *Pediatrics*, *131*(4), 724–731.

Michaud, S., Ménard, S., & Arbeit, R. D. (2005). Role of real-time molecular typing in the surveillance of *Campylobacter enteritis* and comparison of pulsed-field gel electrophoresis profiles from chicken and human isolates. *Journal of Clinical Microbiology*, *43*(3), 1105–1111.

Miyajima, Y., Takahashi, M., Eguchi, H., Honma, M., Tanahashi, S., Matui, Y., et al. (2007). Outbreak of enterohemorrhagic *Escherichia coli* O26 in Niigata city, Japan. *Japanese Journal of Infectious Diseases*, *60*(4), 238–239.

Montes, M., Ardanuy, C., Tamayo, E., Domènech, A., Liñares, J., & Pérez-Trallero, E. (2011). Epidemiological and molecular analysis of *Streptococcus pyogenes* isolates causing invasive disease in Spain (1998-2009): Comparison with non-invasive isolates. *European Journal of Clinical Microbiology & Infectious Diseases*, *30*(10), 1295–1302.

Mukhi, S. N., May-Hadford, J., Plitt, S., Preiksaitis, J., & Lee, B. (2010). DIAL: A platform for real-time laboratory surveillance. *Online Journal of Public Health Informatics, 2*(3). pii: ojphi.v2i3.3041.

National Association of State Public Health Veterinarians, Inc. (NASPHV), & Centers for Disease Control and Prevention (CDC). (2011). Compendium of measures to prevent disease associated with animals in public settings, 2011: National Association of State Public Health Veterinarians, Inc. *MMWR - Recommendations and Reports*, *60*(RR-04), 1–24.

O'Loughlin, R. E., Roberson, A., Cieslak, P. R., Lynfield, R., Gershman, K., Craig, A., et al. (2007). The epidemiology of invasive group A streptococcal infection and potential vaccine implications: United States, 2000-2004. *Clinical Infectious Diseases*, *45*(7), 853–862.

Pabbaraju, K., Wong, S., Lee, B., Tellier, R., Fonseca, K., Louie, M., et al. (2011). Comparison of a singleplex real-time RT-PCR assay and multiplex respiratory viral panel assay for detection of influenza "A" in respiratory specimens. *Influenza and Other Respiratory Viruses*, *5*(2), 99–103.

Pabilonia, K. L., Cadmus, K. J., Lingus, T. M., Bolte, D. S., Russell, M. M., Van Metre, D. C., et al. (2014). Environmental *Salmonella* in agricultural fair poultry exhibits in Colorado. *Zoonoses and Public Health*, *61*(2), 138–144.

Pachkoria, E., Vashakidze, E., Megrelishvili, T., & Tevzadze, L. (2014). Clinical and epidemiological peculiarities of hemorrhagic colitis complicated by hemolytic-uremic syndrome. *Georgian Medical News*, (234), 70–73.

Paine, S., Thornley, C., Wilson, M., Dufour, M., Sexton, K., Miller, J., et al. (2014). An outbreak of multiple serotypes of *Salmonella* in new zealand linked to consumption of contaminated tahini imported from Turkey. *Foodborne Pathogens and Disease*, *11*(11), 887–892.

Painter, J. A., Hoekstra, R. M., Ayers, T., Tauxe, R. V., Braden, C. R., Angulo, F. J., et al. (2013). Attribution of foodborne illnesses, hospitalizations, and deaths to food commodities by using outbreak data, United States, 1998-2008. *Emerging Infectious Diseases*, *19*(3), 407–415.

Park, S. H., & Ricke, S. C. (2014). Development of multiplex PCR assay for simultaneous detection of *Salmonella* genus, *Salmonella* subspecies I, Salm. Enteritidis, Salm. Heidelberg and Salm. Typhimurium. *Journal of Applied Microbiology*, *118*(1), 152–160. http://dx.doi. org/10.1111/jam.12678. [Epub ahead of print] PMID: 25358641.

Parsons, M. B., Cooper, K. L., Kubota, K. A., Puhr, N., Simington, S., Calimlim, P. S., et al. (2007). PulseNet USA standardized pulsed-field gel electrophoresis protocol for subtyping of *Vibrio parahaemolyticus*. *Foodborne Pathogens and Disease*, *4*(3), 285–292.

Paton, J. C., & Paton, A. W. (1998). Pathogenesis and diagnosis of Shiga toxin-producing Escherichia coli infections. *Clinical Microbiology Reviews*, *11*(3), 450–479.

Pavel, A. B., & Vasile, C. I. (2012). PyElph—A software tool for gel images analysis and phylogenetics. *BMC Bioinformatics*, *13*, 9.

Perch, M., Braden, C. R., Bishop, R., Fields, P., Plikaytis, R., & Tauxe, R. V. (2003). *Salmonella surveillance summary, 2003*. Atlanta, GA: U.S. Department of Health and Human Services, Centers for Disease Control and Prevention.

Pérez, L., Apezteguía, L., Piñeyrúa, C., Dabezies, A., Bianco, M. N., Schelotto, F., et al. (2014). Hemolytic uremic syndrome with mild renal involvement due to Shiga toxin-producing *Escherichia coli* (STEC) O145 strain. *Revista Argentina de Microbiología*, *46*(2), 103–106.

Peterson, G., Gerdes, B., Berges, J., Nagaraja, T. G., Frye, J. G., Boyle, D. S., et al. (2010). Development of microarray and multiplex polymerase chain reaction assays for identification of serovars and virulence genes in *Salmonella enterica* of human or animal origin. *Journal of Veterinary Diagnostic Investigation*, *22*(4), 559–569.

Pitout, J. D., Reisbig, M. D., Mulvey, M., Chui, L., Louie, M., Crowe, L., et al. (2003). Association between handling of pet treats and infection with *Salmonella enterica* serotype newport expressing the AmpC beta-lactamase, CMY-2. *Journal of Clinical Microbiology*, *41*(10), 4578–4582.

Popoff, M. Y., Bockemühl, J., & Gheesling, L. L. (2004). Supplement 2002 (no. 46) to the Kauffmann-White scheme. *Research in Microbiology*, *155*(7), 568–570.

Popovich, K. J., Weinstein, R. A., & Hota, B. (2008). Are community-associated methicillin-resistant *Staphylococcus aureus* (MRSA) strains replacing traditional nosocomial MRSA strains? *Clinical Infectious Diseases*, *46*(6), 787–794.

Raguenaud, M. E., Le Hello, S., Salah, S., Weill, F. X., Brisabois, A., Delmas, G., et al. (2012). Epidemiological and microbiological investigation of a large outbreak of monophasic *Salmonella* Typhimurium 4,5,12:i:- in schools associated with imported beef in Poitiers, France, October 2010. *Euro Surveillance*, *17*(40), 20289.

Redziniak, D. E., Diduch, D. R., Turman, K., Hart, J., Grindstaff, T. L., MacKnight, J. M., et al. (2009). Methicillin-resistant *Staphylococcus aureus* (MRSA) in the athlete. *International Journal of Sports Medicine*, *30*(8), 557–562.

Reimer, A. R., Van Domselaar, G., Stroika, S., Walker, M., Kent, H., Tarr, C., et al. (2011). Comparative genomics of *Vibrio cholerae* from Haiti, Asia, and Africa. *Emerging Infectious Diseases*, *17*(11), 2113–2121.

Reuter, S., Ellington, M. J., Cartwright, E. J., Köser, C. U., Török, M. E., Gouliouris, T., et al. (2013). Rapid bacterial whole-genome sequencing to enhance diagnostic and public health microbiology. *JAMA Internal Medicine*, *173*(15), 1397–1404.

Ribot, E. M., Fair, M. A., Gautom, R., Cameron, D. N., Hunter, S. B., Swaminathan, B., et al. (2006). Standardization of pulsed-field gel electrophoresis protocols for the subtyping of *Escherichia coli* O157:H7, *Salmonella*, and *Shigella* for PulseNet. *Foodborne Pathogens and Disease*, *3*(1), 59–67.

Sabat, A. J., Chlebowicz, M. A., Grundmann, H., Arends, J. P., Kampinga, G., Meessen, N. E., et al. (2012). Microfluidic-chip-based multiple-locus variable-number tandem-repeat fingerprinting with new primer sets for methicillin-resistant *Staphylococcus aureus*. *Journal of Clinical Microbiology*, *50*(7), 2255–2262.

Saleh-Lakha, S., Allen, V. G., Li, J., Pagotto, F., Odumeru, J., Taboada, E., et al. (2013). Subtyping of a large collection of historical *Listeria monocytogenes* strains from Ontario, Canada, by an improved multilocus variable-number tandem-repeat analysis (MLVA). *Applied and Environmental Microbiology*, *79*(20), 6472–6480.

Schimmer, B., Ter Schegget, R., Wegdam, M., Züchner, L., de Bruin, A., Schneeberger, P. M., et al. (2010). The use of a geographic information system to identify a dairy goat farm as the most likely source of an urban Q-fever outbreak. *BMC Infectious Diseases*, *10*, 69.

Schwartz, D. C., & Cantor, C. R. (1984). Separation of yeast chromosome-sized DNAs by pulsed field gradient gel electrophoresis. *Cell*, *37*(1), 67–75.

Schwartz, D. C., & Koval, M. (1989). Conformational dynamics of individual DNA molecules during gel electrophoresis. *Nature*, *338*(6215), 520–522.

Schwartz, D. C., Saffran, W., Welsh, J., Haas, R., Goldenberg, M., & Cantor, C. R. (1983). New techniques for purifying large DNAs and studying their properties and packaging. *Cold Spring Harbor Symposia on Quantitative Biology*, *47*(Pt. 1), 189–195.

Seeleang, K., Manning, M. L., Saks, M., & Winstead, Y. (2014). Skin and soft tissue infection management, outcomes, and follow-up in the emergency department of an urban academic hospital. *Advanced Emergency Nursing Journal*, *36*(4), 348–359.

Seemann, T. (2014). Prokka: Rapid prokaryotic genome annotation. *Bioinformatics*, *30*(14), 2068–2069.

Semenza, J. C., Suk, J. E., Estevez, V., Ebi, K. L., & Lindgren, E. (2012). Mapping climate change vulnerabilities to infectious diseases in Europe. *Environmental Health Perspectives*, *120*(3), 385–392.

Shi, C., Singh, P., Ranieri, M. L., Wiedmann, M., & Moreno Switt, A. I. (2015). Molecular methods for serovar determination of Salmonella. *Critical Reviews in Microbiology*, *41*(3), 309–325. PMID: 24228625.

Shopsin, B., Gomez, M., Montgomery, S. O., Smith, D. H., Waddington, M., Dodge, D. E., et al. (1999). Evaluation of protein A gene polymorphic region DNA sequencing for typing of *Staphylococcus aureus* strains. *Journal of Clinical Microbiology*, *37*(11), 3556–3563.

Smith, A. M., Mthanti, M. A., Haumann, C., Tyalisi, N., Boon, G. P., Sooka, A., et al. (2014). Nosocomial outbreak of *Salmonella enterica* serovar Typhimurium primarily affecting a pediatric ward in South Africa in 2012. *Journal of Clinical Microbiology*, *52*(2), 627–631.

Snary, E. L., Munday, D. K., Arnold, M. E., & Cook, A. J. (2010). Zoonoses action plan *Salmonella* monitoring programme: An investigation of the sampling protocol. *Journal of Food Protection*, *73*(3), 488–494.

Supply, P., Allix, C., Lesjean, S., Cardoso-Oelemann, M., Rüsch-Gerdes, S., Willery, E., et al. (2006). Proposal for standardization of optimized mycobacterial interspersed repetitive unit-variable number tandem repeat typing of *Mycobacterium tuberculosis*. *Journal of Clinical Microbiology*, *44*, 4498–4510.

Supply, P., Lesjean, S., Savine, E., Kremer, K., van Soolingen, D., & Locht, C. (2001). Automated high-throughput genotyping for study of global epidemiology of *Mycobacterium tuberculosis* based on mycobacterial interspersed repetitive units. *Journal of Clinical Microbiology*, *39*(10), 3563–3571.

Supply, P., Magdalena, J., Himpens, S., & Locht, C. (1997). Identification of novel intergenic repetitive units in a mycobacterial two-component system operon. *Molecular Microbiology*, *26*(5), 991–1003.

Swain, R. A., Hatcher, J. C., Azadian, B. S., Soni, N., & De Souza, B. (2013). A five-year review of necrotising fasciitis in a tertiary referral unit. *Annals of the Royal College of Surgeons of England*, *95*(1), 57–60.

Swaminathan, B., Gerner-Smidt, P., Ng, L. K., Lukinmaa, S., Kam, K. M., Rolando, S., et al. (2006). Building PulseNet international: An interconnected system of laboratory networks to facilitate timely public health recognition and response to foodborne disease outbreaks and emerging foodborne diseases. *Foodborne Pathogens and Disease*, *3*(1), 36–50.

Sylvester, W. R., Amadi, V., Pinckney, R., Macpherson, C. N., McKibben, J. S., Bruhl-Day, R., et al. (2014). Prevalence, serovars and antimicrobial susceptibility of *Salmonella* spp. from wild and domestic green iguanas (Iguana iguana) in Grenada, West Indies. *Zoonoses and Public Health*, *61*(6), 436–441.

Taboada, E. N., Ross, S. L., Mutschall, S. K., Mackinnon, J. M., Roberts, M. J., Buchanan, C. J., et al. (2012). Development and validation of a comparative genomic fingerprinting method for high-resolution genotyping of *Campylobacter jejuni*. *Journal of Clinical Microbiology*, *50*(3), 788–797.

Tankouo-Sandjong, B., Sessitsch, A., Stralis-Pavese, N., Liebana, E., Kornschober, C., Allerberger, F., et al. (2008). Development of an oligonucleotide microarray method for *Salmonella* serotyping. *Microbial Biotechnology*, *1*(6), 513–522.

Tanner, J., Lin, Y., Kornblum, J., Herzig, C. T., Bystritsky, R., Uhlemann, A. C., et al. (2014). Molecular characterization of methicillin-resistant *Staphylococcus aureus* clinical isolates obtained from the Rikers Island Jail System from 2009 to 2013. *Journal of Clinical Microbiology*, *52*(8), 3091–3094.

Teh, C. S., Chua, K. H., & Thong, K. L. (2011). Genetic variation analysis of *Vibrio cholerae* using multilocus sequencing typing and multi-virulence locus sequencing typing. *Infection, Genetics and Evolution*, *11*(5), 1121–1128.

Terajima, J., Izumiya, H., Iyoda, S., Mitobe, J., Miura, M., & Watanabe, H. (2006). Effectiveness of pulsed-field gel electrophoresis for the early detection of diffuse outbreaks due to Shiga toxin-producing *Escherichia coli* in Japan. *Foodborne Pathogens and Disease*, *3*(1), 68–73.

Thompson, J. D., Gibson, T. J., & Higgins, D. G. (2002). Multiple sequence alignment using ClustalW and ClustalX. *Current Protocols in Bioinformatics*. Chapter 2, Unit 2.

Timbo, B. B., Keys, C., & Klontz, K. (2010). Characterization of *Listeria monocytogenes* recovered from imported cheese contributed to the National PulseNet Database by the U.S. Food and Drug Administration from 2001 to 2008. *Journal of Food Protection*, *73*(8), 1511–1514.

Tonn, K., & Ryan, T. J. (2013). Community-associated methicillin-resistant *Staphylococcus aureus* in college residential halls. *Journal of Environmental Health*, *75*(6), 44–49.

Török, M. E., Reuter, S., Bryant, J., Köser, C. U., Stinchcombe, S. V., Nazareth, B., et al. (2013). Rapid whole-genome sequencing for investigation of a suspected tuberculosis outbreak. *Journal of Clinical Microbiology*, *51*(2), 611–614.

Uhlmann, S., Galanis, E., Takaro, T., Mak, S., Gustafson, L., Embree, G., et al. (2009). Where's the pump? Associating sporadic enteric disease with drinking water using a geographic information system, in British Columbia, Canada, 1996-2005. *Journal of Water and Health*, *7*(4), 692–698.

van Belkum, A. (1999). Short sequence repeats in microbial pathogenesis and evolution. *Cellular and Molecular Life Sciences*, *56*(9-10), 729–734.

van Belkum, A., Sluijuter, M., de Groot, R., Verbrugh, H., & Hermans, P. W. (1996). Novel BOX repeat PCR assay for high-resolution typing of *Streptococcus pneumoniae* strains. *Journal of Clinical Microbiology*, *34*(5), 1176–1179.

Varga, C., Pearl, D. L., McEwen, S. A., Sargeant, J. M., Pollari, F., & Guerin, M. T. (2013). Incidence, distribution, seasonality, and demographic risk factors of *Salmonella* Enteritidis human infections in Ontario, Canada, 2007-2009. *BMC Infectious Diseases*, *13*, 212.

Verghese, B., Schwalm, N. D., 3rd., Dudley, E. G., & Knabel, S. J. (2012). A combined multi-virulence-locus sequence typing and Staphylococcal Cassette Chromosome mec typing scheme possesses enhanced discriminatory power for genotyping MRSA. *Infection, Genetics and Evolution*, *12*(8), 1816–1821.

Versalovic, J., Koeuth, T., & Lupski, J. R. (1991). Distribution of repetitive DNA sequences in eubacteria and application to fingerprinting of bacterial genomes. *Nucleic Acids Research*, *19*(24), 6823–6831.

Wang, X., Taylor, M., Hoang, L., Ekkert, J., Nowakowski, C., Stone, J., et al. (2013). Comparison of clinical and epidemiological features of Shiga toxin-producing *Escherichia coli* O157 and non-O157 infections in British Columbia, 2009 to 2011. *The Canadian Journal of Infectious Diseases & Medical Microbiology*, *24*(4), e102–e106.

Wattiau, P., Van Hessche, M., Schlicker, C., Vander Veken, H., & Imberechts, H. (2008). Comparison of classical serotyping and PremiTest assay for routine identification of common *Salmonella enterica* serovars. *Journal of Clinical Microbiology*, *46*(12), 4037–4040.

Wattiau, P., Weijers, T., Andreoli, P., Schliker, C., Veken, H. V., Maas, H. M., et al. (2008). Evaluation of the Premi Test *Salmonella*, a commercial low-density DNA microarray system intended for routine identification and typing of *Salmonella* enterica. *International Journal of Food Microbiology*, *123*(3), 293–298.

Weniger, T., Krawczyk, J., Supply, P., Niemann, S., & Harmsen, D. (2010). MIRU-VNTRplus: A web tool for polyphasic genotyping of *Mycobacterium tuberculosis* complex bacteria. *Nucleic Acids Research*, *38*(Suppl.), W326–W331.

Wilson, J., Guy, R., Elgohari, S., Sheridan, E., Davies, J., Lamagni, T., et al. (2011). Trends in sources of meticillin-resistant *Staphylococcus aureus* (MRSA) bacteraemia: Data from the national mandatory surveillance of MRSA bacteraemia in England, 2006-2009. *The Journal of Hospital Infection*, *79*(3), 211–217.

Yoshida, C., Lingohr, E. J., Trognitz, F., MacLaren, N., Rosano, A., Murphy, S. A., et al. (2014). Multi-laboratory evaluation of the rapid genoserotyping array (SGSA) for the identification of *Salmonella* serovars. *Diagnostic Microbiology and Infectious Disease*, *80*(3), 185–190. http://dx.doi.org/10.1016/j.diagmicrobio.2014.08.006. pii: S0732-8893 (14)00324-1, PMID: 25219780.

Young, I., Rajić, A., Letellier, A., Cox, B., Leslie, M., Sanei, B., et al. (2010). Knowledge and attitudes toward food safety and use of good production practices among Canadian broiler chicken producers. *Journal of Food Protection*, *73*(7), 1278–1287.

Zhang, W., Jayarao, B. M., & Knabel, S. J. (2004). Multi-virulence-locus sequence typing of *Listeria monocytogenes*. *Applied and Environmental Microbiology*, *70*(2), 913–920. Erratum in: Applied and Environmental Microbiology (2013) *79*(9), 3146.

Molecular Strain Typing and Characterisation of Toxigenic *Clostridium difficile*

9

Tanis C. Dingle*, Duncan R. MacCannell[†],[1]

Icahn School of Medicine at Mount Sinai, New York, USA
†Centers for Disease Control and Prevention (CDC), Atlanta, Georgia, USA
[1]Corresponding author: e-mail address: dmaccannell@cdc.gov

1 INTRODUCTION

Although toxigenic *Clostridium difficile* was first isolated from the stool of asymptomatic infants and young children nearly 80 years ago, the pathogenic importance of this spore-forming, Gram-positive anaerobic rod and its association with antibiotic-associated diarrhoea and colitis were not well understood until the late 1970s (Bartlett, Chang, Gurwith, Gorbach, & Onderdonk, 1978; George, Sutter, Goldstein, Ludwig, & Finegold, 1978; Larson, Price, Honour, & Borriello, 1978). *C. difficile* colonises hosts in which the normal intestinal microbiota has been disturbed, most commonly by broad-spectrum antibiotics. Two large exotoxins, Toxin A (TcdA) and Toxin B (TcdB), are released upon intestinal colonisation and are largely responsible for the watery diarrhoea seen in patients with *C. difficile* infection (CDI) (Poutanen & Simor, 2004; Voth & Ballard, 2005). In some, this can result in pseudomembranous colitis and death. Since the early 2000s, major outbreaks of CDI have occurred worldwide and have been attributed to the emergence of a hyper-virulent strain of *C. difficile* called the restriction endonuclease analysis (REA) type B1/North American pulsed field (NAP) type 1/polymerase chain reaction (PCR) ribotype 027 (B1/NAP1/027) strain (Loo et al., 2005; McDonald et al., 2005; Pepin, Valiquette, & Cossette, 2005). This strain produces a third toxin, known as the *C. difficile* transferase or binary toxin, that is thought to contribute at least in part to its increased virulence, however other factors, both known and unknown, may also play a role (Gerding, Johnson, Rupnik, & Aktories, 2014; Lanis, Heinlen, James, & Ballard, 2013; MacCannell et al., 2006; Warny et al., 2005). The B1/NAP1/027 strain, along with other emerging strain types, is associated with increased severity, increased treatment failures and can occur in patients without the traditional risk factors for CDI (O'Connor, Johnson, & Gerding, 2009).

With this shift in incidence of CDI at the turn of the century, epidemiological strain typing has been critical to our understanding of the pathogenesis and spread of disease. *C. difficile* strain typing has been applied to a number of outbreak investigations (Hardy et al., 2012; Rafferty et al., 1998; Samore et al., 1997), to track transmission patterns (Marsh et al., 2006; Walters, Stafford, Roberts, & Seneviratne, 1982) and to monitor the emergence of new, more virulent strains (Loo et al., 2005; McDonald et al., 2005). The first methods to type *C. difficile* strains were developed in the 1980s and involved phenotypic characterisation by evaluating antibiotic resistance patterns, serotyping by slide agglutination and immunoblotting (Burdon, 1982; Delmee, Homel, & Wauters, 1985; Mulligan, Peterson, Kwok, Clabots, & Gerding, 1988; Nakamura et al., 1981). These methods suffered from low reproducibility and the inability to discriminate between large numbers of strains. In recent decades, epidemiologic studies of *C. difficile*, and indeed most infectious pathogens, have shifted to predominantly genotypic methods to characterise each isolate and assign strain type lineage. These methods, introduced in the 1990s, overcame many of the obstacles associated with phenotypic typing (Cohen, Tang, & Silva, 2001). A number of genotypic methods have been used routinely for *C. difficile* strain typing including pulsed field gel electrophoresis (PFGE), restriction endonuclease analysis (REA), toxinotyping and PCR ribotyping. In the last decade, newer methods such as multilocus variable-number repeat analysis (MLVA) and whole genome sequencing (WGS) have been introduced that offer higher discriminatory power than traditional methods.

Unfortunately, culture-independent enzyme immunoassay and PCR-based diagnostics have largely replaced routine culture for *C. difficile* in most clinical microbiology laboratories. Few facilities are capable of performing advanced molecular characterisation or strain typing to support surveillance or outbreak response. Even among laboratories that strain type *C. difficile* routinely, the historical lack of common protocols, standards or consistent strain typing nomenclature has been an important limitation to the meaningful interchange of strain type data and, perhaps more importantly, to our understanding of the molecular epidemiology of the organism. In the sections that follow, several of the most common molecular approaches for bacterial typing will be discussed in detail, with particular emphasis on their applications and limitations in past and present studies of *C. difficile*. Additionally, a review of recent epidemiological studies and standardisation efforts will be discussed.

2 METHODS FOR MOLECULAR STRAIN TYPING
2.1 RESTRICTION FRAGMENT METHODS
2.1.1 *Restriction endonuclease analysis*
REA is one of the first described and most straightforward genotypic strain typing methods. It uses an enzyme, usually *Hind*III, which cuts the genome sequence frequently leading to a large number of bands that can be resolved by conventional gel electrophoresis (Kuijper, Oudbier, Stuifbergen, Jansz, & Zanen, 1987; Wren & Tabaqchali, 1987). Its use was first described in several *C. difficile* outbreak studies in 1987. Kuijper et al. used REA to show that two patients, who developed

pseudomembranous colitis while hospitalised in the same room, were infected with the same *C. difficile* strain. Additionally, *C. difficile* isolates from the environment surrounding these patients were shown to have identical restriction patterns, suggesting an environmental source of infection (Kuijper et al., 1987). A second study performed around the same period used REA to show that 10 isolates of *C. difficile* isolated during a hospital outbreak were indistinguishable and that the profiles obtained were strain-specific (Wren & Tabaqchali, 1987). Peerbooms and colleagues collected *C. difficile* isolates over a 4-month period and subjected them to REA (Peerbooms, Kuijt, & Maclaren, 1987). They were able to distinguish 22 different restriction types among 38 strains, with the most common being assigned to REA types A–E. These patterns correlated well to antimicrobial susceptibility patterns, though they were less discriminatory than REA typing. In 1993, Clabots et al. were the first to describe the REA method in depth and to catalogue 1965 *C. difficile* isolates into 206 unique REA types (Clabots et al., 1993). This seminal work was the basis for currently used REA typing nomenclature.

REA remains an important epidemiologic tool in the strain typing of *C. difficile* and is used extensively by the Gerding laboratory at Hines, VA. As of 2003, their collection of *C. difficile* isolates, collected longitudinally over more than two decades numbered more than 5000, comprising at least 436 different REA types across 108 REA strain type groupings. The availability of extensive and historical REA strain typing data has been pivotal in our understanding of a number of large-scale epidemiologic studies of *C. difficile*, including the recent outbreaks of NAP1/BI strains in Eastern Canada and the United States (Killgore et al., 2008; McDonald et al., 2005). While REA has been used extensively in epidemiological studies of *C. difficile* and has consistently demonstrated both high discriminatory power and reproducibility, interpretation of the band patterns is difficult, the protocol itself is technically demanding and the portability of resultant data between laboratories is problematic. Because of these challenges, few laboratories report the routine use of REA for *C. difficile* strain typing, although reference typing is available through consultation with an REA expert laboratory.

2.1.2 Pulsed field gel electrophoresis

PFGE is the most commonly used method for bacterial strain typing in support of outbreak investigations, including *C. difficile* typing, and is the gold standard for strain type comparisons in public health laboratories across North America and throughout the world. In this method, an infrequently cutting enzyme, such as *Sma*I, is used to digest genomic DNA that has been purified and suspended intact in an agarose matrix. The large DNA fragments (up to 10 Mb) are often too large to be reliably resolved by conventional gel electrophoresis and are separated by an electric field that repeatedly switches directions over the course of a 15–24 h run (Herschleb, Ananiev, & Schwartz, 2007). PFGE for *C. difficile* typically uses contour-clamped homogenous electric field PFGE, which includes a 120° angular separation, with switch times of 5–40 s, and an overall runtime of 18–22 h (REF). Banding patterns observed by PFGE are assigned to NAP types based on >80% pattern similarity (by the Dice/unweighted pair group arithmetic mean method) (Killgore et al., 2008). There are currently 12 NAP types (NAP1–NAP12) that have been named by the CDC in conjunction with the Public Health Agency of Canada. Since 2009, the Emerging Infections Program (EIP) at the CDC has been characterising both hospital and community-acquired isolates of *C. difficile* from 10 sites in the United States using PFGE (http://www.cdc.gov/hai/eip/Clostridium-difficile.html), which has been further understanding of the epidemiology of disease.

Due to the high resolving power of PFGE, it lends itself to being highly discriminative, while also being very reproducible. It has become central to modern studies of molecular epidemiology and bacterial ecology. Unfortunately, like a number of other recalcitrant bacteria, clinical isolates of *C. difficile* are often non-typeable by standard PFGE protocols due to the extensive and rapid degradation of genomic DNA during sample preparation (Alonso et al., 2005). A number of new and improved protocols have been suggested over the years to overcome this problem, including the use of thiourea in the gel and running buffers, the use of freshly sub-cultured *C. difficile* isolates, increasing lysozyme and detergent concentrations, and adjusting time or temperature conditions for lysis (Alonso et al., 2005; Corkill, Graham, Hart, & Stubbs, 2000; Gal, Northey, & Brazier, 2005). Even under the best conditions, PFGE is laborious, time consuming and requires the use of expensive instrumentation, making it a technology mostly reserved for public health laboratories, such as the CDC. Additionally, despite efforts to standardise protocols for interpretation of PFGE data (Tenover et al., 1995), comparing normalised banding patterns between different laboratories has proven challenging. Given time and resources, however, PFGE still remains a valuable epidemiologic tool for outbreak investigation.

2.1.3 Toxinotyping

Toxigenic strains of *C. difficile* typically encode two principal large glucosylating clostridial toxins, TcdA and TcdB, and carry the corresponding *tcdA* and *tcdB* genes on a 19 kb chromosomal pathogenicity locus (PaLoc). In addition to *tcdA* and *tcdB*, the PaLoc encodes additional genes (*tcdC, tcdR, tcdE*) that are involved in regulating toxin expression, in addition to a putative holin (*tcdE*), which may also have a role in pathogenesis. Most toxigenic strains of *C. difficile* elaborate both TcdA and TcdB (A+B+) toxins, however, some strains have been found to produce only TcdB (A−B+). To date, there have not been verified reports of infections from TcdA+ (A+B−) *C. difficile* strains, and as a result, many molecular diagnostic platforms are currently based on the detection of *tcdB* and other common markers, such as glutamate dehydrogenase. In non-toxigenic strains of *C. difficile*, the PaLoc, a short 115 bp filler sequence occupies the PaLoc site (Braun, Hundsberger, Leukel, Sauerborn, & von Eichel-Streiber, 1996; Cohen, Tang, & Silva, 2000).

It has been noted for some time that *C. difficile* strains can be differentiated or toxinotyped based on the variance (deletions, insertions or point mutations) seen in the PaLoc (Rupnik, 2010), and that these toxinotypes correlate to important molecular epidemiologic characteristics. The currently used toxinotyping method was described in 1998 and is a restriction fragment length polymorphism (RFLP)-PCR-based method (Rupnik, Avesani, Janc, von Eichel-Streiber, & Delmee, 1998). A variable region in *tcdA* (A1) and *tcdB* (B3) is amplified by PCR and then digested using *Eco*RI or *Hinc*II/*Acc*I, respectively (Rupnik, 2008, 2010). The restriction patterns obtained determine the toxinotype, and those strains in the same toxinotype are understood to have similar configurations of the PaLoc cassette (Rupnik et al., 1998). The first study to toxinotype *C. difficile* strains in this manner categorised 219 toxigenic isolates into 10 different toxinotype groups (Rupnik et al., 1998). There are currently 32 known *C. difficile* toxinotypes, I–XXXII, along with toxinotype 0 from the reference strain, VPI 10463. Their genotypic and phenotypic characteristics (including toxin production and cytopathic effect observed on tissue culture cells) of toxinotypes I–XXXI, along with publications describing most toxinotypes in detail can be found in a curated resource maintained by Rupnik and colleagues (www.mf.uni-mb.si/mikro/tox). Toxinotype XXXII was recently described and is lacking the complete TcdA gene and has an unusual organisation of the PaLoc integration site (Janezic, Marin, Martin, & Rupnik, 2015).

Toxinotyping is often considered a relatively low-resolution approach to molecular strain typing, and suffers from many of the same limitations as other fragment-based molecular strain typing methods, namely: subjective interpretation of results, limited availability of consensus standards, intensive time and resource demands and limited throughput. Nevertheless, in an experienced and well-equipped laboratory it is highly reproducible and offers significant and unique advantages that have made it an invaluable asset to many epidemiologic studies of *C. difficile*. Because toxin production is central to the pathogenic process of *C. difficile*, differences in PaLoc structure provide important insight into the epidemiology of circulating strains and may provide important information on the pathogenic potential of different strain types or isolates according to the composition of their toxigenic machinery.

2.2 PCR AMPLIFICATION METHODS

2.2.1 PCR ribotyping

PCR ribotyping or ribospacer PCR is the most widely used method for typing *C. difficile* in Europe. This method uses primers directed at conserved regions of the 16S rRNA and 23S rRNA ribosomal genes to amplify the intergenic spacer (ITS) region between these sequences. Because different strains of *C. difficile* may carry different complements and configurations of the rRNA (*rrn*) operon, amplification of the ITS region results in a highly discriminatory and reproducible banding pattern that may be used for surveillance purposes and outbreak investigations alike. The original primer set developed for PCR ribotyping resulted in fragments too large to differentiate by agarose gel electrophoresis (Gurtler, 1993). O'Neill et al. resolved this problem by designing primers closer to the ITS region. Unfortunately, the 23S primer used by the group was designed off of sequence information from *C. botulinum* rather than *C. difficile* (O'Neill, Ogunsola, Brazier, & Duerden, 1996). One of the first optimised PCR-Ribotyping methods for strain typing of

C. difficile was described by Stubbs, Brazier, O'Neill, and Duerden (1999), with subsequent refinements to primer design by Bidet, Barbut, Lalande, Burghoffer, and Petitt (1999). While all three primer sets are understood to have similar performance characteristics, the Bidet primers result in smaller overall fragment sizes and may have some advantages for accurate and reproducible sizing.

Separation of PCR-Ribotyping fragments originally used agarose gel electrophoresis. This method has a relatively low cost and complexity of implementation, and offers high sensitivity, specificity and single-laboratory reproducibility relative to other amplicon-based techniques. However, the band patterns that it generates are often difficult to interpret consistently between individual laboratories, and it is often easier to exchange isolates directly than compare pattern data between research groups. Another important limitation is lack of a standardised database for comparison. In an attempt to standardise data interpretation and methods, Indra et al. incorporated high-resolution, capillary gel electrophoresis (CGE)-based fragment analysis into the standard PCR-Ribotyping method (Indra, Huhulescu, et al., 2008). CGE PCR ribotyping uses the same primers as agarose gel-based PCR ribotyping except that the 16S rRNA primer is labelled with a fluorescent tag at the 5′ end. Amplicon/fragment size is then determined by software, allowing data to be compared easily between laboratories (Indra, Huhulescu, et al., 2008). CGE-based PCR ribotyping resolved many of the issues related to pattern interpretation compared to traditional PCR-Ribotyping methods, and introduced the notion of a centralised strain typing data repository.

A centralised, web-based, PCR-Ribotyping repository, known as WEBRIBO (webribo.ages.at) analyses CGE-based PCR-Ribotyping data in an attempt to standardise data interpretation. While this effort had the support of the ESCMID Study Group for *C. difficile* (ESGCD), the adoption and effectiveness of WEBRIBO was diminished significantly by the lack of protocol standardisation among system users and limited support for unified data quality standards or process controls. In practice, users of the WEBRIBO system reported significant variation in fragment sizes due to differences in sequencer make and model, differences in capillary/polymer configuration and the selection of reference molecular size standard. Overall, the platform was limited in its ability to accurately categorise and identify submitted isolates, even of well-known strain types, and the lack of a standardised protocol and framework for quality control were clearly required. To address this need, an international collaboration of prominent *C. difficile* research and public health laboratories recently published a standardised PCR-Ribotyping protocol for CGE, which includes validated laboratory procedures, reference quality controls and interpretive criteria (Fawley et al., 2015). When these factors were controlled, the standardised protocol resulted in highly reproducible ribotype patterns between the four participating centres, with an average fragment size discrepancy of less than 4 bp, and an overall accuracy of 97.7% in assigning a set of 50 blinded isolates to the correct ribotype.

Another approach to standardisation was described in a recent innovative report by Westblade et al. which described the use of a standard Agilent bioanalyzer for the separation, detection and documentation of PCR-Ribotype fragment patterns

(Westblade et al., 2013). While this approach has not been broadly explored in the literature, it is nonetheless a relatively turnkey option to resolve PCR-Ribotype patterns on a hardware platform that is relatively inexpensive, accessible to most molecular microbiology laboratories and is both robust and deployable across a range of environments and settings.

2.2.2 Repetitive extragenic palindromic PCR

Repetitive extragenic palindromic PCR (REP-PCR) is a molecular fingerprinting technique that uses heterogeneous PCR primers that amplify outwards from short, repetitive, palindromic (i.e. REP) regions that are interspersed in non-coding regions of most bacterial genomes. PCR amplification using REP consensus sequences was first proposed as a discriminatory epidemiologic tool by Versalovic, Koeuth, and Lupskii (1991). The band sizes produced vary in accordance with the spacing and number of REP sites throughout the bacterial genome. Amplicons of different sizes are then separated by gel electrophoresis, with resulting banding patterns being specific for individual bacterial clones. This method has been used successfully for typing a number of Gram-negative bacteria (Jersek, Tcherneva, Rijpens, & Herman, 1996; Rodriguez-Barradas et al., 1995; Stumpf, Roggenkamp, & Hoffmann, 2005; Woods, Versalovic, Koeuth, & Lupski, 1992).

Several methods to perform REP-PCR in *C. difficile* have been proposed over the years. Spigaglia and Mastrantonio showed that this method is highly reproducible and more highly discriminatory than PCR ribotyping (Spigaglia & Mastrantonio, 2003). In 2005, Rahmati and colleagues demonstrated that REP-PCR could consistently distinguish seven subtypes of *C. difficile* ribotype 001 among 50 isolates from across the UK (Rahmati, Gal, Northey, & Brazier, 2005). A subsequent article by the same group described improvements to this method, with eight subtypes demonstrated among 200 clinical isolates, and good correlation to PFGE. The authors concluded that this improved method was highly reproducible, less laborious and less time consuming than PFGE (Northey, Gal, Rahmati, & Brazier, 2005). Recently, two studies evaluated a commercially available, more rapid, REP-PCR system (DiversiLab, bioMérieux) for *Clostridium* species. In the first study, 181 clinical isolates representing a diverse set of PCR ribotypes were subjected to PFGE, PCR ribotyping and the DiversiLab system (Pasanen et al., 2011). They found that the REP-PCR and PCR-Ribotyping clusters were comparable in 75% of cases. Additionally, all of the ribotype 001 and 027 clustered into their own REP-types (Pasanen et al., 2011). In the second study from Canada, the REP-PCR system was also found to be highly reproducible, however, it did not resolve a selection of reference and clinical strains of *C. difficile* as well as PCR ribotyping (Church, Chow, Lloyd, & Gregson, 2011). One specific concern in this study was the failure of the system to discriminate between ribotype 001 and 027 strains. The authors suggest that the discrepancy observed between the two studies to differentiate ribotypes 001 and 027 might be due to the geographic diversity of the strains and the difference in PCR-Ribotyping methods used (Church et al., 2011).

It would seem that concurrent typing with both PCR ribotyping and the more rapid REP-PCR would offer significant improvements in the granularity of strain type resolution over PCR ribotyping alone, and this indeed seems to be the case among these initial studies described above. However, consideration of added costs and standardisation of both lab developed and commercial REP-PCR methods is still necessary.

2.2.3 Multilocus variable-number tandem repeat analysis

MLVA is a highly discriminatory method for typing bacteria, including *C. difficile*. MLVA exploits and targets short sequences of repetitive DNA in the genome, also known as tandem repeats. These areas arise from slipped strand mispairing during DNA replication and differ between isolates of the same bacterial species (van Belkum, 2007). Therefore, the number and size of tandem repeats at multiple loci in the genome can provide information about the genetic composition of the organism which results in a powerful typing tool (van Belkum, 2007). To perform MLVA, PCR primers are designed to amplify variable-number tandem repeat (VNTR) areas of the genome. Gel electrophoresis and sequencing then determine the size of the resulting products and number of repeats per loci. The Manhattan coefficient is then used to determine the summed tandem repeat difference, with ≤2 being genetically related (Eckert et al., 2011; Marsh et al., 2006; van den Berg, Schaap, Templeton, Klaassen, & Kuijper, 2007).

The two MLVA methods for *C. difficile* described in the literature amplify seven loci containing short tandem repeats spread over the *C. difficile* strain 630 genome (Marsh et al., 2006; van den Berg et al., 2007). One of these studies found 71 unique MLVA types in 86 clinical and reference strains of *C. difficile* from a single institution, which included many of the most commonly encountered REA types (Marsh et al., 2006). The other found that MLVA could correctly classify and discriminate 64 reference strains spanning known toxinotypes, serotypes or ribotype subtypes (van den Berg et al., 2007). Both studies also found that MLVA types are stable over time with strains passaged up to 30 times or paired isolates of *C. difficile* from the same patient collected on different dates resulting in the same MLVA type (Marsh et al., 2006; van den Berg et al., 2007). Recently, Manzoor and colleagues identified eight additional VNTR loci in *C. difficile*. By using these in combination with seven of the previously described loci, they found this extended MLVA (eMLVA) scheme to be in complete concordance with PCR ribotyping (Manzoor et al., 2011).

While the discriminatory power and reproducibility of MLVA are high, this method is not without its limitations. In fact, some have suggested that MLVA may be too highly discriminatory. Tanner et al. typed five individual *C. difficile* colonies cultured from each of 39 stool specimens. All were found to be PCR ribotype 027, but in five of these specimens, multiple MLVA types were identified. The authors suggest that this may confound results of outbreak investigations should only single colonies from each specimen be typed (Tanner, Hardy, & Hawkey, 2010). This led Broukhanski and colleagues to define criteria for distinguishing outbreak isolates from sporadic isolates from a single individual. They found that a 1–3% difference in MLVA types when multiple colonies from a single individual were analysed, and therefore suggest a 3–5% cut-off based on the Manhattan distance-based clustering to identify true differences encountered in outbreak settings (Broukhanski, Simor, & Pillai, 2011).

2.2.4 Arbitrarily primed PCR/random amplified polymorphic DNA

Arbitrarily primed PCR (AP-PCR) and random amplified polymorphic DNA (RAPD) are closely related methods that detect polymorphisms in the genome without the prior knowledge of DNA sequence (Williams, Kubelik, Livak, Rafalski, & Tingey, 1990). Both techniques use short oligonucleotide primers (up to 53 bp for AP-PCR, ~10 bp for RAPD) of arbitrary sequence and permissive thermocycler conditions to amplify genomic DNA. Polymorphisms are detected when DNA is amplified in one genome but not another. By using multiple primers, a genetic map of "RAPD markers" can be determined and compared between strains (Williams et al., 1990).

McMillin and Muldrow were the first to describe AP-PCR for use in molecular typing of *C. difficile*. Six arbitrary primers were used to distinguish six toxigenic *C. difficile* strains (McMillin & Muldrow, 1992). Barbut et al. showed that by using only two RAPD primers, AP4 and AP5, 11 epidemiologically distinct strains could be identified by this method, and the results were both reproducible and comparable to the known serologic profiles of each organism (Barbut, Mario, Delmee, Gozian, & Petit, 1993). In a subsequent clinical study, the authors used the technique to clarify the aetiology of outbreaks among hospitalised patients with AIDS. The same RAPD profiles were identified in 25 isolates from 15 different patients, suggesting horizontal transmission of a single epidemic clone (Barbut et al., 1994). Killgore and Kato used a similar approach to separate 41 clinical isolates from a hospital outbreak of antibiotic-associated diarrhoea into nine groups, with 66% falling within in the same type. In comparison to immunoblotting, they found agreement in 33 out of 34 isolates with AP-PCR capable of distinguishing an additional seven types (Killgore & Kato, 1994).

While the AP-PCR/RAPD method has proven useful in the molecular typing of *C. difficile*, results are often inconsistent, both within and between research groups, and the banding patterns that are produced are often subject to variation according to subtle differences in the thermal profile of the amplification and/or the amount of genomic template used (Bidet et al., 2000; Brazier, 2001; Collier, Stock, DeGirolami, Samore, & Cartwright, 1996). AP-PCR/RAPD is moderately discriminating and comparable to other molecular strain typing approaches. Although attempts to optimise RAPD protocols have been made (Green, Worthington, Hilton, & Lambert, 2011), the use of this method has been primarily restricted to supportive work and method validation.

2.3 SEQUENCE-BASED METHODS

2.3.1 Targeted sequencing approaches

2.3.1.1 Multilocus sequence typing

Multilocus sequence typing (MLST) typically refers to the systematic sequencing of six or seven well-conserved, house-keeping genes or loci within the bacterial genome. Allelic variation at each locus is catalogued, and a sequence type or lineage is assigned by comparing the set of alleles to other isolate profiles in the database. MLST databases for many organisms may be found at http://www.mlst.net or http://www.pubmlst.org; the predominant MLST reference database for *C. difficile* is hosted at http://pubmlst.org/cdifficile/. While MLST is an important tool for understanding the molecular evolution of *C. difficile*, it does not typically have the necessary resolution for outbreak investigations. Moreover, using conventional Sanger sequencing approaches, MLST was often cost prohibitive and difficult to perform at scale. As WGS becomes increasingly ubiquitous, however, it has become relatively straightforward to extract MLST profiles directly from WGS datasets using tools such as SRST2 (Inouye et al., 2014), and the MLST concept has also been extended to include schemes with many hundreds or even thousands of loci whole genome multilocus sequence typing (wgMLST) (Jolley & Maiden, 2014).

2.3.1.2 Surface layer protein A sequencing

Single locus sequence typing using surface layer protein A (*slpA*) was initially developed by Kato et al., with subsequent work by Lemee et al. that described genetic sequence variability in several *C. difficile* adhesins, surface components and flagellar proteins (Kato, Yokoyama, & Arakawa, 2005; Lemee et al., 2005). A 2006 study by Eidhin et al. described *slpA* sequence variability among 14 different PCR ribotypes of *C. difficile* and found limited correlation, with *slpA* sequence variability presumably related to differences in selective pressure (Eidhin, Ryan, Doyle, Walsh, & Kelleher, 2006). Despite this apparent instability, a handful of strain typing assays have been described that target *slpA* for discrimination, including a degenerate multiplex PCR assay to approximate PCR-Ribotypes (Xiao et al., 2013). Recent evidence suggests that *slpA* plays a pivotal role in bacterial adhesion to the host epithelium, and it will likely continue to be an important target for genomic and proteomic characterisation.

2.3.1.3 Tandem repeat sequence typing

Tandem repeat sequence typing (TRST) was first described by Zaiss et al. in 2009 and involves the sequencing of two tandem repeat regions, designated TR6 and TR10 (Zaiss et al., 2009). While the initial validation data suggested that TRST offered useful discrimination and epidemiologic concordance, and the technique was a reasonable compromise to the cost, complexity and interpretive challenges of MLVA and MLST, TRST has not been extensively used in published reports.

2.3.2 *Whole genome sequencing*

With the introduction of commercial next-generation sequencing (NGS) platforms less than a decade ago, high-throughput, WGS has become an increasingly cost-effective and practical option for the characterisation and strain typing of many microbial pathogens, and has provided important new insights into the molecular epidemiology and transmission dynamics of many infectious diseases. NGS platforms present a rapidly evolving technology sector, with regular and significant updates and improvements to the available sequencing hardware, reagents and consumables.

Although different NGS instrument platforms can vary significantly in terms of their overall performance characteristics (output, read length, runtime, cost) and error model, most vendors have adopted common, open file formats (e.g. FASTQ) for initial sample output. Even so, the most important barrier to the implementation of NGS-based methods in many public health and clinical laboratories lies not in the cost and complexity of the laboratory equipment and procedures themselves, but in the management and analysis of the vast amounts of sequence data that they produce. A typical benchtop sequencer, for example, can generate more than 25 GB of raw sequence data in a single 1–2 days run, which represents several hundred-thousand times more data than conventional molecular approaches.

This complexity and abundance of data presents both challenges and opportunities for *C. difficile* surveillance and outbreak investigations, as well as basic research. On the one hand, near-complete genome sequencing allows unprecedented resolution for strain typing and allows rapid and objective interpretation of transmission dynamics. Moreover, comprehensive characterisation is possible from a single dataset, which may ultimately result in faster time-to-answer and reduced costs. On the other hand, Laboratory Information Management Systems (LIMS), High-Performance Computing (HPC) requirements and quality/reliability measures must all be addressed to handle data of this volume and complexity. For most applications, the development of standardised, validated procedures and bioinformatics workflows for sequence analysis, interpretation and reporting have not yet been developed.

There are several general approaches to strain typing bacterial isolates from short-read, NGS data. The applicability of these techniques depends, to a large extent, on the quality and depth of the sequence data available, the applicability of quality reference or comparator sequences, the overall genomic structure, plasticity and population diversity of the bacterium, the context, scope and desired outcome of the analysis, and the overall cost and throughput of sequencing, bioinformatics and data management. Nearly all of these analytical approaches have been applied to *C. difficile* outbreak investigations, surveillance or studies of the molecular evolution and population dynamics of the organism.

2.3.2.1 Single nucleotide polymorphism typing

As its name implies, single nucleotide polymorphism (SNP) typing relies on sets of strain-specific, SNPs to determine the genetic similarity of one or more bacterial genomes. In this approach, a reference genome is used, typically an epidemiologically relevant, high quality reference genome, or in some cases, a high quality assembly of the genome from an epidemiologically important isolate from the investigation, such as an index case or suspected source. The NGS sequencing reads from all query isolates are then mapped to this single reference, and SNPs are identified. The initial determination is typically algorithmic, although SNPs may be filtered or masked according to sequence differences, sequencing coverage/depth and the complexity of the genomic structure. For some organisms, the use of canonical SNP panels may be possible, but in the case of mosaic bacterial genomes, such as *C. difficile*, it is generally necessary to realign and recompute SNP matrices at each application. For this reason, SNPs are well-suited for outbreak investigations, but may not be useful or practical for long-term surveillance studies. Over the course of time, the number of discriminatory loci may change significantly, making longitudinal comparisons and the development of consistent strain type nomenclature challenging.

2.3.2.2 Whole genome multilocus sequence typing

Another approach, wgMLST, uses NGS data to extend the traditional six or seven locus MLST model to encompass thousands of complete genes or marker sequences from across the genome or pangenome of the organism. In this approach, NGS data from a query isolate is assembled, the open reading frames are identified and compared against a database (e.g. BIGSdb http://pubmlst.org/software/database/bigsdb) of allelic variants to develop a comprehensive profile for the organism (Jolley & Maiden, 2014). Subsets of genes may be included in different typing schemes, and may provide additional information about the genotype of the organism, such as its likely virulence or antimicrobial resistance properties. For surveillance purposes, wgMLST-based subtyping of the core genome or pangenome of the organism provides a stable reference frame for strain typing, and allows for reliable comparisons and consistent nomenclature of both alleles and strain type profiles over time. That said, this approach is thought to sacrifice some resolution through its omission of intragenic sequences, and its inability to represent the scale of allelic differences, and may not be ideal for outbreak investigations where fine-grained resolution is required. While wgMLST and other expanded MLST schemes have not been extensively used for *C. difficile* strain typing and characterization to date, the recent incorporation of wgMLST into large-scale molecular surveillance programs, such as PulseNet, will almost certainly extend these techniques to *C. difficile* and other bacterial pathogens over the coming years.

2.3.2.3 k-mer-based comparisons

k-mer-based methods have become increasingly popular tools for the analysis and comparison of bacterial genomes due to their relatively low computational intensity, limited reliance on reference sequence data and rapid time-to-answer. In these approaches, the assembled or raw sequence data are exploded into subsequences of length "k" (typically between 20 and 35 bp), and the presence, absence or relative abundance of each k-mer is used to assess overall similarity. k-mer-based algorithms may be adapted for both reference-based and reference-free applications. While the overall sensitivity and specificity of k-mer-based phylogenetic approaches are generally lower than reference-based methods, such as SNP typing, k-mer-based approaches are often useful in outbreak investigations where rapid typing information is needed, or as part of a tiered approach to guide the selection of reference sequences and more detailed comparisons of putative clusters (Gardner & Hall, 2013).

2.4 **GENOMIC AND METAGENOMIC APPLICATIONS**

WGS, and in particular, high-resolution SNP-based phylogenetics, have been applied to outbreak investigations and evolutionary studies of the strain type diversity and transmission dynamics of *C. difficile*. A recent retrospective survey of 151 027/NAP1/BI isolates from North America, Europe, Australia and Asia, for example, demonstrated the emergence and global dissemination of two separate lineages of the epidemic strain, FQR1 and FQR2, with relatively low sequence diversity and a highly conserved PaLoc (He et al., 2013). Irrespective of strain type, most evolutionary studies of *C. difficile* suggest a relatively low rate of genetic drift, with estimates in the range of 1–2 mutations per genome per year (Didelot et al., 2012; He et al., 2013).

Genomic surveys have been used to detect emerging strains of *C. difficile*, such as PCR ribotype 244, to study human/animal transmission dynamics in rural and agricultural settings and to better understand antimicrobial susceptibility and resistance to fidaxomicin and metronidazole (Eyre et al., 2014; Knetsch et al., 2014; Lim et al., 2014; Lynch et al., 2013). High-quality draft sequencing of *C. difficile* has also been used to demonstrate concordance with MLVA-based typing results and has been proposed as an important tool to improve or approximate the results from PCR ribotyping, MLVA and other fragment-based molecular surveillance methods (Eyre et al., 2012; Gurtler & Grando, 2013).

High-throughput sequencing has also been used for comprehensive metagenomic analysis of *C. difficile*-positive stools from healthcare and community-associated clinical cases, to survey patients with recurrent CDI or to assess their response to antimicrobial therapy and to better understand the transmission dynamics of contaminated environmental surfaces. Both targeted (e.g. 16S) and unbiased sequencing approaches have been used, and for the most part, these studies have been largely consistent with earlier work using denaturing gradient gel electrophoresis or terminal restriction fragment length polymorphisms (t-RFLP) to describe microbial communities and microecologic change (Chang et al., 2008). The application of metagenomics to *C. difficile* is a powerful new tool in our understanding of microbial population dynamics during an infection, and the role of microbiome-associated risk factors (Schubert et al., 2014). It also provides an important means to address the declining availability of culture isolates for pathogen characterisation and public health surveillance.

3 **COMPARISON OF MOLECULAR TYPING METHODS**

Molecular strain typing techniques are often assessed on the basis of several criteria, which include the typeability of the organism within a given method, the stability and reproducibility of the test, discriminatory power and epidemiologic concordance, but also practical considerations, such as cost, ease of use, ease of data interpretation, scalability and sample throughput (Brazier, 2001; Struelens, 1998). Table 1 compares the most common methods based on these parameters. As with most bacteria, the "gold standard" of *C. difficile* strain typing in North America is generally considered to be PFGE, although due to inherent difficulties with non-typeable strains, PCR ribotyping, REA, AP-PCR and toxinotyping are all common methods used among the handful of clinical laboratories that actively culture and study the organism. Unfortunately, due to a lack of standardisation efforts, no one method is used or can be compared at a global level. For this reason, it is not uncommon for study isolates to be referred to by multiple typing designations (e.g. NAP1/BI/027; PFGE/REA/PCR ribotyping).

Over the past several decades, and with the emergence of each new strain typing approach, comparison studies have been published to determine the internal validity of different methods, and to quantitatively determine the relative sensitivity, specificity and predictive value of each technology relative to newer methods.

In 2000, Bidet compared the sensitivity and reproducibility of PFGE, PCR ribotyping and AP-PCR of 99 serogrouped strains of *C. difficile* (Bidet et al., 2000). The typing ability of PCR ribotyping and AP-PCR was 100%, while PFGE was only 90% due to DNA degradation issues among serogroup G strains. The discriminatory power of all methods was greater than 95%. Although reproducibility for PFGE and PCR ribotyping was excellent (100%), AP-PCR reproducibility was only 33–88% depending on the PCR primer set used. The authors concluded that PCR ribotyping offered the best advantages as an initial molecular typing tool due to a straightforward and quick protocol, even though the fingerprints were more difficult to interpret than PFGE (Bidet et al., 2000). Similar conclusions were drawn in a number of other studies. Collier and colleagues found 83% concordance between PFGE and PCR ribotyping of 39 *C. difficile* isolates, while only 44–60% agreement between PFGE and two AP-PCR methods (Collier et al., 1996). In 2003, Wullt and colleagues compared PCR ribotyping to AP-PCR and demonstrated poor overall concordance between the two methods. They found PCR ribotyping to be significantly more consistent and discriminatory for routine experimental use, due to the limited reproducibility of AP-PCR patterns (Wullt, Burman, Laurell, & Akerlund, 2003).

In a more recent study, Killgore et al. compared seven different typing techniques (REA, PFGE, PCR ribotyping, MLST, MLVA, *slpA* typing and amplified fragment length polymorphism, AFLP) for 42 isolates contributed by laboratories in North America and Europe (Killgore et al., 2008). Compared to PFGE, all the strains showed a high concordance of groupings by all methods, with 100% typeability. The highest discrimination indices (Hunter & Gaston, 1988) were seen for those methods that could distinguish subtypes of *C. difficile*, namely MLVA, REA, PFGE and slpA typing. REA, *slpA* typing, MLST and PCR ribotyping all successfully grouped the B1/NAP1/027 strains into a single group. However, only REA and MLVA were capable of distinguishing North American B1/NAP1/027 isolates from European isolates. The authors also noted that while MLVA had the highest discrimination index at 0.964, the method may actually over-discriminate isolates, thereby obscuring strain similarities that are seen by PFGE. Due to the small sample size used in this study, reproducibility analysis was not performed. Additionally, there was a

Table 1 Discriminatory Power, Reproducibility, Cost and Ease of Data Interpretation of Molecular Typing Methods for *Clostridium difficile*

Method Type	Method	Discriminatory Power	Reproducibility	Technical Difficulty	Cost	Time	Ease of Data Interpretation	Ease of Inter-Laboratory Comparison
Restriction fragment methods	REA	+++	+++	++	++	++	+	+
	PFGE	+++	+++	+++	+++	+++	++	+
	Toxinotyping	++	+++	++	++	+++	+	++
PCR amplification methods	Ribotyping (CGE)	+++	+++	+	++	++	++	++
	REP-PCR	+++	++	++	++	++	++	++
	MLVA	+++	+++	+++	++	+++	+	+
	AP-PCR/RAPD	++	+	+	+	+	+	+
Sequence-based methods	slpA	++	+	+	+	+	+++	
	MLST	++	++	+++	+++	+++	++	+++
	Tandem repeat ST	++	++	++	+	++	+	+++
	WGS	++++	+++	+++	+++ / +	+++	+	+++

+, Low; ++, medium; +++, high; ++++, very high.

large predominance of B1/NAP1/027 strains out of the 42 isolates tested, which may have masked the ability of the different methods to differentiate non-B1/NAP1/027 strains.

A large-scale study performed by Tenover et al. compared PCR ribotyping, PFGE and REA for 350 toxigenic isolates collected from seven sites in North America over a 1-year period (Tenover et al., 2011). Of these toxigenic strains, 244 (70%) and 187 (54%) were typeable by PCR ribotyping and PFGE, respectively. 224 (68%) were categorised into one of eight common REA groups. The diversity index (Hunter & Gaston, 1988) for PCR ribotyping (0.8271) was the highest followed by REA (0.7825) then PFGE (0.6530). The authors noted that all methods are useful for strain typing of *C. difficile*, but they each provide different clustering patterns and that more than one method should be used to optimise strain discrimination (Tenover et al., 2011). They also state that of the three techniques compared in this study, PCR ribotyping is the most readily implemented in the clinical laboratory.

More recently, three phenotypic (serogrouping, radio-PAGE, immunoblot) and four genotypic (REA, AP-PCR, PCR ribotyping, RFLP) methods were compared for 100 international toxigenic *C. difficile* isolates representing 39 PCR ribotypes of *C. difficile* (Manzo et al., 2014). The authors found that all methods, particularly REA and PCR ribotyping, showed consistent results both internally and when compared with each other. REA typing had the highest diversity index (0.9957) followed by AP-PCR (0.9618) and PCR ribotyping (0.9541). RFLP was the least discriminatory of all methods (diversity index $= 0.7353$) including the phenotypic methods, as multiple PCR ribotypes were classified within a single RFLP type. Of note, ribotype 027 isolates were noticeably absent from this study presumably due to isolate collection between 1982 and 1991, prior to emergence of the NAP1/B1/027 strain (Manzo et al., 2014).

Several studies have also evaluated newer methods such as MLVA and REP-PCR and compared them to more traditional molecular typing methods. Marsh and colleagues compared MLVA and MLST to REA for 157 *C. difficile* isolates, and found MLVA to be similarly discriminatory to REA, with MLST being the least discriminatory (Marsh et al., 2010). MLVA was also capable of identifying previously unrecognised lineages that REA and MLST could not distinguish. In comparison to PCR ribotyping, eMLVA is more highly discriminatory and accurately clusters isolates into appropriate ribotypes while further dividing the isolates into subtypes at a level that is useful epidemiologically (Manzoor et al., 2011).

4 IMPACT AND APPLICATIONS OF MOLECULAR STRAIN TYPING

A comprehensive understanding of *C. difficile* diversity remains unclear. Nevertheless, it is evident that the epidemiology of CDI has shifted dramatically since the turn of the century. Since the early 2000s, the NAP1/B1/027 strain has emerged worldwide in patients with atypical risk factors and is associated with increased disease severity and mortality rates (Freeman et al., 2010). While it is clear that the NAP1/B1/027 strain has contributed to increasing *C. difficile* rates, other PCR ribotypes have also emerged in recent years. In both North America and Europe, there has been a rise in PCR ribotype 078, a strain highly associated with food animals including calves and pigs (Keel, Brazier, Post, Weese, & Songer, 2007). Similar to the NAP1/B1/027 strain, ribotype 078 produces binary toxin and has been associated with higher morbidity and mortality (Goorhuis et al., 2008). However, compared to NAP1/B1/027, it tends to occur in a younger population and is more frequently associated with community-associated disease. Together with ribotype 001, 002 and 014/020, ribotype 078 and NAP1/B1/027 represent the most frequently isolated ribotypes in North America and Europe (Bauer et al., 2011; Waslawski et al., 2013). Other PCR ribotypes have been the cause of outbreaks in isolated countries (ribotype 106 in England, 017 in Ireland and Asia, 018 in Italy and 053 in Austria), but spread beyond regional outbreaks has yet to occur (Drudy, Harnedy, Fanning, O'Mahony, & Kyne, 2007; Indra, Schmid, et al., 2008; Kim et al., 2010; Spigaglia, Barbanti, Dionisi, & Mastrantonio, 2010).

With this change in the epidemiology of *C. difficile*, molecular typing has become an essential tool for monitoring changes in circulating strains at both regional and global levels and for monitoring hospital outbreaks of *C. difficile*. Numerous studies already mentioned in this review have used molecular typing methods as a critical tool for tracking strains responsible for hospital outbreaks. With this knowledge, infection control measures may be implemented to attempt to halt further spread of the causative strain. Unfortunately, most clinical microbiology laboratories do not routinely culture and type the circulating *C. difficile* strains in their hospitals. In cases where typing is performed by a reference laboratory, turnaround times (TATs) can be so long that implementation of infection control measures in a timely fashion is not possible. Hardy et al. performed a multicenter study in England where the objective was to determine if the number of cases of *C. difficile* could be reduced when using a more rapid and discriminatory strain typing method, MLVA, compared to PCR ribotyping (Hardy et al., 2012). A significant difference in TAT was observed in the eight hospitals using PCR ribotyping (mean TAT of 13.6 days) compared to the eight hospitals using MLVA (mean TAT of 5.3 days). There were no differences in *C. difficile* rates between the two groups over the study period, which the authors stipulate may be due to the wide range of infection control measures instituted by the different hospitals. However, following distribution of a questionnaire to determine if typing resulted in implementation of measures to reduce *C. difficile*, 40.9% of respondents at hospitals using MLVA strongly agreed that the results helped them in management of case clusters, compared to 9.9% at the hospitals using PCR ribotyping (Hardy et al., 2012). The authors indicated that one of the main advantages of the timely availability of typing data is that relevant interventions could be implemented while the patient was still in the hospital (Hardy et al., 2012).

While timely availability of strain typing data is one critical barrier to rapid implementation of measures to control spread of *C. difficile*, it is evident that national surveillance efforts in both North America and Europe have helped control the spread of CDI through the identification of emergent strain types, and the support of antibiotic stewardship, clinical management and infection control practices.

Molecular typing data has also been used to predict outcomes from CDIs though results are conflicting. Some suggest that the NAP1/B1/027 strain is not associated with more severe disease than non-NAP1 strains (Cloud, Noddin, Pressman, Hu, & Kelly, 2009; Morgan et al., 2008). Other, more large-scale, studies have suggested otherwise. In 2010, Miller and colleagues pulled data from the Canadian Nosocomial Infection Surveillance Program (CNISP) to determine if strain type was associated with a more severe outcome (ICU admission, colectomy or death within 30 days of a positive test) (Miller et al., 2010). The most common strains observed in the study were NAP1 (31%) and NAP2/J (28%). Severe outcome was noted in 12.5% of patients infected with the NAP1 strain compared to 5.9% of patients infected with other strain types, which was clinically significant ($p < 0.001$). Additionally, they showed that patients aged 60–90 were two times more likely to experience a severe outcome if infected with the NAP1 strains compared to those infected with non-NAP1 strains (Miller et al., 2010). Another study analysed 2057 cases of CDI with strain typing data to determine if strain type was associated with severe disease (ileus, toxic megacolon, PMC within 5 days or WBC $\geq 15,000$ cells/μl), severe outcome (as defined above) or death within 14 days of a positive test (See et al., 2014). NAP1 (28.4%), NAP4 (10.2%) and NAP11 (9.1%) accounted for the most common strains isolated. The NAP1 strain was associated with severe disease (adjusted odds ratio, AOR = 1.74), severe outcome (AOR = 1.66) and death within 14 days (AOR = 2.12) (See et al., 2014).

Recurrent CDI occurs in 20–30% of patients with CDI, presumably due to the persistence of spores in the gut following a treatment course. Strain typing data has been applied to the study of recurrences, specifically to determine if subsequent episodes were caused by the initial offending strain or by infection with a new strain of *C. difficile*. REA was the first strain typing method to study CDI recurrences (Figueroa et al., 2012; Johnson, Adelmann, Clabots, Peterson, & Gerding, 1989; O'Neill, Beaman, & Riley, 1991). More recently, Figueroa et al. used REA to find that 75 of 90 (83.3%) patients with recurrent CDI were infected with the same REA type of *C. difficile* (Figueroa et al., 2012). Other studies found similar results using AP-PCR (66.6% caused by the initial strain) and a combination of PFGE and PCR ribotyping (80% caused by the initial strain) (Oka et al., 2012; Tang-Feldman, Mayo, Silva, & Cohen, 2003). Strain typing studies have also shown that the B1/NAP1/027 strain has been more commonly associated with relapsing disease than other *C. difficile* strains (Figueroa et al., 2012; Marsh et al., 2012; Petrella et al., 2012). Petrella and colleagues found that the recurrence rates of B1 strains were significantly higher (27.4%, 51 of 186) compared to non-B1 strains (16.6%, 66 of 397) ($p = 0.002$) (Petrella et al., 2012).

5 STANDARDISATION EFFORTS

As strain typing and molecular surveillance of *C. difficile* become increasingly important in many industrialised countries, the lack of an internationally standardised, high-resolution strain typing platform that is both technologically accessible and robust, yet that also results in reproducible and interoperable data, has become a critical gap for both public health and applied research.

As we discussed in an earlier section, one strain typing method that has seen significant efforts towards standardisation has been PCR ribotyping, which is relatively straightforward to implement for most laboratories, and has been widely used for outbreak investigations in a number of different countries and settings. In most cases, however, the protocols are customised by individual laboratories, and as demonstrated by the WebRibo project, the exchange of individual datasets has proven difficult, particularly for the consistent interpretation of similar strains (e.g. 078/126, 014/020, 027/036/075). To address this important gap, and the need for standardised surveillance protocols, a collaborative effort by public health surveillance laboratories from the Centers for Disease Control and Prevention (CDC; USA), the Public Health Agency of Canada (PHAC; Canada), Leiden University Medical Centre (LUMC; NL) and Leeds (UK) recently published a standardised and validated protocol, including interpretive criteria and downloadable reference patterns and other resources for high-resolution PCR ribotyping using CGE (Fawley et al., 2015).

The development of internationally standardised surveillance methods, and consistent strain type nomenclature is critical to our understanding of the transmission dynamics of *C. difficile* and the early identification and response to emerging hyper-virulent or antimicrobial resistant phenotypes (Huber, Foster, Riley, & Paterson, 2013). As sequence-based methods, such as WGS, become increasingly cost-effective, standardisation of laboratory and bioinformatics protocols, reference databases and nomenclature will continue to be critical to enable the rapid and seamless exchange of information between participating public health and clinical laboratories. As the importance of WGS-based methods continues to expand, appropriate quality assurance methods will be needed to describe and validate the reference databases and bioinformatic workflows that were used for analysis and reporting of strain typing and characterisation data from public health and clinical microbiology laboratories. In these efforts, standards organisations or multilateral groups such as the Global Microbial Identifier effort (http://www.g-m-i.org) may be useful in developing the necessary frameworks and standards for reliable information reporting and exchange.

6 FUTURE DIRECTIONS

As clinical laboratories continue to move towards culture-independent diagnostic testing, routine culture and analysis of *C. difficile* and other anaerobic pathogens are likely to continue to decrease. For public health applications, the targeted recovery of isolates for WGS, transcriptomic analysis and proteomic characterisation seems likely to overtake and eventually replace most conventional molecular approaches for *C. difficile* strain typing and characterisation. While metagenomics and target-specific capture or amplification may eventually permit the direct analysis and characterisation of *C. difficile* directly from clinical stool samples, limitations of the current generation of sample processing and enrichment techniques, and the enormous bioinformatic complexity of the data make these approaches largely unfeasible.

In the meantime, some of the current generation of clinical molecular diagnostic platforms (Cepheid GenXpert *C. difficile*/Epi, for example) are able to provide some level of categorical strain type, such as the prediction the hyper-virulent BI/NAP1/027 strain. Although limited to a single strain type, this method has been shown to quickly detect BI/NAP1/027 strains directly from stool specimens at the same time as TcdB without the need to wait days to weeks for molecular typing data from reference laboratories. This rapid identification of BI/NAP1/027 strains may be useful in quickly detecting nosocomial *C. difficile* outbreaks, and implementing appropriate infection control interventions at the outset.

CONCLUSION

The epidemiology of CDIs has changed considerably over the past decade, with the emergence of more virulent strains and with disease occurring in those without traditional risk factors. Molecular strain typing has become an important tool to our understanding of the changing epidemiology of this disease. As typing methods have shifted from phenotypic to genotyping techniques, so have diagnostic methods for *C. difficile*. Unfortunately, without the use of culture-based diagnostic methods in many clinical microbiology laboratories, the ability to perform routine molecular strain typing on isolates has become increasingly limited, and our understanding of global *C. difficile* strain diversity and the molecular epidemiology of these pathogens has been impacted as a result. As NGS becomes increasingly cost-effective, however, genome-scale molecular epidemiology has the potential to give tremendous insight into the pathogenesis, microbial risk factors and molecular ecology of the organism. These same high-throughput sequencing technologies may be leveraged to assess the metagenomic composition of the clinical sample to better understand the infection in the context of the patient's microbiome, and to identify important biomarkers and host risk factors in the progression or outcome of disease. The impact of genomics, proteomics and other high-throughput laboratory technologies is fundamentally changing clinical and public health laboratory practice, and standards for the characterisation and strain typing of *C. difficile* and other anaerobic bacteria is certain to change significantly over the next decade.

REFERENCES

Alonso, R., Martin, A., Pelaez, T., Marin, M., Rodriguez-Creixems, M., & Bouza, E. (2005). An improved protocol for pulsed-field gel electrophoresis typing of *Clostridium difficile*. *Journal of Medical Microbiology, 54*(Pt. 2), 155–157.

Barbut, F., Mario, N., Delmee, M., Gozian, J., & Petit, J. C. (1993). Genomic fingerprinting of *Clostridium difficile* isolates by using a random amplified polymorphic DNA (RAPD) assay. *FEMS Microbiology Letters, 114*(2), 161–166.

Barbut, F., Mario, N., Meyohas, M. C., Binet, D., Frottier, J., & Petit, J. C. (1994). Investigation of a nosocomial outbreak of *Clostridium difficile*-associated diarrhoea among AIDS patients by random amplified polymorphic DNA (RAPD) assay. *The Journal of Hospital Infection, 26*(3), 181–189.

Bartlett, J. G., Chang, T. W., Gurwith, M., Gorbach, S. L., & Onderdonk, A. B. (1978). Antibiotic-associated pseudomembranous colitis due to toxin-producing clostridia. *The New England Journal of Medicine, 298*(10), 531–534. http://dx.doi.org/10.1056/NEJM197803092981003.

Bauer, M. P., Notermans, D. W., van Benthem, B. H., Brazier, J. S., Wilcox, M. H., Rupnik, M., et al. (2011). *Clostridium difficile* infection in Europe: A hospital-based survey. *Lancet, 377*(9759), 63–73. http://dx.doi.org/10.1016/S0140-6736(10)61266-4.

Bidet, P., Barbut, F., Lalande, V., Burghoffer, B., & Petit, J. C. (1999). Development of a new PCR-ribotyping method for *Clostridium difficile* based on ribosomal RNA gene sequencing. *FEMS Microbiology Letters, 175*(2), 261–266.

Bidet, P., Lalande, V., Salauze, B., Burghoffer, B., Avesani, V., Delmee, M., et al. (2000). Comparison of PCR-ribotyping, arbitrarily primed PCR, and pulsed-field gel electrophoresis for typing *Clostridium difficile*. *Journal of Clinical Microbiology, 38*(7), 2484–2487.

Braun, V., Hundsberger, T., Leukel, P., Sauerborn, M., & von Eichel-Streiber, C. (1996). Definition of the single integration site of the pathogenicity locus in *Clostridium difficile*. *Gene, 181*(1–2), 29–38.

Brazier, J. S. (2001). Typing of *Clostridium difficile*. *Clinical Microbiology and Infection, 7*(8), 428–431.

Broukhanski, G., Simor, A., & Pillai, D. R. (2011). Defining criteria to interpret multilocus variable-number tandem repeat analysis to aid *Clostridium difficile* outbreak investigation. *Journal of Medical Microbiology, 60*(Pt. 8), 1095–1100. http://dx.doi.org/10.1099/jmm.0.029819-0.

Burdon, D. W. (1982). *Clostridium difficile*: The epidemiology and prevention of hospital-acquired infection. *Infection, 10*(4), 203–204.

Chang, J. Y., Antonopoulos, D. A., Kalra, A., Tonelli, A., Khalife, W. T., Schmidt, T. M., et al. (2008). Decreased diversity of the fecal microbiome in recurrent *Clostridium difficile*-associated diarrhoea. *The Journal of Infectious Diseases, 197*(3), 435–438. http://dx.doi.org/10.1086/525047.

Church, D. L., Chow, B. L., Lloyd, T., & Gregson, D. B. (2011). Evaluation of automated repetitive-sequence-based PCR (DiversiLab) compared to PCR ribotyping for rapid molecular typing of community- and nosocomial-acquired *Clostridium difficile*. *Diagnostic Microbiology and Infectious Disease, 70*(2), 183–190. http://dx.doi.org/10.1016/j.diagmicrobio.2010.12.024.

Clabots, C. R., Johnson, S., Bettin, K. M., Mathie, P. A., Mulligan, M. E., Schaberg, D. R., et al. (1993). Development of a rapid and efficient restriction endonuclease analysis typing system for *Clostridium difficile* and correlation with other typing systems. *Journal of Clinical Microbiology, 31*(7), 1870–1875.

Cloud, J., Noddin, L., Pressman, A., Hu, M., & Kelly, C. (2009). *Clostridium difficile* strain NAP-1 is not associated with severe disease in a nonepidemic setting. *Clinical Gastroenterology and Hepatology, 7*(8), 868–873. http://dx.doi.org/10.1016/j.cgh.2009.05.018, e862.

Cohen, S. H., Tang, Y. J., & Silva, J., Jr. (2000). Analysis of the pathogenicity locus in *Clostridium difficile* strains. *The Journal of Infectious Diseases, 181*(2), 659–663. http://dx.doi.org/10.1086/315248.

Cohen, S. H., Tang, Y. J., & Silva, J., Jr. (2001). Molecular typing methods for the epidemiological identification of *Clostridium difficile* strains. *Expert Review of Molecular Diagnostics, 1*(1), 61–70. http://dx.doi.org/10.1586/14737159.1.1.61.

Collier, M. C., Stock, F., DeGirolami, P. C., Samore, M. H., & Cartwright, C. P. (1996). Comparison of PCR-based approaches to molecular epidemiologic analysis of *Clostridium difficile*. *Journal of Clinical Microbiology, 34*(5), 1153–1157.

Corkill, J. E., Graham, R., Hart, C. A., & Stubbs, S. (2000). Pulsed-field gel electrophoresis of degradation-sensitive DNAs from *Clostridium difficile* PCR ribotype 1 strains. *Journal of Clinical Microbiology, 38*(7), 2791–2792.

Delmee, M., Homel, M., & Wauters, G. (1985). Serogrouping of *Clostridium difficile* strains by slide agglutination. *Journal of Clinical Microbiology, 21*(3), 323–327.

Didelot, X., Eyre, D. W., Cule, M., Ip, C. L., Ansari, M. A., Griffiths, D., et al. (2012). Microevolutionary analysis of *Clostridium difficile* genomes to investigate transmission. *Genome Biology, 13*(12), R118. http://dx.doi.org/10.1186/gb-2012-13-12-r118.

Drudy, D., Harnedy, N., Fanning, S., O'Mahony, R., & Kyne, L. (2007). Isolation and characterisation of toxin A-negative, toxin B-positive *Clostridium difficile* in Dublin, Ireland. *Clinical Microbiology and Infection, 13*(3), 298–304. http://dx.doi.org/10.1111/j.1469-0691.2006.01634.x.

Eckert, C., Van Broeck, J., Spigaglia, P., Burghoffer, B., Delmee, M., Mastrantonio, P., et al. (2011). Comparison of a commercially available repetitive-element PCR system (DiversiLab) with PCR ribotyping for typing of *Clostridium difficile* strains. *Journal of Clinical Microbiology, 49*(9), 3352–3354. http://dx.doi.org/10.1128/JCM.00324-11.

Eidhin, D. N., Ryan, A. W., Doyle, R. M., Walsh, J. B., & Kelleher, D. (2006). Sequence and phylogenetic analysis of the gene for surface layer protein, slpA, from 14 PCR ribotypes of *Clostridium difficile*. *Journal of Medical Microbiology, 55*(Pt. 1), 69–83. http://dx.doi.org/10.1099/jmm.0.46204-0.

Eyre, D. W., Babakhani, F., Griffiths, D., Seddon, J., Del Ojo Elias, C., Gorbach, S. L., et al. (2014). Whole-genome sequencing demonstrates that fidaxomicin is superior to vancomycin for preventing reinfection and relapse of infection with *Clostridium difficile*. *The Journal of Infectious Diseases, 209*(9), 1446–1451. http://dx.doi.org/10.1093/infdis/jit598.

Eyre, D. W., Walker, A. S., Griffiths, D., Wilcox, M. H., Wyllie, D. H., Dingle, K. E., et al. (2012). *Clostridium difficile* mixed infection and reinfection. *Journal of Clinical Microbiology, 50*(1), 142–144. http://dx.doi.org/10.1128/JCM.05177-11.

Fawley, W. N., Knetsch, C. W., MacCannell, D. R., Harmanus, C., Du, T., Mulvey, M. R., et al. (2015). Development and validation of an internationally-standardized, high-resolution capillary Gel-based electrophoresis PCR-ribotyping protocol for *Clostridium difficile*. *PLoS One, 10*(2), e0118150. http://dx.doi.org/10.1371/journal.pone.0118150.

Figueroa, I., Johnson, S., Sambol, S. P., Goldstein, E. J., Citron, D. M., & Gerding, D. N. (2012). Relapse versus reinfection: Recurrent *Clostridium difficile* infection following treatment with fidaxomicin or vancomycin. *Clinical Infectious Diseases, 55*(Suppl. 2), S104–S109. http://dx.doi.org/10.1093/cid/cis357.

Freeman, J., Bauer, M. P., Baines, S. D., Corver, J., Fawley, W. N., Goorhuis, B., et al. (2010). The changing epidemiology of *Clostridium difficile* infections. *Clinical Microbiology Reviews, 23*(3), 529–549. http://dx.doi.org/10.1128/CMR.00082-09.

Gal, M., Northey, G., & Brazier, J. S. (2005). A modified pulsed-field gel electrophoresis (PFGE) protocol for subtyping previously non-PFGE typeable isolates of *Clostridium difficile* polymerase chain reaction ribotype 001. *The Journal of Hospital Infection, 61*(3), 231–236. http://dx.doi.org/10.1016/j.jhin.2005.01.017.

Gardner, S. N., & Hall, B. G. (2013). When whole-genome alignments just won't work: kSNP v2 software for alignment-free SNP discovery and phylogenetics of hundreds of microbial genomes. *PLoS One*, *8*(12), e81760. http://dx.doi.org/10.1371/journal.pone.0081760.

George, W. L., Sutter, V. L., Goldstein, E. J., Ludwig, S. L., & Finegold, S. M. (1978). Aetiology of antimicrobial-agent-associated colitis. *Lancet*, *1*(8068), 802–803.

Gerding, D. N., Johnson, S., Rupnik, M., & Aktories, K. (2014). *Clostridium difficile* binary toxin CDT: Mechanism, epidemiology, and potential clinical importance. *Gut Microbes*, *5*(1), 15–27. http://dx.doi.org/10.4161/gmic.26854.

Goorhuis, A., Bakker, D., Corver, J., Debast, S. B., Harmanus, C., Notermans, D. W., et al. (2008). Emergence of *Clostridium difficile* infection due to a new hypervirulent strain, polymerase chain reaction ribotype 078. *Clinical Infectious Diseases*, *47*(9), 1162–1170. http://dx.doi.org/10.1086/592257.

Green, L. M., Worthington, T., Hilton, A. C., & Lambert, P. A. (2011). Genetic characterization of clinical isolates of *Clostridium difficile* using an optimized RAPD protocol and PCR ribotyping reveals strain diversity between two tertiary referral Trusts in the West Midlands, UK. *Journal of Medical Microbiology*, *60*(Pt. 9), 1287–1291. http://dx.doi.org/10.1099/jmm.0.030999-0.

Gurtler, V. (1993). Typing of *Clostridium difficile* strains by PCR-amplification of variable length 16S-23S rDNA spacer regions. *Journal of General Microbiology*, *139*(12), 3089–3097.

Gurtler, V., & Grando, D. (2013). New opportunities for improved ribotyping of C. difficile clinical isolates by exploring their genomes. *Journal of Microbiological Methods*, *93*(3), 257–272. http://dx.doi.org/10.1016/j.mimet.2013.02.013.

Hardy, K., Manzoor, S., Marriott, C., Parsons, H., Waddington, C., Gossain, S., et al. (2012). Utilizing rapid multiple-locus variable-number tandem-repeat analysis typing to aid control of hospital-acquired *Clostridium difficile* Infection: A multicenter study. *Journal of Clinical Microbiology*, *50*(10), 3244–3248. http://dx.doi.org/10.1128/JCM.00784-12.

He, M., Miyajima, F., Roberts, P., Ellison, L., Pickard, D. J., Martin, M. J., et al. (2013). Emergence and global spread of epidemic healthcare-associated *Clostridium difficile*. *Nature Genetics*, *45*(1), 109–113. http://dx.doi.org/10.1038/ng.2478.

Herschleb, J., Ananiev, G., & Schwartz, D. C. (2007). Pulsed-field gel electrophoresis. *Nature Protocols*, *2*(3), 677–684. http://dx.doi.org/10.1038/nprot.2007.94.

Huber, C. A., Foster, N. F., Riley, T. V., & Paterson, D. L. (2013). Challenges for standardization of *Clostridium difficile* typing methods. *Journal of Clinical Microbiology*, *51*(9), 2810–2814. http://dx.doi.org/10.1128/JCM.00143-13.

Hunter, P. R., & Gaston, M. A. (1988). Numerical index of the discriminatory ability of typing systems: An application of Simpson's index of diversity. *Journal of Clinical Microbiology*, *26*(11), 2465–2466.

Indra, A., Huhulescu, S., Schneeweis, M., Hasenberger, P., Kernbichler, S., Fiedler, A., et al. (2008). Characterization of *Clostridium difficile* isolates using capillary gel electrophoresis-based PCR ribotyping. *Journal of Medical Microbiology*, *57*(Pt. 11), 1377–1382. http://dx.doi.org/10.1099/jmm.0.47714-0.

Indra, A., Schmid, D., Huhulescu, S., Hell, M., Gattringer, R., Hasenberger, P., et al. (2008). Characterization of clinical *Clostridium difficile* isolates by PCR ribotyping and detection of toxin genes in Austria, 2006–2007. *Journal of Medical Microbiology*, *57*(Pt. 6), 702–708. http://dx.doi.org/10.1099/jmm.0.47476-0.

Inouye, M., Dashnow, H., Raven, L. A., Schultz, M. B., Pope, B. J., Tomita, T., et al. (2014). SRST2: Rapid genomic surveillance for public health and hospital microbiology labs. *Genome Medicine*, *6*(11), 90. http://dx.doi.org/10.1186/s13073-014-0090-6.

Janezic, S., Marin, M., Martin, A., & Rupnik, M. (2015). A new type of toxin A-negative, toxin B-positive *Clostridium difficile* strain lacking a complete tcdA gene. *Journal of Clinical Microbiology*, *53*, 692–695. http://dx.doi.org/10.1128/JCM.02211-14.

Jersek, B., Tcherneva, E., Rijpens, N., & Herman, L. (1996). Repetitive element sequence-based PCR for species and strain discrimination in the genus Listeria. *Letters in Applied Microbiology*, *23*(1), 55–60.

Johnson, S., Adelmann, A., Clabots, C. R., Peterson, L. R., & Gerding, D. N. (1989). Recurrences of *Clostridium difficile* diarrhoea not caused by the original infecting organism. *The Journal of Infectious Diseases*, *159*(2), 340–343.

Jolley, K. A., & Maiden, M. C. (2014). Using MLST to study bacterial variation: Prospects in the genomic era. *Future Microbiology*, *9*(5), 623–630. http://dx.doi.org/10.2217/fmb.14.24.

Kato, H., Yokoyama, T., & Arakawa, Y. (2005). Typing by sequencing the slpA gene of *Clostridium difficile* strains causing multiple outbreaks in Japan. *Journal of Medical Microbiology*, *54*(Pt. 2), 167–171.

Keel, K., Brazier, J. S., Post, K. W., Weese, S., & Songer, J. G. (2007). Prevalence of PCR ribotypes among *Clostridium difficile* isolates from pigs, calves, and other species. *Journal of Clinical Microbiology*, *45*(6), 1963–1964. http://dx.doi.org/10.1128/JCM.00224-07.

Killgore, G. E., & Kato, H. (1994). Use of arbitrary primer PCR to type *Clostridium difficile* and comparison of results with those by immunoblot typing. *Journal of Clinical Microbiology*, *32*(6), 1591–1593.

Killgore, G., Thompson, A., Johnson, S., Brazier, J., Kuijper, E., Pepin, J., et al. (2008). Comparison of seven techniques for typing international epidemic strains of *Clostridium difficile*: Restriction endonuclease analysis, pulsed-field gel electrophoresis, PCR-ribotyping, multilocus sequence typing, multilocus variable-number tandem-repeat analysis, amplified fragment length polymorphism, and surface layer protein A gene sequence typing. *Journal of Clinical Microbiology*, *46*(2), 431–437. http://dx.doi.org/10.1128/JCM.01484-07.

Kim, S. J., Kim, H., Seo, Y., Yong, D., Jeong, S. H., Chong, Y., et al. (2010). Molecular characterization of toxin A-negative, toxin B-positive variant strains of *Clostridium difficile* isolated in Korea. *Diagnostic Microbiology and Infectious Disease*, *67*(2), 198–201. http://dx.doi.org/10.1016/j.diagmicrobio.2010.01.007.

Knetsch, C. W., Connor, T. R., Mutreja, A., van Dorp, S. M., Sanders, I. M., Browne, H. P., et al. (2014). Whole genome sequencing reveals potential spread of *Clostridium difficile* between humans and farm animals in the Netherlands, 2002 to 2011. *Euro Surveillance*, *19*(45), 20954.

Kuijper, E. J., Oudbier, J. H., Stuifbergen, W. N., Jansz, A., & Zanen, H. C. (1987). Application of whole-cell DNA restriction endonuclease profiles to the epidemiology of *Clostridium difficile*-induced diarrhoea. *Journal of Clinical Microbiology*, *25*(4), 751–753.

Lanis, J. M., Heinlen, L. D., James, J. A., & Ballard, J. D. (2013). *Clostridium difficile* 027/BI/NAP1 encodes a hypertoxic and antigenically variable form of TcdB. *PLoS Pathogens*, *9*(8), e1003523. http://dx.doi.org/10.1371/journal.ppat.1003523.

Larson, H. E., Price, A. B., Honour, P., & Borriello, S. P. (1978). *Clostridium difficile* and the aetiology of pseudomembranous colitis. *Lancet*, *1*(8073), 1063–1066.

Lemee, L., Bourgeois, I., Ruffin, E., Collignon, A., Lemeland, J. F., & Pons, J. L. (2005). Multilocus sequence analysis and comparative evolution of virulence-associated genes and housekeeping genes of *Clostridium difficile*. *Microbiology*, *151*(Pt. 10), 3171–3180. http://dx.doi.org/10.1099/mic.0.28155-0.

Lim, S. K., Stuart, R. L., Mackin, K. E., Carter, G. P., Kotsanas, D., Francis, M. J., et al. (2014). Emergence of a ribotype 244 strain of *Clostridium difficile* associated with severe disease and related to the epidemic ribotype 027 strain. *Clinical Infectious Diseases, 58*(12), 1723–1730. http://dx.doi.org/10.1093/cid/ciu203.

Loo, V. G., Poirier, L., Miller, M. A., Oughton, M., Libman, M. D., Michaud, S., et al. (2005). A predominantly clonal multi-institutional outbreak of *Clostridium difficile*-associated diarrhoea with high morbidity and mortality. *The New England Journal of Medicine, 353*(23), 2442–2449. http://dx.doi.org/10.1056/NEJMoa051639.

Lynch, T., Chong, P., Zhang, J., Hizon, R., Du, T., Graham, M. R., et al. (2013). Characterization of a stable, metronidazole-resistant *Clostridium difficile* clinical isolate. *PLoS One, 8*(1), e53757. http://dx.doi.org/10.1371/journal.pone.0053757.

MacCannell, D. R., Louie, T. J., Gregson, D. B., Laverdiere, M., Labbe, A. C., Laing, F., et al. (2006). Molecular analysis of *Clostridium difficile* PCR ribotype 027 isolates from Eastern and Western Canada. *Journal of Clinical Microbiology, 44*(6), 2147–2152. http://dx.doi.org/10.1128/JCM.02563-05.

Manzo, C. E., Merrigan, M. M., Johnson, S., Gerding, D. N., Riley, T. V., Silva, J., Jr., et al. (2014). International typing study of *Clostridium difficile*. *Anaerobe, 28*, 4–7. http://dx.doi.org/10.1016/j.anaerobe.2014.04.005.

Manzoor, S. E., Tanner, H. E., Marriott, C. L., Brazier, J. S., Hardy, K. J., Platt, S., et al. (2011). Extended multilocus variable-number tandem-repeat analysis of *Clostridium difficile* correlates exactly with ribotyping and enables identification of hospital transmission. *Journal of Clinical Microbiology, 49*(10), 3523–3530. http://dx.doi.org/10.1128/JCM.00546-11.

Marsh, J. W., Arora, R., Schlackman, J. L., Shutt, K. A., Curry, S. R., & Harrison, L. H. (2012). Association of relapse of *Clostridium difficile* disease with BI/NAP1/027. *Journal of Clinical Microbiology, 50*(12), 4078–4082. http://dx.doi.org/10.1128/JCM.02291-12.

Marsh, J. W., O'Leary, M. M., Shutt, K. A., Pasculle, A. W., Johnson, S., Gerding, D. N., et al. (2006). Multilocus variable-number tandem-repeat analysis for investigation of *Clostridium difficile* transmission in hospitals. *Journal of Clinical Microbiology, 44*(7), 2558–2566. http://dx.doi.org/10.1128/JCM.02364-05.

Marsh, J. W., O'Leary, M. M., Shutt, K. A., Sambol, S. P., Johnson, S., Gerding, D. N., et al. (2010). Multilocus variable-number tandem-repeat analysis and multilocus sequence typing reveal genetic relationships among *Clostridium difficile* isolates genotyped by restriction endonuclease analysis. *Journal of Clinical Microbiology, 48*(2), 412–418. http://dx.doi.org/10.1128/JCM.01315-09.

McDonald, L. C., Killgore, G. E., Thompson, A., Owens, R. C., Jr., Kazakova, S. V., Sambol, S. P., et al. (2005). An epidemic, toxin gene-variant strain of *Clostridium difficile*. *The New England Journal of Medicine, 353*(23), 2433–2441. http://dx.doi.org/10.1056/NEJMoa051590.

McMillin, D. E., & Muldrow, L. L. (1992). Typing of toxic *Clostridium difficile* using DNA fingerprints generated with arbitrary polymerase chain reaction primers. *FEMS Microbiology Letters, 71*(1), 5–9.

Miller, M., Gravel, D., Mulvey, M., Taylor, G., Boyd, D., Simor, A., et al. (2010). Health care-associated *Clostridium difficile* infection in Canada: Patient age and infecting strain type are highly predictive of severe outcome and mortality. *Clinical Infectious Diseases, 50*(2), 194–201. http://dx.doi.org/10.1086/649213.

Morgan, O. W., Rodrigues, B., Elston, T., Verlander, N. Q., Brown, D. F., Brazier, J., et al. (2008). Clinical severity of *Clostridium difficile* PCR ribotype 027: A case-case study. *PLoS One, 3*(3), e1812. http://dx.doi.org/10.1371/journal.pone.0001812.

Mulligan, M. E., Peterson, L. R., Kwok, R. Y., Clabots, C. R., & Gerding, D. N. (1988). Immunoblots and plasmid fingerprints compared with serotyping and polyacrylamide gel electrophoresis for typing *Clostridium difficile*. *Journal of Clinical Microbiology*, *26*(1), 41–46.

Nakamura, S., Mikawa, M., Nakashio, S., Takabatake, M., Okado, I., Yamakawa, K., et al. (1981). Isolation of *Clostridium difficile* from the feces and the antibody in sera of young and elderly adults. *Microbiology and Immunology*, *25*(4), 345–351.

Northey, G., Gal, M., Rahmati, A., & Brazier, J. S. (2005). Subtyping of *Clostridium difficile* PCR ribotype 001 by REP-PCR and PFGE. *Journal of Medical Microbiology*, *54*(Pt. 6), 543–547. http://dx.doi.org/10.1099/jmm.0.45989-0.

O'Connor, J. R., Johnson, S., & Gerding, D. N. (2009). *Clostridium difficile* infection caused by the epidemic BI/NAP1/027 strain. *Gastroenterology*, *136*(6), 1913–1924. http://dx.doi.org/10.1053/j.gastro.2009.02.073.

Oka, K., Osaki, T., Hanawa, T., Kurata, S., Okazaki, M., Manzoku, T., et al. (2012). Molecular and microbiological characterization of *Clostridium difficile* isolates from single, relapse, and reinfection cases. *Journal of Clinical Microbiology*, *50*(3), 915–921. http://dx.doi.org/10.1128/JCM.05588-11.

O'Neill, G. L., Beaman, M. H., & Riley, T. V. (1991). Relapse versus reinfection with *Clostridium difficile*. *Epidemiology and Infection*, *107*(3), 627–635.

O'Neill, G. L., Ogunsola, F. T., Brazier, J. S., & Duerden, B. I. (1996). Modification of a PCR ribotyping method for application as a routine typing scheme for *Clostridium difficile*. *Anaerobe*, *2*(4), 205–209.

Pasanen, T., Kotila, S. M., Horsma, J., Virolainen, A., Jalava, J., Ibrahem, S., et al. (2011). Comparison of repetitive extragenic palindromic sequence-based PCR with PCR ribotyping and pulsed-field gel electrophoresis in studying the clonality of *Clostridium difficile*. *Clinical Microbiology and Infection*, *17*(2), 166–175. http://dx.doi.org/10.1111/j.1469-0691.2010.03221.x.

Peerbooms, P. G., Kuijt, P., & Maclaren, D. M. (1987). Application of chromosomal restriction endonuclease digest analysis for use as typing method for *Clostridium difficile*. *Journal of Clinical Pathology*, *40*(7), 771–776.

Pepin, J., Valiquette, L., & Cossette, B. (2005). Mortality attributable to nosocomial *Clostridium difficile*-associated disease during an epidemic caused by a hypervirulent strain in Quebec. *CMAJ: Canadian Medical Association Journal*, *173*(9), 1037–1042. http://dx.doi.org/10.1503/cmaj.050978.

Petrella, L. A., Sambol, S. P., Cheknis, A., Nagaro, K., Kean, Y., Sears, P. S., et al. (2012). Decreased cure and increased recurrence rates for *Clostridium difficile* infection caused by the epidemic C. difficile BI strain. *Clinical Infectious Diseases*, *55*(3), 351–357. http://dx.doi.org/10.1093/cid/cis430.

Poutanen, S. M., & Simor, A. E. (2004). *Clostridium difficile*-associated diarrhoea in adults. *CMAJ: Canadian Medical Association Journal*, *171*(1), 51–58.

Rafferty, M. E., Baltch, A. L., Smith, R. P., Bopp, L. H., Rheal, C., Tenover, F. C., et al. (1998). Comparison of restriction enzyme analysis, arbitrarily primed PCR, and protein profile analysis typing for epidemiologic investigation of an ongoing *Clostridium difficile* outbreak. *Journal of Clinical Microbiology*, *36*(10), 2957–2963.

Rahmati, A., Gal, M., Northey, G., & Brazier, J. S. (2005). Subtyping of *Clostridium difficile* polymerase chain reaction (PCR) ribotype 001 by repetitive extragenic palindromic PCR genomic fingerprinting. *The Journal of Hospital Infection*, *60*(1), 56–60. http://dx.doi.org/10.1016/j.jhin.2004.09.034.

Rodriguez-Barradas, M. C., Hamill, R. J., Houston, E. D., Georghiou, P. R., Clarridge, J. E., Regnery, R. L., et al. (1995). Genomic fingerprinting of Bartonella species by repetitive element PCR for distinguishing species and isolates. *Journal of Clinical Microbiology, 33*(5), 1089–1093.

Rupnik, M. (2008). Heterogeneity of large clostridial toxins: Importance of *Clostridium difficile* toxinotypes. *FEMS Microbiology Reviews, 32*(3), 541–555. http://dx.doi.org/10.1111/j.1574-6976.2008.00110.x.

Rupnik, M. (2010). *Clostridium difficile* toxinotyping. *Methods in Molecular Biology, 646,* 67–76. http://dx.doi.org/10.1007/978-1-60327-365-7_5.

Rupnik, M., Avesani, V., Janc, M., von Eichel-Streiber, C., & Delmee, M. (1998). A novel toxinotyping scheme and correlation of toxinotypes with serogroups of *Clostridium difficile* isolates. *Journal of Clinical Microbiology, 36*(8), 2240–2247.

Samore, M., Killgore, G., Johnson, S., Goodman, R., Shim, J., Venkataraman, L., et al. (1997). Multicenter typing comparison of sporadic and outbreak *Clostridium difficile* isolates from geographically diverse hospitals. *The Journal of Infectious Diseases, 176*(5), 1233–1238.

Schubert, A. M., Rogers, M. A., Ring, C., Mogle, J., Petrosino, J. P., Young, V. B., et al. (2014). Microbiome data distinguish patients with *Clostridium difficile* infection and non-C. difficile-associated diarrhoea from healthy controls. *mBio, 5*(3), e01021–01014. http://dx.doi.org/10.1128/mBio.01021-14.

See, I., Mu, Y., Cohen, J., Beldavs, Z. G., Winston, L. G., Dumyati, G., et al. (2014). NAP1 strain type predicts outcomes from *Clostridium difficile* infection. *Clinical Infectious Diseases, 58*(10), 1394–1400. http://dx.doi.org/10.1093/cid/ciu125.

Spigaglia, P., Barbanti, F., Dionisi, A. M., & Mastrantonio, P. (2010). *Clostridium difficile* isolates resistant to fluoroquinolones in Italy: Emergence of PCR ribotype 018. *Journal of Clinical Microbiology, 48*(8), 2892–2896. http://dx.doi.org/10.1128/JCM.02482-09.

Spigaglia, P., & Mastrantonio, P. (2003). Evaluation of repetitive element sequence-based PCR as a molecular typing method for *Clostridium difficile*. *Journal of Clinical Microbiology, 41*(6), 2454–2457.

Struelens, M. J. (1998). Molecular epidemiologic typing systems of bacterial pathogens: Current issues and perspectives. *Memórias do Instituto Oswaldo Cruz, 93*(5), 581–585.

Stubbs, S. L., Brazier, J. S., O'Neill, G. L., & Duerden, B. I. (1999). PCR targeted to the 16S-23S rRNA gene intergenic spacer region of *Clostridium difficile* and construction of a library consisting of 116 different PCR ribotypes. *Journal of Clinical Microbiology, 37*(2), 461–463.

Stumpf, A. N., Roggenkamp, A., & Hoffmann, H. (2005). Specificity of enterobacterial repetitive intergenic consensus and repetitive extragenic palindromic polymerase chain reaction for the detection of clonality within the Enterobacter cloacae complex. *Diagnostic Microbiology and Infectious Disease, 53*(1), 9–16. http://dx.doi.org/10.1016/j.diagmicrobio.2005.04.003.

Tang-Feldman, Y., Mayo, S., Silva, J., Jr., & Cohen, S. H. (2003). Molecular analysis of *Clostridium difficile* strains isolated from 18 cases of recurrent *Clostridium difficile*-associated diarrhoea. *Journal of Clinical Microbiology, 41*(7), 3413–3414.

Tanner, H. E., Hardy, K. J., & Hawkey, P. M. (2010). Coexistence of multiple multilocus variable-number tandem-repeat analysis subtypes of *Clostridium difficile* PCR ribotype 027 strains within fecal specimens. *Journal of Clinical Microbiology, 48*(3), 985–987. http://dx.doi.org/10.1128/JCM.02012-09.

Tenover, F. C., Akerlund, T., Gerding, D. N., Goering, R. V., Bostrom, T., Jonsson, A. M., et al. (2011). Comparison of strain typing results for *Clostridium difficile* isolates from

North America. *Journal of Clinical Microbiology*, *49*(5), 1831–1837. http://dx.doi.org/10.1128/JCM.02446-10.

Tenover, F. C., Arbeit, R. D., Goering, R. V., Mickelsen, P. A., Murray, B. E., Persing, D. H., et al. (1995). Interpreting chromosomal DNA restriction patterns produced by pulsed-field gel electrophoresis: Criteria for bacterial strain typing. *Journal of Clinical Microbiology*, *33*(9), 2233–2239.

van Belkum, A. (2007). Tracing isolates of bacterial species by multilocus variable number of tandem repeat analysis (MLVA). *FEMS Immunology and Medical Microbiology*, *49*(1), 22–27. http://dx.doi.org/10.1111/j.1574-695X.2006.00173.x.

van den Berg, R. J., Schaap, I., Templeton, K. E., Klaassen, C. H., & Kuijper, E. J. (2007). Typing and subtyping of *Clostridium difficile* isolates by using multiple-locus variable-number tandem-repeat analysis. *Journal of Clinical Microbiology*, *45*(3), 1024–1028. http://dx.doi.org/10.1128/JCM.02023-06.

Versalovic, J., Koeuth, T., & Lupski, J. R. (1991). Distribution of repetitive DNA sequences in eubacteria and application to fingerprinting of bacterial genomes. *Nucleic Acids Research*, *19*(24), 6823–6831.

Voth, D. E., & Ballard, J. D. (2005). *Clostridium difficile* toxins: Mechanism of action and role in disease. *Clinical Microbiology Reviews*, *18*(2), 247–263. http://dx.doi.org/10.1128/CMR.18.2.247-263.2005.

Walters, B. A., Stafford, R., Roberts, R. K., & Seneviratne, E. (1982). Contamination and crossinfection with *Clostridium difficile* in an intensive care unit. *Australian and New Zealand Journal of Medicine*, *12*(3), 255–258.

Warny, M., Pepin, J., Fang, A., Killgore, G., Thompson, A., Brazier, J., et al. (2005). Toxin production by an emerging strain of *Clostridium difficile* associated with outbreaks of severe disease in North America and Europe. *Lancet*, *366*(9491), 1079–1084. http://dx.doi.org/10.1016/S0140-6736(05)67420-X.

Waslawski, S., Lo, E. S., Ewing, S. A., Young, V. B., Aronoff, D. M., Sharp, S. E., et al. (2013). *Clostridium difficile* ribotype diversity at six health care institutions in the United States. *Journal of Clinical Microbiology*, *51*(6), 1938–1941. http://dx.doi.org/10.1128/JCM.00056-13.

Westblade, L. F., Chamberland, R. R., MacCannell, D., Collins, R., Dubberke, E. R., Dunne, W. M., Jr., et al. (2013). Development and evaluation of a novel, semiautomated *Clostridium difficile* typing platform. *Journal of Clinical Microbiology*, *51*(2), 621–624. http://dx.doi.org/10.1128/JCM.02627-12.

Williams, J. G., Kubelik, A. R., Livak, K. J., Rafalski, J. A., & Tingey, S. V. (1990). DNA polymorphisms amplified by arbitrary primers are useful as genetic markers. *Nucleic Acids Research*, *18*(22), 6531–6535.

Woods, C. R., Jr., Versalovic, J., Koeuth, T., & Lupski, J. R. (1992). Analysis of relationships among isolates of Citrobacter diversus by using DNA fingerprints generated by repetitive sequence-based primers in the polymerase chain reaction. *Journal of Clinical Microbiology*, *30*(11), 2921–2929.

Wren, B. W., & Tabaqchali, S. (1987). Restriction endonuclease DNA analysis of *Clostridium difficile*. *Journal of Clinical Microbiology*, *25*(12), 2402–2404.

Wullt, M., Burman, L. G., Laurell, M. H., & Akerlund, T. (2003). Comparison of AP-PCR typing and PCR-ribotyping for estimation of nosocomial transmission of *Clostridium difficile*. *The Journal of Hospital Infection*, *55*(2), 124–130.

Xiao, K., Kong, F., Wang, Q., Jin, P., Thomas, L., Xiong, L., et al. (2013). Multiplex PCR targeting slpA: A rapid screening method to predict common *Clostridium difficile* ribotypes among fluoroquinolone resistant clinical strains. *Pathology*, *45*(6), 595–599. http://dx.doi.org/10.1097/PAT.0b013e3283650c37.

Zaiss, N. H., Rupnik, M., Kuijper, E. J., Harmanus, C., Michielsen, D., Janssens, K., et al. (2009). Typing *Clostridium difficile* strains based on tandem repeat sequences. *BMC Microbiology*, *9*, 6. http://dx.doi.org/10.1186/1471-2180-9-6.

Accurate Whole-Genome Sequencing-Based Epidemiological Surveillance of *Mycobacterium Tuberculosis*

10

Hannes Pouseele*, Philip Supply[†,‡,§,¶,∥,#,1]

**Applied Maths, Sint-Martens-Latem, Belgium*
[†]*Center for Infection and Immunity of Lille, Lille, France*
[‡]*INSERM U1019, Lille, France*
[§]*CNRS UMR 8204, Lille, France*
[¶]*Université Lille, Lille, France*
[∥]*Institut Pasteur de Lille, Lille, France*
[#]*Genoscreen, Lille, France*
[1]*Corresponding author: e-mail address: philip.supply@ibl.cnrs.fr*

1 INTRODUCTION

1.1 THE NEED FOR POWERFUL, STANDARDISED MOLECULAR-GUIDED TB EPIDEMIOLOGICAL SURVEILLANCE

Efficient epidemiological surveillance is especially important for tuberculosis (TB) control and prevention, for multiple reasons. This disease causes 9 million new cases and 1.5 million deaths each year (World Health Organisation, 2014). Moreover, an estimated two billion individuals are infected by the etiologic agent *Mycobacterium tuberculosis* or by its closely related variants of the so-called *M. tuberculosis* complex (MTBC, also comprising *Mycobacterium africanum*, *Mycobacterium bovis*, *Mycobacterium microti* and *Mycobacterium pinnipedii*). The risk of progression from infection to active disease is about 10% among immuno-competent individuals, and the risk is several-fold higher among immuno-deficient individuals. The global reservoir of persons at risk of developing TB is thus huge.

The disease also represents a special challenge from prophylactic, diagnostic and therapeutic points of view. The available BCG vaccine only offers a relatively low protection against the infectious, pulmonary forms of TB (Colditz et al., 1994).

Clinical signs of active disease are rarely directly apparent, and risk of TB infection is typically, although not always, associated with socioeconomically underprivileged groups often with reduced access to public health facilities (Fok, Numata, Schulzer, & FitzGerald, 2008). This frequently results in delayed diagnosis and thus prolonged patient infectiousness, even in developed countries. The comprehensive standard treatment regimen, consisting of a 4-month intensive phase combining four first-line drugs followed by a 2-month follow-up phase including two first-line drugs, and its potential side effects can pose problems of compliance, potentially leading to development drug resistance, which ideally necessitates supervised treatment in some cases. Treatment of latent infection likewise requires lengthy chemoprophylaxis, also causing problems of adherence in persons who are otherwise asymptomatic (Dye, 2011).

The TB control challenge further extends to epidemiological aspects. Before infectious cases are identified and diagnosed, infection and disease can insidiously propagate in the population, which can lead to outbreaks even in groups being considered *a priori* at very low risk (Coitinho et al., 2013). As a result of delays in diagnosis or of poor compliance to treatment, a small number of highly infectious cases/active transmitters, also termed super-spreaders, can disproportionally contribute to the total burden of active TB transmission in populations (Small et al., 1994; Walker, Ip, et al., 2013). Especially when involving multidrug resistant (MDR; defined as resistant to the main two first-line drugs isoniazid and rifampicin) TB, uncontrolled epidemic transmission can obviously have major adverse consequences on public health (Bifani et al., 1996). Tracing of TB transmission is additionally complicated by the fact that transmission is not limited to the most frequent household relatives, but extends to social and casual contacts as well, which can be much more difficult to detect by classical epidemiological investigation (Gardy et al., 2011; Walker et al., 2014).

Epidemiological challenges also exist at a global level, while the pathogen is globally distributed, strong disparities exist among and within the different continents in prevalence of TB in general and MDR TB (World Health Organisation, 2014) in particular, and often also among different regions within a country, e.g., between rural and urban areas (Jenkins et al., 2014). The globalisation of human travel and population movement therefore generates important fluxes of (potential) cases between different, often widely distant, geographic regions, which are important to recognise in order to have a clear epidemiological understanding. Tracking of such fluxes is possible as there is an association of strain types with the geographic origins of the human patients that carry them (Gagneux et al., 2006; Hirsh, Tsolaki, DeRiemer, Feldman, & Small, 2004).

Because of these specific features, efficient, molecular-guided epidemiological surveillance based on MTBC strain typing has a higher priority and is clearly necessary for TB more than for almost any other infectious disease. In addition, because of the global aspects and interconnections between different regions described above, international standardisation is indispensable in order to enable trans-regional and -national analyses, which necessarily implies full comparability of MTBC strain molecular data collected across different geographic areas.

1.2 CONSTRAINTS IMPOSED BY BIOLOGICAL SPECIFICITIES OF THE PATHOGEN

The methodological requirements indicated above have to overcome major biological specificities of the pathogen. The tubercle bacilli are characterised by their particular slow growth, and their culture requires bio-safety level 3 facilities. They also show a low level of genetic diversity (Comas et al., 2010; Sreevatsan et al., 1997) and a high degree of clonality (Hirsh et al., 2004; Supply et al., 2003), with the exception of the rare *Mycobacterium canettii* strains (Fabre et al., 2004; Gutierrez et al., 2005; Supply et al., 2013). These characteristics have prevented the use of genotyping techniques that have been established for many other bacterial species, such as pulse-field gel electrophoresis (Arbeit et al., 1990) requiring sufficient cellular material to obtain the required amount of purified DNA, and multi-locus sequence typing (MLST) (Maiden et al., 1998), requiring substantial diversity of house-keeping gene sequences for molecular discrimination at strain level. This has prompted the development of internationally standardised PCR-based, repetitive sequence-based genotyping assays, which still represent the most widely used tools today. As these genetic markers have been the subject of other recent reviews (e.g. Niemann & Supply, 2014; Schurch & van Soolingen, 2012), their main methodological features are summarised below, with a specific section including the latest developments of the current typing standard (MIRU-VNTR typing; see Section 2.3) for molecular tracing of MTBC strains.

As it will undoubtedly become the next standard for typing MTBC strains as well as other bacterial pathogens, emphasis will also be paid to whole-genome sequencing (WGS)-based approaches. Crucial bioinformatics analysis components determining the accuracy of single nucleotide polymorphism (SNP) calling will be overviewed, in relation again with the particular genomic complexity of the MTBC bacteria. In addition to an in-depth discussion of whole-genome (wg) SNP mapping, two main alternative methodologies, based on chromosome alignment or wgMLST, will be presented.

2 CLASSICAL MOLECULAR TYPING FOR MOLECULAR-GUIDED INVESTIGATION

2.1 IS*6110* RFLP

Restriction fragment length polymorphism (RFLP) targeting the insertion sequence IS*6110* was the first standardised molecular typing method for MTBC strains. IS*6110* RFLP uses the variability of the numbers and positions of IS*6110* elements in MTBC genomes to generate strain-associated molecular fingerprints (van Embden et al., 1993; Figure 1). This method includes the use of a reference fingerprint from a specific *M. tuberculosis* strain that is used to normalise the banding patterns obtained from test strains in different experiments and different laboratories, through the use of specific image analysis software packages such as GelCompar (Applied Maths, Belgium) or BioImage Whole Band Analyzer or Advanced Quantifier (Bio Image Systems, MI, USA). This technique combines a relatively high discriminatory power, as reflected by clearly different IS*6110*-RFLP profiles often seen among isolates from unrelated TB cases, with a general stability of the DNA profiles in longitudinal isolates from chronically infected patients or from outbreak cases supported by classical contact tracing. On these grounds, clusters of active TB transmission in a patient population are identified based on the detection of identical IS*6110*-RFLP profiles among the corresponding isolates, which are assumed to reflect infection by a same clone. As such, it has been the preferred approach for more than a decade, for the investigation of outbreaks and nosocomial infection, detection of laboratory cross-contamination and distinguishing between exogenous reinfection and unsuccessful treatment of the initial infection in relapse cases (for review see, e.g., Van Soolingen, 2001).

However, this technique has a number of limitations including being labour intensive, the frequently complex DNA profiles remain difficult to compare despite standardisation and the method requires weeks of *in vitro* culture of the slow-growing mycobacteria before sufficient amounts of purified DNA can be obtained for subsequent typing. Therefore, the method has often been used as a tool for retrospective epidemiological analyses, including identification of TB transmission risk groups, and retrospective outbreak confirmation or detection.

2.2 SPOLIGOTYPING

Spoligotyping was the first PCR-based method developed in an attempt to overcome these problems (Kamerbeek et al., 1997). This method targets the so-called direct repeat locus present in virtually all MTBC strains, which is a member of the large bacterial family of clustered, regularly interspaced short palindromic repeat loci (Makarova et al., 2011). The locus contains a series of variable spacer sequences that are PCR amplified with primers against conserved repeated motives flanking each spacer. The amplicons generated are then compared to a reference set of 43 spacers by a membrane-based, reverse line blot hybridisation assay to generate a portable 43-bit barcode, signalling the presence or absence of each reference spacer in the tested strain (Barnes & Cave, 2003; Kamerbeek et al., 1997; Figure 1). This assay is relatively cheap to perform, although hybridisation requires significant manipulation time. However, the principle of interrogating the reference spacer sequences has been elegantly transposed onto the Luminex multi-analyte profiling system, which enables faster and more automated analysis (Cowan, Diem, Brake, & Crawford, 2004). Despite these relative advantages, spoligotyping does not provide resolution at the strain level, but rather at the strain sub-lineage level (Kremer et al., 1999). This limitation is exasperated for strains that belong to the Beijing lineage, which

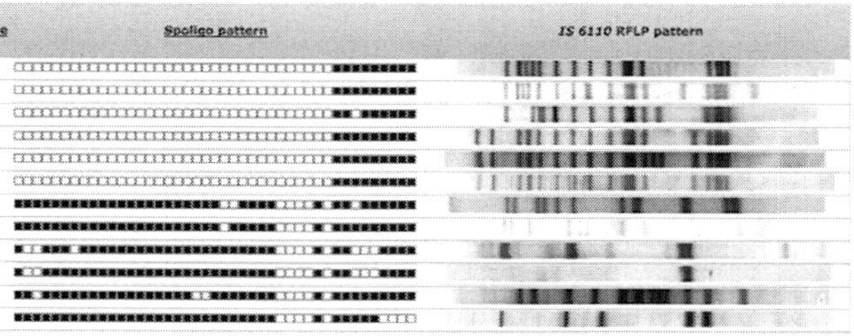

FIGURE 1

Classical typing methods for *M. tuberculosis* complex strains. Genotypes based on standard 24-locus MIRU-VNTR typing (represented by 24-number strings), spoligotyping (represented by digitalised 43-spacer profiles; individual spacers present or absent are shown as black or white boxes, respectively) and IS*6110* RFLP banding patterns are shown for selected representative strains included in MIRU-VNTR*plus* database (www.miru-vntrplus.org). From left to right, information on strain identification number (ID), species, genetic lineage, standard 24-locus MIRU-VNTR nomenclature (MLVA MtbC15-9) and spoligotype designation according to SpolDB4 is also shown. Standard MIRU-VNTR locus names are shown according to locus position on the *M. tuberculosis* H37Rv chromosome (in kbp; i.e. 154, 424, etc.), in addition to alias names according to historical designation (i.e. MIRU02, Mtub04, ETRC, etc.).

ID	Species	Lineage	MLVA MtbC15-9	SpolDB4-Type	Spoligo pattern	IS 6110 RFLP pattern
4445/02	M. tuberculosis	Beijing	95-34	1		
4498/02	M. tuberculosis	Beijing	96-32	1		
4499/02	M. tuberculosis	Beijing	97-32	265		
3243/02	M. tuberculosis	Beijing	98-32	1		
3329/02	M. tuberculosis	Beijing	99-32	1		
4436/02	M. tuberculosis	Beijing	100-32	1		
947/01	M. tuberculosis	EAI	109-40	1390		
7190/03	M. tuberculosis	EAI	110-41	591		
4850/03	M. tuberculosis	EAI	111-42			
1797/03	M. tuberculosis	EAI	112-43	11		
9267/01	M. tuberculosis	EAI	113-44	19		
6006/03	M. tuberculosis	EAI	114-45	138		

represent a major MTBC branch with a global geographic distribution, because most Beijing strains share a single, dominant spoligotype (Kremer et al., 2004; see Beijing strain spoligotypes in Figure 1). Therefore, spoligotyping cannot be used as a sole method to assess and ascertain epidemiological links between TB cases (Barnes & Cave, 2003). It is better suited to phylogenetic exploration of MTBC strain populations, facilitated by the availability of a large, global spoligotype database (Brudey et al., 2006). It has to be noted, however, that this use of spoligotyping also has some limitations, because this marker is subject to a significant level of homoplasy that blurs its phylogenetic signal (Comas, Homolka, Niemann, & Gagneux, 2009).

2.3 MIRU-VNTR TYPING

Because of the spoligotyping limitations indicated above, the use of IS*6110* RFLP as a standard for molecular epidemiological typing of MTBC strains has only been superseded by PCR-based typing of variable numbers of tandem repeats (VNTRs) of genetic elements including mycobacterial interspersed repetitive units (MIRUs), which are mostly intergenic. This method was among the very first multi-locus VNTR analysis (MLVA) (Keim et al., 1999) techniques developed for bacterial typing. After initial identifications of such genetic elements in MTBC genomes (Frothingham, 1995; Supply, Magdalena, Himpens, & Locht, 1997), different typing formats based on different numbers and sets of markers have been used (Gutierrez et al., 2006; Le Fleche, Fabre, Denoeud, Koeck, & Vergnaud, 2002; Magdalena, Supply, & Locht, 1998; Mazars et al., 2001; Roring et al., 2002; Skuce et al., 2002; Smittipat et al., 2005; Smittipat & Palittapongarnpim, 2000; Supply et al., 2000; Warren et al., 2004). A key step in the now generalised use of MIRU-VNTR typing has been its international standardisation using a 24-locus format, selected for a balance between the power to resolve isolates from the main MTBC lineages and clonal stability in longitudinal isolates from chronically infected patients or from human transmission chains (Supply et al., 2006).

Similar to other MLVA methods, MIRU-VNTR typing relies on amplification of the target regions containing DNA tandem repeats, using primers against the unique flanking regions, followed by sizing of the amplicons. The numbers of repeat units in each target region are then deduced, based on the known lengths of the repeat units ranging from about 50 to 110 bp depending on the marker. In the standard format, this generates a portable, 24-number type code, which facilitates direct comparison of results between laboratories (Figure 1).

Another advantage of MIRU-VNTR typing is its adaptability to different analysis platforms. Sizing of the PCR fragments, from which allelic profiles can be performed by electrophoretic separation, either on agarose gels or on multi-capillary systems. Using gels, PCRs and sizing of the amplicons are performed separately for each locus (Frothingham & Meeker-O'Connell, 1998; Mazars et al., 2001). Using automated DNA analysers with up to 96 capillaries coupled to multi-dye chemistry, PCRs are multiplexed and size analysis and allele calling are semi-automated via the use of specific software tools, for high-throughput (Allix, Supply, & Fauville-Dufaux, 2004; Cowan et al., 2005; Supply et al., 2006). Commercially produced kits (Genoscreen) adapted for the Applied Biosystems platforms integrate ready-to-use mixes, allowing the amplification of the 24 standard loci in 6 quadruplex PCRs (Supply, Gaudin, & Raze, 2014) or of 4 hypervariable loci (Allix-Beguec et al., 2014; see Section 2.4) in a single-quadruplex assay. Calibration kits are also available, integrating customised bioinformatic modules for facilitating the analysis and the compatibility with MIRU-VNTR*plus* database (see below), and for calibrating size calling of the different alleles of each MIRU-VNTR marker, based on the use of allelic ladders. This calibration corrects for deviations between the expected sizes based on the allele sequence lengths and the actually observed sizes after an electrophoretic run, which vary according to each marker, the alleles of each marker and individual DNA analysers (Allix et al., 2004; Supply et al., 2006). Single dye-based multi-capillary systems, such as the QiaXcel platform, enabling relatively quick analysis of products from each individually amplified marker, offer an intermediate throughput between those obtained with agarose gels and multi-capillary DNA analysers coupled to multi-dye chemistry (Gauthier et al., 2015; Matsumoto et al., 2013). Separation of individual PCR fragments has also been described based on non-denaturing HPLC (Evans et al., 2004). International proficiency studies have shown that the use of automated DNA analysers with commercial kits provided the highest reproducibility of typing results (de Beer, Kremer, Kodmon, Supply, & van Soolingen, 2012), although initial evaluations and complementary reports indicated also high reproducibility of systems based on agarose gels or automated DNA analysers using in-house PCR conditions (Cowan et al., 2012; Cowan, Mosher, Diem, Massey, & Crawford, 2002; Savine et al., 2002). Further technical optimisation of 24-locus MIRU-VNTR typing compared to initial standard conditions and conditions in commercial kits has recently been claimed (de Beer et al., 2014), but this assertion has been contradicted (Supply et al., 2014).

MIRU-VNTR typing can be performed on purified DNA prepared for IS*6110* RFLP analysis (van Soolingen, Hermans, de Haas, Soll, & van Embden, 1991; Walker, Ip, et al., 2013), or on crude DNA extracts obtained by resuspending mycobacterial colonies from solid cultures or pellets from liquid culture into 100–200 ml 10 mM Tris–HCl, 1 mM EDTA (pH 7.0), followed by incubation at 95 °C for 30 min and centrifugation of the suspension to recover the supernatant containing the DNA (Mazars et al., 2001). In situations where urgent epidemiological questions are to be resolved, MIRU-VNTR typing can even be applied directly on clinical samples, although the degree of typing success depends on the quality of the sample, the bacterial load present and possibly the use of column-based DNA purification (Alonso et al., 2012; Bidovec-Stojkovic, Seme, Zolnir-Dovc, & Supply, 2014; Mokrousov et al., 2009).

The numerical format of MIRU-VNTR types has been exploited to develop databases freely accessible via the Internet, such as MIRU-VNTR*plus* (www.miru-vntrplus.org; Allix-Beguec, Harmsen, Weniger, Supply, & Niemann, 2008) and SITVIT (http://www.pasteur-guadeloupe.fr:8081/SITVIT_ONLINE/; Demay et al.,

2012). MIRU-VNTR*plus* integrates multiple functionalities for easy (non-permanent) uploading and analysis of MIRU-VNTR types (and spoligotypes) of local strains and comparison versus reference strains (see example of genotypes of reference strains from MIRU-VNTR*plus* in Figure 1) that are well characterised by classical phylogenetic markers such as genomic deletions/regions of differences. These functionalities enable the users to perform both cluster analysis for primary molecular epidemiological purposes and phylogenetic prediction of the genetic lineages of local isolates, based on similarity search and/or tree-based analysis with pre-calibrated parameters (Allix-Beguec, Harmsen, et al., 2008). Of note, as an expected consequence of its multi-locus basis, phylogenetic prediction based on standard MIRU-VNTR typing is less prone to homoplasy than spoligotyping (Cardoso Oelemann et al., 2011; Comas et al., 2009; Wirth et al., 2008). The MIRU-VNTR*plus* database also includes a universal MIRU-VNTR nomenclature system, called MLVA MtbC 15-9, reflecting the allelic combinations over the 15 most discriminatory loci and the 9 less discriminatory loci (Weniger, Krawczyk, Supply, Niemann, & Harmsen, 2010). Database users can choose to automatically generate new universal genotypes for their isolates, in case of allelic combinations that are not yet known. As of December 2014, more than 16,000 different types based on the sole 15-locus portion of the standard 24-locus types were registered (http://www.miru-vntrplus.org/MIRU/types15.faces). For comparison, at the same date, the SITVIT database included 2380 12-locus MIRU-VNTR types, as well as some 7000 spoligotype patterns (http://www.pasteur-guadeloupe.fr:8081/SITVIT_ONLINE/description.jsp#). Last but not least, MIRU-VNTR*plus* comprises detailed tutorials and helps functionalities to guide beginners, a downloadable detailed technical protocol (http://www.miru-vntrplus.org/MIRU/files/MIRU-VNTRtypingmanualv6.pdf) on the basics of MIRU-VNTR typing (and protocols on other typing methods), as well as allele calling tables applicable to fragment sizing using electrophoretic separation on agarose gels and 24-locus control types of reference strains.

Compared to MIRU-VNTR*plus*, the SITVIT database is more oriented towards the collection of typing data, as well as the analysis of type frequency and geographic distribution.

2.4 EPIDEMIOLOGICAL INFERENCES BASED ON CLASSICAL TYPING

Different population-based studies showed that the standard 24-locus format, especially when used in combination with spoligotyping, provides a predictive value comparable to IS*6110* RFLP to trace TB transmission in Western European populations (Allix-Beguec, Fauville-Dufaux, & Supply, 2008; Alonso-Rodriguez et al., 2009; Bidovec-Stojkovic, Zolnir-Dovc, & Supply, 2011; Oelemann et al., 2007). The largest demonstration of the utility of this approach was established through a 5-year nationwide survey performed in the Netherlands, comparing MIRU-VNTR typing, IS*6110* RFLP results and data from cluster investigations using almost 4000 MTBC isolates (de Beer et al., 2013). However, standard 24-locus MIRU-VNTR typing lacks resolution for accurately discriminating closely related clones of Beijing strain populations, which are predominant in certain world regions such as Russia or China. IS*6110* RFLP can possibly be used for subtyping Beijing isolates that are indistinguishable using standard 24-locus MIRU-VNTR typing (Roetzer et al., 2011). However, in addition to the other limitations of IS*6110* RFLP indicated above, IS*6110* RFLP profiles of Beijing strains can be especially complex. Therefore, an international consortium recently proposed an additional consensus set of four so-called hypervariable MIRU-VNTR loci for subtyping Beijing clonal complexes and clusters identified by standard typing (see Allix-Beguec et al., 2014; and references therein). This set was defined based on the analysis of the additional resolution power, clonal stability and inter-laboratory reproducibility demonstrated when using these markers. These parameters were evaluated using more than 600 Beijing isolates from different world regions. A corresponding kit based on a single 4-plex assay and its specific companion kit for calibration (Genoscreen) are available. Interestingly, three of these four consensus loci were also found to be important for differentiating unrelated Beijing strains specifically from China, a country where strains of this type are highly prevalent (Luo, Yang, Pang, et al., 2014).

The definition of a cluster suggestive of clonal transmission based on MIRU-VNTR typing has been established based on the measure of the rate of allelic drift of *M. tuberculosis* clones in infected patients, in actual transmission chains and in a large longitudinal outbreak spanning more than 10 years (Allix-Beguec et al., 2014; Savine et al., 2002; Supply et al., 2006). Only rare single-locus variations were observed among clonal isolates in such situations. Therefore, clustering should be based on strict identity of MIRU-VNTR alleles over the 24 standard loci (and over the 4 hypervariable loci, if used in addition), as a proxy to ongoing TB transmission in populations (Allix-Beguec et al., 2014; Savine et al., 2002; Supply et al., 2006). As another consequence of this definition, the simultaneous detection of double alleles in two MIRU-VNTR loci or more in an isolate indicates mixed infection or a mixture of strains resulting from culture cross-contamination (Allix et al., 2004; Garcia de Viedma, Alonso Rodriguez, Andres, Ruiz Serrano, & Bouza, 2005; Shamputa et al., 2006).

As indicated above, spoligotyping can be used as a convenient independent typing method, in addition to standard (and hypervariable) MIRU-VNTR typing. As a reflection of the lower resolution power of spoligotyping relative to MIRU-VNTR typing, it is expected that spoligotype changes among isolates clustered by MIRU-VNTR typing will be infrequent, and thus that the gain in total discriminatory power when using both methods in combination will be limited compared to the sole use of MIRU-VNTR typing (Allix-Beguec, Fauville-Dufaux, et al., 2008; Oelemann et al., 2007). Given that spoligotype changes are virtually never seen in clonal isolates from confirmed epidemiologically linked cases, the definition of a cluster suggestive of recent transmission in TB patient populations is then based on strict identity of both MIRU-VNTR and spoligotype profiles.

The general molecular-guided strategy for investigating TB transmission established on these grounds is described in Figure 2, which also shows the possible connection with WGS-based approaches presented below.

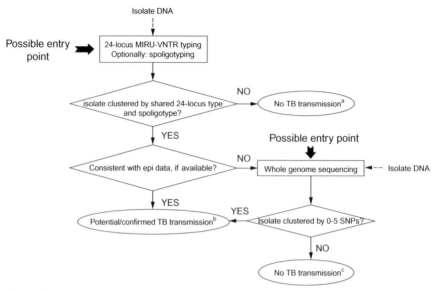

FIGURE 2

Molecular-guided strategy for investigating TB transmission. Superscripted letters indicate possible modulation of conclusions at end points (circled) as follows: a, if the isolate differs by at least two MIRU-VNTR loci relative to isolates of comparison, transmission can be confidently excluded; if the difference consists in a single-locus variation only, the conclusion on no transmission is likely, but should be compared to epidemiological data, if available, for further assessment of a specific link (see text for discussion of rare single-locus variation seen in transmission chains). b, the degree of likelihood of TB transmission will be highest if the conclusion is based on whole-genome sequencing data supported by epidemiological data, if available. c, if the genomic divergence relative to isolates of comparison is in an intermediate range of about 6–15 SNPs, the conclusion on no transmission is likely, but should be compared to epidemiological and clinical data (e.g. information on case with a particularly persistent disease), if available. Note that in the case of clusters defined by standard 24-locus MIRU-VNTR types (and possibly by spoligotypes) but not supported by available epidemiological data, an intermediate step of subtyping based on four consensus hypervariable MIRU-VNTR loci (see text) can be performed before possibly proceeding to whole-genome sequencing.

3 WHOLE-GENOME-BASED TYPING FOR MOLECULAR-GUIDED INVESTIGATION

Despite being the most advanced classical typing system for MTBC strains, MIRU-VNTR typing, as the other classical typing methods, only samples small genomic regions, used as surrogate markers of the corresponding clonal backgrounds. Although these regions were selected for their significant polymorphism, classical typing inherently cannot detect variation that can independently occur in other parts of the genome, potentially signalling larger genetic distances than could be recognised using classical typing results. This suboptimal sensitivity for detecting all possible epidemiologically significant variations may partly explain why only about 50% of the links suggested by molecular clusters based on classical typing (MIRU-VNTR typing or IS*6110* RFLP) are confirmed by classical epidemiological investigations (de Beer et al., 2013). Although it is largely used as a proxy to transmission in TB patient populations, classical typing provides us with an underestimate of this parameter in many cases, the extent of which may depend on various factors, including the diversity of the circulating strains and the actual level of recent TB transmission in the patient populations.

In contrast, by capturing (nearly) all of the genetic information available, WGS-based approaches maximise the discriminatory power and, in principle, permit the detection of all possible epidemiologically significant variation (Koser et al., 2012). In addition, WGS analysis can simultaneously deliver information relevant to patient management such as prediction of drug resistance, based on the analysis of known drug resistance-associated gene sequences. Such approaches take advantage of the rapid improvements and increasing affordability of next-generation sequencing (NGS) technologies.

WGS analysis for epidemiological investigation of TB transmission involves the following steps: (1) extraction of genomic DNA; (2) DNA fragmentation and construction of DNA libraries; (3) massively parallel sequencing; (4) analysis of sequencing results, by wgSNP and/or wgMLST; and (5) epidemiological inferences based on detected genome-wide genetic distances. These steps will be overviewed in the next sections (with steps 1 and 2 combined). Special importance will be paid to step 4, which perhaps represents the most important limiting factor today and is the key determinant of the final NGS result accuracy and of the correctness of the resulting epidemiological inferences.

3.1 DNA EXTRACTION, AND FRAGMENTATION AND CONSTRUCTION OF DNA LIBRARIES

Various DNA extraction protocols can be used, including DNA purification using CTAB (*N*-cetyl-*N,N,N*-trimethyl ammonium bromide)/NaCl similar to IS*6110* RFLP (van Soolingen et al., 1991; Walker, Ip, et al., 2013), or column-based DNA purification (Gardy et al., 2011). We have also successfully performed WGS using crude DNA extracts from heat-inactivated mycobacterial cultures, similar to those used for MIRU-VNTR typing (Mazars et al., 2001). Starting from crude, heat-inactivated extracts, DNA libraries were constructed using enzymatic fragmentation (tagmentation) with the Nextera XT kit (Illumina). While a variety of other commercial kits and physical fragmentation procedures can also be used, the Nextera XT kit has the advantage of requiring small amounts of DNA as low as 1–2 ng, therefore reducing the incubation period of the mycobacterial culture.

Even though such a small amount of DNA is still well above the amount that can in principle be obtained from 10^4 mycobacteria in 1 ml of a clinical sample, corresponding to the classical threshold of detection by microscopy, contamination with large amounts of human DNA is also an obstacle for performing direct WGS analysis on clinical samples. A recent study which included a differential lysis step to reduce contamination by human DNA demonstrated the potential of performing a culture-independent shotgun metagenomics-like WGS analysis for diagnostic detection of MTBC bacilli in sputum samples (Doughty, Sergeant, Adetifa, Antonio, & Pallen, 2014). However, even with the differential lysis step, the reference genome coverages that can be reached (from $0.002 \times$ to $0.7 \times$ in this case) are insufficient for reliable SNP analysis and epidemiological inferences.

Depending on the specifications of the DNA library construction kits used, the quality and yield of DNA libraries are verified by inspecting the DNA fragment size distribution, using electrophoretic systems such as Bioanalyzer (Agilent) or QiaXcel platforms (Qiagen).

3.2 MASSIVE PARALLEL SEQUENCING

The relative performances of the different second-generation sequencing platforms available on the market in terms of, e.g., DNA output, cost per Gb obtained, rates and types of error on read level are now relatively well established and impact on the sequencing step and the choice of the instrument (Harris et al., 2013; Junemann et al., 2013; Loman et al., 2012; Miyamoto et al., 2014). Although some initial studies used GS FLX (Roche/454) platforms to investigate WGS analysis for epidemiological investigation (Sandegren et al., 2011; Schurch et al., 2010), there has been very limited use of this technology subsequently. A single study used a SOLiD 5500XL instrument (Applied Biosystems) (Colangeli et al., 2014); however, most studies have used either the Illumina Hiseq or Miseq instruments (Gardy et al., 2011; Kato-Maeda et al., 2013; Niemann et al., 2009; Roetzer et al., 2013; Smit et al., 2015; Walker, Ip, et al., 2013; Walker et al., 2014). The advantages of using the Illumina instrument rather than Ion Torrent and Ion Proton platforms (Life Technologies) are the higher DNA sequence outputs of Illumina systems, which impact on sample multiplexing capacity, cost per Gb and effective genome coverage depth (Harris et al., 2013; Junemann et al., 2013; Loman et al., 2012; Miyamoto et al., 2014), as well as 75- to 300-bp read length ranges, which make them well adapted for SNP analysis. Read length as well as the choice of using single or paired-end reads has varied between studies that have utilised Illumina instruments (Gardy et al., 2011; Niemann et al., 2009; Roetzer et al., 2013; Smit et al., 2015; Walker, Ip, et al., 2013; Walker et al., 2014). However, reference genome coverage and potential quality of SNP analysis can be excellent even with reads as short as 50 bp as used by Gardy et al. (2011), provided correct analysis parameters are utilised (H. Pouseele & P. Supply, unpublished; see also below). Longer reads are to be preferred when approaches including a *de novo* assembly step are used, such as done in wgMLST (see Section 3.4.3), or especially when completely circularised genomes are targeted. In the latter case, single molecule, real-time (SMRT) sequencing (Pacific Bioscience), producing read lengths that can exceed 15 kbp, appears to be the best method currently (Chin et al., 2013). However, as sequence variation can be captured over more than 92–95% of the *M. tuberculosis* reference genome by classical, short read-based re-sequencing approaches, SMRT sequencing, with its higher cost per Gb of DNA sequence output, is currently used more for generating new reference genomes or investigating research-oriented questions (Miyamoto et al., 2014).

3.3 ANALYSIS OF SEQUENCING RESULTS

The wgSNP approach is the most widely used for the analysis of sequencing results and has been utilised in all studies published so far, except one (Kohl et al., 2014), which also used wgMLST. The wgSNP approach is based on mapping the sequence reads obtained from the studied isolates against a completely assembled *M. tuberculosis* reference genome (either *M. tuberculosis* H37Rv or CDC1551), followed by SNP calling on the mapped positions shared by all the isolates being investigated. However, the key parameters and a standardised pipeline for wgSNP analysis have not been established. Published studies have used different analysis parameters and pipelines, combining diverse and relatively generic mapping and SNP calling tools with home-made scripts. Lack of a standardised method has resulted in very divergent results, demonstrated by the relatively high amount of SNPs detected by WGS among outbreak isolates in some studies (e.g. Gardy et al., 2011) compared to other studies (e.g. Roetzer et al., 2013; Smit et al., 2015; Walker, Ip, et al., 2013). Identifying which parameters to use is not straightforward, because the levels of accuracy of the WGS analysis pipelines have been only partially verified in the different studies. For instance, SNPs identified by Illumina in pre-test or study samples were not verified by independent sequencing (e.g. Sanger) in two studies (Gardy et al., 2011; Walker, Ip, et al., 2013). In other cases (Kato-Maeda et al., 2013; Roetzer et al., 2013), only SNPs pre-identified by Illumina sequencing were verified by Sanger sequencing, which gives an indication of the rate of false-positive SNP calling but does not identify the rate of false-negative SNP calling (i.e. the proportion of SNPs potentially missed).

Therefore, in the following sections, we will present the main types of artefacts that can create noise in SNP analyses, illustrating them with real-life examples. Then, we will show how these artefacts can be identified and removed from the SNP analysis using the appropriate filters. It should be noted that, in contrast to previous reports, we rigorously calibrated the analysis pipeline using the aforementioned filters, to determine both the associated rates of false-positive and false-negative SNP calling. To perform this calibration, we used a unique dataset, combining SMRT sequencing or Sanger sequencing confirmed by Sequenom SNP analysis, with Illumina re-sequencing of *M. tuberculosis* H37Rv and *M. bovis* BCG Pasteur (P. Supply & H. Pouseele, unpublished). Finally, we will more briefly present alternative approaches to wgSNP analysis, including wgMLST.

3.3.1 SNP artefacts

For any analysis of molecular data, it is important to both reduce noise levels as much as possible and maximise the strength of the signal. The low genome mutation rate of MTBC makes this particularly challenging in a SNP analysis, as the signal expected in an outbreak situation, often involving closely related strains, is barely above the detection threshold, and any kind of noise might significantly interfere with the signal. Therefore, a good understanding of the biological and technical processes that adversely affect an SNP analysis is essential, in order to control and reduce their effect.

3.3.1.1 Artefacts caused by synteny breaks

Although MTBC and even *M. canettii* genomes are highly syntenic (i.e. have a highly conserved gene order/are highly collinear) (Supply et al., 2013), the genome of any tested isolate almost inevitably contains some structural rearrangements with respect to the reference genome. There are many different factors underlying such macro-variations between reference and sample genomes, such as differences in repeat structures (e.g. in MIRU-VNTR regions) (Supply et al., 2000), large deletions or insertions (so-called regions of differences or RDs, or large sequence polymorphisms (LSPs) (Brosch et al., 2002; Hirsh et al., 2004)), rearrangements caused by mobile elements such as the IS*6110* element (Brosch et al., 1999) (see Figure 3) or large translocations or inversions (Shitikov et al., 2014).

From a mapping point of view, these types of differences between reference and sample that exceed the length of a single read, or even the total length encompassed by a paired-end read set, cause the reference-based assembly process to break down. To illustrate this, let us consider a rearrangement position, that is, a position on the reference genome for which, on the 3′ side, the reference and sample genome sequences are locally syntenic/collinear (by sharing orthologous sequences), but on the 5′ side, the reference and the sample sequences are not. A read from the sample that contains this position thus captures a genetic content equivalent to that of the reference on its 3′ arm, but a distinct genetic on its 5′ arm. As this read only partially aligns properly with the reference sequence, it will be discarded or retained for mapping, depending on the rearrangement position and the minimal threshold of sequence identity accepted for mapping. Roughly speaking, if a read of length l contains a rearrangement position at position p (with p between 1 and l), then the sequence identity s of this read is approximately

$$s = \frac{p \cdot 100\% + (l-p) \cdot 25\%}{l}$$

For instance, taking a read that contains the rearrangement position as its middle, the sequence identity with the reference sequence is high (say 99%) for its 3′ end, but low (typically 25%) for its 5′ end. This yields an overall sequence identity score of about 62.5%. The use of a sequence identity threshold $s_{\min}$ lower than 100% is of course required to accommodate for sequencing errors and biological variation. With a sequence identity threshold of 90%, the read considered above will be correctly

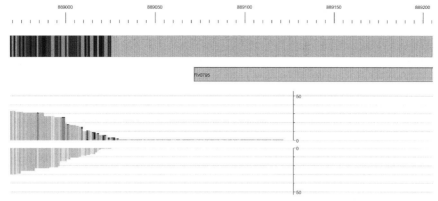

FIGURE 3

Ambiguous base calls and coverage deviations created by a synteny break linked to an IS*6110* element. From top to bottom, numbers and scale indicate genomic positions in the *M. tuberculosis* H37Rv genome, a box with four colour and grey-shaded portions corresponds to the consensus base calls and absent sequence segment for a tested isolate, a purple-coloured box shows the position of Rv0795 CDS corresponding to a transposase fragment in H37Rv and histograms show forward and reverse coverage of the reference region by sequence reads from a tested isolate (with coverage levels shown by scales on the right). In this example, an IS*6110* is located at position starting from 889021 in the *M. tuberculosis* H37Rv genome. The very low coverage observed beyond this position indicates that the mobile element is absent at this position in the tested isolate. The more strongly coloured blocks in the coverage graph from about position 889010 to 889030 indicate mismatches between consensus bases and bases in the individual reads at different positions. These mismatching bases originate from a read that was mapped on a more internal segment of the IS*6110* element, and which partially overlaps but mismatches the flanking region. (See the color plate.)

discarded. However, a grey zone exists when reads contain a rearrangement position close to their 3′ or 5′ ends. To be precise, when

$$p \leq \frac{100\% - S_{min}}{75\%} \cdot l \text{ or } p \geq \frac{s_{min} - 25\%}{75\%} \cdot l$$

the sequence identity between the read and the reference is high enough for inclusion in mapping. For instance, using a same sequence identity threshold of 90%, and read length of 100 bases, reads with rearrangement positions from 1 to 13 (so $1 \leq p \leq 13$) or from positions 87 to 100 (so $87 \leq p \leq 100$) will be accepted. This implies that, around a rearrangement position, a zone of about 25 bp contains two signals: one from correctly aligned read sides and one from incorrectly aligned read sides. This translates into unreliable consensus base calls for such 25-bp regions, usually creating a lot of ambiguous base calls, but possibly also base calls that could be considered an SNP, but are in fact artefacts. An example of such ambiguous calls, created by the presence at a specific position of an IS*6110* element in the *M. tuberculosis* H37Rv reference but not in a tested isolate, is shown in Figure 3.

In addition to high densities of ambiguous base calls, these regions of rearrangement also exhibit other characteristics. As only a fraction of the reads partially matching such regions are actually accepted for mapping, their coverage is typically several-fold lower compared to the coverage of their flanking regions (see the coverage histograms around the IS*6110* position at the bottom part of Figure 3). Assuming a random choice of starting position for the reads, one can expect to have only 25% of the reads containing the rearrangement position to be accepted. Therefore, at a specific position on the reference genome, a drop in coverage down to 25% of the coverage seen on the flanking regions may suggest a possible rearrangement position and thus a potential region with artificial sequence variation.

3.3.1.2 Artefacts caused by repeated elements

Repeated elements, and especially those that exceed the length of a read or even the total length of a paired-end read set, are another cause of uncertainty when mapping reads from a sample on a reference.

First, let us consider the case with an exactly repeated sequence, that is, a sequence that exists as two identical copies on the reference genome. Then for a read that is contained entirely within such a region, there is no way to decide which copy of the genomic region the read truly matches, as the read matches both copies equally well. In a sense, it does not really matter where this read originates from, and choosing randomly between the two possible locations is a defendable strategy.

However, in practice, the multiple copies of repeated elements are rarely completely exact, and a read does not always match a copy exactly. The strategy to randomly choose between the best positions, even if they are only marginally better than the runner-up positions, is then a frequent source for error and ambiguity.

For instance, let us consider a region that appears twice on the reference genome, and a test isolate that has an SNP in one of its corresponding copies. Assuming unbiased sequencing of either copy, half of the reads will have the SNP. Randomly mapping those reads over the two copies will lead, in the best case, to two ambiguous positions, as half of the reads will have a different base call at this position in each of the respective copies. In a worst-case scenario where coverages would by coincidence be unbalanced, the result might be an SNP position falsely identified in one or both of the two copies. The situation becomes even worse when the amount of (relative) variation among the paralogous sequences and the number of copies of such repeated sequences increase. The problem of aligning a read on the right copy in the reference becomes more difficult, as more candidate positions become equally possible. This leads to regions of ambiguous bases and possibly false SNPs.

This issue is particularly crucial for MTBC genomes, where repeated sequence-containing regions such as numerous copies of the PE_PGRS- and PPE_MPTR-encoding gene families account for >8% of the coding capacity (Cole et al., 1998). An example of the treatment of such a PE_PGRS-encoding gene region is presented further below (see Section 3.3.2.1 and Figure 5).

3.3.1.3 Artefacts caused by strand bias

As most of the sequencing errors are random, piling up of many independent reads for the same position provides a reliable way of determining the correct base call. However, problems arise in case of systematic sequencing errors such as strand-specific biases (Guo et al., 2012). Such biases and systematic errors can be caused, e.g., by certain DNA sequence patterns, such as A/T-rich regions present upstream of the sequenced segment (Oyola et al., 2014). Such biases can lead to ambiguous positions or even to artificial variation when the strand where the bias occurs is overly represented. An example of such strand-specific bias, leading to an ambiguous base call if the strand-specific variation is not filtered, is shown in Figure 4 for a read set from a *Campylobacter jejuni* isolate, mapped on a reference from the same species.

In a typical example of an MTBC genome analysed by Illumina sequencing with an overall coverage depth of >50, the number of individual bases covered at a minimum of fivefold, described by the C5 statistics, represents 97.3% of the reference genome. On this fraction, most of the bases that are not covered in both directions are typically found towards the ends of covered region and represent only about 0.6% of the reference genome. Thus, imposing a coverage of at least onefold forward and reverse coverage in addition to a total coverage of fivefold (quantified by the so-called C5C1C1 statistics) to avoid strand bias errors results in a minimal loss of information.

3.3.2 Avoiding SNP artefacts

One possible solution to tackle the problems indicated above is to use what is known about repetitive structures of the reference genome to create a mask to remove unreliable regions from the analysis. A number of studies used such an approach, by excluding from the reference sequence a quite extensive list of >250 complete genes linked to PE, PE_PGRS, PPE and PPE_MPTR family proteins, as well as transposases and phage-related proteins, based on gene annotation information (Comas et al., 2010; Roetzer et al., 2013). As an alternative annotation-based approach, other studies excluded (potential) SNPs found in such regions after mapping, resulting in the elimination of up to 9.3% of the genome (Casali et al., 2014, 2012; Gardy et al., 2011). However, this approach is both overly conservative, as such annotated regions and genes often also comprise non-repetitive parts, and incomplete, as there is a significant number of repetitive regions (including intergenic ones) and genes that are not annotated as such in reference genome databases. A more systematic approach included masking of repetitive regions by use of self-self BLAST, resulting in the exclusion of 7.4% of the genome reference (Walker et al., 2014). However, even this approach does not take into account genomic changes that are specific to the tested isolates.

In contrast, the approach described below is not based on the structure of the reference genome, but instead on the sequencing data from the tested isolates themselves. This has the advantage that the specificities of the tested samples and of the sequencing data are taken into account, maximising genome coverage while at the same time maintaining the precision. Such an approach can for instance profit

FIGURE 4

Example of strand-specific sequence bias. The figure shows an extract from an assembly of reads obtained from a *Campylobacter jejuni* isolate mapped on a reference from the same species. The position of an ambiguous base call resulting from a (not filtered) strand-specific sequence variation is highlighted in orange. Total coverage at this position is $85\times$, consisting of 28 Gs and 57 Ts. Forward coverage is $52\times$, with 28 Gs and 24 Ts. Reverse coverage is $33\times$, and all reverse mapped reads have a T at this position. This implies that the G in the forward mapped reads is a strand-specific bias. (See the color plate.)

from increased read lengths and paired-end insert sizes, which will result in increased genome coverage.

3.3.2.1 Handling ambiguous read positions

Before addressing the problem of ambiguous mapping, some introduction on the notions of mapping degeneracy and mapping discrimination is needed. Given the mapping parameters, and especially the sequence identity threshold between read and reference to accept a certain position on the reference sequence as a possible position for the read, the mapping degeneracy is the number of genomic positions/regions onto which a read can be mapped. Then, the mapping discrimination is defined as follows:

$$1 - \frac{s_2 - s_{min}}{s_1 - s_{min}}$$

where s_{min} is the sequence identity threshold, s_1 is the sequence identity of the best position for the read and s_2 is the sequence identity of the second-best position for the read. When no second-best position is available, the mapping discrimination of a read equals 1. A read with mapping discrimination of 0.5 has a best position that has an excess of sequence identity relative to the sequence identity threshold that is the double of that of the second-best position. With a sequence identity threshold s_{min} of 95%, this could for instance mean that the best position has a sequence identity of 100% and the second-best position a sequence identity of 97.5%.

Using such mapping degeneracy and mapping discrimination allows for a fine-tuned treatment of positioning ambiguous reads. Setting the accepted mapping degeneracy at 1 results in maximal stringency (i.e. no ambiguity is allowed). However, when sequencing data are paired-end, this can be calculated not at individual read level but at paired read level. In this case, the mapping degeneracy of a pair of reads is the number of pairs of positions, separated by a sequence length that is below the expected maximal insert size. The mapping degeneracy of a pair of reads is then calculated on the sum of the sequence identities relative to s_{min} for the different acceptable pairs of mapping positions. This paired-end approach allows for positioning of reads even in difficult genomic regions, when only one of the paired reads is ambiguous, so that the other read mapping on the non-repetitive flanking regions serves as a reliable anchor (Figure 5).

3.3.2.2 Handling genome rearrangements

To detect and handle artefacts linked to genome rearrangements, two complementary methods can be used. First, the fact that coverage significantly decreases around a position of rearrangement can be captured by analysis of local coverage. In our pipeline, for a region around the considered position, an average coverage is calculated over a 50-bp window that is centred on this position. Then, for the considered position, this average coverage is compared to the mean of all average coverages calculated by sliding the window over 1000 bp in either direction of that region. If the windowed average coverage centred on the considered position is too low,

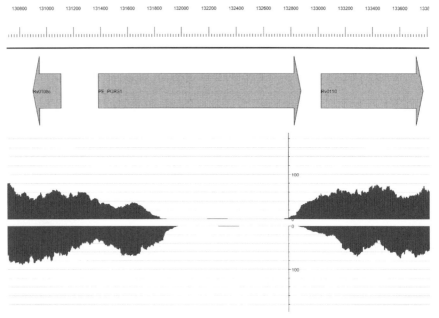

FIGURE 5

Treatment of a repetitive PE_PGRS-encoding gene. From top to bottom, numbers and scale indicate genomic positions in the *M. tuberculosis* H37Rv genome, purple-coloured boxes (grey in the print version) show the position of the PE_PGRS1 CDS as well as those of two flanking CDS and histogram curves in dark purple (dark grey in the print version) show forward and reverse coverage of the reference region by sequence reads from a tested isolate. Note how the beginning and the end of the PE_PGRS1 CDS are covered thanks to external anchoring based on paired-end information, whereas reads cannot be reliably placed on the more internal portions of the CDS (hence a zero coverage) because external anchoring is not possible on positions that are too distant from the non-repetitive flanking regions.

i.e., less than 25% of the mean obtained by sliding the window over the region, the position is labelled as unreliable. This is repeated for every position in the reference genome.

The probability of such a deviation occurring by chance is very small. In a typical example with average base coverage of 75.06 and standard deviation of 22.90, and assuming a normal distribution for the windowed average base coverage, the likelihood of having a 50-bp window with an average coverage below 18.77 is about 0.003%.

In this example, the filtering labels about 8500 bases (0.2%) as unreliable. This gives some order of magnitude of the maximal number of artefacts that could be otherwise potentially created and, at the same time, of the minimal impact of such filtering in terms of loss of information.

Second, the artefacts caused by incorrect alignment of 3′ or 5′ terminal parts of reads comprising rearrangement positions can be detected by searching for ambiguous base calls in the vicinity of a variable position. In our pipeline, such a position is labelled as unreliable when ambiguous bases are detected within a 25-bp window around this position. Moreover, as those ambiguous bases might turn up as variable positions themselves due to random coverage effects, different variable positions co-occurring within a same 25-bp surrounding region are considered unreliable. It has to be noted, however, that the latter parameter should not be used when analysing specific sequence targets that are under positive selection, such as targets associated with drug resistance.

3.3.3 Mapping analysis and SNP calling algorithm

The resulting mapping analysis and SNP calling algorithm proposed here consist of two phases. Given a reference sequence and sequencing data for a set of samples, the first phase determines, for each sample, the positions and the consensus base calls for this sample with respect to the reference. In the second phase, the SNP positions for the set of samples are determined. This SNP calling algorithm is available under Bio-Numerics® 7.6.

Phase 1: reliable base calls

1. Map the reads to the reference sequence. The mapping algorithm, while of course crucial to obtain good results, does not play a specific role in the approach we present here. The only requirement is that the mapping algorithm allows for calculations of mapping degeneracy and discrimination. A mapping algorithm that allows to do so is the BioNumerics® Power Assembler mapping algorithm.

2. Filter the read alignments, removing all reads that have a mapping degeneracy above, or mapping discrimination below, the threshold. In our case, a mapping degeneracy threshold of 1 is used, forcing the mapping discrimination to be equal to 1.

3. Calculate the forward and reverse coverage of every position of the reference genome, using only the reads that passed the filter in step 2, and use this to determine the well-covered positions in the reference genome. Using separate forward and reverse coverages (containing only reads that have been mapped in the forward or the reverse direction, respectively) allows for the exclusion of strand-specific biases. In our case, the minimum total coverage is set to 5, while the forward and reverse coverage thresholds are set to 1.

4. Determine base positions with normal coverage behaviour, by excluding all reference genome positions where the coverage is abnormally low with respect to the surrounding positions. In our algorithm, we exclude every position with windowed average coverage (using a half-window size of 25) below 25% of the average windowed average coverage of the surrounding positions (defined as a 1000-bp region centred on the position under consideration).

5. For every reliable position (based on unbiased coverage threshold (step 3) and normal coverage behaviour (step 4)), determine the consensus base calls. In our case, at a reliable position, a single-base call is made when this base is found in at least 85% of the reads at this position. If this is not the case, the appropriate double or triple base call is made, yielding an ambiguous base call labelled according to the double or triple base call detected.

Phase 2: SNP calling

1. Given, for each sample, the reliable positions and their base calls, determine the positions that are reliable for, and shared by all samples.
2. Among those universally reliable positions, determine the positions that are variable within the sample set. This implies that positions that are constant in the sample set but variable with respect to the reference sequence are not considered.
3. Filter the list of variable positions to exclude those that are too proximal. All variable positions that are within a certain range of ambiguous bases or other variable positions are discarded. In our case, a half-window size of 12 is used to define proximity.

By using the reference dataset indicated above (combining SMRT sequencing or Sanger sequencing confirmed by Sequenom SNP analysis, with Illumina re-sequencing of *M. tuberculosis* H37Rv and *M. bovis* BCG Pasteur), we determined that the use of this analysis pipeline resulted in a rate of false-positive SNP calling below the detection threshold, without a single-false SNP detected on 95.5% of the H37Rv reference effectively covered by Illumina reads. We further determined our false-negative SNP detection rate to be at most 1.6–4.7%, after classical exclusion of probable artefactual SNPs linked to indels and repeated sequences, as defined by a similarity-matching analysis. Most of the undetected SNPs actually correspond to (other) sequence positions inaccessible to mapping (P. Supply & H. Pouseele, unpublished). These results thus demonstrate the accuracy of this analysis pipeline.

3.4 ALTERNATIVE ANALYSIS METHODS

It is clear from the above description that the reference-based SNP analysis procedure, although undoubtedly powerful, has its limitations. The need for a sample to be as syntenic as possible with respect to the reference, as well as the difficulties in mapping repeated regions, reduce to a certain extent the fraction of the genome that can be used for reliable SNP detection. Differences in mobile or repeated elements between samples further reduce the fraction of the genome that can be reliably analysed across all samples.

Moreover, and more importantly, a whole-genome SNP analysis is inherently unstable. The set of positions on the reference genome that vary in a sample set completely depends on the sample set, and adding or removing a single sample can change the variable position set. A sample can introduce new variable positions or can cause variable positions to be lost because of coverage effects or vicinity artefacts. Due to the low MTBC genome mutation rate, a single-variable position being missed can significantly influence any hypothesis on transmission.

In the remainder of this section, we therefore briefly discuss two alternative analysis methods, which can help circumventing these limitations to some extent, and discuss a third most promising alternative more in depth.

3.4.1 The kSNP approach

The kSNP methodology (Gardner & Hall, 2013) is an alternative SNP calling procedure based on a different concept of position. In a reference-based system, position is defined as position with respect to the reference sequence, and the correctness of the mapping procedure that aligns reads with respect to the reference is of utmost importance. In contrast, position in the kSNP approach is defined by local neighbourhood. Two bases are at the same position if and only if a region before and/or a region after both positions are exactly the same. This implies that a kSNP algorithm does not require mapping against a reference, and can do SNP calling directly on raw reads. Moreover, as the kSNP definition of position does not allow for ambiguity in the flanking regions, there is no discussion about correct or incorrect alignment. A set of samples can then be compared by determining the bases at the same position that are present in all the samples (defining a core kSNP approach), which yields an SNP matrix, or directly a similarity matrix. Alternatively, samples can be compared in a pairwise fashion, maximising genome coverage at the cost of providing a similarity matrix only, and no SNP matrix.

The kSNP procedure overcomes a number of issues of reference-based SNP calling. First, due to its lack of ambiguity in the definition of position, there are no issues with repeated regions or mobile elements. Second, set of flanking regions can be determined from the sample itself, or better, within each pair of samples, thus avoiding the restrictions linked to the use of a single reference.

However, there are some caveats. First, an analysis technology that is based directly on raw reads is very prone to sequencing errors. Appropriate checks need to be built in to make sure that a base (and its surroundings) is correct. This can be done through individual base quality scores (for instance, using only bases with quality scores above a certain value). Also a coverage check is absolutely necessary to avoid erroneous base calls. Again, basing this on forward, reverse and total coverage is necessary. Also, an appropriate treatment of ambiguous positions is required.

The main theoretical drawback of this methodology is that it is error prone, despite the suggestions listed above, and the fact that proximal mutations cannot be detected, due to the requirement of conserved flanking regions.

3.4.2 Pairwise SNP calling based on chromosome comparison

An alternative to reference-based SNP calling that is less localised compared to kSNP is the so-called chromosome comparison. In this approach, *de novo* assembled sequences are aligned in a pairwise fashion, and the dissimilarity between two samples is based on the number of variant positions found in the pairwise sequence alignment, resulting in a (dis)similarity matrix. Tools such as MUMmer (Kurtz et al.,

2004), Mauve (Darling, Mau, & Perna, 2010) or BioNumerics' Chromosome Comparison module can be used to perform such an analysis.

Similar to the kSNP approach, this procedure avoids the restrictive use of a reference sequence, but in contrast to the former method, the output is more directly making biological sense, thanks to the intrinsic integration of gene synteny information. However, as with reference-based alignment, this approach is also subject to alignment artefacts linked to repetitive sequences. As an example showing the potential extent of the problem, while the original comparison of the *M. bovis* BCG Pasteur genome with that of *M. bovis* AF2122/97 identified 736 SNPs using classical Sanger sequencing and the diffseq tool from the EMBOSS package (Brosch et al., 2007) (http://emboss.sourceforge.net/), an alignment and subsequent SNP analysis of the same complete chromosome sequences using Mauve yields a total of 974 SNP positions, potentially overestimating the variation between these two genomes by more than 32%.

3.4.3 Whole genome-based MLST

All of the above SNP procedures do not fundamentally address the issue of instability, as the set of variables that are considered when comparing samples are not universally defined beforehand. Such stable upfront definition is possible with approaches extending the principle of MLST (Maiden et al., 1998). In this approach, the analysis is based on a predefined schema of loci and their alleles. A locus is a genomic region that captures a certain biological trait of the organism. On this basis, classical MLST targets sequence segments from six to seven housekeeping genes, assumed to reflect the overall evolutionary history of the genome. Each sequence variant defines a specific gene allele, and the combination of each allele of the six to seven targets defines a sequence type (ST), which constitutes a portable and universal genetic identifier of clones and clonal complexes. Genetic relationships are analysed based on the comparison of allelic identity or differences, regardless of the number of SNPs that differ between two alleles of a gene, which simplifies the analysis. In addition, the influence of an importation event of a distant sequence (differing by multiple SNPs), due to horizontal gene transfer/recombination between two distant strain types, is attenuated when comparing a strain to another, otherwise very closely related strain. Therefore, this system is better at preserving the phylogenetic signal conveyed by genetic variation that was vertically transmitted in the other target genes. However, the latter advantage applies only to bacterial species prone to frequent inter-strain recombination, which is not the case for classical strains of tubercle bacilli (Hirsh et al., 2004; Supply et al., 2013, 2003).

In a genome-wide MLST context, a locus usually is a coding sequence (CDS), but could in principle also be a genomic feature such as a spoligotyping spacer. In any case, an allele is then a sequence variant of the locus. Although synteny and *in silico* flanking primers can play a role, a locus usually is defined by local sequence alignment and similarity: basically, if a genomic sequence closely resembles a known allele of a locus, it is considered as an allele of that locus. For CDS loci, start and stop codons are taken into account to determine the allele, as alignment can be uncertain especially close to the ends of the allele.

The effect using a predefined set of variables to compare samples can best be seen in a core genome approach (i.e. cgMLST). For MTBC, this system has been developed in Kohl et al. (2014) and is available through the publicly available platform BIGSDb (Jolley & Maiden, 2010), and the commercial software packages SeqSphere + (Ridom) and BioNumerics (Applied Maths). The cgMLST schema for MTBC contains in total 3362 CDS loci, defined by their common presence in *M. tuberculosis* H37Rv (used as a starting reference) and in six other MTBC genome references, which are assumed to be largely present in all other MTBC samples. In this schema, all CDSs associated to repetitive regions listed in Comas et al. (2010) were discarded, because they cannot be easily reconstructed. In the proof-of-principle study by Kohl et al., cgMLST based on this set of 3362 loci was shown to be almost as discriminatory as wgSNP to resolve isolates from a selected MTBC outbreak. However, in other cases, outbreak isolates might not be totally resolved, e.g., because of a few possible SNPs in intergenic regions or non-core (accessory) genes that are not captured by such cgMLST, which might prevent possible fine-tuned dissection of transmission chains. Thus, the cgMLST approach intrinsically sacrifices some discriminatory power. Therefore, such MLST approach is more primarily useful for generating universal strain identifiers based on a common reference set of core genes defined at genome-wide level, while wgSNP remains the approach of choice when maximal resolution power is desired for fine-tuned wg-based epidemiological investigation.

However, it is noteworthy that cgMLST nomenclature efforts need to take into account the fact that, although the core gene set has been defined as being shared by selected MTBC genome references, each specific MTBC strain will miss some of the core loci, due to variation in both sequencing coverage and core genomes among strains.

As an alternative to cgMLST, and an extension thereof, a pan-genomic MLST approach (pgMLST) was very recently developed. Under this system, an extended schema of loci can be created based on the cumulative compilation of genes, e.g., both from MTBC and from *M. canettii* (representing ancestral lineages of the tubercle bacilli (Supply et al., 2013)) reference genomes, providing thus more genome coverage and resolution power compared to cgMLST. This pgMLST scheme is available through BioNumerics.

3.5 EPIDEMIOLOGICAL INFERENCES BASED ON WGS

As for classical typing, the relevance of epidemiological inferences based on WGS relies on a properly calibrated definition of a WGS-based molecular cluster. This definition obviously depends on the actual genome mutation rate during infection and human transmission. This parameter has been estimated by different studies. Two initial reports found no more than 4 SNPs (Sandegren et al., 2011; Schurch et al., 2010) between the genomes of isolate pairs from well-characterised transmission chains separated by 9–12 years and several intermediate transmission events. Two subsequent reports studied the genome evolution rate more systematically, based on larger numbers of isolates collected from chronically infected or epidemiologically linked patients. In a first study of a specific longitudinal TB outbreak in Northern Germany, about 60% of the isolates within any of eight well-identified transmission chains showed no SNP differences, while a maximum of 3 SNPs were identified among the remaining isolates that were part of a same transmission chain (Roetzer et al., 2013). Along the same lines, another study investigating longitudinal isolates, from various genetic backgrounds, collected at different time points from individual patients or from different patients in various household outbreaks in the UK Midlands found that over 95% of the corresponding genomes differed by 5 or fewer SNPs over a few-year distance. The few remaining cases, differing by no more than 11 SNPs, essentially corresponded to special cases of pulmonary infection persisting during up to 10 years, linked to development of drug resistance (Walker, Ip, et al., 2013). In accordance, both studies estimated closely matching genome evolution rates of 0.4–0.5 SNP per genome per year. Interestingly, similar values of wg mutation rates were estimated during active disease, as during latency, when using an experimental simian infection model (Ford et al., 2011). Further in accordance with these findings, from 0 to a maximum of 2 SNPs per transmission event were found when other specific TB outbreaks spanning a few years were studied by WGS (Kato-Maeda et al., 2013; Smit et al., 2015). Finally, a recent study used systematic WGS to investigate the epidemiology of TB transmission among 247 TB cases in an unselected population in Oxfordshire (UK) over 6 years. The authors initially defined an upper threshold of 12 SNPs separating two isolates to define a genome-based cluster, a limit that was conservatively set according to the extreme level of divergence noted between epidemiologically related isolates and isolates within hosts, especially those with a persistent disease (Walker, Ip, et al., 2013). Interestingly, the actual differences within the 13 initially so-defined genomic clusters identified all ranged from 0 to 7 SNPs (median 1 SNP), with most of them again not exceeding 4 SNPs.

In this context, the report of Gardy et al. (2011) appears as a clear outlier. In contrast to the above reports, this study found not less than 12 SNPs separating the most closely isolates from an outbreak that occurred over a 3-year period in Canada. We found that these results were most likely due to the analysis parameters used, rather than issues with the primary WGS data. Indeed, upon re-analysis of the same original sequence read set using our pipeline with the calibrated parameters described above, we found no more than 3 SNPs separating the 32 outbreak isolates, with a majority of cases clustered by a single-genome ST (H. Pouseele & P. Supply, unpublished).

Thus, the overall picture that emerges from the studies above converges to a range from 0 to 5 SNPs, as a reasonable threshold for defining a wg-based cluster in prospective studies of TB transmission. Integration of this parameter in the proposed general molecular-guided strategy for investigating TB transmission is shown in Figure 2. This proposed general strategy includes the option to start directly with WGS analysis of MTBC isolates, or to first use MIRU-VNTR typing optionally combined with spoligotyping, before considering WGS in the case of MIRU-VNTR (and spoligotype)-based clusters not supported by epidemiological data, if available.

Although currently available data suggest that the limit of 0–5 SNPs will be valid in many cases, it is important to note that it will not necessarily be universally respected or applicable. As indicated earlier, clonal variants within particular patients with an especially persistent infection might show a higher than average pairwise divergence, of up to 11 SNPs in the study of Walker, Monk, Grace Smith and Peto (2013). A few other cases with clonal variants showing an apparently higher than average divergence were identified by another study, identifying up to 7 SNPs between isolates obtained from a same patient and 14 SNPs between isolates from two epidemiologically linked isolates (although possible uncaptured intermediate cases were not excluded) (Perez-Lago et al., 2014). Conversely, such a limit of 0–5 SNPs might still cluster a number of cases that are actually more distant than what this limited divergence may initially suggest. Like for classical typing, albeit to a lower extent, links suggested by genome-based clustering, even in the case of completely matching genome sequences, are not systematically supported by epidemiological investigation (Walker et al., 2014). Whereas this might reflect the difficulty in linking cases resulting, e.g., from casual and/or trans-regional contacts, such discordances might also result from the intrinsic stochasticity around the slow average mutation rate of 0.4–0.5 SNP per genome per year, which might lead to (a minority of) strains sharing genomes with ≤ 5 SNPs but descending from a common ancestor more than a decade ago. Furthermore, as indirectly suggested by a recent report (Luo, Yang, Peng, et al., 2014), the particular genetic homogeneity of *M. tuberculosis* strains in some world regions dominated by a specific genetic lineage, such as Beijing, might conceivably require the use of a more stringent divergence threshold (i.e. closer to 0 SNP), in order to more specifically distinguish cases directly part of an outbreak from unrelated cases caused by highly endemic strains.

Because of the vertical inheritance and accumulation of mutations and the scarceness of reversions in MTBC, the use of WGS can provide also useful information on the directionality of transmission and on the status or potential presence of particularly contagious cases (so-called super-spreaders) in the population considered. Such super-spreader cases can in principle be identified by detection of star-like patterns in the SNP-based trees, indicating multiple secondary cases from a same common node representing a particularly contagious subject (Gardy et al., 2011; Perez-Lago et al., 2014; Walker, Ip, et al., 2013; Walker et al., 2014; Walker, Monk, et al., 2013). Moreover, the presence of a potentially missing super-spreader can be postulated when using analysis algorithms allowing possible hypothetical nodes, such as minimum spanning and maximum likelihood trees (available, e.g., under BioNumerics), or median joining network (Gardy et al., 2011; Perez-Lago et al., 2014; Walker, Ip, et al., 2013; Walker et al., 2014; Walker, Monk, et al., 2013). However, such analysis aiming at refining the epidemiological linkages within a cluster/an outbreak requires extremely accurate SNP calling pipeline, as even single SNP differences can dramatically affect the structure of connections among very closely related cases in the phylogenetic trees obtained. Moreover, the potential coexistence and differential transmission of clonal variants, particularly in cases of persistent disease with levels of divergence sometimes even exceeding the general limit of ≤ 5 SNPs, might blur inferences made on super-spreader cases (Perez-Lago et al., 2014).

CONCLUDING REMARKS

Clearly, WGS-based approaches hold great promise for improved molecular-guided surveillance and control of TB. Its intrinsic superior resolution power compared to classical typing is expected to result in more sensitive detection of clusters of ongoing transmission. Moreover, bacterial WGS can provide unique molecular information on the directionality of transmission and on the occurrence of particularly infectious cases in the population, which is particularly valuable to guide efforts in outbreak investigation and control. However, in contrast to classical typing systems, methodologies of genome-based typing of MTBC have not been standardised yet and are still evolving, partially as a reflection of the constant progress in NGS technologies. In this context, this chapter aimed to provide the reader with an up-to-date overview of current methods and their limitations, of critical parameters influencing the accuracy of WGS analysis, and proposed a general strategy for WGS-guided epidemiological investigation of TB.

ACKNOWLEDGEMENTS

Parts of this chapter result from research supported by a grant from the European Union's Seventh Framework Program (FP7/2007-2013) under Grant Agreement No. 278864 in the framework of the PathoNGenTrace project.

Competing financial interests: P.S. is a consultant for Genoscreen and H.P. is an employee of Applied Maths.

REFERENCES

Allix, C., Supply, P., & Fauville-Dufaux, M. (2004). Utility of fast mycobacterial interspersed repetitive unit-variable number tandem repeat genotyping in clinical mycobacteriological analysis. *Clinical Infectious Diseases*, *39*, 783–789.

Allix-Beguec, C., Fauville-Dufaux, M., & Supply, P. (2008). Three-year population-based evaluation of standardized mycobacterial interspersed repetitive-unit-variable-number tandem-repeat typing of Mycobacterium tuberculosis. *Journal of Clinical Microbiology*, *46*, 1398–1406.

Allix-Beguec, C., Harmsen, D., Weniger, T., Supply, P., & Niemann, S. (2008). Evaluation and strategy for use of MIRU-VNTRplus, a multifunctional database for online analysis of genotyping data and phylogenetic identification of Mycobacterium tuberculosis complex isolates. *Journal of Clinical Microbiology*, *46*, 2692–2699.

Allix-Beguec, C., Wahl, C., Hanekom, M., Nikolayevskyy, V., Drobniewski, F., Maeda, S., et al. (2014). Proposal of a consensus set of hypervariable mycobacterial interspersed repetitive-unit-variable-number tandem-repeat loci for subtyping of Mycobacterium tuberculosis Beijing isolates. *Journal of Clinical Microbiology, 52*, 164–172.

Alonso, M., Herranz, M., Martinez Lirola, M., Gonzalez-Rivera, M., Bouza, E., & Garcia de Viedma, D. (2012). Real-time molecular epidemiology of tuberculosis by direct genotyping of smear-positive clinical specimens. *Journal of Clinical Microbiology, 50*, 1755–1757.

Alonso-Rodriguez, N., Martinez-Lirola, M., Sanchez, M. L., Herranz, M., Penafiel, T., Bonillo Mdel, C., et al. (2009). Prospective universal application of mycobacterial interspersed repetitive-unit-variable-number tandem-repeat genotyping to characterize Mycobacterium tuberculosis isolates for fast identification of clustered and orphan cases. *Journal of Clinical Microbiology, 47*, 2026–2032.

Arbeit, R. D., Arthur, M., Dunn, R., Kim, C., Selander, R. K., & Goldstein, R. (1990). Resolution of recent evolutionary divergence among Escherichia coli from related lineages: The application of pulsed field electrophoresis to molecular epidemiology. *The Journal of Infectious Diseases, 161*, 230–235.

Barnes, P. F., & Cave, M. D. (2003). Molecular epidemiology of tuberculosis. *The New England Journal of Medicine, 349*, 1149–1156.

Bidovec-Stojkovic, U., Seme, K., Zolnir-Dovc, M., & Supply, P. (2014). Prospective genotyping of Mycobacterium tuberculosis from fresh clinical samples. *PLoS One, 9*, e109547.

Bidovec-Stojkovic, U., Zolnir-Dovc, M., & Supply, P. (2011). One year nationwide evaluation of 24-locus MIRU-VNTR genotyping on Slovenian Mycobacterium tuberculosis isolates. *Respiratory Medicine, 105*(Suppl. 1), S67–S73.

Bifani, P. J., Plikaytis, B. B., Kapur, V., Stockbauer, K., Pan, X., Lutfey, M. L., et al. (1996). Origin and interstate spread of a New York City multidrug-resistant Mycobacterium tuberculosis clone family. *JAMA, 275*, 452–457.

Brosch, R., Gordon, S. V., Garnier, T., Eiglmeier, K., Frigui, W., Valenti, P., et al. (2007). Genome plasticity of BCG and impact on vaccine efficacy. *Proceedings of the National Academy of Sciences of the United States of America, 104*, 5596–5601.

Brosch, R., Gordon, S. V., Marmiesse, M., Brodin, P., Buchrieser, C., Eiglmeier, K., et al. (2002). A new evolutionary scenario for the Mycobacterium tuberculosis complex. *Proceedings of the National Academy of Sciences of the United States of America, 99*, 3684–3689.

Brosch, R., Philipp, W. J., Stavropoulos, E., Colston, M. J., Cole, S. T., & Gordon, S. V. (1999). Genomic analysis reveals variation between Mycobacterium tuberculosis H37Rv and the attenuated M. tuberculosis H37Ra strain. *Infection and Immunity, 67*, 5768–5774.

Brudey, K., Driscoll, J. R., Rigouts, L., Prodinger, W. M., Gori, szA., Al-Hajoj, S. A., et al. (2006). Mycobacterium tuberculosis complex genetic diversity: Mining the fourth international spoligotyping database (SpolDB4) for classification, population genetics and epidemiology. *BMC Microbiology, 6*, 23.

Cardoso Oelemann, M., Gomes, H. M., Willery, E., Possuelo, L., Batista Lima, K. V., Allix-Beguec, C., et al. (2011). The forest behind the tree: Phylogenetic exploration of a dominant Mycobacterium tuberculosis strain lineage from a high tuberculosis burden country. *PLoS One, 6*, e18256.

Casali, N., Nikolayevskyy, V., Balabanova, Y., Harris, S. R., Ignatyeva, O., Kontsevaya, I., et al. (2014). Evolution and transmission of drug-resistant tuberculosis in a Russian population. *Nature Genetics, 46*, 279–286.

Casali, N., Nikolayevskyy, V., Balabanova, Y., Ignatyeva, O., Kontsevaya, I., Harris, S. R., et al. (2012). Microevolution of extensively drug-resistant tuberculosis in Russia. *Genome Research, 22*, 735–745.

Chin, C. S., Alexander, D. H., Marks, P., Klammer, A. A., Drake, J., Heiner, C., et al. (2013). Nonhybrid, finished microbial genome assemblies from long-read SMRT sequencing data. *Nature Methods, 10*, 563–569.

Coitinho, C., Greif, G., Robello, C., Laserra, P., Willery, E., & Supply, P. (2013). Rapidly progressing tuberculosis outbreak in a very low risk group. *European Respiratory Journal, 43*(3), 903–906.

Colangeli, R., Arcus, V. L., Cursons, R. T., Ruthe, A., Karalus, N., Coley, K., et al. (2014). Whole genome sequencing of Mycobacterium tuberculosis reveals slow growth and low mutation rates during latent infections in humans. *PLoS One, 9*, e91024.

Colditz, G. A., Brewer, T. F., Berkey, C. S., Wilson, M. E., Burdick, E., Fineberg, H. V., et al. (1994). Efficacy of BCG vaccine in the prevention of tuberculosis. Meta-analysis of the published literature. *JAMA, 271*, 698–702.

Cole, S. T., Brosch, R., Parkhill, J., Garnier, T., Churcher, C., Harris, D., et al. (1998). Deciphering the biology of Mycobacterium tuberculosis from the complete genome sequence. *Nature, 393*, 537–544.

Comas, I., Chakravartti, J., Small, P. M., Galagan, J., Niemann, S., Kremer, K., et al. (2010). Human T cell epitopes of Mycobacterium tuberculosis are evolutionarily hyperconserved. *Nature Genetics, 42*, 498–503.

Comas, I., Homolka, S., Niemann, S., & Gagneux, S. (2009). Genotyping of genetically monomorphic bacteria: DNA sequencing in Mycobacterium tuberculosis highlights the limitations of current methodologies. *PLoS One, 4*, e7815.

Cowan, L. S., Diem, L., Brake, M. C., & Crawford, J. T. (2004). Transfer of a Mycobacterium tuberculosis genotyping method, Spoligotyping, from a reverse line-blot hybridization, membrane-based assay to the Luminex multianalyte profiling system. *Journal of Clinical Microbiology, 42*, 474–477.

Cowan, L. S., Diem, L., Monson, T., Wand, P., Temporado, D., Oemig, T. V., et al. (2005). Evaluation of a two-step approach for large-scale, prospective genotyping of Mycobacterium tuberculosis isolates in the United States. *Journal of Clinical Microbiology, 43*, 688–695.

Cowan, L. S., Hooks, D. P., Christianson, S., Sharma, M. K., Alexander, D. C., Guthrie, J. L., et al. (2012). Evaluation of mycobacterial interspersed repetitive-unit-variable-number tandem-repeat genotyping as performed in laboratories in Canada, France, and the United States. *Journal of Clinical Microbiology, 50*, 1830–1831, author reply 1832.

Cowan, L. S., Mosher, L., Diem, L., Massey, J. P., & Crawford, J. T. (2002). Variable-number tandem repeat typing of Mycobacterium tuberculosis isolates with low copy numbers of IS6110 by using mycobacterial interspersed repetitive units. *Journal of Clinical Microbiology, 40*, 1592–1602.

Darling, A. E., Mau, B., & Perna, N. T. (2010). ProgressiveMauve: Multiple genome alignment with gene gain, loss and rearrangement. *PLoS One, 5*, e11147.

de Beer, J. L., Akkerman, O. W., Schurch, A. C., Mulder, A., van der Werf, T. S., van der Zanden, A. G., et al. (2014). Optimization of standard in-house 24-locus variable-number tandem-repeat typing for Mycobacterium tuberculosis and its direct application to clinical material. *Journal of Clinical Microbiology, 52,* 1338–1342.

de Beer, J. L., Kremer, K., Kodmon, C., Supply, P., & van Soolingen, D. (2012). First worldwide proficiency study on variable-number tandem-repeat typing of Mycobacterium tuberculosis complex strains. *Journal of Clinical Microbiology, 50,* 662–669.

de Beer, J. L., van Ingen, J., de Vries, G., Erkens, C., Sebek, M., Mulder, A., et al. (2013). Comparative study of IS6110 restriction fragment length polymorphism and variable-number tandem-repeat typing of Mycobacterium tuberculosis isolates in the Netherlands, based on a 5-year nationwide survey. *Journal of Clinical Microbiology, 51,* 1193–1198.

Demay, C., Liens, B., Burguiere, T., Hill, V., Couvin, D., Millet, J., et al. (2012). SITVIT-WEB—A publicly available international multimarker database for studying Mycobacterium tuberculosis genetic diversity and molecular epidemiology. *Infection, Genetics and Evolution, 12,* 755–766.

Doughty, E. L., Sergeant, M. J., Adetifa, I., Antonio, M., & Pallen, M. J. (2014). Culture-independent detection and characterisation of Mycobacterium tuberculosis and M. africanum in sputum samples using shotgun metagenomics on a benchtop sequencer. *PeerJ, 2,* e585.

Dye, C. (2011). Practical preventive therapy for tuberculosis? *The New England Journal of Medicine, 365,* 2230–2231.

Evans, J. T., Hawkey, P. M., Smith, E. G., Boese, K. A., Warren, R. E., & Hong, G. (2004). Automated high-throughput mycobacterial interspersed repetitive unit typing of Mycobacterium tuberculosis strains by a combination of PCR and nondenaturing high-performance liquid chromatography. *Journal of Clinical Microbiology, 42,* 4175–4180.

Fabre, M., Koeck, J. L., Le Fleche, P., Simon, F., Herve, V., Vergnaud, G., et al. (2004). High genetic diversity revealed by variable-number tandem repeat genotyping and analysis of hsp65 gene polymorphism in a large collection of "Mycobacterium canettii" strains indicates that the M. tuberculosis complex is a recently emerged clone of "M. canettii". *Journal of Clinical Microbiology, 42,* 3248–3255.

Fok, A., Numata, Y., Schulzer, M., & FitzGerald, M. J. (2008). Risk factors for clustering of tuberculosis cases: A systematic review of population-based molecular epidemiology studies. *The International Journal of Tuberculosis and Lung Disease, 12,* 480–492.

Ford, C. B., Lin, P. L., Chase, M. R., Shah, R. R., Iartchouk, O., Galagan, J., et al. (2011). Use of whole genome sequencing to estimate the mutation rate of Mycobacterium tuberculosis during latent infection. *Nature Genetics, 43,* 482–486.

Frothingham, R. (1995). Differentiation of strains in Mycobacterium tuberculosis complex by DNA sequence polymorphisms, including rapid identification of M. bovis BCG. *Journal of Clinical Microbiology, 33,* 840–844.

Frothingham, R., & Meeker-O'Connell, W. A. (1998). Genetic diversity in the Mycobacterium tuberculosis complex based on variable numbers of tandem DNA repeats. *Microbiology, 144*(Pt 5), 1189–1196.

Gagneux, S., DeRiemer, K., Van, T., Kato-Maeda, M., de Jong, B. C., Narayanan, S., et al. (2006). Variable host-pathogen compatibility in Mycobacterium tuberculosis. *Proceedings of the National Academy of Sciences of the United States of America, 103,* 2869–2873.

Garcia de Viedma, D., Alonso Rodriguez, N., Andres, S., Ruiz Serrano, M. J., & Bouza, E. (2005). Characterization of clonal complexity in tuberculosis by mycobacterial interspersed repetitive unit-variable-number tandem repeat typing. *Journal of Clinical Microbiology, 43,* 5660–5664.

Gardner, S. N., & Hall, B. G. (2013). When whole-genome alignments just won't work: kSNP v2 software for alignment-free SNP discovery and phylogenetics of hundreds of microbial genomes. *PLoS One, 8*, e81760.

Gardy, J. L., Johnston, J. C., Ho Sui, S. J., Cook, V. J., Shah, L., Brodkin, E., et al. (2011). Whole-genome sequencing and social-network analysis of a tuberculosis outbreak. *The New England Journal of Medicine, 364*, 730–739.

Gauthier, M., Bidault, F., Mosnier, A., Bablishvili, N., Tukvadze, N., Somphavong, S., et al. (2015). High-throughput MIRU-VNTR genotyping for Mycobacterium tuberculosis epidemiological studies. *Journal of Clinical Microbiology, 53*(2), 498–503.

Guo, Y., Li, J., Li, C. I., Long, J., Samuels, D. C., & Shyr, Y. (2012). The effect of strand bias in Illumina short-read sequencing data. *BMC Genomics, 13*, 666.

Gutierrez, M. C., Ahmed, N., Willery, E., Narayanan, S., Hasnain, S. E., Chauhan, D. S., et al. (2006). Predominance of ancestral lineages of Mycobacterium tuberculosis in India. *Emerging Infectious Diseases, 12*, 1367–1374.

Gutierrez, M. C., Brisse, S., Brosch, R., Fabre, M., Omais, B., Marmiesse, M., et al. (2005). Ancient origin and gene mosaicism of the progenitor of *Mycobacterium tuberculosis*. *PLoS Pathogens, 1*, e5.

Harris, S. R., Torok, M. E., Cartwright, E. J., Quail, M. A., Peacock, S. J., & Parkhill, J. (2013). Read and assembly metrics inconsequential for clinical utility of whole-genome sequencing in mapping outbreaks. *Nature Biotechnology, 31*, 592–594.

Hirsh, A. E., Tsolaki, A. G., DeRiemer, K., Feldman, M. W., & Small, P. M. (2004). Stable association between strains of Mycobacterium tuberculosis and their human host populations. *Proceedings of the National Academy of Sciences of the United States of America, 101*, 4871–4876.

Jenkins, H. E., Gegia, M., Furin, J., Kalandadze, I., Nanava, U., Chakhaia, T., et al. (2014). Geographical heterogeneity of multidrug-resistant tuberculosis in Georgia, January 2009 to June 2011. *Euro Surveillance, 19*, 1–10.

Jolley, K. A., & Maiden, M. C. (2010). BIGSdb: Scalable analysis of bacterial genome variation at the population level. *BMC Bioinformatics, 11*, 595.

Junemann, S., Sedlazeck, F. J., Prior, K., Albersmeier, A., John, U., Kalinowski, J., et al. (2013). Updating benchtop sequencing performance comparison. *Nature Biotechnology, 31*, 294–296.

Kamerbeek, J., Schouls, L., Kolk, A., van Agterveld, M., van Soolingen, D., Kuijper, S., et al. (1997). Simultaneous detection and strain differentiation of Mycobacterium tuberculosis for diagnosis and epidemiology. *Journal of Clinical Microbiology, 35*, 907–914.

Kato-Maeda, M., Ho, C., Passarelli, B., Banaei, N., Grinsdale, J., Flores, L., et al. (2013). Use of whole genome sequencing to determine the microevolution of Mycobacterium tuberculosis during an outbreak. *PLoS One, 8*, e58235.

Keim, P., Klevytska, A. M., Price, L. B., Schupp, J. M., Zinser, G., Smith, K. L., et al. (1999). Molecular diversity in Bacillus anthracis. *Journal of Applied Microbiology, 87*, 215–217.

Kohl, T. A., Diel, R., Harmsen, D., Rothganger, J., Walter, K. M., Merker, M., et al. (2014). Whole-genome-based Mycobacterium tuberculosis surveillance: A standardized, portable, and expandable approach. *Journal of Clinical Microbiology, 52*, 2479–2486.

Koser, C. U., Ellington, M. J., Cartwright, E. J., Gillespie, S. H., Brown, N. M., Farrington, M., et al. (2012). Routine use of microbial whole genome sequencing in diagnostic and public health microbiology. *PLoS Pathogens, 8*, e1002824.

Kremer, K., Glynn, J. R., Lillebaek, T., Niemann, S., Kurepina, N. E., Kreiswirth, B. N., et al. (2004). Definition of the Beijing/W lineage of Mycobacterium tuberculosis on the basis of genetic markers. *Journal of Clinical Microbiology, 42*, 4040–4049.

Kremer, K., van Soolingen, D., Frothingham, R., Haas, W. H., Hermans, P. W., Martin, C., et al. (1999). Comparison of methods based on different molecular epidemiological markers for typing of Mycobacterium tuberculosis complex strains: Interlaboratory study of discriminatory power and reproducibility. *Journal of Clinical Microbiology, 37*, 2607–2618.

Kurtz, S., Phillippy, A., Delcher, A. L., Smoot, M., Shumway, M., Antonescu, C., et al. (2004). Versatile and open software for comparing large genomes. *Genome Biology, 5*, R12.

Le Fleche, P., Fabre, M., Denoeud, F., Koeck, J. L., & Vergnaud, G. (2002). High resolution, on-line identification of strains from the Mycobacterium tuberculosis complex based on tandem repeat typing. *BMC Microbiology, 2*, 37.

Loman, N. J., Misra, R. V., Dallman, T. J., Constantinidou, C., Gharbia, S. E., Wain, J., et al. (2012). Performance comparison of benchtop high-throughput sequencing platforms. *Nature Biotechnology, 30*, 434–439.

Luo, T., Yang, C., Pang, Y., Zhao, Y., Mei, J., & Gao, Q. (2014). Development of a hierarchical variable-number tandem repeat typing scheme for Mycobacterium tuberculosis in China. *PLoS One, 9*, e89726.

Luo, T., Yang, C., Peng, Y., Lu, L., Sun, G., Wu, J., et al. (2014). Whole-genome sequencing to detect recent transmission of Mycobacterium tuberculosis in settings with a high burden of tuberculosis. *Tuberculosis (Edinburgh, Scotland), 94*, 434–440.

Magdalena, J., Supply, P., & Locht, C. (1998). Specific differentiation between Mycobacterium bovis BCG and virulent strains of the Mycobacterium tuberculosis complex. *Journal of Clinical Microbiology, 36*, 2471–2476.

Maiden, M. C., Bygraves, J. A., Feil, E., Morelli, G., Russell, J. E., Urwin, R., et al. (1998). Multilocus sequence typing: A portable approach to the identification of clones within populations of pathogenic microorganisms. *Proceedings of the National Academy of Sciences of the United States of America, 95*, 3140–3145.

Makarova, K. S., Haft, D. H., Barrangou, R., Brouns, S. J., Charpentier, E., Horvath, P., et al. (2011). Evolution and classification of the CRISPR-Cas systems. *Nature Reviews. Microbiology, 9*, 467–477.

Matsumoto, T., Koshii, Y., Sakane, K., Murakawa, T., Hirayama, Y., Yoshida, H., et al. (2013). A novel approach to automated genotyping of Mycobacterium tuberculosis using a panel of 15 MIRU VNTRs. *Journal of Microbiological Methods, 93*, 239–241.

Mazars, E., Lesjean, S., Banuls, A. L., Gilbert, M., Vincent, V., Gicquel, B., et al. (2001). High-resolution minisatellite-based typing as a portable approach to global analysis of Mycobacterium tuberculosis molecular epidemiology. *Proceedings of the National Academy of Sciences of the United States of America, 98*, 1901–1906.

Miyamoto, M., Motooka, D., Gotoh, K., Imai, T., Yoshitake, K., Goto, N., et al. (2014). Performance comparison of second- and third-generation sequencers using a bacterial genome with two chromosomes. *BMC Genomics, 15*, 699.

Mokrousov, I., Valcheva, V., Sovhozova, N., Aldashev, A., Rastogi, N., & Isakova, J. (2009). Penitentiary population of Mycobacterium tuberculosis in Kyrgyzstan: Exceptionally high prevalence of the Beijing genotype and its Russia-specific subtype. *Infection, Genetics and Evolution, 9*, 1400–1405.

Niemann, S., Koser, C. U., Gagneux, S., Plinke, C., Homolka, S., Bignell, H., et al. (2009). Genomic diversity among drug sensitive and multidrug resistant isolates of Mycobacterium tuberculosis with identical DNA fingerprints. *PLoS One*, *4*, e7407.

Niemann, S., & Supply, P. (2014). Diversity and evolution of Mycobacterium tuberculosis: Moving to whole-genome-based approaches. *Cold Spring Harbor Perspectives in Medicine*, *4*, a021188.

Oelemann, M. C., Diel, R., Vatin, V., Haas, W., Rusch-Gerdes, S., Locht, C., et al. (2007). Assessment of an optimized mycobacterial interspersed repetitive-unit-variable-number tandem-repeat typing system combined with spoligotyping for population-based molecular epidemiology studies of tuberculosis. *Journal of Clinical Microbiology*, *45*, 691–697.

Oyola, S. O., Manske, M., Campino, S., Claessens, A., Hamilton, W. L., Kekre, M., et al. (2014). Optimized whole-genome amplification strategy for extremely AT-biased template. *DNA Research*, *21*, 661–671.

Perez-Lago, L., Comas, I., Navarro, Y., Gonzalez-Candelas, F., Herranz, M., Bouza, E., et al. (2014). Whole genome sequencing analysis of intrapatient microevolution in Mycobacterium tuberculosis: Potential impact on the inference of tuberculosis transmission. *The Journal of Infectious Diseases*, *209*, 98–108.

Roetzer, A., Diel, R., Kohl, T. A., Ruckert, C., Nubel, U., Blom, J., et al. (2013). Whole genome sequencing versus traditional genotyping for investigation of a Mycobacterium tuberculosis outbreak: A longitudinal molecular epidemiological study. *PLoS Medicine*, *10*, e1001387.

Roetzer, A., Schuback, S., Diel, R., Gasau, F., Ubben, T., di Nauta, A., et al. (2011). Evaluation of Mycobacterium tuberculosis typing methods in a 4-year study in Schleswig-Holstein, Northern Germany. *Journal of Clinical Microbiology*, *49*, 4173–4178.

Roring, S., Scott, A., Brittain, D., Walker, I., Hewinson, G., Neill, S., et al. (2002). Development of variable-number tandem repeat typing of Mycobacterium bovis: Comparison of results with those obtained by using existing exact tandem repeats and spoligotyping. *Journal of Clinical Microbiology*, *40*, 2126–2133.

Sandegren, L., Groenheit, R., Koivula, T., Ghebremichael, S., Advani, A., Castro, E., et al. (2011). Genomic stability over 9 years of an isoniazid resistant Mycobacterium tuberculosis outbreak strain in Sweden. *PLoS One*, *6*, e16647.

Savine, E., Warren, R. M., van der Spuy, G. D., Beyers, N., van Helden, P. D., Locht, C., et al. (2002). Stability of variable-number tandem repeats of mycobacterial interspersed repetitive units from 12 loci in serial isolates of Mycobacterium tuberculosis. *Journal of Clinical Microbiology*, *40*, 4561–4566.

Schurch, A. C., Kremer, K., Kiers, A., Daviena, O., Boeree, M. J., Siezen, R. J., et al. (2010). The tempo and mode of molecular evolution of Mycobacterium tuberculosis at patient-to-patient scale. *Infection, Genetics and Evolution*, *10*, 108–114.

Schurch, A. C., & van Soolingen, D. (2012). DNA fingerprinting of Mycobacterium tuberculosis: From phage typing to whole-genome sequencing. *Infection, Genetics and Evolution*, *12*, 602–609.

Shamputa, I. C., Jugheli, L., Sadradze, N., Willery, E., Portaels, F., Supply, P., et al. (2006). Mixed infection and clonal representativeness of a single sputum sample in tuberculosis patients from a penitentiary hospital in Georgia. *Respiratory Research*, *7*, 99.

Shitikov, E. A., Bespyatykh, J. A., Ischenko, D. S., Alexeev, D. G., Karpova, I. Y., Kostryukova, E. S., et al. (2014). Unusual large-scale chromosomal rearrangements in Mycobacterium tuberculosis Beijing B0/W148 cluster isolates. *PLoS One*, *9*, e84971.

Skuce, R. A., McCorry, T. P., McCarroll, J. F., Roring, S. M., Scott, A. N., Brittain, D., et al. (2002). Discrimination of Mycobacterium tuberculosis complex bacteria using novel VNTR-PCR targets. *Microbiology*, *148*, 519–528.

Small, P. M., Hopewell, P. C., Singh, S. P., Paz, A., Parsonnet, J., Ruston, D. C., et al. (1994). The epidemiology of tuberculosis in San Francisco. A population-based study using conventional and molecular methods. *The New England Journal of Medicine*, *330*, 1703–1709.

Smit, P. W., Vasankari, T., Aaltonen, H., Haanpera, M., Casali, N., Marttila, H., et al. (2015). Enhanced tuberculosis outbreak investigation using whole genome sequencing and IGRA. *European Respiratory Journal*, *45*, 276–279.

Smittipat, N., Billamas, P., Palittapongarnpim, M., Thong-On, A., Temu, M. M., Thanakijcharoen, P., et al. (2005). Polymorphism of variable-number tandem repeats at multiple loci in Mycobacterium tuberculosis. *Journal of Clinical Microbiology*, *43*, 5034–5043.

Smittipat, N., & Palittapongarnpim, P. (2000). Identification of possible loci of variable number of tandem repeats in Mycobacterium tuberculosis. *Tubercle and Lung Disease*, *80*, 69–74.

Sreevatsan, S., Pan, X., Stockbauer, K. E., Connell, N. D., Kreiswirth, B. N., Whittam, T. S., et al. (1997). Restricted structural gene polymorphism in the Mycobacterium tuberculosis complex indicates evolutionarily recent global dissemination. *Proceedings of the National Academy of Sciences of the United States of America*, *94*, 9869–9874.

Supply, P., Allix, C., Lesjean, S., Cardoso-Oelemann, M., Rusch-Gerdes, S., Willery, E., et al. (2006). Proposal for standardization of optimized mycobacterial interspersed repetitive unit-variable-number tandem repeat typing of Mycobacterium tuberculosis. *Journal of Clinical Microbiology*, *44*, 4498–4510.

Supply, P., Gaudin, C., & Raze, D. (2014). Optimization of standard 24-locus variable-number tandem-repeat typing of Mycobacterium tuberculosis isolates: A multicenter perspective. *Journal of Clinical Microbiology*, *52*, 3518–3519.

Supply, P., Magdalena, J., Himpens, S., & Locht, C. (1997). Identification of novel intergenic repetitive units in a mycobacterial two-component system operon. *Molecular Microbiology*, *26*, 991–1003.

Supply, P., Marceau, M., Mangenot, S., Roche, D., Rouanet, C., Khanna, V., et al. (2013). Genomic analysis of smooth tubercle bacilli provides insights into ancestry and pathoadaptation of Mycobacterium tuberculosis. *Nature Genetics*, *45*, 172–179.

Supply, P., Mazars, E., Lesjean, S., Vincent, V., Gicquel, B., & Locht, C. (2000). Variable human minisatellite-like regions in the Mycobacterium tuberculosis genome. *Molecular Microbiology*, *36*, 762–771.

Supply, P., Warren, R. M., Banuls, A. L., Lesjean, S., Van Der Spuy, G. D., Lewis, L. A., et al. (2003). Linkage disequilibrium between minisatellite loci supports clonal evolution of *Mycobacterium tuberculosis* in a high tuberculosis incidence area. *Molecular Microbiology*, *47*, 529–538.

van Embden, J. D., Cave, M. D., Crawford, J. T., Dale, J. W., Eisenach, K. D., Gicquel, B., et al. (1993). Strain identification of Mycobacterium tuberculosis by DNA fingerprinting: Recommendations for a standardized methodology. *Journal of Clinical Microbiology*, *31*, 406–409.

Van Soolingen, D. (2001). Molecular epidemiology of tuberculosis and other mycobacterial infections: Main methodologies and achievements. *Journal of Internal Medicine*, *249*, 1–26.

van Soolingen, D., Hermans, P. W., de Haas, P. E., Soll, D. R., & van Embden, J. D. (1991). Occurrence and stability of insertion sequences in Mycobacterium tuberculosis complex strains: Evaluation of an insertion sequence-dependent DNA polymorphism as a tool in the epidemiology of tuberculosis. *Journal of Clinical Microbiology*, 29, 2578–2586.

Walker, T. M., Ip, C. L., Harrell, R. H., Evans, J. T., Kapatai, G., Dedicoat, M. J., et al. (2013). Whole-genome sequencing to delineate Mycobacterium tuberculosis outbreaks: A retrospective observational study. *The Lancet Infectious Diseases*, 13, 137–146.

Walker, T. M., Lalor, M. K., Broda, A., Saldana Ortega, L., Morgan, M., Parker, L., et al. (2014). Assessment of Mycobacterium tuberculosis transmission in Oxfordshire, UK, 2007–12, with whole pathogen genome sequences: An observational study. *The Lancet. Respiratory Medicine*, 2, 285–292.

Walker, T. M., Monk, P., Grace Smith, E., & Peto, T. E. (2013). Contact investigations for outbreaks of Mycobacterium tuberculosis: Advances through whole genome sequencing. *Clinical Microbiology and Infection*, 19(9), 796–802.

Warren, R. M., Victor, T. C., Streicher, E. M., Richardson, M., van der Spuy, G. D., Johnson, R., et al. (2004). Clonal expansion of a globally disseminated lineage of Mycobacterium tuberculosis with low IS6110 copy numbers. *Journal of Clinical Microbiology*, 42, 5774–5782.

Weniger, T., Krawczyk, J., Supply, P., Niemann, S., & Harmsen, D. (2010). MIRU-VNTRplus: A web tool for polyphasic genotyping of Mycobacterium tuberculosis complex bacteria. *Nucleic Acids Research*, 38(Suppl.), W326–W331.

Wirth, T., Hildebrand, F., Allix-Beguec, C., Wolbeling, F., Kubica, T., Kremer, K., et al. (2008). Origin, spread and demography of the Mycobacterium tuberculosis complex. *PLoS Pathogens*, 4, e1000160.

World Health Organisation. (2014). *Global tuberculosis report 2014*. http://apps.who.int/iris/bitstream/10665/137094/1/9789241564809_eng.pdf?ua=1 (Accessed 15.01.09).

Solid and Suspension Microarrays for Microbial Diagnostics

11

Steve Miller*, Ulas Karaoz†, Eoin Brodie†, Sherry Dunbar‡,1

*Clinical Microbiology Laboratory, University of California, San Francisco, California, USA
†Earth Sciences Division, Lawrence Berkeley National Laboratory, Berkeley, California, USA
‡Luminex Corporation, Austin, Texas, USA
1Corresponding author: e-mail address: sdunbar@luminexcorp.com

1 INTRODUCTION

Microarrays can be most simply defined as a collection of microscopic spots (or features) of biological capture molecules (such as DNA or proteins) spatially arranged in a predefined order on a solid substrate. In DNA microarrays (also known as DNA chips or biochips), each spot contains predefined short (25- to 70-nucleotide) single-stranded DNA oligonucleotides or larger (200- to 800-basepair) double-stranded DNA, which are known as probes. A microarray experiment consists of the hybridisation of a sample of target DNA in solution, whose identity or abundance is being sought, with a typically large number of probes fixed to the substrate. Target DNA molecules are prepared with fluorescent tags, the microarray is exposed to the resulting mixture under conditions that favour hybridisation, and any unbound DNA fragments are washed away. Microarray positions with DNA–DNA hybrids are then detected optically, indicating the components that were present in the starting sample. Each microarray experiment is therefore a large number of hybridisation experiments done in parallel and as such, microarrays can be thought of as 'highly parallel' processors of biological queries. Various microarray methods are available and differ in the type of the probe, the solid support and the methods used for addressing the probe and detecting the target (Miller & Tang, 2009). Microarray technologies commonly used for the detection and identification of microbial targets include solid-state microarrays, electronic microarrays and suspension bead microarrays. Details of these technologies, including probe length, number of possible features (or density) and surface chemistry employed, are described below and compared in Table 1. In this chapter, we review the concepts behind each of these microarray technologies and describe clinical applications of microarray analysis for pathogen detection, genotyping and antimicrobial resistance testing in the microbiology laboratory.

Methods in Microbiology, Volume 42, ISSN 0580-9517, http://dx.doi.org/10.1016/bs.mim.2015.04.002

Table 1 Comparison of Microarray Technologies

Technology	Principle	Probe Length	Features	Solid Support	References
Printed	Pre-synthesised probes are deposited onto solid surface and attached by covalent or electrostatic chemistry	25–80 base	≤30,000	Glass slide	Goldmann and Gonzalez (2000) and Lausted et al. (2004)
In situ synthesised with photolithographic mask	Probes are synthesised in parallel directly on solid surface with phosphoramidite protection/deprotection chemistry using mask-directed photolithography	20–25 base	>1,000,000	Silicon chip	Beier and Hoheisel (2000) and Pease et al. (1994)
In situ synthesised with digital micromirror	Probes are synthesised in parallel directly on solid surface with phosphoramidite protection/deprotection chemistry using maskless photodeprotection with digital micromirror devices	60–100 base	15,000–>1,000,000	Silicon chip	Singh-Gasson et al. (1999)
In situ synthesised with inkjet	Probes are synthesised in parallel directly on solid surface with phosphoramidite protection/deprotection chemistry using acid-mediated chemical deprotection with inkjet printing	60 base	15,000–>1,000,000	Glass slide	Hughes et al. (2001)
Electronic	Pre-synthesised probes are transported to specific locations on the solid surface by electric field and immobilised by streptavidin–biotin bonding	20–30 base	400	Silicon chip	Sosnowski, Tu, Butler, O'Connell, and Heller (1997) and Edman et al. (1997)
Suspension bead array	Pre-synthesised probes are covalently coupled to solid surface	15–100 base	≤500	Polystyrene microspheres	Dunbar (2013) and Dunbar and Jacobson (2007)

2 SOLID-STATE MICROARRAYS

Solid-state (also called solid-phase) microarrays, in particular, use rigid, impermeable substrates (such as glass or silicon wafers) as the solid support (Foglieni et al., 2010; Palmieri, Alessi, Conoci, Marchi, & Panvini, 2008). The microscopic features, each with thousands of identical probes immobilised to the solid surface, are arranged in an orderly fashion at high density. Solid-state microarrays are the technological descendants of 'dot-blotting' techniques (Kafatos, Jones, & Efstratiadis, 1979), which involved the application of macroscopic spots over porous membranes (i.e. nitrocellulose or gel pads). The introduction of solid substrates was critical to the success of DNA microarrays for high-throughput large-scale biological applications (Khrapko et al., 1989). With probes tethered to a solid surface rather than being buried in pores, the rate of hybridisation is significantly increased and the subsequent washing steps are unhampered by diffusion effect. The rigidity of the solid support improves image acquisition and processing by producing sharper images, and high image definition is indispensable to the micron-scale feature sizes achieved on these microarrays.

2.1 PRINTED MICROARRAYS

Printed (spotted) microarrays are built by depositing pre-synthesised probes onto glass slides using fine-pointed pins (Figure 1; Goldmann & Gonzalez, 2000; Lausted et al., 2004). The probes can be short oligonucleotides, cDNA or PCR products, and the surface attachment chemistry can be either covalent or non-covalent. Covalent attachment requires the addition of an aliphatic amine (NH_2) group to the 5′-end of the probe. The probes are then attached to the glass from their 5′-end. Non-covalent attachment involves electrostatic attraction between the amine groups on the glass slide and the phosphate group of the probe's backbone. Non-covalent interactions are not strong enough to tether short oligonucleotide probes to glass, thus non-covalent attachment is more appropriate for cDNA microarrays.

The delivery of probes onto the glass slide is achieved with spotting robots using either contact or non-contact microarray printing. Contact printing technologies have been used since the early days of microarray technology. They are based on pin tools, which are dipped into the sample solution to take up a small volume of sample and then brought in contact with the slide surface to deliver the sample as a small spot. The currently available pins deliver spot volumes in 0.5–12 nl range, resulting in spot diameters from 62.5 to 600 μm. Non-contact printing enables microarray printing without direct contact to the surface. It uses piezoelectric, bubble-generated and micro-solenoid-driven dispensers that work with the same principle as inkjet printers. These systems are capable of dispensing single drops down to a volume of several hundred picolitres.

After the probes are printed, the surface of the array is usually chemically modified (fixed) making the attachment of additional DNA unfeasible. Fixing is desired to avoid non-specific binding of the DNA target from the sample to the glass slide

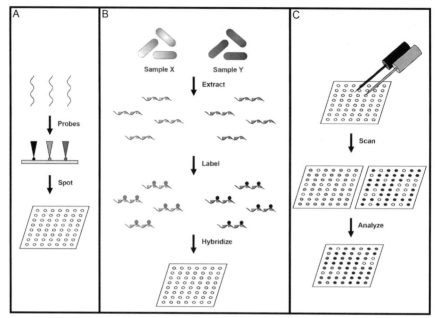

FIGURE 1

Printed microarrays. (A) Pre-synthesised oligonucleotide or DNA probes are spotted onto glass slides using fine-pointed pins. After printing, the probes are chemically fixed to the slide. (B) Samples containing target DNA are processed through extraction, labelling and hybridisation to the microarray. (C) The hybridised microarray is scanned, and the raw images are processed and analysed. (See the color plate.)

during hybridisation. Post-production modification of the surface to make it more hydrophilic is also common so as to facilitate the mixing of the target DNA during hybridisation.

2.2 *IN SITU*-SYNTHESISED MICROARRAYS

Probes for very high-density microarrays are synthesised base-by-base and *in situ* on the solid surface of the microarray. The most straightforward method for synthesising oligonucleotides involves adding nucleotides one at a time to the growing end of an oligonucleotide chain. In this method, the sequence of the oligonucleotide chain is dictated by the order in which dNTPs are added to the reaction. For solid-phase *in situ* chemical synthesis of DNA, an alternative chemical oligonucleotide synthesis based on phosphoramidite (protection/deprotection) chemistry is coupled with a solid support containing covalent linker molecules (Beier & Hoheisel, 2004). Phosphoramidite building blocks (phosphoramidite nucleosides) derived from chemically protected 2′-deoxynucleosides (dA, dC, dG and dT) are used in phosphoramidite protection/deprotection chemistry. The protection of the amine, phosphate and hydroxyl reactive groups ensures the prevention of unwanted side reactions. For the protection of the exocyclic amino group, standard schemes use a benzoyl group for adenosine and cytidine and an isobutyryl group for guanosine. Thymidine does not have a reactive exocyclic amino group and hence does not need any protection.

The *in situ* synthesis of the oligonucleotide chain using protection/deprotection chemistry consists of reaction cycles, where each cycle of reactions results in the addition of a single base. The base added has a protective group on its $5'$-end to prevent the addition of more than one base during each round of synthesis. At the initial stage, the terminal $3'$-base is covalently attached to the solid support. Oligonucleotides are synthesised in the $3'$- to $5'$-direction (as opposed to the $5'$- to $3'$-direction in DNA replication) by repeating the synthesis cycle, as many times as required until the desired length of DNA is achieved. A protection/deprotection synthesis cycle has the following four conceptual steps:

1. *Deprotection*: the protecting group of the $5'$-terminal nucleotide is removed, yielding a reactive $5'$-hydroxyl (OH) ready to bond with the as yet to be added phosphoramidite.
2. *Coupling*: an activated nucleoside phosphoramidite is brought in contact with the support-bound oligonucleotide precursor. The $5'$-OH group reacts with the activated phosphoramidite moiety forming a phosphite triester bond. This is an unstable linkage for the oligonucleotide at the end of the coupling reaction and a small fraction of the $5'$-OH groups remains unreacted. The fraction of the newly added bases that are actually incorporated into the growing DNA chain is called coupling efficiency. One desired property of a high-yield nucleotide synthesis scheme is a high coupling efficiency.
3. *Capping*: the permanent blocking of the unreacted $5'$-OH groups by treating with a chemical catalyst mixture.
4. *Oxidation*: the phosphite triester bond formed after coupling is unstable under the oligonucleotide synthesis conditions. An oxidation step involves the oxidation of the trivalent phosphite to a more stable pentavalent phosphate.

Different methods used in the deprotection step result in three technologies for *in situ*-synthesised microarrays. These three technologies differ in whether they use photo- or acid-labile protecting groups for $5'$-OH group of the nucleoside phosphoramidite and in how they achieve the combinatorial chemistry involved in parallel synthesis of thousands of probes.

2.2.1 Photodeprotection using photolithographic mask

In this method, which forms the basis of Affymetrix® GeneChip® technology, light is used to convert the protective group on the $5'$-terminal nucleotide into an OH group (Figure 2). During each round of synthesis, the selective exposure of specific array locations (features) to light is achieved with a physical chromium mask that allows light to reach only certain areas of the microarray. This technique, known as photolithography, was adapted from the manufacturing of silicon microprocessor chips (Beier & Hoheisel, 2000; Pease et al., 1994). Those features exposed to light become

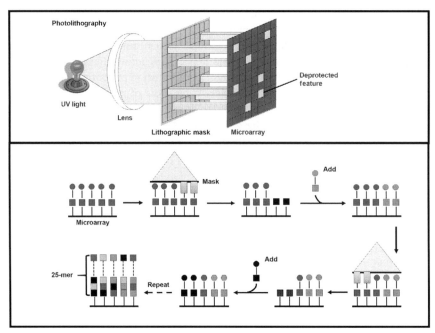

FIGURE 2

In situ-synthesised microarray using mask-assisted photolithography. (Top) A lithographic mask is used to transmit or block UV light from chemically protected nucleotides on the microarray surface. (Bottom) Protecting groups (circles) are removed by UV exposure, allowing addition of a single protected nucleotide (square). The cycle is repeated using different masks sequentially to dictate the order of nucleotide sequence in the oligonucleotide probes. (See the color plate.)

active for further synthesis. The microarray surface is then flushed with a solution containing a single type of nucleotide, but the chemical coupling occurs only in those features that have been deprotected. The newly coupled nucleotides also have a protecting group so the process can be repeated with the application of another mask and subsequent flushing of another nucleotide, until all probes have been fully synthesised. Each probe on the microarray requires four masks per round of synthesis. The probes on an Affymetrix GeneChip microarray are 25 nucleotides long, requiring nearly 100 masks per microarray. The photolithographic masks are expensive to produce but once designed and manufactured, they can be reused to make a large number of identical microarrays with high fidelity. Thus, Affymetrix GeneChip technology is well positioned for making large numbers of standardised microarrays.

2.2.2 Maskless photodeprotection using digital micromirror devices

Maskless photodeprotection is also based on light-directed synthesis but uses a digital micromirror device for selective light exposure of array features (Singh-Gasson et al., 1999). This is the method behind the Roche NimbleGen Maskless Array

Synthesiser (MAS) technology. A digital micromirror device is a solid-state silicon device with an array of aluminium mirrors that can be computer controlled to direct light to desired locations of the solid microarray surface (Figure 3). It effectively creates virtual masks replacing standard photolithographic masks. NimbleGen's digital micromirror device is able to pattern up to 786,000 individual pixels of light which enables minimum feature sizes of 17 μm. Since there are no physical masks to produce, the primary expense of the maskless approach is the chemistry. The flexibility of the virtual masks enables production of small batches of high-density microarrays with different array layouts in a matter of days.

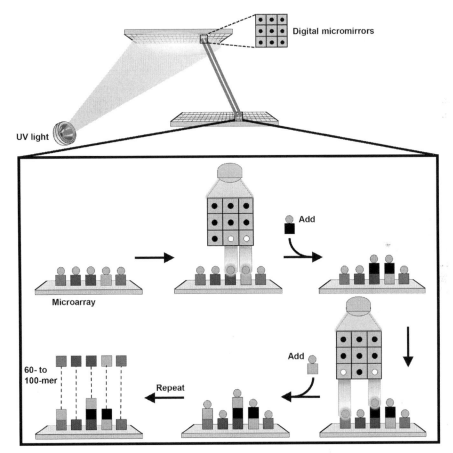

FIGURE 3

Maskless *in situ*-synthesised microarray using digital micromirrors. Digital micromirror technology is used to create virtual masks for photodeprotection. UV light is directed through the mask image onto the microarray surface to remove the protecting group (circles) and allow addition of a single nucleotide (square). The steps are repeated with different virtual masks to direct the synthesis of the desired oligonucleotides in a selected pattern. (See the color plate.)

2.2.3 Chemical deprotection with inkjet-mediated synthesis

This method underlying Agilent Technologies' microarrays couples chemical acid-mediated deprotection with inkjet printing to deposit 60 basepair oligonucleotides one base at a time on standard glass slides (Hughes et al., 2001). The deprotection of the DMT-protected nucleoside phosphoramidite is achieved with the removal of the DMT with trichloroacetic acid. In the coupling step of the synthesis cycle, the inkjet printer uses tiny nozzles for each monomer and chemical activators to place picolitre drops of reagents in specified locations on the microarray (Figure 4).

2.2.4 Comparison of in situ synthesis technologies

Micromirror and inkjet technologies allow computer controlled synthesis of each feature at the time of array production, while photolithography requires physical pre-made masks. This makes the former technologies highly flexible but these technologies are less efficient for making large batches of identical arrays compared to photolithographic masking. Variants of phosphoramidite-based synthesis have different coupling efficiencies. The yield of the desired probes is dependent on the length of the oligonucleotide being synthesised. With low coupling efficiencies and longer probes, eventually the yield of full-length oligos becomes unacceptably low. Photodeprotection and acid-mediated deprotection of dimethoxytrityl protecting group have coupling efficiencies of approximately 95% and 98%, respectively (McGall et al., 1997). The chemical deprotection used in inkjet-mediated synthesis allows the inkjet technology to be more suitable for high-fidelity synthesis of longer oligonucleotides.

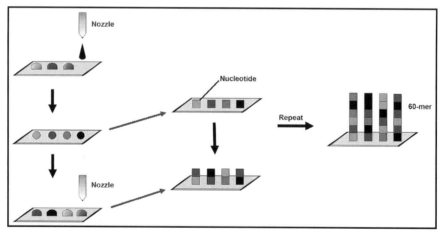

FIGURE 4

In situ-synthesised microarray using chemical deprotection with inkjet-mediated synthesis. This method uses acid-mediated deprotection with inkjet printing for oligonucleotide probe synthesis. The printer uses tiny nozzles to deposit picolitre amounts of nucleotides and reagents in specified locations of the microarray. Repeated rounds of printing extend the probe length to 60 nucleotides. (See the color plate.)

2.3 ELECTRONIC MICROARRAYS

The printed and *in situ*-synthesised microarrays rely on passive transport for the hybridisation of nucleic acids. In contrast, electronic microarrays utilise active hybridisation via electric fields to control nucleic acid transport (Figure 5). Commercially available as NanoChip® XL by Savyon Diagnostics (formerly NanoChip® 400 from Nanogen, Inc.), the core technology of the microelectronic cartridge uses complementary metal oxide semiconductor technology for the electronic addressing of nucleic acids (Sosnowski et al., 1997). The cartridge has 12 connectors that control 400 individual test sites. The microarray is generated by transport of negatively charged nucleic acids to specific sites when an electric field is applied to one or more test sites on the microarray (Edman et al., 1997). The surface of the microarray contains streptavidin, which allows the formation of streptavidin–biotin bonds with biotinylated probes reaching each location through the application of a specific electrical field, thus immobilising the probe onto the surface. The electric field is then removed from the active features, and new test sites can be activated.

Once the probes have been hybridised at discrete test sites, the microarray is ready for the application of fluorescently labelled target DNA. Target DNA

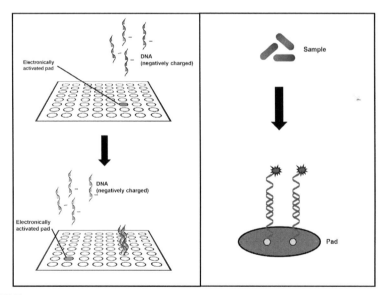

FIGURE 5

Electronic microarray. (Left) Positive current is applied to specific test sites (pads) to assist movement of negatively charged DNA probes to the activated pads. Probes are attached by streptavidin–biotin binding. The site is deactivated and the current applied to a different pad to localise a new probe. The process is repeated until all of the probes are arrayed. (Right) Samples are extracted, amplified and hybridised to the microarray surface. Target-specific secondary probes and fluorescent detector oligonucleotides are used to measure positive hybridisation reactions. (See the color plate.)

passively hybridises with the immobilised probes on the microarray or is concentrated electronically (Barlaan, Sugimori, Furukawa, & Takeuchi, 2005; Kumar et al., 2008). Electronically addressing the capture probe first is the most commonly used assay format, but amplicon first and sandwich assays have also been described (Feng & Nerenberg, 1999). If hybridisation occurs between the probe and the target DNA, fluorescent reporters will be present at the positive test site(s) and will be detected when the electronic microarray is scanned and analysed by an imaging instrument using a charge-coupled device (CCD) camera with data collection and analysis software.

3 SUSPENSION BEAD ARRAYS

Suspension bead microarray technologies are three-dimensional arrays that use microparticles (microspheres or beads) as the solid support for the binding reaction and flow cytometry or CCD imaging for detection of the bead and associated target (Figure 6). The particles can be distinguished into subsets by different classification features, such as internally absorbed fluorophores, size or diameter and surface or composition characteristics, and multiplexing is achieved through the unique features of the particle subsets. The xMAP® Technology platform by Luminex® is a bead-based suspension microarray technology commonly used for detection of nucleic acids and utilises polystyrene microspheres that are internally dyed with two or three spectrally distinct fluorochromes (Dunbar, 2006). Using precise amounts of each of these fluorochromes, an array is created consisting of different bead sets with specific spectral addresses. A specific capture oligonucleotide probe is covalently coupled to the surface of the specific bead sets or bead sets precoupled with unique capture oligonucleotides are commercially available for nucleic acid microarray assay development. The target nucleic acid is captured by hybridisation to the complementary probe sequence and labelled with a reporter molecule (i.e. fluorescent dye) to allow detection and quantitation of the specific binding that has occurred at the particle surface. Multiple readings are made per bead set or target, providing a fluorescent output signal of the hybridisation results.

A variety of assay chemistries can be used for generating the DNA targets for suspension bead hybridisation, including direct capture, competitive hybridisation and solution-based enzymatic chemistries in combination with universal capture probes (Figure 7).

3.1 DIRECT HYBRIDISATION

Direct hybridisation of a labelled PCR-amplified target DNA to bead sets bearing specific oligonucleotide capture probes is a simple and commonly used assay chemistry and typically employs probes of approximately 20 nucleotides in length (Dunbar, 2006; Dunbar & Jacobson, 2007). The probes are complementary in sequence to the labelled strand of the PCR product and the polymorphic nucleotide

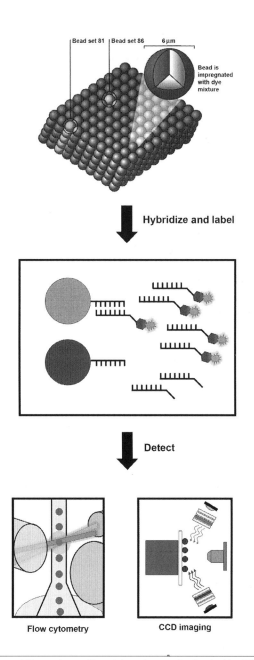

FIGURE 6

Suspension bead array. Microspheres (beads) are internally dyed with different intensities of red and infrared dyes to create 500 beads sets with unique spectral identities. Bead sets coupled with specific oligonucleotide probes are hybridised to amplified targets generated from the starting sample. Hybridised targets are labelled using a fluorescent reporter dye, and the microsphere suspension is analysed by flow cytometry or CCD imaging to identity each bead and quantify the reporter probe-target reaction on the microsphere surface. (See the color plate.)

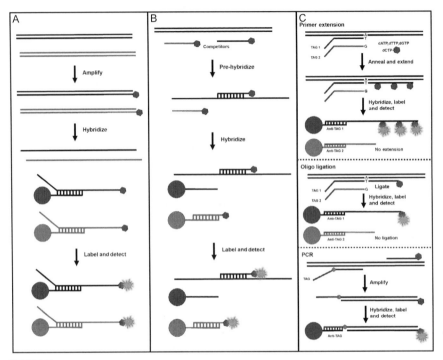

FIGURE 7

Suspension bead assay formats. (A) In direct hybridisation, target DNA is amplified and biotin labelled. The amplified products are denatured, hybridised to specific probe-coupled bead sets and labelled with streptavidin-linked fluorescent reporter. (B) In competitive hybridisation, labelled oligonucleotides containing the target sequences compete with target DNA in the test sample for hybridisation to specific probe-coupled bead sets. The sample target DNA concentration is inversely proportional to the bead-associated reporter signal, which decreases as sample concentration increases. (C) Enzymatic assay chemistries (primer extension, oligonucleotide ligation and PCR) are used to identify the target sequences in the sample and incorporate specific capture tag sequences. The targets are hybridised to bead sets bearing complementary anti-TAG probes, and labelled for detection. (See the color plate.)

(for single-nucleotide discrimination) is located at or near the centre of the probe. The oligo probes are modified with a 5′-terminal amine for covalent attachment to the functional carboxyl group on the microsphere surface and contain a spacer (usually C-12) between the amine and probe sequence to minimise steric hindrance of hybridisation at the microsphere surface. PCR primers are typically designed to amplify 50- to 300-basepair regions of target sequence with one primer of each pair biotinylated at the 5′ end for labelling the target strand of the amplicon. Using a small target DNA minimises the potential for steric hindrance to affect hybridisation efficiency adversely at the bead surface (Dunbar, 2006).

3.2 COMPETITIVE HYBRIDISATION

Competitive hybridisation is similar to direct hybridisation in probe and target design. However, in the competitive assay format, unlabelled double-stranded PCR-amplified targets compete with labelled single-stranded oligonucleotide targets for annealing to the sequence-specific capture probes on the microspheres (Fulton, McDade, Smith, Kienker, & Kettman, 1997).

3.3 ENZYMATIC CHEMISTRIES WITH UNIVERSAL CAPTURE PROBES

Another approach is to use a sequence-specific enzymatic reaction in solution to identify the target DNA, followed by capture onto a universal probe sequence on the bead surface for detection. This format involves the incorporation of a capture sequence during the enzymatic step that allows hybridisation to a complementary sequence attached to the bead surface. Commonly used enzymatic methods for sequence determination rely on the discriminating ability of DNA polymerases and DNA ligases, and include allele-specific or target-specific primer extension, oligonucleotide ligation, single-base chain extension and target-specific PCR (Dunbar, 2013; Taylor et al., 2001; Ye et al., 2001). These chemistries are used in combination with bead reagents precoupled with universal oligonucleotide capture sequences that are chosen to have identical hybridisation parameters with minimal cross-reactivity (Mahony et al., 2007). This approach takes advantage of the speed of solution-phase hybridisation and permits the universal bead sets to be used in many different assays where new sequences can be targeted by adding the appropriate capture sequence to the target-specific oligonucleotide used in the enzymatic step.

4 CLINICAL APPLICATIONS OF MICROARRAY TESTING

The detection and genetic characterisation of microbial pathogens by microarray methods have been developed for several applications in the clinical microbiology laboratory. The large number of microarray probes available in several commercial platforms allows for highly multiplexed assays, enabling detection of a broad range of organisms and/or discrimination of multiple genetic elements within the targeted species (Miller & Tang, 2009). At the current time, the clinical utility of microarray methods is primarily limited by our lack of detailed understanding about how genetic changes affect organism phenotype and outcomes of infections. The rapid expansion of knowledge about multigenic virulence and resistance traits will allow for clinically actionable interpretation of complex microarray data and additional uses for this testing in patient care. While other molecular methods such as next-generation sequencing can yield the same or additional genetic data for interpretation, targeted microarray methods will likely continue to be used in conjunction with high-throughput sequencing, as they can provide more rapid or cheaper tests for particular indications (Sibley, Peirano, & Church, 2012).

The combination of microbial detection and genotypic characterisation in a single assay is one of the attractions of microarray testing in the clinical laboratory. Probes specific for individual pathogens can be utilised, along with genetic markers such as resistance elements, to provide information enabling early optimisation of therapy. For some pathogens such as *Mycobacterium tuberculosis*, phenotypic susceptibility testing can take days or weeks to complete, and empiric treatment must be initiated in the interim (Neonakis, Gitti, Krambovitis, & Spandidos, 2008; Schanne et al., 2008). Often in bacterial infections, phenotypic susceptibility test results are available within 24–48 h, and the utilisation of microarray testing will depend on factors such as turnaround time, cost and the failure rate of empiric antibiotic coverage. It should be noted that the sensitivity may vary for species detection versus resistance markers, so that resistance prediction could be less reliable for organisms present in low titers (Thissen et al., 2014). Guidelines have been developed for validation and implementation of microarrays for diagnosis of infectious disease in clinical laboratories (Clinical and Laboratory Standards Institute, 2014).

4.1 PATHOGEN DETECTION

Most microarrays detect target organisms based on conserved gene targets that are amplified and labelled prior to hybridisation and detection. Therefore, the spectrum of detection depends on both the number of available probes covering the desired species and the ability to amplify the regions of interest using conserved primers. No single target gene or genetic region will allow for detection of all pathogens of interest, so microarrays will typically target one or more classes of organisms. The most commonly used gene targets are listed in Table 2, and include 16S ribosomal DNA (rDNA) (bacteria), internal transcribed spacers for 18S or 28S rDNA (fungi) and *rpoB* (mycobacteria) (Baker et al., 2003; Gingeras et al., 1998; McCabe et al., 1999). Because it is not possible to design a primer pair that is completely conserved among all prokaryotic species for any gene target, PCR may be performed using degenerate primers or under conditions that allow for primer mismatches to still amplify the expected product. Therefore, species with lower sequence identity to primer regions can be identified, but require higher numbers of organisms for detection. More distantly related organisms may not be detectable, and the diversity of detectable species can be predicted based on known sequences or determined experimentally (Baker et al., 2003; McCabe et al., 1999).

4.1.1 Identification of organisms isolated by culture-based methods

Microarrays can be useful for the identification of unusual or difficult to identify organisms isolated by culture-based methods. Arrays are designed for a particular pathogen type, with species diversity dependent on the microarray targets and amplification strategy used. Microarrays can be applied to common bacterial pathogens, but are more often used when alternative techniques such as biochemical tests, MALDI-TOF mass spectrometry or 16S gene sequencing are not available (Yoo et al., 2009).

Table 2 Conserved Gene Targets Used for Microarray Detection

Gene	Function	Organism Class	References
16S rDNA	Protein synthesis	Bacteria, Mycobacteria	McCabe, Zhang, Huang, Wagar, and McCabe (1999), Baker, Smith, and Cowan, (2003) and Gingeras et al. (1998)
23S rDNA	Protein synthesis	Bacteria, Mycobacteria	Yoo et al. (2010) and Lee, Jelfs, Sintchenko, and Gilbert, (2009)
rpoB	RNA polymerase	Bacteria, Mycobacteria	Adekambi, Colson, and Drancourt, (2003) and Gingeras et al. (1998)
gyrB	Topoisomerase	Bacteria, Mycobacteria	Jarvinen et al. (2009) and Fukushima et al. (2003)
16S–23S rRNA internal transcribed spacer	Protein synthesis	Bacteria, Mycobacteria	Kirschner et al. (1996) and Wang et al. (2009)
cpn60	Chaperonin	Bacteria	Maynard et al. (2005)
parE	Topoisomerase	Bacteria	Jarvinen et al. (2009)
recA	DNA repair	Bacteria	Champagne et al. (2011)
wecE	Lipid synthesis	Enterobacteriaceae	Maynard et al. (2005)
hsp65	Chaperonin	Mycobacteria	Ringuet et al. (1999)
dnaJ	Chaperonin	Mycobacteria	Nasr Esfahani, Rezaei Yazdi, Moghim, Ghasemian Safaei, and Zarkesh Esfahani (2012)
IS6110	Promoter	Mycobacteria	Thierry et al. (1990)
sodA	Superoxide dismutase	Mycobacteria	Kooken et al. (2014)
32 kDa protein	Secreted protein	Mycobacteria	Soini, Skurnik, Liippo, Tala, and Viljanen (1992)
18S and 28S rRNA internal transcribed spacers	Protein synthesis	Fungi	Leinberger, Schumacher, Autenrieth, and Bachmann (2005), Campa et al. (2008) and Landlinger et al. (2009)
Genus-specific viral family targets	Various	Viruses	Xiao-Ping et al. (2009)

Bacterial identification by microarray has been demonstrated for potential bio-threat organisms, such as *Bacillus anthracis*, where reliable differentiation from environmental *Bacillus* species is necessary (Burton, Oshota, & Silman, 2006) and for food or waterborne pathogenic bacteria (Huang et al., 2014; Maynard et al., 2005; Zhou et al., 2011). Substantial genotype information can be derived from these isolates at a single-nucleotide polymorphism (SNP) level (Thierry et al., 2013). Comparative genomic arrays can allow for sensitive discrimination of related isolates and detect virulence-associated genes in pathogens such as *Staphylococcus aureus* (Spence et al., 2008; Witney et al., 2005).

Slow growing and fastidious organisms can be efficiently identified by microarrays, including mycobacteria (Fukushima et al., 2003; Tobler, Pfunder, Herzog, Frey, & Altwegg, 2006) and fungi (Campa et al., 2008; Leinberger et al., 2005). A commercially available line probe system uses the 16S–23S mycobacterial spacer for identification of 17 mycobacterial species with high sensitivity and specificity (Mijs et al., 2002; Tortoli, Mariottini, & Mazzarelli, 2003). Microarrays have also been applied directly on patient samples to aid diagnosis, where assay sensitivity is a critical factor (Lazzeri, Santoro, Oggioni, Iannelli, & Pozzi, 2012). Microarrays can also be applied to the identification of viruses isolated in culture, using conserved targets within viral families, which can be applied to atypical or highly pathogenic viruses (Fitzgibbon & Sagripanti, 2006; Xiao-Ping et al., 2009).

4.1.2 Identification of organisms in positive blood cultures

The identification of organisms growing in blood cultures and the detection of resistance mechanisms may facilitate earlier and more targeted antimicrobial therapy. This has the potential to improve patient outcomes while reducing the need for the use of broad-spectrum antibiotic therapy. Microarray technology has been successfully applied to positive blood cultures, where the organism concentration is high enough that successful detection can be demonstrated. Molecular diagnostic techniques and MALDI-TOF mass spectrometry can yield similar information 1–3 days sooner than standard biochemical tests, and there are now several commercial assays available for clinical use (Buchan et al., 2013; Laakso, Kirveskari, Tissari, & Maki, 2011; Liesenfeld, Lehman, Hunfeld, & Kost, 2014).

Microarray targets can include many clinically relevant organisms, or target-specific pathogen types as determined by the initial Gram stain. The differentiation of Gram-positive cocci into virulent *S. aureus* versus the much less pathogenic *Staphylococcus epidermidis* can lead to significant treatment changes, and these panels typically contain resistance elements such as *mecA* for staphylococcal methicillin resistance and *vanA* for vancomycin-resistant enterococci (Palka-Santini, Putzfeld, Cleven, Kronke, & Krut, 2007; Pasko et al., 2012; Samuel et al., 2013). Gram-negative bacterial microarray panels typically identify *Enterobacteriaceae* species and *Pseudomonas aeruginosa*, and detect resistance mechanisms such as beta-lactamases and carbapenemases (Sullivan et al., 2014). The accuracy of these assays in predicting cephalosporin and carbapenemase susceptibility will depend on the extent of the panel targeting various classes of beta-lactamase genes and the local circulating strain epidemiology. Panels can also target fungal pathogens, typically identifying *Candida* species to enable optimised antifungal therapy (Aittakorpi et al., 2012; Yoo et al., 2010).

Data are still emerging about the optimal approach to translate blood culture microarray results into improved clinical care. While the presence of resistance elements such as *mecA* and *bla* beta-lactamase gene influence outcomes, host factors such as age and comorbidities are major factors as well (Rieg et al., 2013). It appears that intervention by an infectious disease or pharmacy practitioner is needed to realise maximum benefit from earlier identification and resistance marker results (Sango et al., 2013). As clinicians become more familiar with these tests and decision algorithms are developed, specialist consultation may become less necessary to achieve their potential improvements in outcomes.

4.1.3 Application of microarrays directly to patient samples
4.1.3.1 Respiratory infections

The use of multiplexed respiratory pathogen detection panels has become routine in many clinical laboratories. Several commercial multiplex panels, including microarray assays, are available and have been approved by regulatory agencies. These typically contain a variety of respiratory virus targets and differ in the scope of detection and ability to subtype (i.e. inclusion of human coronaviruses and subtyping of parainfluenza or adenovirus), as well as the workflow and hands-on time required (Krunic, Merante, Yaghoubian, Himsworth, & Janeczko, 2011; Mahony, Petrich, & Smieja, 2011). Comparative studies have shown some differences in assay sensitivities for particular targets, particularly adenovirus and influenza B (Balada-Llasat, LaRue, Kelly, Rigali, & Pancholi, 2011; Popowitch, O'Neill, & Miller, 2013). Some assays may in addition contain bacterial targets such as *Bordetella pertussis*, *Chlamydophila pneumoniae* and *Mycoplasma pneumoniae*, and they all can be adapted to diagnose lower tract as well as upper tract respiratory infections (Lodes et al., 2007; Ruggiero, McMillen, Tang, & Babady, 2014).

Larger microarray panels allow for the diagnosis of more cases of respiratory infection (Bierbaum et al., 2012; Lin et al., 2006; Simoes et al., 2013), but can be difficult to develop and validate for use in the clinical laboratory. Resequencing microarrays allow for detection of co-infections, strain discrimination and epidemiologic tracking of respiratory pathogens (Lin, Blaney, et al., 2007; Lin, Malanoski, et al., 2007). Microarrays using pan-pathogen detection methods such as random amplification followed by broad-spectrum target hybridisation demonstrate a large amount of diversity in viral flora and have diagnosed atypical infections such as human parainfluenzavirus 4, rubella, measles and influenza C (Chiu et al., 2006; Kistler et al., 2007; Shen et al., 2013).

Fungal respiratory infections are an attractive target for microarray testing, since these pathogens can be difficult to grow and some require relatively long incubation periods. Species-specific identification of yeasts, moulds and dimorphic organisms including *Histoplasma capsulatum* and *Coccidioides immitis* can be detected in a single test and correlate with clinical and microbiological disease diagnosis (Landlinger et al., 2009). Microarrays can be particularly useful for diagnosis of immunocompromised patients with suspected invasive fungal infection, where detection is difficult by conventional methods (Spiess et al., 2007).

Rapid detection of mycobacterial pathogens can also improve treatment decisions, and commercially available assays have been evaluated for direct patient sample testing, yielding high specificity but lower sensitivity than when compared to conventional culture-based methods (Tortoli & Marcelli, 2007).

The clinical interpretation of respiratory microarray testing is relatively straight-forward in the case of viral detection in a symptomatic patient, although the significance of co-infections is less certain. Rapid molecular methods can be useful to guide therapy for pneumonia based on the type of organism detected and resistance genes present, but clinical trial data have not yet been published (Endimiani et al., 2011).

4.1.3.2 Gastrointestinal infections

Conventional methods of identification for gastrointestinal pathogens are typically time-consuming and limited in scope to relatively few organisms. Several different tests are employed (i.e. bacterial culture, microscopy for parasite identification, and viral antigen detection or PCR testing), but additional tests may still be needed to detect less common pathogens. By combining multiple organism types onto a single panel, microarray methods have the potential to simplify testing protocols, while providing high sensitivity and the test panels can include species not commonly tested for by conventional methods (Jin, Wen, Chen, Lin, & Wang, 2006). Mixed infections may also be detected and information regarding the infecting parasite species or bacterial subspecies can be generated and used for epidemiologic studies (Marotta et al., 2013; Wang, Orlandi, & Stenger, 2005). Commercially available microarray tests have been shown to increase the number of pathogens detected compared to conventional methods and may provide a faster turnaround time for bacterial and parasitic pathogens (Halligan et al., 2014; Mengelle et al., 2013; Navidad, Griswold, Gradus, & Bhattacharyya, 2013). Future uses may include identification of the intestinal microbiota associated with certain disease states and manipulation of this through probiotic therapy (Carey, Kirk, Ojha, & Kostrzynska, 2007).

4.1.3.3 Bloodstream infections

The full benefit of rapid molecular diagnosis of sepsis requires use of direct patient blood samples to allow for the shortest time to actionable results. These panels can include fastidious organisms that are difficult to diagnose using conventional methods (Fenollar & Raoult, 2007). However, many septic patients have less than one bacteria cell per millilitre of blood, and the extraction of large amounts of blood is technically difficult (Klouche & Schroder, 2008). Therefore, the sensitivity of direct bloodstream detection is a limiting factor, but these results can still be useful to guide more rapid treatment decisions (Huttunen, Syrjanen, Vuento, & Aittoniemi, 2013; Schrenzel, 2007). Some molecular panels have been shown to increase the number of detected bacterial infections over culture, assisting the diagnosis in immunocompromised and in patients where antimicrobial therapy has already been initiated (Burdino et al., 2014; Negoro et al., 2013).

Several microarray-based assays for the detection of bacteria, yeast and viruses have been described, some with sensitivities of 100% compared to blood culture (Debaugnies et al., 2014; Shang, Chen, Wu, Du, & Zhao, 2005; Wiesinger-Mayr et al., 2007). Optimised DNA extraction methods, primer sets and amplification conditions are needed to achieve sensitivities down to 10^1 bacterial cells per millilitre of whole blood (Palka-Santini, Cleven, Eichinger, Kronke, & Krut, 2009; Wiesinger-Mayr, Jordana-Lluch, Martro, Schoenthaler, & Noehammer, 2011). Microarrays can also be used to interrogate the host transcriptomic profile and generate gene expression patterns that correlate with bacterial sepsis or other inflammatory states (Sutherland et al., 2011). However, validated clinical algorithms using direct bloodstream pathogen detection and confirmatory data from large prospective studies demonstrating improvements in clinical outcomes are not yet available.

4.1.3.4 Urinary tract infections

Microarrays have been applied to the direct detection of uropathogens with high sensitivity, but since they are not quantitative, the ability to differentiate a true pathogen in urine containing mixed flora is limited (Liao et al., 2006). Therefore, most studies have focused on the detection of resistant organisms such as carbapenem-resistant *Enterobacteriaceae* and fluoroquinolone-resistant *Escherichia coli* or nosocomial bacteria such as *P. aeruginosa* and *Acinetobacter baumannii* (Keum et al., 2006; Peter et al., 2012; Yu et al., 2007). Novel approaches including electronic biosensor arrays are in development for rapid detection of uropathogens with minimal sample processing, with the eventual goal of providing point-of-care testing capability (Lam et al., 2013).

4.1.3.5 Genital tract infections

Detection of sexually transmitted disease pathogens through microarray panels has been described for *Neisseria gonorrhoeae*, *Chlamydia trachomatis* and *Ureaplasma urealyticum* using a multiplex primer-amplification approach (Cao, Wang, Zhang, & Gao, 2006; Lee et al., 2013; Shi, Wen, Chen, & Wang, 2005). However, commercial development has focused on human papillomavirus (HPV) detection and genotyping as part of cervical cancer diagnosis and screening. Genotyping is needed to distinguish high-risk HPV types (16 and 18) from other intermediate and low-risk types. Microarrays in a variety of formats have been developed to identify up to 120 known genotypes of HPV infection in cervical cytology and swab specimens (Albrecht et al., 2006; Feng et al., 2009; Shen-Gunther & Rebeles, 2013). For this application, detection sensitivity is less important than the ability to discriminate between low-risk patients and those with carcinomas that are likely to progress clinically, which is typically assessed using abnormal histology at cervical intraepithelial neoplasia (CIN) grade 2+ or higher. Most tests have been compared to the Hybrid Capture II assay and show similar or improved analytical sensitivity (Cho, Do, Kim, Bae, & Ahn, 2011; Park et al., 2013). Currently, only the real-time PCR-based Cobas HPV assay has been evaluated in a large trial and specifically approved by the U.S. Food and Drug Administration for primary cervical cancer screening, which improved sensitivity from 52% to 88% for CIN2 or greater and from 53% to 92% for CIN3 or greater compared to liquid-based cytology (Castle et al., 2011; Wright et al., 2012).

4.1.3.6 Skin, soft tissue and joint infections

Detection microarrays are best suited to identify non-commensal pathogens in nonsterile site infections or a wide range of organisms in sterile sites. They have been reported to detect cutaneous mycobacterial infection, and superficial and invasive mycoses (Campa et al., 2008; Carlos, Tang, Adler, & Kovarik, 2012; Sato et al., 2010). Profiling of the microbiota of healing wounds shows an association of *Acinetobacter* with wound failure, and enteric-associated flora with successful healing (Be et al., 2014). The identification of prosthetic joint infection can be enhanced by molecular techniques such as 16S PCR sequencing, and a microarray-based assay enhances detection in patients previously treated with antibiotics (Metso et al., 2014).

4.1.3.7 Central nervous system infections

Microarrays have been applied to diagnosis of bacterial meningitis, where they have been demonstrated to detect more cases than conventional methods alone, although methods have not been standardised, and there is potential for false-positives through sample contamination (Ben et al., 2008; Liu, Han, Huang, & Zhu, 2005; Rafi, Chandramuki, Mani, Satishchandra, & Shankar, 2010). Identification and differentiation of the fungal pathogens *Cryptococcus neoformans* and *Cryptococcus gattii* are possible in cerebrospinal fluid, but methods need to be optimised for sensitive detection (Bovers et al., 2007). Organisms which cause viral meningitis and encephalitis can be targeted through viral microarrays with probes for human herpesviruses, measles, mumps and enteroviruses (Boriskin et al., 2004; Leveque et al., 2011). Specialised microarrays have also been developed for highly pathogenic viruses such as chikungunya, yellow fever, dengue, hantavirus and emerging zoonotic agents (Xiao-Ping et al., 2009).

4.2 PATHOGEN GENOTYPING

4.2.1 Bacterial genotyping

A large number of research studies have used microarray methods to determine the presence of virulence factors and other genetic elements in bacteria. These techniques are important in investigating pathogens public health significance such as *Salmonella* and *Legionella*, and for outbreak investigations, for example, a multistate *Listeria* outbreak associated with cantaloupe (Gomgnimbou et al., 2014; Gronlund et al., 2011; Laksanalamai et al., 2012). From a clinical diagnostic standpoint, bacterial genotyping is rarely needed. In special circumstances, genotyping can assist investigation of infections in immunodeficient patients, for instance to determine whether an individual with *Streptococcus pneumoniae* infection who has been previously vaccinated has a strain included by the vaccine (Wang et al., 2007). As we acquire more knowledge about the clinical relevance of certain bacterial genotypes and virulence factors in predicting disease severity, genotyping could become useful to track disease epidemiology and indicate more aggressive treatment for certain strains, such as *Clostridium difficile* ribotype 027 (Knetsch et al., 2013).

4.2.2 Mycobacterial genotyping

M. tuberculosis strain typing is important for public health control efforts. PCR-based testing for the presence of variable direct repeat spacer oligonucleotides (spoligotyping) is a widely accepted technique, and has been adopted on microarray platforms as a rapid affordable method (Cowan, Diem, Brake, & Crawford, 2004; Ruettger et al., 2012). Genome deletions and SNPs are stable in *M. tuberculosis* clones, making these potential surveillance methods and approaches to study the pathogenesis of strains with known deletions (Kato-Maeda et al., 2001; Mahasirimongkol et al., 2009). Clinical microarrays often will contain markers for antibiotic resistance as well (see Section 4.3.2).

4.2.3 Fungal genotyping

Like bacteria, microarrays have been used to determine strain epidemiology and investigate virulence factors in fungal pathogens (Balajee, Sigler, & Brandt, 2007; Garaizar, Brena, Bikandi, Rementeria, & Ponton, 2006). For example, gene expression analysis yielded candidate virulence factors in *C. gattii* from outbreak strains (Ngamskulrungroj, Price, Sorrell, Perfect, & Meyer, 2011).

4.2.4 Viral genotyping

With the exception of HPV genotyping (described in Section 4.1.3.5), clinical viral genotyping yields prognostic information that can be used to guide treatment decisions or can be used for epidemiological and research purposes. Microarrays can be targeted to specific viral families or pan-viral using unbiased amplification with hybridisation to conserved sequence regions (Wang et al., 2002). Hepatitis C virus (HCV) genotyping is important to determine the optimal treatment regimen, and microarray-based methods are commercially available (Duarte et al., 2010; Nadarajah, Khan, Miller, & Brooks, 2007; Park et al., 2010). Hepatitis B virus (HBV) genotype can influence treatment response, and microarray methods can differentiate the various genotypes (Gauthier et al., 2010). For other viruses, microarray assays can generate epidemiological information about circulating strains, including segment origin of influenza A or rotavirus genomes, or typing of emerging viruses (Gardner & Jaing, 2013; Honma et al., 2007; Paulin et al., 2014).

4.3 ANTIMICROBIAL RESISTANCE GENE DETECTION

4.3.1 Bacterial resistance testing

The ability to accurately predict phenotypic resistance from bacteria genetic makeup depends on having a large panel of resistance markers, along with prediction algorithms for each antibiotic drug or class of drugs. Microarrays are well suited to detect resistance genes and cassettes, notably for beta-lactamases and carbapenemases, and can also detect SNPs associated with drug resistance. The tests have reasonably high sensitivity and specificity for detection of known resistance elements, but often will not differentiate functional from non-functional genes and fail to identify other mutation types, such as porin mutants, that can affect resistance profiles (Cuzon, Naas,

Bogaerts, Glupczynski, & Nordmann, 2012; Frye et al., 2006; Naas, Cuzon, Truong, Bernabeu, & Nordmann, 2010).

Currently available panels for Gram-negative organisms yield a correlation of approximately 90% for positive microarray results with antimicrobial resistance, but microarray-negative results correlate only about 65% with antimicrobial sensitivity (Card et al., 2013). These predictive rates vary per drug class and are generally better for plasmid-mediated resistance elements than for chromosomal mutations. They are well suited for identification of bacterial producing specific extended-spectrum beta-lactamases or carbapenemases for infection control and resistance monitoring (Naas, Cuzon, Bogaerts, Glupczynski, & Nordmann, 2011). Quinolone resistance in *E. coli* can be identified with known SNPs in *gyrA*, though correlation with phenotypic resistance has been shown for a limited number of isolates (Barl et al., 2008; Yu, Susa, Knabbe, Schmid, & Bachmann, 2004). Microarray targets for genotyping antibiotic resistance from nosocomial Gram-negative pathogens include efflux regulators, DNA gyrase, beta-lactamases, aminoglycoside-modifying enzymes and carbapenemases, with genotype–phenotype correlations of 88% for *P. aeruginosa* and >95% for selected strains of multidrug-resistant *A. baumannii* (Dally et al., 2013; Weile, Schmid, Bachmann, Susa, & Knabbe, 2007). Mutations in *Helicobacter pylori* 23S rRNA gene associated with clarithromycin resistance can also be detected by microarray testing (Chen, Li, & Yu, 2008).

For Gram-positive pathogens such as *S. aureus*, microarrays can be used to detect clinically relevant resistance mutations for several antibiotics (Strommenger et al., 2007). Correlations between array detection and phenotypic resistance are approximately 90% for penicillin, methicillin, aminoglycoside, macrolide and lincosamide resistance by phenotypic testing (Zhu et al., 2007). Other techniques, such as next-generation sequencing, have a high correlation with microarray methods for detection of bacterial resistance mutations. In general, the current data support the ability of these molecular techniques to detect known resistance mutations, but they have not been prospectively tested with large numbers of clinical isolates from a variety of locations to determine the performance characteristics relative to conventional phenotypic susceptibility testing (Veenemans et al., 2014). Assay turnaround time and manual manipulation steps generally have to be reduced for these tests results to be available in a clinically relevant time frame.

4.3.2 Mycobacterial resistance testing

Mycobacterial resistance microarray platforms have focused on *M. tuberculosis*, the most clinically significant pathogen in this group. Phenotypic testing is delayed due to the organism's slow growth, so that molecular techniques can yield rapid detection of resistance mutations, allowing earlier transition to second-line treatment. About 95% of rifampin-resistant strains have certain known point mutations and rearrangements in *rpoB* gene (Caoili et al., 2006; Yue et al., 2004). Approximately 80% of isoniazid-resistant strains have a *katG* mutation and about 20% have *inhA* mutations (Kim et al., 2006; Yao et al., 2010). Pyrazinamide resistance is associated with mutations in the *pncA* gene, though about 40% of resistant strains have no mutation in this gene (Denkin, Volokhov, Chizhikov, & Zhang, 2005). Microarray-detected *embB* mutations are found in about 75% of ethambutol-resistant strains (Moure et al., 2014).

For second-line *M. tuberculosis* therapies, mutations in *rpsL* were found associated with streptomycin resistance, though sensitivity was low due to limited microarray gene coverage. The same study found sensitivities for detection of resistance at 100% for rifampin, 90% for isoniazid and 70% for ethambutol (Linger et al., 2014). Other second-line drugs for which microarrays can yield detection of resistance mutations include fluoroquinolones, kanamycin and capreomycin (Zimenkov et al., 2013). Other than tuberculosis, a microarray has been developed to detect resistance polymorphisms in *Mycobacterium leprae* (Matsuoka et al., 2008). Overall, microarray methods for detecting resistance in mycobacterial species are promising, but still require testing of a large number of geographically diverse isolates to determine their ability to guide therapeutic choices.

4.3.3 Fungal resistance testing

Several mechanisms of antifungal resistance in yeast have been identified, and many of these involve overexpression of genes encoding for drug targets or efflux pump transporters. Gene expression microarrays can identify genes associated with azole resistance in *Candida albicans* and are useful in identifying specific mutations affecting promoter expression (Yan et al., 2008). Increased expression of ergosterol biosynthesis genes is important for amphotericin B resistance in *C. albicans*, while azole resistance in *Candida glabrata* is associated with upregulation of the *CgPDR1* pleiotropic drug resistance network locus (Barker et al., 2004; Tsai et al., 2010). As more knowledge is gained about the specific mutations causing these upregulations, microarrays can be used to identify strains carrying known resistance mutations.

4.3.4 Viral resistance testing

Detection of known resistance mutations in viruses is amenable to microarray testing and can be implemented as a more economical method than full gene sequencing in resource-limited settings. Testing for HIV resistance in protease and reverse transcriptase genes yields >95% concordance with sequencing, with a few discrepant cases due to increased sensitivity of microarray over sequencing methods (Zhang, Cai, et al., 2013). For HBV, microarray targets can accurately identify mutations associated with lamivudine resistance (Chen et al., 2005). Microarrays designed to detect mutations in the neuraminidase and matrix genes of influenza A can identify these strains with high accuracy (Zhang, Liu, Wang, Chen, & Wang, 2013). As more viral resistance mutations conferring resistance to therapy are discovered, microarrays will offer a cost-effective method to detect them in clinical and epidemiological settings.

CONCLUSIONS

Microarrays offer a convenient method for interrogating specific pathogen sequences on scales of tens to hundreds of thousands of oligonucleotide probes. Several technologies and platforms are available, including solid-state, electronic and suspension bead microarrays, each with multiple variations in probe synthesis chemistries, linkage and data collection and analysis formats. Samples are typically amplified using multiplex primer sets, followed by hybridisation, washing and signal detection. Microarrays have been adopted for microbial detection, genotyping and antimicrobial resistance gene detection directly from clinical samples or from isolated organisms. Limits of detection are comparable to other molecular techniques, but still may not be equivalent to large-volume culture techniques such as blood cultures.

Microarray assays are limited by the set of available probe sequences, so they may not detect unusual sequence variants or novel genetic elements of clinical significance. Careful consideration must be taken during assay design and validation to distinguish between closely related species differing in virulence (i.e. *S. pneumoniae* vs. viridans *Streptococcus*), and to ensure that relevant circulating strains containing sequence variations are adequately detected and identified. Ideally, probe sets should be updated to reflect changing genotypes and strain epidemiology. The performance of microarray tests needs to be determined for different specimen types or matrices that may be encountered during clinical testing.

Microarray tests are well suited to offer a panel approach to diagnose specific patient presentations, such as respiratory viral infection or infectious diarrhoea. They can also discriminate between clinical isolates by genotype, which can be useful for tracking epidemiology and for outbreak investigations. As more information has become available regarding specific genes and pathways involved in antimicrobial resistance, we are beginning to be able to predict susceptibility patterns based on sequence interrogation for particular organisms. With further advances in automated microarray processing methods and genotype–phenotype prediction algorithms, these tests will become useful as an adjunct or replacement for conventional susceptibility testing, allowing for more rapid selection of targeted therapy for infectious disease.

REFERENCES

Adekambi, T., Colson, P., & Drancourt, M. (2003). rpoB-based identification of nonpigmented and late-pigmenting rapidly growing mycobacteria. *Journal of Clinical Microbiology*, *41*(12), 5699–5708.

Aittakorpi, A., Kuusela, P., Koukila-Kahkola, P., Vaara, M., Petrou, M., Gant, V., et al. (2012). Accurate and rapid identification of Candida spp. frequently associated with fungemia by using PCR and the microarray-based Prove-it Sepsis assay. *Journal of Clinical Microbiology*, *50*(11), 3635–3640.

Albrecht, V., Chevallier, A., Magnone, V., Barbry, P., Vandenbos, F., Bongain, A., et al. (2006). Easy and fast detection and genotyping of high-risk human papillomavirus by dedicated DNA microarrays. *Journal of Virological Methods, 137*(2), 236–244.

Baker, G. C., Smith, J. J., & Cowan, D. A. (2003). Review and re-analysis of domain-specific 16S primers. *Journal of Microbiological Methods, 55*(3), 541–555.

Balada-Llasat, J. M., LaRue, H., Kelly, C., Rigali, L., & Pancholi, P. (2011). Evaluation of commercial ResPlex II v2.0, MultiCode-PLx, and xTAG respiratory viral panels for the diagnosis of respiratory viral infections in adults. *Journal of Clinical Virology, 50*(1), 42–45.

Balajee, S. A., Sigler, L., & Brandt, M. E. (2007). DNA and the classical way: Identification of medically important molds in the 21st century. *Medical Mycology, 45*(6), 475–490.

Barker, K. S., Crisp, S., Wiederhold, N., Lewis, R. E., Bareither, B., Eckstein, J., et al. (2004). Genome-wide expression profiling reveals genes associated with amphotericin B and fluconazole resistance in experimentally induced antifungal resistant isolates of Candida albicans. *The Journal of Antimicrobial Chemotherapy, 54*(2), 376–385.

Barl, T., Dobrindt, U., Yu, X., Katcoff, D. J., Sompolinsky, D., Bonacorsi, S., et al. (2008). Genotyping DNA chip for the simultaneous assessment of antibiotic resistance and pathogenic potential of extraintestinal pathogenic Escherichia coli. *International Journal of Antimicrobial Agents, 32*(3), 272–277.

Barlaan, E. A., Sugimori, M., Furukawa, S., & Takeuchi, K. (2005). Electronic microarray analysis of 16S rDNA amplicons for bacterial detection. *Journal of Biotechnology, 115*(1), 11–21.

Be, N. A., Allen, J. E., Brown, T. S., Gardner, S. N., McLoughlin, K. S., Forsberg, J. A., et al. (2014). Microbial profiling of combat wound infection through detection microarray and next-generation sequencing. *Journal of Clinical Microbiology, 52*(7), 2583–2594.

Beier, M., & Hoheisel, J. D. (2000). Production by quantitative photolithographic synthesis of individually quality checked DNA microarrays. *Nucleic Acids Research, 28*(4), E11.

Beier, M., & Hoheisel, J. D. (2004). Synthesis of 5′-O-phosphoramidites with a photolabile 3′-O-protecting group. *Current Protocols in Nucleic Acid Chemistry,* 17:12.3:12.3.1–12.3.10.

Ben, R. J., Kung, S., Chang, F. Y., Lu, J. J., Feng, N. H., & Hsieh, Y. D. (2008). Rapid diagnosis of bacterial meningitis using a microarray. *Journal of the Formosan Medical Association, 107*(6), 448–453.

Bierbaum, S., Konigsfeld, N., Besazza, N., Blessing, K., Rucker, G., Kontny, U., et al. (2012). Performance of a novel microarray multiplex PCR for the detection of 23 respiratory pathogens (SYMP-ARI study). *European Journal of Clinical Microbiology & Infectious Diseases, 31*(10), 2851–2861.

Boriskin, Y. S., Rice, P. S., Stabler, R. A., Hinds, J., Al-Ghusein, H., Vass, K., et al. (2004). DNA microarrays for virus detection in cases of central nervous system infection. *Journal of Clinical Microbiology, 42*(12), 5811–5818.

Bovers, M., Diaz, M. R., Hagen, F., Spanjaard, L., Duim, B., Visser, C. E., et al. (2007). Identification of genotypically diverse Cryptococcus neoformans and Cryptococcus gattii isolates by Luminex xMAP technology. *Journal of Clinical Microbiology, 45*(6), 1874–1883.

Buchan, B. W., Ginocchio, C. C., Manii, R., Cavagnolo, R., Pancholi, P., Swyers, L., et al. (2013). Multiplex identification of gram-positive bacteria and resistance determinants directly from positive blood culture broths: Evaluation of an automated microarray-based nucleic acid test. *PLoS Medicine, 10*(7), e1001478.

Burdino, E., Ruggiero, T., Allice, T., Milia, M. G., Gregori, G., Milano, R., et al. (2014). Combination of conventional blood cultures and the SeptiFast molecular test in patients with suspected sepsis for the identification of bloodstream pathogens. *Diagnostic Microbiology and Infectious Disease*, *79*(3), 287–292.

Burton, J. E., Oshota, O. J., & Silman, N. J. (2006). Differential identification of Bacillus anthracis from environmental Bacillus species using microarray analysis. *Journal of Applied Microbiology*, *101*(4), 754–763.

Campa, D., Tavanti, A., Gemignani, F., Mogavero, C. S., Bellini, I., Bottari, F., et al. (2008). DNA microarray based on arrayed-primer extension technique for identification of pathogenic fungi responsible for invasive and superficial mycoses. *Journal of Clinical Microbiology*, *46*(3), 909–915.

Cao, X., Wang, Y. F., Zhang, C. F., & Gao, W. J. (2006). Visual DNA microarrays for simultaneous detection of Ureaplasma urealyticum and Chlamydia trachomatis coupled with multiplex asymmetrical PCR. *Biosensors & Bioelectronics*, *22*(3), 393–398.

Caoili, J. C., Mayorova, A., Sikes, D., Hickman, L., Plikaytis, B. B., & Shinnick, T. M. (2006). Evaluation of the TB-Biochip oligonucleotide microarray system for rapid detection of rifampin resistance in Mycobacterium tuberculosis. *Journal of Clinical Microbiology*, *44*(7), 2378–2381.

Card, R., Zhang, J., Das, P., Cook, C., Woodford, N., & Anjum, M. F. (2013). Evaluation of an expanded microarray for detecting antibiotic resistance genes in a broad range of gram-negative bacterial pathogens. *Antimicrobial Agents and Chemotherapy*, *57*(1), 458–465.

Carey, C. M., Kirk, J. L., Ojha, S., & Kostrzynska, M. (2007). Current and future uses of real-time polymerase chain reaction and microarrays in the study of intestinal microbiota, and probiotic use and effectiveness. *Canadian Journal of Microbiology*, *53*(5), 537–550.

Carlos, C. A., Tang, Y. W., Adler, D. J., & Kovarik, C. L. (2012). Mycobacterial infection identified with broad-range PCR amplification and suspension array identification. *Journal of Cutaneous Pathology*, *39*(8), 795–797.

Castle, P. E., Stoler, M. H., Wright, T. C., Jr., Sharma, A., Wright, T. L., & Behrens, C. M. (2011). Performance of carcinogenic human papillomavirus (HPV) testing and HPV16 or HPV18 genotyping for cervical cancer screening of women aged 25 years and older: A subanalysis of the ATHENA study. *The Lancet. Oncology*, *12*(9), 880–890.

Champagne, J., Diarra, M. S., Rempel, H., Topp, E., Greer, C. W., Harel, J., et al. (2011). Development of a DNA microarray for enterococcal species, virulence, and antibiotic resistance gene determinations among isolates from poultry. *Applied and Environmental Microbiology*, *77*(8), 2625–2633.

Chen, L. Y., Huang, J., Zhang, X. P., Qiao, P., Zhang, W., Yang, N. M., et al. (2005). Clinical evaluation of oligonucleotide microarrays for the detection of HBV mutants associated with lamivudine resistance. *Pharmacogenomics*, *6*(7), 721–730.

Chen, S., Li, Y., & Yu, C. (2008). Oligonucleotide microarray: A new rapid method for screening the 23S rRNA gene of Helicobacter pylori for single nucleotide polymorphisms associated with clarithromycin resistance. *Journal of Gastroenterology and Hepatology*, *23*(1), 126–131.

Chiu, C. Y., Rouskin, S., Koshy, A., Urisman, A., Fischer, K., Yagi, S., et al. (2006). Microarray detection of human parainfluenzavirus 4 infection associated with respiratory failure in an immunocompetent adult. *Clinical Infectious Diseases*, *43*(8), e71–e76.

Cho, E. J., Do, J. H., Kim, Y. S., Bae, S., & Ahn, W. S. (2011). Evaluation of a liquid bead array system for high-risk human papillomavirus detection and genotyping in comparison with Hybrid Capture II, DNA chip and sequencing methods. *Journal of Medical Microbiology*, *60*(Pt. 2), 162–171.

Clinical and Laboratory Standards Institute. (2014). *Microarrays for diagnosis and monitoring of infectious diseases; approved guideline: CLSI document MM22-A* (p. 84). Wayne, PA: Clinical and Laboratory Standards Institute.

Cowan, L. S., Diem, L., Brake, M. C., & Crawford, J. T. (2004). Transfer of a Mycobacterium tuberculosis genotyping method, Spoligotyping, from a reverse line-blot hybridization, membrane-based assay to the Luminex multianalyte profiling system. *Journal of Clinical Microbiology, 42*(1), 474–477.

Cuzon, G., Naas, T., Bogaerts, P., Glupczynski, Y., & Nordmann, P. (2012). Evaluation of a DNA microarray for the rapid detection of extended-spectrum beta-lactamases (TEM, SHV and CTX-M), plasmid-mediated cephalosporinases (CMY-2-like, DHA, FOX, ACC-1, ACT/MIR and CMY-1-like/MOX) and carbapenemases (KPC, OXA-48, VIM, IMP and NDM). *The Journal of Antimicrobial Chemotherapy, 67*(8), 1865–1869.

Dally, S., Lemuth, K., Kaase, M., Rupp, S., Knabbe, C., & Weile, J. (2013). DNA microarray for genotyping antibiotic resistance determinants in Acinetobacter baumannii clinical isolates. *Antimicrobial Agents and Chemotherapy, 57*(10), 4761–4768.

Debaugnies, F., Busson, L., Ferster, A., Lewalle, P., Azzi, N., Aoun, M., et al. (2014). Detection of Herpesviridae in whole blood by multiplex PCR DNA-based microarray analysis after hematopoietic stem cell transplantation. *Journal of Clinical Microbiology, 52*(7), 2552–2556.

Denkin, S., Volokhov, D., Chizhikov, V., & Zhang, Y. (2005). Microarray-based pncA genotyping of pyrazinamide-resistant strains of Mycobacterium tuberculosis. *Journal of Medical Microbiology, 54*(Pt. 12), 1127–1131.

Duarte, C. A., Foti, L., Nakatani, S. M., Riediger, I. N., Poersch, C. O., Pavoni, D. P., et al. (2010). A novel hepatitis C virus genotyping method based on liquid microarray. *PLoS One, 5*(9), e12822.

Dunbar, S. A. (2006). Applications of Luminex xMAP technology for rapid, high-throughput multiplexed nucleic acid detection. *Clinica Chimica Acta, 363*(1–2), 71–82.

Dunbar, S. A. (2013). Bead-based suspension arrays for the detection and identification of respiratory viruses. In Y. W. Tang & C. W. Stratton (Eds.), *Advanced techniques in diagnostic microbiology* (2nd ed., pp. 813–834). New York: Springer.

Dunbar, S. A., & Jacobson, J. W. (2007). Quantitative, multiplexed detection of Salmonella and other pathogens by Luminex xMAP suspension array. *Methods in Molecular Biology, 394*, 1–19.

Edman, C. F., Raymond, D. E., Wu, D. J., Tu, E., Sosnowski, R. G., Butler, W. F., et al. (1997). Electric field directed nucleic acid hybridization on microchips. *Nucleic Acids Research, 25*(24), 4907–4914.

Endimiani, A., Hujer, K. M., Hujer, A. M., Kurz, S., Jacobs, M. R., Perlin, D. S., et al. (2011). Are we ready for novel detection methods to treat respiratory pathogens in hospital-acquired pneumonia? *Clinical Infectious Diseases, 52*(Suppl. 4), S373–S383.

Feng, Q., Cherne, S., Winer, R. L., Balasubramanian, A., Lee, S. K., Hawes, S. E., et al. (2009). Development and evaluation of a liquid bead microarray assay for genotyping genital human papillomaviruses. *Journal of Clinical Microbiology, 47*(3), 547–553.

Feng, L., & Nerenberg, M. (1999). Electronic microarray for DNA analysis. *Gene Therapy and Molecular Biology, 4*, 183–191.

Fenollar, F., & Raoult, D. (2007). Molecular diagnosis of bloodstream infections caused by non-cultivable bacteria. *International Journal of Antimicrobial Agents, 30*(Suppl. 1), S7–S15.

Fitzgibbon, J. E., & Sagripanti, J. L. (2006). Simultaneous identification of orthopoxviruses and alphaviruses by oligonucleotide macroarray with special emphasis on detection of variola and Venezuelan equine encephalitis viruses. *Journal of Virological Methods*, *131*(2), 160–167.

Foglieni, B., Brisci, A., San Biagio, F., Di Pietro, P., Petralia, S., Conoci, S., et al. (2010). Integrated PCR amplification and detection processes on a Lab-on-Chip platform: A new advanced solution for molecular diagnostics. *Clinical Chemistry and Laboratory Medicine*, *48*(3), 329–336.

Frye, J. G., Jesse, T., Long, F., Rondeau, G., Porwollik, S., McClelland, M., et al. (2006). DNA microarray detection of antimicrobial resistance genes in diverse bacteria. *International Journal of Antimicrobial Agents*, *27*(2), 138–151.

Fukushima, M., Kakinuma, K., Hayashi, H., Nagai, H., Ito, K., & Kawaguchi, R. (2003). Detection and identification of Mycobacterium species isolates by DNA microarray. *Journal of Clinical Microbiology*, *41*(6), 2605–2615.

Fulton, R. J., McDade, R. L., Smith, P. L., Kienker, L. J., & Kettman, J. R., Jr. (1997). Advanced multiplexed analysis with the FlowMetrix system. *Clinical Chemistry*, *43*(9), 1749–1756.

Garaizar, J., Brena, S., Bikandi, J., Rementeria, A., & Ponton, J. (2006). Use of DNA microarray technology and gene expression profiles to investigate the pathogenesis, cell biology, antifungal susceptibility and diagnosis of Candida albicans. *FEMS Yeast Research*, *6*(7), 987–998.

Gardner, S. N., & Jaing, C. J. (2013). Bioinformatics for microbial genotyping of equine encephalitis viruses, orthopoxviruses, and hantaviruses. *Journal of Virological Methods*, *193*(1), 112–120.

Gauthier, M., Bonnaud, B., Arsac, M., Lavocat, F., Maisetti, J., Kay, A., et al. (2010). Microarray for hepatitis B virus genotyping and detection of 994 mutations along the genome. *Journal of Clinical Microbiology*, *48*(11), 4207–4215.

Gingeras, T. R., Ghandour, G., Wang, E., Berno, A., Small, P. M., Drobniewski, F., et al. (1998). Simultaneous genotyping and species identification using hybridization pattern recognition analysis of generic Mycobacterium DNA arrays. *Genome Research*, *8*(5), 435–448.

Goldmann, T., & Gonzalez, J. S. (2000). DNA-printing: Utilization of a standard inkjet printer for the transfer of nucleic acids to solid supports. *Journal of Biochemical and Biophysical Methods*, *42*(3), 105–110.

Gomgnimbou, M. K., Ginevra, C., Peron-Cane, C., Versapuech, M., Refregier, G., Jacotin, N., et al. (2014). Validation of a microbead-based format for spoligotyping of Legionella pneumophila. *Journal of Clinical Microbiology*, *52*(7), 2410–2415.

Gronlund, H., Riber, L., Vigre, H., Lofstrom, C., Folling, L., Huehn, S., et al. (2011). Microarray-based genotyping of Salmonella: Inter-laboratory evaluation of reproducibility and standardization potential. *International Journal of Food Microbiology*, *145*(Suppl. 1), S79–S85.

Halligan, E., Edgeworth, J., Bisnauthsing, K., Bible, J., Cliff, P., Aarons, E., et al. (2014). Multiplex molecular testing for management of infectious gastroenteritis in a hospital setting: A comparative diagnostic and clinical utility study. *Clinical Microbiology and Infection*, *20*(8), O460–O467.

Honma, S., Chizhikov, V., Santos, N., Tatsumi, M., Timenetsky Mdo, C., Linhares, A. C., et al. (2007). Development and validation of DNA microarray for genotyping group A rotavirus VP4 (P[4], P[6], P[8], P[9], and P[14]) and VP7 (G1 to G6, G8 to G10, and G12) genes. *Journal of Clinical Microbiology*, *45*(8), 2641–2648.

Huang, A., Qiu, Z., Jin, M., Shen, Z., Chen, Z., Wang, X., et al. (2014). High-throughput detection of food-borne pathogenic bacteria using oligonucleotide microarray with quantum dots as fluorescent labels. *International Journal of Food Microbiology*, *185*, 27–32.

Hughes, T. R., Mao, M., Jones, A. R., Burchard, J., Marton, M. J., Shannon, K. W., et al. (2001). Expression profiling using microarrays fabricated by an ink-jet oligonucleotide synthesizer. *Nature Biotechnology*, *19*(4), 342–347.

Huttunen, R., Syrjanen, J., Vuento, R., & Aittoniemi, J. (2013). Current concepts in the diagnosis of blood stream infections. Are novel molecular methods useful in clinical practice? *International Journal of Infectious Diseases*, *17*(11), e934–e938.

Jarvinen, A. K., Laakso, S., Piiparinen, P., Aittakorpi, A., Lindfors, M., Huopaniemi, L., et al. (2009). Rapid identification of bacterial pathogens using a PCR- and microarray-based assay. *BMC Microbiology*, *9*, 161.

Jin, D. Z., Wen, S. Y., Chen, S. H., Lin, F., & Wang, S. Q. (2006). Detection and identification of intestinal pathogens in clinical specimens using DNA microarrays. *Molecular and Cellular Probes*, *20*(6), 337–347.

Kafatos, F. C., Jones, C. W., & Efstratiadis, A. (1979). Determination of nucleic acid sequence homologies and relative concentrations by a dot hybridization procedure. *Nucleic Acids Research*, *7*(6), 1541–1552.

Kato-Maeda, M., Rhee, J. T., Gingeras, T. R., Salamon, H., Drenkow, J., Smittipat, N., et al. (2001). Comparing genomes within the species Mycobacterium tuberculosis. *Genome Research*, *11*(4), 547–554.

Keum, K. C., Yoo, S. M., Lee, S. Y., Chang, K. H., Yoo, N. C., Yoo, W. M., et al. (2006). DNA microarray-based detection of nosocomial pathogenic Pseudomonas aeruginosa and Acinetobacter baumannii. *Molecular and Cellular Probes*, *20*(1), 42–50.

Khrapko, K. R., Lysov Yu, P., Khorlyn, A. A., Shick, V. V., Florentiev, V. L., & Mirzabekov, A. D. (1989). An oligonucleotide hybridization approach to DNA sequencing. *FEBS Letters*, *256*(1–2), 118–122.

Kim, S. Y., Park, Y. J., Song, E., Jang, H., Kim, C., Yoo, J., et al. (2006). Evaluation of the CombiChip Mycobacteria drug-resistance detection DNA chip for identifying mutations associated with resistance to isoniazid and rifampin in Mycobacterium tuberculosis. *Diagnostic Microbiology and Infectious Disease*, *54*(3), 203–210.

Kirschner, P., Rosenau, J., Springer, B., Teschner, K., Feldmann, K., & Bottger, E. C. (1996). Diagnosis of mycobacterial infections by nucleic acid amplification: 18-month prospective study. *Journal of Clinical Microbiology*, *34*(2), 304–312.

Kistler, A., Avila, P. C., Rouskin, S., Wang, D., Ward, T., Yagi, S., et al. (2007). Pan-viral screening of respiratory tract infections in adults with and without asthma reveals unexpected human coronavirus and human rhinovirus diversity. *The Journal of Infectious Diseases*, *196*(6), 817–825.

Klouche, M., & Schroder, U. (2008). Rapid methods for diagnosis of bloodstream infections. *Clinical Chemistry and Laboratory Medicine*, *46*(7), 888–908.

Knetsch, C. W., Lawley, T. D., Hensgens, M. P., Corver, J., Wilcox, M. W., & Kuijper, E. J. (2013). Current application and future perspectives of molecular typing methods to study Clostridium difficile infections. *Euro Surveillance*, *18*(4), 20381.

Kooken, J., Fox, K., Fox, A., Altomare, D., Creek, K., Wunschel, D., et al. (2014). Identification of staphylococcal species based on variations in protein sequences (mass spectrometry) and DNA sequence (sodA microarray). *Molecular and Cellular Probes*, *28*(1), 41–50.

Krunic, N., Merante, F., Yaghoubian, S., Himsworth, D., & Janeczko, R. (2011). Advances in the diagnosis of respiratory tract infections: Role of the Luminex xTAG respiratory viral panel. *Annals of the New York Academy of Sciences, 1222*, 6–13.

Kumar, S., Wang, L., Fan, J., Kraft, A., Bose, M. E., Tiwari, S., et al. (2008). Detection of 11 common viral and bacterial pathogens causing community-acquired pneumonia or sepsis in asymptomatic patients by using a multiplex reverse transcription-PCR assay with manual (enzyme hybridization) or automated (electronic microarray) detection. *Journal of Clinical Microbiology, 46*(9), 3063–3072.

Laakso, S., Kirveskari, J., Tissari, P., & Maki, M. (2011). Evaluation of high-throughput PCR and microarray-based assay in conjunction with automated DNA extraction instruments for diagnosis of sepsis. *PLoS One, 6*(11), e26655.

Laksanalamai, P., Joseph, L. A., Silk, B. J., Burall, L. S., Tarr, C. L., Gerner-Smidt, P., et al. (2012). Genomic characterization of Listeria monocytogenes strains involved in a multistate listeriosis outbreak associated with cantaloupe in US. *PLoS One, 7*(7), e42448.

Lam, B., Das, J., Holmes, R. D., Live, L., Sage, A., Sargent, E. H., et al. (2013). Solution-based circuits enable rapid and multiplexed pathogen detection. *Nature Communications, 4*, 2001.

Landlinger, C., Preuner, S., Willinger, B., Haberpursch, B., Racil, Z., Mayer, J., et al. (2009). Species-specific identification of a wide range of clinically relevant fungal pathogens by use of Luminex xMAP technology. *Journal of Clinical Microbiology, 47*(4), 1063–1073.

Lausted, C., Dahl, T., Warren, C., King, K., Smith, K., Johnson, M., et al. (2004). POSaM: A fast, flexible, open-source, inkjet oligonucleotide synthesizer and microarrayer. *Genome Biology, 5*(8), R58.

Lazzeri, E., Santoro, F., Oggioni, M. R., Iannelli, F., & Pozzi, G. (2012). Novel primer-probe sets for detection and identification of mycobacteria by PCR-microarray assay. *Journal of Clinical Microbiology, 50*(11), 3777–3779.

Lee, A. S., Jelfs, P., Sintchenko, V., & Gilbert, G. L. (2009). Identification of non-tuberculous mycobacteria: Utility of the GenoType Mycobacterium CM/AS assay compared with HPLC and 16S rRNA gene sequencing. *Journal of Medical Microbiology, 58*(Pt. 7), 900–904.

Lee, S. Y., Jeong, J. S., Ahn, J. J., Lee, S. W., Seo, H., Ahn, Y., et al. (2013). Development of electrochemical microbiochip for the biological diagnosis of Neisseria gonorrhoeae. *Analytical Sciences, 29*(12), 1203–1208.

Leinberger, D. M., Schumacher, U., Autenrieth, I. B., & Bachmann, T. T. (2005). Development of a DNA microarray for detection and identification of fungal pathogens involved in invasive mycoses. *Journal of Clinical Microbiology, 43*(10), 4943–4953.

Leveque, N., Van Haecke, A., Renois, F., Boutolleau, D., Talmud, D., & Andreoletti, L. (2011). Rapid virological diagnosis of central nervous system infections by use of a multiplex reverse transcription-PCR DNA microarray. *Journal of Clinical Microbiology, 49*(11), 3874–3879.

Liao, J. C., Mastali, M., Gau, V., Suchard, M. A., Moller, A. K., Bruckner, D. A., et al. (2006). Use of electrochemical DNA biosensors for rapid molecular identification of uropathogens in clinical urine specimens. *Journal of Clinical Microbiology, 44*(2), 561–570.

Liesenfeld, O., Lehman, L., Hunfeld, K. P., & Kost, G. (2014). Molecular diagnosis of sepsis: New aspects and recent developments. *European Journal of Microbiology & Immunology, 4*(1), 1–25.

Lin, B., Blaney, K. M., Malanoski, A. P., Ligler, A. G., Schnur, J. M., Metzgar, D., et al. (2007). Using a resequencing microarray as a multiple respiratory pathogen detection assay. *Journal of Clinical Microbiology, 45*(2), 443–452.

Lin, B., Malanoski, A. P., Wang, Z., Blaney, K. M., Ligler, A. G., Rowley, R. K., et al. (2007). Application of broad-spectrum, sequence-based pathogen identification in an urban population. *PLoS One*, *2*(5), e419.

Lin, B., Wang, Z., Vora, G. J., Thornton, J. A., Schnur, J. M., Thach, D. C., et al. (2006). Broad-spectrum respiratory tract pathogen identification using resequencing DNA microarrays. *Genome Research*, *16*(4), 527–535.

Linger, Y., Kukhtin, A., Golova, J., Perov, A., Lambarqui, A., Bryant, L., et al. (2014). Simplified microarray system for simultaneously detecting rifampin, isoniazid, ethambutol, and streptomycin resistance markers in Mycobacterium tuberculosis. *Journal of Clinical Microbiology*, *52*(6), 2100–2107.

Liu, Y., Han, J. X., Huang, H. Y., & Zhu, B. (2005). Development and evaluation of 16S rDNA microarray for detecting bacterial pathogens in cerebrospinal fluid. *Experimental Biology and Medicine (Maywood, N.J.)*, *230*(8), 587–591.

Lodes, M. J., Suciu, D., Wilmoth, J. L., Ross, M., Munro, S., Dix, K., et al. (2007). Identification of upper respiratory tract pathogens using electrochemical detection on an oligonucleotide microarray. *PLoS One*, *2*(9), e924.

Mahasirimongkol, S., Yanai, H., Nishida, N., Ridruechai, C., Matsushita, I., Ohashi, J., et al. (2009). Genome-wide SNP-based linkage analysis of tuberculosis in Thais. *Genes and Immunity*, *10*(1), 77–83.

Mahony, J., Chong, S., Merante, F., Yaghoubian, S., Sinha, T., Lisle, C., et al. (2007). Development of a respiratory virus panel test for detection of twenty human respiratory viruses by use of multiplex PCR and a fluid microbead-based assay. *Journal of Clinical Microbiology*, *45*(9), 2965–2970.

Mahony, J. B., Petrich, A., & Smieja, M. (2011). Molecular diagnosis of respiratory virus infections. *Critical Reviews in Clinical Laboratory Sciences*, *48*(5–6), 217–249.

Marotta, F., Zilli, K., Tonelli, A., Sacchini, L., Alessiani, A., Migliorati, G., et al. (2013). Detection and genotyping of Campylobacter jejuni and Campylobacter coli by use of DNA oligonucleotide arrays. *Molecular Biotechnology*, *53*(2), 182–188.

Matsuoka, M., Aye, K. S., Kyaw, K., Tan, E. V., Balagon, M. V., Saunderson, P., et al. (2008). A novel method for simple detection of mutations conferring drug resistance in Mycobacterium leprae, based on a DNA microarray, and its applicability in developing countries. *Journal of Medical Microbiology*, *57*(Pt. 10), 1213–1219.

Maynard, C., Berthiaume, F., Lemarchand, K., Harel, J., Payment, P., Bayardelle, P., et al. (2005). Waterborne pathogen detection by use of oligonucleotide-based microarrays. *Applied and Environmental Microbiology*, *71*(12), 8548–8557.

McCabe, K. M., Zhang, Y. H., Huang, B. L., Wagar, E. A., & McCabe, E. R. (1999). Bacterial species identification after DNA amplification with a universal primer pair. *Molecular Genetics and Metabolism*, *66*(3), 205–211.

McGall, G. H., Barone, A. D., Diggelmann, M., Fodor, S. P. A., Gentalen, E., & Nga, N. (1997). The efficiency of light-directed synthesis of DNA arrays on glass substrates. *Journal of the American Chemical Society*, *119*, 5081–5090.

Mengelle, C., Mansuy, J. M., Prere, M. F., Grouteau, E., Claudet, I., Kamar, N., et al. (2013). Simultaneous detection of gastrointestinal pathogens with a multiplex Luminex-based molecular assay in stool samples from diarrhoeic patients. *Clinical Microbiology and Infection*, *19*(10), E458–E465.

Metso, L., Maki, M., Tissari, P., Remes, V., Piiparinen, P., Kirveskari, J., et al. (2014). Efficacy of a novel PCR- and microarray-based method in diagnosis of a prosthetic joint infection. *Acta Orthopaedica*, *85*(2), 165–170.

Mijs, W., De Vreese, K., Devos, A., Pottel, H., Valgaeren, A., Evans, C., et al. (2002). Evaluation of a commercial line probe assay for identification of mycobacterium species from liquid and solid culture. *European Journal of Clinical Microbiology & Infectious Diseases, 21*(11), 794–802.

Miller, M. B., & Tang, Y. W. (2009). Basic concepts of microarrays and potential applications in clinical microbiology. *Clinical Microbiology Reviews, 22*(4), 611–633.

Moure, R., Espanol, M., Tudo, G., Vicente, E., Coll, P., Gonzalez-Martin, J., et al. (2014). Characterization of the embB gene in Mycobacterium tuberculosis isolates from Barcelona and rapid detection of main mutations related to ethambutol resistance using a low-density DNA array. *The Journal of Antimicrobial Chemotherapy, 69*(4), 947–954.

Naas, T., Cuzon, G., Bogaerts, P., Glupczynski, Y., & Nordmann, P. (2011). Evaluation of a DNA microarray (Check-MDR CT102) for rapid detection of TEM, SHV, and CTX-M extended-spectrum beta-lactamases and of KPC, OXA-48, VIM, IMP, and NDM-1 carbapenemases. *Journal of Clinical Microbiology, 49*(4), 1608–1613.

Naas, T., Cuzon, G., Truong, H., Bernabeu, S., & Nordmann, P. (2010). Evaluation of a DNA microarray, the check-points ESBL/KPC array, for rapid detection of TEM, SHV, and CTX-M extended-spectrum beta-lactamases and KPC carbapenemases. *Antimicrobial Agents and Chemotherapy, 54*(8), 3086–3092.

Nadarajah, R., Khan, G. Y., Miller, S. A., & Brooks, G. F. (2007). Evaluation of a new-generation line-probe assay that detects 5′ untranslated and core regions to genotype and subtype hepatitis C virus. *American Journal of Clinical Pathology, 128*(2), 300–304.

Nasr Esfahani, B., Rezaei Yazdi, H., Moghim, S., Ghasemian Safaei, H., & Zarkesh Esfahani, H. (2012). Rapid and accurate identification of Mycobacterium tuberculosis complex and common non-tuberculous mycobacteria by multiplex real-time PCR targeting different housekeeping genes. *Current Microbiology, 65*(5), 493–499.

Navidad, J. F., Griswold, D. J., Gradus, M. S., & Bhattacharyya, S. (2013). Evaluation of Luminex xTAG gastrointestinal pathogen analyte-specific reagents for high-throughput, simultaneous detection of bacteria, viruses, and parasites of clinical and public health importance. *Journal of Clinical Microbiology, 51*(9), 3018–3024.

Negoro, E., Iwasaki, H., Tai, K., Ikegaya, S., Takagi, K., Kishi, S., et al. (2013). Utility of PCR amplification and DNA microarray hybridization of 16S rDNA for rapid diagnosis of bacteremia associated with hematological diseases. *International Journal of Infectious Diseases, 17*(4), e271–e276.

Neonakis, I. K., Gitti, Z., Krambovitis, E., & Spandidos, D. A. (2008). Molecular diagnostic tools in mycobacteriology. *Journal of Microbiological Methods, 75*(1), 1–11.

Ngamskulrungroj, P., Price, J., Sorrell, T., Perfect, J. R., & Meyer, W. (2011). Cryptococcus gattii virulence composite: Candidate genes revealed by microarray analysis of high and less virulent Vancouver island outbreak strains. *PLoS One, 6*(1), e16076.

Palka-Santini, M., Cleven, B. E., Eichinger, L., Kronke, M., & Krut, O. (2009). Large scale multiplex PCR improves pathogen detection by DNA microarrays. *BMC Microbiology, 9*, 1.

Palka-Santini, M., Putzfeld, S., Cleven, B. E., Kronke, M., & Krut, O. (2007). Rapid identification, virulence analysis and resistance profiling of Staphylococcus aureus by gene segment-based DNA microarrays: Application to blood culture post-processing. *Journal of Microbiological Methods, 68*(3), 468–477.

Palmieri, M., Alessi, E., Conoci, S., Marchi, M., & Panvini, G. (2008). Developments of the in-check platform for diagnostic applications. In W. Wang & C. Vauchier (Eds.), *Proceedings of the SPIE: 6886. Microfluidics, bioMEMs, and medical microsystems VI* (p. 14), 688602.

Park, S., Kang, Y., Kim, D. G., Kim, E. C., Park, S. S., & Seong, M. W. (2013). Comparison of the analytical and clinical performances of Abbott RealTime High Risk HPV, Hybrid Capture 2, and DNA Chip assays in gynecology patients. *Diagnostic Microbiology and Infectious Disease, 76*(4), 432–436.

Park, J. C., Kim, J. M., Kwon, O. J., Lee, K. R., Chai, Y. G., & Oh, H. B. (2010). Development and clinical evaluation of a microarray for hepatitis C virus genotyping. *Journal of Virological Methods, 163*(2), 269–275.

Pasko, C., Hicke, B., Dunn, J., Jaeckel, H., Nieuwlandt, D., Weed, D., et al. (2012). Staph ID/R: A rapid method for determining staphylococcus species identity and detecting the mecA gene directly from positive blood culture. *Journal of Clinical Microbiology, 50*(3), 810–817.

Paulin, L. F., de los, D. S.-D. R. M., Sanchez, I., Hernandez, J., Gutierrez-Rios, R. M., Lopez-Martinez, I., et al. (2014). PhyloFlu, a DNA microarray for determining the phylogenetic origin of influenza A virus gene segments and the genomic fingerprint of viral strains. *Journal of Clinical Microbiology, 52*(3), 803–813.

Pease, A. C., Solas, D., Sullivan, E. J., Cronin, M. T., Holmes, C. P., & Fodor, S. P. (1994). Light-generated oligonucleotide arrays for rapid DNA sequence analysis. *Proceedings of the National Academy of Sciences of the United States of America, 91*(11), 5022–5026.

Peter, H., Berggrav, K., Thomas, P., Pfeifer, Y., Witte, W., Templeton, K., et al. (2012). Direct detection and genotyping of Klebsiella pneumoniae carbapenemases from urine by use of a new DNA microarray test. *Journal of Clinical Microbiology, 50*(12), 3990–3998.

Popowitch, E. B., O'Neill, S. S., & Miller, M. B. (2013). Comparison of the Biofire FilmArray RP, Genmark eSensor RVP, Luminex xTAG RVPv1, and Luminex xTAG RVP fast multiplex assays for detection of respiratory viruses. *Journal of Clinical Microbiology, 51*(5), 1528–1533.

Rafi, W., Chandramuki, A., Mani, R., Satishchandra, P., & Shankar, S. K. (2010). Rapid diagnosis of acute bacterial meningitis: Role of a broad range 16S rRNA polymerase chain reaction. *The Journal of Emergency Medicine, 38*(2), 225–230.

Rieg, S., Jonas, D., Kaasch, A. J., Porzelius, C., Peyerl-Hoffmann, G., Theilacker, C., et al. (2013). Microarray-based genotyping and clinical outcomes of Staphylococcus aureus bloodstream infection: An exploratory study. *PLoS One, 8*(8), e71259.

Ringuet, H., Akoua-Koffi, C., Honore, S., Varnerot, A., Vincent, V., Berche, P., et al. (1999). hsp65 sequencing for identification of rapidly growing mycobacteria. *Journal of Clinical Microbiology, 37*(3), 852–857.

Ruettger, A., Nieter, J., Skrypnyk, A., Engelmann, I., Ziegler, A., Moser, I., et al. (2012). Rapid spoligotyping of Mycobacterium tuberculosis complex bacteria by use of a microarray system with automatic data processing and assignment. *Journal of Clinical Microbiology, 50*(7), 2492–2495.

Ruggiero, P., McMillen, T., Tang, Y. W., & Babady, N. E. (2014). Evaluation of the BioFire FilmArray respiratory panel and the GenMark eSensor respiratory viral panel on lower respiratory tract specimens. *Journal of Clinical Microbiology, 52*(1), 288–290.

Samuel, L. P., Tibbetts, R. J., Agotesku, A., Fey, M., Hensley, R., & Meier, F. A. (2013). Evaluation of a microarray-based assay for rapid identification of Gram-positive organisms and resistance markers in positive blood cultures. *Journal of Clinical Microbiology, 51*(4), 1188–1192.

Sango, A., McCarter, Y. S., Johnson, D., Ferreira, J., Guzman, N., & Jankowski, C. A. (2013). Stewardship approach for optimizing antimicrobial therapy through use of a rapid microarray assay on blood cultures positive for Enterococcus species. *Journal of Clinical Microbiology, 51*(12), 4008–4011.

Sato, T., Takayanagi, A., Nagao, K., Tomatsu, N., Fukui, T., Kawaguchi, M., et al. (2010). Simple PCR-based DNA microarray system to identify human pathogenic fungi in skin. *Journal of Clinical Microbiology*, *48*(7), 2357–2364.

Schanne, M., Bodem, J., Gerhold-Ay, A., Jacob, A., Fellenberg, K., Krausslich, H. G., et al. (2008). Genotypic resistance testing in HIV by arrayed primer extension. *Analytical and Bioanalytical Chemistry*, *391*(5), 1661–1669.

Schrenzel, J. (2007). Clinical relevance of new diagnostic methods for bloodstream infections. *International Journal of Antimicrobial Agents*, *30*(Suppl. 1), S2–S6.

Shang, S., Chen, G., Wu, Y., Du, L., & Zhao, Z. (2005). Rapid diagnosis of bacterial sepsis with PCR amplification and microarray hybridization in 16S rRNA gene. *Pediatric Research*, *58*(1), 143–148.

Shen, H., Shi, W., Wang, J., Wang, M., Li, J., Zhang, C., et al. (2013). Development of a new resequencing pathogen microarray based assay for detection of broad-spectrum respiratory tract viruses in patients with community-acquired pneumonia. *PLoS One*, *8*(9), e75704.

Shen-Gunther, J., & Rebeles, J. (2013). Genotyping human papillomaviruses: Development and evaluation of a comprehensive DNA microarray. *Gynecologic Oncology*, *128*(3), 433–441.

Shi, G., Wen, S. Y., Chen, S. H., & Wang, S. Q. (2005). Fabrication and optimization of the multiplex PCR-based oligonucleotide microarray for detection of Neisseria gonorrhoeae, Chlamydia trachomatis and Ureaplasma urealyticum. *Journal of Microbiological Methods*, *62*(2), 245–256.

Sibley, C. D., Peirano, G., & Church, D. L. (2012). Molecular methods for pathogen and microbial community detection and characterization: Current and potential application in diagnostic microbiology. *Infection, Genetics and Evolution*, *12*(3), 505–521.

Simoes, E. A., Patel, C., Sung, W. K., Lee, C. W., Loh, K. H., Lucero, M., et al. (2013). Pathogen chip for respiratory tract infections. *Journal of Clinical Microbiology*, *51*(3), 945–953.

Singh-Gasson, S., Green, R. D., Yue, Y., Nelson, C., Blattner, F., Sussman, M. R., et al. (1999). Maskless fabrication of light-directed oligonucleotide microarrays using a digital micromirror array. *Nature Biotechnology*, *17*(10), 974–978.

Soini, H., Skurnik, M., Liippo, K., Tala, E., & Viljanen, M. K. (1992). Detection and identification of mycobacteria by amplification of a segment of the gene coding for the 32-kilodalton protein. *Journal of Clinical Microbiology*, *30*(8), 2025–2028.

Sosnowski, R. G., Tu, E., Butler, W. F., O'Connell, J. P., & Heller, M. J. (1997). Rapid determination of single base mismatch mutations in DNA hybrids by direct electric field control. *Proceedings of the National Academy of Sciences of the United States of America*, *94*(4), 1119–1123.

Spence, R. P., Wright, V., Ala-Aldeen, D. A., Turner, D. P., Wooldridge, K. G., & James, R. (2008). Validation of virulence and epidemiology DNA microarray for identification and characterization of Staphylococcus aureus isolates. *Journal of Clinical Microbiology*, *46*(5), 1620–1627.

Spiess, B., Seifarth, W., Hummel, M., Frank, O., Fabarius, A., Zheng, C., et al. (2007). DNA microarray-based detection and identification of fungal pathogens in clinical samples from neutropenic patients. *Journal of Clinical Microbiology*, *45*(11), 3743–3753.

Strommenger, B., Schmidt, C., Werner, G., Roessle-Lorch, B., Bachmann, T. T., & Witte, W. (2007). DNA microarray for the detection of therapeutically relevant antibiotic resistance determinants in clinical isolates of Staphylococcus aureus. *Molecular and Cellular Probes*, *21*(3), 161–170.

Sullivan, K. V., Deburger, B., Roundtree, S. S., Ventrola, C. A., Blecker-Shelly, D. L., & Mortensen, J. E. (2014). Pediatric multicenter evaluation of the Verigene gram-negative blood culture test for rapid detection of inpatient bacteremia involving gram-negative organisms, extended-spectrum beta-lactamases, and carbapenemases. *Journal of Clinical Microbiology, 52*(7), 2416–2421.

Sutherland, A., Thomas, M., Brandon, R. A., Brandon, R. B., Lipman, J., Tang, B., et al. (2011). Development and validation of a novel molecular biomarker diagnostic test for the early detection of sepsis. *Critical Care, 15*(3), R149.

Taylor, J. D., Briley, D., Nguyen, Q., Long, K., Iannone, M. A., Li, M. S., et al. (2001). Flow cytometric platform for high-throughput single nucleotide polymorphism analysis. *Biotechniques, 30*(3), 661–666, 668–669.

Thierry, D., Brisson-Noel, A., Vincent-Levy-Frebault, V., Nguyen, S., Guesdon, J. L., & Gicquel, B. (1990). Characterization of a Mycobacterium tuberculosis insertion sequence, IS6110, and its application in diagnosis. *Journal of Clinical Microbiology, 28*(12), 2668–2673.

Thierry, S., Hamidjaja, R. A., Girault, G., Lofstrom, C., Ruuls, R., & Sylviane, D. (2013). A multiplex bead-based suspension array assay for interrogation of phylogenetically informative single nucleotide polymorphisms for Bacillus anthracis. *Journal of Microbiological Methods, 95*(3), 357–365.

Thissen, J. B., McLoughlin, K., Gardner, S., Gu, P., Mabery, S., Slezak, T., et al. (2014). Analysis of sensitivity and rapid hybridization of a multiplexed Microbial Detection Microarray. *Journal of Virological Methods, 201*, 73–78.

Tobler, N. E., Pfunder, M., Herzog, K., Frey, J. E., & Altwegg, M. (2006). Rapid detection and species identification of Mycobacterium spp. using real-time PCR and DNA-microarray. *Journal of Microbiological Methods, 66*(1), 116–124.

Tortoli, E., & Marcelli, F. (2007). Use of the INNO LiPA Rif.TB for detection of Mycobacterium tuberculosis DNA directly in clinical specimens and for simultaneous determination of rifampin susceptibility. *European Journal of Clinical Microbiology & Infectious Diseases, 26*(1), 51–55.

Tortoli, E., Mariottini, A., & Mazzarelli, G. (2003). Evaluation of INNO-LiPA MYCOBACTERIA v2: Improved reverse hybridization multiple DNA probe assay for mycobacterial identification. *Journal of Clinical Microbiology, 41*(9), 4418–4420.

Tsai, H. F., Sammons, L. R., Zhang, X., Suffis, S. D., Su, Q., Myers, T. G., et al. (2010). Microarray and molecular analyses of the azole resistance mechanism in Candida glabrata oropharyngeal isolates. *Antimicrobial Agents and Chemotherapy, 54*(8), 3308–3317.

Veenemans, J., Overdevest, I. T., Snelders, E., Willemsen, I., Hendriks, Y., Adesokan, A., et al. (2014). Next-generation sequencing for typing and detection of resistance genes: Performance of a new commercial method during an outbreak of extended-spectrum-beta-lactamase-producing Escherichia coli. *Journal of Clinical Microbiology, 52*(7), 2454–2460.

Wang, M., Cao, B., Gao, Q., Sun, Y., Liu, P., Feng, L., et al. (2009). Detection of Enterobacter sakazakii and other pathogens associated with infant formula powder by use of a DNA microarray. *Journal of Clinical Microbiology, 47*(10), 3178–3184.

Wang, D., Coscoy, L., Zylberberg, M., Avila, P. C., Boushey, H. A., Ganem, D., et al. (2002). Microarray-based detection and genotyping of viral pathogens. *Proceedings of the National Academy of Sciences of the United States of America, 99*(24), 15687–15692.

Wang, Z., Orlandi, P. A., & Stenger, D. A. (2005). Simultaneous detection of four human pathogenic microsporidian species from clinical samples by oligonucleotide microarray. *Journal of Clinical Microbiology*, *43*(8), 4121–4128.

Wang, Q., Wang, M., Kong, F., Gilbert, G. L., Cao, B., Wang, L., et al. (2007). Development of a DNA microarray to identify the Streptococcus pneumoniae serotypes contained in the 23-valent pneumococcal polysaccharide vaccine and closely related serotypes. *Journal of Microbiological Methods*, *68*(1), 128–136.

Weile, J., Schmid, R. D., Bachmann, T. T., Susa, M., & Knabbe, C. (2007). DNA microarray for genotyping multidrug-resistant Pseudomonas aeruginosa clinical isolates. *Diagnostic Microbiology and Infectious Disease*, *59*(3), 325–338.

Wiesinger-Mayr, H., Jordana-Lluch, E., Martro, E., Schoenthaler, S., & Noehammer, C. (2011). Establishment of a semi-automated pathogen DNA isolation from whole blood and comparison with commercially available kits. *Journal of Microbiological Methods*, *85*(3), 206–213.

Wiesinger-Mayr, H., Vierlinger, K., Pichler, R., Kriegner, A., Hirschl, A. M., Presterl, E., et al. (2007). Identification of human pathogens isolated from blood using microarray hybridisation and signal pattern recognition. *BMC Microbiology*, *7*, 78.

Witney, A. A., Marsden, G. L., Holden, M. T., Stabler, R. A., Husain, S. E., Vass, J. K., et al. (2005). Design, validation, and application of a seven-strain Staphylococcus aureus PCR product microarray for comparative genomics. *Applied and Environmental Microbiology*, *71*(11), 7504–7514.

Wright, T. C., Jr., Stoler, M. H., Behrens, C. M., Apple, R., Derion, T., & Wright, T. L. (2012). The ATHENA human papillomavirus study: Design, methods, and baseline results. *American Journal of Obstetrics and Gynecology*, *206*(1), 46.e1–46.e11.

Xiao-Ping, K., Yong-Qiang, L., Qing-Ge, S., Hong, L., Qing-Yu, Z., & Yin-Hui, Y. (2009). Development of a consensus microarray method for identification of some highly pathogenic viruses. *Journal of Medical Virology*, *81*(11), 1945–1950.

Yan, L., Zhang, J., Li, M., Cao, Y., Xu, Z., Cao, Y., et al. (2008). DNA microarray analysis of fluconazole resistance in a laboratory Candida albicans strain. *Acta Biochimica et Biophysica Sinica*, *40*(12), 1048–1060.

Yao, C., Zhu, T., Li, Y., Zhang, L., Zhang, B., Huang, J., et al. (2010). Detection of rpoB, katG and inhA gene mutations in Mycobacterium tuberculosis clinical isolates from Chongqing as determined by microarray. *Clinical Microbiology and Infection*, *16*(11), 1639–1643.

Ye, F., Li, M. S., Taylor, J. D., Nguyen, Q., Colton, H. M., Casey, W. M., et al. (2001). Fluorescent microsphere-based readout technology for multiplexed human single nucleotide polymorphism analysis and bacterial identification. *Human Mutation*, *17*(4), 305–316.

Yoo, S. M., Choi, J. Y., Yun, J. K., Choi, J. K., Shin, S. Y., Lee, K., et al. (2010). DNA microarray-based identification of bacterial and fungal pathogens in bloodstream infections. *Molecular and Cellular Probes*, *24*(1), 44–52.

Yoo, S. M., Lee, S. Y., Chang, K. H., Yoo, S. Y., Yoo, N. C., Keum, K. C., et al. (2009). High-throughput identification of clinically important bacterial pathogens using DNA microarray. *Molecular and Cellular Probes*, *23*(3–4), 171–177.

Yu, X., Susa, M., Knabbe, C., Schmid, R. D., & Bachmann, T. T. (2004). Development and validation of a diagnostic DNA microarray to detect quinolone-resistant Escherichia coli among clinical isolates. *Journal of Clinical Microbiology*, *42*(9), 4083–4091.

Yu, X., Susa, M., Weile, J., Knabbe, C., Schmid, R. D., & Bachmann, T. T. (2007). Rapid and sensitive detection of fluoroquinolone-resistant Escherichia coli from urine samples using a genotyping DNA microarray. *International Journal of Medical Microbiology, 297*(6), 417–429.

Yue, J., Shi, W., Xie, J., Li, Y., Zeng, E., Liang, L., et al. (2004). Detection of rifampin-resistant Mycobacterium tuberculosis strains by using a specialized oligonucleotide microarray. *Diagnostic Microbiology and Infectious Disease, 48*(1), 47–54.

Zhang, G., Cai, F., Zhou, Z., DeVos, J., Wagar, N., Diallo, K., et al. (2013). Simultaneous detection of major drug resistance mutations in the protease and reverse transcriptase genes for HIV-1 subtype C by use of a multiplex allele-specific assay. *Journal of Clinical Microbiology, 51*(11), 3666–3674.

Zhang, Y., Liu, Q., Wang, D., Chen, S., & Wang, S. (2013). Simultaneous detection of oseltamivir- and amantadine-resistant influenza by oligonucleotide microarray visualization. *PLoS One, 8*(2), e57154.

Zhou, G., Wen, S., Liu, Y., Li, R., Zhong, X., Feng, L., et al. (2011). Development of a DNA microarray for detection and identification of Legionella pneumophila and ten other pathogens in drinking water. *International Journal of Food Microbiology, 145*(1), 293–300.

Zhu, L. X., Zhang, Z. W., Wang, C., Yang, H. W., Jiang, D., Zhang, Q., et al. (2007). Use of a DNA microarray for simultaneous detection of antibiotic resistance genes among staphylococcal clinical isolates. *Journal of Clinical Microbiology, 45*(11), 3514–3521.

Zimenkov, D. V., Antonova, O. V., Kuz'min, A. V., Isaeva, Y. D., Krylova, L. Y., Popov, S. A., et al. (2013). Detection of second-line drug resistance in Mycobacterium tuberculosis using oligonucleotide microarrays. *BMC Infectious Diseases, 13*, 240.

CHAPTER

Gene Amplification and Sequencing for Bacterial Identification

12

Susanna K.P. Lau*,†, Jade L.L. Teng*, Chi-Chun Ho*, Patrick C.Y. Woo*,†,1

*Department of Microbiology, The University of Hong Kong, Queen Mary Hospital, Hong Kong
†State Key Laboratory of Emerging Infectious Diseases, The University of Hong Kong, Queen Mary Hospital, Hong Kong
1Corresponding author: e-mail address: pcywoo@hku.hk

1 IMPORTANCE OF ACCURATE IDENTIFICATION OF BACTERIA

The fundamental and most important task of the clinical microbiology laboratory is to accurately identify the responsible pathogens, most commonly encountered bacteria, to species level. At the patient level, accurate identification of the bacteria can give clues on whether the microbe is causing genuine infection or is a contaminant or coloniser. For example, if the blood culture isolate from a patient is identified as *Staphylococcus aureus*, the bacterium is always regarded as a pathogen until proven otherwise. On the other hand, if *Staphylococcus epidermidis* or many other coagulase-negative staphylococci are recovered from the blood culture, particularly if it is only recovered from just one out of several blood samples, there is a chance that it is just a contaminant (To et al., 2011; Woo, Leung, Leung, & Yuen, 2001). In addition, accurate identification is also crucial for the determination of the choice and duration of antimicrobial agents for treatment as well as the infection control procedures. For example, identifying a bacterial isolate as an *Actinomyces* species in a case of pelvic inflammatory disease implies that the patient is suffering from pelvic actinomycosis, which requires prolonged antibiotic treatment (Woo, Fung, Lau, Hon, & Yuen, 2002; Woo, Fung, Lau, Teng, et al., 2003). At the population scale, this is important for defining the epidemiology of infectious diseases, antibiotic resistance patterns of bacteria and hence the choice of empiric antimicrobials, as well as outcome and prognosis of the infections.

2 EVOLUTION OF BACTERIAL IDENTIFICATION METHODS

2.1 TRADITIONAL PHENOTYPIC METHODS

Traditionally, identification of bacteria is performed using conventional phenotypic tests. These include appearance of the bacterium under light microscope after Gram staining or using other special stains; growth requirements such as requirement of oxygen, carbon dioxide or other nutrients, or growth on a specific carbon source; and biochemical tests, most importantly sugar metabolism and production of specific enzymes. Companies have incorporated various biochemical tests into commercially available strips or slides so that the biochemical profile of a bacterial isolate can be determined in a user-friendly way. Computer software is also used for matching the biochemical profile of a particular isolate with those of known bacteria in order to determine its identity and associated probability. Although these traditional phenotypic tests are accurate for most commonly encountered bacteria in clinical laboratories, they may fail to work in some circumstances. First, some bacterial isolates may possess atypical biochemical profiles, such as catalase-negative or coagulase-negative *S. aureus*, or H_2S-positive *Escherichia coli*. Second, correct identification requires the presence of the biochemical profile of a particular bacterial species in the database, but such profiles are not always available for bacteria rarely encountered clinically. Third, as most commercial kits require reading the results in 24–48 h and some reactions require growth of the bacterium, such methods are not applicable to slowly growing or non-cultivable bacteria.

2.2 GAS CHROMATOGRAPHY–MASS SPECTROMETRY

Since traditional phenotypic methods are not very useful for identification of anaerobic bacteria and mycobacteria, gas chromatography–mass spectrometry has been used for their identification. However, this method has not gained much popularity in clinical microbiology laboratories because special equipment and the associated expertise are required and the equipment is only used for a small group of bacteria.

2.3 GENE SEQUENCING

Since the 1970s, Carl Woese and others have started to use 16S rRNA gene sequencing for classification of bacteria, exploiting the fact that each bacterial species has its specific 16S rRNA gene sequence. In the 1990s, this technology was put to more routine clinical use for identification of difficult-to-identify bacteria. Not only has 16S rRNA gene sequencing been used as the standard of classifying and identifying bacteria, but it has also led to the discovery of an unprecedented number of novel bacterial genera and species. During the process, it was recognised that for some groups of bacteria, 16S rRNA gene sequencing alone may not be discriminative enough for identifying the bacteria. Therefore, other genes or multiple genes were sequenced for this purpose.

2.4 MATRIX-ASSISTED LASER DESORPTION IONISATION-TIME-OF-FLIGHT MASS SPECTROMETRY

In recent years, matrix-assisted laser desorption ionisation-time-of-flight mass spectrometry (MALDI-TOF MS) has emerged as a new technique for rapid bacterial identification at a low cost and has become widely used for bacterial identification in clinical microbiology laboratories. This technology involves the ionisation of proteins in a bacterium using a laser beam and measuring the time required for each of the ionised proteins to move in an electric field using MS. The resulting mass-to-charge ratio profile of the ionised proteins is specific to the bacterial species, hence useful for its identification.

3 16S rRNA GENE SEQUENCING

Among the various studied gene targets for bacterial identification, 16S rRNA gene has been the most widely used. During the early days where sequencing was not widely available, many studies used polymerase chain reaction (PCR)–restriction fragment length polymorphisms (RFLPs) of 16S rRNA gene for genetic identification. However, RFLP patterns, which depend on nucleotide sequences at the critical restriction sites, may not be reliable even if there is only single nucleotide mutation. Therefore, sequencing of 16S rRNA gene is much more reliable and is now considered the gold standard for identification of most bacterial species. Since the 1990s, a large amount of bacterial 16 rRNA gene sequence data have been generated and deposited in various databases, which have enabled matching of query sequences from any bacterial isolates. In the new genomics era, the availability of bacterial genome sequences has also confirmed the representativeness of 16S rRNA gene in comparison to complete genome for bacterial phylogeny studies (Snel, Bork, & Huynen, 1999).

16S rRNA gene sequencing has played a pivotal role for accurate and rapid identification of bacteria for both research and diagnostic purposes. It is most commonly applied for bacterial isolates of uncertain identities from environmental, veterinary to clinical specimens. In clinical microbiology laboratories, 16S rRNA gene sequences can be used to identify most bacterial species including the common Gram-positive and Gram-negative bacteria. However, its usefulness is best highlighted by the ability to identify bacteria with unusual phenotypic profiles and rare or slow-growing bacteria. As a result, diagnosis of infections caused by unusual, rare or slow-growing bacteria has been made possible, allowing better understanding of their clinical disease associations and epidemiology. In some cases, such accurate identification has also enabled decisions on the appropriate antibiotic regimen and infection control procedures. Another advantage of using molecular techniques is their ability to detect bacteria from direct clinical specimens without pure isolates, as well as non-viable or non-culturable bacteria. Therefore, 16S rRNA gene sequencing has also been found helpful for rapid diagnosis of bacteraemia, culture-negative infections and infections due to non-culturable bacteria.

3.1 16S rRNA GENE SEQUENCING FOR IDENTIFICATION OF BACTERIAL ISOLATES

16S rRNA gene sequencing can offer accurate species identification of various groups of bacteria (Bosshard, Abels, Altwegg, Bottger, & Zbinden, 2004; Bosshard, Abels, Zbinden, Bottger, & Altwegg, 2003; de Melo Oliveira, Abels, Zbinden, Bloemberg, & Zbinden, 2013; Heikens, Fleer, Paauw, Florijn, & Fluit, 2005; Lau, Ng, et al., 2006; Song et al., 2005; Teng et al., 2011; Therese et al., 2009; Woo, Chung, et al., 2007; Woo, Teng, Wu, et al., 2009). However, with its costs and expertise requirements, the use of 16S rRNA gene sequencing in clinical microbiology laboratories is still often reserved for bacterial isolates that are difficult to identify by phenotypic tests, rare or phenotypically aberrant. These are the times where mistakes in identification often occur and 16S rRNA gene sequencing is most useful. For rarely encountered bacteria, their biochemical profiles are often poorly defined or not included in existing commercial identification system databases. Clinical laboratories also sometimes encounter bacterial isolates with atypical phenotypes which can lead to mis-identification. 16S rRNA gene sequencing can provide genus identification in >90% and species identification up to 89% of these circumstances (Bharadwaj et al., 2012; Tang, Ellis, et al., 1998). Many of these bacteria are now better defined in terms of their disease association, pathogenic potential and epidemiology. In many circumstances, accurate identification by 16S rRNA gene sequencing has also allowed wiser choice of antibiotics and decision on treatment duration and led to improved clinical outcomes (Woo, Cheung, Leung, & Yuen, 2001; Woo, Chong, Leung, Que, & Yuen, 2001; Woo, Lau, et al., 2007; Woo, Leung, Leung, & Yuen, 2000; Woo, Tsoi, et al., 2000).

3.1.1 Defining disease associations and epidemiology

With a few exceptions, 16S rRNA gene sequencing is very useful for identification of catalase-negative Gram-positive cocci. Although conventional phenotypic tests can readily identify common *Streptococcus* species, such as *Streptococcus pyogenes* and *Streptococcus agalactiae*, 16S rRNA gene sequencing may help in situations where clinical syndromes do not fit into the microbial characteristics. For example, we have used the technique to define a rare case of psoas abscess due to group A streptococcus which is rarely associated with abscess formation (Lau, Woo, Yim, To, & Yuen, 2003). Moreover, species identification of other catalase-negative Gram-positive cocci is often difficult by phenotypic tests, which has hampered understanding their epidemiology and pathogenic potential. We have previously used the technique for defining bacteraemia caused by Lancefield group G beta-haemolytic streptococci and found that all cases were almost exclusively caused by one of the four species belonging to the group, *Streptococcus dysgalactiae* subsp. *equisimilis* (Woo, Fung, Lau, Wong, & Yuen, 2001). In another study, 16S rRNA gene sequencing was used to define infective endocarditis caused by 'Streptococcus milleri group', and *Streptococcus anginosus* was shown to associate with the highest propensity to cause infective endocarditis among the three species of 'S. milleri group' (Woo, Tse, et al., 2004). 16S rRNA gene sequencing has also been a useful tool for diagnosis of invasive *Streptococcus iniae* infections (Koh, Kurup, & Chen, 2004; Lau, Woo, Luk, et al., 2006; Lau, Woo, Tse, et al., 2003; Sun, Yan, Yeh, Lee, & Lu, 2007). This rare aquatic bacterium was first reported to cause human infection in North America and is difficult to identify by phenotypic tests alone. 16S rRNA gene sequencing has allowed diagnosis of cases in Asia, which has led to a better understanding on its geographical distribution and phenotypic variations between strains of different origins (Weinstein et al., 1997). Other *Streptococcus*-related Gram-positive cocci such as *Helcococcus*, *Gemella*, *Lactococcus* and *Globicatella* and the nutritionally deficient streptococci, *Granulicatella adiacens* and *Abiotrophia defectiva*, are also difficult to identify by phenotypic tests. 16S rRNA gene sequencing has been used successfully for their identification, resulting in better understanding of their prevalence and disease associations (Chan et al., 2011; Lau, Woo, Li, et al., 2006; Lau, Woo, Teng, Leung, & Yuen, 2002; Tang, Ellis, et al., 1998; Woo, Chong, et al., 2001; Woo, Fong, et al., 2009; Woo, Fung, Lau, Chan, et al., 2003; Woo, Lau, Fung, et al., 2003; Woo, Ngan, Lau, & Yuen, 2003; Woo, Tse, Wong, et al., 2005). Similarly for catalase-positive Gram-positive cocci, 16S rRNA gene sequencing has also assisted in defining infections caused by rare species (Chan, Wong, et al., 2012).

16S rRNA gene sequencing is also useful for identification of various Gram-negative bacteria such as *Haemophilus* and *Campylobacter* species which are fastidious and may not be readily identified by conventional phenotypic tests. Among members of the genus *Haemophilus*, *Haemophilus influenzae* is most well studied for its disease associations and pathogenicity. In contrast, other *Haemophilus* species may be difficult to identify and, hence, are relatively poorly understood in terms of clinical significance. 16S rRNA gene sequencing allows accurate identification of the various *Haemophilus* species and has led to better understanding of their prevalence and significance (Chow & Clarridge, 2014; Lau, Woo, Chan, et al., 2002; Lau, Woo, Mok, et al., 2004; Mak, Ho, Tse, Lau, & Wong, 2005; Norskov-Lauritsen, 2014). Using the technique, it has been demonstrated that *Haemophilus segnis* is an important cause of non-influenzae *Haemophilus* bacteraemia (Chow & Clarridge, 2014; Lau, Woo, Chan, et al., 2002; Lau, Woo, Mok, et al., 2004; Mak et al., 2005). 16S rRNA gene sequencing has also facilitated the recognition of thermo-tolerant *Campylobacter fetus* as an important cause of bacteraemia in immunocompromised patients (Woo, Leung, Tsoi, et al., 2002). Cases of limb-threatening or fatal infections caused by *Campylobacter rectus* have also been identified by 16S rRNA gene sequencing, which were rarely reported to cause invasive infections in the past (Lam et al., 2011). The technique has also been used to identify infections caused by rare Gram-negative bacteria, such as *Streptobacillus moniliformis*, *Bordetella* and *Arcobacter*, which may be difficult to diagnose by conventional methods (Chen et al., 2007; Dubois et al., 2008; Lau, Woo, Teng, et al., 2002; Tang, Ellis, et al., 1998; Tang, Hopkins, et al., 1998; Woo, Chong, et al., 2001).

16S rRNA gene sequencing is particularly useful for identifying anaerobes which are often ignored for their clinical significance as a result of difficulties in their isolation and identification. Application of the technique has led to recognition of many previously undescribed or 'rarely encountered' anaerobic Gram-positive rods for their associations with more serious diseases than previously thought (Bosshard, Zbinden, & Altwegg, 2002; Chan, Lau, et al., 2012; Lau, McNabb, et al., 2007; Lau, Teng, et al., 2006; Lau, Woo, Fung, et al., 2004; Lau, Woo, Woo, et al., 2004; Woo, Lau, Chan, et al., 2005; Woo, Teng, et al., 2010). We have previously used 16S rRNA gene sequencing to characterise bacteraemia caused by anaerobic, non-sporulating, Gram-positive rod and found that the former genus *Eggerthella* accounted for a high proportion of clinically significant bacteraemia (Lau, Woo, Fung, et al., 2004; Lau, Woo, Woo, et al., 2004). Moreover, two novel species, *Eggerthella hongkongensis* and *Eggerthella sinensis*, now reclassified under the genus *Paraeggerthella*, were discovered and found to contribute to half of the cases of *Eggerthella* bacteraemia (Lau, Woo, Woo, et al., 2004; Wurdemann et al., 2009). As for bacteraemia caused by the spore-forming *Clostridium* species, cases as defined by 16S rRNA gene sequencing were often found to be associated with gastrointestinal or hepatobiliary diseases, which were also risk factors for increased mortality (Woo, Lau, Chan, et al., 2005). Similarly, 16S rRNA gene sequencing is useful in defining the clinical spectrum of infections caused by various anaerobic Gram-negative bacteria (Lau, Woo, Woo, et al., 2006). These studies have led to a better understanding on the prevalence and potential pathogenicity of the understudied anaerobes.

16S rRNA gene sequencing has also been applied for diagnosis of infections caused by acid-fast or partially acid-fast bacilli which are difficult to identify. Using the technique, we have previously described a novel clinical syndrome, acupuncture mycobacteriosis, caused by relatively alcohol-resistant mycobacteria in patients who have received acupuncture (Woo, Leung, Wong, et al., 2002; Woo, Li, Tang, & Yuen, 2001). Another novel clinical syndrome, *Tsukamurella*-associated conjunctivitis and keratitis, has also been defined using 16S rRNA gene sequencing (Woo, Fong, et al., 2009; Woo, Ngan, et al., 2003).

3.1.2 Guiding antibiotic regimens

16S rRNA gene sequencing can readily offer species identification, which, in turn, may affect choice of antibiotic treatment. An important group of anaerobic, non-sporulating, Gram-positive rods of clinical importance are the *Actinomyces*, which cause actinomycosis, a serious infection often requiring prolonged courses of antibiotic treatment to prevent relapse. However, differentiation of *Actinomyces* and non-*Actinomyces* anaerobic Gram-positive bacilli may not be easily achieved by conventional phenotypic tests, and 16S rRNA gene sequencing is a promising tool in such situation for definitive diagnosis of actinomycosis and guiding antibiotic regimen (Llenas-Garcia et al., 2012; Seo, Yeom, & Ko, 2012; Woo, Fung, Lau, Hon, et al., 2002; Woo, Fung, Lau, & Yuen, 2002).

16S rRNA gene sequencing has been used for diagnosis of a case of empyema thoracis caused by *Enterococcus cecorum*, which differs from other *Enterococcus* species by its susceptibility to cephalosporins. As a result of accurate identification by 16S rRNA gene sequencing, the patient was treated successfully with continuation of cefotaxime (Woo, Tam, Lau, Fung, & Yuen, 2004).

3.1.3 Implications on infection control and public health measures

Apart from establishing the correct microbiological diagnosis and guiding antibiotic treatment, accurate species identification could have important management and public health significance. Some Gram-negative bacteria such as members of *Enterobacteriaceae* are also known to be difficult to identify by phenotypic tests and may benefit from 16S rRNA gene sequencing (Tang, Ellis, et al., 1998; Woo, Cheung, et al., 2001; Woo, Fung, Wong, Tsoi, & Yuen, 2001; Woo, Leung, et al., 2000). Among members of the *Enterobacteriaceae*, identification of *Salmonella* is important because of public health implications. 16S rRNA gene sequencing has been used to differentiate between *Salmonella enterica* serotype Typhi and other members of *Enterobacteriaceae*, which is important in guiding decision on the need of cholecystectomy and eradication of carrier state (Woo, Cheung, et al., 2001; Woo, Fung, Wong, et al., 2001; Woo, Leung, et al., 2000).

Tularemia is a serious systemic disease caused by the highly virulent bacterium, *Francisella tularensis*, which is considered a potential bioterrorism agent. However, *F. tularensis* may be misidentified as *Neisseria meningitidis* or *Actinobacillus actinomycetemcomitans* by phenotypic methods (Clarridge et al., 1996). 16S rRNA gene sequencing should be applied to suspicious isolates of *F. tularensis* isolated from blood cultures, which is crucial for prompt antibiotic treatment and public health decisions.

3.1.4 Reflections on phenotypic and genotypic variations

The accurate species identification by 16S rRNA gene sequencing is also an important means to better define inter- and intraspecies phenotypic and genotypic variations, through studies on atypical bacterial isolates. Unusual strains of various Gram-positive and Gram-negative bacteria have been recognised by 16S rRNA gene sequencing (To et al., 2011; Woo, Cheung, et al., 2001; Woo, Leung, Leung, et al., 2001; Woo, Leung, et al., 2000; Woo, Teng, et al., 2003). For example, using 16S rRNA gene sequencing, a case of endocarditis and pericarditis caused by catalase-negative *S. aureus* was defined, which has led to the finding of a novel nonsense mutation in the *katA* gene encoding the catalase protein (To et al., 2011). 16S rRNA gene sequencing has also been applied to define rare infections such as a case of *Haemophilus quentini* bacteraemia leading to the identification of genetic mechanisms for reduced levofloxacin susceptibility and tetracycline resistance (Mak et al., 2005).

3.1.5 Speeding up laboratory diagnosis of infections

16S rRNA gene sequencing is particularly useful to shorten the time to identify slow-growing bacteria and speed up clinical diagnosis and to guide prompt antibiotic treatment. For example, many *Mycobacterium* species take up to 6–8 weeks to grow in culture and species identification by phenotypic tests can lengthen this process. However, a major limitation of 16S rRNA gene sequencing is the high sequence similarity between some *Mycobacterium* species. In these circumstances, alternative gene targets, such as *hsp65*, may be required for species identification (Lau, Curreem, et al., 2011).

3.2 16S rRNA GENE SEQUENCING FOR DIRECT DETECTION OF BACTERIA FROM CLINICAL SPECIMENS

Besides application to bacterial isolates, 16S rRNA gene sequencing has also been used for bacterial detection and identification from direct clinical specimens or positive cultures (Kotilainen et al., 2006; Li et al., 2009; Miyazato et al., 2012; Qian et al., 2001; Tang, 2009). Another advantage of using molecular techniques such as 16S rRNA gene sequencing is that they can be applied on non-cultivable or non-viable bacteria. First, certain bacteria are difficult to culture or known to be uncultivable. Second, culture-negative infections are notoriously difficult to diagnose with conventional microbiological techniques. The first breakthrough in understanding the pathogenesis of Whipple's disease was only achieved almost a century after its initial description in 1907, with the identification of *Tropheryma whipplei* as the causative agent by 16S rRNA gene sequencing (Fenollar & Raoult, 2001; Marin et al., 2007; Relman, Schmidt, MacDermott, & Falkow, 1992; Wilson, Blitchington, Frothingham, & Wilson, 1991). 16S rRNA gene sequencing has also been successfully used to identify the etiological agents of other infections such as bacillary angiomatosis (caused by *Bartonella henselae* and *Bartonella quintana*; Regnery et al., 1992; Relman, Loutit, Schmidt, Falkow, & Tompkins, 1990) and human ehrlichiosis (caused by bacteria in the genera *Ehrlichia* and *Anaplasma*; Anderson, Dawson, Jones, & Wilson, 1991; Chen, Dumler, Bakken, & Walker, 1994; Maeda et al., 1987), which are difficult to isolate and grow *in vitro*. Direct PCR of the 16S rRNA gene has been used to detect borreliae in sera from patients with suspected Lyme disease (Lee, Vigliotti, Vigliotti, Jones, & Shearer, 2014), *Mycoplasma* and *Chlamydia* in patients with respiratory tract infections (Borel et al., 2008; Touati et al., 2010) and *Mycobacterium* species in patients with mycobacteriosis (Nakano et al., 2010; Syre, Myneedu, Arora, & Grewal, 2009). 16S rRNA gene sequencing is particularly useful for diagnosis of leprosy caused by the non-cultivable mycobacterium, *Mycobacterium leprae*, using skin biopsies of leprosy patients (Kurabachew, Wondimu, & Ryon, 1998; Phetsuksiri et al., 2006).

16S rRNA gene sequencing has also been applied for diagnosis of various culture-negative infections as a result of prior antibiotic therapy or inadequate isolation techniques. For example, PCR and sequencing of 16S rRNA gene using DNA directly extracted from infected heart valves have been used to identify the etiological agents of culture-negative infective endocarditis (Cursons, Jeyerajah, & Sleigh, 1999; Duffett, Missaghi, & Daley, 2012; Harris et al., 2014; Kumar, Menon, Pathipati, & Cherian, 2013). The technique has also been used to diagnose other culture-negative infections, such as meningitis, where cultures may be affected by prior antibiotic treatment (Cursons et al., 1999; Lorino et al., 1999; Pandit, Kumar, Karunasagar, & Karunasagar, 2005; Srinivasan, Pisapia, Shah,

Halpern, & Harris, 2012). However, a potential problem associated with the use of broad-range PCR by universal primer sets in such applications is PCR contamination resulting in false positives (Deutch et al., 2006; Marin et al., 2007; Schabereiter-Gurtner et al., 2008). Therefore, caution should be taken when interpreting any positive PCR results.

3.3 16S rRNA GENE SEQUENCING FOR BACTERIAL DISCOVERY

Besides identification of existing bacterial species, 16S rRNA gene sequencing has also been leading useful tool for bacterial discovery during the past two decades. This has led to the discovery of an enormous diversity of previously undescribed, novel bacterial species. As a result of the easy access to clinical specimens, the oral cavity and the gastrointestinal tract are the most important sources for bacterial discovery in humans (Woo, Lau, Teng, Tse, & Yuen, 2008; Woo, Tse, Lau, et al., 2005; Woo, Wong, et al., 2010; Woo et al., 2014; Yuen et al., 2001). The use of 16S rRNA gene sequencing for detection has also facilitated downstream studies of these novel bacteria to help understand their disease associations, pathogenicity, transmission routes and potential reservoir(s) (Couturier, Slechta, Goulston, Fisher, & Hanson, 2012; Curreem et al., 2011; Elsendoorn, Robert, Culos, Roblot, & Burucoa, 2011; Faibis et al., 2008; Kim et al., 2011; Lau, Fan, et al., 2012, 2011; Lau et al., 2005, 2009; Lau, McNabb, et al., 2007; Lau, Wong, Li, Woo, & Yuen, 2008; Lau, Wong, et al., 2011; Lau, Woo, Fan, Lee, et al., 2007; Lau, Woo, Fan, Ma, et al., 2007; Lau, Woo, Hui, et al., 2003; Ni et al., 2007; Smith, Pandey, & Ussher, 2012; Tang et al., 2013; Teng, Huang, et al., 2014; Teng et al., 2005; Uckay et al., 2007; Woo, Kuhnert, et al., 2003; Woo, Lau, Teng, & Yuen, 2005; Woo, Lau, et al., 2009; Woo, Ma, et al., 2005, 2007; Woo, Tam, et al., 2002; Woo, Teng, Lau, & Yuen, 2006; Woo, Teng, et al., 2004, 2008; Woo, Teng, Ma, Lau, & Yuen, 2007; Woo, Teng, Tsang, et al., 2009; Woo, Wong, et al., 2010; Woo et al., 2014; Xiong et al., 2014; Yuen et al., 2001).

3.4 PERFORMING 16S rRNA GENE SEQUENCING AND ANALYSIS

Because the technique requires expertise and significant labour and has a relatively high cost, 16S rRNA gene sequencing is not routinely used in most clinical laboratories. Instead, it is often used when there are bacterial isolates that are 'unidentifiable' or not confidently identified by conventional phenotypic tests or commercial bacterial identification systems. Laboratories may also consider using 16S rRNA gene sequencing when they are not confident with the identification of a rare bacterial species or when there is mismatch between the species identity and expected phenotypic profiles. The bacterial isolates are then subjected to DNA extraction followed by PCR amplification using universal or group-specific primers targeting the conserved regions of 16S rRNA gene.

The technique for amplification and sequencing of 16S rRNA genes is straightforward and well established in many clinical laboratories. Automated systems such as MicroSeq provide universal or degenerate primer sets that work for all bacterial species, which is convenient to technicians who are unfamiliar with primer design. However, such universal primers may be associated with increased risks of PCR contamination from previously generated PCR products leading to false-positive PCR reactions. Therefore, less degenerate primers for specific groups of bacteria are often preferred to avoid the problems associated with carry-over of PCR products. After purification, the PCR products can be used for DNA sequencing.

In contrast, sequence analysis is the most difficult part when interpreting results and an inexperienced scientist may report a wrong identification from a correct sequence. The accuracy of identification depends on the length and quality of 16S rRNA gene sequences, the choice of appropriate programs and databases for matching, as well as correct interpretation of results from similarity searches against databases. For sequence lengths, a minimum of 500–527 bp, which covers the more variable 5′ region, is usually required for sufficient discrimination between closely related bacterial species. Based on these criteria, databases such as 500-MicroSeq, which incorporated bacterial sequences from the 5′ region of 16S rRNA genes, have been developed. Nevertheless, full 16S rRNA gene sequences are usually preferred for more accurate identification, especially for those bacteria such as *Campylobacter* sharing high sequence similarity between related species in the 5′ regions (Drancourt et al., 2000; Janda & Abbott, 2007).

3.5 DATABASES FOR BACTERIAL IDENTIFICATION

The most widely used database for 16S rRNA gene sequence analysis is the GenBank. Although GenBank provides the largest collection of 16S rRNA gene sequences, it contains a lot of unvalidated sequences so to overcome this problem, curated databases have been developed. Since 2011, we have established a user-friendly and efficient database, named 16SpathDB (http://147.8.74.24/16SpathDB), which comprises all medically important bacteria listed in the most recent version of the *Manual of Clinical Microbiology* (Versalovic, Carroll, Jorgensen, Funke, & Landry, 2011). The database is also updated regularly when new editions of the manual are published (Teng, Ho, et al., 2014; Woo et al., 2011). In addition to providing the closest match to a particular 16S rRNA gene sequence, it also gives other possible identities of the sequence for the user to consider. In contrast to the GenBank which usually takes about 20–30 s to match one sequence, 16SpathDB requires less than 2 s for analysing one sequence. 16SpathDB has been validated using 91 bacteria identified with a polyphasic approach as well as 689 complete bacterial genome sequences and has been demonstrated to be highly accurate (Teng, Ho, et al., 2014; Woo et al., 2011).

4 SEQUENCING OF OTHER GENE TARGETS FOR BACTERIAL IDENTIFICATION

Although 16S rRNA gene sequencing has been successfully used for bacterial identification, sometimes two bacterial species possess highly similar 16S rRNA gene sequences and therefore are indistinguishable by this technique. In such circumstances, other gene targets, such as *groEL*, *rpoB*, and *recA*, which may potentially be able to distinguish between the two species, are sequenced. The criteria for interpretation of such data are similar to 16S rRNA gene sequencing. For example, we have observed that the difference between the 16S rRNA gene sequences of *Burkholderia pseudomallei* and *Burkholderia thailandensis* is only around 1%, making it difficult to distinguish between these two species confidently (Woo, Woo, Lau, Wong, & Yuen, 2002). On the other hand, the difference between the *groEL* gene sequences of these two species is more than 2%, which makes it a much better gene target for identifying them (Woo, Woo, et al., 2002). Other examples of using these gene targets for identification include sequencing *groEL* for identifying atypical mycobacteria (Lau, Curreem, et al., 2011; Ringuet et al., 1999) and coagulase-negative staphylococci (Goh et al., 1996), *rpoB* for *Acinetobacter* species (Lee, Jang, et al., 2014) and *Enterobacteriaceae* (Mollet, Drancourt, & Raoult, 1997), *recA* for *Geobacillus* species (Weng, Chiou, Lin, & Yang, 2009) and *Streptococcus mitis* group (Zbinden, Kohler, & Bloemberg, 2011) and *gltA* for *Ehrlichia* (Inokuma, Brouqui, Drancourt, & Raoult, 2001) and *Rickettsia* species (Inokuma et al., 2001; Roux, Rydkina, Eremeeva, & Raoult, 1997). Since these targets are less frequently used than the 16S rRNA gene, the number of sequences for these genes in GenBank is much lower than that for the 16S rRNA gene. Therefore, one usually needs a more thorough evaluation by sequencing quite a large number of strains in closely related species before concluding whether a particular gene is useful for distinguishing among these closely related bacterial species.

5 BACTERIAL IDENTIFICATION BY MULTILOCUS AND GENOME SEQUENCE ANALYSIS

One major limitation of using single gene targets for species identification is that closely related bacterial species may possess minimal differences between their genomes. Despite the development of alternative gene targets to supplement 16S rRNA gene sequencing for identification of bacteria species that share highly similar 16S rRNA gene sequences, on many occasions, no single gene target is found to be superior or provide sufficient discrimination. In such circumstances, multiple gene targets or multilocus sequencing may be employed to resolve the species identity, because more nucleotide differences are expected to be observed when more genes from the different species are used for comparison. This multilocus approach is now increasingly possible with the availability of enormous numbers of bacterial genome sequences from high-throughput sequencing projects. Using genome sequences from related bacterial species, different gene targets can be studied for their potential role in phylogeny and identification of bacteria. However, there are occasional circumstances that even multilocus sequencing may not resolve species identity. One of the most well-known examples is the difficulty in differentiating *Streptococcus pneumoniae*, *Streptococcus pseudopneumoniae*, *S. mitis* and *Streptococcus oralis*. Although these closely related *Streptococcus* species exhibit considerable differences in their pathogenicity and habitat, they share highly similar sequences in their 16S rRNA gene and most housekeeping genes. In such cases, genome sequencing may offer ultimate molecular resolution which is superior to any phenotypic and genotypic tests.

With the advent of next-generation sequencing technology available at a much lower cost than traditional Sanger sequencing, genome sequencing not only allows rapid microbial identification but has also been increasingly used for characterisation of transmission events of various infectious diseases (Bryant et al., 2013; Harris et al., 2013; Koser et al., 2012; Loman et al., 2013; Snyder et al., 2013; Tse et al., 2012). In particular, draft genome sequencing can be more readily achieved, instead of complete genome sequencing, without the need for the tedious task of closing gaps and complete genome assembly. Recently, we have employed rapid draft genome sequencing for identification of the emerging bacterium, *Elizabethkingia anophelis*, isolated from a cluster of bacteraemic cases involving two neonates with meningitis and one neonate's mother with chorioamnionitis (Lau, Wu, et al., 2014). While *E. anophelis* was first discovered from mosquito gut and has been associated with neonatal meningitis and nosocomial outbreaks, little is known about its mode of transmission. The three *E. anophelis* isolates were misidentified as *Elizabethkingia meningoseptica* by phenotypic methods and MALDI-TOF MS, while results from 16S rRNA gene sequencing were inconclusive due to incorrect species identities of sequences deposited in GenBank (Lau, Wu, et al., 2014). Their species identity was only resolved eventually by intergenomic comparison, which also revealed evidence for perinatal vertical transmission from a mother to her neonate. Moreover, transmission in the labour ward was excluded, because the genome sequence identified the other neonatal isolate was genomically distinct (Lau, Wu, et al., 2014). The findings provided genomic evidence for vertical transmission in *E. anophelis* neonatal meningitis and suggested that mosquitoes were unlikely to be the vehicle of transmission. This highlights the power of genome sequencing in tracking transmission routes of emerging pathogens and guiding infection control measures and emphasises the importance of using a polyphasic approach for identification of those 'difficult-to-identify' bacteria. With the polyphasic approach, different methods in multiple phases are used, e.g., phenotypic tests coupled with sequencing of 16S rRNA gene and/or other gene targets and/or genome sequencing, in order to achieve accurate species identification. With further automation and lower sequencing costs, draft genome sequencing may emerge as a tool for identification of pathogens that are difficult to identify in clinical laboratories, while at the same time typing can be achieved for outbreak investigation purposes.

6 COMPARISON OF GENE SEQUENCING AND MALDI-TOF MS FOR BACTERIAL IDENTIFICATION

In the last few years, MALDI-TOF MS has emerged as a new technique for rapid identification of bacteria at a low cost and has become a widely used technique for bacterial identification in clinical microbiology laboratories (Bizzini & Greub, 2010; Eigner et al., 2009; Seng et al., 2009). Recently, we have used MALDI-TOF MS for identification of bacteria endemic to our locality, such as *B. pseudomallei* and *Laribacter hongkognensis* as well as other difficult-to-identify bacteria (Lau, Tang, et al., 2012, 2014; Tang et al., 2013). It has been demonstrated that the main limitation of MALDI-TOF MS, similar to other bacterial identification platforms, is the content of the database. With a comprehensive database which covers both bacterial species endemic to particular localities and sufficient number of strains for each species to control for intraspecies variation, MALDI-TOF MS should be an efficient, user-friendly and inexpensive method for identification of bacteria in clinical microbiology laboratories. However, for the bacterial species that are identified by MALDI-TOF MS with a low score, which may represent a bacterial species not included in the database or a novel bacterial species, retrieving the mass-to-charge ratio profile of the ionised proteins for a particular bacterial strain for manual interpretation is not possible. This is in contrast to gene sequencing which we can manually interpret the sequence alignment, or even the traditional biochemical tests which we manually interpret the biochemical profile, to arrive at a final decision as to the identification of a particular strain, and therefore represents another important limitation of MALDI-TOF MS. In such circumstances, gene sequencing or a polyphasic approach involving both phenotypic and genotypic tests should be performed for final confirmation of identification.

7 IDENTIFICATION BY SPECIES-SPECIFIC GENE AMPLIFICATION

Synapomorphy, or shared derived state, is the presence of distinguishing and often novel characteristics in an organism and its direct progenies (Hennig, 1950). Such characteristics are important as they form the basis of the current classification of bacteria, in which phylogenetic relationships are often emphasised (Brenner, Staley, & Krieg, 2005). As the definition of bacterial species has required the detection of certain phenotypic characters, identification of bacteria by detecting the genes, chromosomal or extra-chromosomal genomic regions that underpin these phenotypes may be seen as a natural extension of the phenotypic identification process in the molecular era. Species-specific gene amplification as an operational method, therefore, resounds with the phylogenetic ideal (Hennig, 1979; Ho, Lau, & Woo, 2013).

Instead of sequencing an orthologous marker gene, such as the 16S rRNA (Lau, Ren, et al., 2007; Lau et al., 2013; Woo, Teng, et al., 2005; Woo et al., 2014; Yuen et al., 2001), *recA* (Cesarini, Bevivino, Tabacchioni, Chiarini, & Dalmastri, 2009; Costechareyre et al., 2010; Dai, Liu, & Wang, 2012; McDowell, Perry, Lambert, & Patrick, 2008; Owusu-Kwarteng et al., 2012; Payne et al., 2005; Zhu et al., 2013) or *groEL* (Leclerque & Kleespies, 2008; Woo, Lau, Woo, et al., 2003; Woo, Leung, Wong, Ho, & Yuen, 2001) and subsequently determining the phylogenetic position of the isolate in question to achieve identification, species-specific gene amplification represents a more intuitive approach that can be applied in the time-critical setting of a clinical laboratory or the rudimentary setting of field work. While the single-nucleotide resolution of gene sequencing methods enables precise phylogenetic placement and discrimination of closely related clones, species-specific gene amplification typing offers a yes or no, unambiguous answer to the identity of an organism (Ho, Lau, et al., 2011). Surprisingly, this simplistic approach, when suitably multiplexed, has been shown to equal or exceed the resolution of antigenic serotyping, multilocus sequence typing and restriction digestion-based pulsed-field gel electrophoresis (Cornelius, Gilpin, Carter, Nicol, & On, 2010).

Genetic target (probe) discovery is the basis of bacterial identification by species-specific gene amplification. The discovery process is either top down or bottom up. The top-down approach examines phenotypic characters or expressed proteins specific to the species of interest and selects corresponding gene probes to use in identification (Bielaszewska et al., 2011; Kooken, Fox, Fox, & Wunschel, 2014; Wongtrakoongate, Mongkoldhumrongkul, Chaijan, Kamchonwongpaisan, & Tungpradabkul, 2007). This provides coding probes with phenotypic correlation (Bielaszewska et al., 2011), as long as the phenotypic or proteomic survey is comprehensive. The bottom-up approach, conversely, starts with obtaining the genetic information of target and non-target bacteria and uses *in vitro* (Akopyants et al., 1998; Coudeyras, Marchandin, Fajon, & Forestier, 2008; Marenda, Sagne, Poumarat, & Citti, 2005) or *in silico* (Ho, Lau, et al., 2011; Ho et al., 2012; Ho, Yuen, Lau, & Woo, 2011; Ou, Ju, et al., 2007; Qin et al., 2011; Yu et al., 2010) methods to select genomic regions that are only present in the targeted species but not in other ones. This allows the discovery process to reach into intergenic spacers, non-coding regions and genes that are not expressed in experimental conditions, resulting in a theoretically larger repertoire of potential probes. To ensure the empirical sensitivity and specificity of the species-specific gene amplification assay, potential probes generated from either approach must be validated against a panel of target and non-target strains not previously used in the probe screening and selection process (Figure 1).

With the exponential increase in DNA sequencing throughput (Loman et al., 2012) and, correspondingly, the amount of publicly available bacterial genomic information since the mid-2000s (Field, Wilson, & van der Gast, 2006), researchers employing the bottom-up approach can very often directly obtain the required sequence information from public repositories, such as the NCBI Genome Database (www.ncbi.nlm.nih.gov/genome/). They no longer need to perform the complete genome sequencing of the target and non-target organisms as long as an adequate number of their genomes have been published (Ho, Yuen, et al., 2011; Qin et al., 2011). On the other hand, while phenotypic information of bacterial species has been made available in reference texts, laboratory manuals and journal articles since the start of bacteriology, the lack of a central repository and the practical impossibility of having a common, comparable set of phenotypic characters limit the top-down approach to a few extensively sampled species. The antigenic or proteomics informed probe selection approaches are likewise limited by the lack of comprehensive public

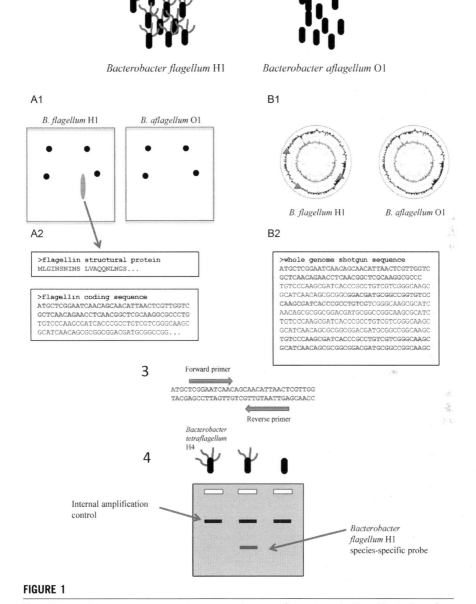

FIGURE 1

Top-down and bottom-up approaches to species-specific probe selection. Illustration of
the top-down (A) and bottom-up (B) approaches to species-specific probe selection using
the hypothetical bacterial species *Bacterobacter flagellum* H1 and *Bacterobacter aflagellum*
O1. In the top-down approach, proteomic profiling of both species is performed and used

(Continued)

databases, standard experimental protocols and computational tools for the meaning-ful comparison of antigenic or proteomic profiles (Taylor et al., 2007). As a result of these issues, the bottom-up, comparative genomics approach is currently preferred.

While individual researchers may develop in-house scripts and programs to uti-lise generic sequence similarity search tools such as BLAST (Altschul et al., 1997), ACT (Carver et al., 2005) and Mauve (Darling, Mau, Blattner, & Perna, 2004), ded-icated software and webservers have been developed to facilitate the comparative genomic analysis essential to the bottom-up approach (Figure 2). These software and web platforms offer enhanced accessibility for biomedical researchers who may not have special training in computer science and programming. The Mobilo-meFINDER (Ou, He, et al., 2007), for example, integrates a set of bioinformatics tools to facilitate the *in silico* identification and characterisation of genomic islands that distinguish a group of bacterial strains and offers automatic PCR primer design. The GenoList webserver (Lechat, Hummel, Rousseau, & Moszer, 2008) is an inte-grated platform with genome browsing and subtractive proteome analysis capabili-ties. The mGenomeSubtractor (Shao et al., 2010) and ssGeneFinder webserver (Ho et al., 2012) are similar and offer BLAST-based comparison of multiple prokary-otic genomes to allow species-specific gene discovery; ssGeneFinder additionally provides a standalone version (Ho, Yuen, et al., 2011) for users who may need to perform their analysis using proprietary or pre-publication data.

Using DNA probes generated from subtractive genome comparison, a number of assays have been designed, with applications ranging from environmental microbiology (Price et al., 2013), industrial quality control (Yu et al., 2010), clinical diagnostics (Ho et al., 2012; Ho, Yuen, et al., 2011; Wang, Sun, & Lu, 2012) and rapid detection and identification of bacterial pathogens with bioterrorism potential (Ho, Lau, et al., 2011; Janse, Hamidjaja, Hendriks, & van Rotterdam, 2013; Lau, Chan, et al., 2014). As the probes generated from the *in silico* analysis can be selected based on their genomic characteristics, e.g., number of copies, highly sensitive assays and quantitative assays may be designed using multi-copy and single-copy probes, respectively. Exhaustive screening to eliminate probe cross reactivity is also made feasible by the availability of numerous genomes from

FIGURE 1—Cont'd to reveal their significant differences (A1), in this case, the flagellant protein (A2). These differences in the extended, proteomic phenotype can then be exploited to look for genomic probes. In the bottom-up approach, the complete genome sequences of the target and non-target species are obtained and analysed by comparative genomic techniques (B1). Genomic regions that distinguish *B. flagellum* from *B. flagellum* can then be identified (B2). Primer design is subsequently performed according to the specific requirements of the amplification techniques, such as isothermal amplification, real-time PCR or end-point PCR (3). Finally, experimental validation of the selected probe should be performed to determine the empirical sensitivity and specificity of the assay (4). As illustrated, additional bacterial isolates are required for validation (i.e. the hypothetical species *Bacterobacter tetraflagellum* H4) and an internal positive control for amplification (i.e. the upper black product bands) may be required to enhance the robustness of the assay. (See the color plate.)

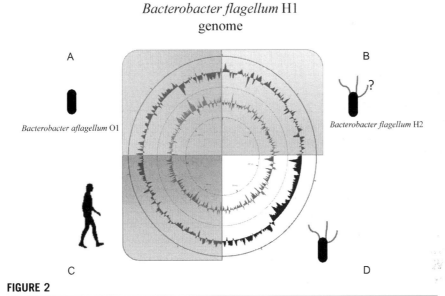

Bacterobacter flagellum H1
genome

A

Bacterobacter aflagellum O1

B

Bacterobacter flagellum H2

C

D

FIGURE 2

Subtractive comparative genomics in probe selection. Subtractive comparative genomics aims to remove three types of sequences not suitable for use as gene probes for species-specific amplification identification. First, sequences with significant similarity to closely related, non-target species have to be excluded (A). Next, sequences with doubtful conservation within the target group have to be eliminated, as probes originating from such regions will not detect other strains of the target group of species (B). Finally, low-complexity regions or sequences sharing inadvertent similarity with host species also need to be precluded from probe selection if the assay is intended for use on mixed or uncultured clinical samples (C). The remaining sequences (D), usually comprising only a small proportion of the complete genome, represent the coding and non-coding genetic regions useful for species-specific probe design. (See the color plate.)

non-target bacteria (Ho, Yuen, et al., 2011; Qin et al., 2011); this does not only improve the theoretical specificity of the probes but also reduce the amount of experimental validation required in the assay design process. The enhanced specificity of the probes additionally allows them to be used on mixed cultures or uncultured samples (Balashov, Mordechai, Adelson, & Gygax, 2014).

In summary, bacterial identification by species-specific gene amplification is a practical approach complementary to the more rigorous typing methods involving gene sequencing, usually applied in a research setting. The availability of bacterial genome data and user-friendly software has allowed the efficient generation of specific probes which are versatile and cost-effective in their application. By focusing on the genomic elements that specifically distinguish a bacterial species instead of comparing core orthologous genes, this approach lends support from the DNA–DNA hybridisation technique used in establishing prokaryotic species since the 1960s;

conversely, the successful application of this approach, especially to closely related bacteria sharing high similarity of their core orthologous genes (e.g. with >97% nucleotide identity in their 16S rRNA genes), substantiates the genomic species concept and reinforces its role in polyphasic taxonomy.

CONCLUSION

In the last few decades, we have witnessed an exponential rise in the number of platforms for bacterial identification: a large number of commercial kits for biochemical testing, commercial databases for 16S rRNA gene sequence interpretation and different models of MALDI-TOF MS. Although these platforms have different importance and serve different purposes in different laboratories along the timeline, due to the objectiveness and large data set, 16S rRNA gene and other gene sequencing still remain the ultimate standard for defining a bacterial species if the other methods cannot give a confident answer. As the technologies for genome sequencing are becoming more widely available and the number of bacterial genomes sequenced is also exponentially increasing, we anticipate a highly comprehensive database for gene sequencing in the near future.

REFERENCES

Akopyants, N. S., Fradkov, A., Diatchenko, L., Hill, J. E., Siebert, P. D., Lukyanov, S. A., et al. (1998). PCR-based subtractive hybridization and differences in gene content among strains of *Helicobacter pylori*. *Proceedings of the National Academy of Sciences of the United States of America*, *95*(22), 13108–13113.

Altschul, S. F., Madden, T. L., Schaffer, A. A., Zhang, J., Zhang, Z., Miller, W., et al. (1997). Gapped BLAST and PSI-BLAST: A new generation of protein database search programs. *Nucleic Acids Research*, *25*(17), 3389–3402.

Anderson, B. E., Dawson, J. E., Jones, D. C., & Wilson, K. H. (1991). *Ehrlichia chaffeensis*, a new species associated with human ehrlichiosis. *Journal of Clinical Microbiology*, *29*(12), 2838–2842.

Balashov, S. V., Mordechai, E., Adelson, M. E., & Gygax, S. E. (2014). Identification, quantification and subtyping of *Gardnerella vaginalis* in noncultured clinical vaginal samples by quantitative PCR. *Journal of Medical Microbiology*, *63*(Pt. 2), 162–175.

Bharadwaj, R., Swaminathan, S., Salimnia, H., Fairfax, M., Frey, A., & Chandrasekar, P. H. (2012). Clinical impact of the use of 16S rRNA sequencing method for the identification of "difficult-to-identify" bacteria in immunocompromised hosts. *Transplant Infectious Disease*, *14*(2), 206–212.

Bielaszewska, M., Mellmann, A., Zhang, W., Kock, R., Fruth, A., Bauwens, A., et al. (2011). Characterisation of the *Escherichia coli* strain associated with an outbreak of haemolytic uraemic syndrome in Germany, 2011: A microbiological study. *The Lancet Infectious Diseases*, *11*(9), 671–676.

Bizzini, A., & Greub, G. (2010). Matrix-assisted laser desorption ionization time-of-flight mass spectrometry, a revolution in clinical microbial identification. *Clinical Microbiology and Infection*, *16*(11), 1614–1619.

Borel, N., Kempf, E., Hotzel, H., Schubert, E., Torgerson, P., Slickers, P., et al. (2008). Direct identification of chlamydiae from clinical samples using a DNA microarray assay: A validation study. *Molecular and Cellular Probes, 22*(1), 55–64.

Bosshard, P. P., Abels, S., Altwegg, M., Bottger, E. C., & Zbinden, R. (2004). Comparison of conventional and molecular methods for identification of aerobic catalase-negative Gram-positive cocci in the clinical laboratory. *Journal of Clinical Microbiology, 42*(5), 2065–2073.

Bosshard, P. P., Abels, S., Zbinden, R., Bottger, E. C., & Altwegg, M. (2003). Ribosomal DNA sequencing for identification of aerobic Gram-positive rods in the clinical laboratory (an 18-month evaluation). *Journal of Clinical Microbiology, 41*(9), 4134–4140.

Bosshard, P. P., Zbinden, R., & Altwegg, M. (2002). *Turicibacter sanguinis* gen. nov., sp. nov., a novel anaerobic, Gram-positive bacterium. *International Journal of Systematic and Evolutionary Microbiology, 52*(Pt. 4), 1263–1266.

Brenner, Don J., Staley, James T., & Krieg, Noel R. (2005). Overview: A Phylogenetic Backbone and Taxonomic Framework for Procaryotic Systematics. In Noel R. Krieg, Don J. Brenner, James T. Staley, & George M. Garrity (Eds.), *Bergey's manual of systematic bacteriology* (pp. 49–66). USA: Springer.

Bryant, J. M., Grogono, D. M., Greaves, D., Foweraker, J., Roddick, I., Inns, T., et al. (2013). Whole-genome sequencing to identify transmission of *Mycobacterium abscessus* between patients with cystic fibrosis: A retrospective cohort study. *The Lancet, 381*(9877), 1551–1560.

Carver, T. J., Rutherford, K. M., Berriman, M., Rajandream, M. A., Barrell, B. G., & Parkhill, J. (2005). ACT: The Artemis comparison tool. *Bioinformatics, 21*(16), 3422–3423.

Cesarini, S., Bevivino, A., Tabacchioni, S., Chiarini, L., & Dalmastri, C. (2009). *RecA* gene sequence and multilocus sequence typing for species-level resolution of *Burkholderia cepacia* complex isolates. *Letters in Applied Microbiology, 49*(5), 580–588.

Chan, J. F., Lau, S. K., Curreem, S. O., To, K. K., Leung, S. S., Cheng, V. C., et al. (2012). First report of spontaneous intrapartum *Atopobium vaginae* bacteremia. *Journal of Clinical Microbiology, 50*(7), 2525–2528.

Chan, J. F., Wong, S. S., Leung, S. S., Fan, R. Y., Ngan, A. H., To, K. K., et al. (2012). First report of chronic implant-related septic arthritis and osteomyelitis due to *Kytococcus schroeteri* and a review of human *K. schroeteri* infections. *Infection, 40*(5), 567–573.

Chan, J. F., Woo, P. C., Teng, J. L., Lau, S. K., Leung, S. S., Tam, F. C., et al. (2011). Primary infective spondylodiscitis caused by *Lactococcus garvieae* and a review of human *L. garvieae* infections. *Infection, 39*(3), 259–264.

Chen, S. M., Dumler, J. S., Bakken, J. S., & Walker, D. H. (1994). Identification of a granulocytotropic *Ehrlichia* species as the etiologic agent of human disease. *Journal of Clinical Microbiology, 32*(3), 589–595.

Chen, P. L., Lee, N. Y., Yan, J. J., Yang, Y. J., Chen, H. M., Chang, C. M., et al. (2007). Prosthetic valve endocarditis caused by *Streptobacillus moniliformis*: A case of rat bite fever. *Journal of Clinical Microbiology, 45*(9), 3125–3126.

Chow, S. K., & Clarridge, J. E., 3rd. (2014). Necessity of 16S rRNA gene sequencing for identifying *Haemophilus parainfluenzae*-like strains associated with opportunistic urinary tract infections. *Journal of Medical Microbiology, 63*(Pt. 6), 805–811.

Clarridge, J. E., 3rd., Raich, T. J., Sjosted, A., Sandstrom, G., Darouiche, R. O., Shawar, R. M., et al. (1996). Characterization of two unusual clinically significant *Francisella* strains. *Journal of Clinical Microbiology, 34*(8), 1995–2000.

Cornelius, A. J., Gilpin, B., Carter, P., Nicol, C., & On, S. L. (2010). Comparison of PCR binary typing (P-BIT), a new approach to epidemiological subtyping of *Campylobacter jejuni*, with serotyping, pulsed-field gel electrophoresis, and multilocus sequence typing methods. *Applied and Environmental Microbiology, 76*(5), 1533–1544.

Costechareyre, D., Rhouma, A., Lavire, C., Portier, P., Chapulliot, D., Bertolla, F., et al. (2010). Rapid and efficient identification of *Agrobacterium* species by *recA* allele analysis: *Agrobacterium recA* diversity. *Microbial Ecology, 60*(4), 862–872.

Coudeyras, S., Marchandin, H., Fajon, C., & Forestier, C. (2008). Taxonomic and strain-specific identification of the probiotic strain *Lactobacillus rhamnosus* 35 within the *Lactobacillus casei* group. *Applied and Environmental Microbiology, 74*(9), 2679–2689.

Couturier, M. R., Slechta, E. S., Goulston, C., Fisher, M. A., & Hanson, K. E. (2012). *Leptotrichia* bacteremia in patients receiving high-dose chemotherapy. *Journal of Clinical Microbiology, 50*(4), 1228–1232.

Curreem, S. O., Teng, J. L., Tse, H., Yuen, K. Y., Lau, S. K., & Woo, P. C. (2011). General metabolism of *Laribacter hongkongensis*: A genome-wide analysis. *Cell & Bioscience, 1*(1), 16.

Cursons, R. T., Jeyerajah, E., & Sleigh, J. W. (1999). The use of polymerase chain reaction to detect septicemia in critically ill patients. *Critical Care Medicine, 27*(5), 937–940.

Dai, J., Liu, X., & Wang, Y. (2012). Genetic diversity and phylogeny of rhizobia isolated from *Caragana microphylla* growing in desert soil in Ningxia, China. *Genetics and Molecular Research, 11*(3), 2683–2693.

Darling, A. C., Mau, B., Blattner, F. R., & Perna, N. T. (2004). Mauve: Multiple alignment of conserved genomic sequence with rearrangements. *Genome Research, 14*(7), 1394–1403.

de Melo Oliveira, M. G., Abels, S., Zbinden, R., Bloemberg, G. V., & Zbinden, A. (2013). Accurate identification of fastidious Gram-negative rods: Integration of both conventional phenotypic methods and 16S rRNA gene analysis. *BMC Microbiology, 13*, 162.

Deutch, S., Pedersen, L. N., Podenphant, L., Olesen, R., Schmidt, M. B., Moller, J. K., et al. (2006). Broad-range real time PCR and DNA sequencing for the diagnosis of bacterial meningitis. *Scandinavian Journal of Infectious Diseases, 38*(1), 27–35.

Drancourt, M., Bollet, C., Carlioz, A., Martelin, R., Gayral, J. P., & Raoult, D. (2000). 16S ribosomal DNA sequence analysis of a large collection of environmental and clinical unidentifiable bacterial isolates. *Journal of Clinical Microbiology, 38*(10), 3623–3630.

Dubois, D., Robin, F., Bouvier, D., Delmas, J., Bonnet, R., Lesens, O., et al. (2008). *Streptobacillus moniliformis* as the causative agent in spondylodiscitis and psoas abscess after rooster scratches. *Journal of Clinical Microbiology, 46*(8), 2820–2821.

Duffett, S., Missaghi, B., & Daley, P. (2012). Culture-negative endocarditis diagnosed using 16S DNA polymerase chain reaction. *The Canadian Journal of Infectious Diseases & Medical Microbiology, 23*(4), 216–218.

Eigner, U., Holfelder, M., Oberdorfer, K., Betz-Wild, U., Bertsch, D., & Fahr, A. M. (2009). Performance of a matrix-assisted laser desorption ionization-time-of-flight mass spectrometry system for the identification of bacterial isolates in the clinical routine laboratory. *Clinical Laboratory, 55*(7–8), 289–296.

Elsendoorn, A., Robert, R., Culos, A., Roblot, F., & Burucoa, C. (2011). *Catabacter hongkongensis* bacteremia with fatal septic shock. *Emerging Infectious Diseases, 17*(7), 1330–1331.

Faibis, F., Mihaila, L., Perna, S., Lefort, J. F., Demachy, M. C., Le Fleche-Mateos, A., et al. (2008). *Streptococcus sinensis*: An emerging agent of infective endocarditis. *Journal of Medical Microbiology, 57*(Pt. 4), 528–531.

Fenollar, F., & Raoult, D. (2001). Molecular techniques in Whipple's disease. *Expert Review of Molecular Diagnostics, 1*(3), 299–309.

Field, D., Wilson, G., & van der Gast, C. (2006). How do we compare hundreds of bacterial genomes? *Current Opinion in Microbiology, 9*(5), 499–504.

Goh, S. H., Potter, S., Wood, J. O., Hemmingsen, S. M., Reynolds, R. P., & Chow, A. W. (1996). *hsp60* gene sequences as universal targets for microbial species identification: Studies with coagulase-negative staphylococci. *Journal of Clinical Microbiology, 34*(4), 818–823.

Harris, S. R., Cartwright, E. J., Torok, M. E., Holden, M. T., Brown, N. M., Ogilvy-Stuart, A. L., et al. (2013). Whole-genome sequencing for analysis of an outbreak of meticillin-resistant *Staphylococcus aureus*: A descriptive study. *The Lancet Infectious Diseases, 13*(2), 130–136.

Harris, K. A., Yam, T., Jalili, S., Williams, O. M., Alshafi, K., Gouliouris, T., et al. (2014). Service evaluation to establish the sensitivity, specificity and additional value of broad-range 16S rDNA PCR for the diagnosis of infective endocarditis from resected endocardial material in patients from eight UK and Ireland hospitals. *European Journal of Clinical Microbiology & Infectious Diseases, 33*(11), 2061–2066.

Heikens, E., Fleer, A., Paauw, A., Florijn, A., & Fluit, A. C. (2005). Comparison of genotypic and phenotypic methods for species-level identification of clinical isolates of coagulase-negative staphylococci. *Journal of Clinical Microbiology, 43*(5), 2286–2290.

Hennig, W. (1950). *Grundzüge einer Theorie der Phylogenetischen Systematik*. Berlin: Deutscher Zentralverlag.

Hennig, W. (1979). *Phylogenetic systematics*. Urbana: University of Illinois Press.

Ho, C. C., Lau, C. C., Martelli, P., Chan, S. Y., Tse, C. W., Wu, A. K., et al. (2011). Novel pangenomic analysis approach in target selection for multiplex PCR identification and detection of *Burkholderia pseudomallei*, *Burkholderia thailandensis*, and *Burkholderia cepacia* complex species: A proof-of-concept study. *Journal of Clinical Microbiology, 49*(3), 814–821.

Ho, C. C., Lau, S. K., & Woo, P. C. (2013). Romance of the three domains: How cladistics transformed the classification of cellular organisms. *Protein and Cell, 4*(9), 664–676.

Ho, C. C., Wu, A. K., Tse, C. W., Yuen, K. Y., Lau, S. K., & Woo, P. C. (2012). Automated pangenomic analysis in target selection for PCR detection and identification of bacteria by use of ssGeneFinder Webserver and its application to *Salmonella enterica* serovar Typhi. *Journal of Clinical Microbiology, 50*(6), 1905–1911.

Ho, C. C., Yuen, K. Y., Lau, S. K., & Woo, P. C. (2011). Rapid identification and validation of specific molecular targets for detection of *Escherichia coli* O104:H4 outbreak strain by use of high-throughput sequencing data from nine genomes. *Journal of Clinical Microbiology, 49*(10), 3714–3716.

Inokuma, H., Brouqui, P., Drancourt, M., & Raoult, D. (2001). Citrate synthase gene sequence: A new tool for phylogenetic analysis and identification of *Ehrlichia*. *Journal of Clinical Microbiology, 39*(9), 3031–3039.

Janda, J. M., & Abbott, S. L. (2007). 16S rRNA gene sequencing for bacterial identification in the diagnostic laboratory: Pluses, perils, and pitfalls. *Journal of Clinical Microbiology, 45*(9), 2761–2764.

Janse, I., Hamidjaja, R. A., Hendriks, A. C., & van Rotterdam, B. J. (2013). Multiplex qPCR for reliable detection and differentiation of *Burkholderia mallei* and *Burkholderia pseudomallei*. *BMC Infectious Diseases, 13*, 86.

Kim, D. S., Wi, Y. M., Choi, J. Y., Peck, K. R., Song, J. H., & Ko, K. S. (2011). Bacteremia caused by *Laribacter hongkongensis* misidentified as *Acinetobacter lwoffii*: Report of the first case in Korea. *Journal of Korean Medical Science*, *26*(5), 679–681.

Koh, T. H., Kurup, A., & Chen, J. (2004). *Streptococcus iniae* discitis in Singapore. *Emerging Infectious Diseases*, *10*(9), 1694–1696.

Kooken, J., Fox, K., Fox, A., & Wunschel, D. (2014). Assessment of marker proteins identified in whole cell extracts for bacterial speciation using liquid chromatography electrospray ionization tandem mass spectrometry. *Molecular and Cellular Probes*, *28*(1), 34–40.

Koser, C. U., Holden, M. T., Ellington, M. J., Cartwright, E. J., Brown, N. M., Ogilvy-Stuart, A. L., et al. (2012). Rapid whole-genome sequencing for investigation of a neonatal MRSA outbreak. *The New England Journal of Medicine*, *366*(24), 2267–2275.

Kotilainen, P., Heiro, M., Jalava, J., Rantakokko, V., Nikoskelainen, J., Nikkari, S., et al. (2006). Aetiological diagnosis of infective endocarditis by direct amplification of rRNA genes from surgically removed valve tissue. An 11-year experience in a Finnish teaching hospital. *Annals of Medicine*, *38*(4), 263–273.

Kumar, N. V., Menon, T., Pathipati, P., & Cherian, K. M. (2013). 16S rRNA sequencing as a diagnostic tool in the identification of culture-negative endocarditis in surgically treated patients. *The Journal of Heart Valve Disease*, *22*(6), 846–849.

Kurabachew, M., Wondimu, A., & Ryon, J. J. (1998). Reverse transcription-PCR detection of *Mycobacterium leprae* in clinical specimens. *Journal of Clinical Microbiology*, *36*(5), 1352–1356.

Lam, J. Y., Wu, A. K., Ngai, D. C., Teng, J. L., Wong, E. S., Lau, S. K., et al. (2011). Three cases of severe invasive infections caused by *Campylobacter rectus* and first report of fatal *C. rectus* infection. *Journal of Clinical Microbiology*, *49*(4), 1687–1691.

Lau, S. K., Chan, S. Y., Curreem, S. O., Hui, S. W., Lau, C. C., Lee, P., et al. (2014). *Burkholderia pseudomallei* in soil samples from an oceanarium in Hong Kong detected using a sensitive PCR assay. *Emerging Microbes & Infections*, *3*, e69.

Lau, S. K., Curreem, S. O., Lin, C. C., Fung, A. M., Yuen, K. Y., & Woo, P. C. (2013). *Streptococcus hongkongensis* sp. nov., isolated from a patient with an infected puncture wound and from a marine flatfish. *International Journal of Systematic and Evolutionary Microbiology*, *63*(Pt. 7), 2570–2576.

Lau, S. K., Curreem, S. O., Ngan, A. H., Yeung, C. K., Yuen, K. Y., & Woo, P. C. (2011). First report of disseminated *Mycobacterium* skin infections in two liver transplant recipients and rapid diagnosis by *hsp65* gene sequencing. *Journal of Clinical Microbiology*, *49*(11), 3733–3738.

Lau, S. K., Fan, R. Y., Lo, H. W., Ng, R. H., Wong, S. S., Li, I. W., et al. (2012). High mortality associated with *Catabacter hongkongensis* bacteremia. *Journal of Clinical Microbiology*, *50*(7), 2239–2243.

Lau, S. K., Fan, R. Y., Wong, G. K., Teng, J. L., Sze, K. H., Tse, H., et al. (2011). Transport genes and chemotaxis in *Laribacter hongkongensis*: A genome-wide analysis. *Cell & Bioscience*, *1*, 28.

Lau, S. K., Ho, P. L., Li, M. W., Tsoi, H. W., Yung, R. W., Woo, P. C., et al. (2005). Cloning and characterization of a chromosomal class C beta-lactamase and its regulatory gene in *Laribacter hongkongensis*. *Antimicrobial Agents and Chemotherapy*, *49*(5), 1957–1964.

Lau, S. K., McNabb, A., Woo, G. K., Hoang, L., Fung, A. M., Chung, L. M., et al. (2007). *Catabacter hongkongensis* gen. nov., sp. nov., isolated from blood cultures of patients from Hong Kong and Canada. *Journal of Clinical Microbiology*, *45*(2), 395–401.

Lau, S. K., Ng, K. H., Woo, P. C., Yip, K. T., Fung, A. M., Woo, G. K., et al. (2006). Usefulness of the MicroSeq 500 16S rDNA bacterial identification system for identification of anaerobic Gram positive bacilli isolated from blood cultures. *Journal of Clinical Pathology*, *59*(2), 219–222.

Lau, K. W., Ren, J., Fung, M. C., Woo, P. C., Yuen, K. Y., Chan, K. K., et al. (2007). *Fangia hongkongensis* gen. nov., sp. nov., a novel gammaproteobacterium of the order *Thiotrichales* isolated from coastal seawater of Hong Kong. *International Journal of Systematic and Evolutionary Microbiology*, *57*(Pt. 11), 2665–2669.

Lau, S. K., Tang, B. S., Curreem, S. O., Chan, T. M., Martelli, P., Tse, C. W., et al. (2012). Matrix-assisted laser desorption ionization-time of flight mass spectrometry for rapid identification of *Burkholderia pseudomallei*: Importance of expanding databases with pathogens endemic to different localities. *Journal of Clinical Microbiology*, *50*(9), 3142–3143.

Lau, S. K., Tang, B. S., Teng, J. L., Chan, T. M., Curreem, S. O., Fan, R. Y., et al. (2014). Matrix-assisted laser desorption ionisation time-of-flight mass spectrometry for identification of clinically significant bacteria that are difficult to identify in clinical laboratories. *Journal of Clinical Pathology*, *67*(4), 361–366.

Lau, S. K., Teng, J. L., Leung, K. W., Li, N. K., Ng, K. H., Chau, K. Y., et al. (2006). Bacteremia caused by *Solobacterium moorei* in a patient with acute proctitis and carcinoma of the cervix. *Journal of Clinical Microbiology*, *44*(8), 3031–3034.

Lau, S. K., Wong, G. K., Li, M. W., Woo, P. C., & Yuen, K. Y. (2008). Distribution and molecular characterization of tetracycline resistance in *Laribacter hongkongensis*. *The Journal of Antimicrobial Chemotherapy*, *61*(3), 488–497.

Lau, S. K., Wong, G. K., Poon, R. W., Lee, L. C., Leung, K. W., Tse, C. W., et al. (2009). Susceptibility patterns of clinical and fish isolates of *Laribacter hongkongensis*: Comparison of the Etest, disc diffusion and broth microdilution methods. *The Journal of Antimicrobial Chemotherapy*, *63*(4), 704–708.

Lau, S. K., Wong, G. K., Tsang, A. K., Teng, J. L., Fan, R. Y., Tse, H., et al. (2011). Virulence determinants, drug resistance and mobile genetic elements of *Laribacter hongkongensis*: A genome-wide analysis. *Cell & Bioscience*, *1*(1), 17.

Lau, S. K., Woo, P. C., Chan, B. Y., Fung, A. M., Que, T. L., & Yuen, K. Y. (2002). *Haemophilus segnis* polymicrobial and monomicrobial bacteraemia identified by 16S ribosomal RNA gene sequencing. *Journal of Medical Microbiology*, *51*(8), 635–640.

Lau, S. K., Woo, P. C., Fan, R. Y., Lee, R. C., Teng, J. L., & Yuen, K. Y. (2007). Seasonal and tissue distribution of *Laribacter hongkongensis*, a novel bacterium associated with gastroenteritis, in retail freshwater fish in Hong Kong. *International Journal of Food Microbiology*, *113*(1), 62–66.

Lau, S. K., Woo, P. C., Fan, R. Y., Ma, S. S., Hui, W. T., Au, S. Y., et al. (2007). Isolation of *Laribacter hongkongensis*, a novel bacterium associated with gastroenteritis, from drinking water reservoirs in Hong Kong. *Journal of Applied Microbiology*, *103*(3), 507–515.

Lau, S. K., Woo, P. C., Fung, A. M., Chan, K. M., Woo, G. K., & Yuen, K. Y. (2004). Anaerobic, non-sporulating, gram-positive bacilli bacteraemia characterized by 16S rRNA gene sequencing. *Journal of Medical Microbiology*, *53*(Pt. 12), 1247–1253.

Lau, S. K., Woo, P. C., Hui, W. T., Li, M. W., Teng, J. L., Que, T. L., et al. (2003). Use of cefoperazone MacConkey agar for selective isolation of *Laribacter hongkongensis*. *Journal of Clinical Microbiology*, *41*(10), 4839–4841.

Lau, S. K., Woo, P. C., Li, N. K., Teng, J. L., Leung, K. W., Ng, K. H., et al. (2006). *Globicatella* bacteraemia identified by 16S ribosomal RNA gene sequencing. *Journal of Clinical Pathology*, *59*(3), 303–307.

Lau, S. K., Woo, P. C., Luk, W. K., Fung, A. M., Hui, W. T., Fong, A. H., et al. (2006). Clinical isolates of *Streptococcus iniae* from Asia are more mucoid and beta-hemolytic than those from North America. *Diagnostic Microbiology and Infectious Disease*, *54*(3), 177–181.

Lau, S. K., Woo, P. C., Mok, M. Y., Teng, J. L., Tam, V. K., Chan, K. K., et al. (2004). Characterization of *Haemophilus segnis*, an important cause of bacteremia, by 16S rRNA gene sequencing. *Journal of Clinical Microbiology*, *42*(2), 877–880.

Lau, S. K., Woo, P. C., Teng, J. L., Leung, K. W., & Yuen, K. Y. (2002). Identification by 16S ribosomal RNA gene sequencing of *Arcobacter butzleri* bacteraemia in a patient with acute gangrenous appendicitis. *Molecular Pathology*, *55*(3), 182–185.

Lau, S. K., Woo, P. C., Tse, H., Leung, K. W., Wong, S. S., & Yuen, K. Y. (2003). Invasive *Streptococcus iniae* infections outside North America. *Journal of Clinical Microbiology*, *41*(3), 1004–1009.

Lau, S. K., Woo, P. C., Woo, G. K., Fung, A. M., Ngan, A. H., Song, Y., et al. (2006). Bacteraemia caused by *Anaerotruncus colihominis* and emended description of the species. *Journal of Clinical Pathology*, *59*(7), 748–752.

Lau, S. K., Woo, P. C., Woo, G. K., Fung, A. M., Wong, M. K., Chan, K. M., et al. (2004). *Eggerthella hongkongensis* sp. nov. and *Eggerthella sinensis* sp. nov., two novel *Eggerthella* species, account for half of the cases of *Eggerthella* bacteremia. *Diagnostic Microbiology and Infectious Disease*, *49*(4), 255–263.

Lau, S. K., Woo, P. C., Yim, T. C., To, A. P., & Yuen, K. Y. (2003). Molecular characterization of a strain of group A *Streptococcus* isolated from a patient with a psoas abscess. *Journal of Clinical Microbiology*, *41*(10), 4888–4891.

Lau, S. K., Wu, A. K., Teng, J. L., Tse, H., Curreem, S. O., Tsui, S. K., et al. (2014). Evidence for *Elizabethkingia anophelis* transmission from mother to infant, Hong Kong. *Emerging Infectious Diseases*, *21*(2), 232–241, Advance online publication.

Lechat, P., Hummel, L., Rousseau, S., & Moszer, I. (2008). GenoList: An integrated environment for comparative analysis of microbial genomes. *Nucleic Acids Research*, *36*(Database issue), D469–D474.

Leclerque, A., & Kleespies, R. G. (2008). 16S rRNA-, GroEL- and MucZ-based assessment of the taxonomic position of 'Rickettsiella melolonthae' and its implications for the organization of the genus *Rickettsiella*. *International Journal of Systematic and Evolutionary Microbiology*, *58*(Pt. 4), 749–755.

Lee, M. J., Jang, S. J., Li, X. M., Park, G., Kook, J. K., Kim, M. J., et al. (2014). Comparison of *rpoB* gene sequencing, 16S rRNA gene sequencing, *gyrB* multiplex PCR, and the VITEK2 system for identification of *Acinetobacter* clinical isolates. *Diagnostic Microbiology and Infectious Disease*, *78*(1), 29–34.

Lee, S. H., Vigliotti, J. S., Vigliotti, V. S., Jones, W., & Shearer, D. M. (2014). Detection of Borreliae in archived sera from patients with clinically suspect Lyme disease. *International Journal of Molecular Sciences*, *15*(3), 4284–4298.

Li, H., Turhan, V., Chokhani, L., Stratton, C. W., Dunbar, S. A., & Tang, Y. W. (2009). Identification and differentiation of clinically relevant mycobacterium species directly from acid-fast bacillus-positive culture broth. *Journal of Clinical Microbiology*, *47*(12), 3814–3820.

Llenas-Garcia, J., Lalueza-Blanco, A., Fernandez-Ruiz, M., Villar-Silva, J., Ochoa, M., Lozano, F., et al. (2012). Primary hepatic actinomycosis presenting as purulent pericarditis with cardiac tamponade. *Infection*, *40*(3), 339–341.

Loman, N. J., Constantinidou, C., Chan, J. Z., Halachev, M., Sergeant, M., Penn, C. W., et al. (2012). High-throughput bacterial genome sequencing: An embarrassment of choice, a world of opportunity. *Nature Reviews. Microbiology*, *10*(9), 599–606.

Loman, N. J., Constantinidou, C., Christner, M., Rohde, H., Chan, J. Z., Quick, J., et al. (2013). A culture-independent sequence-based metagenomics approach to the investigation of an outbreak of Shiga-toxigenic *Escherichia coli* O104:H4. *JAMA*, *309*(14), 1502–1510.

Lorino, G., Lilli, D., Rivanera, D., Guarino, P., Angeletti, S., Gherardi, G., et al. (1999). Polymerase chain reaction, with sequencing, as a diagnostic tool in culture-negative bacterial meningitis. *Clinical Microbiology and Infection*, *5*(2), 92–96.

Maeda, K., Markowitz, N., Hawley, R. C., Ristic, M., Cox, D., & McDade, J. E. (1987). Human infection with *Ehrlichia canis*, a leukocytic rickettsia. *The New England Journal of Medicine*, *316*(14), 853–856.

Mak, G. C., Ho, P. L., Tse, C. W., Lau, S. K., & Wong, S. S. (2005). Reduced levofloxacin susceptibility and tetracycline resistance in a clinical isolate of *Haemophilus quentini* identified by 16S rRNA sequencing. *Journal of Clinical Microbiology*, *43*(10), 5391–5392.

Marenda, M. S., Sagne, E., Poumarat, F., & Citti, C. (2005). Suppression subtractive hybridization as a basis to assess *Mycoplasma agalactiae* and *Mycoplasma bovis* genomic diversity and species-specific sequences. *Microbiology*, *151*(Pt. 2), 475–489.

Marin, M., Munoz, P., Sanchez, M., del Rosal, M., Alcala, L., Rodriguez-Creixems, M., et al. (2007). Molecular diagnosis of infective endocarditis by real-time broad-range polymerase chain reaction (PCR) and sequencing directly from heart valve tissue. *Medicine (Baltimore)*, *86*(4), 195–202.

McDowell, A., Perry, A. L., Lambert, P. A., & Patrick, S. (2008). A new phylogenetic group of *Propionibacterium acnes*. *Journal of Medical Microbiology*, *57*(Pt. 2), 218–224.

Miyazato, A., Ohkusu, K., Tabata, M., Uwabe, K., Kawamura, T., Tachi, Y., et al. (2012). Comparative molecular and microbiological diagnosis of 19 infective endocarditis cases in which causative microbes were identified by PCR-based DNA sequencing from the excised heart valves. *Journal of Infection and Chemotherapy*, *18*(3), 318–323.

Mollet, C., Drancourt, M., & Raoult, D. (1997). *rpoB* sequence analysis as a novel basis for bacterial identification. *Molecular Microbiology*, *26*(5), 1005–1011.

Nakano, N., Wada, R., Yajima, N., Yamamoto, N., Wakai, Y., & Otsuka, H. (2010). Mycobacterial infection of the musculoskeletal tissues: The use of pathological specimens for identification of causative species by PCR-direct sequencing of 16S rDNA. *Japanese Journal of Infectious Diseases*, *63*(3), 188–191.

Ni, X. P., Ren, S. H., Sun, J. R., Xiang, H. Q., Gao, Y., Kong, Q. X., et al. (2007). *Laribacter hongkongensis* isolated from a patient with community-acquired gastroenteritis in Hangzhou City. *Journal of Clinical Microbiology*, *45*(1), 255–256.

Norskov-Lauritsen, N. (2014). Classification, identification, and clinical significance of *Haemophilus* and *Aggregatibacter* species with host specificity for humans. *Clinical Microbiology Reviews*, *27*(2), 214–240.

Ou, H. Y., He, X., Harrison, E. M., Kulasekara, B. R., Thani, A. B., Kadioglu, A., et al. (2007). MobilomeFINDER: Web-based tools for in silico and experimental discovery of bacterial genomic islands. *Nucleic Acids Research*, *35*(Web Server issue), W97–W104.

Ou, H. Y., Ju, C. T., Thong, K. L., Ahmad, N., Deng, Z., Barer, M. R., et al. (2007). Translational genomics to develop a *Salmonella enterica* serovar Paratyphi A multiplex polymerase chain reaction assay. *The Journal of Molecular Diagnostics*, *9*(5), 624–630.

Owusu-Kwarteng, J., Akabanda, F., Nielsen, D. S., Tano-Debrah, K., Glover, R. L., & Jespersen, L. (2012). Identification of lactic acid bacteria isolated during traditional fura processing in Ghana. *Food Microbiology*, *32*(1), 72–78.

Pandit, L., Kumar, S., Karunasagar, I., & Karunasagar, I. (2005). Diagnosis of partially treated culture-negative bacterial meningitis using 16S rRNA universal primers and restriction endonuclease digestion. *Journal of Medical Microbiology*, *54*(Pt. 6), 539–542.

Payne, G. W., Vandamme, P., Morgan, S. H., Lipuma, J. J., Coenye, T., Weightman, A. J., et al. (2005). Development of a *recA* gene-based identification approach for the entire *Burkholderia* genus. *Applied and Environmental Microbiology*, *71*(7), 3917–3927.

Phetsuksiri, B., Rudeeaneksin, J., Supapkul, P., Wachapong, S., Mahotarn, K., & Brennan, P. J. (2006). A simplified reverse transcriptase PCR for rapid detection of *Mycobacterium leprae* in skin specimens. *FEMS Immunology and Medical Microbiology*, *48*(3), 319–328.

Price, E. P., Sarovich, D. S., Webb, J. R., Ginther, J. L., Mayo, M., Cook, J. M., et al. (2013). Accurate and rapid identification of the *Burkholderia pseudomallei* near-neighbour, *Burkholderia ubonensis*, using real-time PCR. *PLoS One*, *8*(8), e71647.

Qian, Q., Tang, Y. W., Kolbert, C. P., Torgerson, C. A., Hughes, J. G., Vetter, E. A., et al. (2001). Direct identification of bacteria from positive blood cultures by amplification and sequencing of the 16S rRNA gene: Evaluation of BACTEC 9240 instrument true-positive and false-positive results. *Journal of Clinical Microbiology*, *39*(10), 3578–3582.

Qin, J., Cui, Y., Zhao, X., Rohde, H., Liang, T., Wolters, M., et al. (2011). Identification of the Shiga toxin-producing *Escherichia coli* O104:H4 strain responsible for a food poisoning outbreak in Germany by PCR. *Journal of Clinical Microbiology*, *49*(9), 3439–3440.

Regnery, R. L., Anderson, B. E., Clarridge, J. E., 3rd., Rodriguez-Barradas, M. C., Jones, D. C., & Carr, J. H. (1992). Characterization of a novel *Rochalimaea* species, *R. henselae* sp. nov., isolated from blood of a febrile, human immunodeficiency virus-positive patient. *Journal of Clinical Microbiology*, *30*(2), 265–274.

Relman, D. A., Loutit, J. S., Schmidt, T. M., Falkow, S., & Tompkins, L. S. (1990). The agent of bacillary angiomatosis. An approach to the identification of uncultured pathogens. *The New England Journal of Medicine*, *323*(23), 1573–1580.

Relman, D. A., Schmidt, T. M., MacDermott, R. P., & Falkow, S. (1992). Identification of the uncultured bacillus of Whipple's disease. *The New England Journal of Medicine*, *327*(5), 293–301.

Ringuet, H., Akoua-Koffi, C., Honore, S., Varnerot, A., Vincent, V., Berche, P., et al. (1999). *hsp65* sequencing for identification of rapidly growing mycobacteria. *Journal of Clinical Microbiology*, *37*(3), 852–857.

Roux, V., Rydkina, E., Eremeeva, M., & Raoult, D. (1997). Citrate synthase gene comparison, a new tool for phylogenetic analysis, and its application for the rickettsiae. *International Journal of Systematic Bacteriology*, *47*(2), 252–261.

Schabereiter-Gurtner, C., Nehr, M., Apfalter, P., Makristathis, A., Rotter, M. L., & Hirschl, A. M. (2008). Evaluation of a protocol for molecular broad-range diagnosis of culture-negative bacterial infections in clinical routine diagnosis. *Journal of Applied Microbiology*, *104*(4), 1228–1237.

Seng, P., Drancourt, M., Gouriet, F., La Scola, B., Fournier, P. E., Rolain, J. M., et al. (2009). Ongoing revolution in bacteriology: Routine identification of bacteria by matrix-assisted laser desorption ionization time-of-flight mass spectrometry. *Clinical Infectious Diseases*, *49*(4), 543–551.

Seo, J. Y., Yeom, J. S., & Ko, K. S. (2012). *Actinomyces cardiffensis* septicemia: A case report. *Diagnostic Microbiology and Infectious Disease*, *73*(1), 86–88.

Shao, Y., He, X., Harrison, E. M., Tai, C., Ou, H. Y., Rajakumar, K., et al. (2010). mGenomeSubtractor: A web-based tool for parallel *in silico* subtractive hybridization analysis of multiple bacterial genomes. *Nucleic Acids Research*, *38*(Web Server issue), W194–W200.

Smith, K., Pandey, S. K., & Ussher, J. E. (2012). Bacteraemia caused by *Catabacter hongkongensis*. *Anaerobe*, *18*(3), 366–368.

Snel, B., Bork, P., & Huynen, M. A. (1999). Genome phylogeny based on gene content. *Nature Genetics*, *21*(1), 108–110.

Snyder, L. A., Loman, N. J., Faraj, L. A., Levi, K., Weinstock, G., Boswell, T. C., et al. (2013). Epidemiological investigation of *Pseudomonas aeruginosa* isolates from a six-year-long hospital outbreak using high-throughput whole genome sequencing. *Euro Surveillance,18*(42), pii: 20611.

Song, Y., Liu, C., Bolanos, M., Lee, J., McTeague, M., & Finegold, S. M. (2005). Evaluation of 16S rRNA sequencing and reevaluation of a short biochemical scheme for identification of clinically significant *Bacteroides* species. *Journal of Clinical Microbiology*, *43*(4), 1531–1537.

Srinivasan, L., Pisapia, J. M., Shah, S. S., Halpern, C. H., & Harris, M. C. (2012). Can broad-range 16S ribosomal ribonucleic acid gene polymerase chain reactions improve the diagnosis of bacterial meningitis? A systematic review and meta-analysis. *Annals of Emergency Medicine:60*(5), 609.e2–620.e2.

Sun, J. R., Yan, J. C., Yeh, C. Y., Lee, S. Y., & Lu, J. J. (2007). Invasive infection with *Streptococcus iniae* in Taiwan. *Journal of Medical Microbiology*, *56*(Pt. 9), 1246–1249.

Syre, H., Myneedu, V. P., Arora, V. K., & Grewal, H. M. (2009). Direct detection of mycobacterial species in pulmonary specimens by two rapid amplification tests, the gen-probe amplified mycobacterium tuberculosis direct test and the genotype mycobacteria direct test. *Journal of Clinical Microbiology*, *47*(11), 3635–3639.

Tang, Y. W. (2009). Duplex PCR assay simultaneously detecting and differentiating *Bartonella quintana*, *B. henselae*, and *Coxiella burnetii* in surgical heart valve specimens. *Journal of Clinical Microbiology*, *47*(8), 2647–2650.

Tang, Y. W., Ellis, N. M., Hopkins, M. K., Smith, D. H., Dodge, D. E., & Persing, D. H. (1998). Comparison of phenotypic and genotypic techniques for identification of unusual aerobic pathogenic Gram-negative bacilli. *Journal of Clinical Microbiology*, *36*(12), 3674–3679.

Tang, Y. W., Hopkins, M. K., Kolbert, C. P., Hartley, P. A., Severance, P. J., & Persing, D. H. (1998). *Bordetella holmesii*-like organisms associated with septicemia, endocarditis, and respiratory failure. *Clinical Infectious Diseases*, *26*(2), 389–392.

Tang, B. S., Lau, S. K., Teng, J. L., Chan, T. M., Chan, W. S., Wong, T. Y., et al. (2013). Matrix-assisted laser desorption ionisation-time of flight mass spectrometry for rapid identification of *Laribacter hongkongensis*. *Journal of Clinical Pathology*, *66*(12), 1081–1083.

Taylor, C. F., Paton, N. W., Lilley, K. S., Binz, P. A., Julian, R. K., Jr., Jones, A. R., et al. (2007). The minimum information about a proteomics experiment (MIAPE). *Nature Biotechnology*, *25*(8), 887–893.

Teng, J. L., Ho, T. C., Yeung, R. S., Wong, A. Y., Wang, H., Chen, C., et al. (2014). Evaluation of 16SpathDB 2.0, an automated 16S rRNA gene sequence database, using 689 complete bacterial genomes. *Diagnostic Microbiology and Infectious Disease*, *78*(2), 105–115.

Teng, J. L., Huang, Y., Tse, H., Chen, J. H., Tang, Y., Lau, S. K., et al. (2014). Phylogenomic and MALDI-TOF MS analysis of *Streptococcus sinensis* HKU4[T] reveals a distinct phylogenetic clade in the genus *Streptococcus*. *Genome Biology and Evolution*, *6*(10), 2930–2943.

Teng, J. L., Woo, P. C., Ma, S. S., Sit, T. H., Ng, L. T., Hui, W. T., et al. (2005). Ecoepidemiology of *Laribacter hongkongensis*, a novel bacterium associated with gastroenteritis. *Journal of Clinical Microbiology*, *43*(2), 919–922.

Teng, J. L., Yeung, M. Y., Yue, G., Au-Yeung, R. K., Yeung, E. Y., Fung, A. M., et al. (2011). *In silico* analysis of 16S rRNA gene sequencing based methods for identification of medically important aerobic Gram-negative bacteria. *Journal of Medical Microbiology*, *60*(Pt. 9), 1281–1286.

Therese, K. L., Bartell, J., Deepa, P., Mangaiyarkarasi, S., Ward, D., Dajcs, J., et al. (2009). DNA sequencing by Microseq kit targeting 16S rRNA gene for species level identification of mycobacteria. *The Indian Journal of Medical Research*, *129*(2), 176–181.

To, K. K., Cheng, V. C., Chan, J. F., Wong, A. C., Chau, S., Tsang, F. H., et al. (2011). Molecular characterization of a catalase-negative *Staphylococcus aureus* subsp. *aureus* Strain collected from a patient with mitral valve endocarditis and pericarditis revealed a novel nonsense mutation in the *katA* gene. *Journal of Clinical Microbiology*, *49*(9), 3398–3402.

Touati, A., Pereyre, S., Bouziri, A., Achour, W., Khaldi, A., Ben Jaballah, N., et al. (2010). Prevalence of *Mycoplasma pneumoniae*-associated respiratory tract infections in hospitalized children: Results of a 4-year prospective study in Tunis. *Diagnostic Microbiology and Infectious Disease*, *68*(2), 103–109.

Tse, H., Bao, J. Y., Davies, M. R., Maamary, P., Tsoi, H. W., Tong, A. H., et al. (2012). Molecular characterization of the 2011 Hong Kong scarlet fever outbreak. *The Journal of Infectious Diseases*, *206*(3), 341–351.

Uckay, I., Rohner, P., Bolivar, I., Ninet, B., Djordjevic, M., Nobre, V., et al. (2007). *Streptococcus sinensis* endocarditis outside Hong Kong. *Emerging Infectious Diseases*, *13*(8), 1250–1252.

Versalovic, J., Carroll, K. C., Jorgensen, J. H., Funke, G., & Landry, M. L. (2011). *Manual of clinical microbiology* (10th ed.). Washington, DC: ASM Press.

Wang, K., Sun, X., & Lu, C. (2012). Development of rapid serotype-specific PCR assays for eight serotypes of *Streptococcus suis*. *Journal of Clinical Microbiology*, *50*(10), 3329–3334.

Weinstein, M. R., Litt, M., Kertesz, D. A., Wyper, P., Rose, D., Coulter, M., et al. (1997). Invasive infections due to a fish pathogen, *Streptococcus iniae*. S. iniae Study Group. *The New England Journal of Medicine*, *337*(9), 589–594.

Weng, F. Y., Chiou, C. S., Lin, P. H., & Yang, S. S. (2009). Application of *recA* and *rpoB* sequence analysis on phylogeny and molecular identification of *Geobacillus* species. *Journal of Applied Microbiology*, *107*(2), 452–464.

Wilson, K. H., Blitchington, R., Frothingham, R., & Wilson, J. A. (1991). Phylogeny of the Whipple's-disease-associated bacterium. *The Lancet*, *338*(8765), 474–475.

Wongtrakoongate, P., Mongkoldhumrongkul, N., Chaijan, S., Kamchonwongpaisan, S., & Tungpradabkul, S. (2007). Comparative proteomic profiles and the potential markers between *Burkholderia pseudomallei* and *Burkholderia thailandensis*. *Molecular and Cellular Probes*, *21*(2), 81–91.

Woo, P. C., Cheung, E. Y., Leung, K., & Yuen, K. (2001). Identification by 16S ribosomal RNA gene sequencing of an *Enterobacteriaceae* species with ambiguous biochemical profile from a renal transplant recipient. *Diagnostic Microbiology and Infectious Disease*, *39*(2), 85–93.

Woo, P. C., Chong, K. T., Leung, K., Que, T., & Yuen, K. (2001). Identification of *Arcobacter cryaerophilus* isolated from a traffic accident victim with bacteremia by 16S ribosomal RNA gene sequencing. *Diagnostic Microbiology and Infectious Disease*, *40*(3), 125–127.

Woo, P. C., Chung, L. M., Teng, J. L., Tse, H., Pang, S. S., Lau, V. Y., et al. (2007). *In silico* analysis of 16S ribosomal RNA gene sequencing-based methods for identification of medically important anaerobic bacteria. *Journal of Clinical Pathology*, *60*(5), 576–579.

Woo, P. C., Fong, A. H., Ngan, A. H., Tam, D. M., Teng, J. L., Lau, S. K., et al. (2009). First report of *Tsukamurella* keratitis: Association between *T. tyrosinosolvens* and *T. pulmonis* and ophthalmologic infections. *Journal of Clinical Microbiology*, *47*(6), 1953–1956.

Woo, P. C., Fung, A. M., Lau, S. K., Chan, B. Y., Chiu, S. K., Teng, J. L., et al. (2003). *Granulicatella adiacens* and *Abiotrophia defectiva* bacteraemia characterized by 16S rRNA gene sequencing. *Journal of Medical Microbiology*, *52*(Pt. 2), 137–140.

Woo, P. C., Fung, A. M., Lau, S. K., Hon, E., & Yuen, K. Y. (2002). Diagnosis of pelvic actinomycosis by 16S ribosomal RNA gene sequencing and its clinical significance. *Diagnostic Microbiology and Infectious Disease*, *43*(2), 113–118.

Woo, P. C., Fung, A. M., Lau, S. K., Teng, J. L., Wong, B. H., Wong, M. K., et al. (2003). *Actinomyces hongkongensis* sp. nov. a novel *Actinomyces* species isolated from a patient with pelvic actinomycosis. *Systematic and Applied Microbiology*, *26*(4), 518–522.

Woo, P. C., Fung, A. M., Lau, S. K., Wong, S. S., & Yuen, K. Y. (2001). Group G beta-hemolytic streptococcal bacteremia characterized by 16S ribosomal RNA gene sequencing. *Journal of Clinical Microbiology*, *39*(9), 3147–3155.

Woo, P. C., Fung, A. M., Lau, S. K., & Yuen, K. Y. (2002). Identification by 16S rRNA gene sequencing of *Lactobacillus salivarius* bacteremic cholecystitis. *Journal of Clinical Microbiology*, *40*(1), 265–267.

Woo, P. C., Fung, A. M., Wong, S. S., Tsoi, H. W., & Yuen, K. Y. (2001). Isolation and characterization of a *Salmonella enterica* serotype Typhi variant and its clinical and public health implications. *Journal of Clinical Microbiology*, *39*(3), 1190–1194.

Woo, P. C., Kuhnert, P., Burnens, A. P., Teng, J. L., Lau, S. K., Que, T. L., et al. (2003). *Laribacter hongkongensis*: A potential cause of infectious diarrhea. *Diagnostic Microbiology and Infectious Disease*, *47*(4), 551–556.

Woo, P. C., Lau, S. K., Chan, K. M., Fung, A. M., Tang, B. S., & Yuen, K. Y. (2005). *Clostridium* bacteraemia characterised by 16S ribosomal RNA gene sequencing. *Journal of Clinical Pathology*, *58*(3), 301–307.

Woo, P. C., Lau, S. K., Fung, A. M., Chiu, S. K., Yung, R. W., & Yuen, K. Y. (2003). *Gemella* bacteraemia characterised by 16S ribosomal RNA gene sequencing. *Journal of Clinical Pathology*, *56*(9), 690–693.

Woo, P. C., Lau, S. K., Lin, A. W., Curreem, S. O., Fung, A. M., & Yuen, K. Y. (2007). Surgical site abscess caused by *Lactobacillus fermentum* identified by 16S ribosomal RNA gene sequencing. *Diagnostic Microbiology and Infectious Disease*, *58*(2), 251–254.

Woo, P. C., Lau, S. K., Teng, J. L., Tse, H., & Yuen, K. Y. (2008). Then and now: Use of 16S rDNA gene sequencing for bacterial identification and discovery of novel bacteria in clinical microbiology laboratories. *Clinical Microbiology and Infection*, *14*(10), 908–934.

Woo, P. C., Lau, S. K., Teng, J. L., & Yuen, K. Y. (2005). Current status and future directions for *Laribacter hongkongensis*, a novel bacterium associated with gastroenteritis and traveller's diarrhoea. *Current Opinion in Infectious Diseases*, *18*(5), 413–419.

Woo, P. C., Lau, S. K., Tse, H., Teng, J. L., Curreem, S. O., Tsang, A. K., et al. (2009). The complete genome and proteome of *Laribacter hongkongensis* reveal potential mechanisms for adaptations to different temperatures and habitats. *PLoS Genetics*, *5*(3), e1000416.

Woo, P. C., Lau, S. K., Woo, G. K., Fung, A. M., Ngan, A. H., Hui, W. T., et al. (2003). Seronegative bacteremic melioidosis caused by *Burkholderia pseudomallei* with ambiguous biochemical profile: Clinical importance of accurate identification by 16S rRNA gene and *groEL* gene sequencing. *Journal of Clinical Microbiology*, *41*(8), 3973–3977.

Woo, P. C., Leung, P. K., Leung, K. W., & Yuen, K. Y. (2000). Identification by 16S ribosomal RNA gene sequencing of an *Enterobacteriaceae* species from a bone marrow transplant recipient. *Molecular Pathology*, *53*(4), 211–215.

Woo, P. C., Leung, A. S., Leung, K. W., & Yuen, K. Y. (2001). Identification of slide coagulase positive, tube coagulase negative *Staphylococcus aureus* by 16S ribosomal RNA gene sequencing. *Molecular Pathology*, *54*(4), 244–247.

Woo, P. C., Leung, K. W., Tsoi, H. W., Wong, S. S., Teng, J. L., & Yuen, K. Y. (2002). Thermo-tolerant *Campylobacter fetus* bacteraemia identified by 16S ribosomal RNA gene sequencing: An emerging pathogen in immunocompromised patients. *Journal of Medical Microbiology*, *51*(9), 740–746.

Woo, P. C., Leung, K. W., Wong, S. S., Chong, K. T., Cheung, E. Y., & Yuen, K. Y. (2002). Relatively alcohol-resistant *Mycobacteria* are emerging pathogens in patients receiving acupuncture treatment. *Journal of Clinical Microbiology*, *40*(4), 1219–1224.

Woo, P. C., Leung, P. K., Wong, S. S., Ho, P. L., & Yuen, K. Y. (2001). *groEL* encodes a highly antigenic protein in *Burkholderia pseudomallei*. *Clinical and Diagnostic Laboratory Immunology*, *8*(4), 832–836.

Woo, P. C., Li, J. H., Tang, W., & Yuen, K. (2001). Acupuncture mycobacteriosis. *The New England Journal of Medicine*, *345*(11), 842–843.

Woo, P. C., Ma, S. S., Teng, J. L., Li, M. W., Kao, R. Y., Lau, S. K., et al. (2005). Construction of an inducible expression shuttle vector for *Laribacter hongkongensis*, a novel bacterium associated with gastroenteritis. *FEMS Microbiology Letters*, *252*(1), 57–65.

Woo, P. C., Ma, S. S., Teng, J. L., Li, M. W., Lau, S. K., & Yuen, K. Y. (2007). Plasmid profile and construction of a small shuttle vector in *Laribacter hongkongensis*. *Biotechnology Letters*, *29*(10), 1575–1582.

Woo, P. C., Ngan, A. H., Lau, S. K., & Yuen, K. Y. (2003). *Tsukamurella* conjunctivitis: A novel clinical syndrome. *Journal of Clinical Microbiology*, *41*(7), 3368–3371.

Woo, P. C., Tam, D. M., Lau, S. K., Fung, A. M., & Yuen, K. Y. (2004). *Enterococcus cecorum* empyema thoracis successfully treated with cefotaxime. *Journal of Clinical Microbiology*, *42*(2), 919–922.

Woo, P. C., Tam, D. M., Leung, K. W., Lau, S. K., Teng, J. L., Wong, M. K., et al. (2002). *Streptococcus sinensis* sp. nov., a novel species isolated from a patient with infective endocarditis. *Journal of Clinical Microbiology*, *40*(3), 805–810.

Woo, P. C., Teng, J. L., Lam, K. K., Tse, C. W., Leung, K. W., Leung, A. W., et al. (2010). First report of *Gordonibacter pamelaeae* bacteremia. *Journal of Clinical Microbiology*, *48*(1), 319–322.

Woo, P. C., Teng, J. L., Lau, S. K., Lum, P. N., Leung, K. W., Wong, K. L., et al. (2003). Analysis of a viridans group strain reveals a case of bacteremia due to lancefield group G alpha-hemolytic *Streptococcus dysgalactiae* subsp *equisimilis* in a patient with pyomyositis and reactive arthritis. *Journal of Clinical Microbiology*, *41*(2), 613–618.

Woo, P. C., Teng, J. L., Lau, S. K., & Yuen, K. Y. (2006). Clinical, phenotypic, and genotypic evidence for *Streptococcus sinensis* as the common ancestor of anginosus and mitis groups of streptococci. *Medical Hypotheses*, *66*(2), 345–351.

Woo, P. C., Teng, J. L., Leung, K. W., Lau, S. K., Tse, H., Wong, B. H., et al. (2004). *Streptococcus sinensis* may react with Lancefield group F antiserum. *Journal of Medical Microbiology*, *53*(Pt. 11), 1083–1088.

Woo, P. C., Teng, J. L., Leung, K. W., Lau, S. K., Woo, G. K., Wong, A. C., et al. (2005). *Anaerospora hongkongensis* gen. nov. sp. nov., a novel genus and species with ribosomal DNA operon heterogeneity isolated from an intravenous drug abuser with pseudobacteremia. *Microbiology and Immunology*, *49*(1), 31–39.

Woo, P. C., Teng, J. L., Ma, S. S., Lau, S. K., & Yuen, K. Y. (2007). Characterization of a novel cryptic plasmid, pHLHK26, in *Laribacter hongkongensis*. *The New Microbiologica*, *30*(2), 139–147.

Woo, P. C., Teng, J. L., Tsang, S. N., Tse, C. W., Lau, S. K., & Yuen, K. Y. (2008). The oral cavity as a natural reservoir for *Streptococcus sinensis*. *Clinical Microbiology and Infection*, *14*(11), 1075–1079.

Woo, P. C., Teng, J. L., Tsang, A. K., Tse, H., Tsang, V. Y., Chan, K. M., et al. (2009). Development of a multi-locus sequence typing scheme for *Laribacter hongkongensis*, a novel bacterium associated with freshwater fish-borne gastroenteritis and traveler's diarrhea. *BMC Microbiology*, *9*, 21.

Woo, P. C., Teng, J. L., Wu, J. K., Leung, F. P., Tse, H., Fung, A. M., et al. (2009). Guidelines for interpretation of 16S rRNA gene sequence-based results for identification of medically important aerobic Gram-positive bacteria. *Journal of Medical Microbiology*, *58*(Pt. 8), 1030–1036.

Woo, P. C., Teng, J. L., Yeung, J. M., Tse, H., Lau, S. K., & Yuen, K. Y. (2011). Automated identification of medically important bacteria by 16S rRNA gene sequencing using a novel comprehensive database, 16SpathDB. *Journal of Clinical Microbiology*, *49*(5), 1799–1809.

Woo, P. C., Tse, H., Chan, K. M., Lau, S. K., Fung, A. M., Yip, K. T., et al. (2004). "*Streptococcus milleri*" endocarditis caused by *Streptococcus anginosus*. *Diagnostic Microbiology and Infectious Disease*, *48*(2), 81–88.

Woo, P. C., Tse, H., Lau, S. K., Leung, K. W., Woo, G. K., Wong, M. K., et al. (2005). *Alkanindiges hongkongensis* sp. nov. A novel *Alkanindiges* species isolated from a patient with parotid abscess. *Systematic and Applied Microbiology*, *28*(4), 316–322.

Woo, P. C., Tse, H., Wong, S. S., Tse, C. W., Fung, A. M., Tam, D. M., et al. (2005). Life-threatening invasive *Helcococcus kunzii* infections in intravenous-drug users and *ermA*-mediated erythromycin resistance. *Journal of Clinical Microbiology*, *43*(12), 6205–6208.

Woo, P. C., Tsoi, H. W., Leung, K. W., Lum, P. N., Leung, A. S., Ma, C. H., et al. (2000). Identification of *Mycobacterium neoaurum* isolated from a neutropenic patient with catheter-related bacteremia by 16S rRNA sequencing. *Journal of Clinical Microbiology*, *38*(9), 3515–3517.

Woo, P. C., Wong, S. S., Teng, J. L., Leung, K. W., Ngan, A. H., Zhao, D. Q., et al. (2010). *Leptotrichia hongkongensis* sp. nov., a novel *Leptotrichia* species with the oral cavity as its natural reservoir. *Journal of Zhejiang University. Science. B*, *11*(6), 391–401.

Woo, P. C., Woo, G. K., Lau, S. K., Wong, S. S., & Yuen, K. (2002). Single gene target bacterial identification. *groEL* gene sequencing for discriminating clinical isolates of

Burkholderia pseudomallei and *Burkholderia thailandensis*. *Diagnostic Microbiology and Infectious Disease*, *44*(2), 143–149.

Woo, P. C., Wu, A. K., Tsang, C. C., Leung, K. W., Ngan, A. H., Curreem, S. O., et al. (2014). *Streptobacillus hongkongensis* sp. nov., isolated from patients with quinsy and septic arthritis, and emended descriptions of the genus *Streptobacillus* and *Streptobacillus moniliformis*. *International Journal of Systematic and Evolutionary Microbiology*, *64*(Pt. 9), 3034–3039.

Wurdemann, D., Tindall, B. J., Pukall, R., Lunsdorf, H., Strompl, C., Namuth, T., et al. (2009). *Gordonibacter pamelaeae* gen. nov., sp. nov., a new member of the *Coriobacteriaceae* isolated from a patient with Crohn's disease, and reclassification of *Eggerthella hongkongensis* Lau et al. 2006 as *Paraeggerthella hongkongensis* gen. nov., comb. nov. *International Journal of Systematic and Evolutionary Microbiology*, *59*(Pt. 6), 1405–1415.

Xiong, L., Teng, J. L., Watt, R. M., Kan, B., Lau, S. K., & Woo, P. C. (2014). Arginine deiminase pathway is far more important than urease for acid resistance and intracellular survival in *Laribacter hongkongensis*: A possible result of *arc* gene cassette duplication. *BMC Microbiology*, *14*, 42.

Yu, S., Chen, W., Wang, D., He, X., Zhu, X., & Shi, X. (2010). Species-specific PCR detection of the food-borne pathogen *Vibrio parahaemolyticus* using the *irgB* gene identified by comparative genomic analysis. *FEMS Microbiology Letters*, *307*(1), 65–71.

Yuen, K. Y., Woo, P. C., Teng, J. L., Leung, K. W., Wong, M. K., & Lau, S. K. (2001). *Laribacter hongkongensis* gen. nov., sp. nov., a novel gram-negative bacterium isolated from a cirrhotic patient with bacteremia and empyema. *Journal of Clinical Microbiology*, *39*(12), 4227–4232.

Zbinden, A., Kohler, N., & Bloemberg, G. V. (2011). *recA*-based PCR assay for accurate differentiation of *Streptococcus pneumoniae* from other viridans streptococci. *Journal of Clinical Microbiology*, *49*(2), 523–527.

Zhu, B., Xiao, D., Zhang, H., Zhang, Y., Gao, Y., Xu, L., et al. (2013). MALDI-TOF MS distinctly differentiates nontypable *Haemophilus influenzae* from *Haemophilus haemolyticus*. *PLoS One*, *8*(2), e56139.

Host-Based Diagnostics for Detection and Prognosis of Infectious Diseases

13

William E. Yang*, Christopher W. Woods[†,‡,§], Ephraim L. Tsalik[‡,§,¶,1]

*Duke University School of Medicine, Center for Applied Genomics and Precision Medicine, Durham, North Carolina, USA
†Medicine Service, Durham VAMC, Durham, North Carolina, USA
‡Center for Applied Genomics & Precision Medicine, Duke University Medical Center, Durham, North Carolina, USA
§Division of Infectious Diseases, Department of Medicine, Duke University Medical Center, Durham, North Carolina, USA
¶Emergency Medicine Service, Durham VAMC, Durham, North Carolina, USA
¹Corresponding author: e-mail address: e.t@duke.edu

1 INTRODUCTION

Laboratory diagnostics have long been a central component of modern medicine. In the realm of infectious diseases, such diagnostics have traditionally focused on pathogen detection and identification. However, pathogen-based diagnostics often have limited sensitivity and specificity, as well as prolonged time to results. Continued innovation in pathogen-focused diagnostics is discussed elsewhere in this book. Here, we will provide an overview of the promising and rapidly evolving state of molecular diagnostics targeting the host response to infection rather than direct pathogen detection.

The host response to infection represents the activation of multiple pathways involved in detecting and eliminating pathogens. These responses occur at the earliest time points in infection, typically preceding clinical manifestations of disease making the host response a logical target when considering novel diagnostic strategies. In addition, clinical symptoms are often the result of the host response itself, rather than the simple presence of a pathogen, facilitating distinctions between actual infection and simple colonisation. Moreover, the host response can have different responses to a given pathogen, varying from person to person and affected by genetics, environment, and circumstance. Defining these differential responses and then developing assays to detect those differences can yield valuable diagnostic and prognostic insights.

The idea of a host-response biomarker is not new. Going back nearly a century, erythrocyte sedimentation rate was identified in 1917 as a marker of tuberculosis

Methods in Microbiology, Volume 42, ISSN 0580-9517, http://dx.doi.org/10.1016/bs.mim.2015.06.001

activity (Waugh, 1923). Shortly thereafter, C-reactive protein (CRP) was described in 1930 in the context of *Streptococcus pneumoniae* infection (Tillett & Francis, 1930). These tests, which indeed reflect the host response, are highly non-specific and, therefore, unable to differentiate infection from no infection, let alone between pathogen types.

More recent advances in our understanding of biological pathways have generated multiple candidate biomarkers, particularly for sepsis, but have yielded mixed results with respect to improving clinical management. Completion of the Human Genome Project in 2003 brought great hope that new diagnostics and therapeutics would quickly follow this monumental scientific achievement. Specifically, knowing the human genome sequence might unlock a new world for scientific and clinical advances. Though these advances were not immediate, the pace of discovery is accelerating. This is due in large part to the development of new technologies and techniques to accompany the vast amounts of information made available by the Human Genome Project. For example, gene expression analysis and the techniques to analyse such big data have made transcriptomics, metabolomics, and proteomics accessible to researchers and more recently, with improved bioinformatics systems, to clinicians (Figure 1).

The first part of this chapter will begin with the commonly used molecular techniques and markers: transcriptomics, proteomics, and metabolomics. We will briefly explore an essential adjunct to the development of these biomarkers, namely computational and analytic tools distilling massive amounts of raw data into clinically useful parameters. The second part of the chapter will explore how these tools are applied to specific pathogen classes and clinical syndromes to define the host response and generate new biomarker candidates.

2 TECHNIQUES AND MARKERS
2.1 TRANSCRIPTOMICS

Advances in the last decade in both sequencing technologies and computational ability have resulted in increased efficiency, decreased cost, and improved capacity to generate and analyse vast amounts of genetic data. The extension of genomic techniques from DNA to RNA opens a window into observing gene expression. Unlike DNA sequencing which captures the largely static genome, the gene expression of cells is subject to change depending on a multitude of factors including the tissue from which it is derived and external stressors. We provide an overview of both the techniques and the targets that have demonstrated utility in host-based diagnostics.

2.1.1 Techniques
2.1.1.1 Whole-genome DNA microarrays
Whole-genome microarrays are large collections of thousands of oligonucleotide probes bound to a substrate, used to detect the presence of complementary sequences in test samples. Substrates are a solid surface, commonly in the form of a chip (e.g. Affymetrix) or microscopic beads (e.g. Illumina). At a fundamental level,

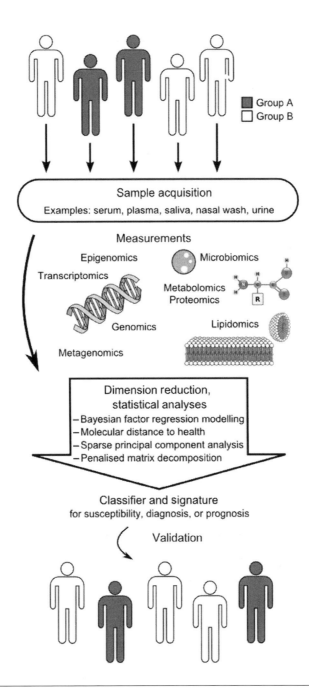

FIGURE 1

Overview of the development process for a host diagnostic biomarker. Beginning with a population that is dichotomised by susceptibility, diagnosis, or prognosis, biological samples are acquired. Omic measurements are run on these samples, which generates large quantities of data. Dimension reduction and statistical analyses generate a classifier or signature that distinguishes the desired characteristic from the original population. The classifier is then validated against a different population to test its generalisability. (See the color plate.)

Table 1 Comparison of Microarrays and RNA-Seq

Microarrays	RNA-Seq
Biases: sequence specific	Biases: longer transcripts, fragmentation
Gene discovery limited to included probes	Can detect novel and splice variants
Requires larger quantity of RNA	Smaller amounts of RNA detectable, but rare transcripts can still be missed
No reference genome needed	Need reference genome
Semi-quantitative	Quantitative
Detect mRNA only	Detect non-coding RNA
Lower cost: $200–800[a]	Higher cost: $300–1600[a]
Small data storage requirements, though normalisation can be computationally intense	Significant computational and data storage burden

[a]Cost estimates based on publicly available data on scienceexchange.com as of 2014. Prices are highly variable based on specific assays or processes and are likely to decline further with continued uptake and optimisation of the technology involved.

whole-genome microarrays are a method by which the entire genome can be assayed for gene expression without bias for specific targets (Table 1).

Real-time PCR is commonly used to validate findings from whole-genome microarrays, which are less sensitive than PCR. In addition, microarrays are semi-quantitative as measurements are in reference to control samples (Chaussabel, Pascual, & Banchereau, 2010). In contrast, the "real-time" component of real-time PCR quantifies the amount of target sequence present during the reaction itself. One important limitation of microarrays is that they are restricted by the probes included on the array. This precludes the detection of splice or sequence variants in most cases.

2.1.1.2 RNA-seq

RNA-seq, also known as whole transcriptome sequencing, is the sequencing of a sample's mRNA content. Notably, it is a method by which a point-in-time snapshot of the transcriptome can be obtained. RNA-seq involves conversion of a sample of RNA to a cDNA library, which is then sequenced and mapped against a reference genome. In addition to the ability to measure the level of gene expression, it provides further information on alternative splicing and non-coding RNA (such as micro-RNA) (Chaussabel et al., 2010). The ability to sequence the transcriptome overcomes some limitations of whole-genome microarrays. Specifically, the measures are quantitative and the spectrum of detectable targets is not limited by the gene probes present on a microarray.

Despite RNA-seq's major advantages, the technique is subject to certain biases inherent to the reverse transcription and PCR amplification processes used. In the former, annealing of primers to RNA is not truly random and results in reduced data from the 5′ and 3′ ends of the strands (Hansen, Brenner, & Dudoit, 2010; Roberts, Trapnell, Donaghey, Rinn, & Pachter, 2011), making it more difficult to identify the starts and ends of transcripts. Meanwhile, PCR tends to amplify long transcripts to a greater degree than short transcripts, resulting in transcript-length bias (Oshlack & Wakefield, 2009).

The cost of equipment is a major impediment to more widespread use. In addition to the molecular sequencing equipment, RNA-seq generates a large quantity of data to analyse; in one particular example, Hiseq2000 (Illumina) can generate 200 million 100-nt reads per lane per sequence run, which is approximately 50 GB (Chu & Corey, 2012). Both computing power and storage are necessary to adequately analyse the sheer volume of data generated.

2.1.2 Nucleic acid markers

2.1.2.1 Messenger RNA

Messenger RNA (mRNA) is well known to be the intermediate between the gene as represented in DNA and translation into proteins. They are the most common targets of measurement when microarrays and RNA-seq are utilised, by way of cDNA libraries, and provide a picture of gene expression. While they do take into account transcriptional regulatory mechanisms, they are limited in only providing a point-in-time snapshot of genetic expression and furthermore do not give information on post-translational modification and regulation (Wong, 2012).

2.1.2.2 microRNA

microRNA (miRNA) is a short (17–22 nucleotide) non-coding RNA molecule involved in post-transcriptional regulation. These small RNAs function to inhibit translation as part of the RNA-induced silencing complex, binding complementary sequences within the 3′ untranslated region (UTR) of mRNA molecules. Although they primarily function intracellularly, miRNAs can also exist stably outside cells. They are remarkably stable, being resistant to boiling, repeated freeze–thaw cycles, or decay over time. As a result, miRNAs make valuable diagnostic targets (Gilad et al., 2008; Mitchell et al., 2008). miRNAs have also been shown to be highly conserved between species, such that discoveries in animal models may inform human disease more than is seen with genomic or transcriptomic analyses (Bartel, 2004; Seok et al., 2013; Stark et al., 2007).

2.1.2.3 Cell-free DNA

Traditionally useful as a tool in prenatal testing (cell-free foetal DNA in the maternal bloodstream), cell-free DNA, or extracellular DNA in blood, has recently become a marker of interest in various acute and chronic disorders. Elevated cell-free DNA has been observed in cancer patients and is hypothesised to be a result of cell apoptosis or necrosis (Jahr et al., 2001). Furthermore, cell-free DNA has been found to correlate with septic shock (Zeerleder et al., 2003), as well as predictive of fatal outcomes in certain populations of sepsis patients (Avriel et al., 2014; Forsblom et al., 2014; Huttunen et al., 2011; Saukkonen et al., 2008). Plasma cell-free DNA has also been correlated with severity of Hantaan virus infection (Yi et al., 2014) and active *Schistosoma japonicum* infection (Kato-Hayashi et al., 2014), though the latter observation involves schistosome cell-free DNA rather than human cell-free DNA.

2.2 PROTEOMICS AND METABOLOMICS

The aim of proteomic and metabolomic analyses is the comprehensive evaluation of proteins and metabolites produced by the host (Figure 2). These analyses include characterisation of modifications, such as post-translational modification of proteins, and interactions with other molecules.

Before discussing specific techniques and markers, it is worth noting that distinctions between proteomic and metabolomic markers are sometimes semantic but useful for organisational purposes. Many but not all metabolites are peptides. As such, the techniques and targets relevant to proteomics and metabolomics overlap considerably.

2.2.1 Techniques
2.2.1.1 Mass spectrometry
Mass spectrometry (MS), commonly used to detect the presence of proteins or metabolites, identifies molecules in a sample by differentiating them based on chemical properties. This includes mass-to-charge ratio, pK, and hydrophobicity. For an overview of the entire analytic process, see Figure 3. While there are some variations in the sample preparation, MS works in principle by ionising gas particles, accelerating them through a vacuum chamber, deflecting them with a magnetic field, and detecting the current produced by the particle (see Figure 3E). The resulting spectra are plotted on a graph to produce a characteristic signature (see Figure 3D) and are then compared to a library of known spectra to determine what molecules are present. Because of the methodology, MS is more limited when used with molecules that have similar spectral signatures, or cannot be ionised. In addition, interpreting the results is highly dependent on the mass spectral library used for comparison. There are many such libraries (e.g. Human Metabolome Library, FiehnLib, Golm Metabolome Database, METLIN, MassBank, and the NIST Standard Reference Database), which improve iteratively over time.

MS technologies continue to evolve, enabling the identification and quantification of a growing number of proteins and metabolites (Cox & Mann, 2011). The accuracy of identification also continues to improve, where measurements of mass-to-charge ratios can now distinguish molecules that differ by only a single neutron (Carin et al., 2012). While this precision provides detailed information on physiological states, it also adds analytical complexity. There also remain limitations in the ability to detect molecules present in low abundance, or that are hydrophobic or alkaline, resulting in incomplete coverage of the proteome and metabolome (Schulze & Usadel, 2010).

2.2.1.2 Immunoassays
Immunoassays, which use antibodies to detect the presence of a specific molecule, have two key characteristics: antigen–antibody binding and antibody labelling for detection. Commonly used labels are colour, fluorescence, and radiation. Common examples include enzyme-linked immunosorbent assays (ELISAs) and radioimmunoassays (RIAs).

A major limitation of immunoassays is specificity and development cost. In order to increase specificity, matching antibodies for multiple epitopes on a given target are

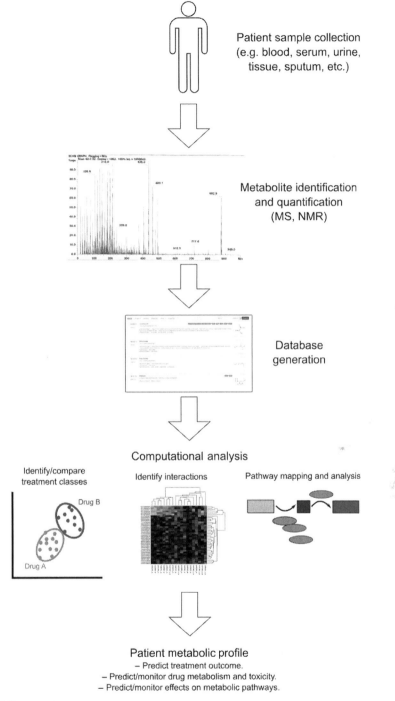

Patient sample collection
(e.g. blood, serum, urine,
tissue, sputum, etc.)

Metabolite identification
and quantification
(MS, NMR)

Database
generation

Computational analysis

Identify/compare
treatment classes

Identify interactions

Pathway mapping and analysis

Drug B

Drug A

Patient metabolic profile
— Predict treatment outcome.
— Predict/monitor drug metabolism and toxicity.
— Predict/monitor effects on metabolic pathways.

FIGURE 2

Sample metabolomics workflow. A biological sample is collected from a patient, metabolomic analyses are matched to a database of known analytes, and computational methods are applied to find associations, interactions, and pathway involvement. MS, mass spectrometry; NMR, nuclear magnetic resonance. (See the color plate.)

required, which greatly increases the cost of development compared to single epitope assays. Even though protein-based immunoassays are challenging to develop, the potential for rapid, low-cost assays for infection diagnostics is compelling. Whereas nucleic acid detection requires amplification, protein and metabolite targets do not

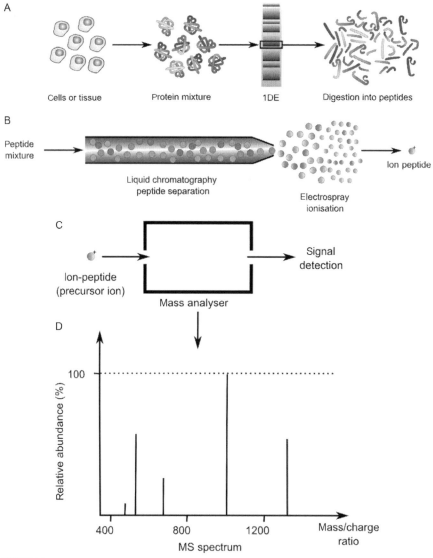

FIGURE 3

Protocol for mass spectroscopy, starting with a biological sample, which is first (A) purified and digested into peptides, which are then (B) separated and ionised. The ions are then run through the (C) mass analyser, which produces (D) characteristic mass spectra.

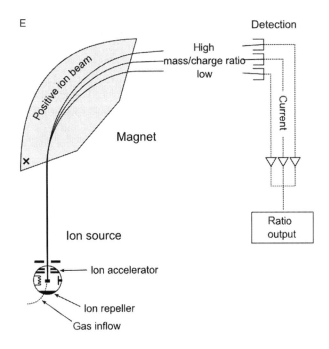

E

Detection

High
mass/charge ratio
low

Current

Positive ion beam

Magnet

Ratio
output

Ion source

Ion accelerator

Ion repeller

Gas inflow

FIGURE 3—Cont'd   (E) A schematic diagram of how the mass analyser separates the ions. The gas sample is accelerated through a magnetic field; the path of particles with a higher mass-to-charge ratio will be affected by the magnetic field to a smaller degree, and this difference in path is detected to produce the plot in (D). (See the color plate.)

Panels (A)–(D): Based on images by Philippe Hupé. Panel (E): Based on an image courtesy of the US Geological Survey.

require it and can be directly detected, as in lateral flow immunochromatography. Such diagnostic assays are well known to manufacturers and users alike, typified by pregnancy screens or rapid Group A *Streptococcus* tests.

2.2.1.3 Nuclear magnetic resonance spectroscopy

Nuclear magnetic resonance (NMR) spectroscopy identifies molecules based on the magnetic properties of certain NMR-active nuclei, the most common being hydrogen-1 and carbon-13. These nuclei are excited by a radiofrequency pulse. The decay response is then detected by sensors, and a Fourier transform is used to obtain structural information about the molecule.

Commonly used in organic chemistry, NMR also has some distinct advantages as a proteomic method as it can simultaneously detect large numbers of metabolites without selecting for specific molecules or pathways. It is inherently quantitative, though tends to have low sensitivity for molecules at low concentration. In those cases, MS remains a better option, though sample preparation is more involved than with NMR. NMR analysis is also non-destructive, in contrast to MS, allowing the same sample to undergo multiple successive analyses (Lutz, Beraud, & Cozzone, 2014). The T2Dx platform manufactured by T2 Biosystems is one example of NMR-based diagnostics for infectious disease (Chung, Castro, Im, Lee, & Weissleder, 2013; Liong et al., 2013; Neely et al., 2013). The T2Dx platform currently offers assays for pathogen detection assays focused on candidaemia and bacteraemia although there are assays targeting host markers in non-infectious disease areas (Agasti et al., 2012; Ling, Pong, Vassiliou, Huang, & Cima, 2011). This demonstrates the potential feasibility of host-based diagnostics in infectious diseases using NMR technology in the future.

2.2.2 Proteomic markers

2.2.2.1 Candidate marker examples

2.2.2.1.1 C-reactive protein. One of the oldest proteomic markers in widespread clinical use, CRP, is an acute-phase reactant and is used as a non-specific serum marker of inflammation. Despite its discovery in 1930 (Tillett & Francis, 1930), CRP continues to be investigated for its diagnostic and prognostic utility. Recent examples include periprosthetic joint infection (Deirmengian et al., 2014), urinary tract infections (Xu, Liu, Liu, & Dong, 2014), tuberculosis in HIV patients (Drain et al., 2014), and as an early predictor of infective complications in laparoscopic colorectal surgery (Nason et al., 2014).

2.2.2.1.2 Procalcitonin. One of the most promising biomarkers in recent years is procalcitonin (PCT), the secreted calcitonin prohormone and acute-phase reactant in bacterial sepsis (Assicot et al., 1993; Moyer, 2012; Muller et al., 2001). PCT has many favourable properties for a diagnostic test as it is easily measured in standard hospital laboratories, is rapidly induced during infection, and has a long half-life (Dandona et al., 1994; Wacker, Prkno, Brunkhorst, & Schlattmann, 2013). Based on observations that viral infections do not stimulate PCT release (Simon, Gauvin, Amre, Saint-Louis, & Lacroix, 2004), this biomarker is well suited to differentiate bacterial from viral aetiologies of acute respiratory infection (ARI). Treatment algorithms that incorporate PCT have safely reduced antibacterial prescription by approximately 50% in ARI (Schuetz et al., 2012).

Despite these encouraging findings, there are still uncertainties around PCT use. PCT has greater specificity for bacterial infection than other biomarkers, but false positives have been described in several clinical conditions such as burns (Carsin et al., 1997) and major trauma (Mimoz et al., 1998), while undergoing cardiopulmonary bypass (Loebe, Locziewski, Brunkhorst, Harke, & Hetzer, 2000), heatstroke (Nylen, Al Arifi, Becker, Snider, & Alzeer, 1997), colonic ischaemia (Nagata, Kobayashi, Nishikimi, & Komori, 2008), acute onset Still's disease (Scire et al., 2006), and cirrhosis with ascites (Attar et al., 2014). In addition, the assay is only FDA approved to identify critically ill patients at the time of admission to the intensive care unit who may be at risk for progression to severe sepsis and septic shock. Nevertheless, PCT is a valuable addition to current diagnostic tests. It underscores the principle that measurement of the host response can differentiate bacterial from viral aetiologies.

2.2.2.1.3 Pattern recognition receptors and toll-like receptors. In addition to acute-phase reactants, there are a number of candidate biomarkers from immune response pathways. Components of the innate immune response, toll-like receptors (TLRs), and other pattern recognition receptors (PRRs) are expressed on cells that bind pathogen-associated molecular patterns (Jenner & Young, 2005). These components grant a certain degree of pathogen class specificity. For example, TLR-3 binds to the double-stranded RNA that certain viruses possess, while TLR-4 binds to lipopolysaccharide found on Gram-negative bacteria. Even more specific than PRRs, *ex vivo* studies of human peripheral blood leukocytes with bacteria have shown that pathogenic processes manipulate the host-immune response in a way that permits discrimination between pathogens (Boldrick et al., 2002; Feezor et al., 2003). Other similar candidates have been described for multiple pathogens and pathogen classes, including viral (Fjaerli et al., 2006; Jenner & Young, 2005; Ramilo et al., 2007), bacterial (Boldrick et al., 2002; Feezor et al., 2003), and fungal infections (Cortez et al., 2006; Huang et al., 2001; Zaas, Aziz, Lucas, Perfect, & Ginsburg, 2010).

2.2.2.1.4 Cytokines. Cytokines are another key component of the immune response, representing a diverse group of signalling molecules such as interleukins and interferons that impact a broad range of target cells and processes. Like PRRs, there is some pathogen class specificity. Tuberculosis and other intracellular pathogens are associated with IL-12 and IFN-gamma production as part of a macrophage and TH1 helper T cell response. Another well-characterised example within the mycobacterial family, *Mycobacterium leprae* (the causative agent of leprosy), can present in either the tuberculoid form or the lepromatous form. These two forms are associated with differential TH1 responses. As might be expected, there are also different cytokine responses. IFN-gamma, along with IFN-inducible protein-10, MCP-1, MIP-1beta, and IL-1beta, has been identified as having potential utility in diagnosing pathogenic versus asymptomatic immune response to *M. leprae* exposure (Geluk et al., 2012). In contrast, extracellular bacterial and fungal pathogens activate TH17 helper T cells, neutrophils, and monocytes mediated by IL-17 and IL-22 (Ouyang, Kolls, & Zheng, 2008). Parasitic infections represent their own pathogen class with a distinct, associated cytokine profile. For example, helminthic parasite infections are associated with IL-4, IL-5, and IL-13 production with TH2 helper T cell, mast cell, and eosinophil activation.

2.2.2.2 Unbiased

An alternative to measuring proteins that are known or suspected to be involved in the host response is an unbiased approach to discovery, which, using the techniques described earlier, aim to analyse the entire proteome. Such an approach has the potential advantage of finding novel response pathways. One example of an unbiased proteomic approach to biomarker discovery was in comparing latent tuberculosis infection (LTBI) patients to healthy controls (Zhang et al., 2014). Among other findings, 14 of the differentially expressed peptides had not previously been implicated in tuberculosis infection and included proteins such as apolipoproteins and transketolase. Three highly accurate models were also developed and evaluated in a blind validation experiment. This is just one example among many studies that have utilised unbiased approaches to achieve results that would not have been possible if only candidate biomarkers had been evaluated.

2.2.3 Metabolomic markers
2.2.3.1 Candidate
2.2.3.1.1 Lactate. Lactate, the conjugate base of lactic acid produced during anaerobic glycolysis, is one of the most commonly used biomarkers for sepsis. Elevated serum lactate levels are associated with higher mortality in ED patients with severe sepsis and shock (Glickman et al., 2010; Howell, Donnino, Clardy, Talmor, & Shapiro, 2007; Mikkelsen et al., 2009; Shapiro et al., 2005; Trzeciak et al., 2007). However, the marker is non-specific as lactate may be elevated for a variety of reasons, including impaired clearance, depressed cellular respiration secondary to insufficient oxygen tissue delivery, impaired microcirculation, and mitochondrial dysfunction (Boulos, Astiz, Barua, & Osman, 2003; Cairns, 2001; Fink, 2002; Shapiro et al., 2009; Trzeciak et al., 2007). However, lactate kinetics may be more useful than a single point-in-time value. Despite its low specificity, lactate clearance has shown improved prognostic value in sepsis (Arnold et al., 2009; Jones, 2013; Jones et al., 2010; Nguyen et al., 2004). This also highlights that biomarker kinetics can in some cases offer higher sensitivity or specificity than a single, isolated measurement.

2.2.3.2 Unbiased
The same unbiased approach to biomarker discovery described above for proteomic analyses applies to metabolomics. A recent study of sepsis patients identified metabolites associated with sepsis diagnosis and sepsis survival (Langley et al., 2013). This was done by defining a population of patients with sepsis, some of whom survived and some who did not. Following the general algorithm demonstrated in Figure 1, samples were obtained from these patients to define the metabolome. Statistical analysis identified acylcarnitines to be particularly important in discriminating survivors from non-survivors, which was validated in an independent dataset. Carnitines, synthesised from lysine and methionine, play an important role in fatty acid oxidation as transporters. When bound to long-chain fatty acids, they are termed acylcarnitines, which are reversibly transported across the mitochondrial membrane. Acylcarnitines have been implicated in experimental models of sepsis (Feingold, Wang, Moser, Shigenaga, & Grunfeld, 2008), though this study revealed their importance in a heterogeneous human population. Specifically, plasma concentrations of multiple carnitine esters differed between sepsis survivors, sepsis non-survivors, and controls with non-infectious systemic inflammatory response syndrome. These results also identified defects in fatty acid β-oxidation as possibly causally related to survival or death, representing a potential therapeutic or prognostic target in sepsis.

It remains unclear if elevations in carnitine esters are unique to sepsis non-survival or are also present in other critical illnesses in the absence of infection; a prospective study of outcomes in such cases would further inform the specificity of these observations. Hypoxia is known to cause increased plasma acylcarnitines (Valkner, Ely, Kerner, Scott, & Bieber, 1985), suggesting a plausible biological association with mitochondrial dysfunction, already known to bear an association with outcomes and survival in sepsis (Brealey et al., 2002; Carre et al., 2010).

The prominence of acylcarnitines identified through an unbiased metabolomic approach reveals the importance of energy homeostasis and metabolism in the host response to infection. These targets expand our traditional view beyond the immune system and further highlight the utility of unbiased approaches in biomarker discovery.

2.3 SYNTHESISING AND INTERPRETING BIG DATA

One of the mixed blessings of modern 'omics is the collection of vast amounts of complex, high-dimensional data on a scale previously unimaginable and not amenable to manual or traditional analytic methods. Also challenging conventional analytics is the "large p, small n" problem, where a small sample size can yield tens of thousands of data points for comparison (West, 2003). Moreover, the traditional single-threaded model of computing lacks the capacity and power to store and process such data. Instead, distributed parallel computing techniques are needed to conduct this massive data processing. The technical details of these methods are outside the scope of this text, but it is worth noting that big data methods have already been employed in other applications. Google Flu Trends is one example of how aggregate analysis of user search queries combined with geospatial data can provide complementary information about the regional incidence of influenza (Ginsberg et al., 2009).

Because these methods are relatively recent developments and may be less familiar to many microbiologists, we provide a brief overview of the commonly used approaches to big data analysis, as well as a few examples of achievable results. Additional details regarding such methods can be found in recent reviews (Aittokallio et al., 2003; Dunkler, Sanchez-Cabo, & Heinze, 2011; He & Yu, 2010).

2.3.1 Data analysis: Reducing dimensionality

Reducing dimensionality is the process of reducing the number of random variables being considered in an analysis. For example, the Affymetrix U133 Plus 2.0 microarray contains probes for over 47,000 transcripts. Using algorithms such as those described below, data from those 47,000 targets can be simplified to hundreds or even dozens of composite variables. There are many approaches to reducing data dimensionality, matching phenotypes to transcriptomic or proteomic observations, and making predictions. Some examples of such algorithms include sparse factor modelling (Carvalho et al., 2008), Bayesian constructions of the elastic net (Chen et al., 2011), sparse principal component analysis (Wu & Liu, 2013), penalised matrix decomposition (Witten, Tibshirani, & Hastie, 2009), modular transcriptional analysis (Chaussabel & Baldwin, 2014), and the molecular distance to health (MDTH) (Pankla et al., 2009). Here we highlight two approaches.

2.3.1.1 Bayesian factor regression modelling

Bayesian factor regression modelling (BFRM), also known as sparse latent factor regression modelling, simplifies analyses of complex high-dimensional data by assuming there are relatively few associations between variables (West, 2003). In the context of genomics, genes with expression patterns that correlate with each other are grouped into "factors" (sometimes also referred to as clusters) (Peng et al., 2013). This reduces the number of variables from tens of thousands of transcripts to dozens or hundreds of factors, allowing for more conventional statistical tests to be subsequently applied. Factors that can make a diagnostic differentiation based on statistical significance are identified, and then the component genes of those factors are individually assessed for significance and biological plausibility. While BFRM makes it feasible to analyse data with hundreds or thousands of variables, reduced data dimensionality always bears the risk of obscuring a significant effect of any single gene if the factor in which it is represented does not reach the same level of significance.

A study of respiratory infections used BFRM to identify a gene expression signature for acute respiratory viral infections (Zaas et al., 2009). The study challenged 56 individuals across three challenge cohorts (rhinovirus, respiratory syncytial virus (RSV), and influenza A virus), approximately half of which become symptomatic. Peripheral blood gene expression was then measured. Data from the three challenge cohorts were combined and used for BFRM analysis. That analysis identified a 30-gene classifier as the most predictive factor of symptomatic versus asymptomatic response upon viral challenge. The classifier was found to have >95% accuracy; in a validation group, the signature had a 100% accuracy distinguishing influenza A versus uninfected, as well as a 93% accuracy distinguishing influenza from pneumococcal respiratory infections.

2.3.1.2 Molecular distance to health

Intuitively, the MDTH is a value that characterises the difference in genetic expression of a given sample (e.g. infected patients) when compared to a control sample (e.g. healthy patients). First, the mean and standard deviation of expression level of each gene being measured in the control sample are calculated. Then, for each gene, the difference is calculated, both in raw expression level and standard deviation units. Finally, the sum of the standard deviations of each gene is used to calculate the MDTH, excluding genes that do not differ significantly in expression. Eliminating such genes that do not differ in expression is one way in which the MDTH achieves dimensional reduction. In addition, the summary score represented by the MDTH enables correlation with clinical metrics. However, as the score is a one-dimensional measure, there remains a risk of obscuring significant effects of any particular gene as might occur with BFRM. Another confounding factor is that changes in gene expression level in standard deviation units are weighted equally, which may not correspond well to the contribution each gene makes.

An example of using MDTH is found in a study of patients with septicaemic melioidosis, a severe disease endemic to Southeast Asia and northern Australia caused by the Gram-negative rod *Burkholderia pseudomallei*. This study used micro-arrays to profile gene transcripts and developed a diagnostic candidate biomarker signature that distinguished melioidosis infection from other causes of sepsis in patients (Pankla et al., 2009). In an intermediate stage of the analysis, a signature that distinguished septic from non-infected patients was developed first, but heterogeneity in the response within septic samples was observed. MDTH scoring was used to quantify the heterogeneity, using non-infected controls as the baseline. The score was applied to 2785 gene probes identified in an unsupervised clustering analysis. Results suggested that the heterogeneity may have been due to disease severity, as the MDTH scores correlated with worse outcomes.

2.3.2 A word of caution: Analytical biases

Analytical advances in biomarker development are accompanied by potential biases, particularly with respect to gene expression-derived classifiers. There are several suggested practices to reduce or avoid biases, which are summarised in a published case study (Lytkin, McVoy, Weitkamp, Aliferis, & Statnikov, 2011). Key points include the observation that multiple predictive signatures may exist. These various signatures are comprised by groups of genes (in the case of gene expression data) with variable overlap between signatures. Focusing on one predictive signature may provide an incomplete picture of the underlying biology, which is better represented by the collection of all predictive signatures. At the same time, "redundant" biomarkers (those that do not improve the predictive value of the model) should only be included to explicitly improve robustness, as inclusion has the potential to reduce signature reproducibility and increase manufacturing cost of the diagnostic assay.

Perhaps of greatest importance is that signatures be derived from the same population as that in which the intended test will be used. As it were, many published studies of host response to infectious disease were instead developed in comparison to healthy controls. This comparison is invaluable to understand host biology, but is not ideal for clinical assay development. The clinical distinction between healthy and sick is not subtle and requires no diagnostic testing. As such, diagnostic targets defined by the host response should be designed to make the desired clinical comparison. For example, a viral pneumonia classifier for clinical diagnostic purposes should be derived from a cohort of respiratory illness patients, rather than from a comparison of viral aetiologies to a healthy population. If the classifier is used in a population from which it was not derived, its results have the potential to mislead.

3 DISEASES

3.1 DIAGNOSTIC APPROACH

When considering disease entities, there are two commonly employed organisational structures. The first of these approaches is by syndromic presentation and is commonly used by clinicians. For example, pneumonias may present with fever, cough, and consolidation on chest X-ray. Treatment can involve respiratory support, empiric antibiotics, and for at-risk populations, intensive care. However, a pneumonia-like presentation may in actuality be due to multiple different or combined aetiologies including bacterial, viral, or non-infectious causes. Making such a determination can be one of the most challenging aspects of clinical care, especially when diagnostic testing is inadequate to support targeted, informed treatment selection.

The second conceptual approach to infectious diseases diagnosis is by aetiologic class. While initial treatment is often empiric and targeted to the most likely aetiologies, diagnostic testing can narrow or alter that treatment by identifying the specific offending pathogen or pathogen class. In the above example, a confirmatory test of viral pneumonia may lead to discontinuation of antibacterials. Other examples where disease syndromes may be due to multiple aetiologic classes include the following: sepsis (bacterial, fungal, non-infectious, or viral in the immunocompromised host); febrile neutropenia (bacterial, fungal, viral, or non-infectious); fever of unknown origin (bacterial, fungal, viral, malignant, or rheumatologic); or fever in the returning traveller (bacterial, fungal, viral, or parasitic).

This chapter focuses on diagnostic approaches and, as such, has organised these considerations by the second conceptual approach—by aetiologic class—while using specific clinical syndromes as examples.

3.2 BACTERIAL PATHOGENS

3.2.1 Sepsis

Sepsis is a common complication of severe infection with substantial associated morbidity, mortality, and cost. In the United States, the incidence of sepsis has increased from 82.7 per 100,000 in 1979 to 240.4 per 100,000 in 2000 (Martin, Mannino, Eaton, & Moss, 2003); meanwhile, the incidence of septic shock has increased from 12.6 per 100,000 in 1998 to 78 per 100,000 in 2009 (Walkey, Wiener, & Lindenauer, 2013). Management of sepsis remains challenging due to its nature, which is heterogeneous not only in causative pathogens but also with respect to the nature and severity of the host response and clinical manifestations. Efforts are ongoing to identify biomarkers for sepsis diagnosis and management.

One of the key clinical questions in managing sepsis patients includes whether or not an infection is even present; that is, distinguishing sepsis from systemic inflammatory response syndrome (SIRS). If a patient indeed has sepsis, then the question of aetiological class, or even specific organism, is important to guide the selection of antimicrobial therapy. Finally, the question of prognosis arises, with such information guiding how intensive the degree of care a patient may require, which is especially important in the appropriate allocation of scarce resources. These are all questions for which there is ongoing host biomarker research. A few examples are detailed below.

PCT has long been a marker of interest for sepsis; elevations of PCT in bacterial sepsis were reported as early as 1993 (Assicot et al., 1993). While PCT is more specific for bacterial infections than other acute-phase reactants, challenges remain in using it to distinguish bacterial from non-bacterial infections, as viral and fungal infections are known to cause PCT elevations that overlap with the range seen in systemic bacterial infection (Assicot et al., 1993). Non-infectious conditions can also result in PCT elevations, as described earlier in this chapter. While establishing different PCT thresholds has been proposed as a solution to the specificity problem, there has also been interest in serial PCT measurement as being more informative than single time-point measurements (Charles et al., 2009; McMaster et al., 2009), as was discussed for lactate. Specifically, decreases in PCT have been correlated with appropriate antibacterial therapy and improved outcomes (Charles et al., 2009).

Soluble CD14 subtype (sCD14-ST), or presepsin, is a more recently discovered biomarker that has demonstrated diagnostic and prognostic utility in sepsis. While in some cases, PCT performed similarly to or better than presepsin for differentiating bacterial from non-bacterial infections (Endo et al., 2012; Ulla et al., 2013), presepsin was generally better able to diagnose sepsis (Liu, Chen, Yin, Zhao, & Li, 2013; Ulla et al., 2013). High presepsin values have also been observed to correlate with worse outcomes during hospitalisation (Cakir Madenci et al., 2014; Masson et al., 2014; Ulla et al., 2013). Pizzolato et al. recently published a thorough review of the role of presepsin in managing sepsis (Pizzolato et al., 2014).

CD64, a marker of neutrophil activation, has also demonstrated utility in both neonatal (Lynema, Marmer, Hall, Meinzen-Derr, & Kingma, 2014) and adult settings (Davis, Olsen, Ahmad, & Bigelow, 2006). It has further demonstrated the ability to distinguish sepsis from severe sepsis and septic shock, as well as predict mortality (Livaditi et al., 2006). Proadrenomedullin has also been of recent interest as a prognostic marker, correlating with sepsis severity and mortality (Akpinar, Rollas, Alagoz, Segmen, & Sipit, 2014; Debiane et al., 2014; Oncel et al., 2012; Suberviola et al., 2012).

In addition to evaluation of candidate biomarkers, unbiased approaches to discovery have also yielded biomarkers of interest. Extensive transcriptomic work has identified a number of gene expression patterns, ranging from a 7-biomarker panel for predicting sepsis in intensive care unit (ICU) patients (Lukaszewski et al., 2008) to a 138-gene signature in peripheral blood mononuclear cells differentiating sepsis from SIRS ICU patients (Tang, McLean, Dawes, Huang, & Lin, 2009). More recently, acylcarnitines, as described earlier in the chapter, have brought to light the role of cellular energy metabolism in sepsis (Langley et al., 2013). An increasingly favoured approach is combining multiple biomarkers to improve the accuracy of diagnostic and prognostic information (Wong et al., 2014), rather than attempting to find any single ideal biomarker.

3.2.2 Tuberculosis
Diagnosis of tuberculosis (TB), caused by *Mycobacterium tuberculosis*, has many well-established algorithms, yet remains practically challenging. Clinically, *M. tuberculosis* causes both latent and active disease, both of which can result in a positive tuberculin skin test (TST). While a chest X-ray is used to distinguish active pulmonary disease from latent disease, it has poor specificity. Further complicating

diagnosis is the use of the BCG vaccine in TB-endemic areas, which commonly cause positive TST results in individuals that have neither latent nor active TB. Fortunately, the recent development of interferon-gamma release assays, such as the QuantiFERON-TB Gold test, improves the ability to distinguish individuals who have TB infection from those that have only received the BCG vaccine, though it remains an imperfect test. Immunocompromised patients may also demonstrate a false-negative TST due to a lack of appropriate immune response. While the diagnostic gold standard for active pulmonary TB remains the sputum smear and culture, it is slow and insensitive. Newly developed PCR-based tests have decreased the length of time to diagnosis but remain less sensitive than smear and culture. Moreover, there remains a relative lack of knowledge regarding the biology of a protective immune response and how to predict progression from latent to active disease.

As identified in a recent review of the immune response to TB (Berry, Blankley, Graham, Bloom, & O'Garra, 2013), proteomic analyses have found several clinically useful signatures. One proteomic study using MS generated a fingerprint that distinguished active TB from latent TB in symptomatic patients presenting to clinic (Sandhu et al., 2012). A different study, also using MS, used a machine-learning approach to construct a classifier differentiating active TB patients from non-TB patients with otherwise similar clinical presentations (Agranoff et al., 2006). Notably, the classifier was unaffected by HIV status. In addition to diagnostic proteomics, proteomics has also been used to define treatment response after 8 weeks of therapy (De Groote et al., 2013).

Metabolomic analyses have also been informative. A 20-metabolite signature distinguished patients with TB from healthy individuals (Weiner et al., 2012). Serum indoleamine 2,3-dioxygenase (IDO) activity has been found to be an independent prognostic marker, with increased activity associated with poorer prognosis (Suzuki et al., 2012). Notably, IDO catalyses the rate-limiting step in tryptophan degradation in the kynurenine pathway, suggesting a target of potential therapeutic interest.

While many of these proteomic and metabolic profiles have yet to be validated in independent cohorts, gene expression studies have demonstrated reproducible signatures. A 393-transcript signature correlated with radiographic extent of disease as well as treatment response and was validated in two independent patient populations (Berry et al., 2010). However, in subsequent studies, this signature was non-specific, demonstrating overlap with other diseases such as sarcoidosis (Dupuy & Simon, 2007; Koth et al., 2011; Maertzdorf et al., 2012) and melioidosis (Koh et al., 2013; Pankla et al., 2009). This limitation aside, the signature was found to distinguish TB from community-acquired pneumonia (CAP) and lung cancer (Bloom et al., 2013), offering new opportunities in these three different clinical syndromes whose initial manifestations can be strikingly similar. Early small-scale studies have also demonstrated a potential for transcriptional signatures to provide information on treatment response after 1–2 weeks of therapy (Cliff et al., 2013; Horne et al., 2010) rather than the current gold standard of sputum acid-fast bacteria (AFB) smear and culture at 2 months, which itself has low sensitivity for detecting relapse.

3.3 VIRAL PATHOGENS

3.3.1 Acute respiratory infection

ARI is a common diagnosis with an easily recognised clinical presentation. However, identifying the causative pathogen is difficult, and as an additional layer of complexity, there may be co-infection with multiple respiratory pathogens in some cases. Two key diagnostic questions arise for ARI. The first is to determine whether the respiratory symptoms are due to an infectious or non-infectious cause (e.g. allergic rhinitis, aspiration pneumonitis). If the symptoms are infectious, then the next question is whether the cause is bacterial or viral, which constitute the two most likely pathogen classes. Their distinction is critical in directing appropriate antimicrobial use considering that 72% of ambulatory care patients with ARI are treated with an antibacterial, even though the majority have viral aetiologies (Cantrell, Young, & Martin, 2002). This excessive and inappropriate antibiotic use contributes to the growing crisis of antimicrobial resistance.

Currently available diagnostics are limited in availability (e.g. multiplex viral PCR), turnaround time (e.g. virus-specific PCR and culture), or performance characteristics. Existing pathogen-based tests are often unable to distinguish between infection and colonisation, which can be as high as 52% in some populations (Singleton et al., 2010). Low pathogen load or poor collection of respiratory samples are also frequent causes of false-negative tests. The host response to infection, however, offers a robust and complimentary source of diagnostic information (Ramilo & Mejias, 2009).

While much of the work surrounding PCT has surrounded sepsis diagnosis, its ability to differentiate bacterial from viral aetiologies is highly valuable for ARI management. A systematic review of 14 randomly controlled trials described the use of PCT to guide initiation and duration of antibiotic treatment resulted in a reduction of the median duration of therapy from 8 to 4 days without an increase in mortality or treatment failure (Schuetz et al., 2013). A review of biomarkers for diagnosing and managing CAP (Christ-Crain & Opal, 2010) also found utility in PCT: low values of PCT make CAP very unlikely (Simon et al., 2004); increased PCT correlated with CAP severity (Muller et al., 2007); PCT kinetics were associated with infection severity; and PCT-guided therapy reduced antibiotic usage (Christ-Crain et al., 2004, 2006) without compromising clinical outcomes.

Though there is currently limited work to support the use of cytokines for ARI diagnosis, they have demonstrated some prognostic value in CAP. In one study, the IL-6/IL-10 ratio has been shown to correlate with CAP mortality (Kellum et al., 2007), where in another study, high IL-6 and IL-10 concentrations as measured at the end of hospitalisation were associated with increased risk of death after discharge (Yende et al., 2008).

A gene expression signature of 854 genes differentiating bacterial and viral ARI in humans was first published by Ramilo et al. (2007). This study included patients with influenza A, *Staphylococcus aureus*, *S. pneumoniae*, or *Escherichia coli* infection. Gene expression was measured using microarrays. Those data led to the identification of host-response genes specific to viral or bacterial infection. For example, genes in the signature included $2',5'$-oligoadenylate synthase (OAS) proteins and a Type I interferon (IFN), both related to the antiviral host response. Furthermore, a 35-gene classifier discriminated infection with influenza A virus from bacterial infection, which was validated in an independent cohort with 94.5% accuracy.

Another example of host response in viral infection (alluded to earlier in the context of BFRM analysis) derives from human viral challenge studies with RSV, influenza A virus, or rhinovirus. Zaas et al. defined gene expression signatures correlating with symptomatic versus asymptomatic response to viral challenge (Zaas et al., 2009). This 28-gene classifier correctly identified 81 of 84 (96.5%) subjects as symptomatic or asymptomatic using leave-one-out cross-validation. The classifier included genes involved in IFN signalling, OAS, and radical S-adenosyl methionine domain (RSAD2) pathways. Classifier performance was consistent across the three viral challenge experiments, highlighting the conserved nature of the host response to viral ARI. The signature was then applied to the aforementioned Ramilo study cohort, where it correctly classified 93% of individuals as viral infection or non-viral (either bacterial infection or healthy), despite several key differences including paediatric versus adult cohort, RNA procurement methodology, and original statistical analysis.

Further demonstrating pathogen-specific host response comparing viral aetiologies, a prospective observational study of children with severe RSV infection identified a 70-gene classifier that discriminated RSV from rhinovirus and influenza infections. The RSV response, when compared with rhinovirus or influenza infection, included overexpression of neutrophil-related genes, with decreased expression of lymphoid lineages and antimicrobial response genes, with the further observation that the magnitude of difference correlated with disease severity (see Figure 4). Validation in an independent cohort revealed 91% accuracy. The magnitude of these changes correlated with disease severity, suggesting prognostic utility of the classifier (Mejias et al., 2013).

miRNA, reflective of post-transcriptional regulation of gene expression, has been studied in response to influenza exposure. Human alveolar and bronchial epithelial cells exposed to influenza A virus *in vitro* showed higher levels of seven miRNAs when compared to unexposed cells (Buggele, Johnson, & Horvath, 2012). The identified miRNAs were determined to target genes such as MAPK3 and interleukin-1 receptor associated kinase 1 (IRAK1), signalling proteins involved in regulating the cellular response to infection. Several other studies developed miRNA classifiers in patients infected with 2009 H1N1 (Song et al., 2013), H7N9 (Zhu et al., 2014), and a mix of H1N1 and H3N2 influenza A strains (Tambyah et al., 2013). Despite the primary difference being influenza strain, the three classifiers exhibited no overlap in miRNAs when compared to each other or the *in vitro* study above, suggesting that methodological differences are of key importance in miRNA analysis.

3.3.2 Dengue

Dengue is a mosquito-borne flavivirus endemic to tropical regions. Upon first exposure to any one of four strains, patients typically develop dengue fever (DF). However, what makes dengue relatively unique from other infectious diseases is that upon

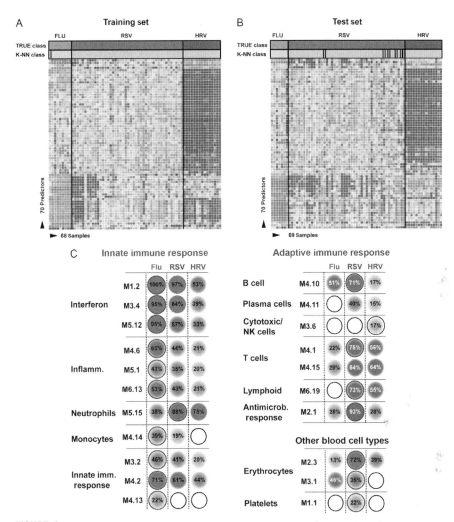

FIGURE 4

An example of gene expression profiling is demonstrated in this analysis of children with influenza, respiratory syncytial virus (RSV), and human rhinovirus (HRV) lower respiratory tract infection (LRTI) (Mejias et al., 2013). (A) A supervised learning algorithm (K-nearest neighbour) was used to identify the 70 top-ranked genes that best discriminated RSV from HRV and influenza LRTI. Using the 70 classifier genes, leave-one-out cross-validation of the training set correctly classified 67 of the 68 samples (influenza [$n=9$, green]; RSV [$n=44$, blue]; HRV [$n=15$, burgundy]) (98% accuracy). Predicted class is indicated by light-coloured rectangles. (B) The 70 classifier genes were cross-validated on an independent set of 69 new patients (test set: influenza, $n=7$; RSV, $n=47$; HRV, $n=15$). The algorithm correctly categorised 63 of the 69 new patient samples (91% accuracy). (C) Mean modular transcriptional fingerprint
(Continued)

a subsequent exposure, usually to a different strain, patients are at increased risk for dengue haemorrhagic fever (DHF) and dengue shock syndrome (DSS). Current pathogen-based diagnostic tests accurately detect viral RNA or NS1 antigen. However, the duration and extent of viraemia may impact the sensitivity of these assays.

DF has been studied extensively with respect to host response, focusing on several key questions. What makes the host response to dengue unique, especially with respect to developing DHF and DSS? One of the prevailing theories, termed the "original antigenic sin" phenomenon in 1983 (Halstead, Rojanasuphot, & Sangkawibha, 1983), proposes that the antibody response during a subsequent infection is misdirected towards the original infection strain rather than the current one, based on observations of neutralising antibody titres in eight children presenting with DSS. While this theory has yet to be definitively established or refuted, characterisations of host response in the past decade have made strides to better understanding the differences between the various clinical manifestations of dengue infection, some of which are described below.

How the host response differs in DF versus DHF or DSS is also of interest for prognostic and potentially therapeutic purposes. Host gene expression responses in 14 adults with acute DHF, when compared to healthy controls, revealed increases in certain immune-associated genes as well as increased expression of cell cycle and endoplasmic reticulum-related genes, suggesting a proliferative component to the response (Simmons et al., 2007). A 24-gene signature in the same population distinguished DSS patients from non-DSS dengue patients, revealing that underexpression of multiple type I IFN-regulated genes may be important in this form of the disease. A different study of host gene expression in DF patients (Fink et al., 2007) found upregulated expression of chemokines IP-10 and I-TAC, both part of the NF-κB immune pathway. These markers were found to be part of the immune response rather than related to viral replication, as dexamethasone administration suppressed IP-10 and I-TAC but did not affect viral replication. A number of IFN response genes were also upregulated, especially viperin, whose expression was

---

FIGURE 4—Cont'd for influenza ($n=16$ and 10 matched controls), RSV ($n=44$ and 14 matched controls), and HRV LRTI ($n=30$ and 14 matched controls). Overall, children with HRV infection demonstrated a milder activation of the innate and adaptive immune responses, compared with children with influenza or RSV infection. Children with influenza displayed a stronger activation of genes related to interferon, inflammation, monocytes, and innate immune response compared with children with RSV or HRV. Neutrophil-related genes were significantly overexpressed in RSV infection, followed by HRV infection and, at a lower level, influenza infection. On the other hand, the suppression of genes related to B cells, T cells, lymphoid lineage, and antimicrobial response observed in RSV infection was significantly milder or not present in children with influenza or HRV LRTI. The outer dark circles highlight the disease group (influenza, RSV, or HRV) with greater (red) or lower (blue) modular activation. (See the color plate.)

correlated with a small but statistically significant reduction in viral replication. Of particular relevance to novel therapeutic discovery was the upregulation of genes in the ubiquitin–proteasome pathway, which prompted testing with proteasome inhibitors. The proteasome inhibitors MG-132 and Ac-Leu-Leu-Nle-al (ALLN) were found to reduce viral replication, demonstrating how host-response profiling can lead to novel therapeutics in addition to diagnostics and prognostics.

3.4 FUNGAL PATHOGENS

3.4.1 Candidaemia

Candida bloodstream infection is a risk to hospitalised patients, those with chronic indwelling vascular catheters, and the immunocompromised. One of the challenges associated with candidaemia is that it can present very similar to bacterial sepsis, which may lead to inappropriate initial treatment with antibacterials instead of antifungals. Blood culture is insensitive and often requires days to result, but remains the usual means of diagnosis. Studies of host response are therefore appealing for their ability to identify response to pathogen. In contrast to ARI, the development of host-based diagnostics for candidaemia is still restricted to murine models rather than derived from actual human disease.

In a murine challenge study with *Candida albicans*, gene expression-derived signatures classified candidaemia with a high degree of accuracy (83–100% sensitivity and 94–98% specificity) (Zaas et al., 2010). These factors distinguished mice with candidaemia from uninfected mice and mice with *S. aureus* sepsis. Implicated genes included those known to be involved in the immune response to *Candida* infections such as interleukin-1β (IL-1β), *Nlrp3*, and genes in the *Tlr4/Myd88*, *Socs3/Il-8*, and *Irak4/Il-1* pathways. These gene expression changes were dynamic, showing evolution through early-, mid-, and late-stage disease (see Figure 5). In another murine challenge study measuring cytokine levels (Chin et al., 2014), immune responses were found to vary among target organs such as the kidney, spleen, and brain, with prognostic and therapeutic implications.

3.4.2 Aspergillosis

Aspergillus is an opportunistic mould responsible for a multitude of clinical diseases including allergic bronchopulmonary aspergillosis, invasive disease in the immunocompromised (including tracheobronchitis, rhinosinusitis, cutaneous, gastrointestinal, and disseminated aspergillosis), allergic fungal sinusitis, and pulmonary aspergillomas. Moreover, *Aspergillus* is a common colonising agent especially in individuals with structural lung disease. Distinguishing between colonisation and infection is a major clinical challenge. The host response presents an improved mechanism to discriminate between these clinical scenarios.

Data from cases of human clinical disease are not yet available. However, *in vitro* exposure of human monocytes to *Aspergillus fumigatus* conidia resulted in altered expression of 1827 genes (Cortez et al., 2006). This included host defence genes such as IL-1β, IL-8, CXCL2, CCL4, CCL3, CCL20, and opsonin long pentraxin 3, as well

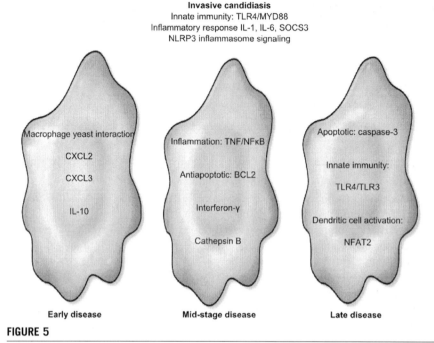

Invasive candidiasis
Innate immunity: TLR4/MYD88
Inflammatory response IL-1, IL-6, SOCS3
NLRP3 inflammasome signaling

Macrophage yeast interaction

CXCL2

CXCL3

IL-10

Inflammation: TNF/NFκB

Antiapoptotic: BCL2

Interferon-γ

Cathepsin B

Apoptotic: caspase-3

Innate immunity:

TLR4/TLR3

Dendritic cell activation:

NFAT2

Early disease   **Mid-stage disease**   **Late disease**

FIGURE 5

The host response to *Candida* bloodstream infection evolves over time. In this depiction of macrophage–yeast interactions, genes expressed in "Early disease" have established roles in early phases of an immune response. Genes expressed in "Mid-stage disease" representing worsening disease burden, revealed a more robust response involving elements of both the innate and the adaptive immune response. "Late disease", representative of the premorbid state, was characterised by upregulation of caspase-3, a protease involved in apoptosis, which is not upregulated in mid-stage disease.

From Zaas et al. (2010). Reprinted with permission from AAAS.

as genes not previously implicated in the host response, such as Hsp40, Hsp110, and connexins 26 and 30. Another study using an *in vitro* model of the human alveolar surface demonstrated a key role for dendritic cells in the immune response, as well as upregulation of some of the same genes, including CXCL2, CXCL5, CCL20, and IL-1β (Morton et al., 2014). These findings lend biological insight into *Aspergillus*–host interactions. Moreover, they offer targets for diagnostic assay development. In a murine challenge study with *S. pneumoniae*, *Pseudomonas aeruginosa*, and *A. fumigatus*, CCL4 robustly distinguished bacterial from non-bacterial infection (Evans et al., 2010). Titin (TTN) was also able to distinguish *A. fumigatus* from sham in 90–100% of the samples. Though promising, these results need to be validated in human clinical samples. In particular, defining the specificity of this response will be key as poor specificity is a particular challenge in *Aspergillus* diagnostic testing. Moreover, many patients affected by *Aspergillus* infections are immunocompromised and so peripheral blood gene expression may not be suitable or reliable as a measure of the host response. Evaluating performance characteristics in such a population will be vital to advancing a host-response diagnostic in cases of suspected *Aspergillus* infections.

3.4.3 Other fungi

There are several other clinically relevant fungal infections including Cryptococci, Mucorales, and endemic dimorphic fungi (e.g. Coccidioides, Blastomyces, and Histoplasma). There is a scarcity of published research exploring the host response as a diagnostic modality in these infections. As the path from basic research to diagnostic assay development becomes better established, host response-derived diagnostics for these pathogens will likely follow.

3.5 PARASITIC PATHOGENS

3.5.1 Malaria

Malarial infection poses a particular diagnostic challenge especially in resource-limited settings where microscopy is often unavailable. Typical clinical management of the febrile patient in malaria-endemic areas involves empiric anti-malarial treatment and if ineffective, then alternative diagnoses are explored. Moreover, the differential diagnosis is particularly broad in locations where malaria is endemic. It includes bacterial infection such as *S. pneumoniae*; viral infection such as dengue or influenza; rickettsial infection such as scrub typhus, typhus fever, and spotted fever; and other parasitic infections such as *Leishmania*. Considering how broad this differential is, pathogen identification is challenging. A host response specific to pathogen class would offer a significant advantage by narrowing the differential considerably.

Of all the parasitic infections, malaria has been the most extensively studied from a host response perspective. Proteomic analyses have identified signatures that distinguish *Plasmodium falciparum* malaria, *Plasmodium vivax* malaria, healthy controls, and leptospirosis with 95% accuracy (Ray et al., 2012). Thirty serum proteins distinguished *P. falciparum* from healthy controls, 31 proteins distinguished *P. vivax* from healthy controls, and 13 were differentially expressed when comparing the two malarial subtypes. The models also performed well in patients with low to moderate parasitaemia. Some of the identified proteins, including serum amyloid A, paraoxonase, apolipoprotein A-1 and E, haptoglobin, haemopexin, and complement C4 were proposed for further investigation based on possible prognostic utility. The advantage to protein targets is the relative ease with which ELISA assays can be generated including lateral flow assays that would be amenable to low-resource settings.

A number of studies have examined gene expression changes in malarial disease, though most focused on the underlying biology rather than diagnostic development. Krupka et al. reported a longitudinal gene profiling study of children with severe malaria, five of whom had a milder malarial reinfection some time later (Krupka et al., 2012). This opportunity to understand disease severity in these five children led to the identification of a 68-gene signature distinguished the mild and severe states. Whereas the signature was derived from a very small cohort, the genes identified were primarily related to interferon or T cell biology, suggesting biological plausibility. Six genes, including thymidine kinase, were associated with severe malaria. These findings also present potential therapeutic targets given that their expression is related to disease severity.

3.5.2 Leishmaniasis

Caused by any number of protozoa in the *Leishmania* genera, this disease is transmitted by the sandfly and classified into cutaneous (CL) or visceral leishmaniasis (VL). Leishmaniasis is diagnosed by a combination of histology, culture, and PCR of skin lesions or affected organs. Serology does not play a role in diagnosis of CL, but can be used for VL, though tests tend to have low specificity for several reasons. Poor test performance can be due to cross-reactivity with other parasite diseases (Reed et al., 1990); positivity with subclinical infection (Bern et al., 2007; Sundar et al., 2006); or continued positive serology after curative treatment (Bern et al., 2007; Houghton et al., 1998).

In a study examining the human macrophage response to *Leishmania* (Lefevre et al., 2013), changes in expression of three C-type lectin receptors (CLRs) were observed; the involved CLRs were identified as part of the antimicrobial response and as a potential target of diagnostic and therapeutic value. Another study employed a proteomic and glycoproteomic comparison between healthy controls and early diagnosed VL patients (Bag et al., 2014), finding 10 differentially expressed proteins, suspected to be involved in parasite survival and with potential utility as early diagnostic markers. However, a major limitation of these studies is the comparison to healthy controls which is not so clinically meaningful, as discussed in Section 2.3.2.

CONCLUSIONS

The currently available infectious disease diagnostics are limited in a number of ways. Pathogen detection assays have highly variable sensitivity and specificity depending on the test in question. Even correct test results rarely provide prognostic information due to high inter-individual variability. Measuring the host response fulfils a complementary role that addresses many limitations of pathogen detection assays. The ability to detect infection at the pathogen class level can guide antibiotic initiation and inform which type of antibiotic is needed. Host-based diagnostics can also distinguish infection from colonisation. With ongoing technical advances, host-based diagnostics have the potential to be rapid, low-cost, and deployable in resource-constrained environments.

There are numerous strategies available to characterise the host response with the goal of developing diagnostic or prognostic assays. Biomarkers may arise as candidates from known biology. Omic technologies have also created opportunities to identify previously unsuspected, unbiased biomarkers as well as biomarker panels drawing from multiple omic analyses. Examples include transcriptomics, proteomics, and metabolomics, which study the respective changes in their namesake. Infection response signatures are developed into classifiers that are then validated in independent cohorts. Industry partnerships are necessary to develop these classifiers into market-ready diagnostic products, deployable in clinical settings. Such development has already begun, setting the stage for a new generation of infectious disease diagnostics.

REFERENCES

Agasti, S. S., Liong, M., Tassa, C., Chung, H. J., Shaw, S. Y., Lee, H., et al. (2012). Supramolecular host-guest interaction for labeling and detection of cellular biomarkers. *Angewandte Chemie International Edition in English*, *51*(2), 450–454.

Agranoff, D., Fernandez-Reyes, D., Papadopoulos, M. C., Rojas, S. A., Herbster, M., Loosemore, A., et al. (2006). Identification of diagnostic markers for tuberculosis by proteomic fingerprinting of serum. *Lancet*, *368*(9540), 1012–1021.

Aittokallio, T., Kurki, M., Nevalainen, O., Nikula, T., West, A., & Lahesmaa, R. (2003). Computational strategies for analyzing data in gene expression microarray experiments. *Journal of Bioinformatics and Computational Biology*, *1*(3), 541–586.

Akpinar, S., Rollas, K., Alagoz, A., Segmen, F., & Sipit, T. (2014). Performance evaluation of MR-proadrenomedullin and other scoring systems in severe sepsis with pneumonia. *Journal of Thoracic Disease*, *6*(7), 921–929.

Arnold, R. C., Shapiro, N. I., Jones, A. E., Schorr, C., Pope, J., Casner, E., et al. (2009). Multicenter study of early lactate clearance as a determinant of survival in patients with presumed sepsis. *Shock*, *32*(1), 35–39.

Assicot, M., Gendrel, D., Carsin, H., Raymond, J., Guilbaud, J., & Bohuon, C. (1993). High serum procalcitonin concentrations in patients with sepsis and infection. *Lancet*, *341*(8844), 515–518.

Attar, B. M., Moore, C. M., George, M., Ion-Nedelcu, N., Turbay, R., Zachariah, A., et al. (2014). Procalcitonin, and cytokines document a dynamic inflammatory state in non-infected cirrhotic patients with ascites. *World Journal of Gastroenterology*, *20*(9), 2374–2382.

Avriel, A., Paryente Wiessman, M., Almog, Y., Perl, Y., Novack, V., Galante, O., et al. (2014). Admission cell free DNA levels predict 28-day mortality in patients with severe sepsis in intensive care. *PLoS One*, *9*(6), e100514.

Bag, A. K., Saha, S., Sundar, S., Saha, B., Chakrabarti, A., & Mandal, C. (2014). Comparative proteomics and glycoproteomics of plasma proteins in Indian visceral leishmaniasis. *Proteome Science*, *12*(1), 48.

Bartel, D. P. (2004). MicroRNAs: Genomics, biogenesis, mechanism, and function. *Cell*, *116*(2), 281–297.

Bern, C., Haque, R., Chowdhury, R., Ali, M., Kurkjian, K. M., Vaz, L., et al. (2007). The epidemiology of visceral leishmaniasis and asymptomatic leishmanial infection in a highly endemic Bangladeshi village. *The American Journal of Tropical Medicine and Hygiene*, *76*(5), 909–914.

Berry, M. P., Blankley, S., Graham, C. M., Bloom, C. I., & O'Garra, A. (2013). Systems approaches to studying the immune response in tuberculosis. *Current Opinion in Immunology*, *25*(5), 579–587.

Berry, M. P., Graham, C. M., McNab, F. W., Xu, Z., Bloch, S. A., Oni, T., et al. (2010). An interferon-inducible neutrophil-driven blood transcriptional signature in human tuberculosis. *Nature, 466*(7309), 973–977.

Bloom, C. I., Graham, C. M., Berry, M. P., Rozakeas, F., Redford, P. S., Wang, Y., et al. (2013). Transcriptional blood signatures distinguish pulmonary tuberculosis, pulmonary sarcoidosis, pneumonias and lung cancers. *PLoS One, 8*(8), e70630.

Boldrick, J. C., Alizadeh, A. A., Diehn, M., Dudoit, S., Liu, C. L., Belcher, C. E., et al. (2002). Stereotyped and specific gene expression programs in human innate immune responses to bacteria. *Proceedings of the National Academy of Sciences of the United States of America, 99*(2), 972–977.

Boulos, M., Astiz, M. E., Barua, R. S., & Osman, M. (2003). Impaired mitochondrial function induced by serum from septic shock patients is attenuated by inhibition of nitric oxide synthase and poly(ADP-ribose) synthase. *Critical Care Medicine, 31*(2), 353–358.

Brealey, D., Brand, M., Hargreaves, I., Heales, S., Land, J., Smolenski, R., et al. (2002). Association between mitochondrial dysfunction and severity and outcome of septic shock. *Lancet, 360*(9328), 219–223.

Buggele, W. A., Johnson, K. E., & Horvath, C. M. (2012). Influenza A virus infection of human respiratory cells induces primary microRNA expression. *The Journal of Biological Chemistry, 287*(37), 31027–31040.

Cairns, C. B. (2001). Rude unhinging of the machinery of life: Metabolic approaches to hemorrhagic shock. *Current Opinion in Critical Care, 7*(6), 437–443.

Cakir Madenci, O., Yakupoglu, S., Benzonana, N., Yucel, N., Akbaba, D., & Orcun Kaptanagasi, A. (2014). Evaluation of soluble CD14 subtype (presepsin) in burn sepsis. *Burns, 40*(4), 664–669.

Cantrell, R., Young, A. F., & Martin, B. C. (2002). Antibiotic prescribing in ambulatory care settings for adults with colds, upper respiratory tract infections, and bronchitis. *Clinical Therapeutics, 24*(1), 170–182.

Carin, L., Hero, A., 3rd, Lucas, J., Dunson, D., Chen, M., Henao, R., et al. (2012). High-dimensional longitudinal genomic data: An analysis used for monitoring viral infections. *IEEE Signal Processing Magazine, 29*(1), 108–123.

Carre, J. E., Orban, J. C., Re, L., Felsmann, K., Iffert, W., Bauer, M., et al. (2010). Survival in critical illness is associated with early activation of mitochondrial biogenesis. *American Journal of Respiratory and Critical Care Medicine, 182*(6), 745–751.

Carsin, H., Assicot, M., Feger, F., Roy, O., Pennacino, I., Le Bever, H., et al. (1997). Evolution and significance of circulating procalcitonin levels compared with IL-6, TNF alpha and endotoxin levels early after thermal injury. *Burns, 23*(3), 218–224.

Carvalho, C. M., Chang, J., Lucas, J. E., Nevins, J. R., Wang, Q., & West, M. (2008). High-dimensional sparse factor modeling: Applications in gene expression genomics. *Journal of the American Statistical Association, 103*(484), 1438–1456.

Charles, P. E., Tinel, C., Barbar, S., Aho, S., Prin, S., Doise, J. M., et al. (2009). Procalcitonin kinetics within the first days of sepsis: Relationship with the appropriateness of antibiotic therapy and the outcome. *Critical Care, 13*(2), R38.

Chaussabel, D., & Baldwin, N. (2014). Democratizing systems immunology with modular transcriptional repertoire analyses. *Nature Reviews. Immunology, 14*(4), 271–280.

Chaussabel, D., Pascual, V., & Banchereau, J. (2010). Assessing the human immune system through blood transcriptomics. *BMC Biology, 8*, 84.

Chen, M., Carlson, D., Zaas, A., Woods, C. W., Ginsburg, G. S., Hero, A., 3rd, et al. (2011). Detection of viruses via statistical gene expression analysis. *IEEE Transactions on Biomedical Engineering*, *58*(3), 468–479.

Chin, V. K., Foong, K. J., Maha, A., Rusliza, B., Norhafizah, M., & Chong, P. P. (2014). Early expression of local cytokines during systemic infection in a murine intravenous challenge model. *Biomedical Reports*, *2*(6), 869–874.

Christ-Crain, M., Jaccard-Stolz, D., Bingisser, R., Gencay, M. M., Huber, P. R., Tamm, M., et al. (2004). Effect of procalcitonin-guided treatment on antibiotic use and outcome in lower respiratory tract infections: Cluster-randomised, single-blinded intervention trial. *Lancet*, *363*(9409), 600–607.

Christ-Crain, M., & Opal, S. M. (2010). Clinical review: The role of biomarkers in the diagnosis and management of community-acquired pneumonia. *Critical Care*, *14*(1), 203.

Christ-Crain, M., Stolz, D., Bingisser, R., Muller, C., Miedinger, D., Huber, P. R., et al. (2006). Procalcitonin guidance of antibiotic therapy in community-acquired pneumonia: A randomized trial. *American Journal of Respiratory and Critical Care Medicine*, *174*(1), 84–93.

Chu, Y., & Corey, D. R. (2012). RNA sequencing: Platform selection, experimental design, and data interpretation. *Nucleic Acid Therapeutics*, *22*(4), 271–274.

Chung, H. J., Castro, C. M., Im, H., Lee, H., & Weissleder, R. (2013). A magneto-DNA nanoparticle system for rapid detection and phenotyping of bacteria. *Nature Nanotechnology*, *8*(5), 369–375.

Cliff, J. M., Lee, J. S., Constantinou, N., Cho, J. E., Clark, T. G., Ronacher, K., et al. (2013). Distinct phases of blood gene expression pattern through tuberculosis treatment reflect modulation of the humoral immune response. *The Journal of Infectious Diseases*, *207*(1), 18–29.

Cortez, K. J., Lyman, C. A., Kottilil, S., Kim, H. S., Roilides, E., Yang, J., et al. (2006). Functional genomics of innate host defense molecules in normal human monocytes in response to *Aspergillus fumigatus*. *Infection and Immunity*, *74*(4), 2353–2365.

Cox, J., & Mann, M. (2011). Quantitative, high-resolution proteomics for data-driven systems biology. *Annual Review of Biochemistry*, *80*, 273–299.

Dandona, P., Nix, D., Wilson, M. F., Aljada, A., Love, J., Assicot, M., et al. (1994). Procalcitonin increase after endotoxin injection in normal subjects. *The Journal of Clinical Endocrinology and Metabolism*, *79*(6), 1605–1608.

Davis, B. H., Olsen, S. H., Ahmad, E., & Bigelow, N. C. (2006). Neutrophil CD64 is an improved indicator of infection or sepsis in emergency department patients. *Archives of Pathology & Laboratory Medicine*, *130*(5), 654–661.

Debiane, L., Hachem, R. Y., Al Wohoush, I., Shomali, W., Bahu, R. R., Jiang, Y., et al. (2014). The utility of proadrenomedullin and procalcitonin in comparison to C-reactive protein as predictors of sepsis and bloodstream infections in critically ill patients with cancer. *Critical Care Medicine*, *42*(12), 2500–2507.

De Groote, M. A., Nahid, P., Jarlsberg, L., Johnson, J. L., Weiner, M., Muzanyi, G., et al. (2013). Elucidating novel serum biomarkers associated with pulmonary tuberculosis treatment. *PLoS One*, *8*(4), e61002.

Deirmengian, C., Kardos, K., Kilmartin, P., Cameron, A., Schiller, K., & Parvizi, J. (2014). Combined measurement of synovial fluid alpha-Defensin and C-reactive protein levels: Highly accurate for diagnosing periprosthetic joint infection. *The Journal of Bone and Joint Surgery. American Volume*, *96*(17), 1439–1445.

Drain, P. K., Mayeza, L., Bartman, P., Hurtado, R., Moodley, P., Varghese, S., et al. (2014). Diagnostic accuracy and clinical role of rapid C-reactive protein testing in HIV-infected individuals with presumed tuberculosis in South Africa. *The International Journal of Tuberculosis and Lung Disease*, *18*(1), 20–26.

Dunkler, D., Sanchez-Cabo, F., & Heinze, G. (2011). Statistical analysis principles for omics data. *Methods in Molecular Biology*, *719*, 113–131.

Dupuy, A., & Simon, R. M. (2007). Critical review of published microarray studies for cancer outcome and guidelines on statistical analysis and reporting. *Journal of the National Cancer Institute*, *99*(2), 147–157.

Endo, S., Suzuki, Y., Takahashi, G., Shozushima, T., Ishikura, H., Murai, A., et al. (2012). Usefulness of presepsin in the diagnosis of sepsis in a multicenter prospective study. *Journal of Infection and Chemotherapy*, *18*(6), 891–897.

Evans, S. E., Tuvim, M. J., Zhang, J., Larson, D. T., Garcia, C. D., Martinez-Pro, S., et al. (2010). Host lung gene expression patterns predict infectious etiology in a mouse model of pneumonia. *Respiratory Research*, *11*, 101.

Feezor, R. J., Oberholzer, C., Baker, H. V., Novick, D., Rubinstein, M., Moldawer, L. L., et al. (2003). Molecular characterization of the acute inflammatory response to infections with gram-negative versus gram-positive bacteria. *Infection and Immunity*, *71*(10), 5803–5813.

Feingold, K. R., Wang, Y., Moser, A., Shigenaga, J. K., & Grunfeld, C. (2008). LPS decreases fatty acid oxidation and nuclear hormone receptors in the kidney. *Journal of Lipid Research*, *49*(10), 2179–2187.

Fink, M. P. (2002). Bench-to-bedside review: Cytopathic hypoxia. *Critical Care*, *6*(6), 491–499.

Fink, J., Gu, F., Ling, L., Tolfvenstam, T., Olfat, F., Chin, K. C., et al. (2007). Host gene expression profiling of dengue virus infection in cell lines and patients. *PLoS Neglected Tropical Diseases*, *1*(2), e86.

Fjaerli, H. O., Bukholm, G., Krog, A., Skjaeret, C., Holden, M., & Nakstad, B. (2006). Whole blood gene expression in infants with respiratory syncytial virus bronchiolitis. *BMC Infectious Diseases*, *6*, 175.

Forsblom, E., Aittoniemi, J., Ruotsalainen, E., Helmijoki, V., Huttunen, R., Jylhava, J., et al. (2014). High cell-free DNA predicts fatal outcome among *Staphylococcus aureus* bacteraemia patients with intensive care unit treatment. *PLoS One*, *9*(2), e87741.

Geluk, A., Bobosha, K., van der Ploeg-van Schip, J. J., Spencer, J. S., Banu, S., Martins, M. V., et al. (2012). New biomarkers with relevance to leprosy diagnosis applicable in areas hyperendemic for leprosy. *Journal of Immunology*, *188*(10), 4782–4791.

Gilad, S., Meiri, E., Yogev, Y., Benjamin, S., Lebanony, D., Yerushalmi, N., et al. (2008). Serum microRNAs are promising novel biomarkers. *PLoS One*, *3*(9), e3148.

Ginsberg, J., Mohebbi, M. H., Patel, R. S., Brammer, L., Smolinski, M. S., & Brilliant, L. (2009). Detecting influenza epidemics using search engine query data. *Nature*, *457*(7232), 1012–1014.

Glickman, S. W., Cairns, C. B., Otero, R. M., Woods, C. W., Tsalik, E. L., Langley, R. J., et al. (2010). Disease progression in hemodynamically stable patients presenting to the emergency department with sepsis. *Academic Emergency Medicine*, *17*(4), 383–390.

Halstead, S. B., Rojanasuphot, S., & Sangkawibha, N. (1983). Original antigenic sin in dengue. *The American Journal of Tropical Medicine and Hygiene*, *32*(1), 154–156.

Hansen, K. D., Brenner, S. E., & Dudoit, S. (2010). Biases in illumina transcriptome sequencing caused by random hexamer priming. *Nucleic Acids Research*, *38*(12), e131.

He, Z., & Yu, W. (2010). Stable feature selection for biomarker discovery. *Computational Biology and Chemistry*, *34*(4), 215–225.

Horne, D. J., Royce, S. E., Gooze, L., Narita, M., Hopewell, P. C., Nahid, P., et al. (2010). Sputum monitoring during tuberculosis treatment for predicting outcome: Systematic review and meta-analysis. *The Lancet Infectious Diseases*, *10*(6), 387–394.

Houghton, R. L., Petrescu, M., Benson, D. R., Skeiky, Y. A., Scalone, A., Badaro, R., et al. (1998). A cloned antigen (recombinant K39) of Leishmania chagasi diagnostic for visceral leishmaniasis in human immunodeficiency virus type 1 patients and a prognostic indicator for monitoring patients undergoing drug therapy. *The Journal of Infectious Diseases*, *177*(5), 1339–1344.

Howell, M. D., Donnino, M., Clardy, P., Talmor, D., & Shapiro, N. I. (2007). Occult hypoperfusion and mortality in patients with suspected infection. *Intensive Care Medicine*, *33*(11), 1892–1899.

Huang, Q., Liu, D., Majewski, P., Schulte, L. C., Korn, J. M., Young, R. A., et al. (2001). The plasticity of dendritic cell responses to pathogens and their components. *Science*, *294*(5543), 870–875.

Huttunen, R., Kuparinen, T., Jylhava, J., Aittoniemi, J., Vuento, R., Huhtala, H., et al. (2011). Fatal outcome in bacteremia is characterized by high plasma cell free DNA concentration and apoptotic DNA fragmentation: A prospective cohort study. *PLoS One*, *6*(7), e21700.

Jahr, S., Hentze, H., Englisch, S., Hardt, D., Fackelmayer, F. O., Hesch, R. D., et al. (2001). DNA fragments in the blood plasma of cancer patients: Quantitations and evidence for their origin from apoptotic and necrotic cells. *Cancer Research*, *61*(4), 1659–1665.

Jenner, R. G., & Young, R. A. (2005). Insights into host responses against pathogens from transcriptional profiling. *Nature Reviews. Microbiology*, *3*(4), 281–294.

Jones, A. E. (2013). Lactate clearance for assessing response to resuscitation in severe sepsis. *Academic Emergency Medicine*, *20*(8), 844–847.

Jones, A. E., Shapiro, N. I., Trzeciak, S., Arnold, R. C., Claremont, H. A., Kline, J. A., et al. (2010). Lactate clearance vs central venous oxygen saturation as goals of early sepsis therapy: A randomized clinical trial. *JAMA*, *303*(8), 739–746.

Kato-Hayashi, N., Leonardo, L. R., Arevalo, N. L., Tagum, M. N., Apin, J., Agsolid, L. M., et al. (2014). Detection of active schistosome infection by cell-free circulating DNA of *Schistosoma japonicum* in highly endemic areas in Sorsogon Province, the Philippines. *Acta Tropica*, *141*(Pt B), 178–183.

Kellum, J. A., Kong, L., Fink, M. P., Weissfeld, L. A., Yealy, D. M., Pinsky, M. R., et al. (2007). Understanding the inflammatory cytokine response in pneumonia and sepsis: Results of the genetic and inflammatory markers of sepsis (GenIMS) study. *Archives of Internal Medicine*, *167*(15), 1655–1663.

Koh, G. C., Schreiber, M. F., Bautista, R., Maude, R. R., Dunachie, S., Limmathurotsakul, D., et al. (2013). Host responses to melioidosis and tuberculosis are both dominated by interferon-mediated signaling. *PLoS One*, *8*(1), e54961.

Koth, L. L., Solberg, O. D., Peng, J. C., Bhakta, N. R., Nguyen, C. P., & Woodruff, P. G. (2011). Sarcoidosis blood transcriptome reflects lung inflammation and overlaps with tuberculosis. *American Journal of Respiratory and Critical Care Medicine*, *184*(10), 1153–1163.

Krupka, M., Seydel, K., Feintuch, C. M., Yee, K., Kim, R., Lin, C. Y., et al. (2012). Mild *Plasmodium falciparum* malaria following an episode of severe malaria is associated with induction of the interferon pathway in Malawian children. *Infection and Immunity*, *80*(3), 1150–1155.

Langley, R. J., Tsalik, E. L., van Velkinburgh, J. C., Glickman, S. W., Rice, B. J., Wang, C., et al. (2013). An integrated clinico-metabolomic model improves prediction of death in sepsis. *Science Translational Medicine, 5*(195), 195ra195.

Lefevre, L., Lugo-Villarino, G., Meunier, E., Valentin, A., Olagnier, D., Authier, H., et al. (2013). The C-type lectin receptors dectin-1, MR, and SIGNR3 contribute both positively and negatively to the macrophage response to *Leishmania infantum*. *Immunity, 38*(5), 1038–1049.

Ling, Y., Pong, T., Vassiliou, C. C., Huang, P. L., & Cima, M. J. (2011). Implantable magnetic relaxation sensors measure cumulative exposure to cardiac biomarkers. *Nature Biotechnology, 29*(3), 273–277.

Liong, M., Hoang, A. N., Chung, J., Gural, N., Ford, C. B., Min, C., et al. (2013). Magnetic barcode assay for genetic detection of pathogens. *Nature Communications, 4*, 1752.

Liu, B., Chen, Y. X., Yin, Q., Zhao, Y. Z., & Li, C. S. (2013). Diagnostic value and prognostic evaluation of Presepsin for sepsis in an emergency department. *Critical Care, 17*(5), R244.

Livaditi, O., Kotanidou, A., Psarra, A., Dimopoulou, I., Sotiropoulou, C., Augustatou, K., et al. (2006). Neutrophil CD64 expression and serum IL-8: Sensitive early markers of severity and outcome in sepsis. *Cytokine, 36*(5–6), 283.

Loebe, M., Locziewski, S., Brunkhorst, F. M., Harke, C., & Hetzer, R. (2000). Procalcitonin in patients undergoing cardiopulmonary bypass in open heart surgery-first results of the Procalcitonin in Heart Surgery study (ProHearts). *Intensive Care Medicine, 26*(Suppl. 2), S193–S198.

Lukaszewski, R. A., Yates, A. M., Jackson, M. C., Swingler, K., Scherer, J. M., Simpson, A. J., et al. (2008). Presymptomatic prediction of sepsis in intensive care unit patients. *Clinical and Vaccine Immunology, 15*(7), 1089–1094.

Lutz, N. W., Beraud, E., & Cozzone, P. J. (2014). Metabolomic analysis of rat brain by high resolution nuclear magnetic resonance spectroscopy of tissue extracts. *Journal of Visualized Experiments*, (91), 51829.

Lynema, S., Marmer, D., Hall, E. S., Meinzen-Derr, J., & Kingma, P. S. (2014). Neutrophil CD64 as a diagnostic marker of sepsis: Impact on neonatal care. *American Journal of Perinatology, 32*(4), 331–336.

Lytkin, N. I., McVoy, L., Weitkamp, J. H., Aliferis, C. F., & Statnikov, A. (2011). Expanding the understanding of biases in development of clinical-grade molecular signatures: A case study in acute respiratory viral infections. *PLoS One, 6*(6), e20662.

Maertzdorf, J., Weiner, J., 3rd, Mollenkopf, H. J., Network, T. B., Bauer, T., Prasse, A., et al. (2012). Common patterns and disease-related signatures in tuberculosis and sarcoidosis. *Proceedings of the National Academy of Sciences of the United States of America, 109*(20), 7853–7858.

Martin, G. S., Mannino, D. M., Eaton, S., & Moss, M. (2003). The epidemiology of sepsis in the United States from 1979 through 2000. *The New England Journal of Medicine, 348*(16), 1546–1554.

Masson, S., Caironi, P., Spanuth, E., Thomae, R., Panigada, M., Sangiorgi, G., et al. (2014). Presepsin (soluble CD14 subtype) and procalcitonin levels for mortality prediction in sepsis: Data from the albumin Italian outcome sepsis trial. *Critical Care, 18*(1), R6.

McMaster, P., Park, D. Y., Shann, F., Cochrane, A., Morris, K., Gray, J., et al. (2009). Procalcitonin versus C-reactive protein and immature-to-total neutrophil ratio as markers of infection after cardiopulmonary bypass in children. *Pediatric Critical Care Medicine, 10*(2), 217–221.

Mejias, A., Dimo, B., Suarez, N. M., Garcia, C., Suarez-Arrabal, M. C., Jartti, T., et al. (2013). Whole blood gene expression profiles to assess pathogenesis and disease severity in infants with respiratory syncytial virus infection. *PLoS Medicine, 10*(11), e1001549.

Mikkelsen, M. E., Miltiades, A. N., Gaieski, D. F., Goyal, M., Fuchs, B. D., Shah, C. V., et al. (2009). Serum lactate is associated with mortality in severe sepsis independent of organ failure and shock. *Critical Care Medicine, 37*(5), 1670–1677.

Mimoz, O., Benoist, J. F., Edouard, A. R., Assicot, M., Bohuon, C., & Samii, K. (1998). Procalcitonin and C-reactive protein during the early posttraumatic systemic inflammatory response syndrome. *Intensive Care Medicine, 24*(2), 185–188.

Mitchell, P. S., Parkin, R. K., Kroh, E. M., Fritz, B. R., Wyman, S. K., Pogosova-Agadjanyan, E. L., et al. (2008). Circulating microRNAs as stable blood-based markers for cancer detection. *Proceedings of the National Academy of Sciences of the United States of America, 105*(30), 10513–10518.

Morton, C. O., Fliesser, M., Dittrich, M., Mueller, T., Bauer, R., Kneitz, S., et al. (2014). Gene expression profiles of human dendritic cells interacting with *Aspergillus fumigatus* in a bilayer model of the alveolar epithelium/endothelium interface. *PLoS One, 9*(5), e98279.

Moyer, M. W. (2012). New biomarkers sought for improving sepsis management and care. *Nature Medicine, 18*(7), 999.

Muller, B., Harbarth, S., Stolz, D., Bingisser, R., Mueller, C., Leuppi, J., et al. (2007). Diagnostic and prognostic accuracy of clinical and laboratory parameters in community-acquired pneumonia. *BMC Infectious Diseases, 7*(1), 10.

Muller, B., White, J. C., Nylen, E. S., Snider, R. H., Becker, K. L., & Habener, J. F. (2001). Ubiquitous expression of the calcitonin-i gene in multiple tissues in response to sepsis. *The Journal of Clinical Endocrinology and Metabolism, 86*(1), 396–404.

Nagata, J., Kobayashi, M., Nishikimi, N., & Komori, K. (2008). Serum procalcitonin (PCT) as a negative screening test for colonic ischemia after open abdominal aortic surgery. *European Journal of Vascular and Endovascular Surgery, 35*(6), 694–697.

Nason, G. J., Barry, B. D., Obinwa, O., McMacken, E., Rajaretnam, N. S., & Neary, P. C. (2014). Early rise in C-reactive protein is a marker for infective complications in laparoscopic colorectal surgery. *Surgical Laparoscopy, Endoscopy & Percutaneous Techniques, 24*(1), 57–61.

Neely, L. A., Audeh, M., Phung, N. A., Min, M., Suchocki, A., Plourde, D., et al. (2013). T2 magnetic resonance enables nanoparticle-mediated rapid detection of candidemia in whole blood. *Science Translational Medicine, 5*(182), 182ra154.

Nguyen, H. B., Rivers, E. P., Knoblich, B. P., Jacobsen, G., Muzzin, A., Ressler, J. A., et al. (2004). Early lactate clearance is associated with improved outcome in severe sepsis and septic shock. *Critical Care Medicine, 32*(8), 1637–1642.

Nylen, E. S., Al Arifi, A., Becker, K. L., Snider, R. H., Jr., & Alzeer, A. (1997). Effect of classic heatstroke on serum procalcitonin. *Critical Care Medicine, 25*(8), 1362–1365.

Oncel, M. Y., Dilmen, U., Erdeve, O., Ozdemir, R., Calisici, E., Yurttutan, S., et al. (2012). Proadrenomedullin as a prognostic marker in neonatal sepsis. *Pediatric Research, 72*(5), 507–512.

Oshlack, A., & Wakefield, M. J. (2009). Transcript length bias in RNA-seq data confounds systems biology. *Biology Direct, 4*, 14.

Ouyang, W., Kolls, J. K., & Zheng, Y. (2008). The biological functions of T helper 17 cell effector cytokines in inflammation. *Immunity, 28*(4), 454–467.

Pankla, R., Buddhisa, S., Berry, M., Blankenship, D. M., Bancroft, G. J., Banchereau, J., et al. (2009). Genomic transcriptional profiling identifies a candidate blood biomarker signature for the diagnosis of septicemic melioidosis. *Genome Biology, 10*(11), R127.

Peng, B., Zhu, D., Ander, B. P., Zhang, X., Xue, F., Sharp, F. R., et al. (2013). An integrative framework for Bayesian variable selection with informative priors for identifying genes and pathways. *PLoS One, 8*(7), e67672.

Pizzolato, E., Ulla, M., Galluzzo, C., Lucchiari, M., Manetta, T., Lupia, E., et al. (2014). Role of presepsin for the evaluation of sepsis in the emergency department. *Clinical Chemistry and Laboratory Medicine*, *52*(10), 1395–1400.

Ramilo, O., Allman, W., Chung, W., Mejias, A., Ardura, M., Glaser, C., et al. (2007). Gene expression patterns in blood leukocytes discriminate patients with acute infections. *Blood*, *109*(5), 2066–2077.

Ramilo, O., & Mejias, A. (2009). Shifting the paradigm: Host gene signatures for diagnosis of infectious diseases. *Cell Host & Microbe*, *6*(3), 199–200.

Ray, S., Renu, D., Srivastava, R., Gollapalli, K., Taur, S., Jhaveri, T., et al. (2012). Proteomic investigation of falciparum and vivax malaria for identification of surrogate protein markers. *PLoS One*, *7*(8), e41751.

Reed, S. G., Shreffler, W. G., Burns, J. M., Jr., Scott, J. M., Orge Mda, G., Ghalib, H. W., et al. (1990). An improved serodiagnostic procedure for visceral leishmaniasis. *The American Journal of Tropical Medicine and Hygiene*, *43*(6), 632–639.

Roberts, A., Trapnell, C., Donaghey, J., Rinn, J. L., & Pachter, L. (2011). Improving RNA-Seq expression estimates by correcting for fragment bias. *Genome Biology*, *12*(3), R22.

Sandhu, G., Battaglia, F., Ely, B. K., Athanasakis, D., Montoya, R., Valencia, T., et al. (2012). Discriminating active from latent tuberculosis in patients presenting to community clinics. *PLoS One*, *7*(5), e38080.

Saukkonen, K., Lakkisto, P., Pettila, V., Varpula, M., Karlsson, S., Ruokonen, E., et al. (2008). Cell-free plasma DNA as a predictor of outcome in severe sepsis and septic shock. *Clinical Chemistry*, *54*(6), 1000–1007.

Schuetz, P., Muller, B., Christ-Crain, M., Stolz, D., Tamm, M., Bouadma, L., et al. (2012). Procalcitonin to initiate or discontinue antibiotics in acute respiratory tract infections. *Cochrane Database of Systematic Reviews*, *9*, CD007498.

Schuetz, P., Muller, B., Christ-Crain, M., Stolz, D., Tamm, M., Bouadma, L., et al. (2013). Procalcitonin to initiate or discontinue antibiotics in acute respiratory tract infections. *Evidence-Based Child Health*, *8*(4), 1297–1371.

Schulze, W. X., & Usadel, B. (2010). Quantitation in mass-spectrometry-based proteomics. *Annual Review of Plant Biology*, *61*, 491–516.

Scire, C. A., Cavagna, L., Perotti, C., Bruschi, E., Caporali, R., & Montecucco, C. (2006). Diagnostic value of procalcitonin measurement in febrile patients with systemic autoimmune diseases. *Clinical and Experimental Rheumatology*, *24*(2), 123–128.

Seok, J., Warren, H. S., Cuenca, A. G., Mindrinos, M. N., Baker, H. V., Xu, W., et al. (2013). Genomic responses in mouse models poorly mimic human inflammatory diseases. *Proceedings of the National Academy of Sciences of the United States of America*, *110*(9), 3507–3512.

Shapiro, N. I., Howell, M. D., Talmor, D., Nathanson, L. A., Lisbon, A., Wolfe, R. E., et al. (2005). Serum lactate as a predictor of mortality in emergency department patients with infection. *Annals of Emergency Medicine*, *45*(5), 524–528.

Shapiro, N. I., Trzeciak, S., Hollander, J. E., Birkhahn, R., Otero, R., Osborn, T. M., et al. (2009). A prospective, multicenter derivation of a biomarker panel to assess risk of organ dysfunction, shock, and death in emergency department patients with suspected sepsis. *Critical Care Medicine*, *37*(1), 96–104.

Simmons, C. P., Popper, S., Dolocek, C., Chau, T. N., Griffiths, M., Dung, N. T., et al. (2007). Patterns of host genome-wide gene transcript abundance in the peripheral blood of patients with acute dengue hemorrhagic fever. *The Journal of Infectious Diseases*, *195*(8), 1097–1107.

Simon, L., Gauvin, F., Amre, D. K., Saint-Louis, P., & Lacroix, J. (2004). Serum procalcitonin and C-reactive protein levels as markers of bacterial infection: A systematic review and meta-analysis. *Clinical Infectious Diseases*, *39*(2), 206–217.

Singleton, R. J., Bulkow, L. R., Miernyk, K., DeByle, C., Pruitt, L., Hummel, K. B., et al. (2010). Viral respiratory infections in hospitalized and community control children in Alaska. *Journal of Medical Virology*, *82*(7), 1282–1290.

Song, H., Wang, Q., Guo, Y., Liu, S., Song, R., Gao, X., et al. (2013). Microarray analysis of microRNA expression in peripheral blood mononuclear cells of critically ill patients with influenza A (H1N1). *BMC Infectious Diseases*, *13*, 257.

Stark, A., Lin, M. F., Kheradpour, P., Pedersen, J. S., Parts, L., Carlson, J. W., et al. (2007). Discovery of functional elements in 12 Drosophila genomes using evolutionary signatures. *Nature*, *450*(7167), 219–232.

Suberviola, B., Castellanos-Ortega, A., Llorca, J., Ortiz, F., Iglesias, D., & Prieto, B. (2012). Prognostic value of proadrenomedullin in severe sepsis and septic shock patients with community-acquired pneumonia. *Swiss Medical Weekly*, *142*, w13542.

Sundar, S., Maurya, R., Singh, R. K., Bharti, K., Chakravarty, J., Parekh, A., et al. (2006). Rapid, noninvasive diagnosis of visceral leishmaniasis in India: Comparison of two immunochromatographic strip tests for detection of anti-K39 antibody. *Journal of Clinical Microbiology*, *44*(1), 251–253.

Suzuki, Y., Suda, T., Asada, K., Miwa, S., Suzuki, M., Fujie, M., et al. (2012). Serum indoleamine 2,3-dioxygenase activity predicts prognosis of pulmonary tuberculosis. *Clinical and Vaccine Immunology*, *19*(3), 436–442.

Tambyah, P. A., Sepramaniam, S., Mohamed Ali, J., Chai, S. C., Swaminathan, P., Armugam, A., et al. (2013). microRNAs in circulation are altered in response to influenza A virus infection in humans. *PLoS One*, *8*(10), e76811.

Tang, B. M., McLean, A. S., Dawes, I. W., Huang, S. J., & Lin, R. C. (2009). Gene-expression profiling of peripheral blood mononuclear cells in sepsis. *Critical Care Medicine*, *37*(3), 882–888.

Tillett, W. S., & Francis, T. (1930). Serological reactions in pneumonia with a non-protein somatic fraction of *Pneumococcus*. *The Journal of Experimental Medicine*, *52*(4), 561–571.

Trzeciak, S., Dellinger, R. P., Chansky, M. E., Arnold, R. C., Schorr, C., Milcarek, B., et al. (2007). Serum lactate as a predictor of mortality in patients with infection. *Intensive Care Medicine*, *33*(6), 970–977.

Trzeciak, S., Dellinger, R. P., Parrillo, J. E., Guglielmi, M., Bajaj, J., Abate, N. L., et al. (2007). Early microcirculatory perfusion derangements in patients with severe sepsis and septic shock: Relationship to hemodynamics, oxygen transport, and survival. *Annals of Emergency Medicine*, *49*(1), 88–98. 98 e81–82.

Ulla, M., Pizzolato, E., Lucchiari, M., Loiacono, M., Soardo, F., Forno, D., et al. (2013). Diagnostic and prognostic value of presepsin in the management of sepsis in the emergency department: A multicenter prospective study. *Critical Care*, *17*(4), R168.

Valkner, K., Ely, S., Kerner, J., Scott, J., & Bieber, L. L. (1985). Effect of hypoxia on pig heart short-chain acylcarnitines. *Comparative Biochemistry and Physiology A, Comparative Physiology*, *80*(1), 123–127.

Wacker, C., Prkno, A., Brunkhorst, F. M., & Schlattmann, P. (2013). Procalcitonin as a diagnostic marker for sepsis: A systematic review and meta-analysis. *The Lancet Infectious Diseases*, *13*(5), 426–435.

Walkey, A. J., Wiener, R. S., & Lindenauer, P. K. (2013). Utilization patterns and outcomes associated with central venous catheter in septic shock: A population-based study. *Critical Care Medicine*, *41*(6), 1450–1457.

Waugh, T. R. (1923). The blood sedimentation test: Its history, technique, nature and clinical application. *Canadian Medical Association Journal*, *13*(8), 604–608.

Weiner, J., 3rd, Parida, S. K., Maertzdorf, J., Black, G. F., Repsilber, D., Telaar, A., et al. (2012). Biomarkers of inflammation, immunosuppression and stress with active disease are revealed by metabolomic profiling of tuberculosis patients. *PLoS One*, *7*(7), e40221.

West, M. (2003). Bayesian factor regression models in the "large p, small n" paradigm. In J. Bernardo, M. Bayarri, J. Berger, A. Dawid, D. Heckerman, A. Smith, & M. West (Eds.), *Bayesian statistics: Vol. 7* (pp. 723–732): Valencia, Spain: Oxford University Press.

Witten, D. M., Tibshirani, R., & Hastie, T. (2009). A penalized matrix decomposition, with applications to sparse principal components and canonical correlation analysis. *Biostatistics*, *10*(3), 515–534.

Wong, H. R. (2012). Clinical review: Sepsis and septic shock—The potential of gene arrays. *Critical Care*, *16*(1), 204.

Wong, H. R., Lindsell, C. J., Pettila, V., Meyer, N. J., Thair, S. A., Karlsson, S., et al. (2014). A multibiomarker-based outcome risk stratification model for adult septic shock*. *Critical Care Medicine*, *42*(4), 781–789.

Wu, Y., & Liu, Y. (2013). Functional robust support vector machines for sparse and irregular longitudinal data. *Journal of Computational and Graphical Statistics*, *22*(2), 379–395.

Xu, R. Y., Liu, H. W., Liu, J. L., & Dong, J. H. (2014). Procalcitonin and C-reactive protein in urinary tract infection diagnosis. *BMC Urology*, *14*, 45.

Yende, S., D'Angelo, G., Kellum, J. A., Weissfeld, L., Fine, J., Welch, R. D., et al. (2008). Inflammatory markers at hospital discharge predict subsequent mortality after pneumonia and sepsis. *American Journal of Respiratory and Critical Care Medicine*, *177*(11), 1242–1247.

Yi, J., Zhang, Y., Zhang, Y., Ma, Y., Zhang, C., Li, Q., et al. (2014). Increased plasma cell-free DNA level during HTNV infection: Correlation with disease severity and virus load. *Viruses*, *6*(7), 2723–2734.

Zaas, A. K., Aziz, H., Lucas, J., Perfect, J. R., & Ginsburg, G. S. (2010). Blood gene expression signatures predict invasive candidiasis. *Science Translational Medicine*, *2*(21), 21ra17.

Zaas, A. K., Chen, M., Varkey, J., Veldman, T., Hero, A. O., 3rd, Lucas, J., et al. (2009). Gene expression signatures diagnose influenza and other symptomatic respiratory viral infections in humans. *Cell Host & Microbe*, *6*(3), 207–217.

Zeerleder, S., Zwart, B., Wuillemin, W. A., Aarden, L. A., Groeneveld, A. B., Caliezi, C., et al. (2003). Elevated nucleosome levels in systemic inflammation and sepsis. *Critical Care Medicine*, *31*(7), 1947–1951.

Zhang, X., Liu, F., Li, Q., Jia, H., Pan, L., Xing, A., et al. (2014). A proteomics approach to the identification of plasma biomarkers for latent tuberculosis infection. *Diagnostic Microbiology and Infectious Disease*, *79*(4), 432–437.

Zhu, Z., Qi, Y., Ge, A., Zhu, Y., Xu, K., Ji, H., et al. (2014). Comprehensive characterization of serum microRNA profile in response to the emerging avian influenza A (H7N9) virus infection in humans. *Viruses*, *6*(4), 1525–1539.

MALDI-TOF Mass Spectrometry in the Clinical Microbiology Laboratory; Beyond Identification

14

Sören Schubert*, Markus Kostrzewa[†],[1]

*Max von Pettenkofer Institute, Ludwig-Maximilians-Universität (LMU), Munich, Germany
[†]Bruker Daltonik GmbH, Bremen, Germany
[1]Corresponding author: e-mail address: markus.kostrzewa@bruker.com

1 MALDI-TOF MS FOR MICROBIAL STRAIN TYPING AND EPIDEMIOLOGY

1.1 INTRODUCTION

Today, there is little doubt that the introduction of MALDI-TOF MS has fundamentally changed the workflow in microbiology diagnostic laboratories. Only a few years after MALDI-TOF MS entered the field as a tool for identification of bacteria and fungi, its speed, accuracy, and cost effectiveness have made this diagnostic system a standard tool in many laboratories around the world. The identification of species relies on the detection of proteins, most of which belong to highly abundant ribosomal proteins. The precise determination of the respective protein masses leads to a species-specific peak pattern, which can rapidly be compared to database entries leading to species identification within a minute (Seng et al., 2010). By providing rapid results regarding the species identification, MALDI-TOF MS stimulated the desire for further fast diagnostic results such as strain typing (Murray, 2010). This is particularly true for microbial forensics or microbial contaminants in drinking water or food, where strain-level resolution is required to track origins of single strains of microorganisms. Further, the identification of virulent strains is of particular interest. But also epidemiologic investigations require rapid identification of a single strain within a single species to determine the origin and spread of an outbreak. In view of the rise of multi-drug-resistant organisms (MDRs), which are increasingly difficult to treat by antimicrobials, the spread of these organisms between patients in health-care settings becomes a major issue. Hygiene interventions with decontamination or separation of patients carrying MDRs are pertinent measures to diminish the risk of transmission of MDRs to other patients. Even more difficult

Methods in Microbiology, Volume 42, ISSN 0580-9517, http://dx.doi.org/10.1016/bs.mim.2015.04.004

is the detection of the epidemiological relationships between bacterial isolates with a "normal" antibiotic resistance pattern. Here, the antibiotic resistance and other phenotypic markers will not help identifying clonally related strains and may delay the detection of an outbreak. On the other hand, the strain typing is also needed to exclude an outbreak. For instance, coexisting "identical" bacterial isolates in a hospital, even of MDRs with identical antibiotic resistance pattern, may reflect the presence of patients with independently acquired phenotypically identical, but genetically unrelated strains.

Until recently, the typing of isolates has mainly been achieved using molecular methods such as fragment pattern-based methods such as pulsed-field gel electrophoresis (PFGE), randomly amplified polymorphic DNA-PCR-based techniques (RAPD-PCR), amplified fragment length polymorphism analysis, or multiple locus variable-number tandem repeat analysis. Sequence-based methods such as single- or multi-locus sequence analysis (MLSA, MLST, *spa*-typing) (de Bruijn, 1992; Killgore et al., 2008) have also been described. Several typing schemes based on these methods have become internationally accepted methods (Harmsen et al., 2003). Each of these approaches has been shown to have sufficiently high resolution to discern microbial strains from one another. However, the use of molecular methods for strain typing generally requires technical expertise and often expensive equipment. Most of these methods are not commercially available and can be time consuming and not suited for high-throughput investigations in possible outbreak situations. Thus, these drawbacks explain why molecular typing methods are only generally performed at reference laboratories.

The advantages of MALDI-TOF MS as a fast, cost-effective, and accurate tool for bacteria species identification together with the wide distribution of the MALDI-TOF MS apparatus in diagnostic laboratories fuelled the idea of using it for strain typing. The species identification using MALDI-TOF MS relies on the detection of protein peaks which are shared by members of the same species. Accordingly, the software algorithms underlying the species identification are developed to spot these common peaks and to ignore to some extent the discriminating peaks. The identification of distinct strains on a subspecies level requires higher resolution approaches and tends to be more challenging, as strains within a single species are quite often extremely similar, genotypically and phenotypically. Thus, several studies have reported the inability to characterise bacteria below the species level using MALDI-TOF MS, mainly due to the fact that members of a single species share remarkably similar MALDI-TOF MS spectra (Gaia, Casati, & Tonolla, 2011; Lasch et al., 2014; Rezzonico, Vogel, Duffy, & Tonolla, 2010). In contrast, a number of reports have demonstrated the suitability of MALDI-TOF MS for typing applications.

In order to perform strain typing on subspecies level, two different approaches have been established: the library-based approach and the bioinformatics-enabled approach (Sandrin, Goldstein, & Schumaker, 2013). The library-based approach is the most commonly used method for the identification of bacteria and fungi on the species level and has been adopted by several researchers to distinguish strains on subspecies level. Three different objectives can be defined for the library-based approach: (i) strain categorisation, which groups together similar strains sharing a particular trait, environmental origin, or subspecies taxon (e.g. serovar); (ii) strain differentiation, which facilitates the discrimination of single strains from another based on similarity levels of their profiles; and (iii) strain identification, based on creation of a reference library consisting of known strains, which enables the identification of yet unidentified strains. The bioinformatics-enabled approach to bacterial profiling is based on the identification of distinct and discriminating peaks in MALDI-TOF MS profiles of microorganisms (biomarkers) corresponding to respective proteins identified from publicly available databases with genome sequence data. Using this approach for strain typing below the species level, the knowledge of discriminating peaks is of great significance. Thus, studies undertaken so far have mainly focused on detecting strain-specific mass peaks (biomarkers), to improve the software algorithms for the detection of strain-specific pattern and to validate the results with reference strains or collections of clinical isolates. As biomarkers represent only a minute portion of the MS spectrum, poor spectrum reproducibility can impede identification of reliable peaks as strain-specific biomarkers. Thus, MS spectrum quality, data richness, reproducibility, and mass accuracy must typically exceed levels necessary for species-level profiling (Dieckmann, Helmuth, Erhard, & Malorny, 2008). The advantage of this bioinformatics-enabled approach is that it does not require construction of libraries of spectra. However, instrumental requirements are sometimes greater with the bioinformatics-enabled approaches than with the library-based approach if protein and peptide sequence data are directly used for biomarker identification as this requires more sophisticated and costly instruments that perform MS–MS (TOF–TOF) analysis (Sandrin et al., 2013).

Published studies indicate that both approaches of MALDI-TOF MS-based typing can successfully be used for microbial typing and identification at the sub-species level (Cherkaoui et al., 2010; Dieckmann et al., 2008; Lartigue et al., 2009, 2011; Williamson et al., 2008). Several authors emphasise the need of distinct sample preparation and data analysis procedures in order to optimise the MALDI-TOF MS method for strain typing (Emonet, Shah, Cherkaoui, & Schrenzel, 2010; Murray, 2010), while the direct utilisation of the identification profiles for typing might sometimes be possible and can be of particular interest (Josten et al., 2013). The limits of the taxonomic resolution of MALDI-TOF MS profiling may largely be due to the nature of the particular bacterium profiled, as some bacteria are remarkably different genetically at the subspecies levels, whereas others are almost indistinguishable (Ghyselinck, Van, Hoste, Heylen, & de, 2011; Tanigawa, Kawabata, & Watanabe, 2010). The focus of this chapter is on the current literature and recent developments of MALDI-TOF MS as a tool for strain typing of Gram-positive and Gram-negative bacteria as well as certain fungi.

1.2 STRAIN TYPING OF GRAM-POSITIVE BACTERIA USING MALDI-TOF MS

Being one of the most common bacterial pathogens, the species *Staphylococcus aureus* has been the target of several MALDI-TOF MS strain typing studies. Besides those aimed at the differentiation of methicillin-sensitive *Staphylococcus aureus* (MSSA) and methicillin-resistant *Staphylococcus aureus* (MRSA) (Du, Yang, Guo, Song, & Wang, 2002; Edwards-Jones et al., 2000), several studies have described reliable subtyping based on MALDI-TOF fingerprints. Walker et al. assessed MALDI-TOF MS for subtyping of MRSA examining a panel of 26 well-characterised staphylococcal isolates (Walker, Fox, Edwards-Jones, & Gordon, 2002). They performed intra- and inter-laboratory reproducibility studies, investigated the effects of culture media, and were able to identify two major MRSA groups, which exhibited differences between certain regions of the MALDI spectra. Majcherczyk et al. explored the discriminatory power of MALDI-TOF MS for detecting subtle differences in isogenic strains of *S. aureus* differing in their expression of resistance to methicillin or teicoplanin. The authors found MALDI-TOF MS to be superior to PFGE profiles or peptidoglycan muropeptide digest patterns (Majcherczyk, McKenna, Moreillon, & Vaudaux, 2006). In the study of Boggs et al., the authors applied MALDI-TOF MS followed by a genetic algorithm model using the ClinProTools software (Bruker Daltonics) to identify the clonal USA300 lineage of MRSA (Boggs, Cazares, & Drake, 2012). This community MRSA strain is endemic in the United States and has become a major cause of skin and soft-tissue infections, community-acquired pneumonia, catheter-related bloodstream infections, and other systemic infections. The authors investigated 47 isolates of USA300 *S. aureus* and 77 non-USA300 *S. aureus* isolates and found three mass/charge peaks (m/z 5932, 6423, and 6592) to be discriminators between the groups of isolates. Using this approach, 197 of 224 test isolates were correctly classified as "USA300 family" or "non-USA300". Wolters et al. described reproducible spectrum differences based on 13 characteristic peaks leading to 15 reproducible bacterial profiles (Wolters et al., 2011). This enabled the robust discrimination of clonal complexes (CC). They concluded that MALDI-TOF MS might be suitable as a first-line tool for inexpensive and rapid typing of MRSA. Josten and colleagues did find peak shifts that differentiate main *S. aureus* clonal complexes CC5, CC22, CC8, CC45, CC30, and CC1 (Josten et al., 2013). They demonstrated that these peak shifts correlate to point mutations in the respective genes. A retrospective study of an MRSA outbreak successfully separated unrelated MSSA, MRSA, and borderline-resistant *S. aureus* strains isolated from health-care workers. This led the authors to the conclusion that MALDI-TOF could act as an identifier and typing tool at the same time.

Giebel and co-workers examined the use of MALDI-TOF-MS for the characterisation of environmental isolates of *Enterococcus* (Giebel, Fredenberg, & Sandrin, 2008). The authors focused on the development of sample preparation protocols for obtaining reproducible MALDI-TOF mass spectra from *Enterococcus* isolates.

They further evaluated methods of data analysis to maximise repeatability of the method and its ability to group isolates according to their respective sources. By including 21 *Enterococcus* isolates from seven unique sources, a repeatability level of 91% as well as grouping by source for isolates could be obtained. MALDI-TOF MS has also been successfully used by Griffin and co-workers to detect *vanB*-positive vancomycin-resistant enterococci (VRE); they further showed its usefulness to investigate the epidemiology of an outbreak (Griffin et al., 2012).

MALDI-TOF MS has been applied for strain typing in different *Streptococcus* species. Moura and colleagues applied a library-based MALDI-TOF approach with random forest analysis and identified unique, strain-specific biomarkers to differentiate strains of *Streptococcus pyogenes* (Moura et al., 2008). A subset of common, characteristic, and reproducible biomarkers in the range of 2000–14,000 Da were detected, which were independent of the culture media. From this, the authors could differentiate invasive group A streptococcus (GAS) strains. The GAS isolates from cases of necrotising fasciitis were clustered together and were distinct from isolates associated with non-invasive infections, despite of sharing the same *emm* type. Almost 30% of the biomarkers detected were tentatively identified as ribosomal proteins.

Lartigue et al. investigated *Streptococcus agalactiae* by MALDI-TOF MS and identified a 6250-Da protein specific to Sequence Type-1 (ST-1) strains and a 7625-Da protein specific to ST-17 strains when used for identification of group B streptococci (Lartigue et al., 2011). The strains of these STs are major causes of meningitis and late-onset disease in neonates. The authors found that the sample preparation was quick and easy, and required only a small amount of biological material such as a few bacterial colonies. Moreover, this method provided highly accurate measurements with a low cost per analysis.

Also, *Streptococcus pneumoniae* (pneumococci, Pnc) have been investigated as a target for MALDI-TOF MS-based strain typing (Williamson et al., 2008). The authors performed a comparative proteomic analysis between non-typeable (NT), non-encapsulated conjunctivitis outbreak strains of *S. pneumoniae* (cPnc), and other known typeable or NT pneumococcal and streptococcal isolates. They compared protein and peptide fragments of the whole-cell bacterial isolate-matrix combinations ranging in size from 2 to 14 kDa. Applying random forest analytical tools and dendrogramic representations, a clustering of the isolates into distinct clonal groups could be achieved. The authors proposed a peak list of protein and peptide masses, which could potentially be used as a biomarker in the rapid diagnosis of pneumococcal conjunctivitis.

But also other clinically relevant Gram-positive bacteria, such as *Clostridium difficile*, have been investigated by MALDI-TOF strain typing. *C. difficile* is a significant cause of nosocomial infections and in recent years, hypervirulent strains belonging to distinct PCR ribotypes have been found in North America and Europe (e.g. ribotypes 078 and 027). These isolates were associated with higher toxin production and increased mortality. Reil and co-workers, (2011) constructed a database consisting of different mass spectra recorded in the SARAMIS software

(AnagnosTec, Zossen, Germany) and validated the database with 355 *C. difficile* strains belonging to 29 different PCR ribotypes. The most frequent PCR ribotypes (types 001, 027, 078/128) were recognised by MALDI-TOF MS.

Moreover, MALDI-TOF MS has been applied to food-borne pathogens such as *Listeria monocytogenes* for identification of distinct isolates. Rapid, accurate discrimination between *Listeria* strains is essential for appropriate therapeutic management and timely intervention for infection control. Barbuddhe et al. investigated 146 strains of different *Listeria* species and serotypes as well as clinical isolates and compared the MALDI-TOF peak spectra with results from PFGE analyses (Barbuddhe et al., 2008). Mass spectra derived from *Listeria* isolates showed characteristic peaks, conserved at both the species and lineage levels.

MALDI-TOF MS typing of strains has also been used for spore-forming aerobic bacteria such as *Bacillus* spp. Ryzhov and colleagues characterised the spores of 14 microorganisms of the *Bacillus cereus* group and obtained unique mass spectra for the different *B. cereus* and *Bacillus thuringiensis* strains, allowing for differentiation at the strain level (Ryzhov, Hathout, & Fenselau, 2000).

Nagy and co-workers investigated MALDI-TOF MS for discrimination of phylotypes I, II, and III of the anaerobic bacterium *Propionibacterium acnes* (Nagy et al., 2013). Phylogroup-specific peaks and peak shifts were then identified for types IA, IB, IC, II, and III. Test of an independent set of further 48 clinical isolates of *P. acnes* revealed agreement with the MLST analysis.

1.3 STRAIN TYPING OF GRAM-NEGATIVE BACTERIA USING MALDI-TOF MS

MALDI-TOF MS has also been applied to Gram-negative bacteria to decipher strain characteristics and to identify strain clusters on the subspecies level. *Escherichia coli* strains can be grouped in different pathotypes according to the repertoire of virulence factors and the respective disease caused. Clark et al. chose 136 *E. coli* isolates representing 8 pathotypes and created MSP database entries specific for each pathotype. Using this novel database enabled the majority of strains to be correctly assigned to the respective pathotype (Clark et al., 2013). In addition, the detection of distinct clones of a pathotype is possible using MALDI-TOF MS. Karger and colleagues describe the successful differentiation of Shiga toxin-producing *E. coli* isolates representing the serotypes O165:H25, O26:H11/H32, and O156:H25 using serotype-specific prototypic mass spectra systematically optimised by filtering out masses that do not contribute to the discrimination of the serotypes (Karger et al., 2011). Christner and co-workers examined isolates of Shiga-toxigenic *E. coli* O104:H4 from a large outbreak in northern Germany in 2011 trying to find a fast and reliable method to distinguish the outbreak strain from other *E. coli* isolates (Christner et al., 2014). Specific peaks in the outbreak strain's spectrum were identified by comparative analysis of archived pre-outbreak spectra that had been acquired for routine species-level identification. Proteins underlying these discriminatory peaks were identified by liquid chromatography-tandem mass spectrometry and validated against publicly available databases. The resulting typing scheme was evaluated against PCR genotyping with 294 *E. coli* isolates from clinical samples collected during the outbreak. From this, the authors identified two characteristic peaks at *m/z* 6711 and 10,883, which correctly classified 292 of 293 study isolates including all O104 outbreak isolates. Besides pathotyping, MALDI-TOF MS has been investigated to determine if it can assign *E. coli* strains to different phylogenetic groups (A, B1, B2, and D), which reflect the evolution of and the relationship within the species *E. coli*. Assignment of phylogenetic groups currently requires time consuming and laborious molecular tests. Sauget et al. studied 656 *E. coli* isolates representing the phylogroups A, B1, B2, and D and identified a single peak shift between isolates of phylogroup B2 and those of groups A, B1, and D (Sauget, Nicolas-Chanoine, Cabrolier, Bertrand, & Hocquet, 2014). This shift accurately classified 89% of the isolates. They further applied the ClinProTools software (Bruker Daltonik) to differentiate strains belonging to the respective phylogroups. The vast majority of *E. coli* strains causing septicaemia, urinary tract infections, and other extra-intestinal infections belong to phylogroup B2. Thus, MALDI-TOF MS could be used to detect the most virulent strains belonging to this phylogroup and to monitor the epidemiology of extra-intestinal pathogenic *E. coli*.

Members of the *Salmonella* spp. are traditionally classified according to the O- and H-antigens by direct agglutination of these polysaccharide antigens based on the White–Kauffmann–Le Minor classification scheme (Guibourdenche et al., 2010). Dieckmann et al. analysed the mass spectral profiles of multiple housekeeping proteins of 126 strains representing *Salmonella enterica* subsp. *enterica* (subspecies I), *S. enterica* subsp. *salamae* (subspecies II), *S. enterica* subsp. *arizonae* (subspecies IIIa), *S. enterica* subsp. *diarizonae* (subspecies IIIb), *S. enterica* subsp. *houtenae* (subspecies IV), and *S. enterica* subsp. *indica* (subspecies VI), and *Salmonella bongori* to obtain a phylogenetic classification of salmonellae (Dieckmann et al., 2008). The authors collected spectra covering a wide mass range (2000–40,000 Da) and were able to assign species-, and subspecies-identifying biomarker peaks on the basis of available genome sequence data for *Salmonella*. Distinct peak patterns of the biomarker profiles of *Salmonella* strains corresponded to single or multiple amino acid changes in housekeeping proteins. From this, a rapid MALDI-TOF MS protocol was established to distinguish *Salmonella* subspecies within minutes. Kuhns et al. (2012) investigated the appropriateness of MALDI-TOF-based intact cell mass spectrometry in order to discriminate *S. enterica* subsp. *enterica* serovar Typhi (*S.* Typhi) from other serovars. A detailed analysis of the mass spectra of 160 isolates of *Salmonella enterica* subsp. *enterica* isolated from 12 different serovars revealed several serovar-specific biomarker ions, allowing the discrimination of *S.* Typhi from others.

MALDI-TOF MS has also been successfully used to determine the relatedness of *Klebsiella pneumoniae* strain of infections in humans that were associated with nosocomial outbreaks. Berrazeg et al. obtained reference mass spectra (MSPs) from 535 *K. pneumoniae* clinical isolates from France and Algeria (Berrazeg et al., 2013). MSP dendrogram revealed five distinct clusters and showed that *K. pneumoniae* strains isolated in Algerian hospitals were significantly associated with respiratory infections and the ESBL phenotype, whereas those from French hospitals were significantly associated with urinary tract infections. Accordingly, Bernaschi et al. were able to distinguish three different clusters of isogenetic or related multi-drug-resistant *K. pneumoniae* strains (Bernaschi et al., 2013). The MALDI-TOF MS results were compared to those of other molecular typing methods such as PFGE and automated repetitive-sequence-based PCR (rep-PCR) genotyping. The results of all of the typing methods correlated; therefore, MALDI-TOF MS proteome profiling may provide a fast and valuable preliminary screening tool to support microbiologists during nosocomial outbreak investigations.

Another important cause of nosocomial infections and outbreaks is *Acinetobacter baumannii*. Typing of these bacteria is an important technique for implementation of control policy measures to stop outbreaks. Mencacci et al. (2013) studied 35 multi-drug-resistant strains of *A. baumannii*, isolated from colonised or infected patients by rep-PCR and MALDI-TOF MS. The statistical analysis of the data showed that the rep-PCR system and MALDI-TOF MS system provided similar results, with a good concordance between the two methods and a high probability of MALDI-TOF MS predicting the rep-PCR results.

MALDI-TOF MS-based bacteria strain typing has further successfully been used for other species, including *Campylobacter jejuni* (Zautner et al., 2013), *Pseudomonas stutzeri* (Scotta, Gomila, Mulet, Lalucat, & Garcia-Valdes, 2013), *Legionella pneumophila* (Fujinami et al., 2011), *Leptospira* spp. (Calderaro et al., 2014), and *Ochrobactrum anthropi* (Quirino et al., 2014).

1.4 STRAIN TYPING OF FUNGI USING MALDI-TOF MS

Very little is known about the potential of MALDI-TOF MS to discriminate fungi isolates (yeasts and mould) below the species level. Pulcrano et al. genotyped 19 strains of *Candida parapsilosis* isolated from the blood cultures of neonates by PCR amplification and subsequent sequencing of eight simple sequence repeats markers as well as by MALDI-TOF MS (Pulcrano et al., 2012). Electrophoretic and spectrometric profile results were compared in order to identify similarities among the isolates and to study micro-evolutionary changes in the *C. parapsilosis* population. Both methods were rapid and effective in identifying indistinguishable strains and studying micro-evolutionary changes in the population. Thus, MALDI-TOF MS appears to be a useful tool both for the identification and for monitoring the spread of *Candida* strains, which is crucial to control nosocomial infections.

2 RESISTANCE TESTING BY MALDI-TOF MS

2.1 INTRODUCTION

In the early days of MALDI-TOF fingerprinting for microbe identification, it was a goal to identify not only the species but also the putative antibiotic resistance of a microorganism. A fast and easy approach was proposed to differentiate resistant and susceptible bacteria by differences in their mass spectral profiles. The "prototype" for this approach was MRSA/MSSA differentiation, which was only partially successful and is still being evaluated today. During recent years, alternative methods have been described which do not use the simple profile differentiation but are directed at the effect of an antibiotic against a microorganism and the function of the respective resistance mechanism.

2.2 MALDI-TOF TYPING FOR RESISTANCE DETECTION

Results indicating the possibility of discrimination between the mass spectral fingerprints obtained from MRSA strains, MSSA strains, and the identification of *S. aureus* to the species level were reported as early as 2000 (Edwards-Jones et al., 2000). These initial results indicated that MALDI-TOF MS can be used for species identification of *S. aureus*, and also for MRSA/MSSA discrimination and strain characterisation (see above). Spectral information for this approach was mainly deduced from the mass range of molecules around 2000 *m/z*. Similar results were obtained with the commercially available MicrobeLynx System (Micromass, UK) in a similar study (Du et al., 2002). In this study, the effect of cultivation conditions, e.g., culture media or cultivation temperature, became clear. In contrast, other workers (Bernardo et al., 2002) did not demonstrate discrimination between MRSA/MSSA in their work. After investigation of several *S. aureus* strains in two different mass ranges (2000–20,000 *m/z*, 800–4000 *m/z*), they stated that a uniform signature profile for MRSA could not be identified although high reproducibility of the spectra was proven. The feasibility of MRSA detection by MALDI-TOF MS fingerprinting remained unconfirmed. In 2012, Szabados and co-workers investigated two strains with the same genetic background, one with and one without the SSCmec cassette harbouring *mecA*, resembling isogenic MRSA and a MSSA strains, respectively. The results clearly demonstrated that there were no characteristic differences between mass fingerprints of MRSA and MSSA strains (Szabados, Kaase, Anders, & Gatermann, 2012). Both strains possessed the same profile spectra, without any peak pattern indicating presence or absence of the SSC*mec* cassette. MALDI-TOF mass spectrometry, although a promising tool for determination of *S. aureus* clonal complexes (see above), seems to be incapable of determining characteristic profiles for MRSA strains. However, in 2014 a German group from Bonn did report that a discriminatory peak at *m/z* 2415 can be found for a group of MRSA strains. This marker peak has a very high specificity for MRSA detection although only limited sensitivity (Josten et al., 2014). The peak had not been detected in earlier studies by the same group (Josten et al., 2013) nor could it be found by Szabados et al. (2012). The simple reason for this was the different preparation method used prior the MALDI-TOF measurement. In the new study, the bacterial cell mass was directly applied to the MALDI target plate, but both older studies had applied a full extraction of bacterial proteins and spotting of the formic acid/acetonitrile extract onto the target position, generally the most reproducible method for detailed spectra analysis. The detected peak mass corresponded to the calculated mass of the singly charged ion of the formylated version of the phenol-soluble modulin (PSM) PSM-*mec*. PSM-*mec* is a small excreted peptide encoded on three SCC*mec* cassettes (type II, III, and VIII) containing the class A *mec* gene complex (Chatterjee et al., 2011). This small peptide, as it is excreted, probably was simply washed away during the extraction procedure before target preparation. In their study collection of *agr*-positive MRSA and MSSA isolates, Josten et al. did find a sensitivity of 0.95 and a specificity of 1.0 for detection of the PSM-*mec* peak (Josten et al., 2014). This finding still has to be confirmed in other studies but show the route for detection of a part of MRSA strains in a very simple and fast way. Also, the evaluation of spectra was done visually/manually in this first study, using a "mass window" of *m/z* 2411–2419. For any routine application, the procedure has to be implemented into a robust automated algorithm. The beauty of this would be the possibility to screen ultrafast from colonies for MRSA, directly by investigation of the routine identification mass spectra automatically after species identification. Although the sensitivity would be relatively low, the specificity of such an approach might be very high. In the reported study, the collection did consist of 89% *agr*-positive strains. This might vary significantly depending on regional distribution of *S. aureus* clonal complexes. For instance, US300 cannot be detected by this method.

Another impressive example of a simple genotypic MALDI-TOF profiling approach has been demonstrated for the anaerobic Gram-negative bacterium *Bacteroides fragilis*. Two different divisions of this important anaerobic pathogen (i.e. division I and division II) exist, and it was demonstrated by two different groups in independent studies that these divisions can also be differentiated by specific peak pattern of their MALDI-TOF profile spectra (Nagy, Becker, Soki, Urban, & Kostrzewa, 2011; Wybo et al., 2011). As only one of the divisions, i.e., division II, does contain a gene encoding for a very powerful metallo-carbapenemase (*cfiA*), the matching of an unknown strain to the respective division is of high practical value to detect the respective carbapenem resistance. The metallo-beta-lactamase activity of these strains inactivates virtually all beta-lactam antibiotics and thereby makes such bacterial infections very difficult to treat.

The differentiation of both divisions is possible by different methods. While Wybo and co-workers (Wybo et al., 2011) found a perfect separation of the two respective clusters after dendrogram calculation with the commercially available MALDI Biotyper software (Bruker Daltonics, Bremen, Germany), Nagy et al. (2011) detected a number of defined peak shifts when they compared several strains of each cluster with each other. The appearance of these peak shifts was then validated with a blinded dataset of 28 further strains (9 *cfiA*+, 19 *cfiA*−). This validation resulted in 100% accuracy of the model. It is important to consider that the expression of the metallo-beta-lactamase is activated by an insertion element if this element is located upstream of *cfiA*. Therefore, a *B. fragilis* strain may appear susceptible to carbapenems initially in resistance testing but will turn to be resistant under the selective pressure of antibiotic treatment.

The third example for such a simple detection of resistant strains of a particular species by profile differentiation comes from results published for VRE. Nowadays, vancomycin-resistant *Enterococcus faecium* outbreaks are one of the big problems for clinical microbiology, in particular those containing either the vancomycin resistance-encoding gene clusters *vanA* or *vanB*. Identification of the different *Enterococcus* species by MALDI-TOF MS is very reliable as shown by Werner et al. (2012). Already, this species information is helpful for some resistance prediction because reliable identification of other, rarer, and intrinsically vancomycin-resistant species is possible. Also, *E. faecium* is securely differentiated from *Enterococcus faecalis*, the second frequently occurring *Enterococcus* species which is less frequently vancomycin resistant.

As described above for *S. aureus* and *B. fragilis*, it would be extremely helpful to detect the resistance directly from the MALDI identification pattern here, too. Recently, Griffin and co-workers described the detection of *vanB* VRE based on MALDI-TOF MS (Griffin et al., 2012) and the creation of a statistical model using the commercial ClinProTools software package (Bruker Daltonics, Bremen, Germany) for differentiation of VRE from susceptible ones. They calculated a respective multivariate statistical model with a support vector machine using whole-cell spectra acquired from locally occurring *E. faecium* strains. Internal cross-validation of the calculated statistical model resulted in 92.4% sensitivity and 85.2% specificity. External validation after incorporation of the statistical model into the daily routine laboratory workflow resulted in an even higher sensitivity and specificity of 96.7% and 98.1%, respectively. These impressive results are in disagreement to the results reported by Lasch et al. (2014) who found that MALDI-TOF patterns are not useful for differentiation of vancomycin-resistant and -susceptible enterococci. The explanation might be in the particular epidemiological situation existing in the New Zealand area, which favoured the *vanB*-positive *E. faecium* detection in the study of Griffin and co-workers but cannot be taken as typical and transferrable to other laboratories. This is supported by the fact that Lasch et al. who reported insufficient discrimination power of MALDI-TOF MS for discrimination of *E. faecium* strains did not identify any statistically relevant marker peaks linked to glycopeptide resistance determinants (*vanA*, *vanB*) in *E. faecium* (Lasch et al., 2014).

2.3 MASS PROFILE CHANGES INDUCED BY ANTIBIOTIC DRUGS

In contrast to such genetically oriented methods, another idea which also is based on standard microorganism profiling was developed for yeast. For this application, the change of a profile spectrum through the action of an antibiotic at a certain concentration is observed. Thereby, the test resembles a phenotypic test similar to classical susceptibility testing but uses MALDI-TOF as the means to "read" the result. In the first report describing such an assay, yeast cells were incubated in a dilution series of fluconazole in liquid medium for 24 h (Marinach et al., 2009). Yeast cells were harvested from the medium and MALDI-TOF profile spectra were measured for those organisms from all antibiotic dilutions. Comparison of the mass spectra of the yeast incubated at different antibiotic concentrations then was performed. It was very obvious that at a certain concentration, the spectral pattern did change significantly. The authors decided to call the lowest antibiotic concentration where a change of mass spectrum was found the "*minimal profile changing concentration*" (MPCC). Comparison of the MPCC with the minimal inhibitory concentration (MIC) determined by the microdilution reference according to the Clinical and Laboratory Standards Institute showed that both values in all investigated cases did not differ more than a single dilution step. A clear drawback at this time was the lack of automated data interpretation. This was overcome in a subsequent study by another group (De Carolis et al., 2012) who applied a similar test scheme with slight modifications to *Candida* species and to *Aspergillus* species and to the antifungal drug caspofungin with results very close to reference data. In their study, the authors did use an algorithm available in a commercial software package which is dedicated to the similarity analysis of mass spectra, the composite correlation index (CCI) analysis (MALDI Biotyper, Bruker Daltonics, Bremen, Germany). This algorithm calculated a similarity matrix between spectra of all categories (fungi grown in different concentrations of caspofungin). The authors did match the spectra from the different dilutions against the spectra of the most extreme dilutions, i.e., against the spectrum at highest concentration of antifungal and the spectrum without caspofungin. The "MPCC" was determined as the concentration at which the derived spectrum is more similar to the one observed at the maximum caspofungin concentration than to the spectrum observed without caspofungin. Although the method has automated the data analysis, an overnight culture was necessary; therefore, the advantage of using the MALDI-TOF MS in comparison to established standard methods is undermined. Subsequently, the same group did improve this approach (Vella et al., 2013) using a simplified and more rapid version of the "MPCC assay". They used breakpoint analysis to determine resistance/susceptibility of fungal strain already after 3 h of incubation time. For this, they incubated yeast in only three concentrations of the drug: without caspofungin, caspofungin at breakpoint concentration, and caspofungin at maximum concentration. Mass spectra of yeast from cells grown at these concentrations were acquired and compared against each other by CCI analysis. If the spectrum at breakpoint concentration was more similar to that acquired without caspofungin than to that at high concentration the strain was considered susceptible, if not the strain was regarded as resistant. In the hands of the authors, this test demonstrated very good results compared to standard routine methods. While this "MALDI-TOF breakpoint assay" may reduce the time required for yeast susceptibility testing, it has only been applied in one routine laboratory. Reproducibility and robustness have to be proven for a broader usage in clinical microbiology as a routine method, and more advanced and adapted bioinformatics tools may be required.

2.4 DETECTION OF DRUG RESISTANCE MECHANISMS: THE BETA-LACTAMASE ASSAY

A different, but also functional approach, which has already been adopted in routine laboratories, is the mass spectrometry-based detection of beta-lactamase activity. This may become the first widely adopted routine resistance detection method using MALDI-TOF mass spectrometry transferring the benefits of the technology from identification to another diagnostic area. The basis of the test is the observation of the enzymatic activity underlying a particular resistance of a bacterium by direct molecular analysis. This approach is effective for the different beta-lactamases, namely ESBL and carbapenemases. Beta-lactam antibiotics are known to be inactivated by hydrolysis of the beta-lactam ring in their chemical structure through enzymes, i.e., beta-lactamases, synthesised, and excreted by resistant microorganisms. Enzymatic cleavage by hydrolysis of this beta-lactam ring leads to an addition of a water molecule to its chemical structure, resulting in an increase of molecular mass about 18 Da. As the resulting molecular structures frequently are not stable, a further mass change of -44 Da happens in many cases. Such mass changes, although too small to be detected by molecular methods, can easily be detected by mass spectrometry. Although usually electrospray mass spectrometry has been used for the analysis of such small molecules, it has been shown that MALDI-TOF MS can be used for a variety of the commonly used beta-lactam antibiotics (Sparbier, Schubert, Weller, Boogen, & Kostrzewa, 2012). Grundt and co-workers employed LC/MS in combination with an HCT Ultra ion trap with an ESI interface (Grundt et al., 2012). An LC/MS system also was used in a study by Peaper et al. (2013). This analytical set-up provides higher sensitivity for small molecules like beta-lactam antibiotics and enables better quantitative evaluations. On the other hand, such instruments are usually not available in microbiological laboratories and liquid chromatography before the MS analysis extends analysis time and introduces new risks including cross-contamination and robustness. Hooff and co-workers applied MALDI-TOF MS as well as MALDI triple quadrupole (QqQ) MS to measure the beta-lactams and their cleavage products (Hooff et al., 2012), with both approaches demonstrating very similar results. The authors did demonstrate MALDI QqQ useful for quantitative analysis, but it has to be questioned if this is of practical advantage and outbalances the lack of availability of such a costly instrument for routine.

Resistance detection using such technology can be very rapid because it is not the growth of microorganism or its inhibition that is being observed, i.e., a secondary effect of the mechanism, but the molecular mechanism itself. Carbapenemases as a significant group of resistance-causing enzymes which can cleave virtually all available beta-lactam antibiotics has led them to them being the focus of assay development published describing this approach (Burckhardt & Zimmermann, 2011; Hrabák, Walkova, Studentova, Chudackova, & Bergerova, 2011). While Hrabák and co-workers used meropenem and an incubation time of 3 h, Burckhardt and Zimmermann used ertapenem and reported a necessary incubation time of 1–2.5 h, depending on the type of carbapenemase. The general assay set-up is very simple and was similar in all published studies. Bacteria, from an overnight agar plate culture, were suspended in a solution of the antibiotic and incubated at between 30 °C and 37 °C for 1–4 h. After this incubation, the bacteria were removed from the solution by a short centrifugation step. About 1 µl of the supernatant containing the antibiotic was spotted onto a MALDI target. The spot was overlaid with MALDI matrix and subsequently measured in a MALDI-TOF instrument. In contrast to profile analysis used for microorganism identification, the settings of the instrument have to be adapted to the analysis. While fingerprinting today is done in a mass range of 2000–20,000 atomic mass units, here a range of about 300–1000 Da has to be applied. Accordingly, other mass spectrometer settings have to be adopted (laser energy, electrode voltages), and this is preferably done by an engineer or product specialist from the instrument manufacturer but only has to be done once. A method containing file for this analysis can be optimised and stored without changing the routine parameters of an instrument.

For most beta-lactamases, activity can be observed after 1–2 h; however, exceptions include the OXA-type lactamases owing to their slower hydrolysis rate, so longer incubation times have been reported and the occurrence of false negatives has also been reported (Carvalhaes et al., 2013). Other workers have also demonstrated excellent results with the MALDI-TOF-based assay for this type of carbapenemase (Hrabák et al., 2012; Lee et al., 2013; Sauget, Cabrolier, Manzoni, Bertrand, & Hocquet, 2014), making it probably the most sensitive technique for their functional detection. A recent publication has shed some light on challenges and possibilities of OXA detection. Hrabák, Chudackova, and Papagiannitsis (2014) did show that the reaction buffer used for carbapenemase detection can have a dramatic effect on the detection time and success at all (Hrabák et al., 2014). As a lysine in the active centre of OXA 48 has to be carboxylated for enhanced activity of the enzyme, carbonate-containing buffers can accelerate the time for cleavage of carbapenems by such enzymes and thereby dramatically impair their detection. Indeed, using imipenem and a hydrogen carbonate-based buffer, the time for incubation of OXA 48-type carbapenemase for their detection can be reduced to 30 min (own, unpublished results, M.K.).

A further opportunity which is offered by MALDI-TOF-based monitoring of the beta-lactam cleavage is the chance to characterise enzymes by their inhibition/non-inhibition through according substances. This has already been described in one of the first publications (Sparbier et al., 2012) where the effect of several inhibitors (clavulanic acid, tazobactam, APBA) and their specificity could be demonstrated. Álvarez-Buylla and co-workers did find that metallo-carbapenemases could not only be detected in *Acinetobacter* species, they further characterised the enzymes using dipicolinic acid as an inhibitor. Dipicolinic acid acts as an inhibitor of metallo-carbapenemases but not of other carbapenemases (Álvarez-Buylla, Picazo, & Culebras, 2013). The same results were demonstrated for *B. fragilis* containing the metallo-beta-lactamase encoded by *cfiA* (Johansson, Nagy, & Soki, 2014).

A most important requirement for novel resistance detection technologies comes from the treatment of sepsis. For several years, MALDI-TOF protocols have been available for the recovery of microbial cells from positive blood cultures and to directly perform MALDI-TOF profile analysis for species identification (Klein et al., 2012; La Scola & Raoult, 2009; Moussaoui et al., 2010; Schubert et al., 2011; Stevenson, Drake, & Murray, 2010). Although this is already regarded to be a significant step forward for sepsis diagnosis and treatment, there is still an urgent requirement for fast resistance determination of the microbes found in a positive blood culture. Initial studies have demonstrated that the MALDI-TOF MS beta-lactamase test can be applied to detect the enzymatic activity from the bacterial cells isolated from blood cultures. The first proof of concept for this was demonstrated by Sparbier et al. (2012). Carvalhaes et al. demonstrated that from 100 randomly selected blood culture vials, 21 of 29 carbapenemase producers could be detected directly in the bacterial pellet obtained from the blood culture (Carvalhaes et al., 2014). This was achieved after 4 h of incubation time for the cleavage reaction. The residual strains, the majority of which were OXA-23 producing *A. baumannii*, were detected the next day from isolated colonies. Jung and co-workers did investigate *Enterobacteriaceae* from 100 consecutive positive blood cultures for resistance against aminopenicillins (*E. coli* only) and third-generation cephalosporins (Jung, Popp, et al., 2014). This allowed discrimination between susceptible and resistant bacteria for the aminopenicillins with a sensitivity and specificity of 100%. The same was true for resistance to third-generation cephalosporins in *Enterobacteriaceae* that constitutively produced class C β-lactamases. Species expressing class A β-lactamases generated some false-positive results leading to a sensitivity and specificity for this group of 100% and 91.5%, respectively. Another interesting study did show that in the case of *B. fragilis*, the bacterial pellet recovered from spiked blood cultures could be used in a sequential "identification, typing, and carbapenemase confirmation" workflow in only 3 h (Johansson et al., 2014). The authors used a software prototype from a manufacturer for the data evaluation and did observe that this did improve result quality. Overall, these results indicate that a MALDI-TOF/beta-lactamase test-based workflow can be established in routine laboratories, which may become a powerful tool for carbapenemase detection, confirmation, and resistance detection in sepsis in the future.

2.5 STABLE ISOTOPE-BASED RESISTANCE DETECTION

While the beta-lactamase monitoring appeared as the first enzyme function-based test using MALDI-TOF MS for resistance detection of bacteria and has a lot of potential for improvement of current practice, its limitations are apparent. For example, not all resistance mechanisms for other antibiotics are related to drug modification by an enzyme. Even for beta-lactam antibiotics, other resistance mechanisms exist, although the beta-lactamase can be considered the most significant today. Recently, a more general test has been described in which the assay directly observes the activity of protein synthesis, not cell growth, using media containing stable isotope-labelled nutrient compounds, thereby making it unique among resistance tests. As cell proliferation is preceded by protein synthesis, the assay has the potential to significantly reduce time for resistance detection and it has been described in two different forms by independent groups. In one publication, the growth of cells was performed in a medium which was totally labelled with ^{13}C (Demirev, Hagan, Antoine, Lin, & Feldman, 2013). The experiment was performed in labelled medium with and without the antibiotic; a control was run without antibiotic in unlabelled medium. The principle is simple with a susceptible cell incubated in labelled medium that is inhibited by the antibiotic drug and therefore will not start protein synthesis to further proliferate; therefore, it will not synthesise proteins with a molecular mass increased by stable isotopes. A resistant microbe will not be inhibited, will start protein synthesis, and thereby quickly synthesises proteins with an increased molecular weight, and in a mass spectrum, the increase of molecular masses easily can be observed. The authors described the calculation of particular peak masses used as biomarkers to determine protein synthesis. This means that for a particular microbial species, conserved biomarker peaks have to be determined which later can be used for a general algorithm specific for this species. A drawback of the method described by Demirev and co-workers is the relatively high cost of the totally ^{13}C-labelled media which are not available commercially currently. Although the authors did have to incubate their generic model (*E. coli* cells and streptomycin) for 6 h, this clearly represents a more rapid detection of resistance compared to current standard methods. This has also been demonstrated by another group using a somewhat different approach, also with observation of stable isotope incorporation by MALDI-TOF MS and called MS-RESIST (*M*ass *S*pectrometry-based *Re*sistance Testing with *S*table *Iso*topes) (Sparbier et al., 2013). In this study, totally isotopically labelled medium was not used and a medium where one amino acid only was substituted by a labelled one (^{13}C$_6$ ^{15}N$_2$-L-lysine). In a small study (10×10), the authors did show the differentiation of MRSA and MSSA using oxacillin as well as cefoxitin in 3 h only. For a similar set-up, only 2 h is sufficient to detect meropenem resistance in *K. pneumoniae* (Kostrzewa, Sparbier, Maier, & Schubert, 2013). For automated data analysis, the authors describe different methods including a correlation analysis between the spectra acquired with and without an antibiotic. Such bioinformatics might facilitate more general data analysis without definition of specific, pre-determined biomarkers. In a second paper (Jung, Eberl, et al., 2014), the same group described the application of the stable isotope method to *Pseudomonas aeruginosa* strains and antibiotics of three different classes (the carbapenem meropenem, the aminoglycoside tobramycin, and the fluoroquinolone ciprofloxacin). Each test was performed with strains with different MICs for the respective antibiotic, for each antibiotic 15 strains in the susceptible and 15 strains in the resistant range according EUCAST, respectively. Incubation times for protein synthesis were 2.5–3 h. All strains were correctly classified including those near to the EUCAST breakpoints. This is a strong indication that the MS-RESIST assay is a very promising technique for a mass spectrometry-based, next-generation resistance test for more rapid diagnostics and therapy guidance. However, the applicability of this approach to positive blood cultures has yet to be established, where the benefit of accelerated resistance determination would be the greatest.

2.6 GROWTH MONITORING BY MALDI-TOF-MS

Recently, another novel technique for resistance and possibly susceptibility detection was published (Lange, Schubert, Jung, Kostrzewa, & Sparbier, 2014), which uses MALDI-TOF MS as a detector for biomass semi-quantification which occurs during microbial growth. This is unusual as MALDI-TOF MS is not the preferred instrument for quantification purposes in the "mass spectrometry community" where quantification is usually the domain of electrospray mass spectrometers. The set-up for the novel assay is quite simple and similar to the MS-RESIST assay, with unmodified standard media used for growth. A control reaction is performed in medium without any antibiotic, and in parallel, from the same start suspension, a second well or microtube with medium plus antibiotic is inoculated. After incubation, the cells are harvested from both reactions and the proteins extracted using a standard method similar to that used for bacterial identification methods. This facilitates the production of standard protein profiles to be acquired in a MALDI-TOF mass spectrometer. However, the extraction solution contains a standard, which is spiked into both samples at a defined concentration. The control (without antibiotic) has to show a "normal" spectrum, only with the addition of the peaks corresponding to the spiked standard. These are two peaks for the standard used in the published work, one corresponding to the single-charged molecule and the other to the double-charged one. Thereby, the control reaction cannot only be used as a growth control but can also be used to control identification. The mass spectrum acquired from the microbes grown in the medium containing the antibiotic is used for resistance detection with the resistant microorganism having almost the same spectrum as the growth control. In contrast, as its growth is inhibited by the antibiotic, a susceptible one will grow only to a very limited extend; therefore, the spectrum will only contain the peaks of the spiked standard and a few small peaks derived from microbial growth. The reason for this is the limited amount of microbial biomass and the suppression effect of the standard protein against this. The publication from Lange and co-workers presented data for a total of 94 *K. pneumoniae* and *Klebsiella oxytoca* isolates obtained from 12 institutions from 6 different countries, with meropenem as the antibiotic drug. The authors demonstrated that a simple visual evaluation of data was possible, and they further present an algorithm which can classify the bacteria automatically. The breakpoint concentration for the meropenem assay was 8 µg/ml, and with only 1 h of incubation time, the assay showed a sensitivity of 97.3% and a specificity of 93.5% for resistance detection. Also in a proof-of-concept experiment using 18 blood cultures spiked with *Klebsiella* strains, 10 with susceptible and 8 with resistant ones, respectively, 17 did result in a correct classification also after only 1 h. The incorrectly classified strain was a susceptible one which did not grow sufficiently in 1 h, as demonstrated by the growth control, but after 2 h growth it was correctly classified.

This novel assay, called MBT–ASTRA by the authors (*M*ALDI *B*io*typer–Antibiotic Susceptibility Test Rapid Assay*), obviously has potential for more rapid resistance testing in the future. As the breakpoint concentration seems to be close to the classical breakpoints determined for laboratory standard assays, this method might obtain very similar results to those of standard assays but more rapidly. MALDI-TOF here acts not as a precise mass detector but as a sensitive detector for semi-quantification of biomass. Further studies with different organisms and other antibiotics are ongoing particularly in areas where currently none or only very limited resistance information can be obtained by the standard methods, e.g., for yeast, mycobacteria, or anaerobic bacteria.

CONCLUDING REMARKS

Besides the identification of bacteria at genus and species level, other fields of application for MALDI-TOF MS are being actively investigated. One field of exceptional interest is strain typing below the species level, which could support epidemiological investigations and may have significant implications in public health and medical care. The accuracy, ease of use, speed, and low cost of the methods may make MALDI-TOF MS a promising alternative to other methods. In the recent years, several encouraging examples of successful strain typing using MALDI-TOF MS have been reported. However, the generalisation of the results obtained in these primarily "proof-of-concept" studies needs to be determined in multi-centre follow-up studies encompassing larger collections of isolates from different regions. Similarly, several very promising approaches have been reported for rapid resistance testing using MALDI-TOF-MS, using the same instruments as used for species identification. In particular, if these new tests can be applied directly to positive blood culture to accelerate the microbiological diagnostic workflow in the case of septicaemia, this will have a high potential to improve clinical microbiological practice; however, this will require more extensive further studies. Finally, the new applications of MALDI-TOF MS that are on the horizon may make MALDI-TOF MS a central and broadly applicable core technology in the clinical microbiology laboratory.

REFERENCES

Álvarez-Buylla, A., Picazo, J. J., & Culebras, E. (2013). Optimized method for *Acinetobacter* species carbapenemase detection and identification by matrix-assisted laser desorption ionization-time of flight mass spectrometry. *Journal of Clinical Microbiology, 51*(5), 1589–1592.

Barbuddhe, S. B., Maier, T., Schwarz, G., Kostrzewa, M., Hof, H., Domann, E., et al. (2008). Rapid identification and typing of listeria species by matrix-assisted laser desorption ionization-time of flight mass spectrometry. *Applied and Environmental Microbiology, 74*(17), 5402–5407.

Bernardo, K., Pakulat, N., Macht, M., Krut, O., Seifert, H., Fleer, S., et al. (2002). Identification and discrimination of *Staphylococcus aureus* strains using matrix-assisted laser desorption/ionization-time of flight mass spectrometry. *Proteomics, 2*(6), 747–753.

Bernaschi, P., Del, C. F., Petrucca, A., Argentieri, A., Ciofi Degli, A. M., Ciliento, G., et al. (2013). Microbial tracking of multidrug-resistant *Klebsiella pneumoniae* isolates in a pediatric hospital setting. *International Journal of Immunopathology and Pharmacology, 26*(2), 463–472.

Berrazeg, M., Diene, S. M., Drissi, M., Kempf, M., Richet, H., Landraud, L., et al. (2013). Biotyping of multidrug-resistant *Klebsiella pneumoniae* clinical isolates from France and Algeria using MALDI-TOF MS. *PLoS One, 8*(4), e61428.

Boggs, S. R., Cazares, L. H., & Drake, R. (2012). Characterization of a *Staphylococcus aureus* USA300 protein signature using matrix-assisted laser desorption/ionization time-of-flight mass spectrometry. *Journal of Medical Microbiology, 61*(Pt. 5), 640–644.

Burckhardt, I., & Zimmermann, S. (2011). Using matrix-assisted laser desorption ionization-time of flight mass spectrometry to detect carbapenem resistance within 1 to 2.5 hours. *Journal of Clinical Microbiology, 49*(9), 3321–3324.

Calderaro, A., Piccolo, G., Gorrini, C., Montecchini, S., Buttrini, M., Rossi, S., et al. (2014). *Leptospira* species and serovars identified by MALDI-TOF mass spectrometry after database implementation. *BMC Research Notes, 7*, 330.

Carvalhaes, C. G., Cayo, R., Assis, D. M., Martins, E. R., Juliano, L., Juliano, M. A., et al. (2013). Detection of SPM-1-producing *Pseudomonas aeruginosa* and class D beta-lactamase-producing *Acinetobacter baumannii* isolates by use of liquid chromatography-mass spectrometry and matrix-assisted laser desorption ionization-time of flight mass spectrometry. *Journal of Clinical Microbiology, 51*(1), 287–290.

Carvalhaes, C. G., Cayo, R., Visconde, M. F., Barone, T., Frigatto, E. A., Okamoto, D., et al. (2014). Detection of carbapenemase activity directly from blood culture vials using MALDI-TOF MS: A quick answer for the right decision. *The Journal of Antimicrobial Chemotherapy, 69*(8), 2132–2136.

Chatterjee, S. S., Chen, L., Joo, H.-S., Cheung, G. Y. C., Kreiswirth, B. N., & Otto, M. (2011). Distribution and regulation of the mobile genetic element-encoded phenol-soluble modulin PSM-mec in methicillin-resistant *Staphylococcus aureus*. *PLoS One, 6*, e28781.

Cherkaoui, A., Hibbs, J., Emonet, S., Tangomo, M., Girard, M., Francois, P., et al. (2010). Comparison of two matrix-assisted laser desorption ionization-time of flight mass spectrometry methods with conventional phenotypic identification for routine identification of bacteria to the species level. *Journal of Clinical Microbiology, 48*(4), 1169–1175.

Christner, M., Trusch, M., Rohde, H., Kwiatkowski, M., Schluter, H., Wolters, M., et al. (2014). Rapid MALDI-TOF mass spectrometry strain typing during a large outbreak of Shiga-toxigenic *Escherichia coli*. *PLoS One, 9*(7), e101924.

Clark, C. G., Kruczkiewicz, P., Guan, C., McCorrister, S. J., Chong, P., Wylie, J., et al. (2013). Evaluation of MALDI-TOF mass spectroscopy methods for determination of *Escherichia coli* pathotypes. *Journal of Microbiological Methods, 94*(3), 180–191.

de Bruijn, F. J. (1992). Use of repetitive (repetitive extragenic palindromic and enterobacterial repetitive intergeneric consensus) sequences and the polymerase chain reaction to fingerprint the genomes of *Rhizobium meliloti* isolates and other soil bacteria. *Applied and Environmental Microbiology, 58*(7), 2180–2187.

De Carolis, E., Vella, A., Florio, A. R., Posteraro, P., Perlin, D. S., Sanguinetti, M., et al. (2012). Use of matrix-assisted laser desorption ionization-time of flight mass spectrometry for caspofungin susceptibility testing of *Candida* and *Aspergillus* species. *Journal of Clinical Microbiology, 50*(7), 2479–2483.

Demirev, P. A., Hagan, N. S., Antoine, M. D., Lin, J. S., & Feldman, A. B. (2013). Establishing drug resistance in microorganisms by mass spectrometry. *Journal of the American Society for Mass Spectrometry, 24*(8), 1194–1201.

Dieckmann, R., Helmuth, R., Erhard, M., & Malorny, B. (2008). Rapid classification and identification of salmonellae at the species and subspecies levels by whole-cell matrix-assisted laser desorption ionization-time of flight mass spectrometry. *Applied and Environmental Microbiology, 74*(24), 7767–7778.

Du, Z., Yang, R., Guo, Z., Song, Y., & Wang, J. (2002). Identification of *Staphylococcus aureus* and determination of its methicillin resistance by matrix-assisted laser desorption/ionization time-of-flight mass spectrometry. *Analytical Chemistry, 74*(21), 5487–5491.

Edwards-Jones, V., Claydon, M. A., Evason, D. J., Walker, J., Fox, A. J., & Gordon, D. B. (2000). Rapid discrimination between methicillin-sensitive and methicillin-resistant *Staphylococcus aureus* by intact cell mass spectrometry. *Journal of Medical Microbiology, 49*(3), 295–300.

Emonet, S., Shah, H. N., Cherkaoui, A., & Schrenzel, J. (2010). Application and use of various mass spectrometry methods in clinical microbiology. *Clinical Microbiology and Infection, 16*(11), 1604–1613.

Fujinami, Y., Kikkawa, H. S., Kurosaki, Y., Sakurada, K., Yoshino, M., & Yasuda, J. (2011). Rapid discrimination of Legionella by matrix-assisted laser desorption ionization time-of-flight mass spectrometry. *Microbiological Research, 166*(2), 77–86.

Gaia, V., Casati, S., & Tonolla, M. (2011). Rapid identification of Legionella spp. by MALDI-TOF MS based protein mass fingerprinting. *Systematic and Applied Microbiology, 34*(1), 40–44.

Ghyselinck, J., Van, H. K., Hoste, B., Heylen, K., & de, V. P. (2011). Evaluation of MALDI-TOF MS as a tool for high-throughput dereplication. *Journal of Microbiological Methods, 86*(3), 327–336.

Giebel, R. A., Fredenberg, W., & Sandrin, T. R. (2008). Characterization of environmental isolates of *Enterococcus* spp. by matrix-assisted laser desorption/ionization time-of-flight mass spectrometry. *Water Research, 42*(4–5), 931–940.

Griffin, P. M., Price, G. R., Schooneveldt, J. M., Schlebusch, S., Tilse, M. H., Urbanski, T., et al. (2012). Use of matrix-assisted laser desorption ionization-time of flight mass spectrometry to identify vancomycin-resistant enterococci and investigate the epidemiology of an outbreak. *Journal of Clinical Microbiology, 50*(9), 2918–2931.

Grundt, A., Findeisen, P., Miethke, T., Jager, E., Ahmad-Nejad, P., & Neumaier, M. (2012). Rapid detection of ampicillin resistance in *Escherichia coli* by quantitative mass spectrometry. *Journal of Clinical Microbiology, 50*(5), 1727–1729.

Guibourdenche, M., Roggentin, P., Mikoleit, M., Fields, P. I., Bockemuhl, J., Grimont, P. A., et al. (2010). Supplement 2003–2007 (No. 47) to the White-Kauffmann-Le Minor scheme. *Research in Microbiology, 161*(1), 26–29.

Harmsen, D., Claus, H., Witte, W., Rothganger, J., Claus, H., Turnwald, D., et al. (2003). Typing of methicillin-resistant *Staphylococcus aureus* in a university hospital setting by using novel software for *spa* repeat determination and database management. *Journal of Clinical Microbiology, 41*(12), 5442–5448.

Hooff, G. P., van Kampen, J. J., Meesters, R. J., van, B. A., Goessens, W. H., & Luider, T. M. (2012). Characterization of beta-lactamase enzyme activity in bacterial lysates using MALDI-mass spectrometry. *Journal of Proteome Research, 11*(1), 79–84.

Hrabák, J., Chudackova, E., & Papagiannitsis, C. C. (2014). Detection of carbapenemases in *Enterobacteriaceae*: A challenge for diagnostic microbiological laboratories. *Clinical Microbiology and Infection, 20*(9), 839–853.

Hrabák, J., Studentova, V., Walkova, R., Zemlickova, H., Jakubu, V., Chudackova, E., et al. (2012). Detection of NDM-1, VIM-1, KPC, OXA-48, and OXA-162 carbapenemases by matrix-assisted laser desorption ionization-time of flight mass spectrometry. *Journal of Clinical Microbiology, 50*(7), 2441–2443.

Hrabák, J., Walkova, R., Studentova, V., Chudackova, E., & Bergerova, T. (2011). Carbapenemase activity detection by matrix-assisted laser desorption ionization-time of flight mass spectrometry. *Journal of Clinical Microbiology, 49*(9), 3222–3227.

Johansson, A., Nagy, E., & Soki, J. (2014). Detection of carbapenemase activities of Bacteroides fragilis strains with matrix-assisted laser desorption ionization–time of flight mass spectrometry (MALDI-TOF MS). *Anaerobe, 26*, 49–52.

Josten, M., Dischinger, J., Szekat, C., Reif, M., Al-Sabti, N., Sahl, H. G., et al. (2014). Identification of *agr*-positive methicillin-resistant *Staphylococcus aureus* harbouring the class A *mec* complex by MALDI-TOF mass spectrometry. *International Journal of Medical Microbiology, 304*(8), 1018–1023.

Josten, M., Reif, M., Szekat, C., Al-Sabti, N., Roemer, T., Sparbier, K., et al. (2013). Analysis of the matrix-assisted laser desorption ionization-time of flight mass spectrum of *Staphylococcus aureus* identifies mutations that allow differentiation of the main clonal lineages. *Journal of Clinical Microbiology, 51*(6), 1809–1817.

Jung, J. S., Eberl, T., Sparbier, K., Lange, C., Kostrzewa, M., Schubert, S., et al. (2014). Rapid detection of antibiotic resistance based on mass spectrometry and stable isotopes. *European Journal of Clinical Microbiology & Infectious Diseases, 33*(6), 949–955.

Jung, J. S., Popp, C., Sparbier, K., Lange, C., Kostrzewa, M., & Schubert, S. (2014). Evaluation of matrix-assisted laser desorption ionization-time of flight mass spectrometry for rapid detection of beta-lactam resistance in Enterobacteriaceae derived from blood cultures. *Journal of Clinical Microbiology, 52*(3), 924–930.

Karger, A., Ziller, M., Bettin, B., Mintel, B., Schares, S., & Geue, L. (2011). Determination of serotypes of Shiga toxin-producing *Escherichia coli* isolates by intact cell matrix-assisted laser desorption ionization-time of flight mass spectrometry. *Applied and Environmental Microbiology, 77*(3), 896–905.

Killgore, G., Thompson, A., Johnson, S., Brazier, J., Kuijper, E., Pepin, J., et al. (2008). Comparison of seven techniques for typing international epidemic strains of *Clostridium difficile*: Restriction endonuclease analysis, pulsed-field gel electrophoresis, PCR-ribotyping, multilocus sequence typing, multilocus variable-number tandem-repeat analysis, amplified fragment length polymorphism, and surface layer protein A gene sequence typing. *Journal of Clinical Microbiology, 46*(2), 431–437.

Klein, S., Zimmermann, S., Kohler, C., Mischnik, A., Alle, W., & Bode, K. A. (2012). Integration of matrix-assisted laser desorption/ionization time-of-flight mass spectrometry in blood culture diagnostics: A fast and effective approach. *Journal of Medical Microbiology, 61*(Pt. 3), 323–331.

Kostrzewa, M., Sparbier, K., Maier, T., & Schubert, S. (2013). MALDI-TOF MS: An upcoming tool for rapid detection of antibiotic resistance in microorganisms. *Proteomics. Clinical Applications, 7*(11–12), 767–778.

Kuhns, M., Zautner, A. E., Rabsch, W., Zimmermann, O., Weig, M., Bader, O., & Groß, U. (2012). Rapid discrimination of *Salmonella enterica* serovar Typhi from other serovars by MALDI-TOF mass spectrometry. *PLoS One, 7*, e40004.

Lange, C., Schubert, S., Jung, J., Kostrzewa, M., & Sparbier, K. (2014). Quantitative matrix-assisted laser desorption ionization-time of flight mass spectrometry for rapid resistance detection. *Journal of Clinical Microbiology, 52*(12), 4155–4162.

Lartigue, M. F., Hery-Arnaud, G., Haguenoer, E., Domelier, A. S., Schmit, P. O., Mee-Marquet, N., et al. (2009). Identification of *Streptococcus agalactiae* isolates from various phylogenetic lineages by matrix-assisted laser desorption ionization-time of flight mass spectrometry. *Journal of Clinical Microbiology, 47*(7), 2284–2287.

Lartigue, M. F., Kostrzewa, M., Salloum, M., Haguenoer, E., Hery-Arnaud, G., Domelier, A. S., et al. (2011). Rapid detection of "highly virulent" Group B Streptococcus ST-17 and emerging ST-1 clones by MALDI-TOF mass spectrometry. *Journal of Microbiological Methods, 86*(2), 262–265.

Lasch, P., Fleige, C., Stammler, M., Layer, F., Nubel, U., Witte, W., et al. (2014). Insufficient discriminatory power of MALDI-TOF mass spectrometry for typing of *Enterococcus faecium* and *Staphylococcus aureus* isolates. *Journal of Microbiological Methods, 100,* 58–69.

La Scola, B., & Raoult, D. (2009). Direct identification of bacteria in positive blood culture bottles by matrix-assisted laser desorption ionisation time-of-flight mass spectrometry. *PLoS One, 4*(11), e8041.

Lee, W., Chung, H. S., Lee, Y., Yong, D., Jeong, S. H., Lee, K., et al. (2013). Comparison of matrix-assisted laser desorption ionization-time-of-flight mass spectrometry assay with conventional methods for detection of IMP-6, VIM-2, NDM-1, SIM-1, KPC-1, OXA-23, and OXA-51 carbapenemase-producing *Acinetobacter* spp., *Pseudomonas aeruginosa,* and *Klebsiella pneumoniae. Diagnostic Microbiology and Infectious Disease, 77*(3), 227–230.

Majcherczyk, P. A., McKenna, T., Moreillon, P., & Vaudaux, P. (2006). The discriminatory power of MALDI-TOF mass spectrometry to differentiate between isogenic teicoplanin-susceptible and teicoplanin-resistant strains of methicillin-resistant *Staphylococcus aureus. FEMS Microbiology Letters, 255*(2), 233–239.

Marinach, C., Alanio, A., Palous, M., Kwasek, S., Fekkar, A., Brossas, J. Y., et al. (2009). MALDI-TOF MS-based drug susceptibility testing of pathogens: The example of *Candida albicans* and fluconazole. *Proteomics, 9*(20), 4627–4631.

Mencacci, A., Monari, C., Leli, C., Merlini, L., De Carolis, E., Vella, A., et al. (2013). Typing of nosocomial outbreaks of *Acinetobacter baumannii* by use of matrix-assisted laser desorption ionization-time of flight mass spectrometry. *Journal of Clinical Microbiology, 51,* 603–606.

Moura, H., Woolfitt, A. R., Carvalho, M. G., Pavlopoulos, A., Teixeira, L. M., Satten, G. A., et al. (2008). MALDI-TOF mass spectrometry as a tool for differentiation of invasive and noninvasive *Streptococcus pyogenes* isolates. *FEMS Immunology and Medical Microbiology, 53*(3), 333–342.

Moussaoui, W., Jaulhac, B., Hoffmann, A. M., Ludes, B., Kostrzewa, M., Riegel, P., et al. (2010). Matrix-assisted laser desorption ionization time-of-flight mass spectrometry identifies 90% of bacteria directly from blood culture vials. *Clinical Microbiology and Infection, 16*(11), 1631–1638.

Murray, P. R. (2010). Matrix-assisted laser desorption ionization time-of-flight mass spectrometry: Usefulness for taxonomy and epidemiology. *Clinical Microbiology and Infection, 16*(11), 1626–1630.

Nagy, E., Becker, S., Soki, J., Urban, E., & Kostrzewa, M. (2011). Differentiation of division I (*cfiA*-negative) and division II (*cfiA*-positive) *Bacteroides fragilis* strains by matrix-assisted laser desorption/ionization time-of-flight mass spectrometry. *Journal of Medical Microbiology, 60*(Pt. 11), 1584–1590.

Nagy, E., Urban, E., Becker, S., Kostrzewa, M., Voros, A., Hunyadkurti, J., et al. (2013). MALDI-TOF MS fingerprinting facilitates rapid discrimination of phylotypes I, II and III of *Propionibacterium acnes. Anaerobe, 20,* 20–26.

Peaper, D. R., Kulkarni, M. V., Tichy, A. N., Jarvis, M., Murray, T. S., & Hodsdon, M. E. (2013). Rapid detection of carbapenemase activity through monitoring ertapenem hydrolysis in *Enterobacteriaceae* with LC-MS/MS. *Bioanalysis, 5*(2), 147–157.

Pulcrano, G., Roscetto, E., Iula, V. D., Panellis, D., Rossano, F., & Catania, M. R. (2012). MALDI-TOF mass spectrometry and microsatellite markers to evaluate *Candida parapsilosis* transmission in neonatal intensive care units. *European Journal of Clinical Microbiology & Infectious Diseases, 31*(11), 2919–2928.

Quirino, A., Pulcrano, G., Rametti, L., Puccio, R., Marascio, N., Catania, M. R., et al. (2014). Typing of *Ochrobactrum anthropi* clinical isolates using automated repetitive extragenic palindromic-polymerase chain reaction DNA fingerprinting and matrix-assisted laser desorption/ionization-time-of-flight mass spectrometry. *BMC Microbiology, 14*, 74.

Reil, M., Erhard, M., Kuijper, E. J., Kist, M., Zaiss, H., Witte, W., et al. (2011). Recognition of *Clostridium difficile* PCR-ribotypes 001, 027 and 126/078 using an extended MALDI-TOF MS system. *European Journal of Clinical Microbiology & Infectious Diseases, 30*, 1431–1436.

Rezzonico, F., Vogel, G., Duffy, B., & Tonolla, M. (2010). Application of whole-cell matrix-assisted laser desorption ionization-time of flight mass spectrometry for rapid identification and clustering analysis of pantoea species. *Applied and Environmental Microbiology, 76*(13), 4497–4509.

Ryzhov, V., Hathout, Y., & Fenselau, C. (2000). Rapid characterization of spores of *Bacillus cereus* group bacteria by matrix-assisted laser desorption-ionization time-of-flight mass spectrometry. *Applied and Environmental Microbiology, 66*(9), 3828–3834.

Sandrin, T. R., Goldstein, J. E., & Schumaker, S. (2013). MALDI TOF MS profiling of bacteria at the strain level: A review. *Mass Spectrometry Reviews, 32*(3), 188–217.

Sauget, M., Cabrolier, N., Manzoni, M., Bertrand, X., & Hocquet, D. (2014). Rapid, sensitive and specific detection of OXA-48-like-producing *Enterobacteriaceae* by matrix-assisted laser desorption/ionization time-of-flight mass spectrometry. *Journal of Microbiological Methods, 105*, 88–91.

Sauget, M., Nicolas-Chanoine, M. H., Cabrolier, N., Bertrand, X., & Hocquet, D. (2014). Matrix-assisted laser desorption ionization-time of flight mass spectrometry assigns *Escherichia coli* to the phylogroups A, B1, B2 and D. *International Journal of Medical Microbiology, 304*(8), 977–983.

Schubert, S., Weinert, K., Wagner, C., Gunzl, B., Wieser, A., Maier, T., et al. (2011). Novel, improved sample preparation for rapid, direct identification from positive blood cultures using matrix-assisted laser desorption/ionization time-of-flight (MALDI-TOF) mass spectrometry. *The Journal of Molecular Diagnostics, 13*(6), 701–706.

Scotta, C., Gomila, M., Mulet, M., Lalucat, J., & Garcia-Valdes, E. (2013). Whole-cell MALDI-TOF mass spectrometry and multilocus sequence analysis in the discrimination of *Pseudomonas stutzeri* populations: Three novel genomovars. *Microbial Ecology, 66*(3), 522–532.

Seng, P., Rolain, J. M., Fournier, P. E., La Scola, B., Drancourt, M., & Raoult, D. (2010). MALDI-TOF-mass spectrometry applications in clinical microbiology. *Future Microbiology, 5*(11), 1733–1754.

Sparbier, K., Lange, C., Jung, J., Wieser, A., Schubert, S., & Kostrzewa, M. (2013). MALDI biotyper-based rapid resistance detection by stable-isotope labeling. *Journal of Clinical Microbiology, 51*(11), 3741–3748.

Sparbier, K., Schubert, S., Weller, U., Boogen, C., & Kostrzewa, M. (2012). Matrix-assisted laser desorption ionization-time of flight mass spectrometry-based functional assay for rapid detection of resistance against beta-lactam antibiotics. *Journal of Clinical Microbiology, 50*(3), 927–937.

Stevenson, L. G., Drake, S. K., & Murray, P. R. (2010). Rapid identification of bacteria in positive blood culture broths by matrix-assisted laser desorption ionization-time of flight mass spectrometry. *Journal of Clinical Microbiology, 48*(2), 444–447.

Szabados, F., Kaase, M., Anders, A., & Gatermann, S. G. (2012). Identical MALDI TOF MS-derived peak profiles in a pair of isogenic SCC*mec*-harboring and SCC*mec*-lacking strains of *Staphylococcus aureus*. *The Journal of Infection, 65*(5), 400–405.

Tanigawa, K., Kawabata, H., & Watanabe, K. (2010). Identification and typing of *Lactococcus lactis* by matrix-assisted laser desorption ionization-time of flight mass spectrometry. *Applied and Environmental Microbiology, 76*(12), 4055–4062.

Vella, A., De Carolis, E., Vaccaro, L., Posteraro, P., Perlin, D. S., Kostrzewa, M., et al. (2013). Rapid antifungal susceptibility testing by matrix-assisted laser desorption ionization-time of flight mass spectrometry analysis. *Journal of Clinical Microbiology, 51*(9), 2964–2969.

Walker, J., Fox, A. J., Edwards-Jones, V., & Gordon, D. B. (2002). Intact cell mass spectrometry (ICMS) used to type methicillin-resistant *Staphylococcus aureus*: Media effects and inter-laboratory reproducibility. *Journal of Microbiological Methods, 48*(2–3), 117–126.

Werner, G., Fleige, C., Fessler, A. T., Timke, M., Kostrzewa, M., Zischka, M., et al. (2012). Improved identification including MALDI-TOF mass spectrometry analysis of group D streptococci from bovine mastitis and subsequent molecular characterization of corresponding *Enterococcus faecalis* and *Enterococcus faecium* isolates. *Veterinary Microbiology, 160*(1–2), 162–169.

Williamson, Y. M., Moura, H., Woolfitt, A. R., Pirkle, J. L., Barr, J. R., Carvalho, M. G., et al. (2008). Differentiation of *Streptococcus pneumoniae* conjunctivitis outbreak isolates by matrix-assisted laser desorption ionization-time of flight mass spectrometry. *Applied and Environmental Microbiology, 74*(19), 5891–5897.

Wolters, M., Rohde, H., Maier, T., Belmar-Campos, C., Franke, G., Scherpe, S., et al. (2011). MALDI-TOF MS fingerprinting allows for discrimination of major methicillin-resistant *Staphylococcus aureus* lineages. *International Journal of Medical Microbiology, 301*(1), 64–68.

Wybo, I., De, B. A., Soetens, O., Echahidi, F., Vandoorslaer, K., Van, C. M., et al. (2011). Differentiation of *cfiA*-negative and *cfiA*-positive *Bacteroides fragilis* isolates by matrix-assisted laser desorption ionization-time of flight mass spectrometry. *Journal of Clinical Microbiology, 49*(5), 1961–1964.

Zautner, A. E., Masanta, W. O., Tareen, A. M., Weig, M., Lugert, R., Gross, U., et al. (2013). Discrimination of multilocus sequence typing-based *Campylobacter jejuni* subgroups by MALDI-TOF mass spectrometry. *BMC Microbiology, 13*, 247.

Next-Generation Sequencing for Pathogen Detection and Identification

15

Kenneth G. Frey*,†, Kimberly A. Bishop-Lilly*,†,1

*Henry M. Jackson Foundation for the Advancement of Military Medicine,
Bethesda, Maryland, USA
†Naval Medical Research Center—Frederick, Fort Detrick, Frederick, Maryland, USA
1Corresponding author: e-mail address: kimberly.a.bishop-lilly.ctr@mail.mil

1 INTRODUCTION TO NEXT-GENERATION SEQUENCING

Next-generation sequencing (NGS) is a general term that describes the class of high-throughput sequencers that have been developed after chemical degradation (Maxam & Gilbert, 1977) and chain termination (Sanger; Sanger, Nicklen, & Coulson, 1977) sequencing. Due to next-generation sequencers' output of millions or billions of sequence reads per instrument run, the process of genome sequencing has been revolutionized such that a single laboratory technician can now single-handedly produce whole-genome sequence (WGS) data for multiple bacterial organisms overnight.

The first next-generation sequencer to be commercially available was the 454 pyrosequencer (Margulies et al., 2005), quickly followed by the Solexa platform (now Illumina; Bennett, 2004), and then the Applied Biosystems SOLiD sequencer (Valouev et al., 2008). The 454, although it offers longer read lengths than Illumina and SOLiD, has been unable to compete in terms of throughput (number of reads), making it generally less cost effective than the other platforms. Although each platform has its own proprietary chemistry, in general, these so-called short-read sequencers share many characteristics in terms of workflows. In comparison to the short-read sequencers, the PacBio sequencer (Eid et al., 2009) is unique both in terms of its extremely long read lengths and its sensitivity, which obviates the need for an amplification step that the other platforms employ (for an overview of NGS chemistries, see Buermans & den Dunnen, 2014). A brief summary of some representative sequencing platforms and their characteristics in terms of read length and throughput is provided in Table 1.

Regardless of the sequencing platform employed for data generation, one of the first steps in data analysis in many cases is genome assembly, which involves fitting together of many individual sequencing reads into larger contiguous blocks of sequence, or 'contigs'. In the case of microbial whole-genome sequencing, a variety

Methods in Microbiology, Volume 42, ISSN 0580-9517, http://dx.doi.org/10.1016/bs.mim.2015.06.004

Table 1 Characteristics of Next-Generation Sequencing Platforms

	Manufacturer	Output	Number of Reads	Read Length
MiSeq/MiSeq Dx	Illumina	15 Gb	25M	2 × 300 bp
NextSeq 500	Illumina	120 Gb	400M	2 × 150 bp
HiSeq 2500	Illumina	1 Tb	4B	2 × 125 bp
HiSeq XTen	Illumina	1.8 Tb	6B	2 × 150 bp
Ion Proton	Life Technologies	10 Gb	80M	400 bp
Ion Torrent	Life Technologies	2 Gb	5.5M	400 bp
PacBio RS II	Pacific Biosciences	6 Gb[a]	800K[b]	Ave. 8.5 kb; max 30 kb

[a]375 Mb/SMRT cell × 16 SMRT cells max.
[b]50K/SMRT cell × 16 SMRT cells max.

of software packages can be applied post-quality filtering to align the reads to a reference genome to produce a reference-guided assembly, or to assemble the reads *de novo* (without using a reference). In general, genomes can be produced to varying levels of completeness, such as standard draft, high-quality draft, improved high-quality draft, and finished (as defined in Chain et al., 2009). As overall throughput of NGS has increased, there has been a trend away from finished genomes and manual genome closure and a trend towards more draft sequencing, whereby the product is assembled contigs, often with gaps between them.

Another current trend in NGS is the decreasing laboratory footprint of sequencers. In addition to the extremely high-throughput options (such as the HiSeq), there are also now scaled-back versions available with smaller footprints (such as the Illumina MiSeq/MiSeqDx and NextSeq 500 as well as Life Technologies Ion Torrent PGM and Ion Proton) that better suit the needs of small academic labs and microbial sequencing work. It is these smaller bench-top sequencers that are currently viewed as having potential for fieldable sequencing (sequencing in remote locations often lacking basic infrastructure) and for clinical applications such as infectious disease diagnostics. Taking the idea of decreased footprint even further is Oxford Nanopore's MinION sequencer (Howorka, Cheley, & Bayley, 2001), which is currently only available through an early access programme. This sequencer has caused much excitement due to its extremely small size (handheld) and long read lengths. However, initial data were recently published that suggest substantial chemistry improvements would be required to support microbial diagnostics in the clinical laboratory or in the field. Despite generating 150 Mb of output per run, the data produced are very error prone, with one report of up to 75% of reads being discarded and those reads that were kept having less than 10% average identity with the reference sequence (Mikheyev & Tin, 2014). A more recent publication reported that the majority of reads obtained mapped to reference and that the proportion of high-quality, informative reads was increased by a chemistry upgrade that was released in

September 2014 (Quick, Quinlan, & Loman, 2014). Further studies and broader access will allow for a more thorough assessment of the technology and its utility, which are likely to improve over time.

Despite the continual improvements in NGS technologies that increasingly make the impossible seem possible, challenges remain. NGS platforms tend to have their own intrinsic error profiles that differ slightly from one another, owing to the differing chemistries. For certain applications, this may necessitate platform-specific quality control (QC) to be performed. Additionally, even for a given platform, error profiles may vary based on genomic context. For instance, homopolymer regions are notorious trouble spots for polymerase-based sequencing platforms. For Ion Torrent sequencing, insertions/deletions are known to be the predominant error type overall (Bragg, Stone, Butler, Hugenholtz, & Tyson, 2013). Interestingly, however, a recent publication showed that as the length of homopolymer tracts increases, deletion rates increase for Illumina, Ion Torrent, and PacBio platforms. Conversely, within these same regions, insertion rates increase for Illumina sequencing, decrease for PacBio, and are unaffected in Ion Torrent sequencing (Ross et al., 2013). Furthermore, while guanosine+cytosine (G+C) content is a known factor that affects sequence coverage and quality, the various NGS platforms show different effects. For instance, a recent study showed that whereas Illumina, PacBio, and Complete Genomics' deletion rates increase in regions of extremely low or extremely high G+C content, PacBio's insertion rate actually decreases, whereas Illumina's increases (Ross et al., 2013). Although the sequencing chemistries continue to improve in terms of not only throughput but also accuracy, for discrimination between sequence errors and true variations, technical replicates and use of multiple complementary chemistries per genome may be required in addition to platform-specific QC protocols (Bragg et al., 2013; Frey et al., 2014).

Increasingly, NGS applications extend beyond simple genome sequencing to include a variety of protocols that have some clinical or diagnostic relevance. For instance, epigenetic sequencing methodologies include methylation sequencing and DNA–protein interaction analysis (ChIP-Seq), both of which are relevant to cancer detection, as aberrant methylation and altered transcription factor binding are two common epigenetic changes in cancer cells. Although each of the sequencers is capable of multiple applications, there are differences among them in how these applications are accomplished. For instance, whereas Illumina offers a specific protocol for methylation sequencing, along with the primary sequence data, the PacBio platform provides methylation information as a by-product of each sequencing run. In addition to epigenetic sequencing, small RNA sequencing is another relevant application compatible with most platforms. MicroRNAs (miRNAs) and other non-coding RNAs can play a role in regulation of gene expression in normal cellular processes such as haematopoietic lineage differentiation (Elton, Selemon, Elton, & Parinandi, 2013) and mammary epithelial differentiation (Yoo et al., 2014), as well as disease states from cancer (Jeong, 2014; Liu & Xiao, 2014; Pickl, Heckmann, Ratz, Klauck, & Sultmann, 2014; Zheng et al., 2014) to infection (Kincaid & Sullivan, 2012; Maudet, Mano, & Eulalio, 2014). NGS has been instrumental in

Table 2 Examples of Infectious Agents for Which Candidate miRNA Biomarkers of Infection Have Been Identified

Organism	Disease	Host/ Pathogen Encoded	Reference
Angiostrongylus cantonensis	Eosinophilic meningoencephalitis	Pathogen	Chen et al. (2014)
Schistosoma mansoni	Schistosomiasis (bilharzia)	Pathogen	Hoy et al. (2014)
Schistosoma japonicum	Schistosomiasis (bilharzia)	Host	He et al. (2013)
BK virus	BK viral nephropathy in renal transplant recipients	Pathogen	Li et al. (2014)
Avian influenza A (H7N9)	Influenza	Host	Zhu et al. (2014)
Enterovirus 71	Hand-foot-and-mouth disease	Host	Cui et al. (2011)
Coxsackievirus 16	Hand-foot-and-mouth disease	Host	Cui et al. (2011)
Human papillomavirus 16 and 18	HPV infection	Host	Wang et al. (2014)
Hepatitis B virus	Hepatitis	Host	Tan et al. (2014)
Mycobacterium tuberculosis	Tuberculosis	Host	Qi et al. (2012), Zhang et al. (2013, 2014)
Francisella tularensis	Tularemia	Host	Cer et al. (manuscript in review)
Burkholderia pseudomallei	Melioidosis	Host	Cer et al. (manuscript in review)

identifying a number of miRNA species that are potential candidates for diagnostic biomarkers (see Table 2 for examples). Finally, metagenomic sequencing, or sequencing from complex samples such as stool or sputum with little or no upfront purification (discussed in detail below), is an NGS application with particular promise for microbial diagnostics.

2 METAGENOMIC SEQUENCING

The term metagenomics was first used by Handelsman and colleagues (Handelsman, Rondon, Brady, Clardy, & Goodman, 1998). Metagenomics was defined as the culture-independent molecular analysis of environmental samples of co-habitating populations. At the time, a bacterial artificial chromosome-based cloning approach was utilized to sequence genomic fragments isolated from soil bacteria. It was noted that this approach had vast potential to facilitate the discovery of novel genes and undescribed metabolic pathways. However, as the extraction protocol was focused towards nucleic acid from bacteria specifically, the sequences obtained would most likely represent the most dominant bacterial species in the soil sample. Thus, the purification method employed can introduce bias in the sequence data (Carrigg, Rice, Kavanagh, Collins, & O'Flaherty, 2007; Delmont, Robe, Clark, Simonet, & Vogel, 2011).

In 2004, metagenomic data were published from a thin biofilm residing in acid rock drainage near an abandoned mine. Due to the extremely low pH (0.83) and high levels of inorganic metal, the number of microbial taxa identified in the community was limited. Using Sanger sequencing, the study was able to recover over 100,000 reads that assembled into over 1000 contigs (Tyson et al., 2004). Genomic scaffolds that resulted from bioinformatic binning of contigs suggested the presence of a previously uncharacterized species of *Ferroplasma*. Further metagenomic sequencing of microbial communities followed, including the well-publicized study of the Sargasso Sea by Venter et al. (2004). Using what was termed a 'whole-genome shotgun sequencing' approach, the group was able to generate over 1 billion non-redundant bases. Assembly and annotation of the sequence reads suggested that the community contained 1.2 million previously unknown genes, including over 700 genes coding for rhodopsin-like proteins.

The above studies highlighted the power of metagenomic sequencing to unveil the incredible genetic diversity present in even a simple environment. However, these studies were performed using clone-based Sanger sequencing. The authors of these studies were limited by the technology of the time, the genomes that were amenable to cloning, and the sampling method employed. Since then, NGS has enabled metagenomic studies of complex communities including the human microbiome (Human Microbiome Jumpstart Reference Strains Consortium et al., 2010). Applications of NGS metagenomic sequencing are far ranging, from epidemiological studies of disease transmission to drug discovery (Parkhill, 2013).

A primary example is a study of honeybee Colony Collapse Disorder (CCD) published in 2007 (Cox-Foster et al., 2007). In this study, a metagenomic 'survey' was conducted, in which numerous healthy colony samples and diseased colony samples were characterized by 454 sequencing and Israeli acute paralysis virus was identified as being correlated with CCD. Then, in early 2008, a study was published in which NGS was used to identify a novel arenavirus in three transplant patients who had died of an unexplained febrile illness (Palacios et al., 2008). In this case, all available microbiological, serological, polymerase chain reaction (PCR), and microarray assays failed to identify a causal agent of the febrile illness. This report highlighted the power of NGS to detect a novel agent from a limited amount of sequence information. Indeed, of over 100,000 sequence reads, only 14 corresponded to the new virus. The sequences were the basis for design of multiple detection assays, including a real-time PCR assay to detect the virus in tissue samples. Using the same experimental approach, a second novel arenavirus was identified from clinical specimens originating from a highly fatal outbreak in South Africa (Briese et al., 2009).

3 NGS FOR RESEARCH DIAGNOSTIC APPLICATIONS

Due to the continuing advances in NGS technology that make sequencing cheaper and easier than ever, sequencing from patient samples for various purposes is becoming more commonplace. In fact, the number of institutions offering NGS services for clinical diagnostics and the wide array of clinical applications that are coming available recently prompted the College of American Pathologists to form an NGS Work Group to define a set of clinical laboratory standards to regulate clinical sequencing. The standards they created, which were published recently in 2014, encompass wet lab and bioinformatic aspects, from documentation and traceability to interpretation of genetic variants. The checklist requirements established by this work group are to be applied to all clinical NGS diagnostic laboratory efforts, from cancer to infectious diseases (Aziz et al., 2015). However, these guidelines as written are somewhat human genome centric, and to govern microbial diagnostics by NGS directly from human specimens (metagenomic sequencing) could conceivably involve development of additional guidelines.

Although clinical use of NGS for microbial diagnostics is currently in its infancy, there is a vast array of diagnostic applications for NGS that provide some precedent for its clinical use. For instance, some of the more established clinical uses for NGS are to characterize cancer-causing mutations (Ku, Cooper, & Roukos, 2013; Loyo et al., 2013; Luthra et al., 2014) and to discover the basis of rare genetic syndromes (Madrigal et al., 2014; Maher, 2007; Ng et al., 2010). Additionally, NGS has been used to perform HLA typing for individuals donating or receiving stem cell transplants (Smith et al., 2014), and in terms of microorganism-related work, to assess stable transfer of gut microbes following faecal transplant (Hamilton, Weingarden, Unno, Khoruts, & Sadowsky, 2013), and to study the relationship between gut microbes and various medical conditions such as Type II diabetes (reviewed in Parkhill, 2013) and inflammatory bowel disease (Alkadhi, Kunde, Cheluvappa, Randall-Demllo, & Eri, 2014). In 2013, Illumina received marketing approval from the U.S. Food and Drug Administration (FDA) for their instrument, MiSeqDx, a bench-top sequencer designed especially with the clinic in mind. MiSeqDx is notable in that it represents the first NGS platform approved by the FDA for diagnostics and it provides the flexibility for researchers to design their own custom amplicon sequencing assays for gene(s) of interest ('FDA-approved NGS system could expand clinical genomic testing: Experts predict MiSeqDx system will make genetic testing more affordable for smaller labs', 2014).

3.1 ADVANTAGES OF NGS FOR INFECTIOUS DISEASE DIAGNOSTICS

As opposed to traditional infectious disease diagnostics, such as those that are based on culture, PCR, and serology, NGS-based diagnostics have certain characteristics that are particularly attractive (Table 3). First, results from NGS-based methods are less biased than other methods. By necessity, traditional diagnostics require some prior knowledge of the agent, whereas NGS requires little or none. For instance, to

Table 3 Characteristics of Conventional Clinical Microbiology Diagnostic Methods Compared to NGS-Based Methods

Method	Unbiased	Rapid Identification of Fastidious Microbes	Identification of Species	Identification of Strains/ Resolve Phylogenies	Identification of Antibiotic Resistance	Identification of Co-infections	Inform Drug Choice	Reference
Culture/ phenotypic	No	No	+/−	−	++	+	+	Hammoudi, Ayoub Moubareck, and Karam Sarkis (2014)
Serological	No	Yes	+/−	−	−	+	+	Oskam, Slim, and Buhrer-Sekula (2003)
Polymerase chain reaction (PCR)	No	Yes	++	+	+	++	+	Leggieri, Rida, Francois, and Schrenzel (2010), Yu et al. (2012)
MALDI-TOF	Yes	Yes	++	++	+++	+++	++	Johansson, Nagy, and Soki (2014), Nagy (2014), Sandrin, Goldstein, and Schumaker (2013)
Multi-locus sequence typing (MLST)	No	Yes	++	++	++	+	+	Foxman, Zhang, Koopman, Manning, and Marrs (2005), Perez-Losada, Cabezas, Castro-Nallar, and Crandall (2013)

Continued

Table 3 Characteristics of Conventional Clinical Microbiology Diagnostic Methods Compared to NGS-Based Methods—cont'd

Method	Unbiased	Rapid Identification of Fastidious Microbes	Identification of Species	Identification of Strains/ Resolve Phylogenies	Identification of Antibiotic Resistance	Identification of Co-infections	Inform Drug Choice	Reference
rRNA gene sequencing[a]	Yes	Yes	++	+	−	+	+	Lane et al. (1985), Wang and Qian (2009), Woo, Lau, Teng, Tse, and Yuen (2008)
NGS[b]	Yes	Yes	+++	+++	+++	+++	+++	Moore et al. (2011), Schaefer et al. (2014), Turingan, Thomann, Zolotova, Tan, and Selden (2013)

(−), not useful; (+), of limited use, additional confirmation required; (++), useful; (+++), extremely useful.
[a]Assumes bacterial agent.
[b]Assumes adequate genome coverage.

amplify a pathogen target via PCR presupposes that the clinician knows what target pathogens to screen for. Though PCR assays can be multiplexed to broaden their ability to detect a range of targets, there are limits to how many targets can be multiplexed together. Additionally, PCR and serological assays have the limitation that as targets evolve or change, be it through natural processes or via genetic manipulation, a false negative may result. PCR assays tend to target conserved regions/loci but even so, their sensitivity can be affected over time by genetic drift and shift. On the other hand, because NGS protocols in general make use of specialized adapter sequences for priming and amplification of DNA fragments, theoretically any DNA sequence that is present in the sample can be sequenced, without any prior knowledge of the organism(s). Furthermore, the unbiased nature of NGS allows for detection of co-infections that might ordinarily be missed by other assays.

As discussed above, NGS requires no prior evidence of the presence of a particular organism. Other sequence-based assays, such as multi-locus sequence typing (MLST), require knowledge about the genus and family of the organism. For instance, in the case of the fastidious organism *Chlamydia trachomatis*, a researcher could design an MLST panel and analyse the data online (http://chlamydia.mlst.net/sql/multiplelocus.asp). However, Sanger sequencing of the resultant amplicons would be necessary and would provide data about only a limited set of loci. By contrast, a single NGS run can provide a much more information-rich data set. Furthermore, the time required for actionable clinical information has been much reduced as compared to using traditional assays (Wilson et al., 2014).

An additional strength of NGS for infectious disease diagnostics is that NGS can detect non-culturable or fastidious organisms. The first step in traditional diagnostic microbiology is generally to isolate the organism from the original sample using some kind of microbiological medium, within cell culture or within embryonated chicken eggs. However, some organisms are very difficult or impossible to grow in the laboratory, while others may take weeks to isolate, delaying subsequent tests such as genotyping or antimicrobial susceptibility testing. Relevant examples include *Mycobacterium tuberculosis* and *Mycobacterium leprae*, the causative agents of tuberculosis and leprosy, respectively. Other difficult to culture organisms include *Chlamydia* sp., which are obligate intracellular parasites. Seth-Smith et al. have shown that NGS can be used in conjunction with immunomagnetic separation and multiple displacement amplification (IMS-MDA; Seth-Smith, Harris, Scott, et al., 2013) to produce WGS data for *C. trachomatis* directly from clinical samples (discarded swabs), obviating the need for time-intensive tissue culture to isolate the organism (Seth-Smith, Harris, Skilton, et al., 2013). Furthermore, Christiansen et al. recently demonstrated that NGS following whole-genome enrichment is even more sensitive than IMS-MDA for detection of *C. trachomatis* and for SNP-based selection of appropriate antibiotic therapy (Christiansen et al., 2014). In the case of co-infections, standard culture-based diagnostics may be biased towards the more readily cultured organism, yielding an incomplete clinical picture and potentially inadequate treatment, whereas NGS may give the clinician a more complete picture and allow for better treatment options to be prescribed. For example, direct sequencing of urine samples identified *Aerococcus urinae* as a component of a mixed infection that was missed by standard culture methods (Hasman et al., 2014).

Perhaps the greatest strength of NGS for infectious disease diagnostics is the simultaneous breadth and granularity of information that is obtainable within a single sequencing reaction. In theory, all the information that is obtained via a myriad of traditional diagnostic tests could be obtained rapidly and within a single step, sequencing, via a standard protocol that does not vary per organism and therefore requires one type of professional expertise versus the expertise required to run many organism-specific protocols (reviewed in Didelot, Bowden, Wilson, Peto, & Crook, 2012). Therefore, when identification and characterization of a strain is required, NGS can potentially provide all the information in one analysis. Characterization may include, for example, prediction of drug resistance, identification of virulence factors, and/or inference of transmission patterns. For instance, it has been demonstrated that NGS can resolve some chains of transmission that traditional molecular methods cannot, as in the case of carbapenemase-encoding Enterobacteriaceae. Recently, Conlan et al. reported PacBio sequencing of carbapenemase-positive patient isolates to discriminate hospital-acquired colonization from acquisition from other sources. Carbapenemase genes are often encoded on plasmids with repetitive sequence content and flanked by transposons; so in this case, the long reads obtained by PacBio sequencing avoided assembly issues typical to other sequencing chemistries and produced contiguous plasmid sequences that allowed for improved discrimination among transmission modes (Conlan et al., 2014). Furthermore, NGS was recently shown to allow for fine-scale strain typing of *Streptococcus pyogenes* isolates from cases of puerperal sepsis that traditional typing did not (Ben Zakour, Venturini, Beatson, & Walker, 2012) and two case reports were recently published in which NGS was used to definitively prove transmission of methicillin-resistant *Staphylococcus aureus* associated with organ transplantation (Altman et al., 2014; Wendt et al., 2014).

3.2 DISADVANTAGES OF NGS FOR MICROBIAL DIAGNOSTICS

Along with the many advantages of NGS come some significant disadvantages. First, NGS is relatively expensive by most laboratories' standards. Not taking into account equipment or labour costs, the reagents alone typically cost a minimum of $1000 per sequencing run. This cost can be mitigated with sample multiplexing if the situation allows, such as following pathogen enrichment or host removal methods, or with amplicon sequencing. However, there are some cases when multiplexing would sacrifice the ability to detect pathogens.

In general, the ability to detect a given pathogen by metagenomic sequencing will depend on the amount of the organism present and the complexity of the total range of organisms in the sample, which determines the contribution of organism-specific reads versus host and commensal reads. This is one main difference of NGS-based diagnostics as compared to specific culture methods. For instance, urine is a less complex sample type than a vaginal swab, and so detection of *C. trachomatis* via

NGS is more sensitive from urine than from vaginal swabs (Christiansen et al., 2014). Additionally, intrinsic characteristics of an organism, such as small genome size, can have a negative effect on the ability to detect by NGS, which is generally not the case for PCR-based detection methods which make use of organism-specific primers. The limit of detection (LoD) of NGS is not only platform specific, and affected by the sample type and therefore complexity of the background sequences, but also affected by the specific nucleic acid extraction protocol or kit and the library preparation protocol used (Greninger et al., 2010).

In general, at or near the LoD, one can expect although NGS may enable detection of a pathogen, NGS's ability to characterize the pathogen would be minimized due to inadequate depth and breadth of genome coverage (Frey et al., 2014). For some sample types, the LoD for metagenomic sequencing is finally beginning to approach that of PCR due to increased throughput (Frey et al., 2014; Greninger et al., 2010), but there are some cases in which PCR or culture-based methods may still be more sensitive for detection of known pathogens. For example, Prachayangprecha et al. conducted NGS on nasopharyngeal aspirates of Thai children with respiratory illness and found that for rhinovirus and human metapneumovirus, NGS and PCR had similar sensitivity. However, for two other viruses detected from the children's samples, NGS was less sensitive than RT-PCR (Prachayangprecha et al., 2014). Loman et al. showed that in the case of an *Escherichia coli* O104:H4 outbreak, NGS of stool samples could be more sensitive than detection by Shiga toxin-specific enzyme-linked immunosorbent assay, but less sensitive than culture (Loman et al., 2013). Table 4 gives some published examples of pathogens detected in various clinical sample types and the relative titres, expressed as genome copies.

In addition to considerations of cost and sensitivity, the quality of nucleic acid required for NGS is another practical consideration that may represent a true obstacle in some situations. Typically, NGS library preparation has more stringent requirements than does PCR, making cold chain critical. When sequencing samples from regions that lack certain basic infrastructure, shipment and/or storage at room temperature or on wet ice rather than dry ice can ruin an NGS experiment even when PCR may still have worked on the specimens. However, as library protocol requirements for quantity of input nucleic acid have been decreasing over time, sequencing these suboptimal samples becomes more feasible.

Time is another factor that can be a disadvantage for using NGS as a diagnostic tool. At minimum, it generally takes at least 24 h from start of library preparation to receipt of raw sequence data, and from that point onwards, the time required to analyse the data can vary greatly depending on sample type (purified organism or complex sample), the granularity of information required (organism identification vs. characterization), and the computational hardware available. For fastidious organisms, NGS may well be the quickest route, but for other organisms it may not. Prachayangprecha et al. discuss the trade-off between time and quantity of information gathered in the context of their viral diagnostic study. Whereas PCR was not only more sensitive in some cases, it also resulted in a faster diagnosis than NGS.

Table 4 Examples of Pathogens Detected via NGS and Pathogen Load

Microorganism	Sample Type	Load	Genome Size	Complete/Partial Genome	Sequencing Platform	Reference
Human astrovirus Puget Sound (HAstV-PS)	Frontal cortex biopsy from encephalitis patient	1.53×10^7 g.c. per RT-PCR	6.6 kb	Partial	Roche 454 GS-FLX	Quan et al. (2010)
Rabies virus	Whole blood spiked with infected cells	1.05×10^4 g.c. per mL	11.9 kb	Partial	Roche 454 GS-FLX Titanium	Cheval et al. (2011)
Influenza virus	Nasal swab from influenza patient	4.74×10^3 g.c. per mL	13.6 kb	Partial	Illumina Genome Analyzer IIx	Greninger et al. (2010)
Influenza virus	Whole blood spiked with virus	Est. 3.2×10^4 to 5.4×10^4 g.c. per mL	13.6 kb	Partial	Illumina MiSeq	Frey et al. (2014)
Leptospira santarosai	CSF from meningitis patient	958 g.c. per mL	3.9 Mb	Partial	Illumina MiSeq	Wilson et al. (2014)
Dolphin polyomavirus 1	FFPE laryngeal tissue from dolphin with tracheobronchitis	2.5×10^6 g.c. per 100 ng DNA	5.2 kb	Complete	Life Technologies Ion Torrent	Anthony et al. (2013)

g.c., genome copies.

On the other hand, NGS resulted in complete strain typing information that PCR did not (Prachayangprecha et al., 2014).

Perhaps the most formidable challenge for adoption of NGS in clinical microbiology is the complexity of the data analysis. Although there have been a succession of technological improvements in sequencing methodologies that have lowered the barrier to enable broader adoption of NGS in many smaller laboratories around the world, for the most part the associated bioinformatics still remain a challenge. These challenges are addressed in detail in Section 4.

Finally, it should be noted that in contrast to traditional culture-based diagnostic practices, identification via metagenomic sequencing is independent of viability, which in the case of diagnostics could theoretically be problematic. For instance, if a patient on antibiotic therapy was to develop a secondary infection and present with symptoms, metagenomic diagnostics may detect presence of the original, now nonviable, pathogen and lead a physician to believe that the original infection had not been treated effectively. However, this same weakness is inherent to all nucleic acid-based diagnostic approaches, including PCR. A potential implementation of NGS that would allow for detection and characterization of only viable, replicating organisms would be transcriptome, rather than genome, sequencing.

3.2.1 Improving the sensitivity of detection of pathogens by NGS

In response to the technical challenges involved in performing NGS directly on clinical samples, which are by nature metagenomic (complex) samples in which the pathogen may not always be the most abundant organism in the sample, a variety of protocols have been developed to increase sensitivity. For instance, host subtraction is one common method for increasing the proportion of reads analysed that are pathogen derived. Host subtraction can be applied prior to sequencing, to physically decrease the proportion of host nucleic acid versus microbial nucleic acid. Additionally, host subtraction can be applied post-sequencing to bioinformatically remove host-derived reads such that one may focus computational resources on the reads of interest, which are generally the microbial-derived reads (except in the case of biomarker-based diagnostic efforts). When sensitivity is a concern, physical reduction of host nucleic acid is of great benefit. In some cases, such as sputum, without a host removal step, as many as 99% of resulting reads may be derived from host nucleic acid (Doughty, Sergeant, Adetifa, Antonio, & Pallen, 2014), which will severely hamper NGS's ability to detect a pathogen that is present at low levels. When sequencing the RNA fraction, as is often the case for viral detection and identification, host ribosomal RNA (rRNA) subtraction can be easily performed via use of commercial kits prior to library construction. Additionally, Illumina recently launched TruSeq™ RNA library preparation kits with rRNA and human globin transcript removal steps built in.

Rather than using probes to select out host nucleic acids via sequence homology as in host removal, a complementary technique is to select for pathogens based on some characteristics such as overall size, or genome sequence, prior to sequencing. There are numerous derivations and examples of this type of technique as applied to environmental and clinical samples in the literature (Ng, Wheeler, et al., 2011; Ng, Willner, et al., 2011; Thurber, Haynes, Breitbart, Wegley, & Rohwer, 2009), and one specific example of a related approach is sequence-independent single-primer amplification (SISPA; Rosseel et al., 2012), which was applied to diagnosis of Schmallenberg virus in lambs and calves. SISPA has also been applied to single plaque sequencing (Depew et al., 2013).

Another method of selecting for pathogens is to use sequence-specific probes to separate pathogens out of the sample and sequence specifically those nucleic acid molecules rather than the host nucleic acid. This technique, commonly referred to as target enrichment, has the advantage of decreasing the number of reads required and therefore allowing for greater multiplexing and lower cost per sample. The main drawback, however, is that the full potential of NGS to detect novel pathogens is not being exploited, as this method relies upon probes designed with previously identified sequences in mind. Still, there are many potential instances in which this kind of diagnostic approach might be worthwhile. For instance, this method was successfully applied in the case of *Streptococcus pneumoniae* detection and typing, whereby essentially one NGS experiment replaced the use of 7 or 8 multiplexed PCR assays (Ip et al., 2014). Similarly, Koehler et al. reported on a panel of Filovirus capture probes used in conjunction with MiSeq sequencing to enable detection of Ebola and Marburg viruses. Their method utilized DxSeq™ probes, which are single-stranded DNA probes of which both ends bind to target sequences followed by filling in of the intervening captured sequence by polymerase and subsequent circularization (Koehler et al., 2014; McCarthy, 2011). Despite the fact that a pre-amplification step was performed prior to sequence capture and subsequent MiSeq sequencing, real-time PCR was still more sensitive for detection than sequencing in this case. Additionally, this NGS approach resulted in some false positives (Koehler et al., 2014).

An additional method which can be used to increase sensitivity of NGS for detection of pathogens was recently published by Seth-Smith et al. This method uses IMS-MDA (Seth-Smith, Harris, Scott, et al., 2013) and is able to generate accurate, whole genomes of difficult to culture microbes from up to 30% of clinical swabs. This sensitivity, however, comes at a cost, as there is a high level of host background and preferential amplification of other than target organism. Additionally, the resulting depth of coverage is highly variable between samples and uneven within samples (Seth-Smith, Harris, Skilton, et al., 2013) which could possibly result in assembly issues and potential implications for detecting copy number variations. Furthermore, this method may not be as sensitive as sequence-based pathogen enrichment (Christiansen et al., 2014).

3.2.2 Enhancing characterization of pathogens using NGS

When isolates can be readily isolated and grown in culture, identification and characterization can be coupled with whole-genome sequencing, producing potentially valuable information regarding an organism's repertoire of drug resistance and virulence genes. However, the full implications of any novel genetic variation are not always easy to predict given genome sequence alone. One approach that may increase the power of NGS to characterize a pathogen is to couple transcriptome sequencing with WGS. For example, Olsen et al. recently demonstrated that when WGS is coupled with transcriptome sequencing of clinical Group A *Streptococcus* isolates, the impact of a given variation on virulence may be more easily discerned. In this case, up-regulation of a hyaluronic acid synthase operon in transcriptomic data from a particular strain drew the group's attention to a small (38 bp) deletion in a non-coding region of the genome. Although this deletion may impact virulence by altering capsular phenotype, it was not noted or flagged for close inspection in the WGS data until the RNA-seq data were analysed (Olsen et al., 2014). This example highlights not only the simultaneous power of genome-wide data and the difficulties involved in extrapolating from it clinically relevant phenotypes but also the synergism of WGS and transcriptome data in correlating genotype with phenotype.

4 CHALLENGES GOING FORTH AND SUCCESSFUL IMPLEMENTATION

There are a variety of potential challenges that may be faced by a clinical lab attempting to implement NGS for routine microbial diagnostics. First is the lack of standards in terms of protocols, reagent choice, and what constitutes a positive or negative result. For instance, there are a myriad of choices in terms of sample preparation kits and protocols, and the choice can influence pathogen detection (Alberti et al., 2014). Although nucleic acid extraction methods can introduce bias in metagenomic sequencing (Carrigg et al., 2007; Delmont et al., 2011; Kennedy et al., 2014; Ó Cuív et al., 2011), there are currently no standards in terms of which methods should be applied to which sample types. Adding to the conundrum are the low titres of pathogens and low yields of pathogen nucleic acid from minimally invasive sample types such as blood, urine, and swabs. Pathogen titre and nucleic acid yield can be expected to vary depending on disease and sampling site. For some sample types, there may be a high variability even within specific disease presentations. Seth-Smith et al. showed that for the fastidious pathogen *C. trachomatis*, total DNA extracted from clinical swabs and urine samples was of highly variable content, with some samples below detection limit. *C. trachomatis* DNA specifically only represented a maximum of 0.6% of the sample (Seth-Smith, Harris, Skilton, et al., 2013). In Hasman et al., 12 of 35 urine samples (34%) did not yield enough total DNA for NGS library construction (Hasman et al., 2014).

Yozwiak et al. performed pooled sequencing of serum samples from acute febrile patients on the Illumina GAII and HiSeq platforms and demonstrated that in positive controls (acute serum from patients whose samples tested positive for Dengue virus), the fraction of reads obtained that were viral in origin ranged from 0.0002% to 2.8% of the total reads. In 37% of previously negative samples, the authors were able to identify viral reads, but these included reads belonging to the Annelloviridae family, which is a family of viruses that can be detected in healthy individuals as well

(Yozwiak et al., 2012). It is important to note that for pathogens present at very low levels, it may be difficult to discriminate these low levels (0.0002%) from run-to-run carry-over and/or sequencing library contamination (published examples of such contamination in Yozwiak et al., 2012 and Cheval et al., 2011). Thus, the challenges involved in discriminating whether there are pathogens truly present at low levels versus reads generated from contamination necessitate stringent laboratory protocols to minimize contamination and carry-over, such as bleach washing of sequencer fluidics, strict rotation of DNA barcodes, and separate spaces for PCR amplification and NGS library preparation.

In cases when reads from a microorganism, especially a novel organism, are present at low levels, it may be difficult to infer causation from sequence data alone. In some cases, the link may be fairly obvious, such as when a known pathogen genome is recovered preferentially from lesions and not from unaffected body sites, or when the same organism's genome is recovered from the majority of patients and not from controls (as in Loman et al., 2013) but in other cases may perhaps require further investigation to determine causality, as in Palacios et al. (2008), for example. With the exception of normally sterile sites, such as cerebrospinal fluid (CSF), in many patient specimen types such as stool and nasal swab, a variety of organisms may be present, though not all cause disease. This challenge is not unique to NGS, but due to the unbiased nature of NGS, NGS is particularly sensitive to it. In addition to run-to-run carry-over and the presence of commensal organisms, laboratory reagent contaminants are another source of microorganism genome sequences that may confound results. For instance, a recent publication reported that parvovirus-like hybrid virus, initially thought to be discovered in various clinical samples, was actually a contaminant introduced into sequencing from silica spin columns used in nucleic acid extraction (Naccache et al., 2013).

Discrimination of pathogens versus commensal flora may be difficult due to a lack of baseline measurements, especially for viruses, though hopefully this will be rectified with future human microbiome-type projects. The vast majority of viruses is likely unsampled and therefore lacks a reference in public databases. Wylie et al. detected 1–15 viral genera in 92% of 102 healthy adults sampled from 5 body sites by sequencing of just the DNA fraction using Illumina technology. This is likely an underestimate of viral complexity as conservative methods were used to identify only those viruses with genomes of double-stranded DNA and with unambiguous matches to viruses in the NCBI NT database (Wylie et al., 2014). Due to the extent of viral diversity that exists on Earth, 80% of viral coding sequences are dissimilar to any pre-existing reference sequences (Greene & Reid, 2013).

In addition to challenges involved in identification of pathogens and discrimination from other sources of microorganism-contributed reads, the clinical lab faces challenges related to using NGS for correlating genotype with phenotype, for instance, predicting virulence or drug resistance from metagenomic sequence data. Though NGS data in theory hold the answers to many questions one could ask about a microbe, in practice, correlating genotype to phenotype in many cases can still be quite difficult. For instance, gaps in assemblies from metagenomic samples can make phenotype predictions difficult—either gaps from missing coverage, low-abundance organisms, or very complex samples, or gaps because of difficulty in assembling the reads even if there is coverage, or maybe difficulty assigning reads to an organism in the mix. For example, though one might detect a particular antibiotic resistance gene that does not necessarily mean it belongs to the pathogen (maybe a commensal) and does not necessarily mean it is expressed. There must be enough genomic context surrounding the gene to be able to have confidence in those assertions. Also, in many cases, it may still be worthwhile to follow-up with or simultaneously run susceptibility testing, as the consequences of falsely reporting susceptibility when in fact the organism is resistant could be grave. Assessing potential virulence using NGS data is still challenging, but in some cases may be feasible. For instance, presence of a capsule can be a clinically relevant virulence factor for some bacterial organisms, such as *Bacillus* species. For other organisms (*Haemophilus influenzae* type b, *S. pneumoniae*, and *Neisseria meningitidis*), virulence is associated with specific capsule types. Compared to making confident calls regarding drug susceptibility, perhaps defining the presence/absence of capsule or defining capsule type from sequence data is a relatively tractable challenge.

In terms of sensitivity of NGS for these applications and the probability of FDA approval, a recent study by Stoesser et al. demonstrated error rates in antimicrobial susceptibility predictions from WGS data of purified clinical isolates to be below the FDA-stipulated levels (Stoesser et al., 2013), suggesting that FDA approval for NGS-based bacterial characterization assays may be attainable. This study, however, employed NGS of only purified *E. coli* and *Klebsiella pneumoniae* isolates, and predictions for seven specific antibiotics (Stoesser et al., 2013); therefore, further research is required to broaden these findings to other organisms and antimicrobial drugs. An additional caveat is that these findings do not necessarily apply to bacterial genome sequences obtained via metagenomic sequencing; further investigation will be required to understand the reliability of resistance predictions from metagenomic data sets. It may make sense to implement NGS for these purposes on the less complex sample types first, as Hasman et al. found that antimicrobial resistance predicted from metagenomic sequencing of urine samples (a metagenomic sample of relatively low complexity) was similar to that predicted from WGS of purified isolates derived from those same urine samples. Unfortunately, despite the fact that results from metagenomic sequencing of urine essentially mirrored those of WGS from purified isolates, for 2 of the 17 samples, prediction of resistance even from WGS data missed resistance to an entire drug class (Hasman et al., 2014), a false-negative result that could have grave consequences.

There are additional challenges relating to analysis of NGS data, many of which are bioinformatic in nature. Due to the unbiased nature of NGS, a typical metagenomic sequencing experiment will likely result in reads being classified as deriving from a variety of microorganisms. These results will vary depending on the sample, the tool that is used to analyse the reads, and the databases employed. The databases in general are skewed towards culturable organisms and organisms 'of interest' (McHardy & Rigoutsos, 2007). The quality of reference genomes varies, which provokes the question of whose job should it be to curate and maintain these sequences should bacterial diagnostics by NGS become routine? There are some efforts aimed at addressing database issues, for instance, the Global Microbial Identifier (GMI) project, an initiative to create a global system of DNA genome databases for microbial and infectious disease identification and diagnostics, and the Genomic Encyclopedia of Bacteria and Archaea (GEBA), which has a broader purpose—to create a comprehensive microbial genomic framework by sampling across the tree of life for every type strain (Kyrpides et al., 2014). However, it is likely that there is still considerable work to be done in terms of reference genome selection and curation in order to support routine, reliable microbial diagnostics via NGS. Indeed, several recent studies have highlighted extensive contamination of several publicly available genomes, including those of *Bos taurus* and *Bacillus cereus* (Longo, O'Neill, & O'Neill, 2011), as well as the human transcriptome (Strong et al., 2014).

Just as database content and quality impact data analysis, selection of bioinformatic software can influence results. Clinical samples that may be interrogated via NGS range in species complexity from blood (relatively non-complex) to stool (very complex). These complex samples are more properly described as metagenomic samples. Bioinformatic tools that were developed for analysis of WGS data often do not befit samples with high microbial complexity. As such, a range of new tools have been developed to overcome this demand (Flicek & Birney, 2009). Several tools initially established for WGS analysis have been updated or improved to dissect metagenomic data. Among these are MetaVelvet, MetAMOS, and MetaRay (Namiki, Hachiya, Tanaka, & Sakakibara, 2012; Treangen et al., 2013). These tools employ two general strategies that both attempt to classify sequence data in taxonomic space: assembly or contig-based or read-based classification. Each approach has advantages and drawbacks. In general, these tools require vast computational resources and computational time and they are error prone (Scholz, Lo, & Chain, 2012). Publications using these tools are sparse and usually serve as only a confirmation of a clinical diagnosis rather than a primary method (Wilson et al., 2014).

4.1 DISCUSSION: LIMITATIONS AND COMMON MISCONCEPTIONS

Often, metagenomic sequencing directly from clinical samples does not yield the depth of coverage necessary for accurate, useful SNP-based analyses of bacterial pathogens (Doughty et al., 2014; Frey et al., 2014). In order to maximize the likelihood of detecting a novel agent, it is critical to maximize the output of the metagenome sequencing, in terms of breadth and depth of coverage. Yet, metagenomic samples yield uneven sequence coverage of differing depth. At present, the community has no agreed-upon standards of adequate genome coverage and depth for species identification. Although there is no published report of this, it remains possible to identify a novel agent from a single novel microbial read present in a complex metagenomic sample. Identification of novel agents has been reported with as few as 14 reads out of over 100,000 (Palacios et al., 2008). Whereas identification of an agent may require detection of only one or more reads, characterization, the crucial next step, is absolutely dependent on complete (100%) representation of the agent's entire genome at adequate depth of coverage, especially in the case of RNA viruses or other microorganisms likely to exhibit functionally relevant minority populations or quasispecies. In this case, it is necessary for follow-on experiments to more fully characterize the genome of the microorganism, such as Sanger sequencing using primers based on the novel fragment(s).

Finally, the field of NGS in general and metagenomics specifically are in their infancy. As such, community-accepted standards, independent and thorough validation, and studies of reproducibility are lacking. The case of microarray-based analysis is instructive. Due to early successes, adaptation of microarray technology was rapid and widespread. However, the field lacked a robust examination and documentation of sources of experimental variability. Indeed, a study of commercially available platforms demonstrated substantial divergence in correlations of gene expression (Tan et al., 2003). This study suggested a critical need for industrial standards as well as thorough validation of the technology. It is clear that NGS requires the same stringency.

4.2 WHEN NGS ADDS VALUE

Although there are some significant challenges that remain pertaining to the use of NGS for infectious disease diagnostics, there are certain situations in which, even in its current state, NGS adds value. These situations are described below.

4.2.1 In the case of immunocompromised patients

Some infections are diagnosed by serology, but immunocompromised patients, including those with genetic disorders and those that have undergone chemotherapy or immune suppression following transplant surgery, may not mount a detectable immune response. Wilson et al. recently reported on the case of an adolescent with severe combined immunodeficiency who reported with fever and headache that progressed to hydrocephalus and seizures, ultimately requiring a medically induced coma. A battery of microbiological tests had failed to produce a diagnosis, and so NGS was applied to CSF and serum, using one multiplexed sequencing run of a MiSeq instrument. From the CSF, majority of bacterial reads (475/589) were identified as belonging to the Leptospiraceae family. The patient was started on an appropriate antibiotic and subsequently recovered (Wilson et al., 2014). This case is notable in being an example in which NGS was used for rapid diagnosis of a patient's infection and resulted in an actionable answer and improved treatment for the patient. The literature is replete with examples where patient samples are sequenced postmortem and provide pathogen identification (Briese et al., 2009; Mi et al., 2014; Palacios et al., 2008; Quan et al., 2010), or when NGS identifies pathogens in samples from living patients (Holtz et al., 2011) but there is no specific treatment that can be applied based on that information. This case is different in that NGS was applied as part of the diagnostic work-up of the patient, and since the identification of a specific bacterial organism was made, the correct antibiotic could be prescribed; therefore, NGS made a positive difference to the patient's treatment (Wilson et al., 2014).

4.2.2 When pathogens are slow growing or fastidious

Another instance in which NGS may add value is when diagnostics are dependent on culture and the organism is slow growing, as in the case of *M. tuberculosis*. In a recent study, differential lysis and metagenomic sequencing were performed on smear-positive sputum samples from patients in the Gambia. Despite very low coverage of the pathogen genome (20–99% of the resulting reads per sample were derived from human DNA), these methods allowed for rapid detection of the organism and assignment to lineages, with implications for treatment. By comparison, culture-based methods would have taken weeks to months and more rapid *M. tuberculosis* tests, though providing results within hours, do not yield the full range of information that NGS did (Doughty et al., 2014). One caveat is that for this study, only smear-positive samples were used as a proof of concept.

In addition to the case of slow-growing or fastidious organisms, another case when NGS may add value for diagnosis of infectious diseases is when small or regional laboratories do not have access to specialized diagnostics and have to send them out to reference laboratories which may have long turn-around time. Reuter et al. recently reported on WGS of isolates from nosocomial infection (vancomycin-resistant *Enterococcus faecium* and carbapenem-resistant *Enterobacter cloacae*) as well as community-acquired meningococcal meningitis. In this case, NGS helped to trace chains of transmission in real time (Reuter et al., 2013).

4.2.3 When more than just identification is required (assuming pathogen load allows for characterization)

Often all the information that one would like to know about a given sample (basic identification followed by strain typing, resistance profiling, presence/absence of virulence factors, and so forth) may necessitate multiple diagnostic tests, some of which may require special expertise or may not be performed routinely in smaller laboratories. In these cases, in which one would like to obtain more than just an organism identification, but also a full characterization, NGS can add value, for it can essentially provide all relevant information for all pathogens, with one, standard assay, even integrating diagnostics with surveillance. In the case of strain typing especially, NGS can be simpler because one assay [sequencing] could replace many different species-specific typing methods/schemes (reviewed in Didelot et al., 2012).

In fact, for some organisms, WGS data have been shown to be superior to other clinical typing methods, such as serological assays, pulsed field gel electrophoresis, or genotyping using a specific gene sequence. For example, Harris et al. performed WGS of pure *C. trachomatis* cultures and showed that traditional typing methods based on the *ompA* gene or the plasmid might be inadequate as they miss recombination events that have occurred (Harris et al., 2012). Another example would be *Salmonella enterica* serovar Enteritidis, a food-borne pathogen which is traditionally typed by PFGE or multi-locus variable-number tandem-repeat analysis, but for which WGS has been shown to provide improved subtype analysis and cluster detection (den Bakker et al., 2014).

4.2.4 In the case of novel pathogens

The situation in which NGS most obviously adds value is in the case of the novel viral pathogen. In this case, a variety of diagnostic tests may be negative because the virus may be divergent enough to evade serology and even PCR assays that target regions of the genome that are conserved within that viral family. By comparison, the unbiased nature of NGS allows for sequencing of the viral pathogen. There are many examples of novel viruses that have been discovered from patient samples by NGS. For instance, Heartland virus is a novel tick-borne virus in the Bunyaviridae family that was found in two febrile patients in northwestern Missouri and characterized via 454 sequencing (McMullan et al., 2012).

4.2.5 In the case of a syndrome with multiple potential causes

Sepsis can be caused by a variety of microorganisms, and fatal if not treated quickly with the appropriate antimicrobial. Furthermore, early sepsis can present similarly as systemic inflammatory response syndrome (SIRS), a condition that is not caused by an infectious agent. Therefore, if NGS can be applied quickly enough, sepsis is another example of a situation in which NGS can add value. NGS may potentially be used either to identify the causative agent by detection of pathogen-derived sequences or to differentiate sepsis from SIRS on the basis of specific miRNA expression (Shankar-Hari & Lord, 2014; Wu et al., 2014).

4.3 WHERE FUTURE EFFORTS SHOULD FOCUS

Owing to some of the current limitations and challenges discussed above, there are some aspects of NGS upon which further research and development should focus. One such aspect would be protocols and/or kits for pathogen enrichment and/or host reduction. While there are commercially available kits that work well for removal of human and other host rRNA, removal of host gDNA is still more of an issue. Methods for removal of host nucleic acid have mostly focused on rRNA (O'Neil, Glowatz, & Schlumpberger, 2013). In the case of RNA-seq studies or detection of RNA viruses, these methods improve coverage of microbial genomes (Bishop-Lilly et al., 2010). However, especially in the case of clinical samples, often host nucleic acid is genomic DNA. Given that the human genome is several orders of magnitude larger than a given bacterium or virus (3 Gb vs. ~5 Mb), NGS data from a clinical sample will be comprised of >90% host reads. At present, there are scant methods for removal of host gDNA (Archer, Long, & Lin, 2010). Instead, researchers depend on bioinformatic removal of sequence reads. This method is dependent on the quality of the database and the specificity of the algorithm. Prior to widespread adoption of NGS in the clinic, robust methods for host gDNA removal should be developed.

Additionally, focusing on faster sequence run time; smaller, lighter instrumentation; and ruggedness for truly fieldable instrumentation could help support diagnostics in military scenarios. Of course, these improvements cannot come at the price of acceptable error rate and throughput. Additionally, focusing on the use of NGS in combination with proteomics might be a useful strategy going forth; for instance, a winning strategy may be to use NGS for identification of candidate biomarkers and mass spectrometry for the assays (Diamandis, 2014). And last but not least, adoption of NGS for routine microbial diagnostics should involve some aspect of training for physicians in terms of interpretation of NGS results.

Disclaimer: Title 17 U.S.C. §105 provides that 'Copyright protection under this title is not available for any work of the United States Government'. Title 17 U.S.C. §101 defines a U.S. Government work as a work prepared by a military service member or employee of the U.S. Government as part of that person's official duties. The opinions or assertions contained herein are the private ones of the author(s) and are not to be construed as official or reflecting the views of either the Department of the Navy or the Department of Defense.

REFERENCES

Alberti, A., Belser, C., Engelen, S., Bertrand, L., Orvain, C., Brinas, L., et al. (2014). Comparison of library preparation methods reveals their impact on interpretation of metatranscriptomic data. *BMC Genomics*, *15*(1), 912.

Alkadhi, S., Kunde, D., Cheluvappa, R., Randall-Demllo, S., & Eri, R. (2014). The murine appendiceal microbiome is altered in spontaneous colitis and its pathological progression. *Gut Pathogens*, *6*, 25.

Altman, D. R., Sebra, R., Hand, J., Attie, O., Deikus, G., Carpini, K. W., et al. (2014). Transmission of methicillin-resistant Staphylococcus aureus via deceased donor liver transplantation confirmed by whole genome sequencing. *American Journal of Transplantation*, *14*(11), 2640–2644.

Anthony, S. J., St Leger, J. A., Navarrete-Macias, I., Nilson, E., Sanchez-Leon, M., Liang, E., et al. (2013). Identification of a novel cetacean polyomavirus from a common dolphin (Delphinus delphis) with tracheobronchitis. *PLoS One*, *8*(7), e68239.

Archer, M. J., Long, N., & Lin, B. (2010). Effect of probe characteristics on the subtractive hybridization efficiency of human genomic DNA. *BMC Research Notes*, *3*, 109.

Aziz, N., Zhao, Q., Bry, L., Driscoll, D. K., Funke, B., Gibson, J. S., et al. (2015). College of American Pathologists' laboratory standards for next-generation sequencing clinical tests. *Archives of Pathology & Laboratory Medicine*, *139*(4), 481–493.

Bennett, S. (2004). Solexa Ltd. *Pharmacogenomics*, *5*(4), 433–438.

Ben Zakour, N. L., Venturini, C., Beatson, S. A., & Walker, M. J. (2012). Analysis of a Streptococcus pyogenes puerperal sepsis cluster by use of whole-genome sequencing. *Journal of Clinical Microbiology*, *50*(7), 2224–2228.

Bishop-Lilly, K. A., Turell, M. J., Willner, K. M., Butani, A., Nolan, N. M., Lentz, S. M., et al. (2010). Arbovirus detection in insect vectors by rapid, high-throughput pyrosequencing. *PLoS Neglected Tropical Diseases*, *4*(11), e878.

Bragg, L. M., Stone, G., Butler, M. K., Hugenholtz, P., & Tyson, G. W. (2013). Shining a light on dark sequencing: Characterising errors in Ion Torrent PGM data. *PLoS Computational Biology*, *9*(4), e1003031.

Briese, T., Paweska, J. T., McMullan, L. K., Hutchison, S. K., Street, C., Palacios, G., et al. (2009). Genetic detection and characterization of Lujo virus, a new hemorrhagic fever-associated arenavirus from southern Africa. *PLoS Pathogens*, *5*(5), e1000455.

Buermans, H. P., & den Dunnen, J. T. (2014). Next generation sequencing technology: Advances and applications. *Biochimica et Biophysica Acta*, *1842*(10), 1932–1941.

Carrigg, C., Rice, O., Kavanagh, S., Collins, G., & O'Flaherty, V. (2007). DNA extraction method affects microbial community profiles from soils and sediment. *Applied Microbiology and Biotechnology*, *77*(4), 955–964.

Cer, R. Z., Herrera-Galeano, J. E., Frey, K. G., Schully, K. L., Luu, T. V., Pesce, J., et al. Human miRNA profiling of human peripheral blood mononuclear cells after pathogenic bacterial exposure. *Submitted to BMC Research Notes*. (manuscript in review).

Chain, P. S., Grafham, D. V., Fulton, R. S., Fitzgerald, M. G., Hostetler, J., Muzny, D., et al. (2009). Genomics. Genome project standards in a new era of sequencing. *Science*, *326*(5950), 236–237.

Chen, X., Li, Z. Y., Maleewong, W., Maleewong, P., Liang, J., Zeng, X., et al. (2014). Serum aca-mir-146a is a potential biomarker for early diagnosis of Angiostrongylus cantonensis infection. *Parasitology Research*, *113*(9), 3221–3227.

Cheval, J., Sauvage, V., Frangeul, L., Dacheux, L., Guigon, G., Dumey, N., et al. (2011). Evaluation of high-throughput sequencing for identifying known and unknown viruses in biological samples. *Journal of Clinical Microbiology*, *49*(9), 3268–3275.

Christiansen, M. T., Brown, A. C., Kundu, S., Tutill, H. J., Williams, R., Brown, J. R., et al. (2014). Whole-genome enrichment and sequencing of Chlamydia trachomatis directly from clinical samples. *BMC Infectious Diseases*, *14*(1), 591.

Conlan, S., Thomas, P. J., Deming, C., Park, M., Lau, A. F., Dekker, J. P., et al. (2014). Single-molecule sequencing to track plasmid diversity of hospital-associated carbapenemase-producing Enterobacteriaceae. *Science Translational Medicine*, *6*(254), 254ra126.

Cox-Foster, D. L., Conlan, S., Holmes, E. C., Palacios, G., Evans, J. D., Moran, N. A., et al. (2007). A metagenomic survey of microbes in honey bee colony collapse disorder. *Science*, *318*(5848), 283–287.

Cui, L., Qi, Y., Li, H., Ge, Y., Zhao, K., Qi, X., et al. (2011). Serum microRNA expression profile distinguishes enterovirus 71 and coxsackievirus 16 infections in patients with hand-foot-and-mouth disease. *PLoS One*, *6*(11), e27071.

Delmont, T. O., Robe, P., Clark, I., Simonet, P., & Vogel, T. M. (2011). Metagenomic comparison of direct and indirect soil DNA extraction approaches. *Journal of Microbiological Methods*, *86*(3), 397–400.

den Bakker, H. C., Allard, M. W., Bopp, D., Brown, E. W., Fontana, J., Iqbal, Z., et al. (2014). Rapid whole-genome sequencing for surveillance of Salmonella enterica serovar Enteritidis. *Emerging Infectious Diseases*, *20*(8), 1306–1314.

Depew, J., Zhou, B., McCorrison, J. M., Wentworth, D. E., Purushe, J., Koroleva, G., et al. (2013). Sequencing viral genomes from a single isolated plaque. *Virology Journal*, *10*, 181.

Diamandis, E. P. (2014). Towards identification of true cancer biomarkers. *BMC Medicine*, *12*(1), 156.

Didelot, X., Bowden, R., Wilson, D. J., Peto, T. E., & Crook, D. W. (2012). Transforming clinical microbiology with bacterial genome sequencing. *Nature Reviews. Genetics*, *13*(9), 601–612.

Doughty, E. L., Sergeant, M. J., Adetifa, I., Antonio, M., & Pallen, M. J. (2014). Culture-independent detection and characterisation of Mycobacterium tuberculosis and M. africanum in sputum samples using shotgun metagenomics on a benchtop sequencer. *PeerJ, 2,* e585.

Eid, J., Fehr, A., Gray, J., Luong, K., Lyle, J., Otto, G., et al. (2009). Real-time DNA sequencing from single polymerase molecules. *Science, 323*(5910), 133–138.

Elton, T. S., Selemon, H., Elton, S. M., & Parinandi, N. L. (2013). Regulation of the MIR155 host gene in physiological and pathological processes. *Gene, 532*(1), 1–12.

FDA-approved next-generation sequencing system could expand clinical genomic testing: Experts predict MiSeqDx system will make genetic testing more affordable for smaller labs. (2014). *American Journal of Medical Genetics. Part A, 164A*(3), x–xi.

Flicek, P., & Birney, E. (2009). Sense from sequence reads: Methods for alignment and assembly. *Nature Methods, 6*(11 Suppl.), S6–S12.

Foxman, B., Zhang, L., Koopman, J. S., Manning, S. D., & Marrs, C. F. (2005). Choosing an appropriate bacterial typing technique for epidemiologic studies. *Epidemiologic Perspectives & Innovations: EP+I, 2,* 10.

Frey, K. G., Herrera-Galeano, J. E., Redden, C. L., Luu, T. V., Servetas, S. L., Mateczun, A. J., et al. (2014). Comparison of three next-generation sequencing platforms for metagenomic sequencing and identification of pathogens in blood. *BMC Genomics, 15,* 96.

Greene, S. E., & Reid, A. (2013). Viruses throughout life & time: Friends, foes, change agents. A report on an American Academy of Microbiology Colloquium.

Greninger, A. L., Chen, E. C., Sittler, T., Scheinerman, A., Roubinian, N., Yu, G., et al. (2010). A metagenomic analysis of pandemic influenza A (2009 H1N1) infection in patients from North America. *PLoS One, 5*(10), e13381.

Hamilton, M. J., Weingarden, A. R., Unno, T., Khoruts, A., & Sadowsky, M. J. (2013). High-throughput DNA sequence analysis reveals stable engraftment of gut microbiota following transplantation of previously frozen fecal bacteria. *Gut Microbes, 4*(2), 125–135.

Hammoudi, D., Ayoub Moubareck, C., & Karam Sarkis, D. (2014). How to detect carbapenemase producers? A literature review of phenotypic and molecular methods. *Journal of Microbiological Methods, 107,* 106–118.

Handelsman, J., Rondon, M. R., Brady, S. F., Clardy, J., & Goodman, R. M. (1998). Molecular biological access to the chemistry of unknown soil microbes: A new frontier for natural products. *Chemistry & Biology, 5*(10), R245–R249.

Harris, S. R., Clarke, I. N., Seth-Smith, H. M., Solomon, A. W., Cutcliffe, L. T., Marsh, P., et al. (2012). Whole-genome analysis of diverse Chlamydia trachomatis strains identifies phylogenetic relationships masked by current clinical typing. *Nature Genetics, 44*(4), 413–419. S411.

Hasman, H., Saputra, D., Sicheritz-Ponten, T., Lund, O., Svendsen, C. A., Frimodt-Moller, N., et al. (2014). Rapid whole-genome sequencing for detection and characterization of microorganisms directly from clinical samples. *Journal of Clinical Microbiology, 52*(1), 139–146.

He, X., Sai, X., Chen, C., Zhang, Y., Xu, X., Zhang, D., et al. (2013). Host serum miR-223 is a potential new biomarker for Schistosoma japonicum infection and the response to chemotherapy. *Parasites & Vectors, 6,* 272.

Holtz, L. R., Wylie, K. M., Sodergren, E., Jiang, Y., Franz, C. J., Weinstock, G. M., et al. (2011). Astrovirus MLB2 viremia in febrile child. *Emerging Infectious Diseases, 17*(11), 2050–2052.

Howorka, S., Cheley, S., & Bayley, H. (2001). Sequence-specific detection of individual DNA strands using engineered nanopores. *Nature Biotechnology*, *19*(7), 636–639.

Hoy, A. M., Lundie, R. J., Ivens, A., Quintana, J. F., Nausch, N., Forster, T., et al. (2014). Parasite-derived microRNAs in host serum as novel biomarkers of helminth infection. *PLoS Neglected Tropical Diseases*, *8*(2), e2701.

Human Microbiome Jumpstart Reference Strains Consortium, Nelson, K. E., Weinstock, G. M., Highlander, S. K., Worley, K. C., Creasy, H. H., et al. (2010). A catalog of reference genomes from the human microbiome. *Science*, *328*(5981), 994–999.

Ip, M., Liyanapathirana, V., Ang, I., Fung, K., Ng, T. K., Zhou, H., et al. (2014). Direct detection and prediction of all pneumococcal serogroups by target enrichment based next generation sequencing. *Journal of Clinical Microbiology*, *52*(12), 4244–4252.

Jeong, H. C. (2014). Clinical aspect of microRNA in lung cancer. *Tuberculosis and Respiratory Diseases*, *77*(2), 60–64.

Johansson, A., Nagy, E., & Soki, J. (2014). Instant screening and verification of carbapenemase activity in Bacteroides fragilis in positive blood culture, using matrix-assisted laser desorption ionization–time of flight mass spectrometry. *Journal of Medical Microbiology*, *63*(Pt. 8), 1105–1110.

Kennedy, N. A., Walker, A. W., Berry, S. H., Duncan, S. H., Farquarson, F. M., Louis, P., et al. (2014). The impact of different DNA extraction kits and laboratories upon the assessment of human gut microbiota composition by 16S rRNA gene sequencing. *PLoS One*, *9*(2), e88982.

Kincaid, R. P., & Sullivan, C. S. (2012). Virus-encoded microRNAs: An overview and a look to the future. *PLoS Pathogens*, *8*(12), e1003018.

Koehler, J. W., Hall, A. T., Rolfe, P. A., Honko, A. N., Palacios, G. F., Fair, J. N., et al. (2014). Development and evaluation of a panel of filovirus sequence capture probes for pathogen detection by next-generation sequencing. *PLoS One*, *9*(9), e107007.

Ku, C. S., Cooper, D. N., & Roukos, D. H. (2013). Clinical relevance of cancer genome sequencing. *World Journal of Gastroenterology*, *19*(13), 2011–2018.

Kyrpides, N. C., Hugenholtz, P., Eisen, J. A., Woyke, T., Goker, M., Parker, C. T., et al. (2014). Genomic encyclopedia of bacteria and archaea: Sequencing a myriad of type strains. *PLoS Biology*, *12*(8), e1001920.

Lane, D. J., Pace, B., Olsen, G. J., Stahl, D. A., Sogin, M. L., & Pace, N. R. (1985). Rapid determination of 16S ribosomal RNA sequences for phylogenetic analyses. *Proceedings of the National Academy of Sciences of the United States of America*, *82*(20), 6955–6959.

Leggieri, N., Rida, A., Francois, P., & Schrenzel, J. (2010). Molecular diagnosis of bloodstream infections: Planning to (physically) reach the bedside. *Current Opinion in Infectious Diseases*, *23*(4), 311–319.

Li, J. Y., McNicholas, K., Yong, T. Y., Rao, N., Coates, P. T., Higgins, G. D., et al. (2014). BK virus encoded microRNAs are present in blood of renal transplant recipients with BK viral nephropathy. *American Journal of Transplantation*, *14*(5), 1183–1190.

Liu, H. S., & Xiao, H. S. (2014). MicroRNAs as potential biomarkers for gastric cancer. *World Journal of Gastroenterology*, *20*(34), 12007–12017.

Loman, N. J., Constantinidou, C., Christner, M., Rohde, H., Chan, J. Z., Quick, J., et al. (2013). A culture-independent sequence-based metagenomics approach to the investigation of an outbreak of Shiga-toxigenic Escherichia coli O104:H4. *JAMA*, *309*(14), 1502–1510.

Longo, M. S., O'Neill, M. J., & O'Neill, R. J. (2011). Abundant human DNA contamination identified in non-primate genome databases. *PLoS One*, *6*(2), e16410.

Loyo, M., Li, R. J., Bettegowda, C., Pickering, C. R., Frederick, M. J., Myers, J. N., et al. (2013). Lessons learned from next-generation sequencing in head and neck cancer. *Head & Neck*, *35*(3), 454–463.

Luthra, R., Patel, K. P., Reddy, N. G., Haghshenas, V., Routbort, M. J., Harmon, M. A., et al. (2014). Next-generation sequencing-based multigene mutational screening for acute myeloid leukemia using MiSeq: Applicability for diagnostics and disease monitoring. *Haematologica*, *99*(3), 465–473.

Madrigal, I., Alvarez-Mora, M. I., Karlberg, O., Rodriguez-Revenga, L., Elurbe, D. M., Rabionet, R., et al. (2014). Efficient application of next-generation sequencing for the diagnosis of rare genetic syndromes. *Journal of Clinical Pathology*, *67*(12), 1099–1103.

Maher, B. (2007). Personal genomics: His daughter's DNA. *Nature*, *449*(7164), 773–776.

Margulies, M., Egholm, M., Altman, W. E., Attiya, S., Bader, J. S., Bemben, L. A., et al. (2005). Genome sequencing in microfabricated high-density picolitre reactors. *Nature*, *437*(7057), 376–380.

Maudet, C., Mano, M., & Eulalio, A. (2014). MicroRNAs in the interaction between host and bacterial pathogens. *FEBS Letters*, *588*(22), 4140–4147.

Maxam, A. M., & Gilbert, W. (1977). A new method for sequencing DNA. *Proceedings of the National Academy of Sciences of the United States of America*, *74*(2), 560–564.

McCarthy, A. (2011). Pathogenica: Diagnosing by DNA. *Chemistry & Biology*, *18*(11), 1343–1344.

McHardy, A. C., & Rigoutsos, I. (2007). What's in the mix: Phylogenetic classification of metagenome sequence samples. *Current Opinion in Microbiology*, *10*(5), 499–503.

McMullan, L. K., Folk, S. M., Kelly, A. J., MacNeil, A., Goldsmith, C. S., Metcalfe, M. G., et al. (2012). A new phlebovirus associated with severe febrile illness in Missouri. *The New England Journal of Medicine*, *367*(9), 834–841.

Mi, Z., Yuan, X., Pei, G., Wang, W., An, X., Zhang, Z., et al. (2014). High-throughput sequencing exclusively identified a novel Torque teno virus genotype in serum of a patient with fatal fever. *Virologica Sinica*, *29*(2), 112–118.

Mikheyev, A. S., & Tin, M. M. (2014). A first look at the Oxford Nanopore MinION sequencer. *Molecular Ecology Resources*, *14*(6), 1097–1102.

Moore, R. A., Warren, R. L., Freeman, J. D., Gustavsen, J. A., Chenard, C., Friedman, J. M., et al. (2011). The sensitivity of massively parallel sequencing for detecting candidate infectious agents associated with human tissue. *PLoS One*, *6*(5), e19838.

Naccache, S. N., Greninger, A. L., Lee, D., Coffey, L. L., Phan, T., Rein-Weston, A., et al. (2013). The perils of pathogen discovery: Origin of a novel parvovirus-like hybrid genome traced to nucleic acid extraction spin columns. *Journal of Virology*, *87*(22), 11966–11977.

Nagy, E. (2014). Matrix-assisted laser desorption/ionization time-of-flight mass spectrometry: A new possibility for the identification and typing of anaerobic bacteria. *Future Microbiology*, *9*(2), 217–233.

Namiki, T., Hachiya, T., Tanaka, H., & Sakakibara, Y. (2012). MetaVelvet: An extension of velvet assembler to de novo metagenome assembly from short sequence reads. *Nucleic Acids Research*, *40*(20), e155.

Ng, S. B., Buckingham, K. J., Lee, C., Bigham, A. W., Tabor, H. K., Dent, K. M., et al. (2010). Exome sequencing identifies the cause of a mendelian disorder. *Nature Genetics*, *42*(1), 30–35.

Ng, T. F., Wheeler, E., Greig, D., Waltzek, T. B., Gulland, F., & Breitbart, M. (2011). Meta-genomic identification of a novel anellovirus in Pacific harbor seal (Phoca vitulina richard-sii) lung samples and its detection in samples from multiple years. *The Journal of General Virology*, *92*(Pt. 6), 1318–1323.

Ng, T. F., Willner, D. L., Lim, Y. W., Schmieder, R., Chau, B., Nilsson, C., et al. (2011). Broad surveys of DNA viral diversity obtained through viral metagenomics of mosquitoes. *PLoS One*, *6*(6), e20579.

Ó Cuív, P., Aguirre de Carcer, D., Jones, M., Klaassens, E. S., Worthley, D. L., Whitehall, V. L., et al. (2011). The effects from DNA extraction methods on the evaluation of microbial diversity associated with human colonic tissue. *Microbial Ecology*, *61*(2), 353–362.

Olsen, R. J., Fittipaldi, N., Kachroo, P., Sanson, M. A., Long, S. W., Como-Sabetti, K. J., et al. (2014). Clinical laboratory response to a mock outbreak of invasive bacterial infections: A preparedness study. *Journal of Clinical Microbiology*, *52*(12), 4210–4216.

O'Neil, D., Glowatz, H., & Schlumpberger, M. (2013). Ribosomal RNA depletion for efficient use of RNA-seq capacity. *Current Protocols in Molecular Biology*, (Chapter 4, Unit 4.19).

Oskam, L., Slim, E., & Buhrer-Sekula, S. (2003). Serology: Recent developments, strengths, limitations and prospects: A state of the art overview. *Leprosy Review*, *74*(3), 196–205.

Palacios, G., Druce, J., Du, L., Tran, T., Birch, C., Briese, T., et al. (2008). A new arenavirus in a cluster of fatal transplant-associated diseases. *The New England Journal of Medicine*, *358*(10), 991–998.

Parkhill, J. (2013). What has high-throughput sequencing ever done for us? *Nature Reviews. Microbiology*, *11*(10), 664–665.

Perez-Losada, M., Cabezas, P., Castro-Nallar, E., & Crandall, K. A. (2013). Pathogen typing in the genomics era: MLST and the future of molecular epidemiology. *Infection, Genetics and Evolution*, *16*, 38–53.

Pickl, J. M., Heckmann, D., Ratz, L., Klauck, S. M., & Sultmann, H. (2014). Novel RNA markers in prostate cancer: Functional considerations and clinical translation. *BioMed Research International*, *2014*, 765207.

Prachayangprecha, S., Schapendonk, C. M., Koopmans, M. P., Osterhaus, A. D., Schurch, A. C., Pas, S. D., et al. (2014). Exploring the potential of next-generation sequencing in detection of respiratory viruses. *Journal of Clinical Microbiology*, *52*(10), 3722–3730.

Qi, Y., Cui, L., Ge, Y., Shi, Z., Zhao, K., Guo, X., et al. (2012). Altered serum microRNAs as biomarkers for the early diagnosis of pulmonary tuberculosis infection. *BMC Infectious Diseases*, *12*, 384.

Quan, P. L., Wagner, T. A., Briese, T., Torgerson, T. R., Hornig, M., Tashmukhamedova, A., et al. (2010). Astrovirus encephalitis in boy with X-linked agammaglobulinemia. *Emerging Infectious Diseases*, *16*(6), 918–925.

Quick, J., Quinlan, A. R., & Loman, N. J. (2014). A reference bacterial genome dataset generated on the MinION portable single-molecule nanopore sequencer. *GigaScience*, *3*, 22.

Reuter, S., Ellington, M. J., Cartwright, E. J., Koser, C. U., Torok, M. E., Gouliouris, T., et al. (2013). Rapid bacterial whole-genome sequencing to enhance diagnostic and public health microbiology. *JAMA Internal Medicine*, *173*(15), 1397–1404.

Ross, M. G., Russ, C., Costello, M., Hollinger, A., Lennon, N. J., Hegarty, R., et al. (2013). Characterizing and measuring bias in sequence data. *Genome Biology*, *14*(5), R51.

Rosseel, T., Scheuch, M., Hoper, D., De Regge, N., Caij, A. B., Vandenbussche, F., et al. (2012). DNase SISPA-next generation sequencing confirms Schmallenberg virus in Belgian field samples and identifies genetic variation in Europe. *PLoS One*, 7(7), e41967.

Sandrin, T. R., Goldstein, J. E., & Schumaker, S. (2013). MALDI TOF MS profiling of bacteria at the strain level: A review. *Mass Spectrometry Reviews*, 32(3), 188–217.

Sanger, F., Nicklen, S., & Coulson, A. R. (1977). DNA sequencing with chain-terminating inhibitors. *Proceedings of the National Academy of Sciences of the United States of America*, 74(12), 5463–5467.

Schaefer, E., Helms, P., Marcellin, L., Desprez, P., Billaud, P., Chanavat, V., et al. (2014). Next-generation sequencing (NGS) as a fast molecular diagnosis tool for left ventricular noncompaction in an infant with compound mutations in the MYBPC3 gene. *European Journal of Medical Genetics*, 57(4), 129–132.

Scholz, M. B., Lo, C. C., & Chain, P. S. (2012). Next generation sequencing and bioinformatic bottlenecks: The current state of metagenomic data analysis. *Current Opinion in Biotechnology*, 23(1), 9–15.

Seth-Smith, H. M., Harris, S. R., Scott, P., Parmar, S., Marsh, P., Unemo, M., et al. (2013). Generating whole bacterial genome sequences of low-abundance species from complex samples with IMS-MDA. *Nature Protocols*, 8(12), 2404–2412.

Seth-Smith, H. M., Harris, S. R., Skilton, R. J., Radebe, F. M., Golparian, D., Shipitsyna, E., et al. (2013). Whole-genome sequences of Chlamydia trachomatis directly from clinical samples without culture. *Genome Research*, 23(5), 855–866.

Shankar-Hari, M., & Lord, G. M. (2014). How might a diagnostic microRNA signature be used to speed up the diagnosis of sepsis? *Expert Review of Molecular Diagnostics*, 14(3), 249–251.

Smith, A. G., Pyo, C. W., Nelson, W., Gow, E., Wang, R., Shen, S., et al. (2014). Next generation sequencing to determine HLA class II genotypes in a cohort of hematopoietic cell transplant patients and donors. *Human Immunology*, 75(10), 1040–1046.

Stoesser, N., Batty, E. M., Eyre, D. W., Morgan, M., Wyllie, D. H., Del Ojo Elias, C., et al. (2013). Predicting antimicrobial susceptibilities for Escherichia coli and Klebsiella pneumoniae isolates using whole genomic sequence data. *The Journal of Antimicrobial Chemotherapy*, 68(10), 2234–2244.

Strong, M. J., Xu, G., Morici, L., Splinter Bon-Durant, S., Baddoo, M., Lin, Z., et al. (2014). Microbial contamination in next generation sequencing: Implications for sequence-based analysis of clinical samples. *PLoS Pathogens*, 10(11), e1004437.

Tan, P. K., Downey, T. J., Spitznagel, E. L., Jr., Xu, P., Fu, D., Dimitrov, D. S., et al. (2003). Evaluation of gene expression measurements from commercial microarray platforms. *Nucleic Acids Research*, 31(19), 5676–5684.

Tan, Y., Ge, G., Pan, T., Wen, D., Chen, L., Yu, X., et al. (2014). A serum microRNA panel as potential biomarkers for hepatocellular carcinoma related with hepatitis B virus. *PLoS One*, 9(9), e107986.

Thurber, R. V., Haynes, M., Breitbart, M., Wegley, L., & Rohwer, F. (2009). Laboratory procedures to generate viral metagenomes. *Nature Protocols*, 4(4), 470–483.

Treangen, T. J., Koren, S., Sommer, D. D., Liu, B., Astrovskaya, I., Ondov, B., et al. (2013). MetAMOS: A modular and open source metagenomic assembly and analysis pipeline. *Genome Biology*, 14(1), R2.

Turingan, R. S., Thomann, H. U., Zolotova, A., Tan, E., & Selden, R. F. (2013). Rapid focused sequencing: A multiplexed assay for simultaneous detection and strain typing of Bacillus anthracis, Francisella tularensis, and Yersinia pestis. *PLoS One*, 8(2), e56093.

Tyson, G. W., Chapman, J., Hugenholtz, P., Allen, E. E., Ram, R. J., Richardson, P. M., et al. (2004). Community structure and metabolism through reconstruction of microbial genomes from the environment. *Nature*, *428*(6978), 37–43.

Valouev, A., Ichikawa, J., Tonthat, T., Stuart, J., Ranade, S., Peckham, H., et al. (2008). A high-resolution, nucleosome position map of C. elegans reveals a lack of universal sequence-dictated positioning. *Genome Research*, *18*(7), 1051–1063.

Venter, J. C., Remington, K., Heidelberg, J. F., Halpern, A. L., Rusch, D., Eisen, J. A., et al. (2004). Environmental genome shotgun sequencing of the Sargasso Sea. *Science*, *304*(5667), 66–74.

Wang, Y., & Qian, P. Y. (2009). Conservative fragments in bacterial 16S rRNA genes and primer design for 16S ribosomal DNA amplicons in metagenomic studies. *PLoS One*, *4*(10), e7401.

Wang, X., Wang, H. K., Li, Y., Hafner, M., Banerjee, N. S., Tang, S., et al. (2014). MicroRNAs are biomarkers of oncogenic human papillomavirus infections. *Proceedings of the National Academy of Sciences of the United States of America*, *111*(11), 4262–4267.

Wendt, J. M., Kaul, D., Limbago, B. M., Ramesh, M., Cohle, S., Denison, A. M., et al. (2014). Transmission of methicillin-resistant Staphylococcus aureus infection through solid organ transplantation: Confirmation via whole genome sequencing. *American Journal of Transplantation*, *14*(11), 2633–2639.

Wilson, M. R., Naccache, S. N., Samayoa, E., Biagtan, M., Bashir, H., Yu, G., et al. (2014). Actionable diagnosis of neuroleptospirosis by next-generation sequencing. *The New England Journal of Medicine*, *370*(25), 2408–2417.

Woo, P. C., Lau, S. K., Teng, J. L., Tse, H., & Yuen, K. Y. (2008). Then and now: Use of 16S rDNA gene sequencing for bacterial identification and discovery of novel bacteria in clinical microbiology laboratories. *Clinical Microbiology and Infection*, *14*(10), 908–934.

Wu, Y., Li, C., He, Y., Li, Q., Wang, G., Wen, P., et al. (2014). Relationship between expression of microRNA and inflammatory cytokines plasma level in pediatric patients with sepsis. *Zhonghua Er Ke Za Zhi*, *52*(1), 28–33.

Wylie, K. M., Mihindukulasuriya, K. A., Zhou, Y., Sodergren, E., Storch, G. A., & Weinstock, G. M. (2014). Metagenomic analysis of double-stranded DNA viruses in healthy adults. *BMC Biology*, *12*(1), 71.

Yoo, H. K., Kang, K., Feuermann, Y., Jin Jang, S., Robinson, G. W., & Hennighausen, L. (2014). The STAT5-regulated miR-193b locus restrains mammary stem and progenitor cell activity and alveolar differentiation. *Developmental Biology*, *395*(2), 245–254.

Yozwiak, N. L., Skewes-Cox, P., Stenglein, M. D., Balmaseda, A., Harris, E., & DeRisi, J. L. (2012). Virus identification in unknown tropical febrile illness cases using deep sequencing. *PLoS Neglected Tropical Diseases*, *6*(2), e1485.

Yu, A. C., Vatcher, G., Yue, X., Dong, Y., Li, M. H., Tam, P. H., et al. (2012). Nucleic acid-based diagnostics for infectious diseases in public health affairs. *Frontiers of Medicine*, *6*(2), 173–186.

Zhang, X., Guo, J., Fan, S., Li, Y., Wei, L., Yang, X., et al. (2013). Screening and identification of six serum microRNAs as novel potential combination biomarkers for pulmonary tuberculosis diagnosis. *PLoS One*, *8*(12), e81076.

Zhang, H., Sun, Z., Wei, W., Liu, Z., Fleming, J., Zhang, S., et al. (2014). Identification of serum microRNA biomarkers for tuberculosis using RNA-seq. *PLoS One*, *9*(2), e88909.

Zheng, G., Du, L., Yang, X., Zhang, X., Wang, L., Yang, Y., et al. (2014). Serum microRNA panel as biomarkers for early diagnosis of colorectal adenocarcinoma. *British Journal of Cancer*, *111*(10), 1985–1992.

Zhu, Z., Qi, Y., Ge, A., Zhu, Y., Xu, K., Ji, H., et al. (2014). Comprehensive characterization of serum microRNA profile in response to the emerging avian influenza A (H7N9) virus infection in humans. *Viruses*, *6*(4), 1525–1539.

Virology: The Next Generation from Digital PCR to Single Virion Genomics

16

Richard Allen White III*[,†,1], Jessica N. Brazelton de Cárdenas[‡], Randall T. Hayden[‡]

*Department of Microbiology and Immunology at the University of British Columbia, Vancouver, British Columbia, Canada

[†]Pacific Northwest National Laboratory, Richland, Washington, USA

[‡]St. Jude Children's Research Hospital, Memphis, Tennessee, USA

[1]Corresponding author: e-mail address: raw937@gmail.com

1 INTRODUCTION

The first written record of the term 'Virus', was coined by the Roman encyclopaedist Aulus Cornelius Celsus in ca. 50 A.D, in association with the rabies virus in dogs (Creager, 2002). The term 'Virus' is of Latin origin and means poison, venom, or slime, in direct association with a disease that cannot be confirmed to be bacteriological. Virology has come a long way from the days of 'unfilterable agents', of Ivanoviskii's and Beijerinck's Tobacco Mosaic virus, to Hershey and Chase's experiments leading to the discovery that genes are composed of DNA (Creager, 2002; Hershey & Chase, 1952). Studies of bacterial viruses dominated until the 1980s, until they were discovered in freshwater and marine systems in the 1990s. Virology, mainly phage biology, has given insight into genetic code, transcription, and genetic exchange, leading to the molecular revolution (Brenner, Jacob, & Meselson, 1961; Crick, Barnett, Brenner, & Watts-Tobin, 1961; Hershey & Chase, 1952).

Just as the Middle Ages led into the Renaissance Age, so too the molecular revolution led to the biotechnology industry, leaving its footprints in virology. Viral genes and proteins are so commonly used in genetic engineering that the work to establish them has long been forgotten. From the genes themselves (e.g. phage/viral promoters: T7/cytomegalovirus), to novel enzymes which catalyse molecular reactions (T4-ligase or Moloney murine leukaemia virus reverse transcriptase), have led to advancements in everything from the genomics of human health, to understanding the biosphere. The discovery of viruses has led to advancements in technology needed to study them, which has had a broader impact on science and society as a whole. Viruses were first identified by electron microscopy and genomic sequencing of DNA and RNA by Sanger sequencing. MS2 (RNA phage) and phiX (DNA phage) were the first RNA and DNA genomes sequenced, but widely used techniques

Methods in Microbiology, Volume 42, ISSN 0580-9517, http://dx.doi.org/10.1016/bs.mim.2015.09.001

such as SDS-PAGE gels and whole-genome amplification with multiple displacement amplification (MDA), have been key technologies allowing the study of other life forms and the biosphere at large (Dean et al., 2002; Dean, Nelson, Giesler, & Lasken, 2001; Fiers et al., 1976; Luria, Delbruck, & Anderson, 1943; Sanger et al., 1977; Weber & Osborn, 1969).

It is generally accepted that viruses are the most abundant biological organism on the planet. Their estimated population of 10^{31} is at least an order of magnitude higher than the 10^{30} estimate for bacteria (Suttle, 2005). Furthermore, the clinical impact of viral infection is indisputable. Viral infections manifest as a broad range of diseases, ranging from asymptomatic carriage to severe, fulminant, and sometimes fatal processes. While diagnostic tools used to detect and characterise infection were initially based on culture, electron microscopy, and antigen detection, these have been increasingly supplanted by molecular techniques, largely endpoint and real-time PCR (Ko et al., 2015; Lipkin & Anthony, 2015; Pang & Lee, 2015). Quantitative applications of these techniques have become integral to clinical care (Dioverti & Razonable, 2015; Tan, Waggoner, & Pinsky, 2015).

2 DIGITAL PCR

Digital PCR (dPCR) is a departure from previous quantitative PCR (qPCR) methodologies, in that it relies on neither rate-based measurements (cycle threshold values) nor calibration curves. Originally developed to detect rare mutations, dPCR was an approach designed to overcome detection difficulties encountered by DNA sequencing, and the difficulty of quantitating a small fraction of mutant molecules within a starting population (Sidransky et al., 1992; Vogelstein & Kinzler, 1999). Further work has demonstrated its potential role in clinical diagnostic virology. dPCR builds on traditional endpoint PCR amplification and fluorescent probe-based detection methods, with the option of adding an RT step for analysis of RNA-based targets. The basic premise of dPCR is centred on limiting dilution, in which single DNA molecules are amplified by PCR in reaction partitions created by methods of separation, including microfluidic chambers, capillaries, or small emulsion droplets. This physical partitioning and separation allows for positive PCR amplifications to be counted directly as the number of positive microreactions (positive partitions) at the reaction endpoint. The limiting dilution factor is chosen such that a high proportion of single reaction partitions (up to 35%) contain no template molecules per partition, giving a '0' (negative) result, with the balance producing a '1' (positive) result, indicating one molecule per positive well (White, Blainey, Fan, & Quake, 2009; White, Quake, & Curr, 2012).

Limiting dilution PCR, the forefather of dPCR, was first described by Sykes et al. (1992) for the quantitation of rearranged immunoglobulin heavy chain (IgH) in leukaemic and non-leukaemic cells, by a limiting dilution-based nested PCR. The Kalinina et al. paper was the next major advancement leading to dPCR. This paper coupled real-time PCR quantitation using $5'$ exonuclease (TaqMan) chemistry and scaled the limiting dilution down to a manageable nanolitre level (Kalinina, Lebedeva, Brown, & Silver, 1997). This facilitated massively parallel reactions by setting the partitioning of these reactions in a capillary (Kalinina et al., 1997). The term 'digital PCR' was coined by Vogelstein and Kinzler (1999), in a study quantifying mutation frequency in colorectal cancer cell lines. dPCR has since been used in a wide variety of biological applications, including maize genotyping and quantitation of copy number in genetically modified corn (Corbisier, Bhat, Partis, Xie, & Emslie, 2010), quantitation of next-generation sequencing libraries (White et al., 2009, 2012), foetal diagnosis of Down's syndrome (Fan, Blumenfeld, Chitkara, Hudgins, & Quake, 2008), diagnosis of organ transplant rejection (Snyder, Khush, Valantine, & Quake, 2011), determination of copy number variation in autism (Sanders, Ercan-Sencicek, et al., 2011), to predict relapse in leukaemia (Mori et al., 2015), response to anti-EGFR therapies in colon cancer (Laurent-Puig et al., 2015), and many more applications too numerous to mention.

The workflow for dPCR is similar to most other PCR methods and is illustrated in Figure 1 (Mazaika & Homsy, 2014).

A typical real-time PCR mastermix, consisting of buffer, deoxynucleotide tri-phosphate solution mix, primers, DNA polymerase, and DNA template material, is made, with TaqMan probes used for detection of amplified product. Usually, one primer probe is specific for the region of interest, while another is targeted to

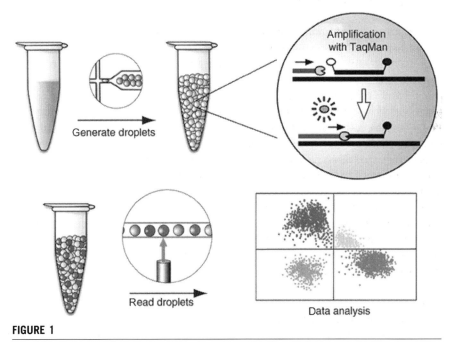

FIGURE 1

Overview of digital PCR workflow.

From Mazaika and Homsy (2014).

a standard reference (internal control). The reaction is partitioned into many separate reactions by either droplet emulsion or physical separation, each partition containing a separate amplification. After endpoint amplification, fluorescent signal in each partition is read and those containing primary target or control will fluoresce in their corresponding channel, whereas those without target will not. The number of positive versus negative reactions is counted, and Poisson statistics are used to directly calculate the number of DNA molecules in the original sample. Because some reactions may contain more than one target molecule, simply counting may lead to an underestimation of the actual concentration of the sample—this is corrected using the Poisson equation, which is used to calculate the average number of molecules per reaction from the observed proportion of positive reactions within the sample (Mazaika & Homsy, 2014; Sedlak & Jerome, 2013).

Currently, a handful of different dPCR platforms are commercially available, differing primarily by their method of partitioning and the number of partitions produced. The plate format, a microfluidics-based system utilised by Fluidigm and Life Technologies, physically separates the reactions into individual reaction wells. These systems allow for up to several thousand partitions per chip or plate. The droplet format, utilised by Bio-Rad Laboratories and RainDance, creates an oil-in-water emulsion in which each droplet represents a single reaction. The droplet format systems allow for tens of thousands of droplets per sample to over a million droplets per sample, for Bio-Rad and RainDance, respectively (Sedlak & Jerome, 2013). Currently, most dPCR systems only allow for two colour multiplexing. Despite limitations in fluorophore number and choice, some groups have been able to create multiplexed assays. By manipulating the concentration of the fluorogenic probes, it is possible to differentiate multiple targets based on resulting fluorescence intensity (Zhong et al., 2011). With the rapid rate of technological advances, an accurate comparison of dPCR systems with regard to pricing and dynamic range is impossible to discuss. There are pros and cons to each platform, and the choice in platform should be inherently linked to the type of research and study design for which it is purchased. With that statement, the growing interest and use in dPCR has led to the introduction of the minimum information for publication of quantitative digital PCR experiments guidelines, in the hope that it will assist researchers in independent evaluation of experimental data and proper reporting of dPCR data (Huggett et al., 2013). Several items specific to dPCR experiments should be included in the publication, namely, mean copies per partition, partition number, template structural information, individual partition volume, total volume of the partitions measured, comprehensive details and appropriate use of controls, examples of positive and negative experimental results as supplemental data, and experimental variance or CI (Huggett et al., 2013). Regardless of the platform used in the experiments, the addition of these guidelines will allow for more reproducible data and reliable scientific reporting.

As stated before, dPCR has been used in a wide variety of biological applications including oncology, genetics, and environmental testing (Corbisier et al., 2010; Fan et al., 2008; Sanders, Huggett, et al., 2011; Snyder et al., 2011; White et al., 2009, 2012), and it has more recently been gaining traction in the field of clinical virology.

An early report described environmental viruses in single bacterial cells, with a subsequent study illustrating quantitation of a cryptic occult human virus known as GB virus C (GBV-C) (Tadmor, Ottesen, Leadbetter, & Phillips, 2011; White et al., 2009). An important discovery was the demonstration that single bacterial cell hosts could be linked to corresponding infectious phage particles, previously a major challenge in environmental virology (Clokie, Millard, Letarov, & Heaphy, 2011; Wagner & Waldor, 2002; Williams, 2013). Classical phage studies have used a susceptible host to probe for new viruses (Tadmor et al., 2011). A remaining weakness for this application is that universal markers for phages do not exist (Tadmor et al., 2011). A given degenerate viral primer may not necessarily detect targets representative of recent infection of the host, but remnants of previous viral infection(s) in which virus integrated into the host genome. White et al. (2009) demonstrated application of dPCR for the quantification of human RNA virus, with potential clinical relevance for HIV-1-positive patients. The GBV-C virus used in this study had been associated with inference of the HIV-1 viral cycle in co-infected patients (Simons, Desai, & Mushahwar, 2000). This study showed a detection limit of three viral genome equivalents in GBV-C infected native peripheral blood mononuclear cells (PBMCs) without the use of a standard curve (White et al., 2009). Further studies might be designed to show dynamic changes in viral load in relation to HIV-1/GBV-C co-infections in cell lines, PBMCs, or in patient samples, or to quantitate quasi-species formed during GBV-C viral replication (White et al., 2009).

These studies represent some of the first steps in the application of this technology in the field of virology. dPCR technology may also prove very useful in understanding environmental virology and clinical pathophysiologic aspects of viral infection in humans. In particular, dPCR has a benefit over qPCR in that it has been shown to be more resistant to PCR inhibition (Coudray-Meunier et al., 2015; Sedlak, Kuypers, & Jerome, 2014). In a clinical setting, many sample types have high concentrations of PCR-interfering substances, rendering PCR inhibition a continuing problem. In a study targeting DNA viruses associated with gastrointestinal disease, it was shown that dPCR was more resistant to PCR inhibitors compared to qPCR (Sedlak, Kuypers, et al., 2014). Additional advantages of dPCR include rare variant detection and the potential to perform precise, low-level quantification in relation to a high background of cellular nucleic acid. It also offers absolute quantification without the need for calibration curves. However, dPCR is still influenced by variance and bias, and study design and optimisation is of utmost importance, particularly for applications involving RNA viruses. Reviews on the advantages and potential pitfalls of dPCR as a molecular diagnostic tool are well summarised elsewhere (Gullett & Nolte, 2015; Hall Sedlak & Jerome, 2014; Huggett, Cowen, & Foy, 2015; Sedlak & Jerome, 2013).

Because of its ability to detect low levels of viral nucleic acid, dPCR has been used extensively in detection of various viral infections, including hepatitis B (Boizeau et al., 2014; Huang et al., 2015), hepatitis C (Mukaide et al., 2014), HIV (Kiselinova et al., 2014; Malatinkova et al., 2014, 2015; Ruelle, Yfantis,

Duquenne, & Goubau, 2014), human Herpesvirus 6A and 6B (Leibovitch et al., 2014; Sedlak, Cook, Huang, et al., 2014), Enterovirus 71 (Lui & Tan, 2014), and cytomegalovirus (CMV) (Hayden et al., 2013, 2015; Nixon et al., 2014; Sedlak, Cook, Cheng, Magaret, & Jerome, 2014), among others. Many of these studies compare dPCR to qPCR, and with varying results. In a study comparing dPCR to real-time PCR for the detection of low levels of hepatitis B virus DNA, it was determined that dPCR detected fewer positive samples and seemed to under-report the mean viral load range, compared to both a commercial HBV real-time assay and a laboratory-developed qPCR assay (Boizeau et al., 2014). A second study using dPCR for quantitative detection of foodborne enteric viruses, such as norovirus and hepatitis A virus, echoed these findings, showing that the number of genomic copies of hepatitis A virus detected by dPCR was lower than the expected numbers calculated from qPCR (Coudray-Meunier et al., 2015). Some, however, have shown that dPCR has linearity and good quantitative correlation with real-time methods when used to test CMV standard material (Hall Sedlak & Jerome, 2014; Hayden et al., 2013; Huggett et al., 2015; Sedlak & Jerome, 2013), while others have found it to be an accurate method for measuring HIV DNA targets in clinical specimens (Strain et al., 2013). In the latter case, existing qPCR assays were used without modification or optimisation for the study, and dPCR was found to have improved accuracy and precision compared to qPCR (Strain et al., 2013).

3 VIRAL LOAD TESTING

Arguably, one of the most common clinical applications of qPCR in the field of infectious diseases is for viral load testing. For many illnesses, even low levels of viraemia are clinically relevant, and changes in viral load can have significance in prognosis and outcome (Sedlak & Jerome, 2013). Additionally, quantitative values are used to determine the efficacy of antiviral therapies or as a trigger for pre-emptive treatment prior to symptomatic infection (Hayden et al., 2013). Studies on CMV have indicated that while both digital and qPCR provide accurate viral load data over a wide dynamic range, dPCR may have a reduced variability (Hayden et al., 2013). Other studies have indicated that increased precision in CMV detection by dPCR compared to qPCR may not always be clinically relevant. For instance, one study found increased precision at viral loads greater than 4 $\log_{10}$, however, that increased precision did not hold true for lower viral loads. As CMV monitoring is important in a transplant setting, where CMV viral loads are often below 3 $\log_{10}$, the clinical relevance of the increased precision in dPCR over qPCR may not be as useful (Sedlak, Cook, Cheng, et al., 2014). Evaluations of dPCR for HIV viral load testing have focused on efforts to characterise latent HIV reservoirs and eradication interventions (Ruelle et al., 2014; Strain et al., 2013). One study reports a limit of quantification of 7 copies/mL of plasma for a HIV-2 viral load assay using dPCR, with increased sensitivity and reproducibility compared to qPCR (Ruelle et al., 2014). Studies of Enterovirus 71 viral load also show a good correlation to qPCR within a range of 2.5×10^0 to 2.5×10^3 copies and show promising use of dPCR for investigation

of low viral count diagnostics (Lui & Tan, 2014). The demonstrated use of dPCR in the analysis of common quantitative standards may also serve to improve agreement among qPCR viral load tests. By its use as a reference standard, particularly for DNA viral quantitation, there may be significant value in facilitating normalisation among qPCR calibration materials; dPCR may become critical to increasing uniformity across platforms and laboratories (Hayden et al., 2015).

4 SINGLE VIRION GENOMICS

The use of dPCR to screen infected cells could be coupled to single cell genomics, then to high-throughput sequencing, to show whether the viral genome is a remnant of a previous infection or a new infection. dPCR may also be used to measure virulence factors in the environment, relating to both human infection and infections that affect the biosphere at large, such as massive fish die offs or viral lysis within the phytoplankton community. In general, dPCR is a novel technology that could be useful in the characterisation, classification, and the biological role that viruses have on the human host and the biosphere.

The link between dPCR and single virion genomics starts with a new application to dPCR, which is the random amplification of high-molecular-weight DNA, driven by Φ29 DNA polymerase (DNAP), referred to as MDA (Blainey & Quake, 2011; Dean et al., 2001). MDA allows circular DNA templates to be amplified over 10,000-fold within a few hours and is the most common method of whole-genome amplification used in single cell genomics (Blainey & Quake, 2011; Dean et al., 2001). Digital MDA (dMDA) combines traditional MDA methods with a digital platform for increased sensitivity in quantifying nucleic acid fragments of an unknown sequence. Using dMDA, rapid whole-genome amplification of single cells or possibly single virions could be accomplished. In theory, dMDA could be used to amplify the viral genome for sequencing on a high-throughput platform.

Currently, the study of single virion genomics has many technological challenges not shared by single cell genomics. Viruses are much smaller than their host, meaning that trapping and capturing unknown members is very challenging. Random amplification methods such as MDA have problems for both single cell genomics and single virion genomics, in that DNA contamination (from mixes) and chimeric genome amplification bias further weaken the possibility of single virion genomics as common place as the study of single cell genomics (Mazaika & Homsy, 2014). In some studies, up to 50% of cloned MDA products were chimeric inserts prior to enzymatic treatments, and up to 6% remained after treatment (Binga, Lasken, & Neufeld, 2008; Lasken & Stockwell, 2007; Zhang et al., 2006). As viruses in nature have chimeric genomes due to random recombination, higher mutation rates, and genetic transfer from host to host, the addition of chimeric genome amplification bias through MDA makes the possibility of single virion genomics a tough prospect (Marine et al., 2014).

Only one paper illustrates a hypothetical design for single virion genomics, linking flow cytometry attempts to sort single virions into low melting agarose and then amplifying the genome with Φ29-based MDA (Allen et al., 2011). In principle, this method illustrates that a single virion could be isolated and the most difficult part of this method is ensuring that MDA provides enough coverage of the single virus to *de novo* assemble the whole genome. This has been a limitation in single cell genomics as well. Additionally, this method does not include a nuclease step to eliminate any viral DNA that could have been mobilised in the agarose prior to the MDA amplification, and was not checked by any quantitation of genome copy number to validate whether it is a problem or not. The authors also state that 75% of agarose reaction wells had a range of viral particles from 1 to >1 (as high as five particles), once again another major problem that plagues this technique (Allen et al., 2011). However, depending on the number of virions that are sorted per well, if the genetic richness is low or high, *de novo* assembly should be relatively easy due to the higher throughput of Illumina sequencing and Sanger read lengths (>700 bp), as long as MDA can provide uniform coverage of the genome without fusing genes/genomes together through the ramp effects of amplification (Lasken & Stockwell, 2007).

Single virion genomics as a technique is still in its infancy, but may be used to define new viral pathogens, catalogue the genetic potential of viruses, and possibly provide insight into their roles in the environment, genetic transfer, and novel host–viral interactions. The cost of sequencing unknown viral genomes is significantly lower compared to bacteria or the human genome. However, with Illumina and other platforms promising a complete human genome at $1000 USD or less, the human genome sequencing push will provide virology with fresh insights that were previously unthinkable 10 years earlier.

CONCLUSIONS

dPCR and single virion genomics are both novel technologies that can be seen as a full circle effect on virology. These methodologies can be used to catalogue and classify the unknown genetic pool that has affected biology from the very beginning. Additionally, their usefulness in the clinical realm, including molecular diagnostics and viral load testing, has already been shown through many studies and across several virus types. While dPCR is still considered a 'new' technique, the implications for its use in viral load testing could have significant impact on treatment, prognosis, and outcome for many patients. The coupling of next-generation sequencing with dPCR and single virion genomics could be the next generation of virology, leading discoveries in all fields, as these novel biological identities have done in the past.

REFERENCES

Allen, L. Z., Ishoey, T., Novotny, M. A., McLean, J. S., Lasken, R. S., & Williamson, S. J. (2011). Single virus genomics: A new tool for virus discovery. *PLoS One*, 6(3), e17722. http://dx.doi.org/10.1371/journal.pone.0017722.

Binga, E. K., Lasken, R. S., & Neufeld, J. D. (2008). Something from (almost) nothing: The impact of multiple displacement amplification on microbial ecology. *ISME Journal, 2*(3), 233–241. http://dx.doi.org/10.1038/ismej.2008.10.

Blainey, P. C., & Quake, S. R. (2011). Digital MDA for enumeration of total nucleic acid contamination. *Nucleic Acids Research, 39*(4), e19. http://dx.doi.org/10.1093/nar/gkq1074.

Boizeau, L., Laperche, S., Desire, N., Jourdain, C., Thibault, V., & Servant-Delmas, A. (2014). Could droplet digital PCR be used instead of real-time PCR for quantitative detection of the hepatitis B virus genome in plasma? *Journal of Clinical Microbiology, 52*(9), 3497–3498. http://dx.doi.org/10.1128/JCM.01306-14.

Brenner, S., Jacob, F., & Meselson, M. (1961). An unstable intermediate carrying information from genes to ribosomes for protein synthesis. *Nature, 190*, 576–581.

Clokie, M. R., Millard, A. D., Letarov, A. V., & Heaphy, S. (2011). Phages in nature. *Bacteriophage, 1*(1), 31–45. http://dx.doi.org/10.4161/bact.1.1.14942.

Corbisier, P., Bhat, S., Partis, L., Xie, V. R., & Emslie, K. R. (2010). Absolute quantification of genetically modified MON810 maize (Zea mays L.) by digital polymerase chain reaction. *Analytical and Bioanalytical Chemistry, 396*(6), 2143–2150. http://dx.doi.org/10.1007/s00216-009-3200-3.

Coudray-Meunier, C., Fraisse, A., Martin-Latil, S., Guillier, L., Delannoy, S., Fach, P., et al. (2015). A comparative study of digital RT-PCR and RT-qPCR for quantification of Hepatitis A virus and Norovirus in lettuce and water samples. *International Journal of Food Microbiology, 201*, 17–26. http://dx.doi.org/10.1016/j.ijfoodmicro.2015.02.006.

Creager, N. H. A. (2002). *The life of a virus: Tobacco mosaic virus as an experimental model, 1930-1965.* Chicago: University of Chicago Press.

Crick, F. H., Barnett, L., Brenner, S., & Watts-Tobin, R. J. (1961). General nature of the genetic code for proteins. *Nature, 192*, 1227–1232.

Dean, F. B., Hosono, S., Fang, L., Wu, X., Faruqi, A. F., Bray-Ward, P., et al. (2002). Comprehensive human genome amplification using multiple displacement amplification. *Proceedings of the National Academy of Sciences of the United States of America, 99*(8), 5261–5266. http://dx.doi.org/10.1073/pnas.082089499.

Dean, F. B., Nelson, J. R., Giesler, T. L., & Lasken, R. S. (2001). Rapid amplification of plasmid and phage DNA using Phi 29 DNA polymerase and multiply-primed rolling circle amplification. *Genome Research, 11*(6), 1095–1099. http://dx.doi.org/10.1101/gr.180501.

Dioverti, M. V., & Razonable, R. R. (2015). Clinical utility of cytomegalovirus viral load in solid organ transplant recipients. *Current Opinion in Infectious Diseases, 28*(4), 317–322. http://dx.doi.org/10.1097/QCO.0000000000000173.

Fan, H. C., Blumenfeld, Y. J., Chitkara, U., Hudgins, L., & Quake, S. R. (2008). Noninvasive diagnosis of fetal aneuploidy by shotgun sequencing DNA from maternal blood. *Proceedings of the National Academy of Sciences of the United States of America, 105*(42), 16266–16271. http://dx.doi.org/10.1073/pnas.0808319105.

Fiers, W., Contreras, R., Duerinck, F., Haegeman, G., Iserentant, D., Merregaert, J., et al. (1976). Complete nucleotide sequence of bacteriophage MS2 RNA: Primary and secondary structure of the replicase gene. *Nature, 260*(5551), 500–507.

Gullett, J. C., & Nolte, F. S. (2015). Quantitative nucleic acid amplification methods for viral infections. *Clinical Chemistry, 61*(1), 72–78. http://dx.doi.org/10.1373/clinchem.2014.223289.

Hall Sedlak, R., & Jerome, K. R. (2014). The potential advantages of digital PCR for clinical virology diagnostics. *Expert Review of Molecular Diagnostics, 14*(4), 501–507. http://dx.doi.org/10.1586/14737159.2014.910456.

Hayden, R. T., Gu, Z., Ingersoll, J., Abdul-Ali, D., Shi, L., Pounds, S., et al. (2013). Comparison of droplet digital PCR to real-time PCR for quantitative detection of cytomegalovirus. *Journal of Clinical Microbiology*, *51*(2), 540–546. http://dx.doi.org/10.1128/JCM.02620-12.

Hayden, R. T., Gu, Z., Sam, S. S., Sun, Y., Tang, L., Pounds, S., et al. (2015). Comparative evaluation of three commercial quantitative cytomegalovirus standards by use of digital and real-time PCR. *Journal of Clinical Microbiology*, *53*(5), 1500–1505. http://dx.doi.org/10.1128/JCM.03375-14.

Hershey, A. D., & Chase, M. (1952). Independent functions of viral protein and nucleic acid in growth of bacteriophage. *Journal of General Physiology*, *36*(1), 39–56.

Huang, J. T., Liu, Y. J., Wang, J., Xu, Z. G., Yang, Y., Shen, F., et al. (2015). Next generation digital PCR measurement of hepatitis B virus copy number in formalin-fixed paraffin-embedded hepatocellular carcinoma tissue. *Clinical Chemistry*, *61*(1), 290–296. http://dx.doi.org/10.1373/clinchem.2014.230227.

Huggett, J. F., Cowen, S., & Foy, C. A. (2015). Considerations for digital PCR as an accurate molecular diagnostic tool. *Clinical Chemistry*, *61*(1), 79–88. http://dx.doi.org/10.1373/clinchem.2014.221366.

Huggett, J. F., Foy, C. A., Benes, V., Emslie, K., Garson, J. A., Haynes, R., et al. (2013). The digital MIQE guidelines: Minimum Information for Publication of Quantitative Digital PCR Experiments. *Clinical Chemistry*, *59*(6), 892–902. http://dx.doi.org/10.1373/clinchem.2013.206375.

Kalinina, O., Lebedeva, I., Brown, J., & Silver, J. (1997). Nanoliter scale PCR with TaqMan detection. *Nucleic Acids Research*, *25*(10), 1999–2004.

Kiselinova, M., Pasternak, A. O., De Spiegelaere, W., Vogelaers, D., Berkhout, B., & Vandekerckhove, L. (2014). Comparison of droplet digital PCR and seminested real-time PCR for quantification of cell-associated HIV-1 RNA. *PLoS One*, *9*(1), e85999. http://dx.doi.org/10.1371/journal.pone.0085999.

Ko, E. R., Yang, W. E., McClain, M. T., Woods, C. W., Ginsburg, G. S., & Tsalik, E. L. (2015). What was old is new again: Using the host response to diagnose infectious disease. *Expert Review of Molecular Diagnostics*, *15*, 1143–1158. http://dx.doi.org/10.1586/14737159.2015.1059278.

Lasken, R. S., & Stockwell, T. B. (2007). Mechanism of chimera formation during the Multiple Displacement Amplification reaction. *BMC Biotechnology*, *7*, 19. http://dx.doi.org/10.1186/1472-6750-7-19.

Laurent-Puig, P., Pekin, D., Normand, C., Kotsopoulos, S. K., Nizard, P., Perez-Toralla, K., et al. (2015). Clinical relevance of KRAS-mutated subclones detected with picodroplet digital PCR in advanced colorectal cancer treated with anti-EGFR therapy. *Clinical Cancer Research*, *21*(5), 1087–1097. http://dx.doi.org/10.1158/1078-0432.CCR-14-0983.

Leibovitch, E. C., Brunetto, G. S., Caruso, B., Fenton, K., Ohayon, J., Reich, D. S., et al. (2014). Coinfection of human herpesviruses 6A (HHV-6A) and HHV-6B as demonstrated by novel digital droplet PCR assay. *PLoS One*, *9*(3), e92328. http://dx.doi.org/10.1371/journal.pone.0092328.

Lipkin, W. I., & Anthony, S. J. (2015). Virus hunting. *Virology*, *479–480*, 194–199. http://dx.doi.org/10.1016/j.virol.2015.02.006.

Lui, Y. L., & Tan, E. L. (2014). Droplet digital PCR as a useful tool for the quantitative detection of Enterovirus 71. *Journal of Virological Methods*, *207*, 200–203. http://dx.doi.org/10.1016/j.jviromet.2014.07.014.

Luria, S. E., Delbruck, M., & Anderson, T. F. (1943). Electron microscope studies of bacterial viruses. *Journal of Bacteriology*, *46*(1), 57–77.

Malatinkova, E., Kiselinova, M., Bonczkowski, P., Trypsteen, W., Messiaen, P., Vermeire, J., et al. (2014). Accurate episomal HIV 2-LTR circles quantification using optimized DNA isolation and droplet digital PCR. *Journal of the International AIDS Society*, *17*(4 Suppl. 3), 19674. http://dx.doi.org/10.7448/IAS.17.4.19674.

Malatinkova, E., Kiselinova, M., Bonczkowski, P., Trypsteen, W., Messiaen, P., Vermeire, J., et al. (2015). Accurate quantification of episomal HIV-1 two-long terminal repeat circles by use of optimized DNA isolation and droplet digital PCR. *Journal of Clinical Microbiology*, *53*(2), 699–701. http://dx.doi.org/10.1128/JCM.03087-14.

Marine, R., McCarren, C., Vorrasane, V., Nasko, D., Crowgey, E., Polson, S. W., et al. (2014). Caught in the middle with multiple displacement amplification: The myth of pooling for avoiding multiple displacement amplification bias in a metagenome. *Microbiome*, *2*(1), 3. http://dx.doi.org/10.1186/2049-2618-2-3.

Mazaika, E., & Homsy, J. (2014). Digital droplet PCR: CNV analysis and other applications. *Current Protocols in Human Genetics*, *82*, 7.24.1–7.24.13. http://dx.doi.org/10.1002/0471142905.hg0724s82.

Mori, S., Vagge, E., le Coutre, P., Abruzzese, E., Martino, B., Pungolino, E., et al. (2015). Age and dPCR can predict relapse in CML patients who discontinued imatinib: The ISAV study. *American Journal of Hematology*. http://dx.doi.org/10.1002/ajh.24120.

Mukaide, M., Sugiyama, M., Korenaga, M., Murata, K., Kanto, T., Masaki, N., et al. (2014). High-throughput and sensitive next-generation droplet digital PCR assay for the quantitation of the hepatitis C virus mutation at core amino acid 70. *Journal of Virological Methods*, *207*, 169–177. http://dx.doi.org/10.1016/j.jviromet.2014.07.006.

Nixon, G., Garson, J. A., Grant, P., Nastouli, E., Foy, C. A., & Huggett, J. F. (2014). Comparative study of sensitivity, linearity, and resistance to inhibition of digital and nondigital polymerase chain reaction and loop mediated isothermal amplification assays for quantification of human cytomegalovirus. *Analytical Chemistry*, *86*(9), 4387–4394. http://dx.doi.org/10.1021/ac500208w.

Pang, X., & Lee, B. E. (2015). Laboratory diagnosis of noroviruses: Present and future. *Clinics in Laboratory Medicine*, *35*(2), 345–362. http://dx.doi.org/10.1016/j.cll.2015.02.008.

Ruelle, J., Yfantis, V., Duquenne, A., & Goubau, P. (2014). Validation of an ultrasensitive digital droplet PCR assay for HIV-2 plasma RNA quantification. *Journal of the International AIDS Society*, *17*(4 Suppl. 3), 19675. http://dx.doi.org/10.7448/IAS.17.4.19675.

Sanders, S. J., Ercan-Sencicek, A. G., Hus, V., Luo, R., Murtha, M. T., Moreno-De-Luca, D., et al. (2011). Multiple recurrent *de novo* CNVs, including duplications of the 7q11.23 Williams syndrome region, are strongly associated with autism. *Neuron*, *70*(5), 863–885. http://dx.doi.org/10.1016/j.neuron.2011.05.002.

Sanders, R., Huggett, J. F., Bushell, C. A., Cowen, S., Scott, D. J., & Foy, C. A. (2011). Evaluation of digital PCR for absolute DNA quantification. *Analytical Chemistry*, *83*(17), 6474–6484. http://dx.doi.org/10.1021/ac103230c.

Sanger, F., Air, G. M., Barrell, B. G., Brown, N. L., Coulson, A. R., Fiddes, C. A., et al. (1977). Nucleotide sequence of bacteriophage phi X174 DNA. *Nature*, *265*(5596), 687–695.

Sedlak, R. H., Cook, L., Cheng, A., Magaret, A., & Jerome, K. R. (2014a). Clinical utility of droplet digital PCR for human cytomegalovirus. *Journal of Clinical Microbiology*, *52*(8), 2844–2848. http://dx.doi.org/10.1128/JCM.00803-14.

Sedlak, R. H., Cook, L., Huang, M. L., Magaret, A., Zerr, D. M., Boeckh, M., et al. (2014b). Identification of chromosomally integrated human herpesvirus 6 by droplet digital PCR. *Clinical Chemistry*, *60*(5), 765–772. http://dx.doi.org/10.1373/clinchem.2013.217240.

Sedlak, R. H., & Jerome, K. R. (2013). Viral diagnostics in the era of digital polymerase chain reaction. *Diagnostic Microbiology and Infectious Disease*, *75*(1), 1–4. http://dx.doi.org/10.1016/j.diagmicrobio.2012.10.009.

Sedlak, R. H., Kuypers, J., & Jerome, K. R. (2014). A multiplexed droplet digital PCR assay performs better than qPCR on inhibition prone samples. *Diagnostic Microbiology and Infectious Disease*, *80*(4), 285–286. http://dx.doi.org/10.1016/j.diagmicrobio.2014.09.004.

Sidransky, D., Tokino, T., Hamilton, S. R., Kinzler, K. W., Levin, B., Frost, P., et al. (1992). Identification of ras oncogene mutations in the stool of patients with curable colorectal tumors. *Science*, *256*(5053), 102–105.

Simons, J. N., Desai, S. M., & Mushahwar, I. K. (2000). The GB viruses. *Current Topics in Microbiology and Immunology*, *242*, 341–375.

Snyder, T. M., Khush, K. K., Valantine, H. A., & Quake, S. R. (2011). Universal noninvasive detection of solid organ transplant rejection. *Proceedings of the National Academy of Sciences of the United States of America*, *108*(15), 6229–6234. http://dx.doi.org/10.1073/pnas.1013924108.

Strain, M. C., Lada, S. M., Luong, T., Rought, S. E., Gianella, S., Terry, V. H., et al. (2013). Highly precise measurement of HIV DNA by droplet digital PCR. *PLoS One*, *8*(4), e55943. http://dx.doi.org/10.1371/journal.pone.0055943.

Suttle, C. A. (2005). Viruses in the sea. *Nature*, *437*(7057), 356–361. http://dx.doi.org/10.1038/nature04160.

Sykes, P. J., Neoh, S. H., Brisco, M. J., Hughes, E., Condon, J., & Morley, A. A. (1992). Quantitation of targets for PCR by use of limiting dilution. *Biotechniques*, *13*(3), 444–449.

Tadmor, A. D., Ottesen, E. A., Leadbetter, J. R., & Phillips, R. (2011). Probing individual environmental bacteria for viruses by using microfluidic digital PCR. *Science*, *333*(6038), 58–62. http://dx.doi.org/10.1126/science.1200758.

Tan, S. K., Waggoner, J. J., & Pinsky, B. A. (2015). Cytomegalovirus load at treatment initiation is predictive of time to resolution of viremia and duration of therapy in hematopoietic cell transplant recipients. *Journal of Clinical Virology*, *69*, 179–183. http://dx.doi.org/10.1016/j.jcv.2015.06.006.

Vogelstein, B., & Kinzler, K. W. (1999). Digital PCR. *Proceedings of the National Academy of Sciences of the United States of America*, *96*(16), 9236–9241.

Wagner, P. L., & Waldor, M. K. (2002). Bacteriophage control of bacterial virulence. *Infection and Immunity*, *70*(8), 3985–3993.

Weber, K., & Osborn, M. (1969). The reliability of molecular weight determinations by dodecyl sulfate-polyacrylamide gel electrophoresis. *Journal of Biological Chemistry*, *244*(16), 4406–4412.

White, R. A., 3rd, Blainey, P. C., Fan, H. C., & Quake, S. R. (2009). Digital PCR provides sensitive and absolute calibration for high throughput sequencing. *BMC Genomics*, *10*, 116. http://dx.doi.org/10.1186/1471-2164-10-116.

White, R. A., 3rd, Quake, S. R., & Curr, K. (2012). Digital PCR provides absolute quantitation of viral load for an occult RNA virus. *Journal of Virological Methods*, *179*(1), 45–50. http://dx.doi.org/10.1016/j.jviromet.2011.09.017.

Williams, H. T. (2013). Phage-induced diversification improves host evolvability. *BMC Evolutionary Biology*, *13*, 17. http://dx.doi.org/10.1186/1471-2148-13-17.

Zhang, K., Martiny, A. C., Reppas, N. B., Barry, K. W., Malek, J., Chisholm, S. W., et al. (2006). Sequencing genomes from single cells by polymerase cloning. *Nature Biotechnology*, *24*(6), 680–686. http://dx.doi.org/10.1038/nbt1214.

Zhong, Q., Bhattacharya, S., Kotsopoulos, S., Olson, J., Taly, V., Griffiths, A. D., et al. (2011). Multiplex digital PCR: Breaking the one target per color barrier of quantitative PCR. *Lab on a Chip*, *11*(13), 2167–2174. http://dx.doi.org/10.1039/c1lc20126c.

Artificial Nucleic Acid Probes and Their Applications in Clinical Microbiology

17

Alon Singer*,1, Yi-Wei Tang†,‡

*HelixBind, Inc., Marlborough, Massachusetts, USA
†Department of Laboratory Medicine, Memorial Sloan-Kettering Cancer Center, New York, USA
‡Department of Pathology and Laboratory Medicine, Weill Medical College of Cornell University, New York, USA
1Corresponding author: e-mail address: asinger@helixbind.com

1 INTRODUCTION

Few would argue that nucleic acids are remarkable molecules—inherently carrying within them the instructions required for life itself. Most recently, a push has been made to harness these molecules, namely, RNA and DNA, for the purpose of improving the quality of life through advancements in medical diagnostics. It is our understanding of the biological physics of these molecules: their interactions, both with one another and with other types of biomolecules, which has enabled us to develop novel technologies which utilise their unique capabilities. Despite this, DNA and RNA did not evolve for these purposes and have, at times, proven to be rather inefficient or at the very least not perfectly suited for a myriad of reasons.

Recently, significant efforts have been made to engineer artificial nucleic acids (NAs) that would be better suited for the purposes of medical diagnostics. While a number of different classes or types of artificial NAs have been proposed; two classes, locked nucleic acids (LNA) and peptide nucleic acids (PNA) have become more widely used in the field and the diversity of applications that they have made an impact on (Egholm, Buchardt, Nielsen, & Berg, 1992; Koshkin & Wengel, 1998; Obika, Nanbu, Hari, & Morio, 1997). Interestingly, the two classes of artificial NAs that have become the most dominant were originally intended not for diagnostic applications, but rather for therapeutic ones and were both first synthesised in roughly the same time period (Briones & Moreno, 2012). Though the approach taken by the inventors of both of these artificial NAs was quite different, the problem they addressed was similar: How do we increase the binding affinity and sequence specificity of NA probes?

Methods in Microbiology, Volume 42, ISSN 0580-9517, http://dx.doi.org/10.1016/bs.mim.2015.05.003

Groups led by Wengel and Imanishi designed and created artificial nucleotides which had a rigidity or inflexibility within its backbone, specifically at the sugar moiety (Koshkin & Wengel, 1998; Obika et al., 1997). This rigidity 'locks' the oligonucleotide into an A-conformation, mimicking the structure of double-stranded (ds) RNA (Suresh & Priyakumar, 2013) which should be more energetically favourable to, among other things, remain in the ds-form as opposed to the free, single-stranded, or un-hybridised form. This new nucleotide which can replace standard, or natural, nucleotides within an oligomer was rightfully termed LNA. Similarly, a group led by Nielsen proposed an alternative strategy of replacing the backbone of the DNA altogether with one that mimics that of a polypeptide, hence this oligonucleotide was termed PNA (Egholm et al., 1992; Nielsen, Egholm, Berg, & Buchardt, 1993). Their focus was to remove the energetic 'penalty' which single-stranded nucleic acids (ssNAs) pay when they bind to one another due to electrostatic interactions. Beyond the increase in binding affinity, both LNAs and PNAs resisted enzymatic degradation as they are completely synthetic, highly advantageous for *in vivo* applications. The development of these two engineered molecules has enabled researchers to design artificial NAs with specific performance criteria.

Although initially considered for therapeutic applications, given their potential, a number of groups began strategically replacing natural NAs in common diagnostic methodologies including, among others, polymerase chain reactions (PCR), microarrays, and *in situ* hybridisations (ISH). From these early beginnings, researchers demonstrated the inherent attributes of these new synthetic molecules: faster reaction kinetics, higher sensitivity, and improved specificity/accuracy.

In this chapter, we will review LNA and PNA technologies describing how they function, their flexibility, and ability to be further user modified. We will likewise describe the various diagnostic applications where these artificial NAs have made an impact and where future applications may be. We will conclude by describing clinical applications of these probes and demonstrate how they already have and will continue to impact the clinical microbiology laboratory.

2 ARTIFICIAL NUCLEIC ACIDS: THE UNMET NEED DRIVES INNOVATION

While LNAs and PNAs were developed with the aim of solving similar sets of problems, the approaches to solving these problems were quite different. In the case of LNAs, an emphasis was placed on the point replacement of one or more NA monomers with an artificial one, while PNAs emphasised replacing the entire oligomer altogether with an artificial one. Although chronologically PNAs arrived prior to LNAs, given our intended emphasis on PNAs in this chapter, LNAs will be discussed initially.

The elegance in LNAs revolves around the modification made to the sugar constituent of the NA backbone. In LNAs, the chemical modification $(-O-CH_2-)$ is introduced, connecting the $C2'$ and $C4'$ positions of the sugar ring (Figure 1).

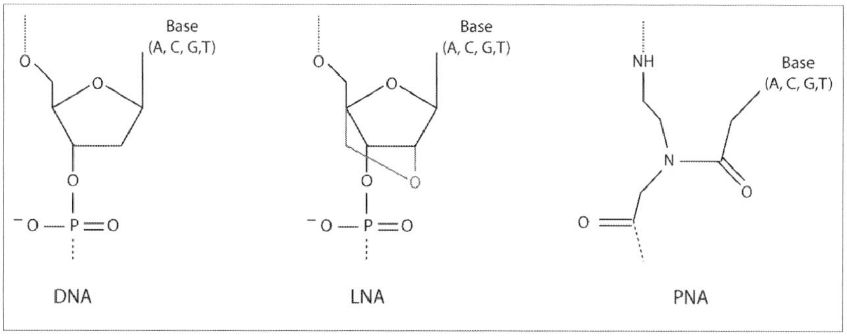

FIGURE 1

Although highly similar to natural NAs, artificial NAs are designed with a specific purpose in mind: LNAs are to be rigid than standard nucleotides and PNAs are designed to be neutral rather than charged. (See the color plate.)

This chemical modification forces the sugar ring into a single conformational state, in essence becoming a rigid structure—hence the term 'Locked NA' (Briones & Moreno, 2012). This added rigidity reduces the entropic loss when the oligonucleotide makes the transition from the single-stranded to the double-stranded state, leading to a more energetically favourable, and hence more stable, duplex. The net result is that the affinity of an LNA-modified hybrid oligomer to its complementary natural NA oligomer is greater than that of a natural oligomer to its complementary natural oligomer, thus enabling the formation of hybrid duplexes through standard Watson–Crick base-pairing rules with increased thermal stability. Numerous studies have quantified this effect where the thermal stability (T_M) increase for a typical RNA/LNA hybrid duplex is 3–10 °C higher for per LNA monomer (Briones & Moreno, 2012; Campbell & Wengel, 2011; Nielsen, Singh, Wengel, & Jacobsen, 2000) over that of a standard duplex RNA.

Although initially intended as an RNA analogue, efforts were made at understanding the capabilities of LNAs for diagnostic applications; which commonly requires hybridisation with DNA. To this end, further studies were conducted, and while the increase in thermal stability of the LNA/DNA hybrid is less pronounced than that of LNA/RNA hybrids, it is nonetheless apparent and is typically quantified at 2–7 °C per LNA monomer (Braasch & Corey, 2001; Briones & Moreno, 2012; Campbell & Wengel, 2011; Petersen & Wengel, 2003). The first published diagnostic study was conducted in 1999 by the Borre group utilising surface-immobilised LNA probes, in a manner similar to ELISA, to detect point mutations in PCR-amplified Factor V Leiden genomic fragments (Ørum, Jakobsen, Koch, Vuust, & Borre, 1999) utilising the affinity of strategically placed LNA stretches within the DNA probe to achieve the required sequence specificity and thermal stability. Since then, LNA-modified oligomers have been used increasingly in diagnostic applications including the reduction of off-target amplification in PCR through elongation inhibition (Hummelshoj, Ryder, Madsen, & Poulsen, 2005), primer clamping

(Thiede et al., 2006), and microarrays (Castoldi et al., 2006). However, further diagnostic applications in clinical microbiology have been limited.

LNAs have a number of disadvantages that need to be taken into account when designing a probe, and when designing an assay utilising that probe. First, the significant increase in T_M and associated increase in binding affinity will inevitably lead to higher levels of non-specific binding. As such, LNAs cannot be employed as 'LNA-only' oligomers but rather as replacement nucleotides *within* a natural oligomer such as a single-stranded DNA. For example, a typical design criteria for PCR clamping probe (discussed later in this chapter) limits the number of LNA nucleotides to around seven, while one should strive to limit the number of consecutive LNA nucleotides to no more than three to four (shorter oligonucleotides favour even less LNA monomers). Moreover, attention needs to be made to dimer formation which is more likely to occur due to the enhanced binding affinity (thermal stability) of LNAs not only to natural NAs but also to themselves, as LNA/LNA duplexes are extremely difficult to disrupt. Lastly, LNA presence in palindrome sequences and to LNA stretches of cytosine or guanine need to be prevented. Despite a number of limitations in oligonucleotide design which reduce flexibility, LNA-modified probes are a useful technology and should be considered in any diagnostic application where NA probes are used.

PNAs were first described, similarly to LNAs, as potential agents for therapeutics but likewise carried the potential for modulating gene expression at the DNA level (Buchardt, Egholm, Berg, & Nielsen, 1993; Egholm et al., 1992). In envisioning and developing PNAs, the strategy was to develop a neutral backbone to replace the negatively charged sugar–phosphate backbone found in all natural NAs. The rationale was clear, double-stranded nucleic acids (dsNAs) continuously play an energetic 'tug-of-war': Combined with entropy, the negatively charged backbone on each strand of a duplex NA tries to break apart the biopolymer while the two single strands are held together by both the hydrogen bonds across the nucleobases as well as by base-pair stacking interactions (Yakovchuk, Protozanova, & Frank-Kamenetskii, 2006). Therefore, by replacing the negatively charged backbone with one that is inherently neutral (Figure 1), one would expect a significantly higher binding affinity to a complementary strand. And as was demonstrated, this is indeed the case with PNA/DNA hybrids being markedly more stable than those of dsDNA with an increase in T_M ranging from 2 to 4 °C/per PNA monomer (Briones & Moreno, 2012). Further, it was shown that PNAs are not only capable of forming duplex hybrids but likewise triplex hybrids by invading dsDNA (Egholm et al., 1993; Peffer et al., 1993) creating what is now termed the (PNA)$_2$–DNA triplex. Thus in contrast to LNAs which primarily strive to increase the T_M of a hybrid duplex, PNAs were additionally advantageous in providing a means of working with both ssNAs and dsNAs as shown in Figure 2. This invasion capability enabled numerous potential applications and was further enhanced by the finding that this triplex structure was indeed sequence specific, potentially down to a single base pair.

However, this PNA triplex-forming capability is not without its constraints and is dependent on a specific range of electrolyte concentrations and pH. This

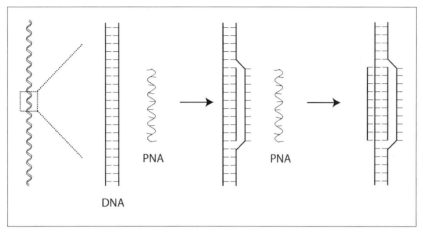

FIGURE 2

PNAs present a unique opportunity as they can bind not only to single-stranded NAs but likewise to double-stranded NAs through a combination of Watson–Crick and Hoogsteen base-pairing rules.

triplex-forming capability is initiated when one of the PNA oligomers binds to the displaced DNA via standard Watson–Crick base-pairing rules, while a second PNA strand, which serves to stabilise this structure, binds to that same displaced strand via Hoogsteen base-pairing rules. Due to this requirement, strand invasion, or triplex formation, is typically limited to homopyrimidine PNAs which naturally require homopurine stretches of DNA (commonly 7–10 bases long), directly limiting the sequences that can be detected (Egholm et al., 1995). Further, given that the $(PNA)_2$–DNA triplex requires the breathing, or opening of the DNA duplex in addition to the formation of Hoogsteen interactions, the rate of triplex formation is quite slow and other factors are required to facilitate this process. Common components found in either biological buffers or media such as monovalent and divalent ions hinder this process to the point that triplex invasion simply does not occur at practical timescales while working at physiological electrolyte concentrations, hence only specific (<20 mM monovalent electrolyte strength) systems are typically considered feasible. Lastly, and once again due to the requirement for Hoogsteen interactions for efficient triplex formation, pH must be controlled as to ensure that the cytosines nucleobases in the PNAs are positively charged facilitating an increased interaction with the negatively charged DNA strand.

The limitations with PNA use are not restricted only due to their use in triplex formation, as other limitations exist, unrelated to triplex formation, where the primary concern is that of solubility. DNA and RNA are naturally charged, making them extremely hydrophilic molecules, whereas PNAs are inherently neutral creating a rather hydrophobic or insoluble molecule in aqueous systems. This typically limits their length to under 12 bases and low GC compositions (<60%) (Nielsen,

1999) with restricted usage of chemical moieties such as dyes or haptens which tend to aggravate this issue. Moreover, even when targeting ssNAs electrolyte concentrations are of importance; as it is PNA's lack of electrostatic repulsion with its complementary ssDNA strand that give it its remarkable affinity and specificity properties. However, this lack of an electrostatic repulsion also leaves the T_M of such hybrid duplexes independent of salt concentrations making PNA/DNA interactions favourable over DNA/DNA interactions only in the low-electrolyte strength regime, from a practical perspective, when [NaCl] < 100 mM.

Despite these concerns or limitations, and in direct contrast to LNAs, diagnostic applications utilising the capabilities of PNAs are common including those depicting PCR clamping through competitive primer binding and elongation arresting (Orum, 2000; Ørum et al., 1993; Takiya et al., 2004; von Wintzingerode, Landt, Ehrlich, & Göbel, 2000). In addition, PNAs have found diagnostic uses as fluorescent probes used in PCR in place of ssDNAs (Choi et al., 2012) and as probes for *in situ* assays (Bonvicini et al., 2006; Wilson, Hall, & Procop, 2010), and microarrays (Choi, Kim, & Park, 2009; Raymond et al., 2005) proving that PNAs have application in range of assays and offer unique advantages in the next generation of diagnostic platforms.

3 THE MANY FLAVOURS OF ARTIFICIAL NUCLEIC ACIDS; DIFFERENT STRUCTURES, INTERNAL MODIFICATIONS, AND CHARACTERISTICS

Although we have described two main classes of artificial NAs, significant effort has been directed to circumvent some of the limitations of the standard model LNA or PNA, and develop more superior probes which could facilitate their use in new applications. Given the breadth of work completed with PNAs in comparison to LNAs, and the general applicability of PNAs to diagnostic approaches in clinical microbiology, modifications to the standard PNA model will be described in more detail.

From a diagnostic perspective, one of the most interesting features of PNAs is their ability to invade dsDNA, which presents multiple opportunities for exploitation from simplifying the sample preparation processes to attaining single-base specificity. As such, additional binding strategies, and chemical modifications and even PNA structures were aimed at improving on the invasion capabilities of PNAs.

Initially, dsDNA invasion was accomplished by incubating two PNA strands, both homoprymidines, with a target DNA containing the complementary homopurine stretch. Under the right conditions, this would induce the formation of a hybrid triplex helix containing two PNAs, and a DNA where one PNA would be bound to the DNA through standard Watson–Crick interactions, where the second PNA would

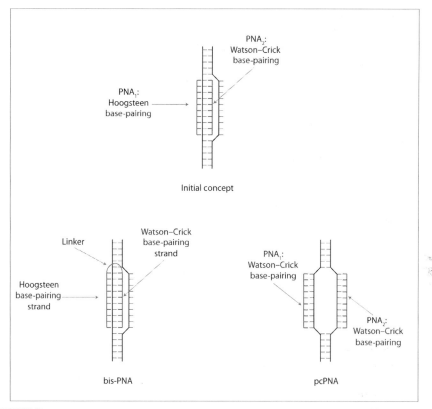

FIGURE 3

Two concepts aimed at improving the kinetics and stability of PNA invasion; either through a tandem Watson–Crick/Hoogsteen base-pairing with a single PNA probe (bis-PNAs) or through binding to both DNA strands within a single region, simultaneously (pcPNA). (See the color plate.)

be bound through Hoogsteen interactions (Figure 3, top). This triplex formation proved highly advantageous in stabilising the overall structure. A method first described by Egholm and co-workers was to join together two homoprymidines PNAs via a flexible linker—in that case, the flexible linker comprised of three adjoining units of 8-amino-3,6-dioxaoctanoic acid. This new PNA, termed *bis-PNA*, proved highly capable of improving the efficiency of the formation of the triplex complex (Figure 3, bottom left) as demonstrated through binding assays (Abibi, Protozanova, Demidov, & Frank-Kamenetskii, 2004; Griffith et al., 1995). Moreover it was shown that not only are the binding kinetics improved, while still meeting single base-pair specificity, but likewise the thermal stability of the bis-PNA is likewise improved (Kuhn, Demidov, Frank-Kamenetskii, & Nielsen, 1998). In order to circumvent the strong pH dependence of the reaction, due to the requirement for protonation of the cytosines in the Hoogsteen strand, Egholm and co-workers demonstrated that through the introduction of an artificial cytosine, pseudoisocytosine (termed a 'J' nucleobase), a pH-independent reaction was feasible. One issue that remains however, is the requirement for homoprymidine stretches (typically 7–10mer), obviously limiting the number of sequences that can be targeted as well as the conditions under which invasion occurs, namely, <20 mM monovalent salt concentrations.

An alternative approach, aimed at removing the sequence restriction described above, targets simultaneously both strands of a prospective target region through Watson–Crick base-pairing rules (Figure 3, bottom right). This approach, utilises what are essentially two complementary PNA strands, termed pseudo-complimentary PNAs or pcPNAs, both targeting a single, specific region (Lohse, Dahl & Nielsen, 1999). In order to overcome the obvious difficulty where the two PNA strands would bind to one another (and bind very effectively with extremely high thermal stability), artificial nucleobases were introduced, taking the place of adenines and thymines; specially 2,6-diaminopurines which recognises thymine more efficiently than does adenine (thereby replacing adenines), and 2-thiouracils (thereby replacing thymines)—natural cytosines and guanines are retained (Kutyavin et al., 1996). The concept of this design scheme relies on 2,6-diaminopurines binding to thymines, and 2-thiouracils binding to adenines while not binding to one another due to steric hindrance. From an energetic perspective, it was deemed that the required free energy of strand displacement (dsDNA invasion) would be met if both strands were targeted simultaneously (Demidov et al., 2002; Lohse et al., 1999). However a drawback of this method is that while it in theory enables sequence-unrestricted invasion of PNAs, in practice, the thermal stability of the overall structure was reduced considerably (Lohse et al., 1999).

Beyond the desire to improve the invasion capabilities of PNAs, their lack of solubility and their tendency to self-aggregate needed to be addressed (Efimov, Choob, Buryakova, Kalinkina, & Chakhmakhcheva, 1998). Although this effect is largely discussed in terms of a reduction in cellular uptake given the therapeutic lineage of PNAs, for diagnostics, this lack of solubility leads to a reduction in the concentration range which is suitable for *in vitro* applications as well as they may display high levels of non-specific binding (Abibi et al., 2004).

In order to address this issue, two simple components are typically added to the PNA oligomer. First, lysine residues are commonly added at both the C-terminus and N-terminus. One of the advantages with lysine lies in its side chain, which is terminated with an amino group (NH_3^+). This amino group (owing to its pK_a value of 10.5) is positively charged at typical working conditions leading to two positive traits: first, this positive charge induces a decrease in the hydrophobicity of the PNA oligomer, and second is that it is beneficial to the kinetics of PNA/DNA hybridisation—while not leading to significant non-specific interactions (Abibi et al., 2004). Although this does not resolve the issue of lack of solubility completely, this simple addition/modification is indeed significant and is considered standard practice in PNA oligomer design. An additional design approach to increase solubility is to incorporate during the synthesis process hydrophilic, uncharged polymers such as poly(ethylene glycol) into the oligomer, such as described in bis-PNAs (Millili, Yin, Fan, Naik, & Sullivan, 2010). The addition of this flexible additive has been found to not adversely affect PNA/DNA stability while increasing solubility significantly.

One of the first attempts to directly address the issue of solubility was described by Peyman and co-workers where they attempted to synthesise a PNA analogue, termed phosphonic-ester nucleic acid or pPNA (Peyman et al., 1996) without compromising the binding properties of PNA. Here, the standard achiral and neutral N-(2-aminoethyl)-glycine PNA backbone is replaced with an achiral and negatively charged backbone where the nucleobases are attached through carboxymethylene linkers. Given that this new structure is isosteric to PNA, yet charged, it was hypothesised that this would produce improved solubility. Despite the marked increase in solubility, binding affinity was significantly lower than that of 'ordinary' PNA. Along those lines, Efimov and co-workers proposed that it might be favourable to develop a scaffold which combines stretches of pPNA with stretches of 'ordinary' PNA. As both these artificial NAs are isosteric, it was hypothesised that PNA–pPNA chimeras might display the excellent binding characteristics of PNAs with the high solubility of pPNAs (Efimov et al., 1998). Although not reaching the same level of thermal stability displayed by PNAs, the hybrid PNA–pPNA did achieve higher levels of thermal stability compared to that of pPNAs by themselves, where the solubility was significantly increased. Due to the reduced thermal stability of the hybrid complex, in addition to the complicated synthesis process, PNA–pPNA hybrids have not become commonplace, but are highly suggestive of the types of modifications possible to achieve the desired characteristics.

More recently, Dragulescu-Andrasi and co-workers described an innovative method to attain sequence-unrestricted invasion of PNAs to dsDNA. Previously, PNA invasion was hindered by being energetically unfavourable, thus requiring alternative strategies to stabilise the invaded hybrid. Dragulescu-Andrasi and co-workers hypothesised that given that the free energy of binding, ΔG, is comprised both of an enthalpic, ΔH, and by an entropic contribution, $T\Delta S$, by minimising the entropic penalty the single-stranded PNA pays due to invasion (which can be described as a reduction in freedom of movement or an increase in rigidity), it might be possible to facilitate stable dsDNA invasion by creating a 'rigid' PNA (Dragulescu-Andrasi et al., 2006). In order to achieve this, they added a (S)-Me stereogenic centre at the γ-position along the backbone (Figure 4). This stereogenic centre, they found, induces the PNA oligomer to form a right-handed helical structure, similar to that found in dsDNA. Given that this PNA, in its free form, is in a helical form prior to invasion, post-invasion when the PNA/DNA hybrid maintains a similar helical structure the relative loss of entropy between the two states is reduced, making the invasion complex energetically favourable. In essence, this PNA structure, termed γPNA, is the first PNA oligomer to have sequence-unrestricted invasion capabilities while maintaining single base-pair specificity (Chenna et al., 2008; He, Rapireddy, Bahal, Sahu, & Ly, 2009). In addition, in order to address the issue of solubility using γPNAs, Sahu et al. demonstrated that

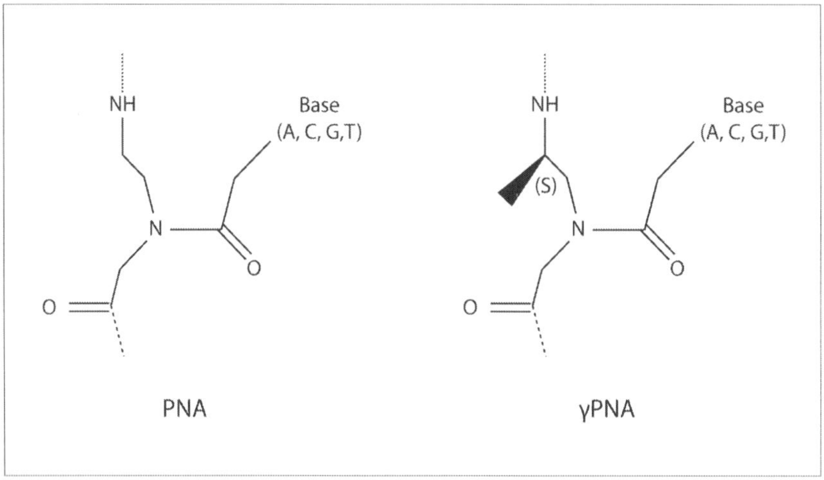

FIGURE 4

Gamma-modified PNAs differ from ordinary PNAs through the creation of a stereogenic centre at the γ-site of the backbone. This simple modification creates a chiral PNA which is far more likely to create a stable dsDNA/PNA duplex than ordinary PNAs.

by incorporating, at the γ-position of the backbone a diethylene glycol moiety, this results in a significant increase in solubility while not adversely affecting affinity or sequence specificity (Bahal, Sahu, Rapireddy, Lee, & Ly, 2012; Sahu et al., 2011).

4 ARTIFICIAL NAs IN USE: VERSATILE, FLEXIBLE, AND OUTPERFORMING DNA/RNA

Although relatively nascent, artificial NAs have demonstrated their diagnostic potential by replacing natural NAs in conventional diagnostic assays, as well as paving the way for more sophisticated diagnostic assays that have been specifically designed to use the capabilities of these synthetic oligomers. In this section, we will describe a number of these diagnostic assays and investigate their improved performance over conventional methods.

4.1 MICROARRAYS

Microarrays (and bead arrays), perhaps better than any other assay types, have demonstrated the utility of artificial NAs. In microarrays, single-stranded oligomers are immobilised or spotted onto a solid surface with a predesigned macro-pattern. Through the addition and subsequent capture of a labelled target to the surface, these oligomers capture the target and immobilise it to the surface. Subsequent imaging of the surface and analysis of the location where a marked increase in optical signature

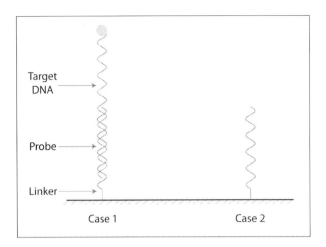

FIGURE 5

Artificial NA capture probes operate in a very similar fashion to standard NAs in microarrays. The presence of a captured target which was pre-amplified with fluorescent molecule allows the optical detection of that capture target and its quantification. Elucidation of its origin is accomplished by comparing the location of the detected signal back to the map of the different capture probes. (See the color plate.)

occurs, allows the user to elucidate the presence, and quantify the amount, of a specific sequence (Figure 5). Though most commonly used to measure gene expression levels, and single nucleotide polymorphisms (SNPs) rather than applications in clinical microbiology, such as identifying the presence or absence of a particular pathogen, artificial NAs are indeed of interest here. The main advantage that artificial NAs were intended to provide was that of specificity and sensitivity. Artificial NAs should provide an improved method to achieve both highly sensitive and high-stringent hybridisation assays.

LNA-modified oligomers have also been investigated to assess their usefulness to replace DNA oligos in microarrays. Although, to the best of our knowledge no studies have been published regarding the use of LNA-modified probes targeting the genomic material from microorganisms, they have been investigated in other applications. Castoldi and co-workers were one of the first to hypothesise that because LNA-modified oligomers demonstrated superior performance in other assays, such as *in situ* hybridisation (discussed below), they may also provide superior results in microarrays (Castoldi et al., 2006). In terms of sensitivity of detection, their findings indicated that LNA-modified oligomers were more sensitive than typical DNA-only oligomers as measured by the signal detected for a given input with the two different probe types. On average, LNA-modified oligomers produced a signal that was 2–2.5 × higher than that of DNA. Although a direct comparison to DNA on sequence specificity was not determined, LNA-modified oligomers proved to have single-base specificity when the mismatch was located in the centre of the probe, emphasising the strategic placement of the LNA nucleotides. Further studies by Fang and co-workers demonstrated that with proper optimisation, attomole sensitivity (10^{-18} moles) was possible using LNA-modified oligomer probes (Fang, Lee, Wark, & Corn, 2006). It should be noted however that few studies have directly compared LNA-modified oligomers with DNA in microarrays and some studies have demonstrated that despite the increase in sensitivity exhibited by microarrays complemented with LNA-modified oligomers, excessive non-specific binding was observed (Diercks, Gescher, Metfies, & Medlin, 2009).

In contrast to LNA-modified oligomers, PNAs have been used more extensively in microarray applications, including those directly related to clinical microbiology. One advantage of PNAs compared to natural NAs is their peptide-like backbone which enables the detection of unlabelled samples through detection of physico-chemical/biochemical signatures which occur during natural NA hybridisation; such as the detection of phosphates and/or sugars (PNAs have neither) through an increase in the net negative charge, as PNAs are inherently neutral (Brandt & Hoheisel, 2004; Liu & Bazan, 2005; Sun, Gaylord, Hong, Liu, & Bazan, 2007). A recent study comparing the capabilities of DNA- and PNA-based microarray technology demonstrated that PNA microarrays can achieve greater sensitivities than those of their DNA-based counterparts owing to both the greater affinity of PNAs to ssDNAs (which directly increases the signal attained for a given concentration) and the lower background due to a reduced level of off-target binding (Choi, Jang, Kim, & Park, 2010) as PNAs displayed a 4–7 × increase in specificity when comparing the effects of mismatches on the attained signal PNAs are a simple tool used to attain higher levels of specificity. Song and co-workers demonstrated that through the use of a PNA-based microarray chip for both the detection and genotyping of the human papillomavirus (HPV), while sensitivity was only marginally improved over a DNA-based microarray chip, the frequency of both Type-I and -II errors is significantly reduced (Choi et al., 2009; Song et al., 2010). Jane and co-workers demonstrated the potential of these probes to investigate antiviral resistance (Jang, Kim, Choi, Son, & Park, 2010) using microarray technology. In their study, PNA probes were designed to identify the mutation in the hepatitis B viral (HBV) genome known to confer resistance to common antiviral drugs: lamivudine, adefovir, and entecavir. Results demonstrated a sensitivity (post-PCR) of 100 viral copies/ml being sufficient to yield a positive signature indicating HBV was present in the serum sample, and the identification of resistance-conferring mutations was identified correctly in 98% of samples tested. These initial studies have demonstrated the superior properties of PNA-based microarrays which have higher levels of specificity and sensitivity than DNA-based microarrays, indicating PNA-based microarrays facilitate significant performance with limited modification in the assay (Singh, Oh, & Choi, 2010).

4.2 *IN SITU* HYBRIDISATION

In situ hybridisation or ISH is a well-established technique to identify the presence or absence of a particular genomic sequence within a cell. In diagnostic assays, DNA or RNA-based fluorescent probes, are commonly used (termed FISH; fluorescent *in situ* hybridisation) which facilitate cell imaging using a standard fluorescence microscope. As described earlier, the advantage of using artificial NAs such as LNA-modified oligomers and PNAs is the inherent thermal stability of these probes when compared to natural NAs, with the higher thermal stability facilitating the utilisation of shorter probes yielding more specific hybridisation (Silahtaroglu, Pfundheller, Koshkin, Tommerup, & Kauppinen, 2004; Silahtaroglu, Tommerup, & Vissing, 2003). In addition, and unlike natural NA probes, when these probes are to be introduced into live cells, they are highly resistant to enzymatic degradation and can be further modified as needed to increase cellular uptake (Amann & Fuchs, 2008). Unlike microarray technology, there has been an increased interest in developing FISH assays for the use in clinical microbiology particularly targeted to pathogen identification whereas the majority of research work with LNAs has been directed at the detection of human miRNA and mRNA (Doné & Beltcheva, 2014; Nielsen, 2012; Singh et al., 2014) or viral targets (Cerqueira et al., 2008; Robertson, Verhoeven, Thach, & Chang, 2010; Shiogama et al., 2013).

One of the first studies using LNA-modified oligomers for rRNA-targeted FISH assays was described by Kubota and co-workers in 2006. They described that poor hybridisation efficiency was a significant limitation of FISH methods due to the low affinity of DNA probes (Kubota, Ohashi, Imachi, & Harada, 2006). This study demonstrated that by substituting 2–4 DNA bases within an 18–21mer DNA probe with LNAs, it was possible to produce a signal enhancement ranging from $2 \times$ to over $22 \times$ indicating that simple modifications can result in significant signal enhancement. They also demonstrated that discriminating among targets with two mismatches is feasible with LNA-modified oligomers which is not possible with DNA probes. This corroborates previous studies demonstrating the LNA-modified oligomers outperform natural NAs as FISH probes (Thomsen, Nielsen, & Jensen, 2005). Montone and co-workers further demonstrated this convincingly in formalin-fixed, paraffin-embedded (FFPE) tissue sections by detecting the 18s rRNA of *Aspergillus*. They demonstrated that standard DNA probes had a significantly weaker signal in the hybridisation reaction than that observed in LNA-modified probes, demonstrating the higher affinity of LNAs to rRNA. Additionally, LNA-modified probes were able to rule out culture-positive cases of *Fusarium*, where the DNA probes failed as they produced numerous false positives (Montone, 2009; Montone & Feldman, 2009).

Unlike LNA-modified oligomers, PNAs have been utilised more extensively in FISH assays and have made a direct impact in the clinical setting. PNAs for FISH-based assays were introduced shortly after PNAs were first developed, and there have been numerous studies published which describe the optimal PNA oligomer lengths to be between 13 and 18mer, in contrast to DNA probes which range from 20 to 25mer. As a result, PNA-FISH assays display remarkable levels of sequence specificity, potentially at the single-base mismatch level, far superior to that possible with DNA probes (Cerqueira et al., 2008). As the majority of PNA-FISH applications are directed at live cells, PNAs are advantageous as they can be designed to promote cellular uptake (given their relatively hydrophobic nature) and are very resistant to the action of nucleases and proteases. Though not limited to, PNA FISH-based assays have mainly been described for the detection of bacterial pathogens either in FFPE, media, or blood cultures. Many studies have described assays using PNAs for detecting and identifying numerous bacterial and fungal species including of *Mycobacterium* sp., *Staphylococcus* sp., *E. coli*, *Pseudomonas* sp., *Klebsiella* sp., *Salmonella* sp., *Listeria* sp., *Streptococcus* sp., *Acinetobacter* sp., *Proteus* sp., and others (Cerqueira et al., 2008; Forrest et al., 2006; González et al., 2004; Lefmann et al., 2006; Montague, Cleary, Martinez, & Procop, 2008; Oliveira, Procop, Wilson, Coull, & Stender, 2002; Oliveira et al., 2003; Peleg et al., 2009; Perry-O'Keefe et al., 2001; Reller, Mallonee, Kwiatkowski, & Merz, 2007; Stender et al., 1999). The majority of assays target the rDNA due to its relative abundance, and in the majority of cases, both sensitivity and sensitivity of over 90% were demonstrated. FISH assays using PNAs have been demonstrated to be superior to DNA-based approaches and are likely to be utilised in many more applications in the near future.

4.3 PCR

Artificial NAs have also been applied in PCR-based approaches where their high sequence specificity has been demonstrated to increase the stringency of a target template in relation to an off-target template that may be present within the sample.

4.3.1 Primer clamping

Primer clamping is the more common of the two methods, where the artificial NA is utilised to compete with a primer (Figure 6). This approach is reliant on not adversely affecting the primer-based extension of the target, while blocking the extension of the off-target. This can be difficult if the two templates are highly similar and vary at only one to two bases between the perfect primer match of the target and the non-specific binding to the off-target.

To address this, two probes are designed, one, DNA-based targeting the correct template sequence (the primer) and another artificial NA probe to be a perfect match for the off-target template(s) (the clamp). The rationale is simple, with the perfectly matched artificial NA having a significantly higher T_M than a mismatched DNA primer it should bind preferentially to the off-target template blocking the binding of the primer. Moreover, given the high sequence specificity of artificial NAs, they should not bind to the target template (which would lower sensitivity). In the case of LNA-modified probes, given that they are a mixed sequence comprising of both DNA/LNA nucleotides, the LNA bases are placed strategically at the mismatches, and they tend to be centred, maximising the ΔT_M, and are additionally placed at the 3′ end of the probe (or at the very least, the 3′ end should be phosphorylated) to prevent polymerase extension. In the case of PNAs, centring the mismatches is typically all that is required, from a design perspective. The presence of an LNA base at the 3′ end, or the presence of a PNA probe prevents the polymerase from extending nucleobase sequence, effectively blocking the amplification of the off-target template.

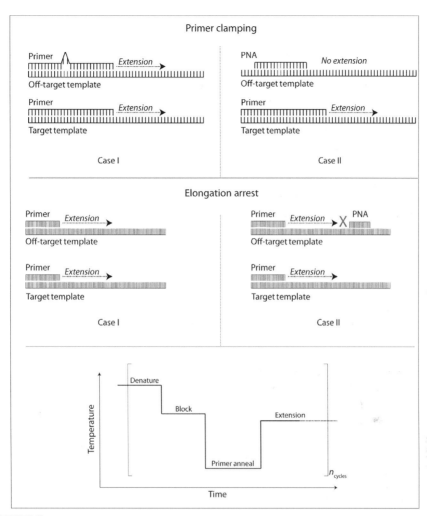

FIGURE 6

Artificial NAs can be used to develop more sensitive amplification processes either by blocking off-priming events (primer clamping—top) or by preventing successful amplification of an unwanted amplicon (elongation arrest—middle). Though not necessarily required, these techniques are assisted by the addition of a 'blocking' step between the 'denaturing' and 'primer annealing' step during thermal cycling allowing the artificial NAs to bind without restrictions. (See the color plate.)

4.3.2 Elongation arrest

A less-commonly used approach is to prevent the extension process itself and not the binding of a primer. This method is favourable where the target and off-target may require the use of identical primers and the key difference lies somewhere between the forward and reverse primers. Here, the aim is to design a blocking probe, specific to the off-target. The rational is straightforward, although extension will be initiated given that the primers will bind to the off-target with the same yield as to the target template, the amplification process cannot be completed given the presence of a blocking probe bound to the template. Given that polymerases which may have either a 5′–3′ exonuclease activity or displacement activity tend to not to be able to act on these artificial NA probes, this is considered an efficient method (Di Giusto & King, 2004; Hummelshoj et al., 2005; Slaitas, Ander, Foldes-Papp, Rigler, & Yeheskiely, 2003). As with clamping methodology, if LNA-modified probes are used, the 3′ end should either contain an LNA nucleotide or be phosphorylated to prevent elongation. This blocker oligo might also be used to prevent the binding of a molecular beacon-type probe or a TaqMan probe in real-time PCR-based assays.

It should be noted that there are multiple design approaches, in both the primer clamping and elongation arrest approach when it has been found favourable to add an additional 'blocking step' during the thermal cycling. Given that artificial probes have an elevated T_M when compared to primers or molecular beacons, effective blocking or clamping is achieved by the incorporation of a short intermediate step between the high-temperature denaturing step and the lower-temperature annealing step.

These two approaches have been demonstrated to produce highly favourable results with minimal modification to the assay itself. The use of LNA-modified oligomers for PCR in clinical microbiology has been limited with applications in clinical microbiology utilising PNA-based approaches. Though used sparingly, LNA-modified oligomers in PCR assays have produced effective results, demonstrating their utility in combination with real-time PCR can increase assay sensitivity significantly, facilitating more effective multiplexing within assays, and support target enrichment (Ikenaga & Sakai, 2014; Ren et al., 2009; Thiede et al., 2006).

von Wintzingerode and co-workers demonstrated that employing PNA clamping of the highly conserved bacterial 16s rDNA at key pinch-points, could selectively enrich a mixed sample containing predominantly off-target template despite only 1–2 bp differences in target/off-targets (von Wintzingerode et al., 2000). More recently, Takiya and co-workers demonstrated that by combining multiplex PCR with PNA clamping, they were able to identify stains of O157:H7 *E. coli* based on a single-base mutation. In their approach, PNA clamping effectively prevented off-target amplification despite the similarity of the two stains (Takiya et al., 2004). Similarly, Iwamoto and co-workers demonstrated that PNA-based clamping is useful for the detection of rifampin resistance in the *rpoB* gene in *M. tuberculosis* (Iwamoto & Sonobe, 2004). This study demonstrated that using an amplification and clamping protocol, they were able to reach a limit of detection of approximately 100 fg of genomic DNA (approximately 25 cells) and were able to suppress amplification from as much as 100 ng of off-target genomic DNA (approximately 2.5×10^7 cells). Combining the sensitivity of the PCR assay with the suppression capabilities of PNA clamping, Iwamoto and co-workers found that they were able to enrich a sample where the genomic load of the target was minimal.

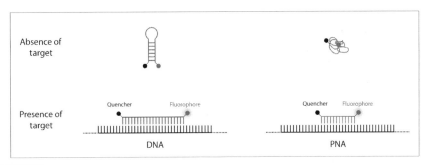

FIGURE 7

Standard DNA-based molecular beacons present an elegant manner to detect, in real-time, the presence of an amplicon. PNA-based molecular beacons not only present an opportunity to gain sequence specificity but likewise ease the design of the molecular beacons given to the ability to remove the stem-loop design criteria. (See the color plate.)

In addition to preventing elongation or priming in PCR/RT-PCR-based assays, PNAs have been demonstrated to be efficient fluorescent beacons. Choi and co-workers developed a highly specific assay that utilised fluorescent PNAs to detect and discriminate among *M. tuberculosis* and non-tuberculous mycobacteria in clinical specimens. Application of the assay to 531 patient specimens, demonstrated no Type-I or Type-II errors and a specificity of 100% and 99.6%, for non-tuberculous *Mycobacteria* and *M. tuberculosis*, respectively (Choi et al., 2012). Though conceptually similar to standard real-time PCR-based approaches using molecular beacons, there are differences. First, in contrast to DNA-based probes, PNAs are not degraded during the elongation process; meaning that the signal from the PNA-based molecular beacon is equivalent to the amount of the amplicon product present in that cycle, where in standard DNA-based molecular beacon assays, the signal is comprised both of the fluorophores released from the probes degraded in the previous cycles, and the amount of the amplicon product present at that cycle, overall, the PNA-based signal is lower but more representative of the actual amount of amplicon present (Stender, Fiandaca, Hyldig-Nielsen, & Coull, 2002). Lastly, it should be noted that beyond the affinity/specificity of PNAs, oligomer design is simplified when compared to DNAs due to the lack of a stem-loop design criteria (Figure 7) (Petersen et al., 2004).

4.4 NOVEL APPROACHES

The use of artificial NAs have opened up new and exciting applications in diagnostics with PNAs being most commonly utilised due to their duplex invasion capabilities which facilitate simplified sample preparation methods and increased sequence specificity. In the following section, we describe two approaches specifically designed to utilise PNA's invasion properties.

Smolina and co-workers described a method to increase the sensitivity and expand upon the capabilities of FISH-based assays. Their novel approach

(Smolina & Frank-Kamenetskii, 2014; Smolina, Miller, & Frank-Kamenetskii, 2010) employed the two bis-PNA molecules to 'open-up' or 'pry-open' a short and specific region of the duplex DNA leaving the exposed strand (in this case, roughly 20 bases in length called a PD-loop) upon which a rolling circle amplification process was performed using standard techniques. The output of the rolling circle amplification, dependent on the displaced strand, enabled the subsequent binding of different coloured fluorescent probes for the identification of different pathogens and potential resistance traits; namely, the discrimination of *S. aureus* from coagulase-negative staphylococcal species and the discrimination of methicillin-sensitive (MSSA) and methicillin-resistant (MRSA) strains. This PD-loop structure is not limited to post-processing amplification, but serves as a method to enable the highly sequence-specific labelling of a dsDNA requiring only a brief heating step (Smolina, Kuhn, Lee, & Frank-Kamenetskii, 2008) and a simple molecular beacon for FISH (Figure 8).

An alternative approach recognises that at the invasion site of a DNA/PNA complex, the conformation of the natural double-stranded DNA changes where a bulge and/or kink becomes apparent, in a highly localised and sequence-specific manner (Cherny et al., 1998). Singer and co-workers demonstrated that by using solid-state nanopores, which can be considered as *single-molecule* Coulter counters, it is possible to detect the PNA invasion site (Figure 9). Solid-state nanopores are novel biosensors capable of detecting DNA molecules as they pass, one-by-one, through them. By fabricating single sub-5 nm pores (only slightly larger than the diameter of the dsDNA helix) in a 30 nm thick silicon-nitride membrane, and by tagging (invading) the duplex structure with bis-PNAs (Singer et al., 2010) they found that as

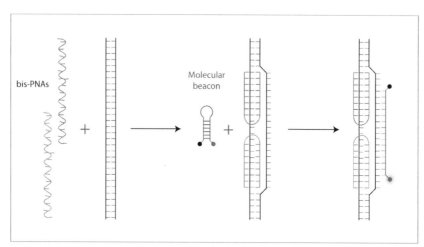

FIGURE 8

The PD-loop design employing two bis-PNA probes with a standard molecular beacon enable the detection of target sequences in a highly sequence-specific manner as not one, but two PNAs are required to 'open-up' or expose the molecular beacon's target region. (See the color plate.)

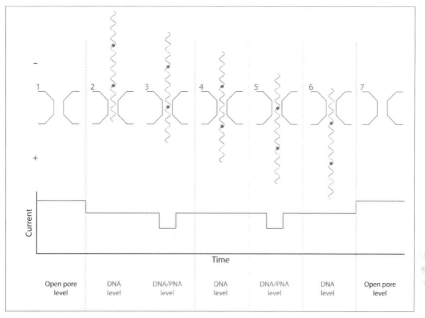

FIGURE 9

Nanopore technology presents a novel approach to using dsDNA invading DNAs due to the change in conformation that are apparent at the PNA invasion site which are detected via a secondary drop in the current (frames 3 and 5) compared to the DNA-only current levels (frames 2, 4, and 6). The two markers within the DNA molecule indicate the point of the DNA/PNA invasion complex. (See the color plate.)

the DNA molecule moves across and through the nanopores, distinct blockade levels were evident, depicting either the blockade level of bare DNA or depicting the blockade level of a bis-PNA/DNA complex. Singer and co-workers expanded on this work, and developed a barcoding strategy employing γPNAs and demonstrated that this process could be used efficiently to discriminate among viral subtypes in a quick and efficient manner simply by counting the number of PNA/DNA complex sites and the time lapse (correlating distances) between them owing to the distinct blockade levels (Singer, Rapireddy, Ly, & Meller, 2012).

5 ARTIFICIAL PROBES AND DIAGNOSTIC CLINICAL MICROBIOLOGY

The emergence of artificial NAs as potential improvements for molecular diagnostic assays targeting infectious diseases is gaining momentum. As we highlighted previously, artificial NAs were initially developed with therapeutic applications in mind, and the transition to clinical microbiology will take time.

South Korea-based Panagene, who specialise in synthesising PNAs, have developed two platforms, an HPV genotyping array-chip, and a qPCR kit for detecting *M. tuberculosis* (TB) and non-tuberculous *Mycobacteria* sp. (NTM). Although both products are not cleared for use in the United States, they have received the European CE mark. The PANarray HPV chip provides the ability to identify up to 32 genotypes of HPV including both high- and low-risk types and can be completed using standard laboratory equipment in under 5 h (Song et al., 2010). The PANarray HPV chip can achieve comparable results to DNA-based arrays in terms of sensitivity, where it does outperform DNA probes is in specificity; it provides a significantly more accurate method to genotype the detected pathogen. Their second product, the PANA qPCR TB/NTM Detection kit, is a simple, one-stop kit which extracts the microbial DNA from several different clinical sample types and using standard qPCR methods, where the molecular beacon has been replaced with a PNA probe, enables the discrimination of TB from NTM in roughly 3 hours while achieving >99% specificity for both targets. However, in one study, sensitivity reached only 69% for NTM but overall, this kit demonstrated similar performance characteristics to other nucleic acid amplification-based assays (Choi et al., 2012). A more recent study comparing the Cobas TaqMan MTN assay (Roche Diagnostics) with Panagene's system found that the performance of the PNA-based assay is superior to the TaqMan assay (Kim et al., 2013). Given the novelty of this approach, further studies are warranted to ascertain and the performance enhancement of using PNAs in place of DNA-based molecular beacons.

AdvanDx is an established leader in commercialising PNA-based assays for detecting pathogens directly from human samples having received FDA clearance for over 10 of their products. AdvanDx employs a PNA FISH-based assay for the identification of the causative microorganism in cases of sepsis enabling a more targeted intervention roughly a day faster than using standard approaches. Starting off with a positive blood culture samples are fixed onto a microscope slide, and stained with as many as three different PNA probes (each labelled with a different fluorophore) targeting the cell's rRNA. Post-staining, the slide can be visualised, using a standard fluorescence microscope, where cells are then identified based on their colouring (Figure 10). Currently, AdvanDx offer an array of rapid assays (QuickFISH) requiring between 20 and 30 min post-positive culture (Tang & Peterson, 2014). Laboratory evaluation of their *Staphylococcus* QuickFISH test demonstrated a high specificity (approximately 99%) for both *S. aureus* and coagulase-negative Staphylococci individually and 90% when both are present together (Deck et al., 2012). Similarly, their Yeast 3-color Traffic Light assay for *Candida* species displayed >96% sensitivity and >95% specificity (Hall, Le Febre, Deml, Wohlfiel, & Wengenack, 2012) while their assay for detecting *Enterococcus* achieve >97% sensitivity and 100% specificity (Deck et al., 2014). Despite the promising results of the evaluation of the AdvanDx's assays, there are limitations: (1) the approach requires a positive culture, thus a key bottleneck—waiting for the culture to turn positive has not been eliminated; (2) the multiplexing abilities are limited to three distinct targets or groups, and given the range of targets that may induce sepsis,

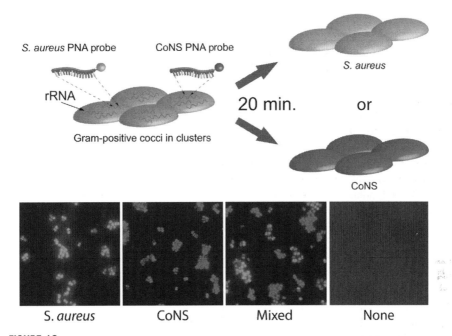

S. aureus PNA probe   CoNS PNA probe

S. aureus

rRNA

20 min.   or

Gram-positive cocci in clusters

CoNS

S. *aureus*   CoNS   Mixed   None

FIGURE 10

Overview of the QuickFISH test. (Top) QuickFISH technology utilises highly specific PNA probes which hybridise to the target's rRNA allowing the user to visualise the presence of a specific pathogen, in this case differentiating among *S. aureus* and coagulase-negative *Staphylococci*. (Bottom) Representative results of a *Staphylococcus* QuickFISH test, as viewed through a fluorescent microscope displaying the difference among the two infections. (See the color plate.)

Figure courtesy of AdvanDx.

more than one assay is required to achieve sufficient coverage; and (3) the assay is an open system, prone to cross contamination, and although very simple, is also prone to human errors and inconsistencies.

CONCLUSION

Significant advances in diagnostics have been made using NA probes and the use of artificial NAs will facilitate the development of improved clinical diagnostic tests. So far, aside from a small number of studies describing the use of PNAs, they have not become common in today's clinical laboratory. However, artificial NAs are now gaining acceptance and with this appreciation comes advancement; both in their chemistry and in the innovative manner in which they can be used.

It is interesting that there is a clear lack of applications with which LNA-modified probes have been employed, especially given that the assay modifications required are rather straightforward and that oligonucleotide synthesis companies already supply LNA probes. However, researchers must first become familiar with them, allowing them to learn how to adapt an assay to take advantage of their capabilities, while addressing or overcoming their limitations. PNAs have without doubt been used more frequently, though they are not standard practice despite being relatively common in the clinical setting. PNAs, unlike LNAs, facilitate the development of a new type of molecular assay; one that utilises dsDNA and not ssDNA. However this pathway has, until now, seldom been used and similarly to LNAs, the difficulties in designing an assay around this capability have perhaps outweighed its benefits. And like LNAs, as more studies are described demonstrating their enhanced performance, more applications will be developed.

Although relatively new players to the established NA probe methodologies, interest in artificial NAs has increased as demonstrated by the increasing number of scientific publications describing the use of these probes. In the future, we foresee improved diagnostics, both conventional assays such as microarrays, FISH, and real-time PCR with improved performance, as well as novel diagnostic approaches specifically designed to utilise the unique attributes of LNAs and PNAs.

REFERENCES

Abibi, A., Protozanova, E., Demidov, V. V., & Frank-Kamenetskii, M. D. (2004). Specific versus nonspecific binding of cationic PNAs to duplex DNA. *Biophysical Journal*, *86*(5), 3070–3078.

Amann, R., & Fuchs, B. M. (2008). Single-cell identification in microbial communities by improved fluorescence in situ hybridization techniques. *Nature Reviews Microbiology*, *6*(5), 339–348.

Bahal, R., Sahu, B., Rapireddy, S., Lee, C. M., & Ly, D. H. (2012). Sequence-unrestricted, Watson-Crick recognition of double helical B-DNA by (R)-miniPEG-gammaPNAs. *Chembiochem*, *13*(1), 56–60.

Bonvicini, F., Filippone, C., Manaresi, E., Gentilomi, G. A., Zerbini, M., Musiani, M., et al. (2006). Peptide nucleic acid-based in situ hybridization assay for detection of parvovirus B19 nucleic acids. *Clinical Chemistry*, *52*(6), 973–978.

Braasch, D. A., & Corey, D. R. (2001). Locked nucleic acid (LNA): Fine-tuning the recognition of DNA and RNA. *Chemistry & Biology*, *8*(1), 1–7.

Brandt, O., & Hoheisel, J. D. (2004). Peptide nucleic acids on microarrays and other biosensors. *Trends in Biotechnology*, *22*(12), 617–622.

Briones, C., & Moreno, M. (2012). Applications of peptide nucleic acids (PNAs) and locked nucleic acids (LNAs) in biosensor development. *Analytical and Bioanalytical Chemistry*, *402*(10), 3071–3089.

Buchardt, O., Egholm, M., Berg, R. H., & Nielsen, P. E. (1993). Peptide nucleic acids and their potential applications in biotechnology. *Trends in Biotechnology*, *11*(9), 384–386.

Campbell, M. A., & Wengel, J. (2011). Locked vs. unlocked nucleic acids (LNA vs. UNA): Contrasting structures work towards common therapeutic goals. *Chemical Society Reviews*, *40*(12), 5680–5689.

Castoldi, M., Schmidt, S., Benes, V., Noerholm, M., Kulozik, A. E., Hentze, M. W., et al. (2006). A sensitive array for microRNA expression profiling (miChip) based on locked nucleic acids (LNA). *RNA*, *12*(5), 913–920.

Cerqueira, L., Azevedo, N. F., Almeida, C., Jardim, T., Keevil, C. W., & Vieira, M. J. (2008). DNA mimics for the rapid identification of microorganisms by fluorescence in situ hybridization (FISH). *International Journal of Molecular Sciences*, *9*(10), 1944–1960.

Chenna, V., Rapireddy, S., Sahu, B., Ausin, C., Pedroso, E., & Ly, D. H. (2008). A simple cytosine to G-clamp nucleobase substitution enables chiral gamma-PNAs to invade mixed-sequence double-helical B-form DNA. *Chembiochem*, *9*(15), 2388–2391.

Cherny, D. I., Fourcade, A., Svinarchuk, F., Nielsen, P. E., Malvy, C., & Delain, E. (1998). Analysis of various sequence-specific triplexes by electron and atomic force microscopies. *Biophysical Journal*, *74*(2), 1015–1023.

Choi, J. J., Jang, M., Kim, J., & Park, H. (2010). Highly sensitive PNA array platform technology for single nucleotide mismatch discrimination. *Journal of Microbiology and Biotechnology*, *20*(2), 287–293.

Choi, J.-j., Kim, C., & Park, H. (2009). Peptide nucleic acid-based array for detecting and genotyping human papillomaviruses. *Journal of Clinical Microbiology*, *47*(6), 1785–1790.

Choi, Y. J., Kim, H. J., Shin, H. B., Nam, H. S., Lee, S. H., Park, J. S., et al. (2012). Evaluation of peptide nucleic acid probe-based real-time PCR for detection of *Mycobacterium tuberculosis* complex and nontuberculous mycobacteria in respiratory specimens. *Annals of Laboratory Medicine*, *32*(4), 257–263.

Deck, M. K., Anderson, E. S., Buckner, R. J., Colasante, G., Coull, J. M., Crystal, B., et al. (2012). Multicenter evaluation of the Staphylococcus QuickFISH method for simultaneous identification of *Staphylococcus aureus* and coagulase-negative staphylococci directly from blood culture bottles in less than 30 minutes. *Journal of Clinical Microbiology*, *50*(6), 1994–1998.

Deck, M. K., Anderson, E. S., Buckner, R. J., Colasante, G., Davis, T. E., Coull, J. M., et al. (2014). Rapid detection of *Enterococcus* spp. direct from blood culture bottles using Enterococcus QuickFISH method: A multicenter investigation. *Diagnostic Microbiology and Infectious Disease*, *78*(4), 338–342.

Demidov, V. V., Protozanova, E., Izvolsky, K. I., Price, C., Nielsen, P. E., & Frank-Kamenetskii, M. D. (2002). Kinetics and mechanism of the DNA double helix invasion by pseudocomplementary peptide nucleic acids. *Proceedings of the National Academy of Sciences of the United States of America*, *99*(9), 5953–5958.

Di Giusto, D. A., & King, G. C. (2004). Strong positional preference in the interaction of LNA oligonucleotides with DNA polymerase and proofreading exonuclease activities: Implications for genotyping assays. *Nucleic Acids Research*, *32*(3), e32.

Diercks, S., Gescher, C., Metfies, K., & Medlin, L. (2009). Evaluation of locked nucleic acids for signal enhancement of oligonucleotide probes for microalgae immobilised on solid surfaces. *Journal of Applied Phycology*, *21*(6), 657–668.

Doné, S., & Beltcheva, O. (2014). In situ hybridization detection of miRNA using LNA™ oligonucleotides. In M. L. Alvarez, & M. Nourbakhsh (Eds.), *RNA mapping: Vol. 1182* (pp. 57–71). New York: Springer.

Dragulescu-Andrasi, A., Rapireddy, S., Frezza, B. M., Gayathri, C., Gil, R. R., & Ly, D. H. (2006). A simple gamma-backbone modification preorganizes peptide nucleic acid into a helical structure. *Journal of the American Chemical Society*, *128*(31), 10258–10267.

Efimov, V. A., Choob, M. V., Buryakova, A. A., Kalinkina, A. L., & Chakhmakhcheva, O. G. (1998). Synthesis and evaluation of some properties of chimeric oligomers containing PNA and phosphono-PNA residues. *Nucleic Acids Research, 26*(2), 566–575.

Egholm, M., Buchardt, O., Christensen, L., Behrens, C., Freier, S. M., Driver, D. A., et al. (1993). PNA hybridizes to complementary oligonucleotides obeying the Watson-Crick hydrogen-bonding rules. *Nature, 365*(6446), 566–568.

Egholm, M., Buchardt, O., Nielsen, P. E., & Berg, R. H. (1992). Peptide nucleic acids (PNA). Oligonucleotide analogs with an achiral peptide backbone. *Journal of the American Chemical Society, 114*(5), 1895–1897.

Egholm, M., Christensen, L., Dueholm, K. L., Buchardt, O., Coull, J., & Nielsen, P. E. (1995). Efficient pH-independent sequence-specific DNA binding by pseudoisocytosine-containing bis-PNA. *Nucleic Acids Research, 23*(2), 217–222.

Fang, S., Lee, H. J., Wark, A. W., & Corn, R. M. (2006). Attomole microarray detection of microRNAs by nanoparticle-amplified SPR imaging measurements of surface poly-adenylation reactions. *Journal of the American Chemical Society, 128*(43), 14044–14046.

Forrest, G. N., Mehta, S., Weekes, E., Lincalis, D. P., Johnson, J. K., & Venezia, R. A. (2006). Impact of rapid in situ hybridization testing on coagulase-negative staphylococci positive blood cultures. *Journal of Antimicrobial Chemotherapy, 58*(1), 154–158.

González, V., Padilla, E., Giménez, M., Vilaplana, C., Pérez, A., Fernández, G., et al. (2004). Rapid diagnosis of *Staphylococcus aureus* bacteremia using *S. aureus* PNA FISH. *European Journal of Clinical Microbiology and Infectious Diseases, 23*(5), 396–398.

Griffith, M. C., Risen, L. M., Greig, M. J., Lesnik, E. A., Sprankle, K. G., Griffey, R. H., et al. (1995). Single and bis peptide nucleic acids as triplexing agents: Binding and stoichiometry. *Journal of the American Chemical Society, 117*(2), 831–832.

Hall, L., Le Febre, K. M., Deml, S. M., Wohlfiel, S. L., & Wengenack, N. L. (2012). Evaluation of the yeast traffic light PNA FISH probes for identification of Candida species from positive blood cultures. *Journal of Clinical Microbiology, 50*(4), 1446–1448.

He, G., Rapireddy, S., Bahal, R., Sahu, B., & Ly, D. H. (2009). Strand invasion of extended, mixed-sequence B-DNA by gammaPNAs. *Journal of the American Chemical Society, 131*(34), 12088–12090.

Hummelshoj, L., Ryder, L. P., Madsen, H. O., & Poulsen, L. K. (2005). Locked nucleic acid inhibits amplification of contaminating DNA in real-time PCR. *BioTechniques, 38*(4), 605–610.

Ikenaga, M., & Sakai, M. (2014). Application of locked nucleic acid (LNA) oligonucleotide—PCR clamping technique to selectively PCR amplify the SSU rRNA genes of bacteria in investigating the plant-associated community structures. *Microbes and Environments, 29*(3), 286–295.

Iwamoto, T., & Sonobe, T. (2004). Peptide nucleic acid-mediated competitive PCR clamping for detection of rifampin-resistant *Mycobacterium tuberculosis*. *Antimicrobial Agents and Chemotherapy, 48*(10), 4023–4026.

Jang, H., Kim, J., Choi, J.-j., Son, Y., & Park, H. (2010). Peptide nucleic acid array for detection of point mutations in hepatitis B virus associated with antiviral resistance. *Journal of Clinical Microbiology, 48*(9), 3127–3131.

Kim, J., Choi, Y., Kim, H., Park, J., Nam, H., Hwangbo, Y., et al. (2013). Comparison of PNA probe-based real-time PCR and Cobas TaqMan MTB for detection of MTBC. *BioChip Journal, 7*(2), 85–88.

Koshkin, A. A., & Wengel, J. (1998). Synthesis of novel 2′,3′-linked bicyclic thymine ribonucleosides. *The Journal of Organic Chemistry, 63*(8), 2778–2781.

Kubota, K., Ohashi, A., Imachi, H., & Harada, H. (2006). Improved in situ hybridization efficiency with locked-nucleic-acid-incorporated DNA probes. *Applied and Environmental Microbiology, 72*(8), 5311–5317.

Kuhn, H., Demidov, V. V., Frank-Kamenetskii, M. D., & Nielsen, P. E. (1998). Kinetic sequence discrimination of cationic bis-PNAs upon targeting of double-stranded DNA. *Nucleic Acids Research, 26*(2), 582–587.

Kutyavin, I. V., Rhinehart, R. L., Lukhtanov, E. A., Gorn, V. V., Meyer, R. B., & Gamper, H. B. (1996). Oligonucleotides containing 2-aminoadenine and 2-thiothymine act as selectively binding complementary agents. *Biochemistry, 35*(34), 11170–11176.

Lefmann, M., Schweickert, B., Buchholz, P., Göbel, U. B., Ulrichs, T., Seiler, P., et al. (2006). Evaluation of peptide nucleic acid-fluorescence in situ hybridization for identification of clinically relevant mycobacteria in clinical specimens and tissue sections. *Journal of Clinical Microbiology, 44*(10), 3760–3767.

Liu, B., & Bazan, G. C. (2005). Methods for strand-specific DNA detection with cationic conjugated polymers suitable for incorporation into DNA chips and microarrays. *Proceedings of the National Academy of Sciences of the United States of America, 102*(3), 589–593.

Lohse, J., Dahl, O., & Nielsen, P. E. (1999). Double duplex invasion by peptide nucleic acid: A general principle for sequence-specific targeting of double-stranded DNA. *Proceedings of the National Academy of Sciences of the United States of America, 96*(21), 11804–11808.

Millili, P. G., Yin, D. H., Fan, H., Naik, U. P., & Sullivan, M. O. (2010). Formulation of a peptide nucleic acid based nucleic acid delivery construct. *Bioconjugate Chemistry, 21*(3), 445–455.

Montague, N. S., Cleary, T. J., Martinez, O. V., & Procop, G. W. (2008). Detection of group B streptococci in Lim broth by use of group B streptococcus peptide nucleic acid fluorescent in situ hybridization and selective and nonselective agars. *Journal of Clinical Microbiology, 46*(10), 3470–3472.

Montone, K. T. (2009). Differentiation of Fusarium from Aspergillus species by colorimetric in situ hybridization in formalin-fixed, paraffin-embedded tissue sections using dual fluorogenic-labeled LNA probes. *American Journal of Clinical Pathology, 132*(6), 866–870.

Montone, K. T., & Feldman, M. D. (2009). In situ detection of aspergillus 18s ribosomal RNA sequences using a terminally biotinylated locked nucleic acid (LNA) probe. *Diagnostic Molecular Pathology Part B, 18*(4), 239–242.

Nielsen, P. E. (1999). Applications of peptide nucleic acids. *Current Opinion in Biotechnology, 10*(1), 71–75.

Nielsen, B. (2012). MicroRNA in situ hybridization. In J.-B. Fan (Ed.), *Next-generation microRNA expression profiling technology: Vol. 822* (pp. 67–84). Springer Protocols: Humana Press.

Nielsen, P. E., Egholm, M., Berg, R. H., & Buchardt, O. (1993). Sequence specific inhibition of DNA restriction enzyme cleavage by PNA. *Nucleic Acids Research, 21*(2), 197–200.

Nielsen, K. E., Singh, S. K., Wengel, J., & Jacobsen, J. P. (2000). Solution structure of an LNA hybridized to DNA: NMR study of the d(CTLGCTLTLCTLGC):d(GCAGAAGCAG) duplex containing four locked nucleotides. *Bioconjugate Chemistry, 11*(2), 228–238.

Obika, S., Nanbu, D., Hari, Y., Morio, K.-i., In, Y., Ishida, T., et al. (1997). Synthesis of 2′-O,4′-C-methyleneuridine and -cytidine. Novel bicyclic nucleosides having a fixed C3, -endo sugar puckering. *Tetrahedron Letters, 38*(50), 8735–8738.

Oliveira, K., Brecher, S. M., Durbin, A., Shapiro, D. S., Schwartz, D. R., De Girolami, P. C., et al. (2003). Direct identification of staphylococcus aureus from positive blood culture bottles. *Journal of Clinical Microbiology*, *41*(2), 889–891.

Oliveira, K., Procop, G. W., Wilson, D., Coull, J., & Stender, H. (2002). Rapid identification of *Staphylococcus aureus* directly from blood cultures by fluorescence in situ hybridization with peptide nucleic acid probes. *Journal of Clinical Microbiology*, *40*(1), 247–251.

Orum, H. (2000). PCR clamping. *Current Issues in Molecular Biology*, *2*(1), 27–30.

Ørum, H., Jakobsen, M. H., Koch, T., Vuust, J., & Borre, M. B. (1999). Detection of the factor V Leiden mutation by direct allele-specific hybridization of PCR amplicons to photoimmobilized locked nucleic acids. *Clinical Chemistry*, *45*(11), 1898–1905.

Ørum, H., Nielsen, P. E., Egholm, M., Berg, R. H., Buchardt, O., & Stanley, C. (1993). Single base pair mutation analysis by PNA directed PCR clamping. *Nucleic Acids Research*, *21*(23), 5332–5336.

Peffer, N. J., Hanvey, J. C., Bisi, J. E., Thomson, S. A., Hassman, C. F., Noble, S. A., et al. (1993). Strand-invasion of duplex DNA by peptide nucleic acid oligomers. *Proceedings of the National Academy of Sciences of the United States of America*, *90*(22), 10648–10652.

Peleg, A. Y., Tilahun, Y., Fiandaca, M. J., D'Agata, E. M. C., Venkataraman, L., Moellering, R. C., et al. (2009). Utility of peptide nucleic acid fluorescence in situ hybridization for rapid detection of *Acinetobacter* spp. and *Pseudomonas aeruginosa*. *Journal of Clinical Microbiology*, *47*(3), 830–832.

Perry-O'Keefe, H., Rigby, S., Oliveira, K., Sørensen, D., Stender, H., Coull, J., et al. (2001). Identification of indicator microorganisms using a standardized PNA FISH method. *Journal of Microbiological Methods*, *47*(3), 281–292.

Petersen, K., Vogel, U., Rockenbauer, E., Vang Nielsen, K., Kølvraa, S., Bolund, L., et al. (2004). Short PNA molecular beacons for real-time PCR allelic discrimination of single nucleotide polymorphisms. *Molecular and Cellular Probes*, *18*(2), 117–122.

Petersen, M., & Wengel, J. (2003). LNA: A versatile tool for therapeutics and genomics. *Trends in Biotechnology*, *21*(2), 74–81.

Peyman, A., Uhlmann, E., Wagner, K., Augustin, S., Breipohl, G., Will, D. W., et al. (1996). Phosphonic ester nucleic acids (PHONAs): Oligonucleotide analogues with an achiral phosphonic acid ester backbone. *Angewandte Chemie, International Edition in English*, *35*(22), 2636–2638.

Raymond, F., Ho, H.-A., Peytavi, R., Bissonnette, L., Boissinot, M., Picard, F., et al. (2005). Detection of target DNA using fluorescent cationic polymer and peptide nucleic acid probes on solid support. *BMC Biotechnology*, *5*(1), 10.

Reller, M. E., Mallonee, A. B., Kwiatkowski, N. P., & Merz, W. G. (2007). Use of peptide nucleic acid-fluorescence in situ hybridization for definitive, rapid identification of five common Candida species. *Journal of Clinical Microbiology*, *45*(11), 3802–3803.

Ren, X. D., Lin, S. Y., Wang, X., Zhou, T., Block, T. M., & Su, Y.-H. (2009). Rapid and sensitive detection of hepatitis B virus 1762T/1764A double mutation from hepatocellular carcinomas using LNA-mediated PCR clamping and hybridization probes. *Journal of Virological Methods*, *158*(1–2), 24–29.

Robertson, K. L., Verhoeven, A. B., Thach, D. C., & Chang, E. L. (2010). Monitoring viral RNA in infected cells with LNA flow-FISH. *RNA*, *16*(8), 1679–1685.

Sahu, B., Sacui, I., Rapireddy, S., Zanotti, K. J., Bahal, R., Armitage, B. A., et al. (2011). Synthesis and characterization of conformationally preorganized, (R)-diethylene glycol-containing gamma-peptide nucleic acids with superior hybridization properties and water solubility. *The Journal of Organic Chemistry*, *76*(14), 5614–5627.

Shiogama, K., Inada, K., Kohara, M., Teramoto, H., Mizutani, Y., Onouchi, T., et al. (2013). Demonstration of hepatitis C virus RNA with in situ hybridization employing a locked nucleic acid probe in humanized liver of infected chimeric mice and in needle-biopsied human liver. *International Journal of Hepatology*, *2013*, 249535.

Silahtaroglu, A., Pfundheller, H., Koshkin, A., Tommerup, N., & Kauppinen, S. (2004). LNA-modified oligonucleotides are highly efficient as FISH probes. *Cytogenetic and Genome Research*, *107*(1–2), 32–37.

Silahtaroglu, A. N., Tommerup, N., & Vissing, H. (2003). FISHing with locked nucleic acids (LNA): Evaluation of different LNA/DNA mixmers. *Molecular and Cellular Probes*, *17*(4), 165–169.

Singer, A., Rapireddy, S., Ly, D., & Meller, A. (2012). Electronic barcoding of a viral gene at the single-molecule level. *Nano Letters*, *12*(3), 1722–1728.

Singer, A., Wanunu, M., Morrison, W., Kuhn, H., Frank-Kamenetskii, M., & Meller, A. (2010). Nanopore based sequence specific detection of duplex DNA for genomic profiling. *Nano Letters*, *10*(2), 738–742.

Singh, U., Keirstead, N., Wolujczyk, A., Odin, M., Albassam, M., & Garrido, R. (2014). General principles and methods for routine automated microRNA in situ hybridization and double labeling with immunohistochemistry. *Biotechnic & Histochemistry*, *89*(4), 259–266.

Singh, R. P., Oh, B.-K., & Choi, J.-W. (2010). Application of peptide nucleic acid towards development of nanobiosensor arrays. *Bioelectrochemistry*, *79*(2), 153–161.

Slaitas, A., Ander, C., Foldes-Papp, Z., Rigler, R., & Yeheskiely, E. (2003). Suppression of exonucleolytic degradation of double-stranded DNA and inhibition of Exonuclease III by PNA. *Nucleosides, Nucleotides & Nucleic Acids*, *22*(5-8), 1603–1605.

Smolina, I., & Frank-Kamenetskii, M. (2014). PNA openers and their applications for bacterial DNA diagnostics. In P. E. Nielsen, & D. H. Appella (Eds.), *Peptide nucleic acids: Vol. 1050* (pp. 121–130). Methods in Molecular Biology: Humana Press.

Smolina, I. V., Kuhn, H., Lee, C., & Frank-Kamenetskii, M. D. (2008). Fluorescence-based detection of short DNA sequences under non-denaturing conditions. *Bioorganic & Medicinal Chemistry*, *16*(1), 84–93.

Smolina, I., Miller, N. S., & Frank-Kamenetskii, M. D. (2010). PNA-based microbial pathogen identification and resistance marker detection: An accurate, isothermal rapid assay based on genome-specific features. *Artificial DNA, PNA & XNA*, *1*(2), 76–82.

Song, H., Lee, J., Kim, B., Song, S., Bae, D., & Kim, D. (2010). Comparison of the performance of the PANArray™ HPV test and DNA chip test for genotyping of human papillomavirus in cervical swabs. *BioChip Journal*, *4*(3), 167–172.

Stender, H., Fiandaca, M., Hyldig-Nielsen, J. J., & Coull, J. (2002). PNA for rapid microbiology. *Journal of Microbiological Methods*, *48*(1), 1–17.

Stender, H., Mollerup, T. A., Lund, K., Petersen, K. H., Hongmanee, P., & Godtfredsen, S. E. (1999). Direct detection and identification of in smear-positive sputum samples by fluorescence in situ hybridization (FISH) using peptide nucleic acid (PNA) probes. *The International Journal of Tuberculosis and Lung Disease*, *3*(9), 830–837.

Sun, C., Gaylord, B. S., Hong, J. W., Liu, B., & Bazan, G. C. (2007). Application of cationic conjugated polymers in microarrays using label-free DNA targets. *Nature Protocols*, *2*(9), 2148–2151.

Suresh, G., & Priyakumar, U. D. (2013). Structures, dynamics, and stabilities of fully modified locked nucleic acid (beta-D-LNA and alpha-L-LNA) duplexes in comparison to pure DNA and RNA duplexes. *The Journal of Physical Chemistry B*, *117*(18), 5556–5564.

Takiya, T., Futo, S., Tsuna, M., Namimatsu, T., Sakano, T., Kawai, K., et al. (2004). Identification of single base-pair mutation on *Escherichia coli* O157:H7 by peptide nucleic acids (PNA) mediated PCR clamping. *Bioscience Biotechnology and Biochemistry*, *68*(2), 360–368.

Tang, Y. W., & Peterson, L. R. (2014). Molecular identification of staphylococcal bacteraemia. *The Lancet Infectious Diseases*, *14*(2), 94–96.

Thiede, C., Creutzig, E., Illmer, T., Schaich, M., Heise, V., Ehninger, G., et al. (2006). Rapid and sensitive typing of NPM1 mutations using LNA-mediated PCR clamping. *Leukemia*, *20*(10), 1897–1899.

Thomsen, R., Nielsen, P. S., & Jensen, T. H. (2005). Dramatically improved RNA in situ hybridization signals using LNA-modified probes. *RNA*, *11*(11), 1745–1748.

von Wintzingerode, F., Landt, O., Ehrlich, A., & Göbel, U. B. (2000). Peptide nucleic acid-mediated PCR clamping as a useful supplement in the determination of microbial diversity. *Applied and Environmental Microbiology*, *66*(2), 549–557.

Wilson, D. A., Hall, G. S., & Procop, G. W. (2010). Detection of group B streptococcus bacteria in LIM enrichment broth by peptide nucleic acid fluorescent in situ hybridization (PNA FISH) and rapid cycle PCR. *Journal of Clinical Microbiology*, *48*(5), 1947–1948.

Yakovchuk, P., Protozanova, E., & Frank-Kamenetskii, M. D. (2006). Base-stacking and base-pairing contributions into thermal stability of the DNA double helix. *Nucleic Acids Research*, *34*(2), 564–574.

Index

Note: Page numbers followed by *f* indicate figures and *t* indicate tables.

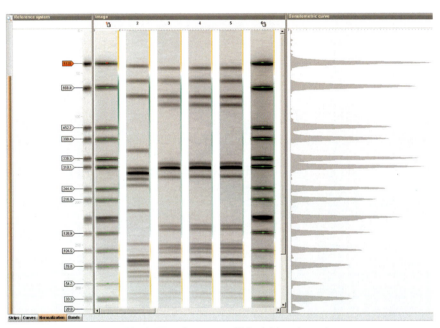

PLATE 1 (Fig. 2 on page 292 of this volume.)

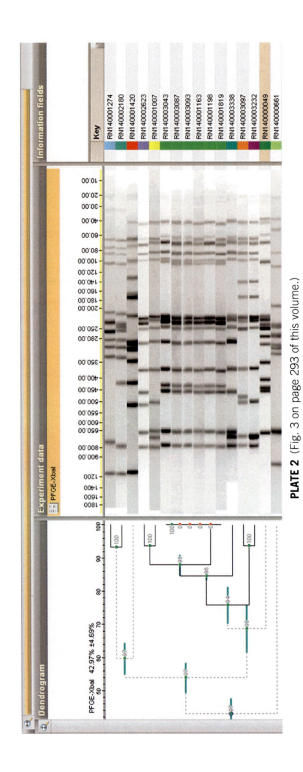

PLATE 2 (Fig. 3 on page 293 of this volume.)

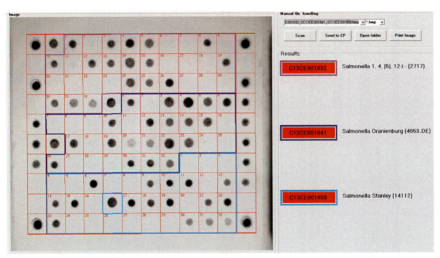

PLATE 3 (Fig. 4 on page 295 of this volume.)

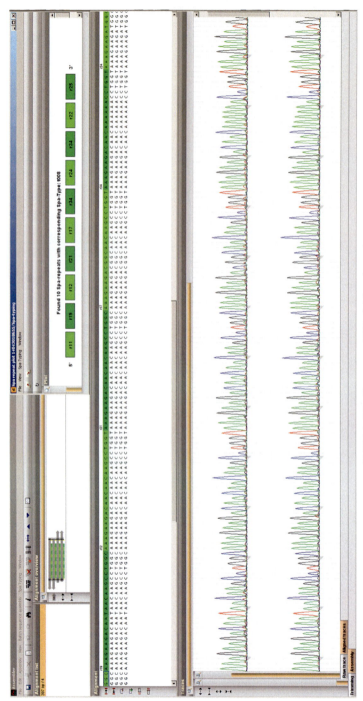

PLATE 4 (Fig. 5 on page 298 of this volume.)

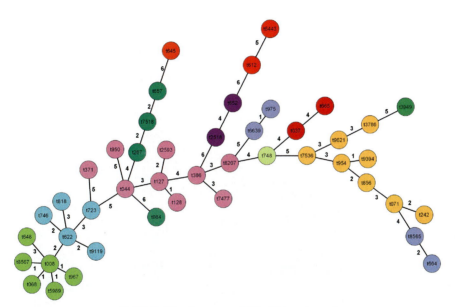

PLATE 5 (Fig. 6 on page 299 of this volume.)

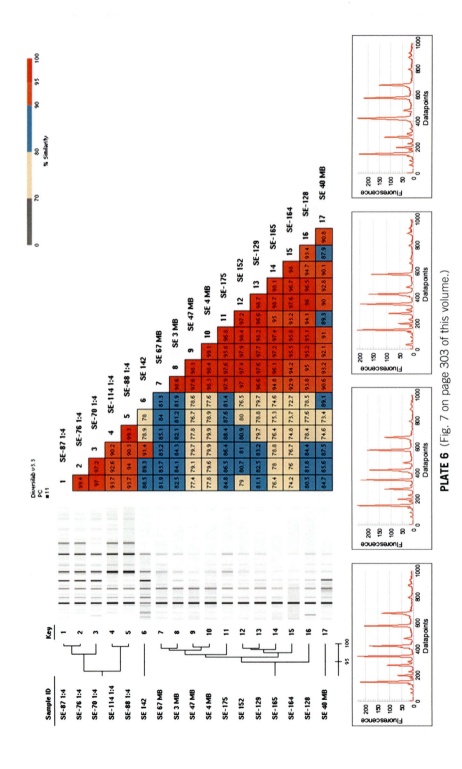

PLATE 6 (Fig. 7 on page 303 of this volume.)

Isolate	Cluster number 100%	MP8					MP7					MP6					MP5					MP4					MP3					MP2					MP1				
		g1294 (160 bp)	g1551 (241 bp)	g0307 (347 bp)	g1324 (440 bp)	g0035 (541 bp)	g1134 (152 bp)	g1552 (222 bp)	g1439 (307 bp)	g1721 (415 bp)	g1679 (529 bp)	g0755 (101 bp)	g0736 (205 bp)	g0967 (301 bp)	g1141 (413 bp)	g1136 (510 bp)	g0421c (127 bp)	g0033 (206 bp)	g0486 (301 bp)	g0569 (399 bp)	g0625 (498 bp)	g1550c (188 bp)	g1329 (307 bp)	g0177 (399 bp)	g1334 (462 bp)	g0566 (558 bp)	g0297c (300 bp)	g1727c (369 bp)	g0264 (406 bp)	g0008 (486 bp)	g1585 (630 bp)	g0057 (175 bp)	g0860 (282 bp)	g1431c (307 bp)	g0733 (441 bp)	g1427c (613 bp)	g0298c (198 bp)	g0728 (296 bp)	g0570 (405 bp)	g0181 (486 bp)	g0483 (612 bp)
1	1	1	0	1	1	1	0	0	0	0	1	0	0	1	0	0	1	0	0	0	1	0	0	0	0	0	1	0	1	0	0	0	1	0	0	0	1	0	1	0	0
2	1	1	0	1	1	1	0	0	0	0	1	0	0	1	0	0	1	0	0	0	1	0	0	0	0	0	1	0	1	0	0	0	1	0	0	0	1	0	1	0	0
3	1	1	0	1	1	1	0	0	0	0	1	0	0	1	0	0	1	0	0	0	1	0	0	0	0	0	1	0	1	0	0	0	1	0	0	0	1	0	1	0	0
4	1	1	0	1	1	1	0	0	0	0	1	0	0	1	0	0	1	0	0	0	1	0	0	0	0	0	1	0	1	0	0	0	1	0	0	0	1	0	1	0	0
5	2	1	0	0	0	1	0	0	0	0	1	0	0	1	0	0	1	0	0	0	1	0	0	0	0	0	1	0	0	0	0	0	1	0	0	0	1	0	1	0	0
6	3	1	0	1	1	1	1	0	0	0	0	0	1	1	0	0	1	0	0	0	1	0	1	0	1	0	1	0	1	0	0	0	1	0	1	0	1	1	1	0	0
7	4	1	0	1	0	1	0	0	0	0	0	1	1	1	0	0	1	0	1	0	1	0	0	1	0	1	1	0	0	0	0	1	1	0	1	0	1	1	1	1	1
8	5	1	0	1	1	1	1	0	0	0	1	0	1	1	0	0	1	0	0	0	1	0	1	1	1	1	1	1	1	1	1	1	0	0	1	1	1	1	1	1	0
9	5	1	0	1	1	1	1	0	0	0	1	0	1	1	0	0	1	0	0	0	1	0	1	1	1	1	1	1	1	1	1	1	0	0	1	0	1	1	1	1	0
10	6	1	0	1	0	1	0	0	0	0	0	1	1	1	0	0	1	0	1	0	1	0	1	1	1	1	1	1	0	0	1	1	0	0	1	1	1	1	1	1	1
11	6	1	1	1	1	1	0	0	0	0	0	1	1	1	0	0	1	1	1	0	1	0	1	1	1	0	1	1	0	0	1	1	0	0	1	1	1	1	1	1	1
NCTC 11168	7	1	1	1	1	1	1	1	1	1	1	1	1	1	1	1	1	1	1	1	1	1	1	1	1	1	1	1	1	1	1	1	1	1	1	1	1	1	1	1	1

PLATE 7 (Fig. 9 on page 306 of this volume.)

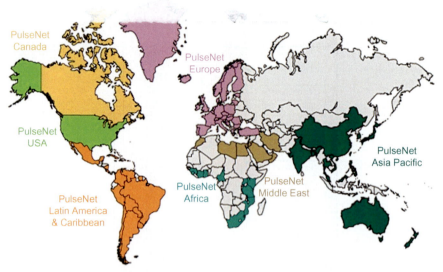

PLATE 8 (Fig. 10 on page 308 of this volume.)

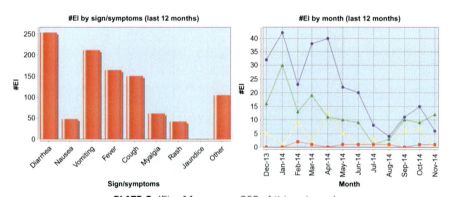

PLATE 9 (Fig. 11 on page 308 of this volume.)

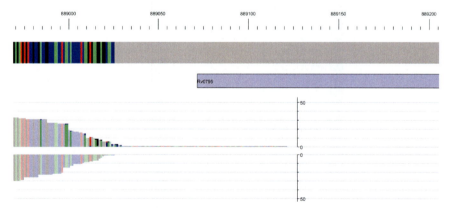

PLATE 10 (Fig. 3 on page 373 of this volume.)

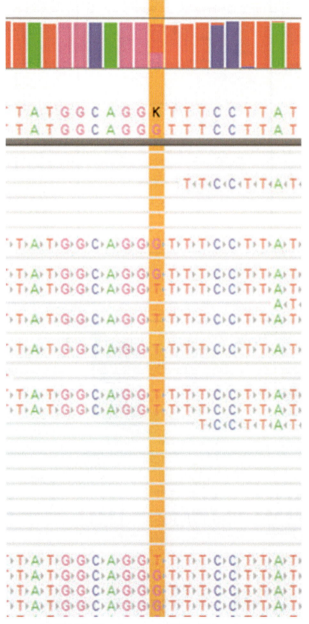

PLATE 11 (Fig. 4 on page 376 of this volume.)

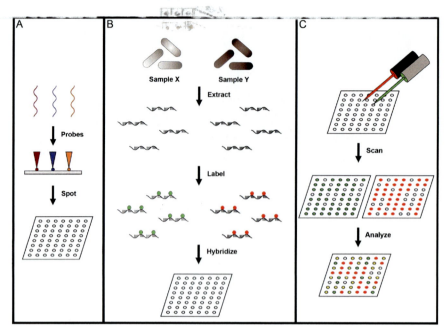

PLATE 12 (Fig. 1 on page 398 of this volume.)

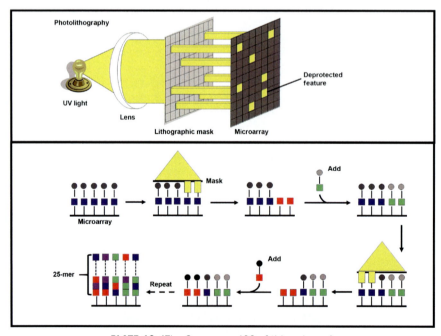

PLATE 13 (Fig. 2 on page 400 of this volume.)

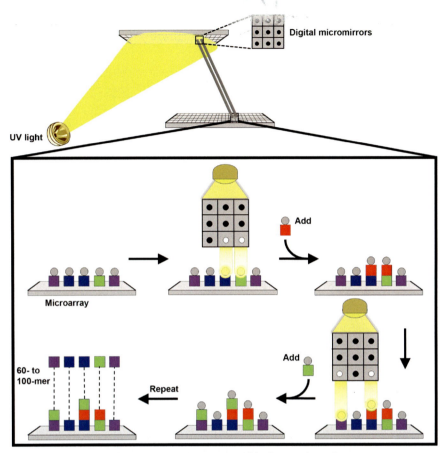

PLATE 14 (Fig. 3 on page 401 of this volume.)

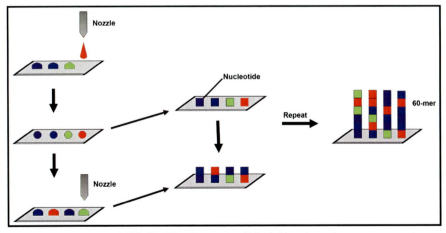

PLATE 15 (Fig. 4 on page 402 of this volume.)

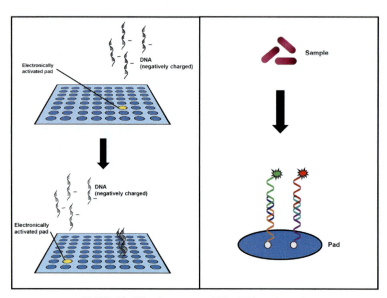

PLATE 16 (Fig. 5 on page 403 of this volume.)

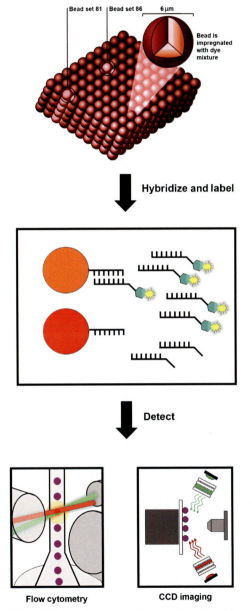

Bead set 81 Bead set 86 6 μm

Bead is
impregnated
with dye
mixture

Hybridize and label

Detect

Flow cytometry   CCD imaging

PLATE 17 (Fig. 6 on page 405 of this volume.)

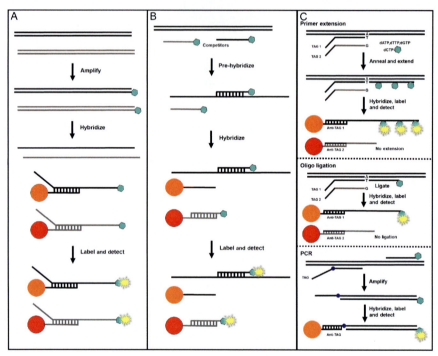

PLATE 18 (Fig. 7 on page 406 of this volume.)

Bacterobacter flagellum H1   *Bacterobacter aflagellum* O1

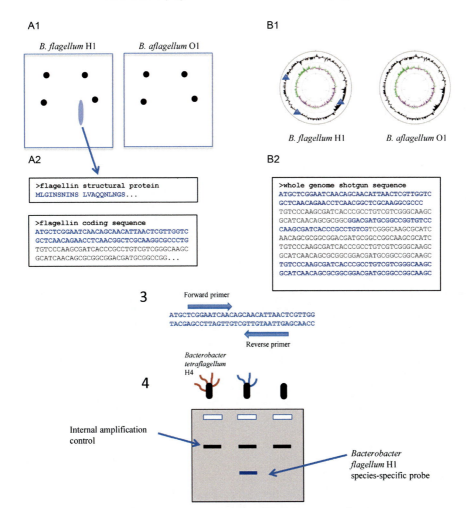

A1

B. *flagellum* H1   B. *aflagellum* O1

A2

>flagellin structural protein
MLGINSNINS LVAQQNLNGS...

>flagellin coding sequence
ATGCTCGGAATCAACAGCAACATTAACTCGTTGGTC
GCTCAACAGAACCTCAACGGCTCGCAAGGCGCCCTG
TGTCCCAAGCGATCACCCGCCTGTCGTCGGGCAAGC
GCATCAACAGCGCGGCGGACGATGCGGCCGG...

B1

B. *flagellum* H1   B. *aflagellum* O1

B2

>whole genome shotgun sequence
ATGCTCGGAATCAACAGCAACATTAACTCGTTGGTC
GCTCAACAGAACCTCAACGGCTCGCAAGGCGCCC
TGTCCCAAGCGATCACCCGCCTGTCGTCGGGCAAGC
GCATCAACAGCGCGGCGGACGATGCGGCCGGTGTCC
CAAGCGATCACCCGCCTGTCGTCGGGCAAGCGCATC
AACAGCGCGGCGGACGATGCGGCCGGCAAGCGCATC
TGTCCCAAGCGATCACCCGCCTGTCGTCGGGCAAGC
GCATCAACAGCGCGGCGGACGATGCGGCCGGCAAGC
TGTCCCAAGCGATCACCCGCCTGTCGTCGGGCAAGC
GCATCAACAGCGCGGCGGACGATGCGGCCGGCAAGC

3   Forward primer

ATGCTCGGAATCAACAGCAACATTAACTCGTTGG
TACGAGCCTTAGTTGTCGTTGTAATTGAGCAACC

Reverse primer

Bacterobacter tetraflagellum H4

4

Internal amplification control

Bacterobacter flagellum H1 species-specific probe

PLATE 19 (Fig. 1 on page 447 of this volume.)

Bacterobacter flagellum H1
genome

A

Bacterobacter aflagellum O1

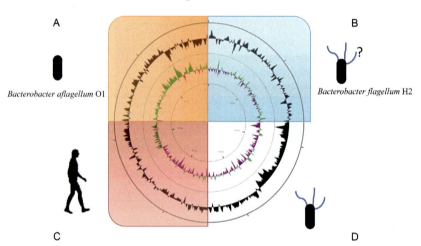

B

Bacterobacter flagellum H2

C

D

PLATE 20 (Fig. 2 on page 449 of this volume.)

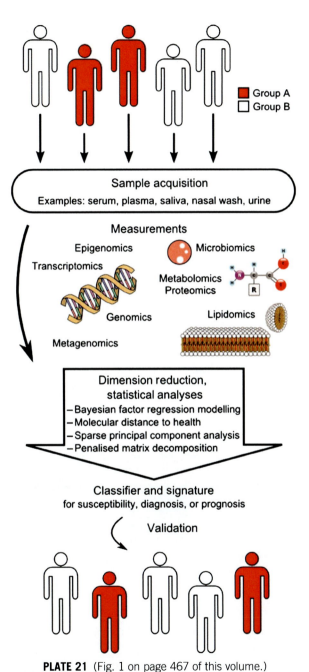

PLATE 21 (Fig. 1 on page 467 of this volume.)

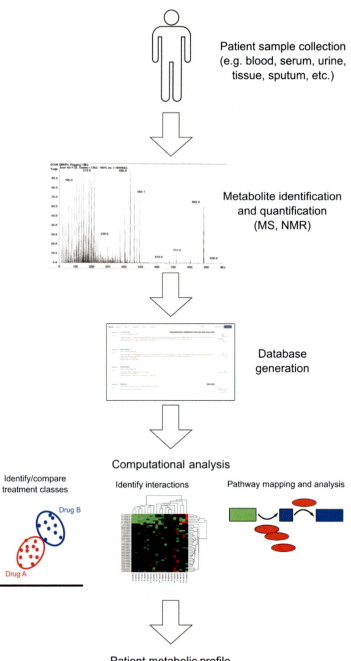

Patient sample collection
(e.g. blood, serum, urine,
tissue, sputum, etc.)

Metabolite identification
and quantification
(MS, NMR)

Database
generation

Computational analysis

Identify/compare
treatment classes

Identify interactions

Pathway mapping and analysis

Drug B

Drug A

Patient metabolic profile
– Predict treatment outcome.
– Predict/monitor drug metabolism and toxicity.
– Predict/monitor effects on metabolic pathways.

PLATE 22 (Fig. 2 on page 471 of this volume.)

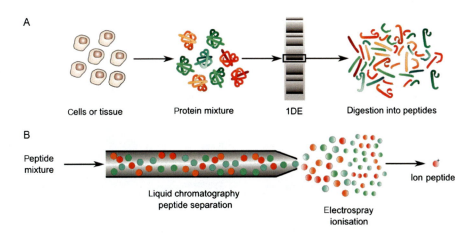

A

Cells or tissue   Protein mixture   1DE   Digestion into peptides

B

Peptide
mixture

Liquid chromatography
peptide separation

Electrospray
ionisation

Ion peptide

PLATE 23 (Fig. 3 on page 472 of this volume.)

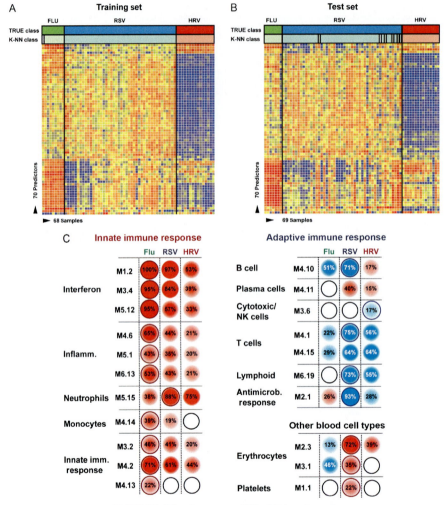

PLATE 24 (Fig. 4 on page 485 of this volume.)

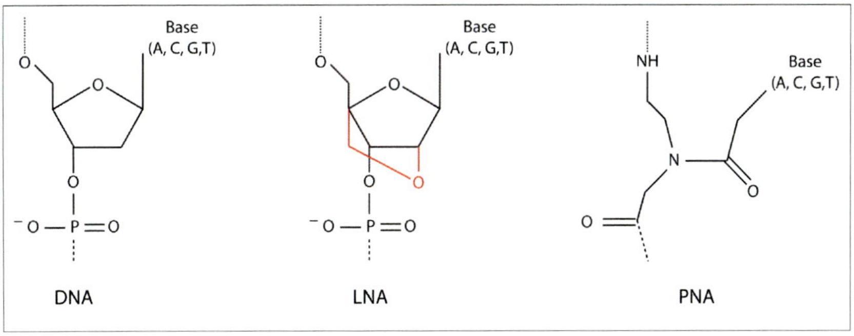

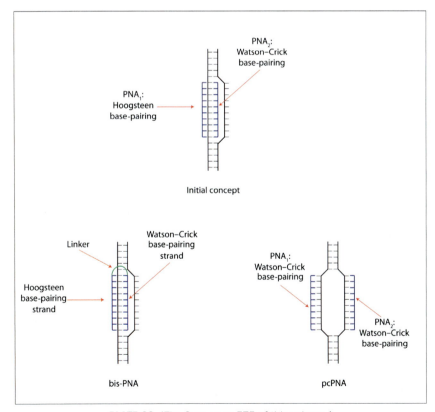

PLATE 25 (Fig. 1 on page 571 of this volume.)

PLATE 26 (Fig. 3 on page 575 of this volume.)

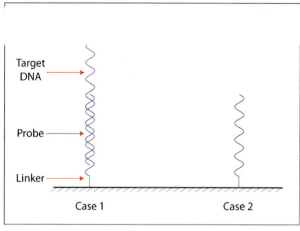

PLATE 27 (Fig. 5 on page 579 of this volume.)

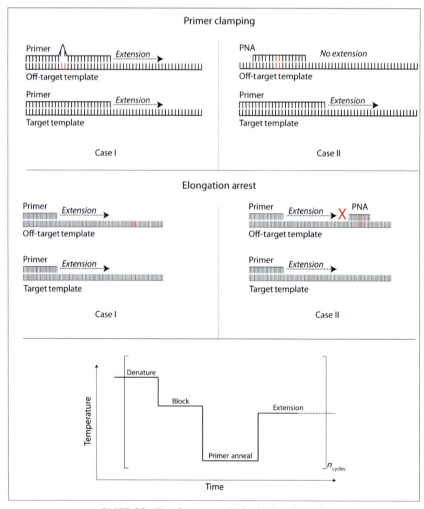

PLATE 28 (Fig. 6 on page 583 of this volume.)

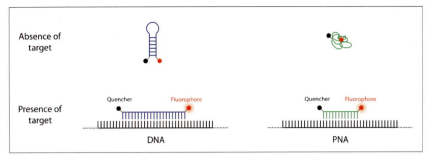

PLATE 29 (Fig. 7 on page 585 of this volume.)

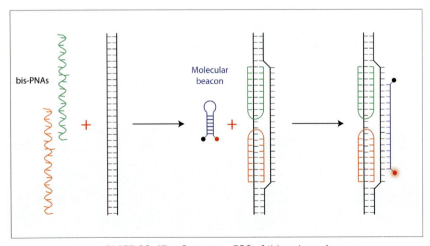

PLATE 30 (Fig. 8 on page 586 of this volume.)

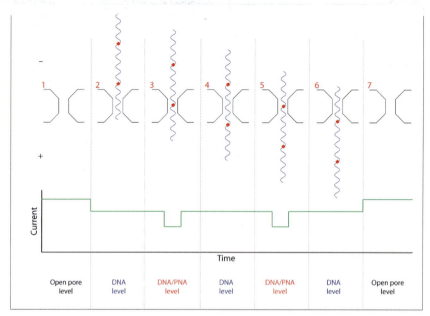

PLATE 31 (Fig. 9 on page 587 of this volume.)

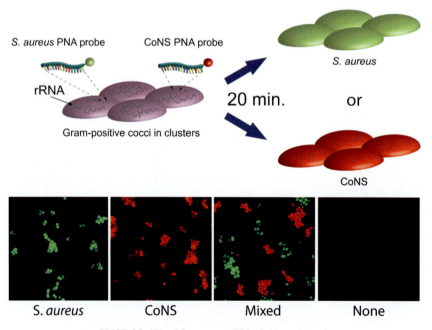

PLATE 32 (Fig. 10 on page 589 of this volume.)

CPI Antony Rowe
Eastbourne, UK
November 20, 2015